DISLOCATION DYNAMICS

DISLOCATION DYNAMICS

Edited by

ALAN R. ROSENFIELD
Metal Science Group, Battelle Memorial Institute,
Columbus Laboratories

GEORGE T. HAHN
Metal Science Group, Battelle Memorial Institute,
Columbus Laboratories

ARDEN L. BEMENT, Jr.
Metallurgy Research Section, Battelle Memorial Institute,
Northwest Laboratories

ROBERT I. JAFFEE
Metal Science Group, Battelle Memorial Institute,
Columbus Laboratories

BATTELLE INSTITUTE
MATERIALS SCIENCE COLLOQUIA
*Seattle, Washington, U. S. A. and
Harrison, British Columbia, Canada
May 1–6, 1967*
Robert I. Jaffee, Chairman

McGRAW-HILL BOOK COMPANY

New York   San Francisco   Toronto   London   Sydney

DISLOCATION DYNAMICS

53807

1234567890 MAMM 7543210698

To Professors Peierls and Nabarro,

whose names have become synonymous with the subject of these proceedings.

The Original Manuscript of the Peierls Dislocation Model*

(kindly made available by E. Orowan)

*A commentary on the Peierls-Nabarro Force is given by R. E. Peierls on page xiii and the original calculation is given in full on page xvii.

PARTICIPANTS

GEORG ALEFELD *KFA Institut für Reaktorwerkstoffe, Jülich, West Germany*

G. A. ALERS *Scientific Laboratory, Ford Motor Company, Dearborn, Michigan, U.S.A.*

R. W. ARMSTRONG *Division of Engineering, Brown University, Providence, Rhode Island, U.S.A.*

Z. S. BASINSKI *National Research Council, Ottawa, Ontario, Canada*

A. L. BEMENT, Jr. *Metallurgy Research Section, Battelle Memorial Institute, Richland, Washington, U.S.A.*

WALTER BOLLMANN *Battelle Memorial Institute, Geneva-Carouge, Switzerland*

NORMAN BROWN *School of Metallurgical Engineering, University of Pennsylvania, Philadelphia, Pennsylvania, U.S.A.*

RONALD BULLOUGH *Theoretical Physics Division, Atomic Energy Research Establishment, Harwell, U.K.*

B. M. BUTCHER *Sandia Corporation, Albuquerque, New Mexico, U.S.A.*

J. D. CAMPBELL *University of Oxford, Oxford, U. K.*

HANS CONRAD *Materials Science and Engineering Division, The Franklin Institute, Philadelphia, Pennsylvania, U.S.A.*

J. E. DORN *Inorganic Materials Research Division, University of California, Berkeley, California, U.S.A.*

B. ESCAIG *Faculté des Sciences, Service de Physique des Solides, Université de Paris, Orsay, France*

J. J. GILMAN *Department of Mining, Metallurgy, and Petroleum Engineering, University of Illinois, Urbana, Illinois, U.S.A.*

A. V. GRANATO *Department of Physics, University of Illinois, Urbana, Illinois, U.S.A.*

PETER HAASEN *Institut für Metallphysik der Universität Göttingen, Göttingen, West Germany*

G. T. HAHN *Metal Science Group, Battelle Memorial Institute, Columbus, Ohio, U.S.A.*

P. B. HIRSCH   *Department of Metallurgy, University of Oxford, Oxford, U.K.*

J. P. HIRTH   *Department of Metallurgical Engineering, The Ohio State University, Columbus, Ohio, U.S.A.*

R. H. HOBART, Jr.   *Fundamental Physics Division, Battelle Memorial Institute, Columbus, Ohio, U.S.A.*

R. I. JAFFEE   *Metal Science Group, Battelle Memorial Institute, Columbus, Ohio, U.S.A.*

A. S. KEH   *Edgar C. Bain Laboratory for Fundamental Research, U.S. Steel Corporation, Monroeville, Pennsylvania, U.S.A.*

J. M. KRAFFT   *Mechanics Division, U.S. Naval Research Laboratory, Washington, D. C., U.S.A.*

J. C. M. LI   *Edgar C. Bain Laboratory for Fundamental Research, U.S. Steel Corporation, Monroeville, Pennsylvania, U.S.A.*

JENS LOTHE   *Fysiks Institut, Blindern Universitet, Oslo, Norway*

J. R. LOW, Jr.   *Research and Development Center, General Electric Company, Schenectady, New York, U.S.A.*

W. P. MASON   *Department of Civil Engineering and Engineering Mechanics, Columbia University, New York, New York, U.S.A.*

F. R. N. NABARRO   *Department of Physics, University of the Witwatersrand, Johannesburg, South Africa*

W. S. OWEN   *Department of Materials Science and Engineering, Cornell University, Ithaca, New York, U.S.A.*

R. E. PEIERLS   *Department of Theoretical Physics, University of Oxford, Oxford, U.K.*

A. R. ROSENFIELD   *Metal Science Group, Battelle Memorial Institute, Columbus, Ohio, U.S.A.*

ALFRED SEEGER   *Institut für Physik, Max Planck Institut für Metallforschung, Stuttgart, West Germany*

A. W. SLEESWYK   *Natuurkundig Laboratorium der Rijks-Universiteit, Groningen, The Netherlands*

F. A. SMIDT, Jr.   *Metallurgy Research Section, Battelle Memorial Institute, Richland, Washington, U.S.A.*

JOHN SPREADBOROUGH   *Battelle Memorial Institute, Geneva-Carouge, Switzerland*

D. F. STEIN   *Research and Development Center, General Electric Company, Schenectady, New York, U.S.A.*

HIDEJI SUZUKI   *Department of Physics, University of Tokyo, Tokyo, Japan*

TAIRA SUZUKI   *The Institute for Solid State Physics, University of Tokyo, Tokyo, Japan*

J. W. TAYLOR   *Los Alamos Scientific Laboratory, University of California, Los Alamos, New Mexico, U.S.A.*

THAD VREELAND, Jr.   *W. M. Keck Laboratory of Engineering Materials, California Institute of Technology, Pasadena, California, U.S.A.*

JOHANNES WEERTMAN   *The Technological Institute, Department of Materials Science, Northwestern University, Evanston, Illinois, U.S.A.*

F. W. YOUNG, Jr.   *Solid State Division, Oak Ridge National Laboratory, Oak Ridge, Tennessee, U.S.A.*

PREFACE

This book presents the proceedings of the Second Battelle Materials Science Colloquium, "Dislocation Dynamics," held in Seattle, Washington, and Harrison, British Columbia, May 1 to 7, 1967.

The purpose of the Colloquium was to assess present knowledge of dislocation dynamics of crystalline materials with emphasis on metals and alloys. The objectives were to identify rate controlling mechanisms, examine the extent to which these mechanisms can be formulated theoretically, review experiments that offer insights into these processes, and describe progress in quantitatively describing macroscopic deformation behavior. The subjects included descriptions of the dislocation core and the Peierls' barrier and interactions of moving dislocation with lattice defects, impurities, other dislocations, and boundaries. The participants were selected from those throughout the world who have made historically significant as well as current advances in the field. Consistent with the wide range of subjects, they represented a breadth of interests and disciplines.

The format for the meeting was patterned after the highly successful First Battelle Materials Science Colloquium, "Phase Stability in Metals and Alloys," held in Geneva and Villars, Switzerland in March, 1966. The first day's meeting was devoted to introductory lectures by five distinguished scientists who have contributed importantly for many years and have greatly influenced the field. These introductory lectures appropriately inaugurated the Battelle-Seattle Science Center and were attended by invited representatives of the Pacific Northwest scientific community.

Four subsequent days at the Harrison Hotel, British Columbia, were devoted consecutively to:

Preliminary considerations
Low speed dislocations
High speed dislocations
Critical problems

In general, in a morning session of three hours, six formal papers were presented. As time allowed, there was a discussion specific to the papers, following such presentations. In these proceedings, these discussions are appended to the respective papers. After an afternoon of relaxation, the participants reconvened for a three-hour Agenda Discussion on the subject of the day. Scientists eminent in the field were selected as the Agenda Discussion Leaders and were given responsibility for outlining critical points and developing the discussion of each point. The Agenda Discussions were tape-recorded, and a more-or-less literal transcription was made by the Agenda Discussion Secretaries. The Agenda Discussion Leaders have condensed, rearranged, organized, added commentary to, and interpreted these discussions, each at his own discretion, in order to contribute a compendium of current thought on each subdivision of the Colloquium. These Agenda Discussions are included in these proceedings following the formal papers of the topic of the day. An Agenda Discussion on "Critical Problems" summed up the status of the field and provided the final paper in these proceedings.

At the end of the Colloquium, an informal banquet was held in honor of Professors Peierls and Nabarro at which each was invited to describe the circumstances surrounding his contributions to the development of the force to move a dislocation and the subsequent application of this concept to dislocation theory. A dramatic surprise during the banquet was the production by Professor Nabarro of the manuscript in which Professor Peierls originally transmitted his calculations of the now-famous "Peierls' force" to Professor Orowan at the latter's request. Because of the historical significance of this calculation, and, despite errors which have since come to light, the manuscript is presented as the frontispiece of these proceedings. Furthermore, autobiographical sketches of Professors Peierls and Nabarro, which provide an interesting perspective into the development of dislocation dynamics are presented in the following section.

The Battelle Materials Science Colloquia, of which this was the second, are made possible through the support of the Battelle

Institute and the continued encouragement of its president, Dr. B. D.
Thomas, and its director of research in the physical sciences,
Dr. F. J. Milford.

The Battelle Northwest members of the organizing committee
were Dr. Spencer H. Bush and Dr. John L. Brimhall. Dr. Sherwood
F. Fawcett, director of the Battelle Northwest Laboratory and
Mr. Kenneth B. Hobbs, Battelle-Seattle Research Center, gen-
erously provided all needed services. The assistance of our
charming and efficient secretaries, Kathryn Drake, Berta Hull,
Rebecca Martin, and Jackie Munson is gratefully acknowledged.

Alan R. Rosenfield
George T. Hahn
Arden L. Bement, Jr.
Robert I. Jaffee

COMMENTARY ON THE
"PEIERLS-NABARRO FORCE"

Rudolph Peierls

I am glad to have this opportunity of making a statement that may help to set the record straight, though the correct facts have already been set out by Orowan.*

I did not have in 1939, nor do I have today, any close knowledge or any deep understanding of the problems of dislocations. In 1939 Orowan asked me for help in the formulation and solution of an approximate model for a dislocation. He had a clear picture of the approximation to be used; namely, to assume that the interaction of the plane of atoms on either side of the slip plane with the adjacent half-space is given by the equations for an elastic continuum, and that the force across the slip plane is dominated by a sinusoidal term. The derivation of the integral equation resulting from this model is straightforward, and I was greatly surprised to find that the simplest function with the expected qualitative behavior turned out to be an exact solution of this integral equation. Orowan says that it would have taken him "days or weeks" to study this problem, and this may be a generous estimate; in any event there is no doubt he could have found the solution without difficulty.

* Orowan, E.: "The Sorby Centennial Symposium on the History of Metallurgy," Gordon and Breach, Science Publ., New York, 1965.

When the result of the calculation which had been reported to a conference in Bristol, was published,* I would have preferred to have this appear as a joint paper with Orowan, or perhaps as an appendix to a paper by him. However, he was not willing to agree to this, and at the time the matter did not seem of great importance. If I had foreseen the attention this paper would receive, and the extent to which it would be quoted even today (which I did not fully realize until I heard some of the lectures at the Battelle Symposium), I would probably have pressed the point more strongly.

In 1947 Nabarro generalized, and for this purpose rederived the formula.[†] It was then discovered that my calculation contained an error of a factor 2. Orowan mentions a factor 2π, but in the interest of historical accuracy, I must point out that this is an exaggeration. Actually, this error occurs in a large exponent, so that even the factor 2 changes the magnitude of the critical stress by several orders of magnitude. It is a sobering thought that it could easily have been a factor 2π. Perhaps it is as well, in view of this, that the paper was published under my name, so that the responsibility for this slip can be correctly assigned; indeed this factor 2 would seem to be my only really original contribution to the subject.

It would evidently be much more satisfactory if this force was known as the "Orowan-Nabarro Force." However, from one's general experience of the way in which the use of names becomes immutable one cannot feel very optimistic about the prospects of this happening.

---

*Peierls, R. E.: *Proc. Phys. Soc.* **52**, 34 (1940).
[†]Nabarro, F. R. N.: *Proc. Phys. Soc.* **59**, 236 (1947).

AUTOBIOGRAPHICAL SKETCH
OF FRANK NABARRO

I went up to Oxford in 1934 to read physics. In those days, physicists did a solid year of mathematics before being allowed into the Clarendon Laboratory. Once I was there, T. C. Keeley, who was in charge of the practical classes, soon recognized that my strength lay on the theoretical side. There was not then much theoretical physics in Oxford, so I decided to ask to work in Bristol under Mott, who was my external examiner. He advised me to read mathematics for an extra year before starting research. Once in Bristol, I worked for a short time with Mott on the theory of magnetic coercivity, until we discovered that our ideas had already been developed in Germany and published in one of those Conference Reports that no one ever reads. We then devised a preliminary theory of solution and precipitation hardening, which we have both been interested in ever since. I was rather disappointed at not being given a problem in classical physics, and Mott let me take some time off to do a proper highbrow problem with Frohlich.

During the War, I was (unlike Hume-Rothery) fortunate to find stimulating problems in the Army Operational Research Group under Schonland. I first helped Mott on problems of diffraction affecting the accuracy of anti-aircraft radar. We then got interested in the effectiveness of anti-aircraft shells, and, after Mott had gone to another job, I extended this to the effectiveness of anti-personnel weapons.

After the War, I rejoined Mott in Bristol. Here he had built up a group (Frank, J. W. Mitchell, Thompson, Eshelby, Cabrera) which made a major contribution to our theoretical and experimental knowledge of dislocations. During a trip to Germany, I met Doris Kuhlmann, and Mott invited her to join the group.

It was soon after I got back to Bristol that I read the paper of Peierls on the core of a dislocation. I realized its importance, but had great difficulty in reproducing its results, because the methods were sketched only very briefly. After I had made some progress, I went to see Peierls in Birmingham. He was very helpful, and told me how he had approached the problem. Thus strengthened, I went back and found the factor of 2 which Peierls had discussed. I also considered the theory of diffusional creep, which, rather to my surprise, has turned out to be a real and even occasionally important phenomenon. After four years in Bristol, I moved to Birmingham in Hanson's metallurgy department, where Cottrell and Bilby were working. Cahn soon came, and Raynor was leading a team on another group of problems. Peierls in his own department was giving for the first time the course of lectures which is published in his *Quantum Theory of Solids*, and his student, Hunter, became my first Ph.D. student.

After four years in Birmingham, I went to Johannesburg as head of the department of physics in the University of Witwatersrand. Apart from Doris Kuhlmann-Wilsdorf, who went to America soon after I arrived, none of my colleagues had extensive experience of the sort of solid state research we were trying to build up. Research in South African universities does not receive governmental support on the scale which is customary in England, and industrial support such as is common in America was then very limited, so it was a hard job to build up the conditions for productive research. But it had its excitements and its rewards as well. Africa needs to build up more centers where good work is done, both on problems of general interest and on problems and materials specially relevant to the local economy, and I like to think that I have made some contribution to this process.

Frank Nabarro

THE PEIERLS DISLOCATION MODEL

The Peierls formulation of the dislocation is given below. The steps in the model are as set down by R. E. Peierls in his original 1939 manuscript illustrated in the Frontispiece. Self-compensating errors in sign and the highly abbreviated form of the calculation are preserved for historical interest.

Half–space $z > 0$

$$\nabla^2 u = 0$$

Surface force per unit area, $F(x) = -G\left(\dfrac{\partial u}{\partial z}\right)_{z=0}$

Let: $u(x, 0) = \phi(x)$

$$\phi(x) = \int_{-\infty}^{\infty} dk\, a_k e^{ikx}$$

$$u(x, z) = \int_{-\infty}^{\infty} dk\, a_k e^{ikx - |k| z}$$

$$a_k = \frac{1}{2\pi} \int_{-\infty}^{\infty} dx'\, \phi(x')\, e^{-ikx'}$$

$$u(x, z) = \frac{1}{2\pi} \int_{-\infty}^{\infty} dk \int e^{ik(x - x') - |k| z} \phi(x')\, dx'$$

$$\int_{0}^{\infty} e^{ik(x - x') - kz}\, dk = \frac{1}{z - i(x - x')}$$

$$u(x, z) = \frac{1}{2\pi} \int_{-\infty}^{\infty} dx' \, \phi(x') \left\{ \frac{1}{z - i(x - x')} + \frac{1}{z + i(x - x')} \right\}$$

$$\frac{\partial u(x, z)}{\partial z} = \frac{i}{2\pi} \int_{-\infty}^{\infty} dx' \, \phi(x') \frac{\partial}{\partial x'} \left\{ \frac{1}{z - i(x - x')} - \frac{1}{z + i(x - x')} \right\}$$

$$= \frac{i}{2\pi} \int_{-\infty}^{\infty} dx' \, \frac{\partial \phi(x')}{\partial x'} \left\{ \frac{1}{z - i(x - x')} - \frac{1}{z + i(x - x')} \right\}$$

$$= -\frac{2}{2\pi} \int_{-\infty}^{\infty} dx' \left(\frac{\partial \phi}{\partial x} \right)_{x'} \frac{(x - x')}{z^2 + (x - x')^2}$$

For $z = 0$: $F(x) = \frac{1}{\pi} G \int_{-\infty}^{\infty} \left(\frac{\partial \phi}{\partial x} \right)_{x'} \frac{1}{(x - x')} \, dx'$

Here the integral has a logarithmic singularity, but from its derivation it is evident that the way in which the integral is to be understood is as the

$$\lim_{\epsilon \to 0} \left\{ \int_{-\infty}^{x-\epsilon} + \int_{x+\epsilon}^{+\infty} \right\}$$

On the other hand, $F(x) = B \sin \frac{2\pi\phi}{a}$.

The connection between B and G can be obtained by working out the curve of a small displacement $u = \alpha z$ in a homogeneous crystal. Then the force is on one hand

$$-G\alpha$$

On the other hand

$$-B \frac{2\pi\alpha}{a} \cdot a$$

Thence

$$B = \frac{G}{2\pi}$$

The equation becomes

$$\frac{1}{\pi} \int_{-\infty}^{\infty} \left(\frac{\partial \phi}{\partial x} \right)_{x'} \frac{1}{x - x'} \, dx' = \frac{1}{2\pi} \sin \frac{2\pi\phi}{a}$$

Introduce the variables:

$$\frac{2\pi\phi}{a} = \psi$$

$$\frac{2\pi x}{a} = \xi$$

Then

$$2\int_{-\infty}^{\infty}\left(\frac{d\psi}{d\xi}\right)_{\xi'}\frac{1}{\xi - \xi'}\,d\xi' = \sin\psi$$

Or if we take ψ as the independent variable,

$$\sin\psi = 2\int_{-\pi}^{\pi}\frac{d\psi'}{\xi(\psi) - \xi(\psi')} = \sin\psi$$

The boundary conditions are:

$$\xi = \begin{cases} +\infty & \psi = \pi \\ -\infty & \psi = -\pi \end{cases}$$

Try as solution

$$\xi = c\tan\frac{\psi}{2}$$

Then

$$\xi(\psi) - \xi(\psi') - c\left(\frac{\sin(\psi/2)}{\cos(\psi/2)} - \frac{\sin(\psi'/2)}{\cos(\psi'/2)}\right) = c\,\frac{\sin[(\psi - \psi')/2]}{\cos(\psi/2)\cos(\psi'/2)}$$

Equation:

$$\sin\psi = \frac{2}{c}\int_{-\pi}^{\pi}d\psi'\cos\frac{\psi}{2}\frac{\cos(\psi'/2)}{\sin[(\psi - \psi')/2]}$$

Write $\cos\dfrac{\psi'}{2} = \cos\dfrac{\psi}{2}\cos\dfrac{\psi - \psi'}{2} + \sin\dfrac{\psi}{2}\sin\dfrac{\psi - \psi'}{2}$

And thus:

$$\sin\psi = \frac{2}{c}\cos\frac{\psi}{2}\int_{-\pi}^{\pi}d\psi'\left\{\cos\frac{\psi}{2}\cot\frac{\psi - \psi'}{2} + \sin\frac{\psi}{2}\right\}$$

$$= \frac{2}{c}\cos\frac{\psi}{2}\cos\frac{\psi}{2}\int_{-\pi}^{\pi}d\psi'\cot\frac{\psi - \psi'}{2} + \frac{2}{c}\cos\frac{\psi}{2}\sin\frac{\psi}{2}\int_{-\pi}^{\pi}d\psi'$$

Of the integrals, the first vanishes, and the second is equal to 2π,

$$\sin\psi = \frac{2}{c} \cdot 2\pi \cdot \cos\frac{\psi}{2} \sin\frac{\psi}{2}$$

Which is O.K. for $c = 2\pi$.

Hence

$$\xi = 2\pi \tan\frac{\psi}{2}$$

$$\frac{x}{a} = \tan\frac{\pi}{a}\phi$$

The point where the force is a maximum, i.e., where

$$\sin\psi = 1, \text{ is}$$
$$\xi = 2\pi, \text{ i.e., } x = a$$

CONTENTS

——————————————————— *Part One*

INTRODUCTORY
LECTURES

THE PLASTIC RESPONSE OF SOLIDS*

J. J. Gilman

*University of Illinois,
Urbana, Illinois*

ABSTRACT

Past developments of the microdynamical description of plastic flow are outlined and the successes of the theory are mentioned. Some extensions in scope are suggested and discussed. These include: application to polymeric solids; collective dislocation behavior; and dynamics of precipitation hardening. The problem of the difference between the flowing-state and the post-flowing-state is emphasized. It is proposed that dislocation lines exist in noncrystalline solids, and provide a convenient description for local shear correlations. The possibility of flutter being induced in

* The work reviewed here has been supported by grants from the U. S. Atomic Energy Commission, the Office of Naval Research, and the Army Research Office — Durham.

moving dislocations by turbulent phonon-gas flow is considered. Finally, an atomic model for dislocation mobility in covalent crystals is presented.

1   INTRODUCTION

Perhaps the most important of all transport phenomena for solids is mechanical transport, that is, plastic flow. It is active in such natural phenomena as the flow of rocks and the movements of proteins; and also in determining the strengths of construction materials, for example, metals, ceramics, and high polymers. Mechanical transport occurs as a result of the motion of dislocation lines within a solid. Orowan[1] was the first to point out explicitly that the rate depends on the number of dislocations moving and their average velocity, as well as how much displacement is carried by each one. Therefore, the basic transport equation for a given glide system has the following form:

$$\dot{\epsilon} = bNv \qquad (1)$$

where $\dot{\epsilon}$ is the shear strain-rate; b is the displacement carried by the dislocation lines; N is the mean flux of lines; and v is their mean velocity. For particular conditions of stressing and material, each of the parameters on the right-hand side of this equation can be dependent on time, and hence are all involved in the discussion of "dislocation dynamics." Mura[2] has shown how the flow equation can be written in general tensor form.

For some time after the initial presentation of ideas about dislocation motions and their relation to plastic flow, there were no sufficiently direct experimental measurements to guide further development of the ideas. Then, about a decade ago, it became possible to measure dislocation velocities directly, and it was learned that substantial internal drag stresses act on moving dislocations. Under many conditions they balance the applied stresses, which results in stable steady-state motion.[3] For nearly perfect pure metals and salts this is not the case; here, the drag is large when they are imperfect. It is also large in covalent crystals.

Experimental verification of the mechanical transport equation was first provided by Johnston and Gilman,[3] but various subsequent verifications have been introduced. A particularly careful study is reported by Haasen at this meeting.

Another dynamic feature of dislocation behavior that became clear through experimentation is that dislocations breed extensively by means of a mechanism proposed by Koehler[4] as a development of the Frank-Read regeneration mechanism. A consequence of breeding is that although individual dislocation motions may be steady, macroscopic plastic flow can become unstable if the initial dislocation density is low. This leads to the interpretation of discontinuous yielding that was given by Gilman and Johnston,[5] and developed by Johnston,[6] Hahn,[7] Patel and Chadhuri,[8] and Gillis and Gilman.[9] So many applications of this interpretation have been made, it is impossible to review all of them here. I can only offer apologies to neglected authors.

Discontinuous yielding is a characteristically dynamic effect and its interpretation requires that the dynamics of the testing machine be included in the analysis. This can be done for constant displacement-rate machines by writing a "machine equation" as suggested by Gilman.[10] For other cases, such as that of plane shock waves, the machine equation can be replaced by an elastic-plastic wave equation as done first by Taylor.[11]

An implication of discontinuous yielding is that the flow instability leads to the formation of propagating plastic fronts (Luders band-fronts). This is discussed in some detail by Campbell in this volume.

As an application of the microdynamic theory, the creep phenomenon is particularly suited for analysis. Therefore, it is not surprising that several authors have discussed this case, for example, Dew-Hughes,[12] Peissker, Haasen, and Alexander,[13] Li,[14] Akulov,[15] Garafalo,[16] Gilman,[17] and Webster.[18] Again I offer apologies to neglected authors.

A most important practical case is that of fracture, especially since a dynamic treatment is the only type that can lead to the large local stress which is necessary to cause parting of a material. A preliminary attempt to treat fracture in dynamical terms has been made elsewhere.[19]

Various means for relating the microscopic quantities in Eq. (1) to the macroscopic shear stress σ_s, and the plastic strain ϵ_p, have been proposed. The one I consider to be most general for low-temperature deformation (other authors will be quick to emphasize its limitations) is the following:[20]

$$\dot{\epsilon}_p = \Phi b v^* (N_o + M\epsilon_p) \exp \frac{-(D + H\epsilon_p)}{\sigma_s} \tag{2}$$

where

Φ = orientation factor
b = Burger's displacement

v^* = terminal dislocation velocity
N_o = initial mobile dislocation density
M = multiplication parameter (may be a function of stress)
D = characteristic drag stress
H = strain-hardening coefficient

The value of the general equation (1) and more specific forms, such as Eq. (2), for interpreting plastic phenomena is now well-established. Since recent developments will be discussed here by other authors, I shall not dwell on them. However, the initial successes of the microdynamic treatment of plastic flow make it appropriate to attempt to broaden its scope of application, and to develop its foundations in atomic theory. Generalization of its formalism is one possible extension.[2] Another is its application to the theory of noncrystalline solids such as polymers, and some progress made by Dey[21] in this direction will be described. Soils and rocks are still other materials that might be treated.

Applications of the theory to increasingly complex stress-states continue to have great potential for expansion. Some examples include spherical elasto-plastic wave propagation, cracks in plastic bodies, the cyclic stress situations that lead to fatigue failures, and metal forming and cutting operations.

An intriguing possibility is the interpretation of the behavior of active media such as muscular tissues. Here the driving force is associated with a change in chemical potential rather than applied stress, and the system is a mechano-chemical engine which employs dislocations as part of its mechanism.

There is a considerable lack of experimental and theoretical knowledge concerning the dynamics of dislocations that lie along the interfaces of two crystals or phases. Yet such knowledge is important for the understanding of high-temperature creep, composite materials, and phase transformation kinetics.

In crystals with high dislocation mobilities, e.g., zinc (basal plane) and copper, the interaction stresses between dislocations can easily exceed the drag stresses. Then dislocations will behave collectively, and traffic-flow theory may be applicable to some situations. This theme is developed in a later section.

Dislocation motions in two-phase materials have been considered by Copley and Kear[22] and the dynamic behavior of nickel-base alloys that contain coherent precipitates has been clarified.

One approach to an atomic theory of plastic flow can be made by considering the basic electronic process that determines the velocity of a kink on a dislocation line in a covalently bonded crystal. Using the formalism of tunneling theory, an expression for the kink velocity can be derived which relates it to the electronic energy gap, the atomic volume, and the applied stress.[23]

2 MACROSCOPIC FLOW

A general review will be omitted here. Instead, we will discuss some new results and ideas. These are concerned with the following: the plastic behavior of Nylon; collective dislocation effects; the lack of stability of the flowing state; and the dynamics of precipitation hardening.

A Propagation of Plastic Band-Fronts in Polymers

The velocities at various stess levels of the interfaces between oriented and nonoriented sections of Nylon fibers have been studied by Dey[21] to determine whether the stress dependence of dislocation motion in Nylon is similar to, or different from that found for crystals.

Since the macroscopic configuration of a reorientation front remains rather constant as it propagates (except when large velocity changes occur), its velocity V_B is proportional to the mean dislocation velocity within the material. Therefore, the stress dependence of dislocation velocities within solid polymers can be investigated by this means. Some of Dey's results are shown in Fig. 1 where it may be seen that indeed the dynamical behavior is similar to that of crystals.

Furthermore, Dey was able to find a self-consistent (although not necessarily unique) set of microscopic parameters to describe the behavior not only of his specimens, but those of previous workers. For this purpose he used an analytic description of a plastic bandfront that was previously derived for crystals.[16]

It may be concluded that the flow of polymers is phenomenologically similar to that of metals and ceramics and is based on dislocation motions. This provides considerable encouragement for the idea that a unified approach to plastic flow phenomena is feasible.

B Collective (Many-Body) Effects in Dislocation Dynamics

The dislocations inside a flowing crystal are often spaced so close together, and therefore interact so much, that they cannot be

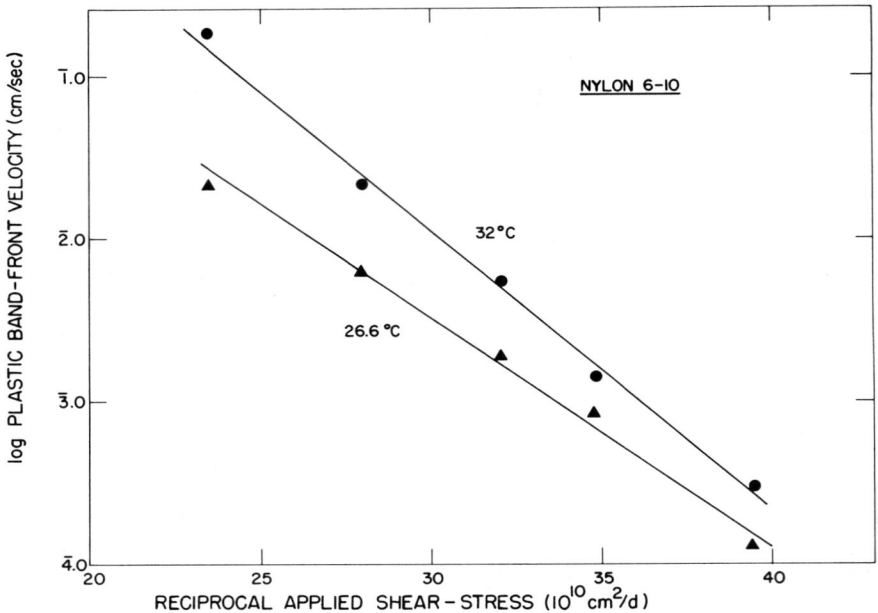

Fig. 1 Showing that the stress dependence of the flow-rate in Nylon 6-10 at constant dislocation density is of the form: $V = V^* \exp(-D/\sigma_s)$.

considered as a set of independent entities.[24] Instead, their collective behavior must be considered. It is difficult to identify the effects of collective behavior when the plastic flow is steady (or nearly so). However, transient effects caused by various kinds of perturbations can be observed. It is believed that some of these can be identified as results of many-body scattering interactions. No detailed discussion will be attempted here, but some characteristics of collective behavior will be discussed and a specific experimental effect will be interpreted as a many-body scattering effect.

Since the effective masses are usually small compared with elastic forces that act on them, acceleration effects are not important for individual dislocations. However, when they behave collectively, the effective mass of a group can be large in comparison to the coupling forces within the group. Then it is necessary to reconsider Eq. (1), and, in particular, its time derivative:

$$\ddot{\epsilon} = b(Na + v\dot{N}) \tag{3}$$

This strain-acceleration equation describes what happens when the conditions of a flowing material are suddenly changed.

In many transient situations, the second term in the equation (which gives the strain acceleration caused by changes of the dislocation density) is the dominant one. However, the first term provides the possibility of purely accelerative effects when the total line-length is constant. One experimental observation that appears to correspond with this is that of transient upper-yield-points. Sometimes it is seen that when a flowing specimen is temporarily unloaded and then reloaded, the stress rises transiently above the previous flow curve, but then returns asymptotically to the projection of the original flow curve as shown in Fig. 2. This latter factor is important because it indicates that the stress over-shoot is a true dynamical transient rather than a result of changes in $\dot{N}$. If the term in $\dot{N}$ were the important one, the stress would not be expected to return to the extension of the original flow curve, but instead to a level either above or below it.

Another factor is that the negative slope of the second half of the stress transient reflects unstable flow. This in turn implies the propagation of a plastic front of some kind. Such a wavelike propagation is more likely to be associated with acceleration effects than with dislocation density changes in a crystal that has already undergone substantial plastic strain.

A set of parallel dislocations moving along a glide plane is analogous to a single-lane traffic flow. Therefore, the dislocations might be expected to obey a "car-following law"; especially one proposed by Gazis, Herman, and Potts.[25] In the present context this states that the acceleration $\ddot{x}_{n+1}$ of the $(n + 1)$th dislocation in a

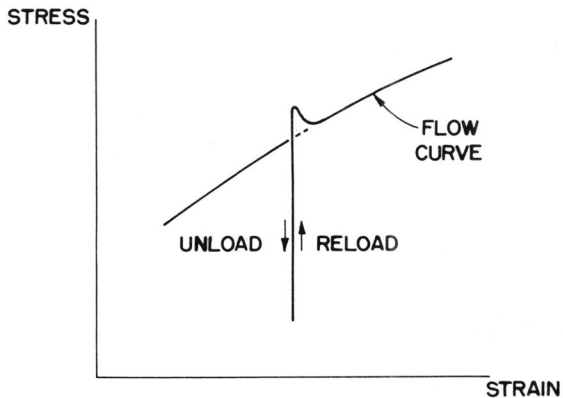

Fig. 2 Schematic drawing of a transient "yield-point" that sometimes appears when a specimen is unloaded and then reloaded.

group is proportional to the difference between its velocity and that of the nth dislocation; in addition, it is inversely proportional to the difference between them (because this determines the interaction force). This may be stated analytically as follows:

$$\ddot{X}_{n+1} = A_0 \left(\frac{\dot{X}_n - \dot{X}_{n+1}}{X_n - X_{n+1}} \right) \tag{4}$$

where A_0 is a characteristic velocity. For Newtonian viscosity this transforms into a standard wave equation since $\dot{X}_n = \Phi\sigma_n$ where Φ is the fluidity. Then Eq. (4) becomes:

$$\ddot{X}_{n+1} = A_0\Phi \left(\frac{\sigma_n - \sigma_{n+1}}{X_n - X_{n+1}} \right) \tag{5}$$

and it may seem that kinematic waves will propagate up and down the chain if it is locally perturbed.

Consider a set of dislocations that has stopped moving because the specimen it is in has been unloaded. When the load is replaced, the whole set will not immediately start moving any more than a line of cars does when the traffic light turns green. Instead, a pulse of motion will begin at the front of the set and propagate back into it. If the damping forces that act on the dislocations are small, then the pulse will propagate at nearly the velocity of sound. However, if substantial damping is present, the response will be slower, which will tend to make the stress overshoot in a constant displacement-rate test as indicated in Fig. 2.

Prigogine's discussions[26] of the statistical dynamics of traffic flow also suggest some interesting collective effects for dislocations. Let the dislocation velocity distribution function be $f(v, t)$. Then for a "dilute" set of dislocations, changes in the distribution will be described by a Boltzmann-like relaxation equation:

$$\frac{\partial f}{\partial t} = -\left(\frac{f - f_0}{\tau} \right) \tag{6}$$

where f_0 is the "free" distribution. This equation indicates that regardless of the initial velocity distribution, the free distribution f_0, is approached in a characteristic time τ. For a "concentrated" set of interacting dislocations, the relaxation equation becomes:

$$\frac{\partial f}{\partial t} = -\left(\frac{f - f_0}{\tau} \right) + \left(\frac{\partial f}{\partial t} \right)_I \tag{7}$$

where $(\partial f/\partial t)_I$ describes changes in the distribution caused by interactions.

For a simple interaction law that causes both colliding dislocations to acquire the lower of the two velocities possessed by them, the relaxation equation becomes (with C = concentration, and β = collision cross section):

$$\frac{\partial f}{\partial t} = -\left(\frac{f - f_0}{\tau}\right) + C\beta f\left[1 - 2\int_0^v f(v')\,dv'\right] \tag{8}$$

The time-independent solution of this is given by the solution of the following integral equation:

$$f = f_0 + C\beta\tau f\left[1 - 2\int_0^v f(v')\,dv'\right] \tag{9}$$

For $v = 0$, the integral in brackets becomes zero and

$$f = \frac{f_0}{1 - C\beta\tau} \tag{10}$$

Therefore $C\beta\tau < 1$ and a critical concentration is reached when

$$C_{\text{crit}} = \frac{1}{\beta\tau}$$

At this concentration, a discontinuous "phase transition" occurs and the average velocity drops almost to zero because of intense scattering interactions.

Explicit solutions of the integral equation will not be discussed here, but they have the following general form:

$$f(\tau) = \frac{f_0(v)}{[1 + C\beta\tau g(v)]^{\frac{1}{2}}} \tag{11}$$

Therefore, the average velocity always decreases as the concentration builds up, and then drops off precipitously when the critical concentration is reached (see Fig. 3).

C The Flowing-State Versus the Post-Flowing State

Critical experimental investigation of the "flowing-state" of a plastic material is difficult because the internal structure may change drastically when the applied stress is removed. Thus, studies of the "post-flowing-state" may reveal very little about

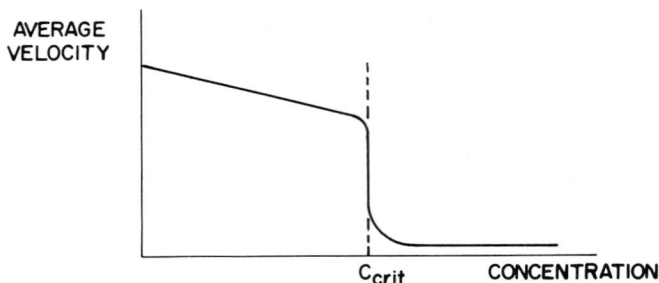

Fig. 3 Schematic behavior of the velocity of a set of interacting parallel dislocations moving along a glide plane as it depends on the concentration.

the actual "flowing-state." The situation is analogous to that of gaseous discharge where the plasma-state exists only during the time that an electric field is applied.

The instability of the flowing state is greatest when the viscous forces on dislocations are small (as in pure metals) because the internal stresses can then cause considerable motion even though no applied stress is present. Also, it is apparent that flowing-states with high dislocation densities will be less stable than those with low densities because the driving forces are larger and the relaxation distances smaller. Finally, patterns of collective motion that arise as a result of dynamical interactions will tend to dissolve when the dynamical conditions are changed.

Crump and Young[2] recently provided experimental evidence of differences between the flowing and post-flowing states. Earlier, less-direct evidence was obtained from observations such as those of "glide polygonization" in LiF crystals.[5] In a flowing crystal the dislocations cannot exist in neat vertical arrays (nor only in cell walls) because this would not allow continuous flow. Therefore, condensation into polygon walls (or cell walls) must occur after the applied stress is removed. Direct observation of this has not been accomplished, but it may be deduced from other observations by means of simple analysis.

Let n be the volume concentration (#/cm^3) of mean-sized dislocation loops (or some other equivalent measure of the dislocation density). Then the rate of change $\dot{n}$ is given by the difference between the rate of multiplication αn (where α is the multiplication coefficient) and the rate of attrition γn^2, which depends on interactions between the dislocations (γ is the attrition coefficient):[3]

$$\dot{n} = \alpha n - \gamma n^2 \tag{12}$$

For steady-state flow, the concentration is fixed so $\dot{n} \to 0$ and $n_{ss} = \alpha/\gamma$.

Suppose that the applied stress is suddenly switched off at time $t = t_s$. Then since multiplication requires input work to create added dislocation lines, it must stop. That is, $\alpha \to 0$. But for at least a short time, pairs of dislocations can mutually interact so γ remains finite; then $\dot{n} = -\gamma n^2$ and upon integration

$$n = \frac{1}{\gamma(t - t_s) + \gamma/\alpha} \simeq \frac{1}{\gamma(t - t_s)} \tag{13}$$

since $\alpha \gg \gamma$; and the recovery rate is given by

$$\dot{n}_r \simeq -\frac{1}{\gamma(t - t_s)^2} \tag{14}$$

which is initially very large. Due to this high recovery rate, it is not valid to infer the flowing-state from the much different post-flowing state.

D Dynamic Treatment of Precipitation Hardening

In an alloy that is hardened by coherent ordered particles, the dislocations move as strongly coupled pairs in the ordered phase because anti-phase-boundary (APB) lies between them. When they move out of the ordered phase they tend to remain as pairs. Each member of a pair may further split into partials but this is neglected in the discussion that follows. Because of the necessity for creating APB when a dislocation pair enters the ordered phase a resistive force arises which delays the entry. Copley and Kear[22] have investigated the dynamics of this situation in some detail.

The schematic kinematics of the motion of a dislocation pair through an ordered particle is shown in Fig. 4 where the ordered particle is designated ω, whereas the surrounding matrix is μ.

At t_1 the leading member of the pair reaches the μ-ω interface. The leading member makes some APB and enters the ω-phase at t_2 while the trailing member lags behind in the μ-phase. The pair is completely inside ω at t_3, and has reached its steady-state width at t_4. The leading member of the pair leaves ω to enter μ at t_5.

The dynamics of this process was studied[22] by means of the following dislocation velocity-stress equation:

$$v = v^* \exp \frac{-D}{\sigma_s} = v^* \exp \frac{-F}{\Sigma f_i} \tag{15}$$

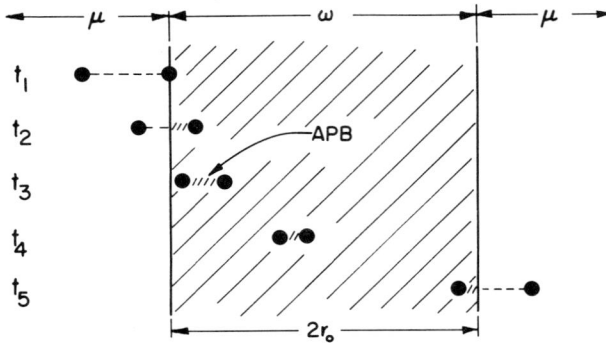

Fig. 4 Sequence of positions for a dislocation as it passes
through an ordered precipitate particle.

where the drag stress D, and the applied stress σ_s, have been replaced by the drag force F, and the sum of other forces f_i caused
by the applied stress, the APB, the second member of a pair, etc.
Thus, if the leading member of a pair is designated by (1) and the
trailing member by (2), their velocities at t_i may be written as
follows:

$$V_1 = V_\omega^* \exp\left[\frac{-F_\omega}{\sigma_s b - A + K/w + \theta_1(x)}\right]$$

$$V_2 = V_\mu^* \exp\left[\frac{-F_\mu}{\sigma_s b - K/w + \theta_2(x)}\right]$$

(16)

where F_ω and F_μ are the drag stresses in the two phases, $\sigma_s b$ is the
stress-induced force, A is the force caused by the APB, K/w (with
w being the separation width and K an elastic coefficient) is the
force caused by the other dislocation, and θ is the line tension which
is a function of the lateral coordinate x. We determined the velocities and displacements by means of stepwise integration of these
functions.

It was found that a dislocation spends most of its time in the
process that is in progress at t_2 of Fig. 4. Hence an approximate
expression for the average velocity can be written

$$\overline{V} = \exp\frac{-(F_\omega + F_\mu)}{2\sigma_s b - A + Gb/r_0}$$

where Gb/r_0 is the effect of the line tension with r_0 being the particle
radius and G the shear modulus.

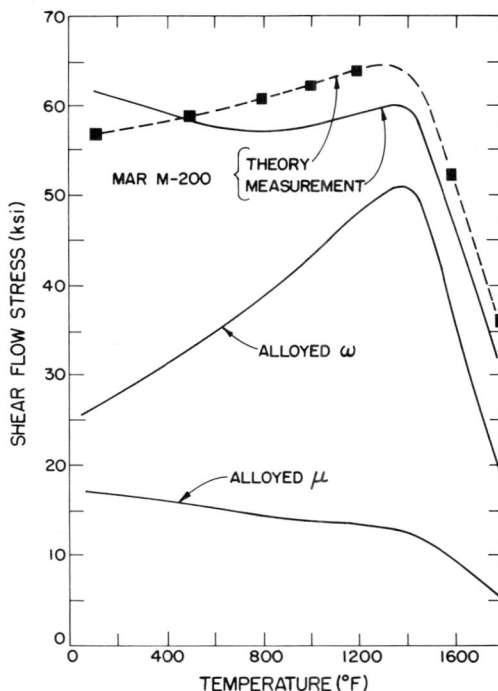

Fig. 5 Comparison of observed flow stresses of a nickel-base alloy (MAR M-200) and calculation of dynamical behavior (after Copley and Kear).

The results of numerical computations for a Nickel-base alloy (MAR M-200) yielded the following average velocities:

$$\text{velocity in pure matrix} = 1.7 \times 10^3 \text{ cm/sec};$$
$$\text{velocity in ordered phase} = 3.3 \times 10^2 \text{ cm/sec};$$
$$\text{velocity in mixture} = 10^{-3} \text{ cm/sec};$$

which show the large effect of the time spent in entering the ordered phase on the average velocity. A comparison between the dynamical calculations and the observed behavior of the alloy is shown in Fig. 5, where it may be seen that quite reasonable agreement is obtained.

3 ATOMIC MECHANISMS

The atomic structures of materials have strong effects on mechanical transport through their effects on: Burgers displacement vector; the primary velocity that a given stress produces;

and the rate of cross-glide. The latter determines how fast dislocations multiply. In crystals the behavior of dislocations is
greatly simplified because the Burgers vector is fixed for a given
line and is restricted to relatively few values that are determined
by the translation vectors of the structure. However, the concept
of a dislocation line remains useful even if Burgers vector does
not have a fixed value; as may be the case for such noncrystalline
solids as glasses.

A   Dislocation Lines in Noncrystalline Solids

In a noncrystalline solid (silica glass being the prototype) a
dislocation line will have a somewhat variable Burger vector
along its length, as suggested in Fig. 6.

The drawings are projections onto the plane of the drawing
of the positions of the silicon atoms of a single sheet in the structure. The oxygen atoms are not shown, but each silicon atom is
bonded to an oxygen atom that lies just above it plus three that lie
below it, parallel to the plane of the drawing. If parts of the upper
oxygen layers are translated while the rest are not, dislocation
lines can be formed at positions indicated by the dashed lines in
Fig. 6. The arrows represent the translations that move oxygen
atoms from initial sites to equivalent final sites in the next higher
layer during an elementary motion of the dislocation line.

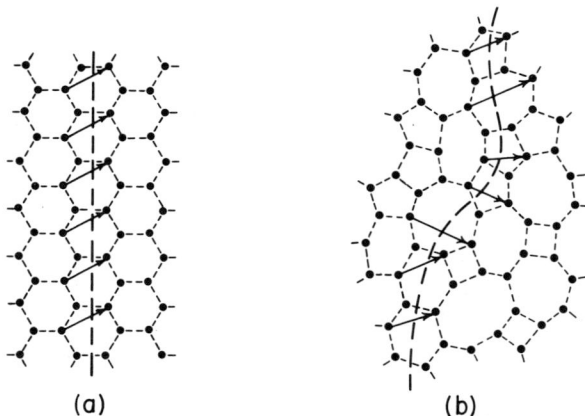

(a)   (b)

Fig. 6 Dislocation lines in crystalline and vitreous silica
compared. Only the silicon atoms (solid dots) are shown;
each is surrounded by an oxygen tetrahedron. (a) Dislocation
line has fixed Burgers vector. (b) Fluctuations in the length
and direction of the Burgers vector occur in the glass structure.

It may be seen that by allowing the magnitude and direction of Burgers vector to fluctuate about mean values, the concept of a dislocation line can be retained for glassy structures. This is not just a forced analogy but is desirable because it allows the flow properties of these structures to be discussed in a more organized way than is otherwise possible. That is, it provides a simple means for describing the correlations that must exist between adjacent elementary shear processes.

For small fluctuations of b the added self-energy of a dislocation is small. Suppose a unit length of dislocation has a Burgers displacement of $b + \delta$ along one-half of its length, and $b - \delta$ along the other half, where δ is a small increment. Then the ratio of its elastic energy to that of the same length without the fluctuations is: $1 + \delta^2/b^2$. Thus fluctuations as large as 30% cause only a 10% energy increase.

In the noncrystalline case of Fig. 6b, the mean Burger displacement has a definite value that is determined by the network dimensions, but there are fluctuations in both its magnitude and its direction along the line. In order to minimize the energy of such a dislocation, it is necessary for the mean b to be conserved over long distances; thus although the local b may fluctuate, there are long-range correlations (occasional large energy densities may cause this condition to be relaxed). Furthermore, there will be little tendency for the line to lie on a single plane, and its local structure will change as it moves. Nevertheless, it is expected that such dislocations will exist in noncrystalline solids, especially under flow conditions. When they are viewed with a somewhat fuzzy microscope (resolution of approximately 10A), their behavior should resemble that of dislocations in crystals.

B   Possibility of Flutter

Mason[28] has pointed out that crystal dislocations flow through a "viscous gas of phonons" at finite temperatures. A rough analogy exists between this and the motion of a cylindrical airfoil through an ordinary gas.

As the temperature increases, it is observed that the motions of screw dislocations become more irregular. This is usually interpreted as thermally activated cross-glide, but this interpretation has some conceptual difficulties. Another possibility is that the flow of the phonon gas past a moving dislocation becomes turbulent above some critical velocity. The turbulence leads to "vortex shedding" which buffets the dislocation line causing it to make excursions from its mean glide plane.

A preliminary test of the feasibility of this process can be made by considering the Reynolds number of the system. The Reynolds number is defined by

$$R = \frac{\rho V s}{\eta} \tag{17}$$

where ρ is the mass density of the gas, V is the relative gas velocity, s is a characteristic dimension of the system, and η is the gas viscosity. The lower critical limit of R which forms the boundary between laminar and turbulent flow is ~ 2000.

At 200°C in copper, Mason has estimated the viscosity of the phonon gas to be $\sim 3 \times 10^{-3}$ poise; a reasonable speed for fast dislocations is 10^5 cm/sec; the characteristic distance between them might be 10^{-4} cm; and the effective mass density of the phonon gas should be about 1 g/cm^3. The resulting value for R is ~ 3000. Therefore it appears that the flow might well be turbulent, and that the turbulence might increase with temperature as the effective density of the phonon gas increases.

C   An Atomic Model for Low-temperature Dislocation Mobility

At high temperatures, dislocation motion can be activated thermally but at very low temperatures it must be induced by applied stresses alone, with negligible assistance from thermal oscillations. Therefore, it is important to conceive of stress–activated processes that allow motion. Dislocation motion is usually described in terms of mechanical mechanisms, but often the mobility is closely correlated with electronic parameters of a crystal; and since mechanical forces have electrical origins, it is possible for the motion to be described as an electronic process. Such a description provides insight to atomic flow mechanisms, and may lead to a quantitative theory as suggested here.

It is proposed that dislocation motion can be thought to consist initially of the transfer of a bonding electron from a site that lies adjacent to a kink into the kink site, followed by repositioning of the atoms. A potential barrier separates the two states, and, therefore, the transition can be described formally as a tunneling process. The electric field that induces the tunneling is provided by the applied shear stress through the local piezoelectric coupling that exists because the local region does not possess a center of symmetry. The treatment is approximate, but can be developed through the use of polaron theory.

The discussion here will concentrate on the behavior of co-valently bonded crystals for which the concept of localized electron-pair bonds is a good approximation. However, with appropriate modifications the model can be applied to impurity centers in salts and metals.

In covalently bonded crystals, dislocation lines move as a result of motion of sharply defined kinks along their lengths. This is confirmed by the strong temperature-dependence of the velocity, the characteristic shapes of dislocation loops, the close correlation between chemical bonding energy and glide activation energy, and the strong stress dependence of the velocity. In addition, a detailed calculation of the configuration by Labusch[29] has yielded approximately 0.4b as the kink width.

At low temperatures, the concentration of kinks is determined by a balance between creation and annihilation rates.[30] If N is the number of kinks per unit length, and α is the number of pairs created per second, then since the pair annihilation rate will be $v_k N^2$, the kink density changes at the following rate:

$$\frac{dN}{dt} = 2\left(\alpha - v_k N^2\right) \tag{18}$$

The kink density saturates at $dN/dt = 0$; hence the saturation density is

$$N^* = \left(\frac{\alpha}{v_k}\right)^{1/2} \tag{19}$$

and the dislocation velocity is

$$v_d = b(\alpha v_k)^{1/2} \tag{20}$$

The quantities α and v_k can be further resolved. The pair-creation rate equals the product of the number of possible creation sites (which is $1/d$ where d is the kink length), and the pair-creation rate at any site. This latter equals the creation attempt frequency ν_c, times the success probability P_c. Thus α is given by

$$\alpha = \frac{\nu_c P_c}{d} \tag{21}$$

The kink velocity depends on the jump rate (which is the jump attempt frequency ν_m, times the success probability P_m), and also on the jump distance d. Hence

$$v_k = \nu_m P_m d \tag{22}$$

Substitution of Eqs. (21 and (22) into (20) yields

$$v_d = b(\nu_c \nu_m)^{\frac{1}{2}} (P_c P_m)^{\frac{1}{2}} \tag{23}$$

Since atomic dimensions are involved, ν_c and ν_m must approximately equal the Debye cut-off frequency. Then the first two terms of Eq. (23) can be replaced by a terminal kink velocity v_k^*, which approximately equals the elastic shear-wave velocity.

The production of a kink pair and the motion of a kink is very nearly the same process in the present model. To create a kink pair, an electron must be transferred from a pair-bond to an adjacent site. To move a kink, an electron must be transferred from a pair-bond into the adjacent kink. Thus the probabilities P_c and P_m are approximately equal and Eq. (23) becomes

$$v_d = v_k^* P_m \tag{24}$$

The next step is to calculate P_m.

In the absence of an applied stress (and other external fields) no net fields exist near a kink. But when a stress σ (causing a strain ϵ) is applied, a local electric potential $\Phi(\sigma) = p\sigma$ is induced through piezoelectric coupling, and the corresponding electric field is given approximately by: $F(\sigma) \simeq p\sigma/b$. The proportionality constant p, can be large since there is no center of symmetry. The local field tends to cause an electron to move to a new position which in turn moves the kink forward by an elementary distance d (see Fig. 7). The temperature is low so that thermal excitations are negligible.

Initially, the electron is bound to its partner. However, to acquire mobility, it must become "ionized" by tunneling into the conduction bond. This is the rate-determining step because the reformation of an electron pair at the kink should be fast. Further approximate development of this idea requires the assumption that local bands exist near kinks. Also, although the electronic energy levels will be more different near a kink than elsewhere, it is assumed that (by analogy with the situation near a free surface) they change in parallel, thereby leaving the band gap essentially unaffected.

KINK

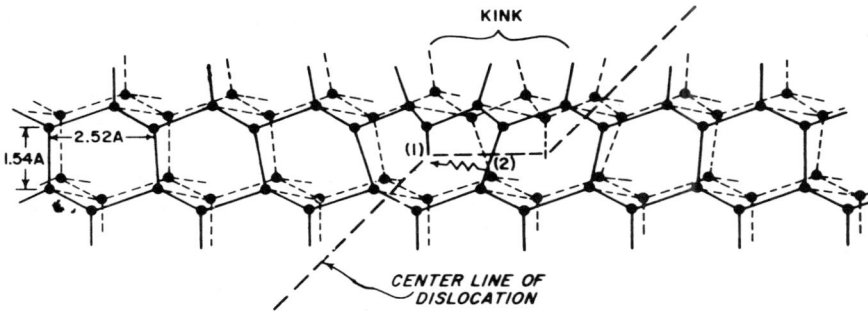

CENTER LINE OF
DISLOCATION

Fig. 7 Configuration at kink in dislocation line that passes through the diamond crystal structure. Tunneling of an electron from bond (2) to bond (1) allows the kink to advance by one atomic distance. The glide plane is {111}; and the glide direction is <110>.

Since the tunneling process is taken to be rate-determining, its probability equals P_m. Also, it must be synchronized with a suitable atomic oscillation so that the net attempt frequency is atomic rather than electronic oscillation. An approximate tunneling probability can be obtained from Zener's band-to-band theory[31,32] which yields:

$$\exp \frac{-\pi^2 (mE_k{}^3)^{\frac{1}{2}}}{heF} \tag{25}$$

where m = effective electron-hole mass; E_k = energy gap at kink; h = Planck's constant; e = electron charge; and F = local electric field. Note that the same form results from Frenkel's ionization theory.[33] Therefore, the assumption of local bands is not crucial to the argument.

Next the local electric field must be related to the applied stress. This is difficult to do directly, but for approximate purposes, a simple thermodynamic argument can be used. Regardless of the details of the process, the net electrical work done must equal the mechanical work because they are the same entity viewed differently. The work done by the mean electrical force during an advance of a kink is eFb. Similarly, for the mean stress force, the work is $\sigma_s b^2 d$; therefore the electrical force may be written as follows:

$$eF = \sigma_s bd \tag{26}$$

To simplify Eq. (25) as much as possible, the following uncertainty relation between the distance b, and the characteristic momentum $(mE_k)^{\frac{1}{2}}$ relation is used to eliminate the effective mass:

$$b(mE_k)^{1/2} \simeq \frac{h}{2\pi} \tag{27}$$

Thus when Eqs. (26) and (27) are substituted into Eq. (25), the expression for the kink velocity becomes:

$$v_k = v_k^* \exp \frac{-\pi E_k}{4\sqrt{2}\sigma_s V} \tag{28}$$

where V = an atomic volume = $b^3/2\sqrt{2}$ and b = d for a 60° dislocation in the diamond structure. This equation may also be written as:

$$v_k = v_k^* \exp \frac{-D}{\sigma_s} \tag{29}$$

where D is the "characteristic drag stress."[20]

Therefore, for rapid kink motion, the applied shear stress must have the following value:

$$\sigma_s^* \simeq D \simeq \frac{E_g}{2V} \tag{30}$$

This is also the stress required for a substantial plastic flow rate; and a simple, although approximate, measure of this at low temperatures is an indentation hardness value. For a comparison with it, σ_s^* is multiplied by two to convert shear stress to compressive stress. Table 1 compares hardness values[34] for Ge and Si with those given by Eq. (30). It may also be seen that the numerical agreement is good. The actual numbers may not be valid because of the approximations, but their consistency tends to support the proposed model.

Two important consequences of this model are: (1) the analytic form of Eq. (29) which makes the velocity an exceedingly strong

TABLE 1 Comparison of Calculated Flow Stresses and Observed Hardnesses for Germanium and Silicon

Crystal	E_g (eV)	b (A)	V (10^{-24} cm^3)	$2\sigma_s^*$ (10^{10} d/cm^2)	Hardness number (10^{10} d/cm^2)
Ge	0.80	3.99	22.7	5.6	6.6
Si	1.14	3.83	20.0	9.1	8.0

function of the stress as would be expected for stress activation; and (2) the fact that strong applied fields of various kinds might influence kink motion by influencing the energy gap. Some of these are strong electric fields (10^5 V/cm or greater), magnetic fields, hydrostatic pressure, and temperature. Data are available only for the last.

At low temperatures the gap decreases linearly with increasing temperature:

$$E_g = E_g^0 - \epsilon T \tag{31}$$

thus the stress needed to cause a given flow rate should decrease linearly with increasing temperature.* Table 2 compares the calculated temperature-dependence of $(2\sigma_s^*)$ with observed hardness changes.[34] The numerical agreement is satisfactory since the hardness data scatter considerably.

TABLE 2 Comparison of Observed and Calculated
Temperature-Dependence of Flow Stress

Crystal	$\partial E_g / \partial T$ $(10^{-16} \text{ ergs} /^\circ\text{K})$	$\Delta(2\sigma_s^*)/\Delta T$ $(10^7 \text{ } d/\text{cm}^2 - {}^\circ\text{K})$	$\Delta H / \Delta T$ $(10^7 \text{ } d/\text{cm}^2 - {}^\circ\text{K})$
Ge	-5.9	-2.6	-1.3
Si	-3.7	-1.9	-4.4

The quantity (E_g/V) in Eq. (30) is an estimate of the binding energy of a kink to a particular location. To average over the complex forces that are involved, it might be better to use an appropriate mechanical energy density, e.g., the elastic shear strength which is known to equal about 10% of the elastic shear stiffness. On the glide plane {111} and in the glide direction <110> the stiffness will be called C_{gd}. Then, to the same degree of approximation as in the previous arguments, the kink velocity is given by

$$v_k = v_k^* \exp \frac{-C_{gd}}{\kappa \sigma_s} \tag{32}$$

where κ is a number of order ten. Consequently, the stress for rapid flow (hardness number) should be proportional to C_{gd} which

---

* The direct effect of the specific volume change is small by comparison, and thus is neglected here.

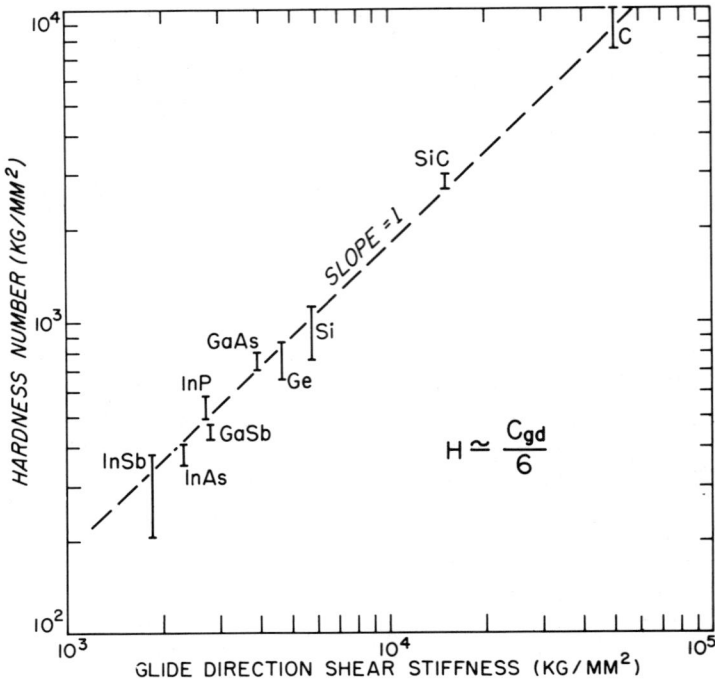

Fig. 8 Correlation of flow stress and glide-plane shear-stiffness for tetrahedrally bonded crystals.

is related to measured elastic stiffness by the following:

$$C_{gd}^{-1} = \frac{1}{3}[S_{44} + 4(S_{11} - S_{12})] \tag{33}$$

and Fig. 8 shows that the expected proportionality is observed with $\kappa \simeq 12$.

ACKNOWLEDGMENTS

Some of the ideas presented here evolved from conversations with colleagues, in particular, Prof. N. Thompson, Dr. D. Koss, and Dr. S. Ben-Abraham, to whom the author is indebted.

REFERENCES

1. Orowan, E.: *Proc. Phys. Soc.* **52**:8 (1940)
2. Mura, T.: *Phil. Mag.* **8**:843 (1963); also, *Int. J. Engr. Sci.* **1**:371 (1963).
3. Johnston, W. G., and J. J. Gilman: *J. Appl. Phys.* **30**:129 (1959).
4. Koehler, J. S.: *Phys. Rev.* **86**:129 (1959).
5. Gilman, J. J., and W. G. Johnston: "Dislocations and Mechanical Properties of Crystals, p. 116, Fisher et al. (ed.), John Wiley & Sons, Inc., New York, 1957.
6. Johnston, W. G.: *J. Appl. Phys.* **33**:2716 (1962).
7. Hahn, G. T.: *Acta Met.* **10**:727 (1962).
8. Patel, J. R., and A. R. Chaudhuri: *J. Appl. Phys.* **34**:2788 (1963)
9. Gillis, P. P., and J. J. Gilman: *J. Appl. Phys.* **36**:3370 (1965).
10. Gilman, J. J.: *Trans. AIME* **206**:1326 (1956).
11. Taylor, J. W.: *J. Appl. Phys.* **36**:3146 (1965).
12. Dew-Hughes, D.: *IBM J. Res. Develop.* **5**:279 (1961).
13. Peissker, E., P. Haasen, and H. Alexander: *Phil. Mag.* **7**:1279 (1962).
14. Li, J. C. M.: *Acta Met.* **11**:1269 (1963).
15. Akulov, N. S.: *Acta Met.* **12**:1195 (1964).
16. Gilman, J. J.: *J. Appl. Phys.* **36**:2772 (1965).
17. Garafalo, F.: "Fundamentals of Creep and Creep-Rupture in Metals," The Macmillan Company, New York, 1965.
18. Webster, G. A.: *Phil. Mag.* (1966).
19. Gilman, J. J.: *Proc. Int. Conf. on Fracture*, vol. 2, p. 733, Sendai, Japan (1966a).
20. Gilman, J. J.: *Proc. 5th U.S. Nat. Congr. Appl. Mech.*, ASME, New York (1966b).
21. Dey, B. N.: *J. Appl. Phys.* **38**: 4144 (1967).
22. Copley, S. M., and B. H. Kear: *Trans. AIME* **239**: 984 (1967)
23. Gilman, J. J.: *J. Met.* p. 1171 (1966).
24. Vreeland, T.: private communication.
25. Gazis, D. C., R. Herman, and R. B. Potts: *Operations Research* **7**:499 (1959).
26. Prigogine, I.: "Theory of Traffic Flow," p. 158, Elsevier Publishing Company, New York, 1961.
27. Crump, J. C., and F. W. Young: *Bull. Amer. Phys. Soc.* **12**:369 (1967).
28. Mason, W. P.: "Physical Acoustics," vol. IIIB, chap. VI, W. P. Mason (ed.), Academic Press, Inc., New York, 1965.
29. Labusch, R.: Berechnung des Peierlspotentials in Diamantgitter, *Phys. Stat. Sol.* **10**: 645 (1965).
30. Celli, V., M. Kabler, T. Nimomiya, and R. Thomson, Theory of Dislocation Mobility in Semiconductors, *Phys. Rev.* **131**:58 (1963).
31. Zener, C.: A Theory of the Electrical Breakdown of Solid Dielectrics, *Proc. Roy. Soc.* (London) **145A**: 523 (1934).
32. Kane, E. O.: Zener Tunneling in Semi-Conductors, *J. Phys. Chem. Solids* **12**: 181 (1959).
33. Frenkel, J.: The Theory of Electric Breakdown of Dielectrics and Electronic Semiconductors, *Tech. Phys.* USSR **5**:685 (1938).
34. Westbrook, J. H., and J. J. Gilman: An Electromechanical Effect in Semiconductors, *J. Appl. Phys.* **33**:2360 (1962).

LOW-TEMPERATURE DISLOCATION MECHANISMS

John E. Dorn

Inorganic Materials Research Division,
University of California,
Berkeley, California

ABSTRACT

A summary of some of the progress already made in rationalizing the macroscopic plastic behavior of metals and alloys in terms of dislocation theory is presented. Emphasis is on temperatures below those at which diffusion-controlled mechanisms become operative. In this range the effects of temperature and strain rate frequently permit identification of the rate-controlling dislocation mechanisms. A unified approach for characterizing the thermally activated mechanisms is adopted.

    Plastic deformation of crystalline materials depends on the motion of dislocations. When such motion can be stimulated by thermal fluctuations, the flow stress becomes dependent not only on the dislocation substructure but also on the temperature and strain rate. Since a basic understanding of these relationships has clear practical importance, considerable effort has been expended in identifying and characterizing the various responsible dislocation mechanisms.

INTRODUCTION

The purpose of this paper is: (1) to place in perspective the present state of knowledge concerning the plastic behavior of metals and alloys; (2) to point out some of the weaknesses in current theories; and (3) to suggest some interesting areas for future research. To confine this presentation to tractable dimensions, certain important and very interesting topics, such as, mechanisms for strain hardening, and mechanisms for high-temperature, diffusion-controlled flow, etc., have been omitted. Since it will not be possible to recount all details of interest, even for the low-temperature mechanisms emphasized here, condensation of this vast subject will be accomplished by adopting a simple classification of dislocation mechanisms. This will provide a frame of reference within which the details for different mechanisms of the same class involve only variations on the same theme.

CLASSIFICATION OF DISLOCATION MECHANISMS

The yield strength of imperfect crystals is determined by the resolved shear stress that is needed to move glide dislocations across their slip planes. If there were no obstacles present, dislocations would sweep through crystals at infinitesimally low stresses. However, real crystals contain obstacles. It is the nature and distribution of such obstacles that determines the plastic behavior of metals and alloys.

The fact that glide dislocations are line imperfections that move on slip planes of a three-dimensional crystal demands that the obstacles they encounter must perforce also have geometrical characteristics. Consequently, obstacles might be classified as localized, linear, and volumetric as suggested in the first column of Table 1. Typical examples of each major type of obstacle are listed in the third column: (1) Localized obstacles serve to arrest dislocations over limited lengths between which the dislocations bow out under applied stresses. (2) Linear obstacles arrest entire dislocation segments along a line. (3) Volumetric obstacles involve energy-dissipative mechanisms resulting from interactions of stress fields of moving dislocations with various lattice phenomena over the volume of the lattice.

The virtue of the proposed classification extends beyond its geometrical origin: Although each major class of obstacles exhibits somewhat different dislocation mechanisms, individual mechanisms within one class have a common basis. Distinctions between mechanisms in any one class appear as interesting variations of a common theme. Mechanisms must be classified, not only in terms of the geometry of the obstacles that dislocations must bypass or

TABLE 1 Classification of Mechanisms

Types of obstacles	Type of mechanism	Representative examples (incomplete listing)	Reference
1. Localized	Thermally activated cutting $$\dot{\gamma} = \dot{\gamma}_0 \exp \frac{-U}{kT} \{\tau^*, \text{struct.}, T\}$$ $-U$ depends on statistics	Repulsive dislocation trees Solute atom stress fields Tetragonal strain centers Guinier-Preston zones	[1–6] [7–12] [9–12] [13–16]
	Either or both	Radiation damage Coherent precipitates Attractive junctions	[17–20] [15–16] [21–23]
	Athermal $$\tau_G = \tau_{Go} \frac{G}{G_o}$$	Incoherent precipitates Long-range stress fields	[24–26] [27–29]
2. Linear	Thermally activated $$\dot{\gamma} = \dot{\gamma}_0 \exp \frac{-U}{kT} \{\tau^*, T\}$$ U not dependent on statistics	Peierls mechanism Cross-slip Recombination Pseudo-Peierls Mechanism Cottrell-Lomer dissociation	[30–35] [36, 37] [38] [39]
	Either	Fisher unlocking Suzuki unlocking	[40] [41]
	$\tau_A \equiv \tau_A \text{ (order)}$	Short-range order Long-range order	[42, 43] [44]
3. Volumetric	Athermal $\dot{\gamma} = \beta\tau$	Thermoelastic Phonon scattering Phonon viscosity Electron viscosity	[45–48] [49–50] [51–54] [55–56]
		Relativistic	[57]

surmount, but also relative to their response to thermal fluctuations: Thermally activated mechanisms are facilitated by thermal fluctuations in energy, whereas athermal mechanisms are much more highly resistant to the effects of such fluctuations. All mechanisms are athermal at the absolute zero temperature since, here, the probability for any thermal fluctuation is zero. By their very nature as volumetric, energy-dissipative processes for moving dislocations, Class III mechanisms are intrinsically athermal and remain so under all environmental conditions. These mechanisms suggest that the velocity v, of a dislocation is linearly related to the force τb acting on the dislocation according to

$$v = B\tau b$$

where τ is the applied stress, b is Burgers vector and B is the mobility.

Class I and Class II mechanisms can further be grouped into two major subclasses—thermally activatable, and athermal. Thermally activatable mechanisms are those in which thermal fluctuations can assist the applied stress in nucleating the forward motion of a segment of the dislocation. Each unit event in a thermally activated process takes place with a frequency dictated by the Boltzmann expression, namely,

$$\nu' = \nu \exp \frac{-U\{\tau, \text{ structure, } T\}}{kT} \tag{2}$$

where ν is a fundamental frequency of the mechanism in question, U is the additional energy that must be supplied by a thermal fluctuation to cause the dislocation to surmount the obstacle, and kT is the Boltzmann constant times the absolute temperature. The applied stress does work on the dislocation as it surmounts an obstacle, thus the energy U that must be supplied by a thermal fluctuation decreases as the applied stress is increased. The activation energy is always mildly sensitive to T as dictated by the variation of the shear modulus of elasticity and, therefore, U varies with temperature. For the localized obstacle mechanisms, U also depends on substructural details. The frequency of the reverse reaction is negligibly small at low temperatures. This is due to the fact that excessively high thermal fluctuations would be required to move dislocations against the force acting on them due to the applied stress. The plastic shear strain rate $\dot{\gamma}$, for a single, isolated, thermally activated mechanism, is given by the following well-known expression:

$$\dot{\gamma} = NAb\nu' = NAb\nu \exp\left(-\frac{U}{kT}\right) \tag{3}$$

where N is the number of points per unit volume where thermal fluctuations can stimulate nucleation of slip; and A is the average area swept out by the dislocation per successful event. The terms N, A, and ν also depend on the details of the mechanism. Thermally activated mechanisms are characterized by a rapidly decreasing stress with increasing temperature for constant strain-rate tests, and by an increasing stress with increasing rate for constant temperature tests.

Athermal mechanisms of Class I and Class II fall into two groups. One group is inherently athermal since the energy for nucleating forward slip never reaches a maximum value. This occurs in short-range order hardening. As a segment of a dislocation in a short-range ordered alloy bows out the energy continuously increases. This is due principally to the disordering that is induced across the slip plane. Therefore, for this type of mechanism, deformation must be induced exclusively mechanically by a sufficiently high stress to cause disordering. Here the yield stress is insensitive to the strain rate and decreases with increasing temperature in proportion to the product of the degree of order and the ordering energy. At high temperatures, however, fluctuations in short-range order take place by diffusional processes which permit local advances of the dislocation and thus lead to thermally activated creep. This subject is beyond the intended scope of the present report.

A second group of mechanisms is environmentally athermal. Typical examples are the surmounting of long-range stress fields and the breaking of attractive junctions. These mechanisms are inherently thermally activatable since the energy to surmount such obstacles has a maximum value. However, at low temperatures and stresses, thermal fluctuations of sufficient energy are so infrequent that they are ineffectual in assisting the nucleation process. Consequently, deformation here must be induced almost exclusively by mechanical means. Hence, the yield stress is insensitive to the strain rate and decreases very modestly with increasing temperature in proportion to the shear modulus of elasticity. The breaking of attractive junctions, etc., are, however, termally activatable at higher stresses and temperatures.

Some mechanisms are thermally activatable under some conditions and athermal under others. For example, dilute and weak Cottrell-atmosphere, solute-atom, locked dislocations are thermally activatable; but more concentrated and stronger Cottrell-atmosphere solute-atom locked dislocations are athermally activated and in many instances the dislocations are so tightly bound that they cannot be torn away from their atmosphere even at very high stresses. Suzuki locking constitutes a second example. Weak Suzuki locking of partials having a narrow stacking fault ribbon are thermally activatable but under otherwise similar conditions; in alloys that have low stacking fault energies, the unlocking mechanism is athermal. The transition of Cottrell locking from a thermally activatable to an athermal mechanism is due primarily to the height of the activation energy barrier. However, this transition for Suzuki locking arises from the fact that high thermal fluctuations in energy can only occur over small volumes of the crystal.

An example of some of the types of phenomena that are observed is illustrated by the experimental data summarized in Fig. 1a.[58] Over the low temperature range (l_T) a thermally activated mechanism, namely intersection of dislocations, is observed. Above T_c an a thermal mechanism that is insensitive to temperature and strain rate is operative. Although the details of the athermal mechanism are still under discussion, it must consist largely of bowing out of dislocations from entanglements,[59,60] breaking of attractive junctions,[21-23] and perhaps surmounting some long-range stress fields.

The stress at absolute zero for the thermally activated range represents that required to surmount obstacles without help from thermal fluctuations. At higher stresses, Class III types of of viscous athermal mechanisms (Fig. 1b), are the only ones that remain effective for determining the velocity of dislocations and the strain rate.[61]

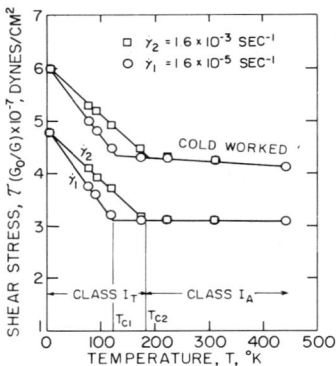

Fig. 1a Relationship between stress, strain-rate and temperature for aluminum single crystals.

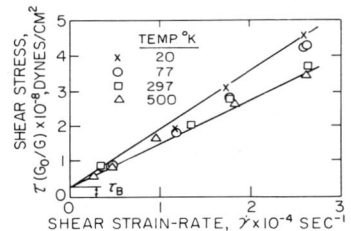

Fig. 1b Shear stress versus shear strain-rate for dynamic shear in aluminum single crystals.

The major concern of investigations on dislocation mechanisms is: (1) to provide a basic understanding of the varied plastic behavior metals and alloys undertake; (2) to formulate plastic behavior so that these concepts may be utilized in engineering applications; and (3) to provide a basis for development of new and superior alloys having special desirable properties. The achievement of even the first objective is a major undertaking primarily because a number of mechanisms of each class of Table 1 is always simultaneously operative. Furthermore, in many instances it is not always clear how to treat the effects of several mechanisms satisfactorily in a unified way. Only the most modest introduction[62] of statistics for simple cases of single mechanisms has been attempted in spite of the fact that the need for statistical approaches is so obvious. On the basis of the inherent complexity of the whole problem, it is indeed remarkable that any progress could have been made. Fortunately, for certain restrictive conditions and over limited ranges of conditions, one identifiable mechanism often seems to predominate. Before proceeding to a discussion of these phenomena, we must first describe and characterize prototype examples of each of the two classes of thermally activated mechanisms and some of their interesting variants.

CUTTING LOCALIZED OBSTACLES

A prototype example for cutting localized obstacles is given by Seeger's approximation for the intersection mechanism.[5] Statistics are neglected and the obstacles are assumed to form a regular "square" array (subscript "s") of l_s on a side as shown in Fig. 2a. Under a resolved shear stress τ_s^* the dislocations contact obstacles and arc out between them as given by $\tau_s^* = \Gamma/bR$ where Γ is the average line energy and R is the radius of curvature. Consequently, a force F_s acts on the obstacle, which is assumed to be rigid. For the present, the force–displacement diagram for cutting Fig. 2b will be assumed, where the strength of the obstacle is given in terms of the line energy Γ, times a strength factor α, namely, $F_{sc} = \alpha\Gamma$. If $F_s \geq \alpha\Gamma$ the obstacle will be cut athermally; but when $F_s \leq \alpha\Gamma$, cutting will take place only with the aid of a thermal fluctuation in energy of magnitude $U = \alpha\Gamma D - \tau_s^* b l_s D$ or greater. Then by Eq. (3)

$$\dot{\gamma} = \frac{\rho}{l_s} l_s^2 b \left(\frac{\nu_0 b}{l_s}\right) \exp \frac{-(\alpha\Gamma D - \tau_s^* b l_s D)}{kT} \tag{4}$$

$$\tau_s^* = \frac{\Gamma}{bR}$$

$$F_s = 2\,\Gamma \sin \phi_s = \frac{2\Gamma \ell_s}{2R} = \tau_s^* \, b\ell_s$$

(a)

$$U = \alpha \Gamma D - \tau_s^* \, b \ell_s D$$

(b)

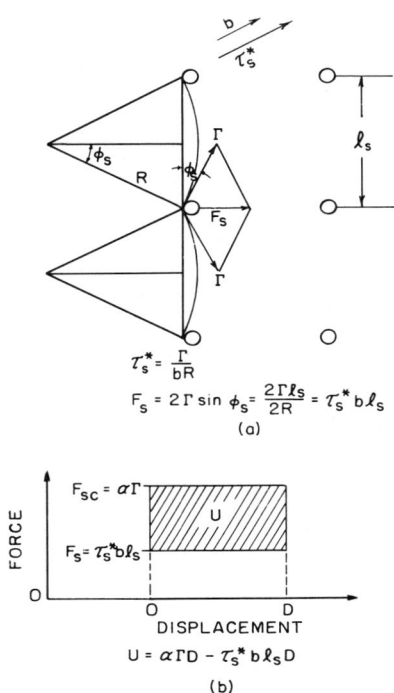

Fig. 2 Prototype for cutting.

where $N \simeq \rho/l_s$, ρ being the density of glide dislocations, $A \simeq l_s^2$ and $\nu \simeq \nu_0 b/l_s$ where ν_0 is the Debye frequency. If, following Seeger's approach, it be assumed that the athermal behavior arises exclusively from long-range stress fields of dislocations, a very simple analysis is obtained: Let, for example, the long-range stress fields exhibit a maximum average amplitude τ_G over a fraction f of the slip plane. Then the effective stress promoting thermal fluctuations is given by $\tau_s^* = \tau_s - \tau_G$ where τ_s is the externally applied stress and the dislocations will glide freely over the fraction $(1 - f)$ of the slip plane. On this basis Eq. (4) reduces to

$$\tau_s = \tau_G + \frac{\alpha \Gamma D}{bl_s D} - \frac{kT}{bl_s D} \ln \frac{\rho b^2 \nu_0}{f\dot{\gamma}}$$

$$0 \leq T \leq T_c \qquad (5a)$$

$$\tau_s = \tau_G \qquad T_c \leq T \qquad (5b)$$

where T_c defined by

$$\alpha \Gamma D = kT_c \ln \frac{\rho b^2 \nu_0}{f\dot{\gamma}} \qquad (5c)$$

is the critical temperature above which thermal fluctuations in energy greater than $\alpha \Gamma D$ occur as often as required to maintain the imposed strain rate even when $\tau^* \simeq 0$.

For this approximation, the mechanical behavior for cutting a simple type of localized obstacle might be interpreted as shown in Fig. 3. The stress to induce flow at the absolute zero is given by $\tau_0 = \tau_{G_0} + \alpha \Gamma_0/bl_s$ and increases with τ_{G_0} and the reciprocal of the mean spacing of obstacles, namely, $1/l_s$. The effects of $\dot{\gamma}$ and T on τ (Fig. 3a) appear to agree, at least qualitatively, with the experimental data recorded in Fig. 1.

On the basis of this model, higher-strength alloys may be made as follows: (1) by increasing τ_G which uniformly elevates the T vs τ curve, (2) by increasing the density of obstacles, i.e., decreasing l_s (Fig. 3b). If only this is done strengthening will be limited to temperatures below T_c; (3) by introducing stronger obstacles, i.e., increasing α (Fig. 3c). For this case the strengthening is extended to higher temperatures since T_c is linearly related to α. However, when α is increased above 2, dislocations will bow through obstacles athermally by the Orowan process[24] at about the stress $\tau_B = \Gamma/bR = 2\Gamma/bl_s$ thus limiting the maximum achievable stress at low temperatures (see Fig. 3c).

VARIATIONS FROM THE IDEALIZED PROTOTYPE

Equations (5) rarely depict the experimentally observed trends with satisfactory accuracy. Four factors serve to contribute to deviations from the idealized case: (A) usually, the force-displacement diagram differs from the simple case assumed in Fig. 2b; (B) obstacles never form a regular array as assumed in Fig. 2a. Occasionally they are clustered as, for example, when forest dislocations in entanglements are cut. Generally, they are more or less randomly distributed as in the case of tetragonal strain centers or Guinier-Preston zones; (C) almost always several kinds of obstacles are present at one time. Even in the simple example of cutting forest dislocations in single crystals of pure f.c.c. and b.c.c. metals, repulsive trees and attractive junctions must be considered simultaneously; (D) the interactions of the same obstacles with dislocations is often greatly dependent on orientation, size and morphology of the

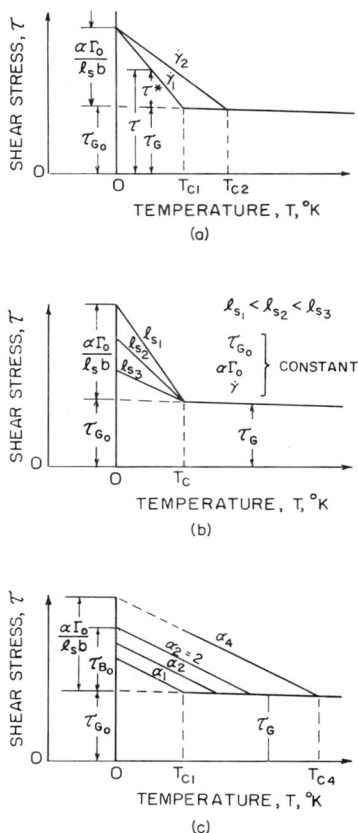

Fig. 3 Simple cutting. (a) Effect of $\dot{\gamma}$. (b) Effect of l_s with no effect of τ_G. (c) Increasing obstacle strengths.

obstacle, and also on whether the dislocation is in screw or edge orientation as is the case for cutting through stress fields due to substantial alloy strengthening. Although these complexities do not change the general format of the approach to the cutting mechanisms given in the preceding section, they modify, and often drastically so, the expected trends.

A   Force-Displacement Diagrams for Cutting Localized Obstacles

The force-displacement diagram (Fig. 2b) for cutting localized obstacles is approximated only for the simple case of intersection of undissociated basal-glide dislocations in h.c.p. systems with unreactive forest dislocations. For all other types of obstacles more complicated diagrams are obtained. The case of cutting non-interacting undissociated forest dislocations by dissociated glide dislocations will serve as an example: Before a pair of jogs can be produced, the dissociated glide dislocation must first be constricted (Fig. 4a). Although the details of how the constriction and

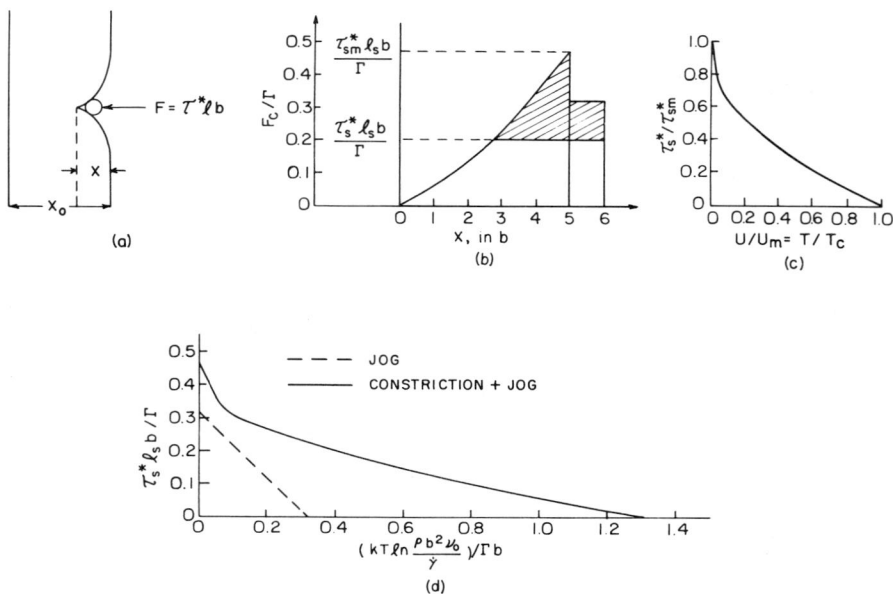

Fig. 4 Cutting of nonreactive forest dislocations by dissociated glide dislocations (a) Partially formed constriction. (b) Force-displacement diagram. (c) Stress versus activation energy. (d) Effect of force-displacement diagram on shape of $\tau^* T$ curve.

jogging sections of the force–displacement diagram might be merged is now well known, an approximate representation given in Fig. 4b cannot be seriously in error. A crude line–energy model for the force-displacement diagram for constriction of modestly dissociated dislocations   suggests that

$$F_c \simeq 2/3\Gamma \sqrt{1 - \left[1 - 0.18\left(\frac{-x}{x_0} + \ln\frac{x_0}{x_0 - x}\right)\right]^2} \qquad (x_0 - x) \geq b \quad (6a)$$

$$F_c \simeq 2/3\Gamma \sqrt{1 - \left[1 - 0.18\left(\frac{-x}{x_0} + 1 - \frac{x_0 - x}{b} + \ln\frac{x_0}{b}\right)\right]^2} \qquad (x_0 - x) \leq b$$

$$(6b)$$

The force to produce two jogs is estimated to be

$$F_{2j} = \frac{\Gamma}{\pi} \tag{6c}$$

Consequently, the force displacement diagram for glide dislocations having an equilibrium dissociation of $x_0 = 5b$ for cutting a repulsive forest dislocation is shown in Fig. 4b. Since the total energy U_m is the total area under the curve, the energy U that must be supplied by a thermal fluctuation to cause cutting when a stress τ_s^* is applied, is denoted by the cross-hatched area. Numerical integration of Fig. 4b gives Fig. 4c which illustrates that for these cases free energy of activation no longer decreases linearly with the stress τ_s^* as was the case for the simple prototype. The strain rate is given by

$$\dot{\gamma} = \rho b^2 \nu_0 \exp -\frac{U}{kT} \tag{7}$$

which reveals that U increases linearly with T for given values of $\dot{\gamma}$ and ρ. Since $U = 0$ and τ_s^* has its maximum values τ_{sm}^* at the absolute zero of temperature, U has its maximum value of U_m, where $\tau_s^* = 0$, at a critical temperature T_c defined by

$$\dot{\gamma} = \rho b^2 \nu_0 \exp -\frac{U_m}{kT_c} \tag{8}$$

T_c increases linearly with U_m and logarithically with $\dot{\gamma}/\rho$. For tests at constant values of $\dot{\gamma}$ and ρ,

$$\frac{U}{U_m} \simeq \frac{T}{T_c} \tag{9}$$

Consequently, Fig. 4c represents a normalized τ_s^* vs. T curve for cutting a regular square distribution of localized obstacles for constant values of $\dot{\gamma}$ and ρ. The value of U_m can be determined from the experimental values of T_c for two strain rates according to

$$\frac{\dot{\gamma}_1}{\dot{\gamma}_2} = \frac{\exp\left(-U_m\{T_{c1}\}/kT_{c1}\right)}{\exp\left(-U_m\{T_{c2}\}/kT_{c2}\right)} \tag{10}$$

where the bracketed terms designate that U_m should be corrected for the differences in the shear modulus of elasticity with temperature. The effects of modification of the force–displacement diagram for cutting obstacles on the τ^* vs. T relation is illustrated in Fig. 4c. Estimates of the force–displacement diagrams for cutting a wide variety of localized obstacles are now available.

B   Randomly Distributed Localized Obstacles

Localized obstacles never present themselves in regular square arrays. Occasionally they are clustered as in the case of dislocation intersection when entanglements develop; nevertheless, these obstacles are somewhat randomly distributed in the entanglements. In other cases, e.g., in the presence of tetragonal strain centers, the dispersion of localized obstacles approaches a random distribution. If a square array of obstacles gave trends that closely agreed with the more realistic random distribution, the issue would be unimportant. However, enough progress has been made on statistical treatments of the problem to suggest that the differences are not always trivial.

As the applied stress is increased, dislocations bow to smaller radii of curvature causing the average link length $\bar{l}$, between the obstacles to decrease. Friedel[62] estimated the effect of τ^* on $\bar{l}$ by assuming steady–state cutting, such that for each obstacle that was cut the average area l_s^2, Fig. 5a, was swept out. For weak obstacles, $\bar{l}$ decreases with increasing τ^* according to

$$\bar{l} = R\,2^{1/3}\left(\frac{l_s}{R}\right)^{2/3} = \left(\frac{2\Gamma l_s^2}{\tau^* b}\right)^{1/3} \tag{11}$$

The athermal yield stress for the Friedel statistical model at the absolute zero, given by $\tau^*\bar{l}b = \alpha\Gamma$, is compared with that predicted for a square array in Fig. 5b. The points obtained from the computerized experiments of Foreman and Makin[64] on cutting randomly distributed obstacles are shown on the same graph. The two approaches lead to similar results which reveal that the athermal

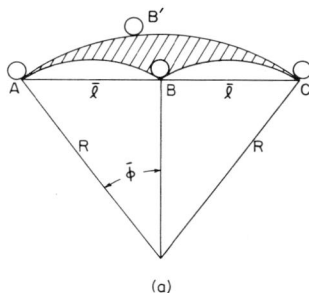

(a)

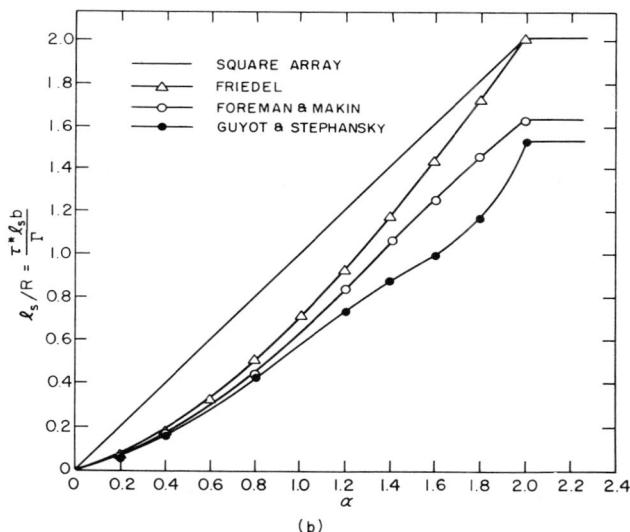

(b)

Fig. 5 Effect of random distribution of localized obstacles. (a) Area swept out per cutting. (b) Athermal yield stress.

flow stress for a random distribution of obstacles is substantially lower than that predicted for a square array.

The distribution of obstacles also has a pronounced influence on the τ^* vs. T relationship deduced for the thermally activated cutting mechanism. Both the average number of obstacles contacted by the dislocation $N = \rho/\bar{l}$ and the average frequency of vibration $\nu = \nu_0 b/l$, are functions of the stress for the random distribution. For weak obstacles the shear strain rate becomes

$$\dot{\gamma} = \left(\frac{\rho}{l}\right) l_s^2 \left(\frac{\nu_0 b}{l}\right) \exp\frac{-U}{kT} = \frac{\rho\nu_0 b^2}{2^{2/3}} \left(\frac{\tau^* b l_s}{\Gamma}\right)^{2/3} \exp\frac{-U}{kT} \qquad (12)$$

where the average area swept out per activation is taken to be l_s^2 and where the activation energy (assuming the force–displacement diagram of Fig. 2b) is given by

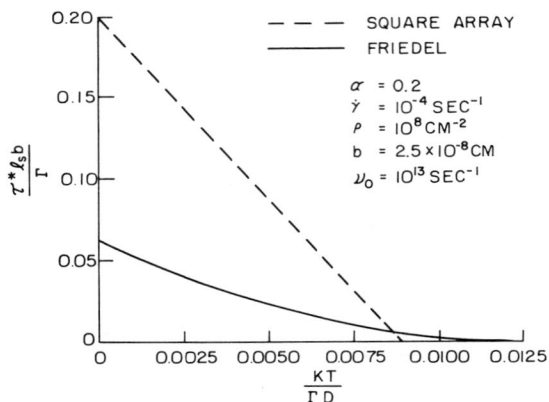

Fig. 6 Comparison of predictions based on Friedel's model with those for a square array.

$$U = \alpha \Gamma D - \tau^* \bar{l} b D = \Gamma D \left\{ \alpha - 2^{1/3} \left(\frac{\tau^* b l_s}{\Gamma} \right)^{2/3} \right\} \tag{13}$$

The bracketed stress term in the preexponential expression of Eq. (12) demands that for any finite value of the strain rate $\dot{\gamma}$, the stress τ^* can never be zero. Therefore, in contrast to the square array model, T_c is infinite. A comparison (see Fig. 6) of the τ^* vs. T relationship predicted by Friedel's model with that deduced for a square array of obstacles, reveals the importance that must be ascribed to a consideration of the statistics of the problem: Randomly dispersed obstacles give much lower stresses over the lower temperature range than those obtained from square arrays. Over the higher temperature range, however, the stress τ^*, for the random distribution of obstacles lies above that predicted for a square array model and decreases very slowly with increasing temperature.

Recently, a detailed statistical approach to the problem of cutting randomly distributed obstacles was presented by Kocks.[65] His method of approach has been adopted by Guyot and Stephansky.[66] They consider an obstacle at 0 of Fig. 7 contacted by an

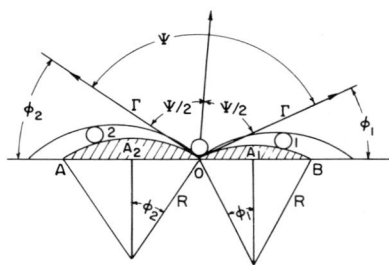

$$A_1 = R^2 (\phi_1 - \sin \phi_1 \cos \phi_1) \tag{14}$$

$$dA_1 = 2R^2 \sin^2 \phi_1 \, d\phi_1 \tag{15}$$

$$F = 2 \Gamma \cos \frac{\psi}{2} = 2 \Gamma \sin(\frac{\phi_1 + \phi_2}{2}) \tag{16}$$

$$F_c = \alpha \Gamma = 2 \Gamma \cos \frac{\psi_c}{2} = 2 \Gamma \sin \phi_c \tag{17}$$

Fig. 7 Force on an obstacle.

originally straight dislocation AB. Under a stress τ^*, R makes an angle ϕ with its original configuration provided there are no obstacles in area A_1 and some in dA_1. The force on the central obstacle is given by Eq. (16) and reaches the cutting force at a critical value of ϕ_c, given by Eq. (17), where ϕ_c is the complement of $\psi_c/2$. The important feature is that the force on the obstacle depends neither directly on the length l_s, as assumed in the square array model, nor on the average link length l, as assumed by Friedel, but rather on the angles ϕ_1 and ϕ_2. Therefore the probability that an obstacle will be cut is equal to the probability that there are no obstacles in A_1 and A_2 or any larger area when $(\phi_1 + \phi_2)/2$ exceeds ϕ_c. The probability that there are no obstacles in A_1 and A_2 is given by the well-known expression $P = \exp\{-(A_1 + A_2)/l_s^2\}$. The probability for cutting an obstacle P_c, deduced from this approach is given in Fig. 8 as a function of $l_s/R = \tau^* b l_s/\Gamma$. Obstacles for which $\alpha > 2$, are not cut but may be by-passed by the Orowan mechanism.

Although it was assumed that a single cutting would permit the dislocation to move only over the average area l_s^2 in the square array and in Friedel's models, it has been demonstrated by Kocks and by the computerized experiments of Foreman and Makin, that a

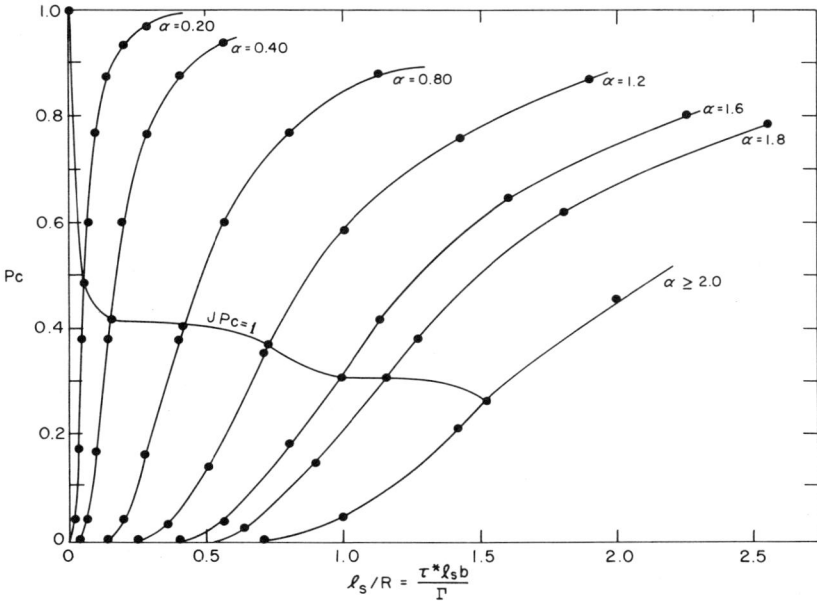

Fig. 8 Probability for cutting and condition for macroscopic yielding.

larger area is swept out. This arises because once a cutting has been achieved there is a certain probability the dislocation can unzip past the next neighbors, etc. An approximate calculation suggests that the area swept out per cutting is given by

$$A_i = \frac{i l_s^2}{i - j P_c} \tag{18}$$

where i equals one plus the number of obstacles that were cut by unzipping, and j is the number of new obstacles contacted by an advance of the dislocation segment thus released. Since j is greater than unity, A_i becomes infinite at some critical stress. This stress is conveniently defined as the athermal yield stress, on which, as detailed analyses have shown, i and j depend. The calculated values of $j P_c$, where yielding takes place, are also recorded in Fig. 8. These data provide the basis for determining the athermal yield stress. The results are in good agreement with the Foreman-Makins curve (Fig. 5), up to $\alpha \simeq 1.2$ and fall somewhat below this curve for still higher values of α.

These results emphasize the need for more complete and accurate statistical treatments for cutting localized obstacles.

C   Effects of Additional Factors

The general problem of plastic deformation arising from the cutting of localized obstacles is complicated by the intrusion of several factors. One concerns the fact that almost invariably several kinds of obstacles having quite different force-displacement diagrams are present simultaneously.

The original concept on intersection in single crystals of pure f.c.c. metals suggested that their plastic behavior could be ascribed to the cutting of repulsive trees imbedded in a long-range internal stress field. Under these circumstances Seeger's superposition principle that $\tau = \tau^* + \tau_G$ is applicable; but only τ can be measured directly and often it is difficult to separate accurately τ^* from τ_G. Furthermore, when Seeger's superposition principle is applied, special assumptions are required in order to account for the observed constancy of the Cottrell-Stokes[67] ratio over Stages II and III.

An alternate suggestion has been made that the crystal cannot support long-range stress fields due to relaxation of stresses by motion of dislocations on secondary slip planes.[68] Therefore, it has been proposed that intersection involves mainly the cutting of repulsive trees and attractive junctions, the long-range stress fields being negligibly small and the attractive junction cutting being largely responsible for the apparent athermal behavior. It is obvious that in this event Seeger's superposition principle cannot rigorously be applied. The yield stress over the higher temperature range will still be influenced by the presence of some repulsive trees located near the attractive junctions and will thus affect the apparent athermal stress level. Furthermore, such an apparent athermal stress level τ_A, cannot be extrapolated into the lower temperature range to give a meaningful $\tau^* = \tau - \tau_A$ because if the weaker repulsive trees are not instantly penetrable, most of the stronger attractive junctions must still remain unbroken. In fact, in this model, the motion of glide dislocations must proceed first by thermally activated cutting some of the repulsive trees which then releases sufficiently long dislocation segments to facilitate activated cutting of attractive junctions. A simple nonstatistical model for this mechanism reveals that the predicted trends of τ vs. T for the repulsive tree—attractive junction model do not differ greatly from those for the repulsive tree—long-range stress field model. Furthermore, the Cottrell-Stokes ratio is inherent to the model. On the other hand, the interpretation of the data are uniquely different for each of the two possibilities.

Because entanglements form which lower the energy of the crystal, it follows that thermally activated cutting of repulsive trees as well as attractive junctions, probably involves motion of the dislocation against the attractive stress field of the entire entanglement. This effect introduces an athermal long-range stress component that warrants additional detailed consideration.

Rather good progress has been made in qualitatively rationalizing some of the effects of solute atoms,[7-12] tetragonal defects,[9-12] and precipitates and dispersed phases[15, 16, 24-26] on the yield strengths of alloys. Since there has been no recent major advance in these areas, their details will not be reviewed here. It is appropriate, however, to reconsider the validity of some of the simplifications that have been made to facilitate analyses. Although some improvement in the accuracy of the force-displacement diagrams for cutting stress fields due to isolated strain centers is needed, the major issues seem to be concerned more with the statistical features of the problem. The stress fields due to lattice strain centers decrease very rapidly in height over larger areas of the slip plane as their distance from the slip plane increases. Therefore, it has been customary to neglect the effects of all strain centers lying a greater distance away than one atomic plane on either side of the slip plane. On this basis, $l_s \simeq b/\sqrt{2c}$, where c is the concentration of strain centers. In a number of examples[11] this simplification is confirmed experimentally relative to the variation of τ^* with $\sqrt{c}$. At higher temperatures, and, therefore, correspondingly lower values of τ^*, however, the weaker and broader stress fields of more distant strain centers must be felt by the dislocations. The very difficult statistics of this problem have been almost completely neglected. Superposition of stress fields due to clusters of strain centers near the slip plane can give rise to local regions with quite different force-displacement than those of the isolated centers. Clusters more remotely displaced from the slip plane, can contribute to long-range stress field interactions with dislocations. In addition to these is the further need for considering the effects of dislocations-localized obstacle interactions on the preexponential expression in the strain-rate equation. In terms of preceding remarks, it is quite understandable why the radiation damage problem[17-20] is so difficult to cope with in detail.

In general, the presence of repulsive trees and attractive junctions have been neglected in problems of alloy strengthening. For cases of strong tetragonal strain centers where $c \geq 10^{-7}$, this omission is justified over the lower temperature range of the thermally activated region since the distance l_s between the repulsive trees and the attractive junctions is much greater than that for the strain centers. Nevertheless, the presence of attractive junctions should have significant effects on the yield stress at higher temperatures.

The problem is somewhat more complicated for isotropic strain centers (e.g., substitutional alloy strengthening) since they interact appreciably only with the edge components of dislocations. Thus the motion of screw components is resisted effectively only by forest dislocations, whereas the edge components are held up principally by the stress fields of the isotropic strain centers. As yet, no theory has been formulated that considers these factors. Furthermore, alloying often alters the stacking fault energy and thus modifies the force-displacement diagrams for cutting trees. In addition, tendencies toward short-range ordering or clustering influence the athermal stress leve. These individual factors are not easily isolated experimentally. There is indeed need for a comprehensive theory that takes these significant factors pertaining to substitutional solid solution strengthening into detailed consideration.

LINEAR OBSTACLES

A Prototype Model

Here Peierls' mechanism will serve as the prototype model for thermal activation of dislocation motion past linear obstacles. A rather complete review of this mechanism[35] has been given recently and therefore only major features are presented here as the basis for discussion.

The energy of a dislocation line is least when it lies parallel to rows of atoms on the slip plane (Fig. 9a). As it moves from one valley to the next, the core energy and hence the line energy change periodically as shown. Peierls' stress τ_p is that which is required to push the dislocation mechanically past the steepest part of the hill whereupon it will advance from one valley to the next, etc. Thus τ_p is the yield stress at the absolute zero. Under a stress $\tau^* < \tau_p$, the dislocation will move part way up the hill to $A_1 A_2$. At the absolute zero no further motion can take place, but at higher temperatures, thermal fluctuations in energy can push the dislocation locally (e.g., B of Fig. 9a) toward the next valley. The fluctuation in energy needed to nucleate a pair of kinks is due to the increased energy needed to nucleate a pair of kinks is due to the increased energy of the now longer dislocation line less the work supplied by the stress τ^* in sweeping out the additional area. For a sufficiently vigorous fluctuation (e.g., D of Fig. 9a), the energy reaches a maximum value. All fluctuations in energy equal to or greater than this value will nucleate a pair of kinks. These then move apart under the action of the applied stress, thereby advancing the dislocation segment of length l, a distance a to the next valley in the slip plane. The energy required to nucleate a

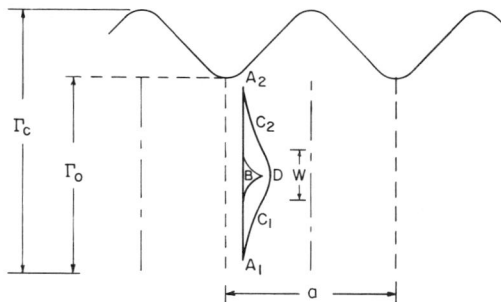

Fig. 9a Peierls mechanism. Nucleation of kinks.

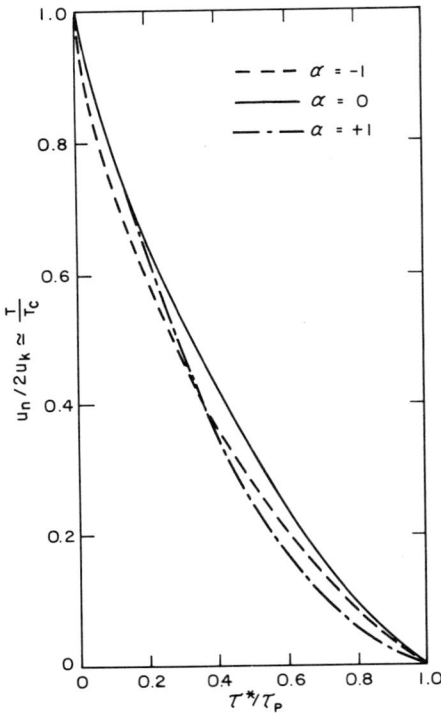

Fig. 9b Peierls mechanism. Activation energy.

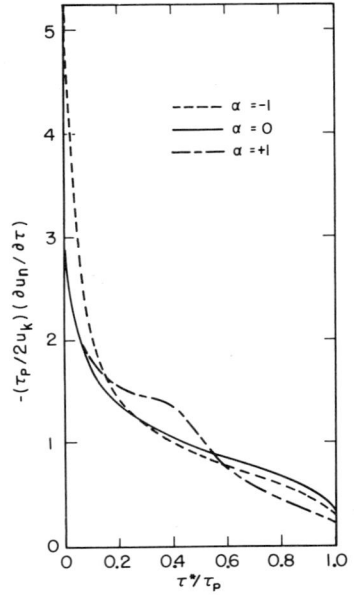

Fig. 9c Peierls mechanism. Activation volume.

pair of kinks U_n divided by twice the energy of a single complete kink U_k in this way, is a simple function of τ^*/τ_p as shown in Fig. 9b. A simple line energy calculation shows that the kink energy is given by

$$\frac{2\pi U_K}{a\Gamma_0} = 5.7 \left(\frac{\tau_p ab}{\Pi\Gamma_0}\right)^{1/2} \tag{19}$$

illustrating that the line energy of a dislocation can be deduced from experimental determination of Peierls' stress and the kink energy.

The strain rate is readily formulated as

$$\dot{\gamma} \simeq \left(\frac{\rho}{l}\right)(la)\, b\left(\frac{\nu_0 b}{w}\right)\left(\frac{l}{2w}\right)\exp\frac{-U_n}{kT} = \frac{\rho lab^2\nu_0}{2w^2}\exp\frac{-U_n}{kT} \tag{20}$$

where (ρ/l) are the number of segments of dislocations that can be activated, (la) is the area swept out, $(\nu_0 b/w)$ is the frequency of vibration of a critical loop (Fig. 9a), and $(l/2w)$ are the number of segments along l at which nucleation can take place. The quantity w has been shown to be almost independent of τ^* and can be approximated by

$$w \simeq \frac{\pi}{2}\left(\frac{2a\Gamma_0}{b\tau_p}\right)^{1/2} \tag{21}$$

As T increases, U_n increases and reaches a value of $2U_k$ at T_c where $\tau^*/\tau_p = 0$. Thus

$$\dot{\gamma} = \frac{\rho l a b^2 \nu_0}{2w^2} \exp \frac{-2U_k}{kT_c} \tag{22}$$

Consequently for a given strain rate

$$\frac{U_n}{2U_k} = \frac{T}{T_c}\frac{G_c}{G} \tag{23}$$

where G_c/G corrects for the change in the shear modulus of elasticity with temperature. The kink energy can be deduced by determining T_c for two different strain rates, namely,

$$\frac{\dot{\gamma}_1}{\dot{\gamma}_2} = \frac{\exp\left(-2U_k/kT_{c1}\right)}{\exp\left(-2U_k/kT_{c2}\right)} \tag{24}$$

The experimentally determined activation volume is

$$v_a = kT\left\{\frac{\partial \ln \dot{\gamma}}{\partial \tau^*}\right\}_T = kT\left\{\frac{\partial \ln(\rho l)}{\partial \tau^*}\right\}_T - \left\{\frac{\partial U_n}{\partial \tau^*}\right\}_T \tag{25}$$

where the theoretical activation volume (Fig. 9c) is given by

$$-\left(\frac{\partial U_n}{\partial \tau^*}\right) = v^* = \frac{-2U_k\,\partial\,(U_n/2U_k)}{\tau_p\,\partial\,(\tau^*/\tau_p)} \tag{26}$$

Thus far, all investigations that seem to confirm Peierls' mechanism have shown that $v_a \simeq v^*$.

A typical example of the confirmation of Peierls' mechanism is given in Fig. 10 where it has been assumed that $\tau = \tau^* + \tau_A$. The experimental data are shown as points and the solid lines represent

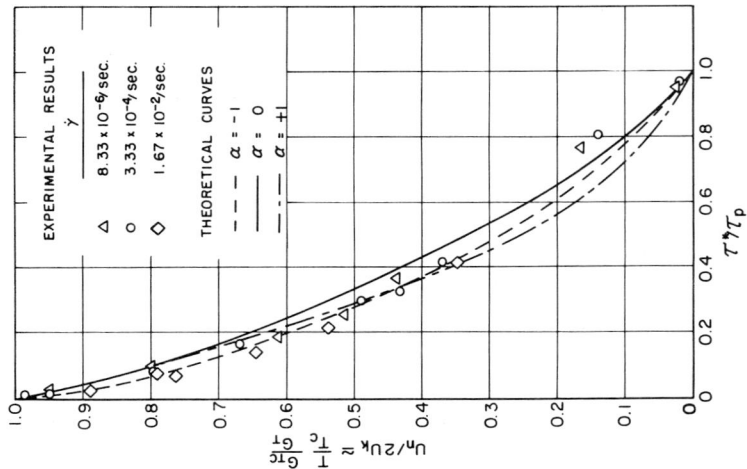

Fig. 10b The thermally activated component of the flow stress versus temperature in dimensionless units.

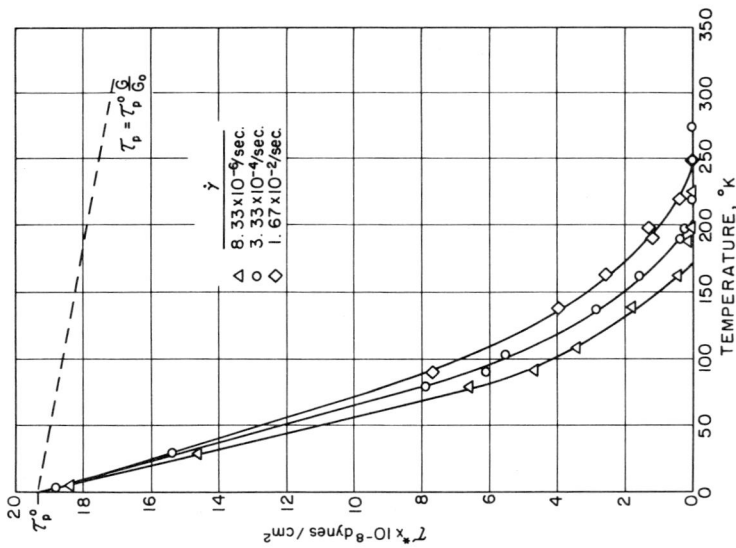

Fig. 10a The thermally activated component of the flow stress versus temperature for different strain rates.

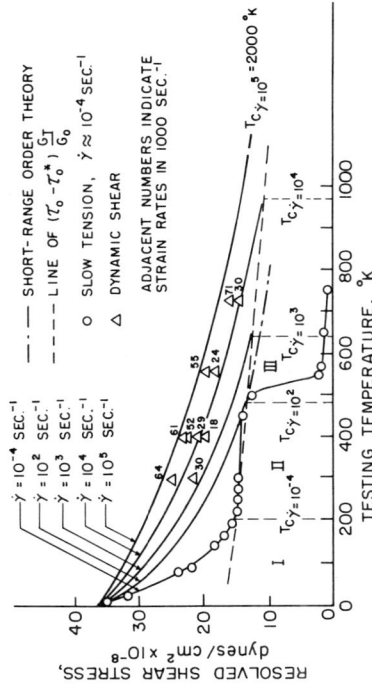

Fig. 10d Resolved shear stress versus temperature for prismatic slip in Ag$_2$-Al.

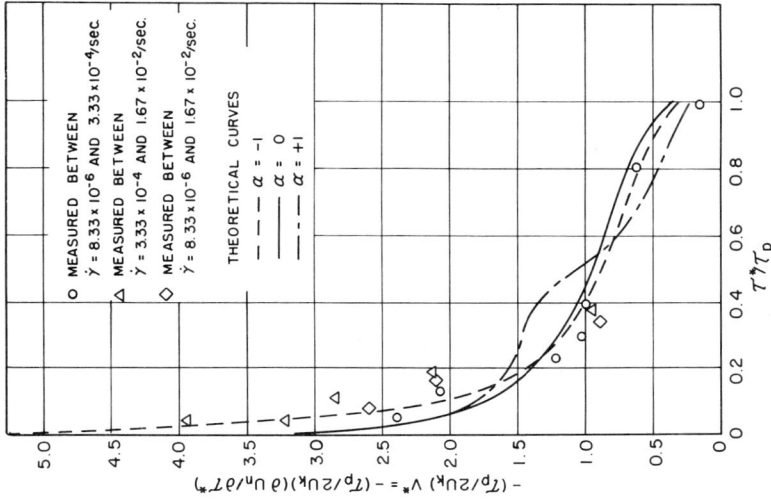

Fig. 10c The thermally activated component of the flow stress versus the activation volume in dimensionless units.

the theoretical curves. Whereas τ_p was deduced from the value of τ^* at $0°K$, $2U_k$ was obtained from T_{c1} and T_{c2}. The line energy deduced from Eq. (19) is $\Gamma_0 \simeq (2.5)\,Gb/2$, i.e., only slightly higher than the crude theoretical estimate. In addition, other details for the model seem to be satisfied quite well.

B Variants of the Prototype Linear Obstacle Model

Variants of the linear obstacle model differ from each other with respect to two features: (1) the effect of stress on the activation energy; and (2) the effect of stress on the preexponential term in the strain rate expression. For example, Fig. 11 shows Friedel's[37] well-formulated mechanism of slip resulting from cross slip of dissociated dislocations on the basal plane to perfect glide dislocations on the prism plane. On the basis of this model

$$\dot\gamma = \frac{NL_s Ab^4 \tau^{*2} \nu_0}{8\Gamma R_e} \exp\frac{-U_c}{kT} \exp\frac{-\left[2^{3/2}(\Gamma R_e^{\,3})^{1/2}\right]}{\tau^* bkT} \tag{27}$$

where N is the number of screw segments of dislocations on the basal plane per unit volume each having a mean length L_s, A is the area swept out per activation, and

$$R_e \simeq \frac{\Gamma}{16} \ln \frac{x_0}{b} \tag{28a}$$

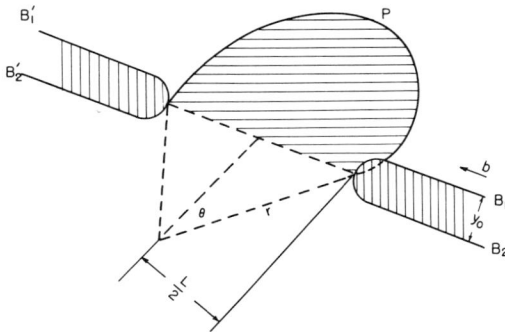

Fig. 11 Nucleation of cross slip as a result of re-combination of the partials B_1 and B_2 or the basal plane along length L and bowing out of the recombined section P on the prism plane.

$$U_c \simeq \frac{\Gamma}{15} x_0 \left(\ln \frac{x_0}{b} \right)^{1/2} \tag{28b}$$

where R_e is the recombination energy per unit length and U_c is the constriction energy for an equilibrium separation x_0, of the partial dislocations on the basal planes. This relation is uniquely different from that for cutting localized obstacles and also that for the Peierls mechanism. Friedel's theory infers, in agreement with several experimental results, that when the resolved shear stress on the basal plane is zero, prismatic slip can only take place at rather high temperatures. As the temperature decreased, the yield stress for prismatic slip increases rapidly and becomes infinitely high even somewhat above the absolute zero. However, the values of U_c and R_e can be reduced by the effects of shear stresses acting on blocked screw dislocations on the basal plane. This illustrates why prismatic slip may be observed in poly-crystalline aggregates even at quite low temperatures.

C   General Comparisons

The above-mentioned formulation of Friedel's cross-slip mechanism has validity only when the two partials of the dissociated dislocation are distinctly separated. Equations (28a) and (28b) are no longer appropriate for cases where the stacking fault energy is so high that $x_0 \simeq 0.5$ to $2b$. In this event it seems more appropriate to imagine that the core energy of a screw dislocation is decreased slightly by this minor separation of its partials rather than as-suming that a true stacking fault is present. This strongly implies that before such a mildly dissociated screw dislocation can be moved on the prism of a h.c.p. metal, its core energy must be in-creased somewhat. Such a screw dislocation will then exhibit a periodic variation in energy as it is displaced from row to row of atoms in the prismatic plane. In this case the plastic behavior for prismatic slip can be expected to agree reasonably well with that for the Peierls mechanism where the potential energy hill may not be too dissimilar from the quasi-parabolic hill already analyzed by Guyot and Dorn.[35]

The above-described pseudo-Peierls mechanism provides an acceptable interpretation of the effects of alloying with Li on the prismatic slip in Mg as illustrated in Fig. 12. There is an ap-parent transition from the Friedel cross-slip mechanism in pure Mg and Mg + 6 at .% Li to a Peierls type of mechanism for Mg + 10

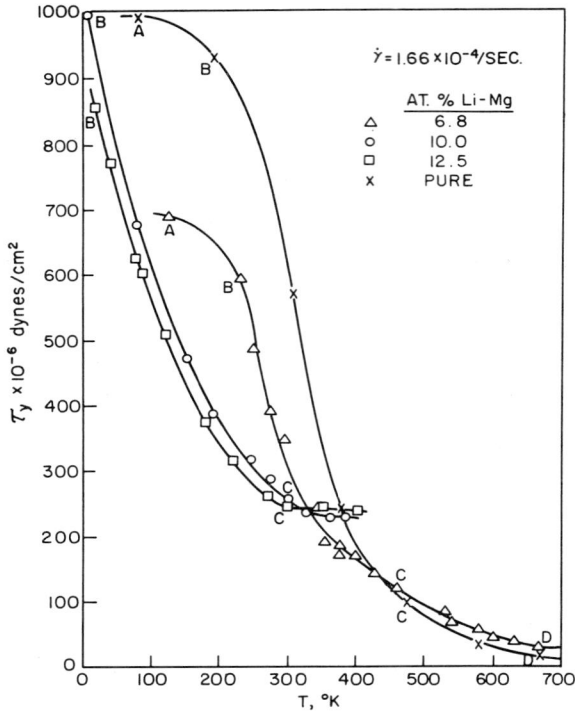

Fig. 12 Critical resolved shear stress versus temperature for prismatic slip in pure Mg and Li-Mg alloys.

at .% Li and Mg + 12 at .% Li. This suggests that Li additions serve to increase the stacking fault energy in Mg.

There are three major points of difference between the true Peierls mechanism and the pseudo-Peierls mechanism: (1) Whereas the true Peierls mechanism applies to any dislocation in any orientation parallel to close-packed rows of atoms on the slip plane, the pseudo-Peierls mechanism applies only to dislocations in screw orientation that dissociate very little on secondary planes; (2) whereas the true Peierls mechanism is expected to be operative in the early stages of straining, some prestraining is required to place dislocations in screw orientation before the pseudo-Peierls mechanism can be controlling. The rate of strain hardening is expected to be high in the microstrain region; (3) whereas the activation energy for the true Peierls mechanism should depend only on the shear stress on the slip plane, that for the pseudo-Peierls mechanisms will depend also on the shear stresses on the secondary planes of dissociation.

The picture concerning the plastic behavior of b.c.c. metals at low temperatures is now rapidly being resolved. There are three principal schools of thought. The first contends that the plastic behavior of b.c.c. metals arises from tetragonal strain centers due to interstitial impurities; a second that b.c.c. metals obey the Peierls mechanism; and a third that their deformation is dependent on recombination of mildly dissociated dislocations, i.e., by a recombination mechanism. Although the validity of the mechanism first mentioned has been demonstrated experimentally in a number of other examples, it still has not been shown that τ^* for b.c.c. metals increases linearly with $\sqrt{c}$ as should be the case if impurities were controlling the deformation. Furthermore, it alone cannot account for the observed effects of orientation on yielding of b.c.c. metals. In many cases there is good experimental confirmation for the validity of the Peierls mechanism, but again the effects of orientation cannot be explained on the basis of this mechanism alone. At present it appears that the recombination mechanism will eventually prove to be the most reliable because it can account for orientation effects, high initial rates of strain hardening and for any given orientation τ^* vs. T relationships that are Peierls-like. It is evident, however, that the more macroscopic formulation of the mechanism in terms of stacking-fault, recombination and constriction energies will have to be replaced by a more detailed atomistic picture of cores of slightly dissociated screw dislocations and the changes in core configurations and energies under general states of stress.

The activation energy for cutting localized obstacles depends on the statistics of their distribution. In contrast, that for nucleating slip past line obstacles is determined completely by the line configuration. Consequently the strain rate for dislocation motion past a series of different kinds of line obstacles depends essentially on the rate of nucleation past the most difficult surmountable obstacle. Usually, it is assumed that the superposition $\tau = \tau^* + \tau_A$ holds as it should if τ_A arises from long-range stress fields. Up to now, however, no adequate theoretical treatment of cases involving both localized and linear obstacles has been reported. On occasions, the preexponential term is not strongly dependent on the stress (e.g., the Peierls mechanism) and then a finite T_c results for linear obstacles at which $\tau^* = 0$. But in other cases a finite T_c is not obtained either because of the stress dependency of the activation energy (e.g., cross slip) or the stress dependency of the preexponential term.

ACKNOWLEDGMENTS

The author expresses his sincere appreciation to Dr. P. Guyot and Mr. T. Stephansky for their numerous invaluable suggestions and contributions to this paper. He also thanks Mr. C. Cheng for his assistance in the proofreading.

This research was conducted as part of the activities of the Inorganic Materials Research Division of the Lawrence Radiation Laboratory of the University of California, Berkeley, and was done under the auspices of the U.S. Atomic Energy Commission.

REFERENCES

1. Mott, N.: *Phil. Mag.* 43:1151 (1952).
2. Cottrell, A.: *J. Mech. and Phys. Sol.* 1:53 (1952).
3. Cottrell, A.: "Dislocations and Plastic Flow in Crystals," p. 170, Oxford University Press, 1953.
4. Friedel, "Dislocations," pp. 121, 221, Pergamon Press, Oxford, 1964.
5. Seeger, A.:*Phil. Mag.* 46:1194 (1955).
6. Basinski, F.: *Phil. Mag.* 4:393 (1959).
7. Cottrell, A., and B. Bilby, *Proc. Phys. Soc.* A62:49 (1949).
8. Mott, N., and F. R. N. Nabarro, *Proc. Phys. Soc.* 52:84 (1940).
9. Cochardt, A., G. Schoeck, and H. Wiedersich, *Acta Met.* 3:533 (1955).
10. Fleischer, R., *Acta Met.* 10:835 (1962).
11. Fleischer, R., and W. Hibbard, "The Relation between the Structure and Mechanical Properties of Metals," p. 262, Her Majesty's Stationery Office, London, 1963.
12. Friedel, J.: "Dislocations," p. 351, Pergamon Press, Oxford, 1964.
13. Kelly, A., and M. Fine, *Acta Met.* 5:365 (1957).
14. Fleischer, R., *Acta Met.* 8:32 (1961).
15. Fine, M.: "The Relation between the Structure and Mechanical Properties of Metals," p. 299, Her Majesty's Stationary Office, London, 1963.
16. Friedel, J.: "Dislocations," p. 376, Pergamon Press, Oxford, 1964.
17. Seeger, A.: *Proc. 2nd U.N. Inter. Conf. PUAE* 6:250 (1958).
18. Seeger, A.: Radiation Damage in Solids, *IAGA* 1:101 (1962).
19. Diehl, J., and W. Shilling, *Proc. 3rd U.N. Inter. Conf. PUAE* 9:72 (1964).
20. Diehl, J., G. Seidel, and L. Nieman,*Phys. Stat. Sol.* 11:339 (1965); *Phys. Stat. Sol.* 12: 405 (1965).
21. Saada, G.: *Acta Met.* 8:841 (1960).
22. Saada, G.: *Acta Met.* 9:2 and 160 (1961).
23. Guyot, P.:*Acta Met.* 14:955 (1966).
24. Orowan, E.: *Symp. on Internal Stresses in Met. and Alloys*, p. 451, Inst. of Met., London, 1948.
25. Orowan, E.:*Disl. in Met.*, p. 131, AIME, New York, 1954.
26. Fisher, J., E. Hart, and R. Pry: *Acta Met.* 336: (1953).
27. Taylor, G.:*Proc. Roy. Soc. L.* 145A:362 (1934).
28. Mott, N.: *Phil. Mag.* 43:1151 (1952).
29. Seeger, A., J. Diehl, S. Mader and H. Rebstock, *Phil. Mag.* 2:323 (1957).
30. Peierls, R.:*Proc. Phys. Soc.* 51:34 (1940).
31. Seeger, A.: *Phil. Mag.* 1:651 (1956).
32. Friedel, J.: "Electron Microscopy and Strength," p. 605, Interscience Publishers, 1963.

33. Celli, V., M. Kabler, T. Nimoiya, and R. Thomson, *Phys. Rev.* 131:58 (1963).
34. Dorn, J., and S. Rajnak: *Trans. AIME* 230:1052 (1964).
35. Guyot, P., and J. Dorn: *Cana. J. of Phys.* 45:983 (1967).
36. Seeger, A.: *Phil. Mag.* 45:771 (1954).
37. Friedel, J.: "Dislocations," p. 264, Pergamon Press, Oxford, 1964.
38. Escaig, B.: *Proc. Phys. des Disl.* Toulouse (1966).
39. Cottrell, A.: *Phil. Mag.* 43:645 (1952).
40. Fisher, J.: *Trans. ASM* 47:451 (1955).
41. Suzuki, H.: *Sci. Rep. Res. Inst. Tohoku Univ.* A4:455 (1952).
42. Fisher, J.: *Acta Met.* 2:9 (1954).
43. Flinn, P.: *Phys. Rev.* 104:350 (1956).
44. Cottrell, A.: *Seminar on Relation of Properties of Microstructure*, p. 151, Amer. Soc. Metals, Cleveland, 1955.
45. Zener, C.: "Elasticity and Anelasticity of Metals," p. 89, Univ. of Chicago Press, 1958.
46. Eshelby, J.: *Proc. Roy. Soc. L.* A197:789 (1956).
47. Granato, A., and K. Lucke, *J. Appl. Phys.* 27:789 (1956).
48. Eshelby, J.: *Proc. Roy. Soc. L.* A197:396 (1957).
49. Liebfried, G.: *Z. Physik* 127:344 (1950).
50. Eshelby, J.: *Proc. Roy. Soc.* A266:222 (1962).
51. Akheiser, A.: *J. Phys. USSR* 1:277 (1939).
52. Mason, W.: *J. Acous. Soc. Amer.* 32:458 (1960).
53. Mason, W., and T. Bateman: *J. Acous. Soc. Amer.* 36:644 (1964).
54. Mason, W.: *J. Appl. Phys.* 35:2779 (1964).
55. Mason, W.: *Phys. Rev.* 97:557 (1955).
56. Mason, W.: *Phys. Rev.* 143:339 (1966).
57. Frank, F.: *Proc. Phys. Soc.* A62:131 (1949).
58. Mukherjee, A., J. Mote, and J. Dorn: *Trans. AIME* 233:1559 (1965).
59. Wilsdorf, H., and D. Kuhlman-Wilsdorf: *Phys. Rev. Ltrs.* 3:170 (1959).
60. Keh, A., and S. Weissman, "Electron Microscopy and Strength of Crystals," p. 231, Interscience Publishers, 1963.
61. Ferguson, W., A. Kumar, and J. Dorn: *J. Appl. Phys.*, to be published, 1967.
62. Friedel, J.: "Dislocations," p. 224, Pergamon Press, Oxford, 1964.
63. Dorn, J., and J. Mitchell: "High-Strength Materials," p. 510, John Wiley & Sons, Inc., New York, 1965.
64. Foreman, A., and M. Makin: *Phil. Mag.* 14:911 (1966).
65. Kocks, U.: *Phil. Mag.* 13:541 (1966).
66. Guyot, P., and T. Stephansky: unpublished research.
67. Cottrell, A., and R. Stokes: *Proc. Roy. Soc.* A233:17 (1955).
68. Wilsdorf, D.: *Trans. AIME* 224:1047 (1962).

EFFECT OF CORE STRUCTURE ON DISLOCATION MOBILITY WITH SPECIAL REFERENCE TO B.C.C. METALS

M. S. Duesbery and P. B. Hirsch

University of Oxford, U.K.

ABSTRACT

The yield stresses for prism slip in certain hexagonal metals, and for slip in b.c.c. metals, have been interpreted either in terms of a Peierls force, or of a model in which screw dislocations are dissociated in planes which are not the slip planes for the particular slip modes considered. The dissociation model is discussed in detail for the case of screw dislocations in b.c.c. metals dissociated partly on {112} and {110} planes. It is shown that at high temperatures and low stresses the conventional constriction model is replaced by a double-kink model. The theory is applied mainly to the deformation of Nb, and it is shown that the temperature dependence of the yield stress, the activation volume, the difference in the yield stress for slip on {110} and {112} planes, and the asymmetry of the yield stress for slip on {112} planes can be explained satisfactorily on the model. The model is compared critically with the Peierls model, and it is concluded that, at present, it is difficult to decide between the applicability of the two on the basis of the experimental evidence available.

1   INTRODUCTION

The mobility of a dislocation can be controlled by interactions with many types of defects, e.g., other dislocations, impurities, solute atoms, precipitates, point defects, etc. However, one of the important factors controlling mobility is the structure of the dislocation itself. In the absence of any interaction with other defects, the structure of the core can affect dislocation motion in at least two ways. (1) If the energy of the misfit across the slip plane is high and the dislocation core is narrow (e.g., in Si or Ge), the energy of the dislocation may be a sensitive, periodic, function of the dislocation position in the lattice. A stress, the Peierls-Nabarro stress, will then be required to move the dislocation from one low energy trough to the next. This process can be aided by thermal activation at finite temperatures by the formation of kinks (Seeger,[1] Friedel,[2] Dorn and Rajnak[3]). (2) The low energy form of the dislocation may be sessile, and before the dislocation can move, it must be changed into a higher energy glissile configuration.

A possible example of this type of mechanism is prismatic slip in certain hexagonal metals. In the hexagonal lattice, dislocations with the basal-plane Burgers vector can dissociate in the basal plane, thereby lowering their energy. Slip on prism planes with the basal Burgers vector can be effected by cross-slip of the screw dislocations onto the prism planes. Before this cross-slip can take place, the dislocation must be constricted into a higher energy configuration which is glissile on the prism plane. Figure 1 (after Friedel[4] illustrates the essential mechanism. The dislocation is constricted and bows out in the prism plane. The energy of the configuration (ii) consists of several terms:

1) the energy of the two half-constrictions;
2) the energy of recombination over the length l;
3) the increase in length of the dislocation line as a result of bowing out on the cross-slip plane;
4) the work done by the applied stress during this process.

The energy of this configuration goes through a maximum as a function of l; this critical value of l_c determines the activation energy for the process for a given stress τ. For lengths l which

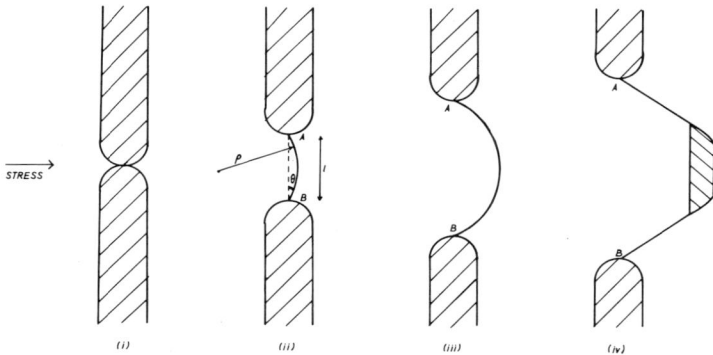

Fig. 1 Cross-slip mechanism for hexagonal close-packed metals (after Friedel[4]).

are large in comparison to the dissociated width of the dislocation, the activation energy H is of the form

$$H = U_c + \frac{K_1}{\tau} \tag{1}$$

where U_c is the energy of the two half-constrictions, and K_1 is a constant which depends on the difference in energy between the constricted and dissociated configurations. The strain rate $\dot{\epsilon}$ from a single mechanism of this type is given by

$$\dot{\epsilon} = \rho b v = \rho b v_0 \frac{b}{l_c} \frac{L}{l_c} d \exp\frac{-H}{KT} \tag{2}$$

where ρ is the screw dislocation density, b is the Burgers vector, v is the velocity of the screw determined by the constriction mechanism, v_0 is the Debye frequency, $v_0 b/l_c$ is the frequency of vibration of the dislocation during activation, l_c is the critical length of the configuration, L is the free length of screw, L/l_c is the number of sites of activation on the screw, and d is the distance traveled by the screw after activation. The preexponential factor depends on stress through l_c (l_c is inversely proportional to stress) and possibly through d. Friedel[5] puts $d = L$, i.e., the mesh length of the dislocation network. More precisely, he assumes that once the dislocation is glissile it does not return to its low-energy configuration. This point will be discussed further in Sec. 4.3. In addition, L is assumed independent of τ or T. This point will also be considered in Sec. 4.3. With these assumptions the preexponential factor varies as r^2; for small dislocation widths the stress dependence of the activation energy H [see Eq. (1)] is more important, and writing the preexponential factor as ν

$$U_c + \frac{K_1}{r} = KT \log\left(\frac{\nu}{\dot{\epsilon}}\right) \tag{3}$$

so that if U_c is small compared with K_1/r, and the variation of $\log \nu$ with r is neglected, r varies inversely as the temperature T and reaches a high value at $T = 0°K$. This theory was developed by Friedel [4,5] to account for the high stresses observed experimentally at low temperatures and the rapid decrease of r with increasing temperature for prismatic slip in Mg, Be and other hexagonal metals.

Another possible example of this type of mechanism is the mobility of dissociated jogs in screw dislocations in the f.c.c. system. As shown by Hirsch,[6] certain types of jogs may be dissociated into sessile configurations, which have to be constricted before conservative glide can take place. This type of dissociation and constriction mechanism is related to prismatic dislocations which can dissociate in one or more planes which are not the slip plane of the total dislocation. The most famous example of this type of dislocation is, of course, the Lomer-Cottrell dislocation which must be constricted before it becomes glissile (Lomer,[7] Cottrell[8]).

Recently, it has been suggested that the plasticity of b.c.c. metals is controlled by the mobility of screw dislocations, whose low-energy configuration is sessile (Hirsch,[9] Mitchell et al.,[10] Escaig,[11] Vitek,[12] Vitek and Kroupa,[13] Foxall et al.[14]). This particular problem will be discussed in detail since it illustrates the possible importance of the complex structure of the dislocation core.

2   MODELS OF DISSOCIATED DISLOCATIONS IN B.C.C. METALS

Various modes of dissociation have been suggested for dislocations with Burgers vector $\frac{1}{2} a \langle 111 \rangle$ in the b.c.c. metals. On $\{112\}$ planes the dislocation can dissociate according to the reaction

$$\frac{1}{2} a[1\bar{1}1] = \frac{1}{3} a[1\bar{1}1] + \frac{1}{6} a[1\bar{1}1] \tag{4}$$

with a twin fault between the partials. Further dissociation according to

$$\tfrac{1}{2}\,a[1\bar{1}1] \;=\; \tfrac{1}{6}\,a[1\bar{1}1] \;+\; \tfrac{1}{6}\,a[1\bar{1}1] \;+\; \tfrac{1}{6}\,a[1\bar{1}1] \tag{5}$$

on the same {112} plane leads to the formation of a high-energy fault involving very close distances of approach of some of the atoms. Figure 2(1) illustrates this type of dissociation. However, screw dislocations coincide with the axis of threefold symmetry in this lattice, and it follows that screws could dissociate on several equivalent crystallographic planes intersecting in this common direction, forming low-energy twin faults on each. Hirsch[9] suggested that the screw would dissociate on all three intersecting {112} planes; Sleeswyk[15] showed that this mode of dissociation is unstable, and that in the stable configuration the dislocation dissociates into three screw partials on two planes as shown in Fig. 2(2). It

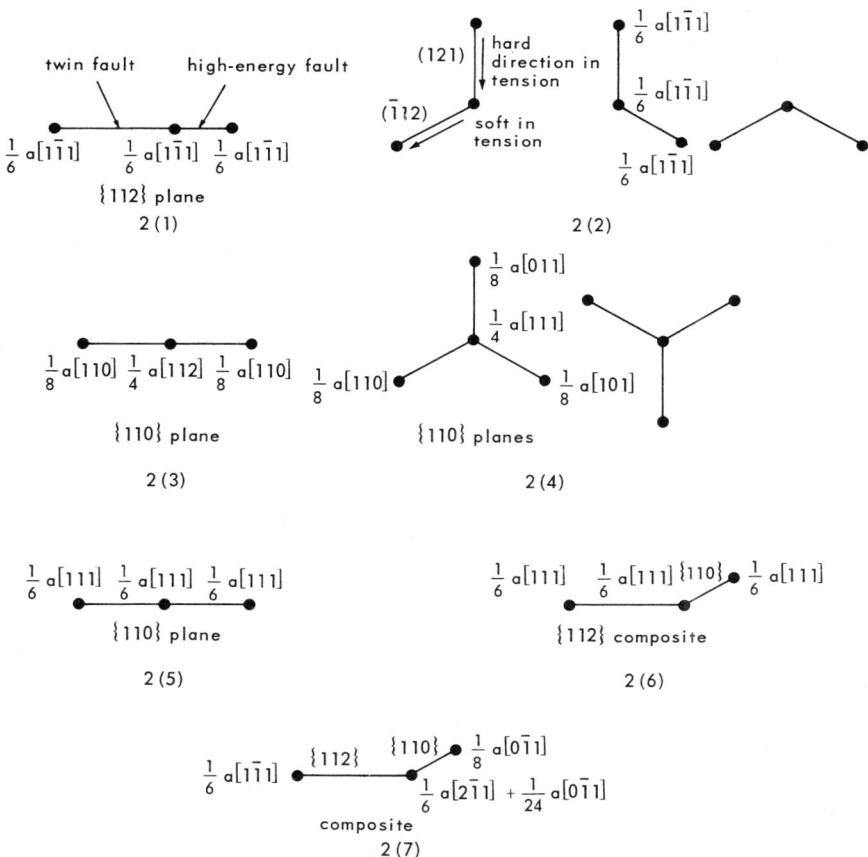

Fig. 2 Possible configurations of dissociated dislocations in the body-centered cubic lattice.

should be noted that only three possible configurations exist as shown in Fig. 2(2). Crussard[16] and Cohen et al.[17] suggested that on {110} planes the dislocation could dissociate according to

$$\tfrac{1}{2}\,a[1\bar{1}1] \;=\; \tfrac{1}{8}\,a[0\bar{1}1] \;+\; \tfrac{1}{4}\,a[2\bar{1}1] \;+\; \tfrac{1}{8}\,a[0\bar{1}1] \tag{6}$$

forming two faults which would be expected on a hard sphere model [Fig. 2(3)]. Kroupa[18] showed that screw dislocations could dissociate on three intersecting {110} planes [see Fig. 2(4)] according to the reaction

$$\tfrac{1}{2}\,a[1\bar{1}1] \;=\; \tfrac{1}{8}\,a[101] \;+\; \tfrac{1}{8}\,a[1\bar{1}0] \;+\; \tfrac{1}{8}\,a[0\bar{1}1] \;+\; \tfrac{1}{4}[1\bar{1}1] \tag{7}$$

where the partial $\tfrac{1}{4}\,a[1\bar{1}1]$ lies at the intersection of the three {110} planes of dissociation. Thus, as shown in Fig. 2(4), there are two possible configurations. Dissociation into three screw partials on {110} planes has also been suggested [see Fig. 2(5)] by Wasilewski.[19]

If faults on {112} and {110} are possible, then it is also possible to find composite dissociations of screws partly on {112} and partly on {110} [see Figs. 2(6) and 2(7)] (Foxall et al.,[14] Kroupa and Vitek[20]).

The energies of these configurations can be compared for various values of the stacking fault energies. In particular, Figs. 2(2), 2(4), and 2(7) involve only the stacking fault energies of the twin fault on {112} (γ_{112}) and of the "hard sphere" fault on {110} (γ_{011}) [we assume that the energy of the high-energy fault on {112} is infinite]. Figure 3 shows the range of stacking fault ratios $\Gamma_0 = \gamma_{011}/\gamma_{112}$ and the values of γ_{011} for which these three types of dislocations are stable; the calculations were performed using isotropic elasticity, and the elastic constants appropriate to niobium. For a certain range of values of Γ_0 and γ_{011} the composite configuration 2(7) appears to be the stable one. Its energy might be lowered further by a dissociation of the dislocation at the intersection of the faults, which has a rather large Burgers vector $(\tfrac{1}{24}\,a[8\bar{5}5])$. It is clear that relatively small variations in the values of the stacking fault energies can change the type of configuration which is stable. Furthermore, the boundaries in Fig. 3 are sensitive to the value of the cut-off radius chosen. Therefore, it cannot at all be claimed that the configuration of the dislocations at this stage can be predicted, since the stacking fault energies are not well known. However, using reasonable estimates of stacking fault energies, of the order as some calculated values (e.g., Hartley[21]), the dislocations would be expected to be dissociated by perhaps 2 – 3 Burgers vectors. Thus, assuming, for example, $\gamma_{011} = 90\,\text{ergs/cm}^2$

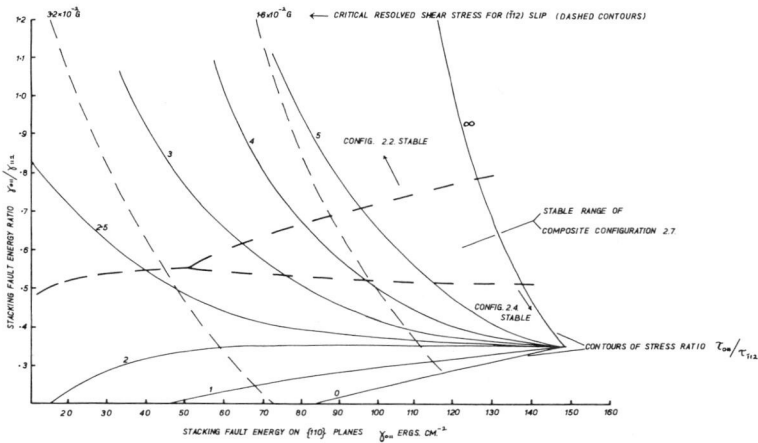

Fig. 3 Stability of configurations 2(2), 2(4) and 2(7) as a function of stacking fault energy on {110} and {112} planes. Configuration 2(7) is stable within the dashed contour; the stable configurations above and below this contour are 2(2), 2(4), respectively. The solid curves are contours of constant critical resolved shear stress ratio $\tau_{011}/\tau_{\bar{1}12}$ at 0°K; the dashed curves are contours of constant critical resolved shear stress on ($\bar{1}12$) at 0°K.

and γ_{112} = 150 ergs/cm^2 for Nb, the distances between the two end partials from the line of intersection of the fault plane in Fig. 2(7) are found to be 1.41b, 2.02b, and on the (011) and ($\bar{1}12$) planes, respectively. Assuming then that such dissociations occur, we shall now examine in some detail the consequences of this type of model.

3   QUALITATIVE CONSEQUENCES OF THE MODEL

1) Before any of the dislocations with the complex configurations 2(2), 2(4), 2(6), and 2(7) can move, the configurations must be changed into the higher-energy glissile configurations 2(1) or 2(3). This can be done by first constricting the dislocations and then changing into the glissile configuration. This is essentially the same mechanism as that proposed by Friedel for slip on prism planes in hexagonal metals (see Sec. 1). This mechanism leads to a high stress at T = 0°K, and also to a rapid temperature dependence. This is in qualitative agreement with experiments on b.c.c. metals (e.g., Conrad,[22] Christian and Masters[23]).

2) This mechanism is confined to screws and predicts that screw dislocations in b.c.c. metals should be much less mobile than edge

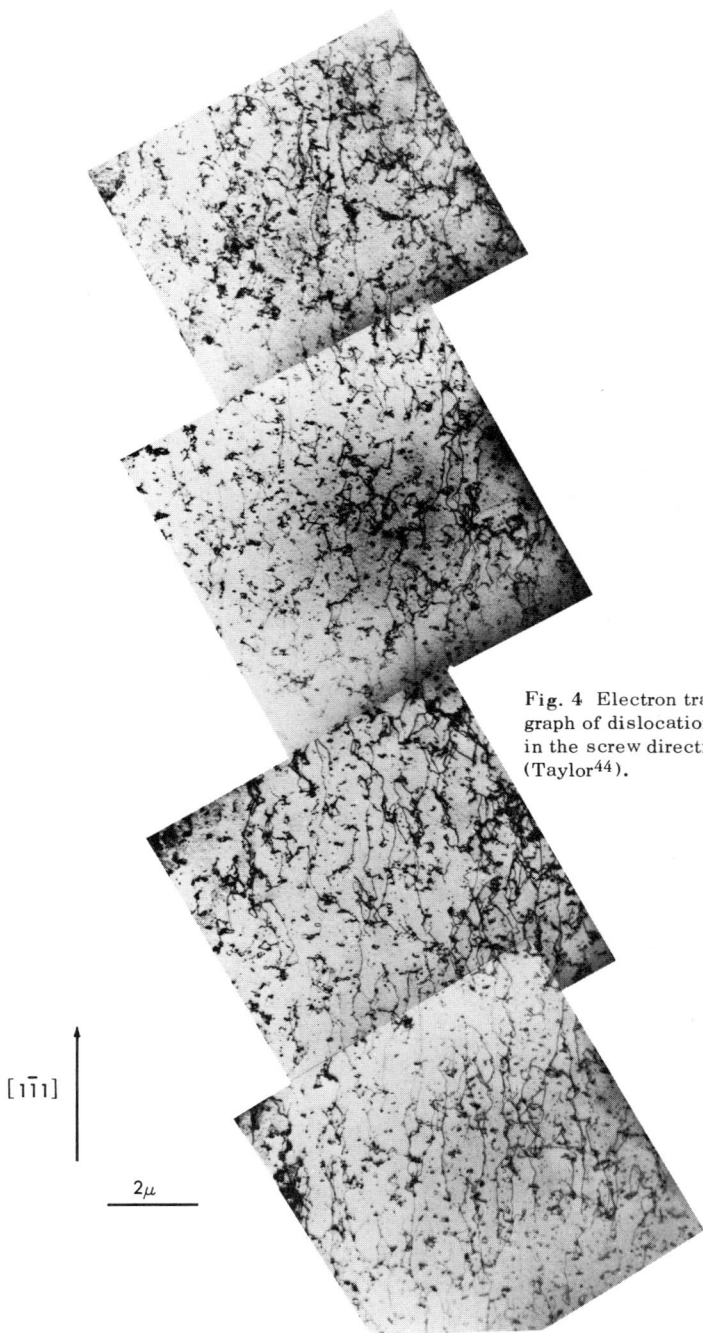

Fig. 4 Electron transmission micrograph of dislocation loops elongated in the screw direction in niobium (Taylor[44]).

$[1\bar{1}1]$

2μ

dislocations. Electron microscopy has provided considerable evidence that this is indeed the case for many b.c.c. metals and alloys (Fe-Si alloys: Low and Guard,[24] Low and Turkalo[25]; Vanadium: Smerd[26]; Niobium: Taylor and Christian[27]). Figure 4 shows an example of dislocation loops elongated in the screw direction in Niobium.

3) Consider the composite configuration 2(7). The six possible dislocations are shown in Fig. 5. For slip on $(\bar{1}12)$ in the direction EF, denoted by the arrow, with the stress direction along b, the constriction D to E necessary to produce slip is aided by the stress. Hence slip in this sense on $(\bar{1}12)$ is therefore relatively easy. On the other hand, slip in the reverse direction is difficult. This is equivalent to producing slip in the "hard" direction for dislocation ABC on (121). If the stress is acting along a, slip on (121) in the "hard" direction (see arrow) can occur if the dislocation C is constricted towards B against the stress or if a complete

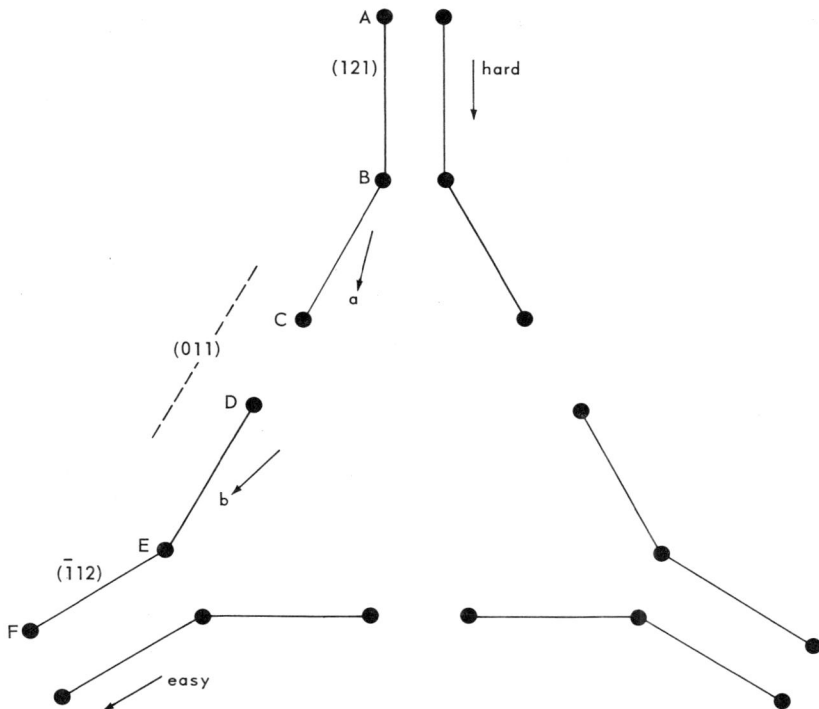

Fig. 5 The six possible dislocations of configuration 2(7). The partials lying on {112} planes have Burgers vector $\frac{1}{6}\ a\langle 111\rangle$; those on {110} planes, $\frac{1}{8}\ a\ \langle 101\rangle$; those at the intersections, $\frac{1}{24}\ a\ \langle 855\rangle$.

constriction of the dislocation is formed. The model therefore immediately predicts an asymmetry in the shear stress for yield on {112} planes. It turns out that slip on {112} planes is always easy if the twinning partial (i.e., dislocations F, A, Burgers vector $\frac{1}{6}a[1\bar{1}1]$) leads, and, conversely, is hard in the reverse sense when twinning partial trails. The dependence of the yield stress on {112} planes on the sense of slip was discovered by Taylor[28] for slip in β-brass. This asymmetry has recently been found in several b.c.c. metals and alloys, in niobium (Duesbery et al.[29]); Bowen et al.,[45] Reid et al.,[46] Taoka et al.,[37] Sestak and Zarubova[30]) and in tungsten (Argon and Maloof[31]). Figure 6 taken from Argon and Maloof shows the differences in yield stress in tension and compression for single crystals of W.

4) The composite configuration 2(7) can give rise to slip on {112} or {110} planes. In general, the activation energies for slip on these planes are different and consequently the critical yield stresses for slip on these planes are also different. This effect has been observed experimentally (Nb: Duesbery et al.,[29] Foxall et al.,[14] Taylor and Christian[27]).

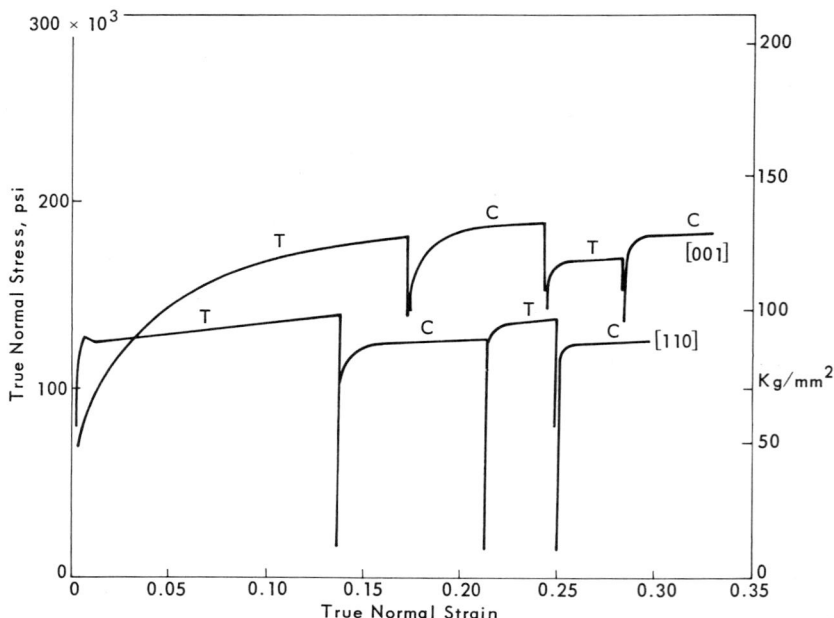

Fig. 6 The effect of reversal of the sense of slip on {112} planes on the stress-strain curve in tungsten: the [001] curve corresponds to the easy sense of slip in tension (T) and the hard sense in compression (C); the [110] curve corresponds to the hard sense of slip in tension and the easy sense in compression (from Argon and Maloof[31]).

5) Since constrictions are necessary for the dislocations to glide, the activation energies for cross-slip are not much larger than for slip; hence cross-slip is relatively easy in b.c.c. metals.

4 THEORY

4.1 General

Calculations on this type of model have been carried out by Escaig,[11] Vitek,[12] Vitek and Kroupa,[13] and Duesbery.[32] Escaig considered the thermally activated slip process for configurations 2(2) and 2(4), but not the orientation dependence or the asymmetry. Vitek also carried out calculations for 2(2) and 2(4), and by assuming that the dislocation has equal possibility of being dissociated in either form, derived an orientation dependence of the slip process and the average slip plane of the dislocation as a function of the orientation of the stress axis. His results explained the asymmetry observed by Sestak and Zarubova in Fe–Si alloys satisfactorily.

In this paper we shall consider the behavior of the composite dislocation 2(7), which has the advantage that it can quite naturally slip on {110} or {112}. We shall discuss two models of the sessile-glissile transformation, the former according to the treatment of Escaig and Vitek.

4.2 Stress at Absolute Zero

At absolute zero, the dislocation can be constricted with the aid of stress to produce slip on {110} or {112} in the easy direction. Figure 3 shows contours of constant stress ratio $\tau_{011}/\tau_{\bar{1}12}$ as a function of stacking fault energies for Nb. The dotted lines show two curves of constant critical resolved shear stress for $(\bar{1}12)$ slip. For the range of stacking fault energies that may occur in practice, the critical resolved shear stress at $T = 0°K$ for slip on $(\bar{1}12)$ is about $2 \times 10^{-2} G$, and for slip on (011) about $6 \times 10^{-2} G$.

4.3 Thermally Activated Slip (Model 1)

Figures 7(i)–(iv) show the sequence of events. The rear part of the dislocation is constricted over a length AB, and produces constrictions at A and B as shown in (iii). For slip on {011} planes the

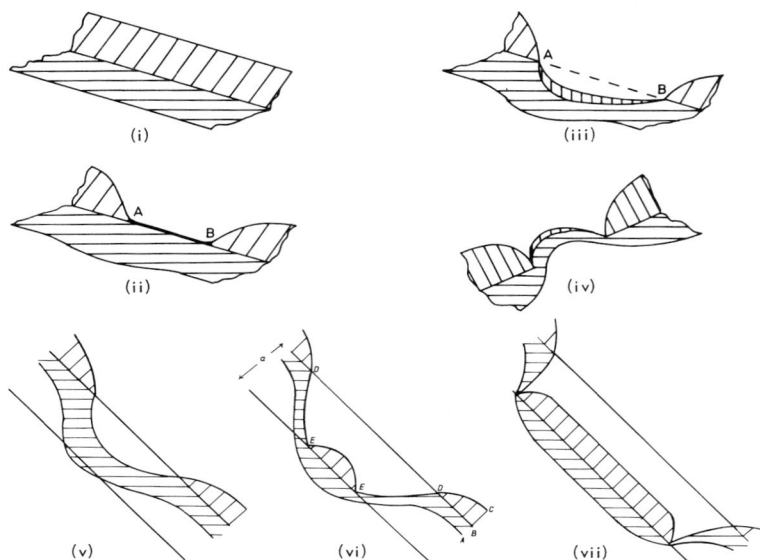

Fig. 7 Two models of slip for configuration 2(7). (i)–(iii). Slip on {110} planes or in the easy sense on {112} planes by constriction and elastic bowing-out in the slip plane. (Model 1). (iv) Slip in the hard sense on {112} planes by constriction against the applied stress and elastic bowing-out in the slip plane. (v)–(vii) Slip by constriction and formation of kinks DE, after redissociation over the length EE into the sessile form; the parallel solid lines represent adjacent low-energy sites in the lattice for the central partial B.

dislocation can redissociate into the glissile configuration 2(3); for slip on {112} planes such dissociation can also take place if the energy of the high-energy fault is not infinite. For slip in the hard direction on {112} the dislocation must constrict against the stress, as shown in Fig. 7(iv).

The energy of configuration (iii) consists of the constriction energies U_c, U_c', of the two types of constriction in each configuration, the recombination energy $l U_R G b^2 / 2\pi$ where l = length AB, the redissociation energy $l U_G G b^2 / 2\pi$, the increase in dislocation length due to bowing out, and the work done by the stress. As in the calculation for prism slip in hexagonal metals, the activation energy is obtained by maximizing this energy with regard to l. The critical length is

$$l_c = \frac{Gb}{\pi\tau}(U_R - U_G)^{1/2} \tag{8}$$

where τ is the effective stress, i.e., the applied stress minus the long-range internal stress.

The activation energy H is

$$H = 2(U_c + U_c') + \frac{G^2 b^3}{3\pi^2 \tau} (U_R - U_G)^{3/2} \tag{9}$$

The maximum distance moved forward by the dislocation at the saddle point is

$$d = \frac{\pi\tau}{2Gb} l_c^2 = \frac{Gb}{2\pi\tau} (U_R - U_G) \tag{10}$$

For the particular case of slip on {110} planes when the stress axis is such that the slip plane is the one bearing the highest resolved shear stress, calculation of the unknown quantities in the above equation gives

$$U_R = \frac{1}{6} \left\{ \ln\left(\frac{t}{t_0}\right) - 1 + \frac{t_0}{t} \right\}$$

where

$$t = \frac{1 + k_0 \tau b / \gamma_{112}}{1 + 0.2887\, \tau b / \gamma_{112}}$$

$$t_0 = t(\tau_0)$$

$$k_0 = \frac{0.1457}{1 + 0.858\,\Gamma_0} - \frac{0.2087}{1 + 0.480\,\Gamma_0}$$

$$\frac{\tau_0 b}{\gamma_{112}} \simeq \frac{(1 + 0.858\,\Gamma_0) - 31.4(1 + 0.480\,\Gamma_0)\gamma_{112}/Gb}{9.06(1 + 0.480\,\Gamma_0)\gamma_{112}/Gb - k_0(1 + 0.858\,\Gamma_0)}$$

$$\text{for} \quad \tau_0 b \ll \gamma_{112}$$

$$2U_c = 0.0337\,Gb^3\,\frac{t}{t_0}\left(1 - \frac{t_0}{t}\right)^2$$

$$2U_c' = 0.0144\,Gb^3\,\frac{x}{b}\left(1 - \frac{b}{x}\right)^2 \tag{11}$$

where

$$\frac{x}{b} = 0.0178\,\frac{Gb}{\gamma_{011}}$$

$$U_G = -0.335 + 5.23\,\frac{\gamma_{011}}{Gb} - 0.0684\,\ln\frac{\gamma_{011}}{Gb} - 0.0492\,\ln\left(1 + 0.0124\,\frac{Gb}{\gamma_{011}}\right)$$

For slip in the easy sense on the $(\bar{1}12)$ plane

$$U_R = 0.062 \left\{ \ln\left(\frac{t}{t_0}\right) - 1 + \frac{t_0}{t} \right\}$$

$$t = \frac{1 + k_0 \tau b / \gamma_{112}}{(1 + 0.1442 \, \tau b / \gamma_{011}) \Gamma_0}$$

$$t_0 = t(\tau_0)$$

$$k_0 = \frac{0.2194}{1 + 0.480 \, \Gamma_0} - \frac{0.2968}{1 - 0.056 \, \Gamma_0} \tag{12}$$

$$\frac{\tau_0 b}{\gamma_{112}} = \frac{1.108 \, (1 - 0.056 \, \Gamma_0) \gamma_{112} / Gb - \Gamma_0 (1 + 0.480 \, \Gamma_0)}{0.1442 (1 + 0.480 \, \Gamma_0) - 1.108 \, k_0 \, (1 - 0.056 \, \Gamma_0) \gamma_{112} / Gb}$$

$$\text{for} \qquad \tau_0 b \ll \gamma_{112}$$

$$2U_c = 0.0137 \, Gb^3 \, \frac{t}{t_0} \left(1 - \frac{t_0}{t}\right)^2$$

$$U_c' = U_G = 0$$

4.4   Stress Dependence of H

For $\tau \ll \tau_0$, the stress at absolute zero, it follows from Eq. (9) that the activation energy is given by

$$H = A + \frac{B}{\tau} \tag{13}$$

where the constants A, B depend on the slip plane, the orientation of the crystal, and the sense of slip on {112}. Figure 8 shows a plot of H vs. τ for slip in {110} and {112} planes in the hard and easy directions. The elastic constants and other parameters apply to Nb; the stacking fault energies were assumed to be

$$\gamma_{011} = 100 \text{ ergs/cm}^2, \qquad \gamma_{112} = 200 \text{ ergs/cm}^2$$

These curves can be used to determine the ratios of critical resolved yield stress for the three types of slip (see Sec. 4.10). Figure 9 shows some theoretical curves of the activation energy for {110} slip for various values of γ_{011} and experimental results for polycrystalline Nb (Christian and Masters,[23] Conrad and Hayes[33]). The general decrease of H with increasing stress is in accordance

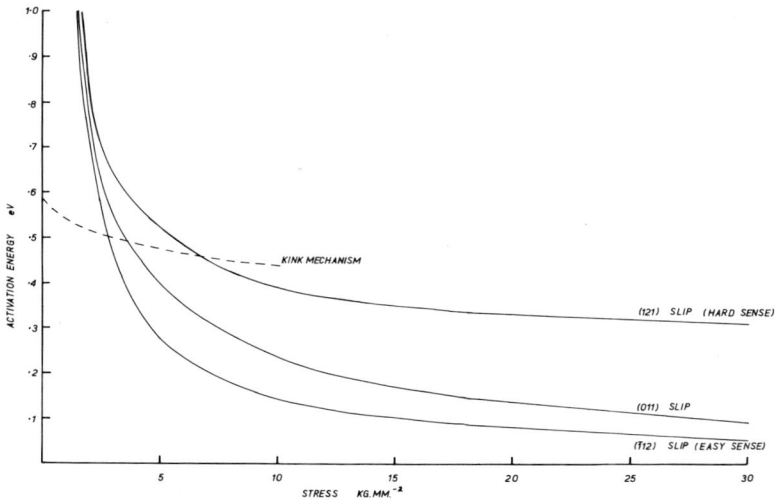

Fig. 8 Theoretical curves of activation energy against stress for slip in niobium on {110} planes and on {112} planes in the hard and easy senses.

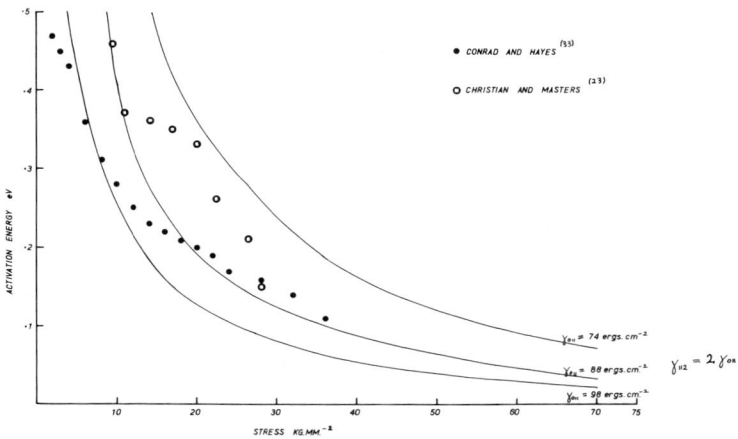

Fig. 9 Experimental variation of activation energy against stress for niobium, and theoretical curves for different stacking fault energies: experimental values from Conrad and Hayes[33] and Christian and Masters[23].

with theory, although the scatter in the experimental results is large; detailed agreement is poor, particularly at low stresses when the theoretical value of H increases much faster than the experimental value.

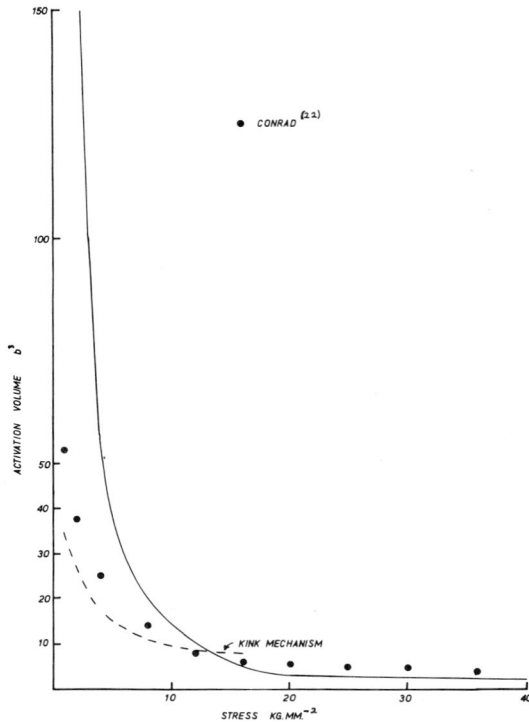

Fig. 10 Theoretical and experimental variation of activation volume with stress for niobium: experimental values from Conrad[22].

4.5 Stress Dependence of the Activation Volume v^*

The activation volume v^* is defined as

$$v^* = -\frac{\partial H}{\partial \tau} = \frac{B}{\tau^2} \quad \text{for} \quad \tau \ll \tau_0 \tag{14}$$

Thus for Nb at 4°K and a stress of $1.7 \times 10^{-2} G$, v^* comes out to $\sim 2b^3$, while at 200°K and a stress of $2.3 \times 10^{-3} G$, $v^* \sim 20 b^3$. Figure 10 shows a plot of v^* as a function of τ and experimental results for Nb taken from Conrad.[22] The rapid decrease of v^* observed with incrasing stress is in general agreement with theory although not very well over the whole range. In particular, at very low stresses the theoretical values of v^* increase much faster than the experimental ones.

4.6 Velocity of Dislocation and Strain Rate

The velocity of a screw dislocation is given approximately by

$$v = v_0 \frac{b}{l_c} \cdot \frac{L}{l_c} \cdot x \, \exp \frac{-H}{kT} \tag{15}$$

where v_0 is the Debye frequency, $v_0 b/l_c$ is the frequency of vibration of the dislocation with wavelength l_c, L/l_c is the number of activation sites in a free length L of dislocation, and x is the distance travelled by the screw before becoming again sessile. The entropy factor has been neglected. (For further discussion see Seeger, Donth and Pfaff,[34] Jøssang, Skylstad and Lothe,[35] and Granato.[36]) Vitek[12] has pointed out that for the glissile-to-sessile transformation 2(4) to 2(3), a constriction must first be formed, and the distance x is therefore affected by an activation process. For the model considered in this paper, however, the reverse transformation should occur with negligible activation since U_G is small or zero and therefore $x \sim d$. It should be pointed out that the same consideration should apply to the case of prismatic slip in hexagonal metals (see Sec. 1). When the dislocation dissociates again the configuration is as shown in Fig. 1(iv), and the area slipped by the dislocation then depends on the distance L traveled by the kinks.

If the time taken for the kinks to travel between pinning centers, e.g., impurities and other dislocations, is small compared with that between successive activations of the screw, in this length, L is given by L_i, the distance between impurities, etc., where the kinks are held up. If on the other hand, the kinks move with finite velocity, determined either by the lattice interaction or by overcoming impurities with the aid of thermal activation, L is determined by the condition that the time taken for one screw activation to take place within L is equal to the time taken by the kink to travel this distance. As shown by Dorn and Rajnak,[3] if v_k is the kink velocity, then

$$L^2 = \frac{v_k l_c^2}{v_0 b} \exp \frac{H}{kT}$$

and hence

$$v = v_0^{1/2} b^{1/2} d \, v_k^{1/2} \exp \frac{-H}{2kT} \tag{16}$$

If v_k itself is determined by an activated mechanism, with activation energy H_k, writing

$$v_k = v_0 bf \exp \frac{-H_k}{kT}$$

we find
$$v = v_0 b d f^{1/2} \exp \frac{-(H + H_k)}{2kT} \tag{17}$$

Thus, it should be emphasized that the velocity of the screw may be affected not only by the activation energy for kink formation, but also by that for kink movement.

In discussing further the effect of kink formation we shall assume that L is determined by defects and is essentially a constant. Using Eqs. (8) and (10) we find

$$v = v_0 \cdot \frac{\pi}{2} \cdot \frac{\tau L_i}{G} \exp \frac{-H}{kT} \tag{18}$$

This relation holds provided $d > a$, where a is the distance between one atomic row and the next. For $d < a$ at very high stress, the validity of the model is doubtful, but in any case the dislocation will travel a minimum distance a before becoming sessile. For $d < a$, therefore,

$$v = v_0 \pi^2 \left(\frac{a}{b}\right) \cdot \frac{L}{(U_R - U_G)} \cdot \left(\frac{\tau}{G}\right)^2 \exp \frac{-H}{kT} \tag{19}$$

For $\tau \ll \tau_0$, the stress dependence of U_R is relatively small, so that the preexponential factor varies as τ or τ^2 depending on the stress. This stress dependence is relatively small compared to that of H. The strain-rate $\dot{\epsilon}$ is given by

$$\dot{\epsilon} = \rho b v \equiv v \exp \frac{-H}{kT} \tag{20}$$

where ρ is the dislocation density.

4.7   The Temperature Dependence of H

From Eq. (20)

$$H = KT \log \frac{v}{\dot{\epsilon}} \tag{21}$$

and neglecting the variation of v with τ, at constant $\dot{\epsilon}$,

$$H = CT$$

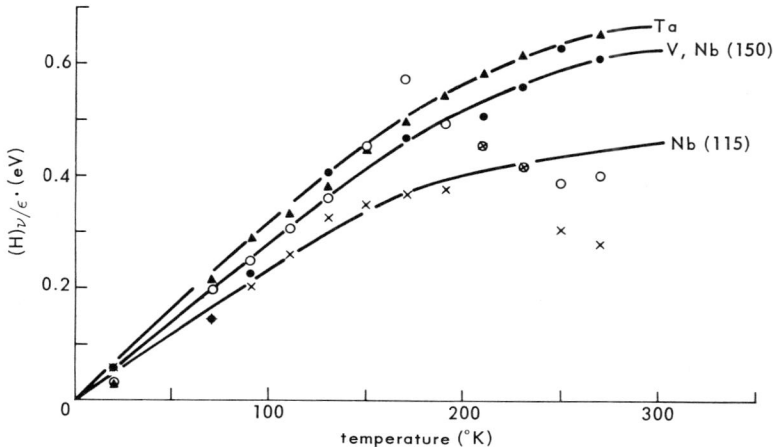

Fig. 11 Experimental variation of activation energy against temperature for various B.C.C. metals (Christian and Masters[23]).

where C is a constant. Figure 11 shows experimental plots for two types of Nb taken from Christian and Masters.[23] Although $H \propto T$ at low temperatures, it tends to a constant value at high temperatures. At high temperatures, according to Eq. (13), it should tend to infinity at τ goes to zero. Clearly, the model is unsatisfactory in this respect, and it will be shown by another mechanism that there is, in fact, a natural limit to the activation energy.

4.8   The Preexponential Factor

From Eqs. (19), (20),

$$\log \frac{\nu}{\dot{\epsilon}} = C = \frac{\rho \nu_0 \pi^2 aL}{(U_R - U_G)} \cdot \left(\frac{\tau}{G}\right)^2 \cdot \frac{1}{\dot{\epsilon}}$$

Inserting typical values for $\rho \sim 10^8$ cm^{-2}, $\nu_0 \sim 10^{13}$ sec^{-1}, $a \sim 2.5$ Å, $L \sim 10,000$ Å, $(U_R - U_G) \sim 1/60$, $\tau/G \sim 10^{-3}$, $\dot{\epsilon} \sim 10^{-4}$ sec^{-1}, gives $\log(\nu/\dot{\epsilon}) \sim 24$, which is of the correct order of magnitude (see, e.g., Christian and Masters,[23] and Conrad[22]).

4.9   Temperature Dependence
of the Yield Stress

It follows from Eqs. (13) and (21) that for $\tau \ll \tau_0$,

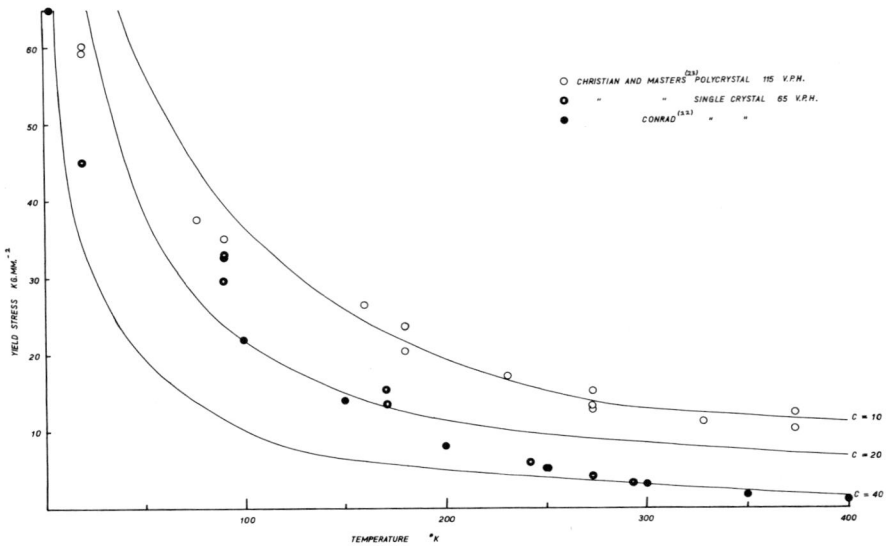

Fig. 12 Experimental variation of yield stress with temperature for niobium, and theoretical curves for various values of the pre-exponential factor: experimental values from Conrad[22] and Christian and Masters[23].

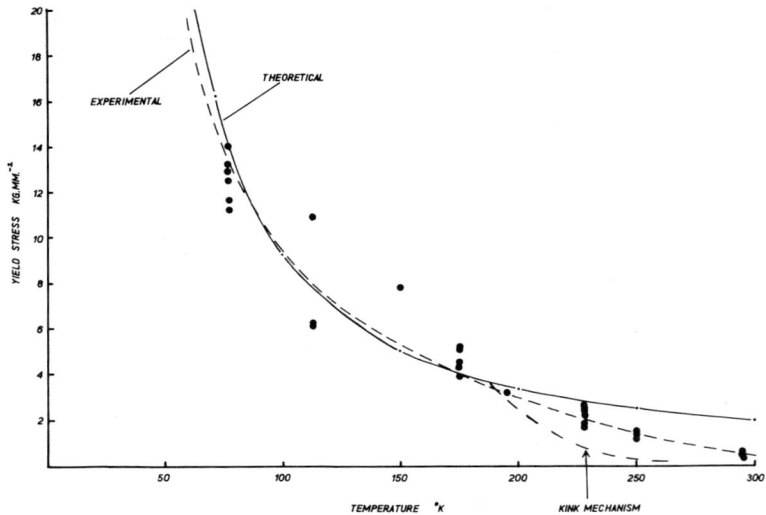

Fig. 13 Experimental and theoretical variation of yield stress with temperature for very pure niobium; the full line shows the theoretical curve predicted by model (1); and the lower dashed curve indicates the modifications expected for model (2).

$$\tau = \frac{B}{CT - A}$$

but for larger stresses the stress dependence B, A and even C must be taken into account. Figure 12 shows theoretical curves for three different values of C (corresponding to $\log(\nu/\dot{\epsilon}) = 10, 20, 40$), and experimental points for Nb from Christian and Masters,[23] and Conrad.[22] While there is agreement in the general trend, overall the experimental values tend to fall below the theoretical curve at higher temperatures. This discrepancy is associated with the tendency for H to reach a constant value. Figure 13 shows Duesbery's results[32] for very pure Nb, which show the same trend. Most b.c.c. metals behave in a similar manner; the theoretical curve can fit the experimental curve only over a certain range, and the nature of the discrepancy is always of the same type; that is, if the fit is made at low temperatures, the experimental values at high temperature fall below the theoretical curve, and if the fit is made at high temperatures, the experimental values at intermediate temperatures are too high.

4.10   Critical Shear Stress for Slip on $\{110\}$ and $\{112\}$, the Slip Plane Boundaries, and Asymmetry

The total velocity of the dislocation, and the mean path is given by Σv, summed over the various modes of slip. The velocities are determined primarily by the activation energies, and these are sufficiently different for the three modes of slip for slip to take place either on $\{110\}$ or on $\{112\}$ in the easy or hard directions, depending on the crystal orientation. The relative values of the yield stresses on these planes can be deduced from Fig. 8 by reading off the stresses for a constant activation energy. Table 1

TABLE 1*

Temperature °K	Ratio $\tau_{011}/\tau_{\bar{1}12}$	Ratio τ_{011}/τ_{121}
295	1.07 (1.06)	0.99 (0.94)
228	1.12 (1.12)	0.91 (0.89)
175	1.45 (1.47)	0.56 (<0.87)

*Experimental values are given in parentheses.

shows the predicted ratios of the yield stresses at various temperatures. In addition, the experimental results are also shown which were obtained by determining the operative slip planes at various temperatures for various crystal orientations (Duesbery,[32] Foxall et al.[14]).

The good agreement between theory and experiment is not of great significance, since slight changes in the stacking fault energies could change the ratios considerably. However, the theoretical curves do predict the observed increase in the ratio $\tau_{011}/\tau_{\bar{1}12}$ with decreasing T, and also the asymmetry for slip on {112} planes.

A detailed study of the orientation of the effective slip plane (on the average noncrystallographic) in Fe-Si alloys as a function of the orientation of the maximum resolved shear stress plane has been carried out by Taoka et al.[37] and Sestak and Zarubova.[30] The orientations of the maximum resolved shear stress plane and of the operative slip plane are determined by the angles Φ, α which they form with the nearest {110} plane [see Fig. 14(i)]. Figure 14(ii) shows the dependence of Φ as a function of α for Fe-3%Si at room temperature. The dotted line at 45° corresponds to the case when the slip plane is always the maximum resolved shear stress plane. The experimental points show deviations from this line, and an asymmetry which is such that if the resolved shear stress acts in the sense that the twinning partial on {112} would lead, then the slip plane follows the maximum resolved shear stress plane. Basically, the explanation is that for $-30° < \alpha < 0$, slip on $(1\bar{1}2)$ in tension is easy, and therefore if slip on (011) is also easy, the mean slip plane is the maximum resolved shear stress plane. On the other hand, for $0 < \alpha < +30°$, slip on (121) in tension is difficult and thus the tendency is for slip to occur preferentially on (011). Figure 14(iii) shows the experimental results for Fe-6%Si; in this case the tendency is for slip to occur preferentially on (011) for orientations $0 < \alpha < 30°$ and $-30° < \alpha < 0$. The solid lines in Figs. 14(ii) and 14(iii) show the theoretical dependence of the slip plane on orientation, expected from the composite configuration 2(7), calculated on the assumption that the direction of the slip plane is given by the vector sum of the dislocation velocities on {110} and {112} planes; the agreement is satisfactory. The stacking fault energies corresponding to the theoretical curves are $\gamma_{011} = 120$ ergs/cm^2, $\gamma_{112} = 270$ ergs/cm^2 for Fe-3%Si, and $\gamma_{011} = 85$ ergs/cm^2, $\gamma_{112} = 170$ ergs/cm^2, for Fe-6%Si. Satisfactory agreement is found also by Vitek and Kroupa[13] on the model based on configurations 2(2) and 2(4), but in this case for stacking fault energies,

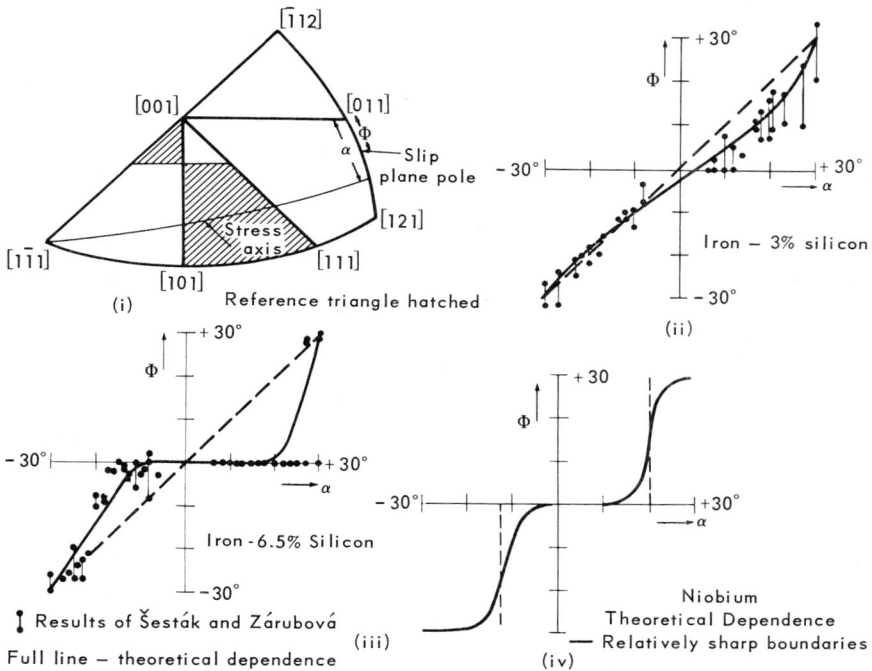

Fig. 14 Theoretical and experimental variation of slip plane with orientation. (i) Definition of the angles Φ, α defining the orientations of the slip plane and stress axis, respectively. (ii) Iron–3% silicon. The experimental points are taken from Sesták and Zárubová[30]; the dashed line shows the variation expected for slip on the maximum resolved shear stress plane. The full line shows the theoretical variation for $\gamma_{011} = 120$ ergs/cm^2, γ_{112} $= 270$ ergs/cm^2; the separation of the partials of configuration 2(7) are 1.25b, 1.21b, on (011) and ($\bar{1}$12) planes respectively. (iii) As (ii), for iron–6.5% silicon: $\gamma_{011} = 85$ ergs/cm^2., $\gamma_{112} = 170$ ergs/cm^2. The separations of the partials are 1.73b, 1.92b, on the (011) and ($\bar{1}$12) planes respectively. (iv) Niobium. The dashed curve shows the experimental values of Table 1; the full line shows the theoretical variation for $\gamma_{011} = 100$ ergs/cm^2., $\gamma_{112} = 200$ ergs/cm^2. The separations of the partials are 1.47b, 1.63b on the (011) and ($\bar{1}$12) planes respectively.

$\gamma_{011} = $ 265 ergs/cm^2, $\gamma_{112} = $ 440 ergs/cm^2, for Fe-3%Si, and $\gamma_{011} = $ 240 ergs/cm^2, $\gamma_{112} = $ 380 ergs/cm^2, for Fe-6%Si.

For comparision, the operative slip planes for Nb appropriate for the results of Table 1 are plotted in Fig. 14(iv). The relatively abrupt change from one slip plane to another is due to the larger differences in the activation energies for slip on {112} and {110}.

While the model explains the difference in τ for slip on different planes and the asymmetry satisfactorily, the orientation dependence of the actual values of yield stress is much larger than predicted on the model (Duesbery et al.,[29] Bowen et al.[45]). This additional orientation dependence also found for other metals (W: Rose et al.,[47] Argon and Maloof;[31] Fe: Stein and Low[48]) has been attributed to jog hardening (Guiu and Pratt[38]), but the detailed explanation is not yet clear.

4.11 A Double Kink Model for Slip

The formation of double kinks is an alternative mechanism, analogous to the Peierls-stress mechanism, by which the screw can move.

Consider the composite configuration 2(7). The intermediate partial lies at the intersection of two fault planes, and this line of intersection repeats at distance a on the slip plane. The minimum distance of advance of a dislocation is therefore from one line of intersection to the next. Figures 7(v)–(vii) illustrate the mechanism [Model (2)]. The dislocation forms a double kink; the intermediate partial transfers from DD to EE. Between these two lines, the partial dislocation must be constricted (here we are neglecting the further glissile dissociation possible). Constrictions are formed at D and E. The energy of the configuration is

$$H' = U_{DK} + 4U_c$$

where U_{DK} is the double kink energy and U_c is the constriction energy. It is clear that this configuration is more advantageous than the completely glissile "kink" considered in Model (1), in that the length over which the dislocation must be combined is reduced; however, the constriction energy is larger since twice the number of constrictions must be formed. There is no doubt that Model (2) must be effective at small stresses or high temperatures when l_c and the activation energy H for Model (1) are very large. For low temperatures the constriction energy may be sufficiently important to make Model (1) preferable. The critical temperature is difficult to calculate since the constriction energy itself is stress-dependent, and the interaction energy between the constrictions is not well known. If we assume the elastic model (1) to hold when $d \lesssim a$, then, as a rough guide, we would expect Model (1) to apply for temperatures below this point, and Model (2) to apply above some slightly higher temperature. From Eq. (10) we find that $d = a$ when

$$\tau = \frac{Gb}{2\pi a} (U_R - U_G)$$

Putting $a \sim b$ and using the same parameters for Nb as in Sec. 4.10,

$$(U_R - U_G) = 1/60$$

so that
$$\tau_{011} \sim 2.4 \times 10^{-3} G$$

which corresponds to a temperature of about $130\,°K$.

This double kink model is essentially a Peierls stress model. The dislocation moves from one energy valley to the next; the difference between the two models is simply that the energy barrier is considered to be due to the necessity of constricting the dislocation running between neighboring energy valleys, so that it can glide in the slip plane rather than to the actual slipping of the core atoms over each other. Formally, however, the kink models developed for a general Peierls stress barrier (e.g., Dorn and Rajnak,[3] Guyot and Dorn[39]) should apply. The constriction energies must, of course, be considered and the barrier model should tend to the rectangular barrier type. To a first approximation, the energy of the dislocation lying between the valleys is independent of position.

The double kink model shows at once that both the need for constriction, and the slipping of the core atoms can be taken into account simultaneously. The energy of the kink dislocation will then involve both types of energy increase. The Peierls energy of the leading partial can also be taken into account in this way. This shows clearly that the two energy terms are additive, and do not correspond separately to alternative mechanisms. However, it does not seem valid to analyze experimental data in terms of a dissociation model in one temperature range and the Peierls model in another (e.g., Guyot and Dorn[39]). It is clear that since the models are essentially similar, it will be rather difficult to decide from experimental data which of the two types of energy barrier is most important. In b.c.c. metals the fact that screw dislocations move more slowly than edges is consistent with the view that although the normal Peierls mechanism may operate for both screws and edges, the additional difficulty in the mobility of the screws is due at least to the constriction mechanism. There is, of course, always the counter argument that the lack of mobility of screws is due to the formation of jogs (see Sec. 5).

In the following section, a very simple approximate treatment of the double kink model is given.

4.12   Simplified Double Kink Model

We shall assume that the kink configuration is independent of stress [Fig. 1(iv)], and that the interaction energy of the constrictions can be neglected. Then the double-kink self-energy is

$$U_{DK} = 2\,a(2E_0\,\Delta E)^{1/2}$$

where E_0 is the line tension, $\Delta E = (Gb^2/2\pi)(U_R - U_G)$, and the single kink length $l_K = a(E_0/2\Delta E)^{1/2}$. Using the values for Nb, $\Delta E \simeq Gb^2/120\pi$ and taking $E_0 \simeq \frac{1}{2}Gb^2$, $l_K \sim 10\,a$.

The interaction energy is

$$U_{INT} \sim \frac{Gb^2}{8\pi}\left(\frac{1 + \nu}{1 - \nu}\right)\frac{a^2}{r}$$

(Kroupa and Brown,[40] Seeger and Schiller[41]) where r is the distance between the two kinks.

The activation energy is

$$H_K = 4U_c + U_{DK} - \left[\frac{G}{2\pi}\left(\frac{1 + \nu}{1 - \nu}\right)\right]^{1/2}a^{3/2}\,b^{3/2}\,\tau^{1/2}$$

This approximation is reasonable for large r/l_K, i.e., for very small stresses. The essential difference between this model and type (1) is that H_K reaches a limiting value

$$H_K = 4U_c + U_{DK}$$

when τ goes to zero. Using the values appropriate to Nb, we find $U_{DK} \simeq 0.37$ eV and $4U_c \simeq 0.22$ eV so that H should reach an upper limit of about 0.59 eV. Conrad's results reach a limit of about 0.56 eV.

The stress below which the kink mechanism takes over is given approximately by

$$\frac{B}{\tau} = 2U_c + U_{DK} - \left[\frac{G}{2\pi}\left(\frac{1 + \nu}{1 - \nu}\right)\right]^{1/2}a^{3/2}\,b^{3/2}\,\tau^{1/2}$$

For Nb, the critical stress is 4.5 kg/mm^2, corresponding to a temperature of about 180°K. This is almost the region in temperature above which τ decreases more rapidly than predicted on Model (1). We also note that the activation volume varies at $\tau^{-1/2}$, and therefore more slowly with stress than on Model (1); this also is observed at high temperatures.

The modifications expected from this model to the theoretical curves of the activation energy and volume against stress, and to the curve of yield stress against temperature, are denoted by the dashed lines in Figs. 8, 10, and 13, respectively.

It is expected that in a fuller treatment of the kink model, in which the optimum shape of the kink and the interaction between the constrictions are taken into account, the discontinuity in $\partial H/\partial \tau$ (see Fig. 8), at the point where one model takes over from the other, will largely disappear.

5 CONCLUSIONS AND COMPARISON WITH THE PEIERLS MODEL

The dissociation model, which goes over to a kink model at high temperatures, explains with reasonable satisfaction the temperature dependence of the yield stress, the variation of activation volume with stress, the difference in yield stress for different slip planes, and the asymmetry effects. The widths of dissociation of the dislocations are very small, and the applicability of the elastic model used is doubtful. Calculations on the lines of Schoek and Seeger[42] for cross-slip in f.c.c. crystals should now be carried out.

The Peierls and dissociation models are similar in that the activation energy depends strongly on and decreases with increasing stress, and the activation volume decreases rapidly with increasing stress. Both models give a limiting activation energy at high temperatures. In addition, both models are physically similar in that the dislocation must pass through a high-energy configuration before slip can take place. In the dissociation model the increase in energy is due to constricting the dislocation to make it narrower; but even in the Peierls model the increase in energy between the energy valleys may imply a narrowing of the dislocation width. The real difference between the models is that on the dissociation model the misfit region is extended in directions out of the slip plane, whereas on the Peierls model the misfit region is confined to the slip plane.

There are, of course, differences in the functional dependence of H on τ for the two models, but it will be very difficult to differentiate between the models on the basis of experimental measurements of the stress dependence of H and v^*. It is not surprising that the values of ΔE deduced on the dissociation model (by determining parameters to fit experimental data) are of the same order as those deduced using the Peierls model (e.g., Christian and Masters,[23] Dorn and Rajnak[3]). This dissociation model, of course, explains the asymmetry effect, but as pointed out by Hirth and Lothe,[42] so could the Peierls model (qualitatively so far) since the dislocation core on {112} is asymmetric. The fact that the screws are relatively immobile is a direct consequence of the dissociation model, but could perhaps be attributed to jogs or the greater line tension and kink energy of the screw. At this stage, therefore, it does not seem possible to decide between the two mechanisms.

ACKNOWLEDGMENT

This research has been sponsored in part by the Air Force Materials Laboratory Research and Technology Division AFSC through the European office of Aerospace Research (OAR).

REFERENCES

1. Seeger, A.: *Phil. Mag.* 1:651 (1956).
2. Friedel, J.: "Electron Microscopy and Strength of Crystals," p. 605, Interscience Publishers, Inc., New York, 1963.
3. Dorn, J. E., and S. Rajnak: *Trans. AIME* 230:1052 (1964).
4. Friedel, J.: "Dislocations," Pergamon Press, 1964.
5. Friedel, J.: "Internal Stresses and Fatigue in Metals," North Holland, Amsterdam, 1959.
6. Hirsch, P. B.: *Phil. Mag.* 7:67 (1962).
7. Lomer, W. M.: *Phil. Mag.* 42:1327 (1951).
8. Cottrell, A. H.: *Phil. Mag.* 43:645 (1952).
9. Hirsch, P. B.: *Fifth International Congress of Crystallography,* Cambridge, 1960.
10. Mitchell, T. E., R. A. Foxall, and P. B. Hirsch: *Phil. Mag.* 8:1895 (1963).
11. Escaig, B.: *J. Phys.* 27:C3-205 (1966).
12. Vitek, V.: *Phys. Stat. Sol.* 18:687 (1966).
13. Vitek, V., and F. Kroupa: *Phys. Stat. Sol.* 18:703 (1966).
14. Foxall, R. A., M. S. Duesbery, and P. B. Hirsch: *Can. J. Phys.* 45:607 (1967).
15. Sleeswyk, A. W.: *Phil. Mag.* 8:1467 (1963).
16. Crussard, C.: *C. R. Acad. Sc.* 252:273 (1961).
17. Cohen, J. B., R. Hinton, K. Lay, and S. Sass: *Acta Met.* 10:894 (1962).
18. Kroupa, F.: *Phys. Stat. Sol.* 3:K-391 (1963).
19. Wasilewski, R. J.: *Acta Met.* 13:40 (1965).
20. Kroupa, F., and V. Vitek: *Can. J. Phys.* 45:945 (1967).
21. Hartley, C. S.: *Phil. Mag.* 14:7 (1966).
22. Conrad, H.: *Symposium on the Relation between the Structure and Properties of Metals* p. 475, N.P.L., Teddington, Her Majesty's Stationery Office, London, 1963.

23. Christian, J. W., and B. C. Masters: *Proc. Roy. Soc.* A28 1: 223 (1964).
24. Low, J. R., and R. W. Guard: *Acta Met.* 7:171 (1959).
25. Low, J. R., and A. M. Turkalo: *Acta Met.* 10:215 (1962).
26. Smerd, P.: private communication (1967).
27. Taylor, G., and J. W. Christian: *Phil. Mag.* 15:873, 893 (1967).
28. Taylor, G. I.: *Proc. Roy. Soc.* A1 18:1 (1928).
29. Duesbery, M. S., R. A. Foxall, and P. B. Hirsch: *Journal de Physique* 27:C3-193 (1966).
30. Sestak, B., and N. Zarubova: *Phys. Stat. Sol.* 10:239 (1965).
31. Argon, A. S., and S. R. Maloof: *Acta Met.* 14:1449 (1966).
32. Duesbery, M. S.: Ph.D. Dissertation, University of Cambridge, 1967.
33. Conrad, H., and W. Hayes: *Trans. ASM* 56:249 (1963).
34. Seeger, A., H. Donth, and P. Pfaff: *Trans. Farad. Soc.* 23:19 (1957).
35. Jøssang, T., K. Skylstad, and L. Lothe: N.P.L. Conf., Teddington, Her Majesty's Stationery Office, London, 1963.
36. Granato, A., and K. Lucke: *J. Appl. Phys.* 32: 327 (1956).
37. Taoka, T., S. Takeuchi, and E. Furukayashi: *J. Phys. Soc. Japan* 19:701 (1964).
38. Guiu, F., and P. L. Pratt: *Phys. Stat. Sol.* 15:539 (1966).
39. Guyot, P., and J. E. Dorn: *Can. J. Phys.* 45: 983 (1967).
40. Kroupa, F., and L. M. Brown: *Phil. Mag.* 6:1267 (1961).
41. Seeger, A., and P. Schiller: *Acta Met.* 10:348 (1962).
42. Schock, G., and A. Seeger: *Acta Met.* 7:469 (1959).
43. Hirth, P., and J. Lothe: *Phys. Stat. Sol.* 15:487 (1966).
44. Taylor, G.: private communication (1967).
45. Bowen, D. K., J. W. Christian, and G. Taylor: *Can. J. Phys.* 45:903 (1967).
46. Reid, C. N., A. Gilbert, and G. T. Hahn: *Acta Met.* 14:975 (1966).
47. Rose, R. M., D. P. Ferris, and J. Wulff: *Trans. AIME* 224:981 (1962).
48. Stein, D. F., and J. R. Low: *Acta Met.* 14:1183 (1966).

KINETICS AND DYNAMICS
IN DISLOCATION PLASTICITY

J. C. M. Li

Edgar C. Bain Laboratory for Fundamental Research,
U. S. Steel Corporation, Research Center,
Monroeville, Pennsylvania

ABSTRACT

The rate equation of dislocation mobility is considered to contain three variables—temperature, pressure, and the effective shear stress. Thermodynamic relations involving the activated state are discussed by means of a Jacobian table. The effect of hydrostatic pressure on creep is reviewed, and the results indicate, in general, a vacancy mechanism. For Zn, a jog mechanism along screw dislocations is suggested. The stress dependence of creep rate can be interpreted in terms of a stress-dependent activation area. This dependence gives several velocity-stress relations, one of which is the power law. The determination of the velocity-stress exponent in this law, from stress relaxation experiments, is summarized and compared with direct etch pit measurements. The consequences of the assumption of zero entropy of activation are discussed.

Low-temperature deformation of potassium is investigated and found to be consistent with a Peierls mechanism for all b.c.c. metals. The problem of the athermal component of flow stress is discussed in terms of dislocation dynamics in a periodic internal stress field. The apparent activation parameters introduced by a sinusoidal internal stress are calculated. A modification of the Mott-Nabarro theory of solid-solution hardening is suggested as a result of considering periodic internal stresses.

1   INTRODUCTION

The concept of thermally activated plastic deformation was introduced as early as 1925 when Becker[1] applied the Boltzmann principle to the nucleation of a slip region. Later, Kanter[2] proposed a self-diffusion theory of secondary creep using the Dushman-Langmuir[3] equation for diffusion. After the introduction of the absolute reaction rate theory of Eyring,[4] Kauzmann[5] formulated a general chemical rate theory of plasticity. Similar equations were derived by Seitz and Read[6] based on thermally activated dislocation motion, and by Nowick and Machlin[7] based on thermally activated dislocation generation. Later efforts concentrated on specific models and on details of the mechanism implied from experimental activation parameters. Questions often asked concern the stress dependence of activation enthalpy,[8,9] the stress and temperature dependence of activation entropy,[9-11] and whether the temperature dependence of shear modulus[11,12] should be considered.

As an aid in answering these questions, an attempt is made in this paper to present a more general formulation making no assumption concerning the activation parameters. In addition to the effect of shear stress and temperature, the effect of hydrostatic pressure or the hydrostatic component of the stress tensor is included. Thermodynamic relations among the activation parameters are discussed so that the consequences of any assumption can be examined. In particular, as an example, the assumption of zero entropy of activation is investigated. Some activation parameters are calculated from recent creep data and their implications on the mechanism of creep are discussed. Recent attempts to obtain velocity-stress relations from stress relaxation experiments are summarized. The relation between the velocity-stress relations and the activation parameters is shown.

Low-temperature deformation of b.c.c. metals is discussed in the light of recent data on potassium. Most of the evidence seems to favor the Peierls-Nabarro barrier as the rate-controlling mechanism. An important remaining problem is the understanding of the athermal component. In this paper an attempt is made to study dislocation dynamics in a periodic internal stress field. An average velocity is calculated as well as the effective internal stress (the athermal component) and its variation with strain rate and temperature. The apparent activation parameters as a result of a sinusoidal internal stress field also are calculated and compared with experimental data near the athermal region. A modified Mott-Nabarro theory of solid-solution hardening is presented as a result of considering the periodic internal stresses.

2 RATE EQUATIONS

The average velocity of a dislocation traveling in an impure, imperfect crystal, such as that measured by etch-pitting techniques, can be considered as a thermally activated process.

$$v = v_c \exp \frac{-\Delta F^{\ddagger}}{kT} \tag{1}$$

where $\Delta F^{\ddagger}$ is the standard free energy of activation, k is the Boltzmann constant, T is the absolute temperature, and v_c is the velocity when $\Delta F^{\ddagger}$ is zero. The term v_c may contain the mean distance the dislocation moves per activation event, a fundamental frequency such as kT/h with h being the Planck constant, and a possible geometric factor. On the other hand, v_c can simply be regarded as the maximum attainable velocity such as the shear wave velocity in the crystal. Dislocations that can travel at such a velocity are in the fully activated state which has a maximum free energy $\Delta F_0^{\ddagger}$ higher than the normal state per activation event.

2.1 Activation Area

If a shear stress τ^* is applied in the slip plane so that τ^* does positive work when the dislocation moves forward, then the free energy of activation is decreased for forward motion and increased for backward motion by $\tau^* b A^*$, where b is the Burgers vector of the dislocation and A^* is the area swept by the dislocation during an activation event (activation area). This indicates that external stress alone may fully activate the dislocation. The stress that can achieve this is τ_c^* which is defined as the friction stress. Let the activation area be A_0^* at $\tau^* = 0$; a consideration of the reversible process shows

$$\Delta F_0^{\ddagger} = b \int_0^{A_0^*} \tau^* \, dA^* \tag{2}$$

assuming that a relation exists between τ^* and A^* during the activation event. Hence at an applied stress τ^*, the activation free energy for the forward motion is

$$\Delta F_f^{\ddagger} = \Delta F_0^{\ddagger} - b\tau^* A^* - b \int_{A^*}^{A_0^*} \tau^* \, dA^*$$

$$= \Delta F_0^{\ddagger} - b \int_0^{\tau^*} A^* \, d\tau^* \tag{3}$$

Similarly, the activation free energy for the backward motion is

$$\Delta F_b^{\ddagger} = \Delta F_0^{\ddagger} + b \int_0^{\tau^*} A^* \, d\tau^* \tag{4}$$

Equations (3) and (4) give τ^* the average velocity of the dislocation:[13,14]

$$v = 2 v_c \exp \frac{-\Delta F_0^{\ddagger}}{kT} \sinh \frac{b}{kT} \int_0^{\tau^*} A^* \, d\tau^* \tag{5}$$

which at small τ^* gives

$$v = 2 v_c \frac{A_0^* \tau^* b}{kT} \exp \frac{-\Delta F_0^{\ddagger}}{kT} \tag{6}$$

a linear relation between stress and velocity. At large τ^* the velocity becomes

$$v = v_c \exp \frac{-\left(\Delta F_0^{\ddagger} - b \int_0^{\tau^*} A^* \, d\tau^* \right)}{kT} \tag{7}$$

A comparison with Eq. (1) shows

$$A^* = -\frac{1}{b} \left(\frac{\partial \Delta F^{\ddagger}}{\partial \tau^*} \right)_{T,P} = \frac{kT}{b} \left[\frac{\partial \ln(v/v_c)}{\partial \tau^*} \right]_{T,P} \tag{8}$$

It is to be noted that Eq. (8) is valid only if the hyperbolic sine function in Eq. (5) can be approximated by an exponential function. In the literature the quantity A^*b is sometimes called the "activation volume." To avoid confusion with the activation volume defined later as the pressure derivative of the standard free energy of activation, the term "activation area" is introduced here and defined by Eq. (5).

2.2 The Activation Enthalpy

The activation enthalpy is defined in thermodynamics by

$$\Delta H^\ddagger = \left[\frac{\partial(\Delta F^\ddagger/T)}{\partial(1/T)}\right]_{\tau^*, P} = kT^2\left[\frac{\partial \ln(v/v_c)}{\partial T}\right]_{\tau^*, P} \tag{9}$$

No correction of the variation of shear modulus μ with temperature is necessary if τ^* is the effective stress exerted on the dislocation. Sometimes the ratio of τ^*/μ is recommended as an independent variable rather than τ^* itself. However, at present, this is also unnecessary since our understanding of the fundamental mechanism is still limited. When τ^* is large, Eq. (8) can be combined with Eq. (9) to give the following:

$$\Delta H^\ddagger = -bA^*T\left(\frac{\partial \tau^*}{\partial T}\right)_{P, v/v_c} \tag{10}$$

Both Eqs. (9) and (10) are widely accepted definitions of activation enthalpy. However, in view of the preceding considerations, they are exact only at high stresses.

2.3 The Activation Volume

Again from thermodynamics the activation volume is defined by

$$\Delta V^\ddagger = \left(\frac{\partial \Delta F^\ddagger}{\partial P}\right)_{\tau^*, T} = -kT\left[\frac{\partial \ln(v/v_c)}{\partial P}\right]_{\tau^*, T} \tag{11}$$

where P is hydrostatic pressure. At large τ^*, Eq. (8) can be combined with Eq. (11) to give

$$\Delta V^\ddagger = bA^*\left(\frac{\partial \tau^*}{\partial P}\right)_{T, v/v_c} \tag{12}$$

Here again no correction of the variation of shear modulus with pressure is necessary since, at present, our understanding of the mechanism is still limited. Similarly, there is no need to use the ratio of pressure to the bulk modulus instead of the pressure itself as an independent variable.

2.4 The Activation Entropy

The activation entropy is defined by

$$\Delta S^{\ddagger} = -\left(\frac{\partial \Delta F^{\ddagger}}{\partial T}\right)_{\tau^*, P} = k\left[\frac{\partial T \ln(v/v_c)}{\partial T}\right]_{\tau^*, P} = \frac{\Delta H^{\ddagger} - \Delta F^{\ddagger}}{T} \tag{13}$$

It is seen that $\Delta H^{\ddagger}$ is not equal to $\Delta F^{\ddagger}$ unless $\Delta S^{\ddagger}$ is zero. Since a linear relation between $\Delta F^{\ddagger}$ and T is expected from Eq. (1) for constant v/v_c, no such linear relation between $\Delta H^{\ddagger}$ and T should be expected (as is often the case) unless $\Delta S^{\ddagger}$ is constant. The possibility of $\Delta S^{\ddagger}$ approaching zero at $0°K$ has been examined for the case of silicon-iron[14] and potassium[15-16] and shown to be very likely. At other temperatures, $\Delta S^{\ddagger}$ can be positive or negative at least from the data of silicon-iron.[14] The validity of the assumption that $\Delta S^{\ddagger}$ should depend only on the variation of shear modulus with temperature[12] is therefore doubtful.

2.5 Thermodynamic Relations

Equations (8), (9), and (11) indicate that the following is a total differential of three independent variables:

$$d \ln \frac{v}{v_c} = \frac{A^* b}{kT} d\tau^* + \frac{\Delta H^{\ddagger}}{kT^2} dT - \frac{\Delta V^{\ddagger}}{kT} dP \tag{14}$$

For the convenience of obtaining relations among the various quantities, we prepare a Jacobian table (Table 1). Six new symbols are introduced for the following partial derivatives:

$$C = \left(\frac{\partial \Delta H^{\ddagger}}{\partial T}\right)_{\tau^*, P} = T\left(\frac{\partial \Delta S^{\ddagger}}{\partial T}\right)_{\tau^*, P} \tag{15}$$

$$\alpha = \left(\frac{\partial \Delta V^{\ddagger}}{\partial T}\right)_{\tau^*, P} = -\left(\frac{\partial \Delta S^{\ddagger}}{\partial P}\right)_{T, \tau^*} \tag{16}$$

TABLE 1 First Partial Derivatives of Dislocation Mobility with Temperature, Pressure, and Effective Shear Stress as Independent Variables

	$\partial/\partial T$	$\partial/\partial P$	$\partial/\partial \tau^*$
T	1	0	0
P	0	1	0
τ^*	0	0	1
$\ln(v/v_c)$	$H^\ddagger/kT^2$	$-\Delta V^\ddagger/kT$	$A^* b/kT$
$\Delta F^\ddagger$	$-\Delta S^\ddagger$	$\Delta V^\ddagger$	$-bA^*$
$\Delta H^\ddagger$	C	$\Delta V^\ddagger - \alpha T$	$b(\gamma T - A^*)$
$\Delta V^\ddagger$	α	ϕ	$-b\beta$
A^*	γ	β	θ
$\Delta S^\ddagger$	C/T	$-\alpha$	$b\gamma$

$$\beta = \left(\frac{\partial A^*}{\partial P}\right)_{T,\tau^*} = -\left(\frac{1}{b}\right)\left(\frac{\partial \Delta V^\ddagger}{\partial \tau^*}\right)_{T,P} \tag{17}$$

$$\gamma = \left(\frac{\partial A^*}{\partial T}\right)_{P,\tau^*} = \left(\frac{1}{b}\right)\left(\frac{\partial \Delta S^\ddagger}{\partial \tau^*}\right)_{T,P} \tag{18}$$

$$\theta = \left(\frac{\partial A^*}{\partial \tau^*}\right)_{T,P} \tag{19}$$

$$\phi = \left(\frac{\partial \Delta V^\ddagger}{\partial P}\right)_{T,\tau^*} \tag{20}$$

As an example, the slope of $\Delta H^\ddagger$ vs. T plot at constant velocity and pressure can be calculated as follows:

$$\left(\frac{\partial \Delta H^\ddagger}{\partial T}\right)_{P,v/v_c} = \frac{J(\Delta H^\ddagger, P, v/v_c)}{J(T, P, v/v_c)} = \frac{\Delta H^\ddagger}{T} + C - \frac{\gamma}{A^*}\Delta H^\ddagger \tag{21}$$

It is seen that $\Delta H^\ddagger$ will be proportional to T if and only if $C = \gamma \Delta H^\ddagger/A^*$ or $\Delta H^\ddagger/A^*$ is independent of temperature at constant τ^* and P. Sometimes the quantity $\Delta H^\ddagger/T$ at constant strain rate is taken as k times the logarithm of the preexponential factor or of the density of mobile dislocations. It is seen here that even if the density of mobile dislocations remains constant, $\Delta H^\ddagger/T$ can still vary with temperature:

$$\left[\frac{\partial(\Delta H^\ddagger/T)}{\partial T}\right]_{P,v/v_c} = \frac{\Delta H^\ddagger}{T}\left[\frac{\partial \ln(\Delta H^\ddagger/A^*)}{\partial T}\right]_{\tau^*,P} \tag{23}$$

Another popular plot is $\Delta H^{\ddagger}$ vs. $\ln \tau^*$ at constant velocity and pressure. The advantage of this plot is the possibility of obtaining $\Delta H^{\ddagger}$ at $\tau^* = 0$ and τ_c^* at $\Delta H^{\ddagger} = 0$. The slope of this plot is as follows:

$$\left(\frac{\partial \Delta H^{\ddagger}}{\partial \ln \tau^*}\right)_{P,\,v/v_c} = b\tau^*\left(\gamma T - A^* - \frac{A^*CT}{\Delta H^{\ddagger}}\right) \tag{23}$$

which should approach zero at small τ^* and a limiting value of $-b\tau_c^* A_c^*$ at τ_c^* near $0°K$.

3   THE ACTIVATION VOLUME AND THE ACTIVATION AREA FOR CREEP

3.1   The Activation Volume

Recent data on the effect of hydrostatic pressure on the creep rate are summarized in Table 2 in terms of the activation volume. It is seen that in almost all cases, the activation volume for creep is similar to that for self-diffusion, indicating a creep mechanism involving vacancies. This is consistent with the fact that the activation enthalpy for creep is also similar to that for self-diffusion.[17]

In Zn the activation volume for self-diffusion parallel to the c axis is different from that perpendicular to it, although the values approach each other at high pressures. The activation volume for

TABLE 2 Activation Volume for Creep of Various Crystals

Crystal	Activation Volumes cc/g-atom		Atomic Volume V cc/g-atom	$\Delta V^{\ddagger}/V$
	Creep	Self-diffusion		
K	$25^{(a)}(28°C)$	—	45.5	0.55
Na	$12.6^{(j)}$ (-20 to 90°C)	$9.6^{(c,d)}$ (-45°C, 90°C)	23.7	0.53
P	$30^{(e)}$ (27°C, 41°C)	$30^{(b)}$ (30°C)	68.2	0.44
Pb	$14^{(f)}$ (70°C)	$13^{(g)}$ (301°C)	18.2	0.77
		$15^{(g)}$ (253°C)		
AgBr	$38^{(b)}$ (300–400°C)	$42^{(i)}$ (350°C)	29	1.31
Sn	$5.1^{(k)}$ (0-57°C)	—	16.3	0.31
Zn	$6^{(l)}$ (27-57°C)	$16^{(m)} \parallel c$ axis $\quad$ (307°C) $5^{(m)} \perp c$ axis	9.6	0.65
Cd	$3.2^{(l)}$ (0-57°C)	—	13	0.63
Al	$13.6^{(n)}$ (270°C)	—	10	1.36
LiF	$40^{(0)}$ (25°C)	—	9.7	4.1

creep agrees better with that for self-diffusion along the basal plane. This observation favors a jog mechanism in which the mobility of screw dislocations in the basal plane is hindered by jogs which climb also in the basal plane. On the other hand, the climb of edge dislocations out of the basal plane would involve diffusion of vacancies parallel to the c axis. The activation volume for the creep of Zn is therefore inconsistent with a mechanism involving the climb of edge dislocations.

3.2 The Activation Area

Although in low-temperature deformation the activation area (under the name "activation volume" which is bA^*) has been used extensively to determine the force-distance (or the stress-area) curve for the activation process,[13, 19-21] it is seldom used in high-temperature creep deformation. As an example the creep data of Christy[22] for AgBr single crystals compressed in the [001] direction are replotted in Fig. 1 showing a linear relation between the logarithm of creep rate and the logarithm of stress at each temperature. The slope m is smaller at higher temperatures. From these straight lines, the activation area is calculated from $A^* = mkT/b\tau^*$ where the shear stress τ^* is taken as one-half of the compressive stress for $(011)[01\bar{1}]$ slip. The activation areas are shown in Fig. 2. Similar to the case of low-temperature deformation, the activation area is more sensitive to stress than to temperature. The shape of the activation area-stress curve indicates a reasonable force-distance relation for the activation process.

The large magnitudes of a few thousand b^2 for the activation area shown in Fig. 2 are not typical for creep. The activation area

References for Table 2

(a) Kohler, C. R. and A. L. Ruoff: J. Appl. Phys. 36: 2444 (1965).

(b) Nachtrieb, N. H. and A. W. Lawson: J. Chem. Phys. 23: 1193 (1955).

(c) Nachtrieb, N. H., J. A. Weil, E. Catalano and A. W. Lawson: J. Chem. Phys. 20: 1189 (1952); see Ref. (a).

(d) Hultsch, R. A. and R. G. Barnes: Phys. Rev. 125: 1832 (1962).

(e) DeVries, K. L., P. Gibbs, H. Miles and H. S. Staten: J. Appl. Phys. 35: 536 (1964).

(f) Butcher, B. M. and A. L. Ruoff: J. Appl. Phys. 32: 2036 (1961).

(g) Nachtrieb, N. H., H. A. Resing and S. A. Rice: J. Chem. Phys. 31: 135 (1959).

(h) Christy, R. W.: Acta Met. 2: 284 (1954).

(i) Tannhauser, D. S.: J. Phys. Chem. Solids 5: 224 (1958).

(j) Ruoff, A. L., R. H. Cornish and B. M. Butcher: Bull. Am. Phys. Soc. 6: 420 (1961).

(k) DeVries, K. L., G. S. Baker and P. Gibbs: J. Appl. Phys. 34: 2258 (1963).

(l) DeVries, K. L. and P. Gibbs: J. Appl. Phys. 34: 3119 (1963).

(m) Liu, T. and H. G. Drickamer: J. Chem. Phys. 22: 312 (1954).

(n) Butcher, B. M., H. Hutto and A. L. Ruoff: Appl. Phys. Letters 7: 34 (1965).

(o) Hanafee, J. E. and S. V. Radcliffe: J. Appl. Phys. 38: 4284 (1967).

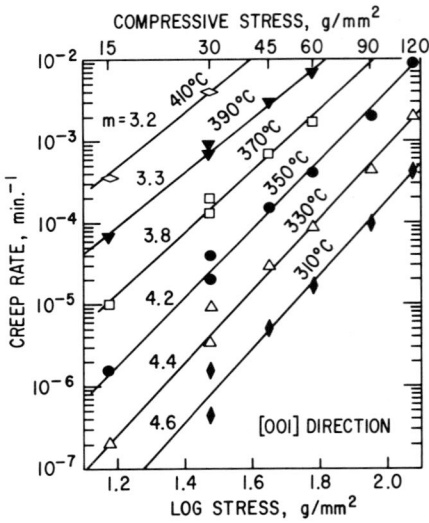

Fig. 1 Strain rate-stress exponent for the
creep of AgBr single crystals. (Data of
R. W. Christy)

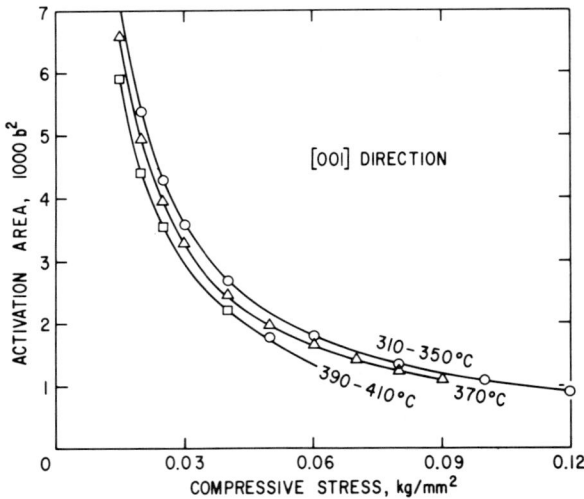

Fig. 2 Activation area for the creep of AgBr single crystals.
(Data of R. W. Christy)

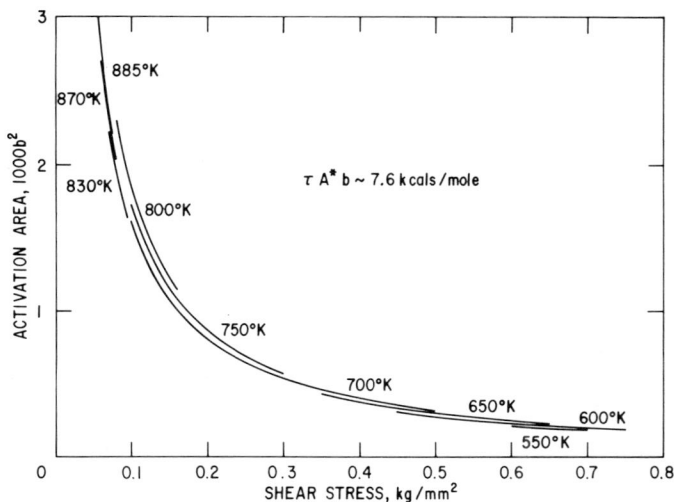

Fig. 3 Activation area for the creep of single crystal Mg-12% Li alloy oriented for basal slip. (Data of Chirouze, Schwartz, and Dorn)

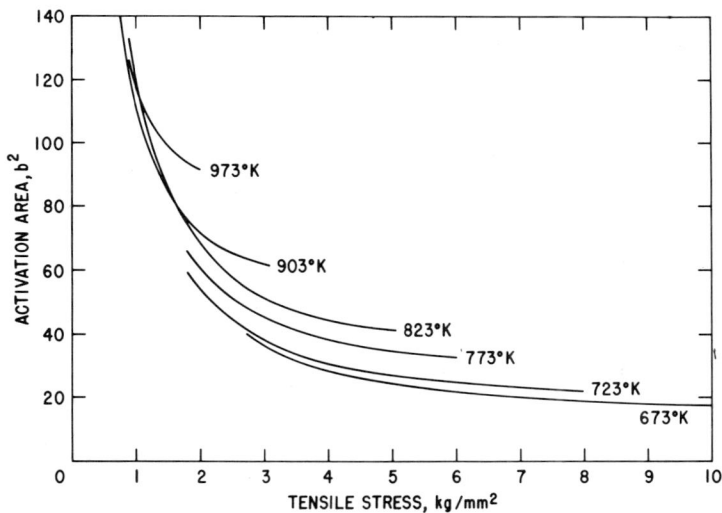

Fig. 4 Activation area for the creep of polycrystalline copper. (Data of Feltham and Meakin)

is smaller at larger stresses. For example, the activation area for the creep of Mg-12 a/o Li alloy single crystals[23] are shown in Fig. 3. The activation area now ranges from 3000 to 200 b^2. Here again the activation area is almost independent of temperature but

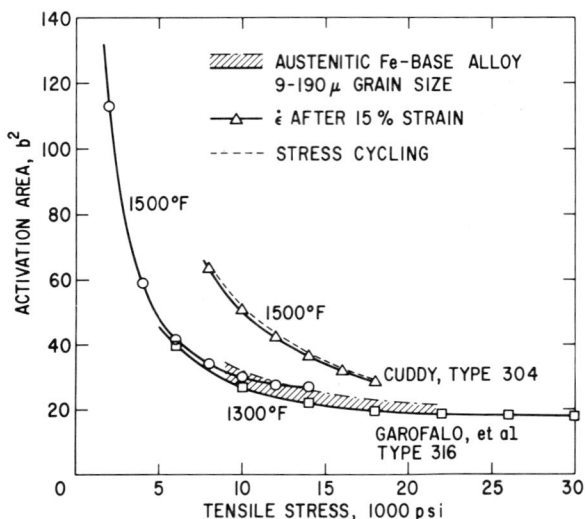

Fig. 5 Activation area for the creep of stainless steel.

depends strongly on stress. With further increase in stress such as that shown in Fig. 4 for copper,[24] the activation area can be as small as 20 b^2. Such small activation area is found also in stainless steel[25-28] as shown in Fig. 5. In the figure, the activation area is shown to be almost independent of grain size for an austenitic Fe base alloy and to be almost the same either by stress cycling on the same specimen or by using different specimens for the 304 stainless steel. The results on both steels indicate that the activation area is structure-independent.

3.3   Dislocation Mechanism for Creep

Figures 2 through 5 can be combined to form an almost continuous curve between the activation area and the shear stress. This observation plus the fact that both activation enthalpy and activation volume agree with those for self-diffusion, indicates that under these conditions, a single dislocation mechanism operates. Several dislocation mechanisms have been suggested in the literature (see Garofalo[18] for a recent review on these mechanisms). Unlike the dislocation mechanisms proposed for low-temperature deformation, none of those proposed for high-temperature creep allow a variation of activation area with stress.

Instead, the stress dependence of creep rate is interpreted by mechanisms other than the dislocation mobility itself.

For example, Weertman[29] interpreted part of the stress dependence of creep rate by using a specific arrangement of edge dislocations. He proposed that edge dislocations of opposite sign gliding on parallel slip planes would interact and pile up when a critical distance between the two slip planes was not exceeded. These two pileups then annihilate each other by climb. Unfortunately, as pointed out by Head[30] and confirmed later by Li,[31] such pileups never form. Even without applied stress, parallel dislocations of opposite sign generated by two sources on two parallel slip planes would meet, pass each other, and form dipoles. The annihilation of any of these dipoles does not seem to be a necessary step for the continuation of creep.

As another example, Barrett and Nix[32] interpreted part of the stress dependence of creep rate by suggesting that the density of mobile dislocations changes with the third power of stress. The activation area of dislocation motion was assumed independent of stress. Although stress dependence of dislocation density during steady-state creep was measured by etch-pitting techniques on Si-Fe, it was difficult to determine whether these were mobile dislocations. Furthermore, the stress dependence of creep rate is about the same[23, 28] either from steady-state measurements on different specimens or from stress-change experiments on the same specimen. Unless the density of mobile dislocations respond immediately to the applied stress, the stress dependence of creep rate cannot have a contribution from the density of mobile dislocations. So far, no great discrepancy between the activation areas obtained from dislocation velocity measurements and those from strain-rate change experiments has been noted. This seems to suggest that the density of mobile dislocations does not change significantly in stress-change or strain-rate change experiments. Therefore, it is likely that the stress dependence of creep rate arises from the stress dependence of dislocation mobility.

In view of the above considerations, it may not be impossible for the activation area to be actually stress-dependent, which is similar to the case of low-temperature deformation. Consider, for example, an interstitial-producing jog on a screw dislocation. The slip process depends on the successive dissolution of an extra row of atoms by either producing interstitials or absorbing vacancies. The applied stress actually exerts a force on the terminal atom to push it into an interstitial position. Because of the distorted core structure at the jog, the distance through which the terminal atom has to move in order to become an interstitial may be several Burgers vectors. Depending on the applied stress, the terminal atom can move only so much by stress alone and the energy associated with the rest of movement has to be supplied thermally, for example, by introducing a vacancy. The distance of movement during the thermally activated step depends therefore on the applied stress. In other words, it is possible to have a stress-dependent activation area.

One of the early arguments against the stress-enhanced activation process is the apparent independence of activation enthalpy with stress. It can easily be shown from Table 1 that

$$\left(\frac{\partial \Delta H^{\ddagger}}{\partial \tau^*}\right)_{T,P} = bA^* \left[\left(\frac{\partial \ln A^*}{\partial \ln T}\right)_{\tau^*,P} - 1\right] \tag{24}$$

It is seen that even if the activation enthalpy is independent of stress, it does not follow that A^* is zero. When A^* is proportional to the absolute temperature, $\Delta H^{\ddagger}$ will be stress-independent no matter how large A^* is. The A^* vs. T proportionality is easily realized when the velocity-stress exponent m^*

$$m^* = \left(\frac{\partial \ln v}{\partial \ln \tau^*}\right)_{T,P} = \frac{\tau^* bA^*}{kT} \tag{25}$$

is independent of temperature. Physically, A^* may change with temperature when the core structure or the distance between jogs or nodes changes with temperature.

4 THE VELOCITY STRESS RELATION

As shown in Figs. 2 through 5 for creep and also for low-temperature deformation, the activation area decreases when the stress increases. When the activation area can be approximated by an inverse proportionality to the stress, a velocity-stress relation

$$v = B(\tau^*)^{m^*} \tag{26}$$

results. In this equation B and m^* are independent of stress but may be functions of both temperature and pressure. Of course, within any small range of stress, the inverse proportionality can always be approximated. Then the m^* is defined by Eq. (25) and may vary with τ^*.

Similarly, when the activation area can be approximated by an inverse proportionality to the square of τ^*, a velocity stress relation

$$v = v_c \exp\left(\frac{-\tau_0}{\tau^*}\right) \tag{27}$$

results. In this equation v_c is the velocity at $\tau^* \to \infty$ and τ_0 is a drag stress. Both v_c and τ_0 may be functions of temperature and pressure.

Both Eqs. (26) and (27) have been used in the literature. When the stress range is small, it is difficult to differentiate between them. A relation exists between τ_0 and m^*:

$$\tau_0 = m^* \tau^* \tag{28}$$

Therefore, it is only necessary to use one of the equations. The following illustrates an application of Eq. (26) to stress-relaxation experiments.

In a stress-relaxation experiment, a specimen is deformed and then the machine is stopped. The stress in the specimen is now relaxed. The elastic strain of the system (specimen and the machine) is replaced by the plastic strain of the specimen:

$$\dot{\epsilon}_p = -\dot{\epsilon}_e = -\frac{1}{E}\dot{\sigma} = \rho b B (\sigma - \sigma_i)^{m^*} \tag{29}$$

where $\dot{\epsilon}_p$ is the plastic strain rate of the specimen, $\dot{\epsilon}_e$ is the elastic strain rate of the system measured on the specimen, E is the combined Young's modulus measured on the load-elongation chart, $\dot{\sigma}$ is the stress rate, ρ is the density of mobile dislocations, and σ_i is the internal stress. By assuming constant ρ Eq. (29) can be integrated to give the stress-time relation:

$$\sigma - \sigma_i = K(t + a)^{-n} \tag{30}$$

with $n = 1/(m^* - 1)$. Differentiating Eq. (30) gives,

$$\dot{\sigma} = -nK(t + a)^{-n-1} \tag{31}$$

The application of these equations to experimental data has already been illustrated [15] and complete accounts of these experiments will be published later.[34] The velocity-stress exponent m^*, determined by this method is compared in Table 3 with that from dislocation

TABLE 3 The Velocity-stress Exponent at Room Temperature

Crystal	m^*	
	Etch pitting	Stress relaxation (purification technique or material grade)
NaCl	$3.9^{(a)}$	4–5 (optical grade)
Nb	$6.7^{(b)}$	5–7 (three zone pass)
Nb	$15^{(c)}$	12–16 (one zone pass)
W	$4.8^{(d)}$	5–9 (zone refined)
LiF	$6^{(e)}$	5–7 (optical grade)
Si–Fe	$35^{(f)}$	20–30 (ZrH_2 purified)

(a) Padawer, G. E.: ScD Thesis, MIT (1966).
(b) Guberman, H. D.: to be published.
(c) Guberman, H. D.: *Radiation Metallurgy Section Solid State Div., Progr. Rept.* ORNL-4020, p. 41 (1966).
(d) Schadler, H. W.: *Acta Met.* 12, 861 (1964).
(e) Sankaran, S. and J. C. M. Li: unpublished results.
(f) Stein, D. F. and J. R. Low, Jr.: *J. Appl. Phys.* 31, 362. (1960).

velocity measurements using etch–pitting techniques. Except for W and Si–Fe, the two methods agree satisfactorily. For W, our material (from *Matl. Res. Corp.*) may not be as pure as that used by Shadler. For Si–Fe we have to remove interstitial C so as to avoid strain raging during relaxation.

5   IDEAL BEHAVIOR WHEN $\Delta S^{\ddagger} = 0$

Equation (1) is written very often with $\Delta H^{\ddagger}$ instead of $\Delta F^{\ddagger}$. Implicitly, it is assumed that $\Delta S^{\ddagger} = 0$. However, seldom is time taken to examine the consequences of this assumption. It is seen from Table 1 that when $\Delta S^{\ddagger}$ is always zero, the quantities C, α, and γ would always be zero and therefore $\Delta H^{\ddagger}$, $\Delta V^{\ddagger}$, and A^* are all temperature-independent at constant τ^* and P. It follows from Eqs. (25) and (28) that the quantities m^*T and $\tau_0 T$ are both temperature independent or the quantities m^* and τ_0 both vary inversely with temperature.

In a strain–rate cycling experiment, if $\Delta\sigma/\Delta \ln\dot{\epsilon}$ can be approximated by σ/m or by σ^2/σ_0

$$\Delta\sigma = \frac{\Delta \ln\dot{\epsilon}}{(mT)}\sigma T, \quad \text{or} \quad \frac{\Delta \ln\dot{\epsilon}}{(\sigma_0 T)}\sigma^2 T \tag{32}$$

and if (mT) or $(\sigma_0 T)$ is constant, Eq. (32) shows that $\Delta\sigma$ is zero at $T = 0$ and, when T increases, σ decreases so that σT or $\sigma^2 T$ first increases then decreases. Hence, at least qualitatively, Eq. (32) shows a $\Delta\sigma - T$ plot with a maximum. Quantitative comparison with experimental results would depend on the validity of the approximation and the assumption just mentioned.

As another consequence of $\Delta S^{\ddagger} = 0$, it is seen from Table 1 that $(\partial\Delta H^{\ddagger}/\partial\tau^{\ddagger})_{T,P} = -m^* kT/\tau^*$ because $\gamma = 0$. Then, when $m^* T$ is shown experimentally to be constant between τ_c^* and τ^*, the $\Delta H^{\ddagger}$ at τ^* is

$$\Delta H^{\ddagger} = m^* kT \ln\left(\frac{\tau_c^*}{\tau^*}\right) \tag{33}$$

which indicates a linear plot between $\Delta H^{\ddagger}$ and $\ln \tau^*$ at constant pressure. Also from Table 1, $\partial\tau^*/\partial T$ at constant v/v_c and pressure is equal to $-\tau^*\Delta H^{\ddagger}/m^* kT^2$. In view of Eq. (33), the stress-temperature relation under these conditions is

$$\ln\left(\frac{\tau^*}{\tau_c^*}\right) = -\left(\frac{\Delta H^{\ddagger}}{m^* kT^2}\right)_T \tag{34}$$

which indicates a linear relation between $\ln \tau^*$ and T. The slope is constant because both $\Delta H^{\ddagger}/T$ and $m^* T$ are constants at constant v/v_c and pressure.

Instead of m^*, the foregoing consideration can be repeated for τ_0 using Eq. (28). The enthalpy-stress relation is now

$$\Delta H^{\ddagger} = \tau_0 kT\left(\frac{1}{\tau^*} - \frac{1}{\tau_c^*}\right) \tag{35}$$

assuming constant τ_0 between τ^* and τ_c^*. The stress-temperature relation becomes

$$\ln\left(\frac{1}{\tau^*} - \frac{1}{\tau_c^*}\right) = \left(\frac{\Delta H^{\ddagger}}{\tau_0 kT^2}\right) T \tag{36}$$

which is a linear relation between $\ln (\tau_c^* - \tau^*)/\tau^*$ and T.

Although these consequences are not exactly found by experiment, they are approximately true. Furthermore, almost all the theoretical models proposed for dislocation mobility ignore the entropy contribution. Therefore it is important to examine the experimental results in the light of these consequences of $\Delta S^{\ddagger} = 0$ before a meaningful comparison with the theoretical models is possible.

6   LOW-TEMPERATURE DISLOCATION MECHANISMS IN B.C.C. METALS

6.1   Low-Temperature Deformation of Potassium

Recent work on potassium[16,35,36] shows that its deformation characteristics are similar to those of other b.c.c. metals. Strong temperature dependence of yield stress is observed below $\sim 25°$K. Extrapolated yield stress at $0°$K is about $0.002\,\mu$ (μ: shear modulus), which is somewhat lower than that of other b.c.c. metals (about $0.01\,\mu$). Activation enthalpy[36] at about $25°$K is about $0.08\,\mu b^3$ similar to other b.c.c. metals (about $0.1\,\mu b^3$)[37] as shown in Fig. 6. Activation area at $10^{-3}\mu$ is about $30\ b^2$ which is also similar to other b.c.c. metals ($20\ b^2$). Impure potassium has a higher yield stress than pure potassium at all temperatures. The difference is either independent of, or increasing with temperature. This seems to rule out impurity as the cause of the thermal component of yield stress. It was concluded[36] that low-temperature deformation of b.c.c. metals is most probably controlled by the Peierls-Nabarro barrier.

6.2   The Problem of the Athermal Component

Although the thermal component of the yield stress is due to a thermally activated process, the cause for the athermal component is not yet clear. After considering all the possible mechanisms, it seems that the internal stress that a dislocation may experience during its journey must have positive and negative regions since the space average of the internal stress must be zero. For example, an edge dislocation passing by another parallel edge dislocation on a parallel slip plane experiences a force which changes sign three times. A flexible dislocation passing through a collection of pinning points must have regions of positive and negative curvature and hence is subjected to forces of different signs at different times and places. In view of this it seems questionable to assume a constant internal stress as the athermal component. Hence it was decided to study the effect of a periodic internal stress field on the average dislocation velocity and, from which, deduce the apparent activation parameters. This will be discussed in Sec. 7.

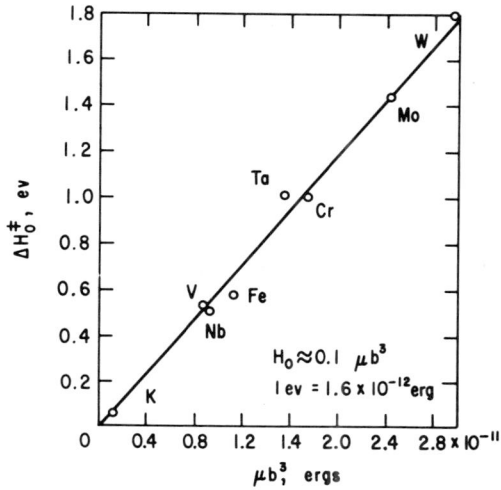

Fig. 6 Relation between the activation enthalpy at $\tau^* = 0$ and the shear modulus for bcc metals.

7 DISLOCATION DYNAMICS IN A PERIODIC INTERNAL STRESS FIELD

As mentioned in Sec. 6.2, the internal stress field experienced by any dislocation during its journey must have both positive and negative regions; a sinusoidal periodic internal stress field will be used as a first approximation:

$$\tilde{\tau}_i = \tau_i \sin\left(\frac{2\pi \xi}{L}\right) \tag{37}$$

where $\tilde{\tau}_i$ is the internal stress at a distance ξ, τ_i is the maximum internal stress and L is the period. The instantaneous velocity $\tilde{v}$ will be assumed to follow a power relation with the instantaneous effective stress

$$\tilde{v} = B(\tau - \tilde{\tau}_i)^{m^*} \tag{38}$$

An average velocity can be calculated from Eqs. (37) and (38) by integrating $1/\tilde{v}$ over the period L and by dividing into L:

$$\bar{v} = \frac{v}{x^{m^*} P_{m^* - 1}(x)} \tag{39}$$

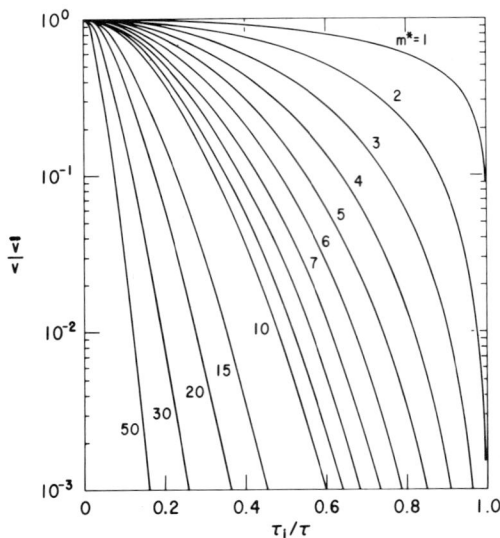

Fig. 7 Average dislocation velocity in a periodic
internal stress field.

where v is the velocity at τ when $\tau_i = 0$, x is given by

$$\frac{1}{x^2} = 1 - \left(\frac{\tau_i}{\tau}\right)^2 \tag{40}$$

and $P_{m^* - 1}(x)$ is the Legendre function of argument x and of degree $m^* - 1$. Chen et al.[38] also calculated the average velocity in a periodic internal stress field by using a velocity-stress relation shown by Eq. (27). However, since the integration cannot be expressed by known analytic functions, it is difficult to examine the apparent activation parameters. Equation (39) is more convenient in this respect.

Equation (39) is plotted in Fig. 7 to show the average velocity as affected by the ratio of internal stress to applied stress and by the velocity-stress exponent. It is seen that the average velocity is always less than the velocity without the internal stress. The decrease is more pronounced when the velocity-stress exponent is large.

7.1   The Effective Internal Stress

An effective stress τ^*, can be defined as that required to produce the same average velocity $\bar{v}$ when the internal stress is

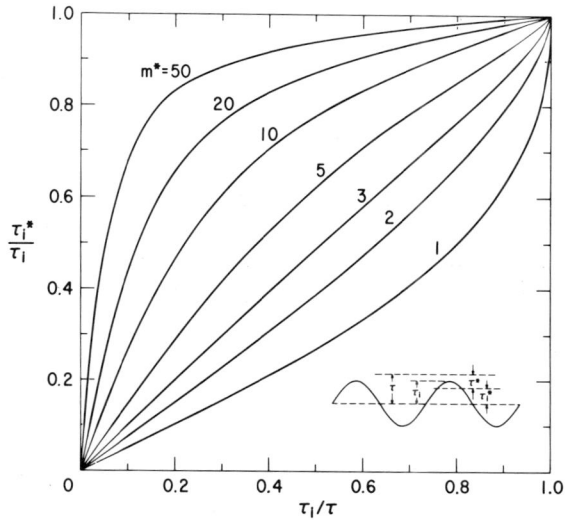

Fig. 8 The effective internal stress in a periodic internal stress field.

everywhere zero:

$$\bar{v} = B(\tau^*)^{m^*} \tag{41}$$

Substituting Eq. (41) into Eq. (39) gives

$$\tau^* = \frac{\tau}{x[P_{m^* - 1}(x)]^{1/m^*}} \tag{42}$$

The effective internal stress, which should be subtracted from the applied stress to obtain the effective stress, is defined by

$$\tau_i^* = \tau - \tau^* = \tau\left[1 - \left(\frac{\bar{v}}{v}\right)^{1/m^*}\right] \tag{43}$$

which is, from Eq. (42),

$$\frac{\tau_i^*}{\tau_i} = \frac{\tau}{\tau_i}\left\{1 - x^{-1}[P_{m^* - 1}(x)]^{-1/m^*}\right\} \tag{44}$$

Equation (44) is plotted in Fig. 8. It is seen that the effective internal stress is always smaller than τ_i, and approaches τ_i only if

τ approaches τ_i. However, the ratio τ_i^*/τ_i deviates appreciably from unity only if both τ_i/τ and m^* are small.

Near the athermal region, τ^* is very small so that τ_i/τ is close to unity. The effective internal stress is then almost the same as the maximum internal stress. At low temperatures when τ^* is large so that τ_i/τ is small, it so happens, at least for b.c.c. metals, that m^* is large. The effective internal stress is then still almost the same as the maximum internal stress. This seems to justify the usual practice of subtracting a constant athermal component from the applied stress. However, as just shown, this is possible only if m^* is large when τ^* is large.

7.2   The Apparent Activation Enthalpy

Equation (39) can be used to calculate an apparent enthalpy of activation. For simplicity, the lattice behavior is assumed ideal, namely, $\Delta S^{\ddagger} = 0$. The apparent enthalpy is

$$\Delta H_a = \Delta H_\tau + k(m^*T) \left\{ \ln x + \left[\frac{\partial \ln P_{m^*-1}(x)}{\partial m^*} \right]_x \right\} \tag{45}$$

where ΔH_τ is the activation enthalpy when the applied stress is τ and the internal stress field is everywhere zero. However, since ΔH_τ depends strongly on τ, a question arises whether ΔH_τ is the appropriate quantity to be compared with ΔH_a. In view of the fact that at low temperatures, both pure and impure metals are usually tested at the same strain rate rather than at the same stress, it is the ΔH at τ^* which should be compared with ΔH_a. By using Eq. (33),

$$\Delta H_a = \Delta H_{\tau^*} + k(m^*T) \left[\left(\frac{\partial \ln P_{m^*-1}(x)}{\partial m^*} \right)_x - \frac{1}{m^*} \ln P_{m^*-1}(x) \right] \tag{46}$$

Equation (46) is shown in Fig. 9. It is seen that ΔH_a is always larger than ΔH_{τ^*}. The difference increases with internal stress for all m^* but is not appreciable until τ_i/τ is near unity for large m^*.

Arsenault and Li[39] discussed the curvature in the $\Delta H - T$ plot at constant strain rate in neutron irradiated copper. They assumed a periodic internal stress field caused by large defect clusters superimposed on a single thermally activated process of overcoming small defects. Roberts and Owen[40] found that the activation enthalpy for martensite at constant strain rate increases

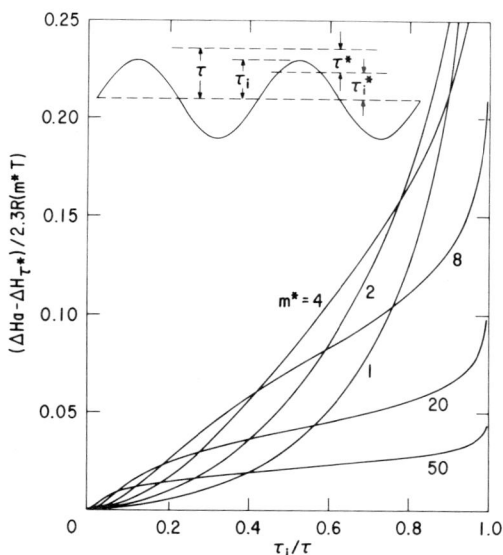

Fig. 9 The effect of periodic internal stress on the enthalpy of activation.

linearly with temperature until τ approaches τ^*. Then the activation enthalpy increases much more rapidly with temperature.

7.3   The Apparent Activation Area

Equation (39) can be used to calculate an apparent activation area

$$\frac{A_a^*(\tau)}{A^*(\tau^*)} = \frac{\tau^* x}{\tau} \left[\frac{x(2m^* - 1)}{m^*} - \frac{m^* - 1}{m^*} \frac{P_{m^* - 2}(x)}{P_{m^* - 1}(x)} \right] \tag{47}$$

where $A^*(\tau^*)$ is the activation area at τ^* without the internal stress field. Equation (47) is plotted in Fig. 10. It is seen that the apparent activation area is always larger than the activation area at the same effective stress. In many studies,[41-43] it was found that the activation area of impure iron was larger than that of pure iron, and some researchers used this as evidence of an impurity mechanism. It is seen that this could be explained by a periodic internal stress field. The effect diminishes with increasing τ and m^* and hence becomes negligible at low temperatures.

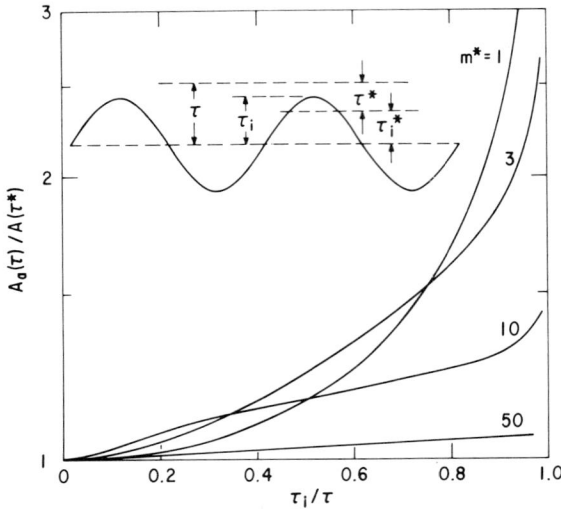

Fig. 10 The effect of periodic internal stress on the area of activation.

7.4 The Apparent Velocity-Stress Exponent

Equation (39) or (47) can be used to calculate an apparent velocity-stress exponent defined by Eq. (25)

$$m_a^* = x^2(2m^* - 1) - x(m^* - 1)\frac{P_{m^*-2}(x)}{P_{m^*-1}(x)} \tag{48}$$

Equation (48) is plotted in Fig. 11. It is seen that m_a^* is always larger than m^*. This effect could be used to explain the general observation that m^* is larger when the impurity content is higher. For large m^*, the effect of a periodic internal stress field is almost equivalent to a uniform back stress τ_i. When this is so, the apparent m_a^* can be approximated by,[14]

$$m_a^* = \frac{m^*}{1 - \dfrac{\tau_i}{\tau}} \tag{49}$$

which can be verified in Fig. 11 for $m^* = 50$.

An example is shown in Fig. 12 for the case of dispersed Al_2O_3 in iron.[44] The strain rate-stress exponent which has been extrapolated to zero strain is used instead of the velocity-stress exponent.

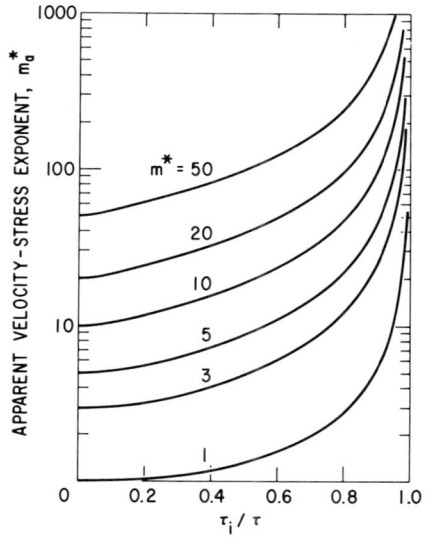

Fig. 11 Apparent velocity-stress exponent in a periodic internal stress field.

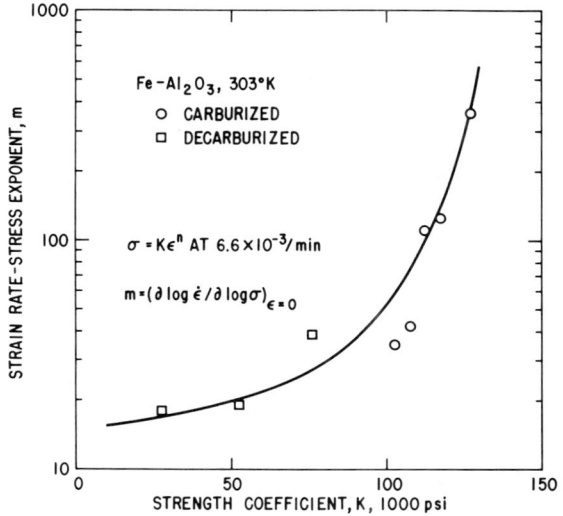

Fig. 12 The effect of dispersed particles on the strain rate-stress exponent. (Data of Felberbauer, Lautenschlager, and Brittain)

These dispersed alloys do not show a yield point but obey a power relation between stress and strain, $\sigma = K\epsilon^n$. The strength coefficient K is used to represent the extent of internal stress field. As expected, the strain rate-stress exponent increases with the internal stress introduced by the dispersed particles. A more quantitative comparison can be made for the case of C and N in iron. The strain rate-stress exponent at zero strain is shown in Fig. 13 using data from many sources.[41,43,45-47] For the same effective stress $\tau^* = \tau - \tau_i$, Eq. (49) can be rewritten as

$$m_a^* = m^* \left(1 + \frac{\tau_i}{\tau^*}\right) \tag{50}$$

which suggests a linear relation between m_a^* and τ_i and, if τ_i is linear with the concentration f (atom fraction), a linear relation between m_a^* and concentration. In view of this observation, a straight line is drawn in Fig. 13. The slope of this line is 24,000 and the intercept is 7. The slope-intercept ratio is 3400. This ratio can be compared with that indicated in Eq. (50) which gives $(1/\tau^*)(d\tau_i/df)$. By using $\tau^* = 1.4$ kg/mm^2 and $d\tau_i/df = 4500$ kg/mm^2 as shown in Fig. 14, this ratio is calculated to be 3200. The agreement seems satisfactory and supports the contention that the apparent strain rate-stress exponent is caused by the periodic internal stresses.

8 SOLID-SOLUTION HARDENING

At least in b.c.c. metals, interstitial solid-solution hardening appears to be an athermal component of flow stress. As shown in Fig. 14, the concentration dependence of flow stress in Fe-N alloys[43] appears to be linear. The slope agrees with that of Heslop and Petch[48] if the tensile strength of polycrystalline specimens is divided by 2.7 to change to the shear stress. Since the slope appears temperature-independent,[48] the contribution of solid-solution hardening is athermal in nature.

The only theory of solid-solution hardening which is temperature-independent and linear with concentration is that of Mott and Nabarro.[49] The give a stress of $\mu\epsilon^2 f$ for f between 0.01 and 0.2 where μ is the shear modulus, $\epsilon = d \ln a/df$ is a misfit parameter, a is lattice dimension, and f is atom fraction. Taking $\epsilon = 0.3$, $\mu = 7900$ kg/mm^2, the predicted slope is 710 kg/mm^2 while the experimental value of Fig. 14 is 4500 kg/mm^2. Mott[50] and Cracknell and Petch[51] modified the Mott-Nabarro theory, and succeeded in obtaining a larger slope.

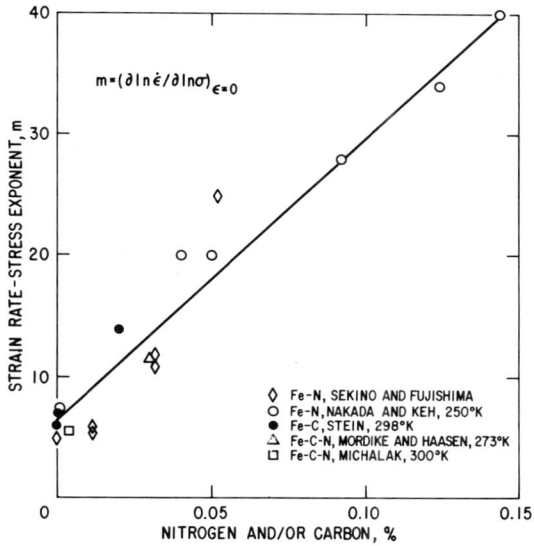

Fig. 13 The effect of interstitial solute on the strain rate-stress exponent.

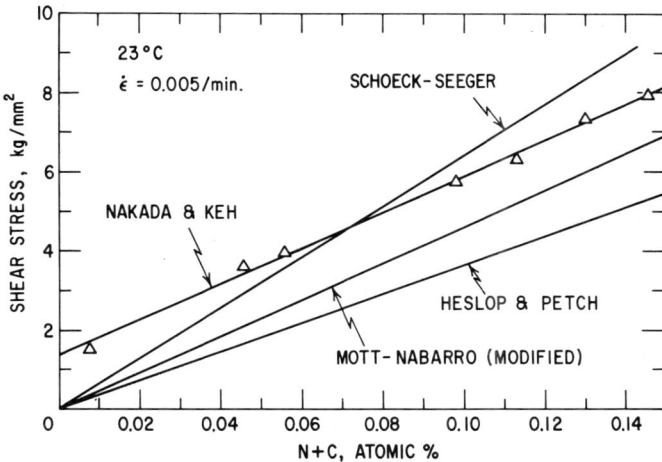

Fig. 14 Solid solution hardening in Fe-N alloys.

In the Mott–Nabarro theory, a partial volume average of the internal stress is used. In view of the dislocation dynamics in the periodic internal stress field just discussed, the effective internal stress is very nearly the maximum internal stress when $\tau_i/\tau \to 1$ (room temperature), or when m^* is large (low temperature). It is therefore expected that the hardening should be more than that predicted by a volume average of the internal stress.

In view of the preceding consideration, a partial area average of the shear stress in the slip plane near the solute atom is calculated as follows. The solute atom is considered as an elastic inclusion of radius α and of misfit ϵ situated at the origin of an (x, y, z) Cartesian or (r, θ, ϕ) spherical coordinate system. The shear stress σ_{xz} along the plane $z = h$ at a point (r, θ, ϕ) is given by

$$\sigma_{xz} = -\frac{6\mu a^3 \epsilon}{h^3} \cos\phi \, \sin^4\phi \, \cos\theta \tag{51}$$

The average of this stress within a quarter circle of radius $\alpha/\sqrt{f}$ in the slip plane is $2\mu\epsilon f\alpha/h(1 - f)$ which has a maximum at $h = \alpha$. For small f, the solid-solution hardening is thus approximately

$$\tau = 2\mu\epsilon f \tag{52}$$

which gives a slope of 4700 kg/mm^2 for N in Fe and compares favorably with the experimental results of Nakada and Keh.[43] It is to be noted that because of the one-dimensional approach to the problem of nonuniform internal stresses, an average over the area around the solute atom is still necessary. More exact calculations may be performed when a two-dimensional approach to the internal stress problem can be formulated.

The Schoeck-Seeger theory[51] predicts a slope slightly higher than the experimental value shown in Fig. 14. The main difficulty with the Schoeck-Seeger theory is that it is a variable strain rate theory. In order to obtain maximum hardening it is necessary to operate at a strain rate which changes with the temperature of the test so as to match the jumping rate of the interstitial atoms. The hardening is not independent of temperature if the strain rate is fixed. Therefore the Schoeck-Seeger theory cannot be used to explain the athermal component of flow stress caused by the interstitial solute atoms.

9   CONCLUSIONS

To analyze dislocation mobility data without a specific model, it is advisable to use general thermodynamic relations without any assumption. This will help in obtaining accurate activation parameters and in testing the self-consistency of experimental data. An example is the entropy of activation. It is essential not to assume zero entropy of activation unless it is indicated from all the consequences of this assumption.

The effect of hydrostatic pressure on creep rate indicates a vacancy mechanism along the slip plane. This suggests a creep mechanism involving screw dislocations with jogs which produce or absorb vacancies. The stress dependence of creep rate indicates a stress-dependent activation area for this mechanism.

Stress-relaxation techniques are believed to be capable of determining the velocity-stress exponent. All the available results are consistent with those from direct etch-pitting techniques.

Recent results on the low-temperature deformation of potassium show a strong temperature-dependence of yield stress below 25°K. A comparison of the extrapolated stress at 0°K, the activation area-stress relation, and the activation enthalpy with all other b.c.c. metals indicates a Peierls mechanism for all of them.

The problem of athermal component in b.c.c. metals and alloys is considered from the viewpoint of dislocation dynamics in a periodic internal stress field. The apparent activation area, activation energy, and the velocity-stress exponent are all larger than those without the internal stress field. However, the effect is pronounced only when the effective stress is small.

From the viewpoint of dislocation dynamics in a periodic internal stress field, the effective internal stress, which should be subtracted from the applied stress to obtain the effective stress, is approximately the maximum internal stress in all the b.c.c. metals and alloys for which the velocity-stress exponent increases with decreasing temperature.

In view of the large effect of the periodic internal stress field, the Mott-Nabarro theory of solid-solution hardening is modified to explain the athermal contribution from interstitial impurities C + N in iron.

REFERENCES

1. Becker, R.: *Physikalische Zeitschrift* 26:919 (1925); *Z. Tech. Phys.* 7:547 (1926).
2. Kanter, J. J.: *Trans. AIME* 131:385 (1938).
3. Dushman, S., and I. Langmuir: *Phys. Rev.* 20:113 (1922).
4. Eyring, H.: *J. Chem. Phys.* 4:283 (1936).
5. Kauzmann, W.: *Trans. AIME* 143:57 (1941).
6. Seitz, F., and T. A. Read: *J. Appl. Phys.* 12:100, 170, 470, 538 (1941).
7. Nowick, A. S., and E. S. Machlin: *J. Appl. Phys.* 18:79 (1947).
8. Sherby, O. D., J. L. Lytton, and J. E. Dorn: *Acta Met.* 5:219 (1957).
9. Basinski, Z. S.: *Acta Met.* 5:684 (1957).
10. Conrad, H., and H. Wiedersich: *Acta Met.* 8:128 (1960).

11. Gibbs, G. B.: *Phys. Stat. Sol.* 5: 693 (1964).
12. Schoeck, G.: *Phys. Stat. Sol.* 8:499 (1965).
13. Christian, J. W., and B. C. Masters: *Proc. Roy. Soc.* A281: 240 (1964).
14. Li, J. C. M.: *Trans. AIME* 233:219 (1965).
15. Li, J. C. M.: *Can. J. Phys.* 45:493 (1967).
16. Bernstein, I. M., J. C. M. Li, and M. Gensamer: *Acta Met.* 15:801 (1967).
17. Dorn, J. E.: "Creep and Recovery," p. 255, American Society for Metals, 1957.
18. Garofalo, F., "Fundamentals of Creep and Creep-Rupture in Metals," p. 87, Macmillan, 1965.
19. Basinski, Z. S.:*Phil. Mag.* 4:393 (1959).
20. Conrad, H., L. Hays, G. Schoeck, and H. Wiedersich: *Acta Met.* 9:367 (1961).
21. Mitra, S. K., P. W. Osborne, and J. E. Dorn: *Trans. AIME* 221:1206 (1961).
22. Christy, R. W.: *Acta Met.* 2: 284 (1954).
23. Chirouze, B. Y., D. M. Schwartz, and J. E. Dorn: *Trans. ASM* 60:51 (1967).
24. Felthan, P., and J. D. Meakin: *Acta Met.* 7:614 (1959).
25. Garofalo, F.: *Trans. AIME* 227:351 (1963).
26. Garofalo, F., O. Richmond, W. F. Domis, and F. von Gemmingen, "Joint International Conference on Creep," p. 1, The Institution of Mech. Eng., London, 1963.
27. Garofalo, F., W. F. Domis, and F. von Gemmingen: *Trans. AIME* 230:1460 (1964).
28. Cuddy, L.: to be published.
29. Weertman, J.: *J. Appl. Phys.* 28:196 (1957).
30. Head, A. K.: *Phil. Mag.* 4:295 (1959).
31. Li, J. C. M.: *Disc. Faraday Soc.*, No. 38, 138 (1964).
32. Barrett, C. R., and W. D. Nix: *Acta Met.* 13:1247 (1965).
33. Hirsch, P. B., and D. H. Warrington: *Phil. Mag.* 6:735 (1961).
34. Gupta, I., and J. C. M. Li: to be published.
35. Hull, D., and H. M. Rosenberg, *Proc. X Int. Congr. Refrigeration*, p. 58, Copenhagen, 1959.
36. Bernstein, I. M., and J. C. M. Li: *Phys. Stat. Sol.* 23:539 (1967).
37. Conrad, H., and W. Hayes: *Trans. ASM* 56:249 (1963).
38. Chen, H. S., J. J. Gilman, and A. K. Head: *J. Appl. Phys.* 35:2502 (1964).
39. Arsenault, R. J., and J. C. M. Li: to be published in *Phil. Mag.*
40. Roberts, M. J., and W. S. Owen: This colloquium, p. 357.
41. Stein, D. F.: *Acta Met.* 14: 99 (1966).
42. Burbach, R., B. L. Mordike, and P. Haasen: *J. Iron Steel Inst.* 204: 390 (1966).
43. Nakada, Y., and A. S. Keh: to be published.
44. Felberbauer, F., E. P. Lautenschlager, and J. O. Brittain: *Trans. AIME* 230: 1596 (1964).
45. Mordike, B. L., and P. Haasen: *Phil. Mag.* 7:459 (1962).
46. Michalak, J. T.: *Acta Met.* 13: 213 (1965).
47. Sekino, S., and T. Fujishima: *Trans. Japan Inst. Met.* 7:142 (1966).
48. Heslop, J., and N. J. Petch: *Phil. Mag.* 1:866 (1956).
49. Mott, N. F., and F. R. N. Nabarro:*Report of Conference on Strength of Solids*, p. 1, Bristol (Phys. Soc, 1948).
50. Mott, N. F.:: "Imperfections in Nearly Perfect Crystals," p. 173, ed. by W. Shcokley, John Wiley, 1952.
51. Cracknell, A., and N. J. Petch: *Acta Met.* 3:186 (1955).
52. Schoeck, G., and A. Seeger:*Acta Met.* 7: 469 (1959).

INTERNAL FRICTION STUDIES
OF DISLOCATION MOTION

A. V. Granato

University of Illinois

ABSTRACT

A review of the type of information concerning dislocation dynamics obtainable from internal friction measurements is given. The particular suitability of such measurements for the study of dislocation interactions with other defects, such as phonons, electrons, point defects, other dislocations, and the lattice itself is stressed. Recent studies of the dependence of dislocation velocities on stress in LiF, NaCl, and Cu, are given particular attention.

1 INTRODUCTION

Internal friction techniques are well suited to the study of dislocation dynamics. In fact, only the dynamics of dislocation behavior can be studied by this method since immobile dislocations give no internal friction. A sound wave traveling in a crystal containing dislocations is attenuated, and the attenuation (and velocity change) depends sensitively on the mobility of the dislocations. The mobility, in turn, depends on the interactions of dislocations with point defects, phonons, electrons, other dislocations, and the lattice. Thus measurements of internal friction, or ultrasonic wave propagation in crystals, are especially suited for the study of dislocation interactions with other defects. Results obtained by this method will be described within a framework in which defect interactions are considered from the "particle" viewpoint.

Some advantages of internal friction measurements over other direct and indirect types of observations derive from: (1) the sensitivity of the method; (2) the selectivity of the measurements; (3) the fact that the measurements are nondestructive; and (4) the fact that the results can be made quantitative.

The sensitivity of the measurements arises from the fact that the damping depends strongly (on the fourth power over a wide frequency range) on the length of the dislocation segment able to vibrate so that a relatively small number of point defects acting as anchoring points can greatly reduce the damping. If the total length of dislocation line per cm^3 is $\Lambda \approx 10^7$ and the average length of segments free to move is $L \approx 10^{-4}$ cm (typical values for unworked pure crystals), we have $\Lambda/L \approx 10^{11}$ pinning points per cm^3. Thus, when the same number of point defects arrive at dislocations, L is reduced by a factor of 2 and the damping by more than a power of ten, so that as few as 10^9 point defects per cm^3 can be detected. This corresponds to a concentration of about 10^{-14}, whereas measurements of resistivity or lattice parameter permit detection of point defects in concentration of the order of 10^{-9} and 10^{-5}, respectively.

The selectivity of the method results from the fact that only those point defects which arrive at the dislocations, pinning them down and restricting their motion, are detected. This is in contrast, for example, with resistivity measurements, where point defects are detected wherever they may be. Thus the internal friction measurements permit distinguishing between different mechanisms of point-defect annihilation.

Due to the nondestructive nature of internal friction measurements, the same specimen may be used repeatedly under different conditions, and also for measurements of other properties. This is a decisive advantage in studies of structure-sensitive properties of materials.

Because of the existence of a quantitative theory of dislocation damping, quantitative information about dislocations and their interactions with other defects can be deduced from the measurements. It will be noted that some of these quantities are difficult to obtain by other methods. Despite the fact that dislocation internal friction effects appear in a rich variety of forms involving many parameters, it has been possible to rationalize the many observations to a high degree in terms of a very simple model: the vibrating string model of a dislocation. There have been many contributions to the development of this model. Mott[1] and Friedel[2] used it to calculate modulus changes at low frequencies. Eshelby[3] and Koehler[4] introduced damping mechanisms. Granato and Lücke[5] connected many of the elements previously introduced separately and worked out the quantitative predictions of the model as a function of frequency and strain amplitude.

The original comparison of the measurements with the predictions of the vibrating string model by Granato and Lücke[6] gave surprisingly good agreement in the area of the strain amplitude-dependent ("dislocation breakaway") loss whereas only scant evidence could be brought in favor of the dislocation resonance loss. At present, the situation appears somehow reversed. Most of the measurements of the strain amplitude–dependent loss have been made at room temperature or higher, thus lying far outside the range of validity of the original zero-point temperature theory, with which such good agreement has been found. The comparison with the extended (finite temperature) theory, however, has not yet been completed so that there remains some doubt about the meaning of the observed agreement. On the other hand, strong evidence has been accumulated that the high-frequency (megacycle) low-amplitude damping is caused by dislocation resonance. The verification of the maximum of the decrement as a function of frequency and the study of this frequency dependence after different types of treatments, have particularly contributed to this evidence.

In Sec. 2, the model will be described, and compared with measurements at low and high strain amplitudes. In Sec. 3, a number of examples will be selected to illustrate the use of the method in the study of various defect interactions.

2 THE MODEL

2.1 Strain Amplitude Independent Effects

There are many ways in which dislocations can contribute to internal friction. A dislocation segment oscillating between two pinning points (vibrating string model) gives one characteristic type of damping (low-amplitude resonance). Dislocations which break away from pinning points under stress lead to another type (amplitude-dependent damping). Dislocations which move by overcoming Peierls barriers are assumed to give rise to the low-temperature Bordoni peaks. There are also other mechanisms, for which models have not yet been developed. In particular, a suitable model still does not exist for the highly deformed state. However, a single dislocation segment model which neglects interactions between dislocations works surprisingly well for moderate deformation (up to about 4%). In the following, we discuss only the vibrating string model since its predictions are definite, and since many experimental checks of these predictions are now available.

The basis for internal friction effects consists in the fact that dislocation motion contributes to the total strain developed in a specimen under stress. For a given applied stress, a solid containing dislocations has a larger strain than a perfect crystal; thus the elastic modulus appears to be lower. Under the action of an alternating stress, the dislocation component of the strain may lag behind the applied stress. This leads then not only to a reduction in modulus, but also to a damping of the applied stress. Simple estimates of the magnitude of the expected effect for typical dislocation densities lead to much larger values than the observed effects, if it is assumed that the dislocations are perfectly mobile with no restrictions on their motion. Therefore, it may be concluded that there must be impediments to the motion of dislocations. Generally speaking, obstacles similar to those assumed in yield stress theories where a similar problem is faced, have also been assumed here. These are, for example, atomic pinning points, network points, jogs, other dislocations, etc. But even with such restrictions, an explanation must be given as to why the dislocation motion lags behind the applied stress. For smooth dislocation motions, it may be imagined that the dislocation is viscously damped as it moves through the electron or phonon gas. Also, impediments which cause a jerky motion of the dislocation will lead to a phase lag. Examples of the latter type of effect are provided by motion over Peierls barriers at low stresses, and catastrophic unpinning of dislocations at high stresses.

In the vibrating string model, advantage is taken of the fact that a dislocation has an effective mass per unit length and an effective tension. Thus an equation of motion for small oscillations of a dislocation may be written as[4]

$$\frac{A\partial^2 y}{\partial t^2} + \frac{B\partial y}{\partial t} - \frac{C\partial^2 y}{\partial x^2} = b\sigma \tag{1}$$

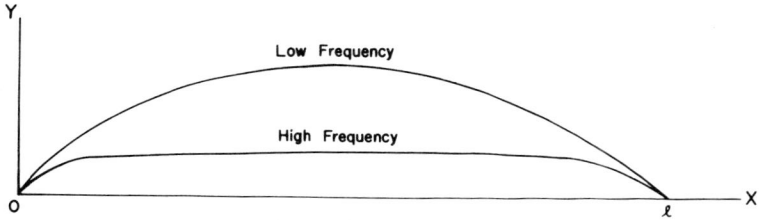

Fig. 1 Schematic dislocation displacement $y(x)$ as a function of coordinate x for low frequencies and high frequencies. At low frequencies the displacement is limited by tension forces. At high frequencies, the displacement is limited by viscous forces.

where

A = effective mass per unit length,
y = dislocation displacement measured from the equilibrium position as indicated in Fig. 1
B = viscous damping constant,
C = tension,
b = Burgers vector
σ = applied stress,
t = time, and
x = a coordinate along the dislocation.

The solution of Eq. (1) together with the boundary conditions $y(o)$ and $y(l) = 0$, where pinning points are placed at $x = 0$ and l gives the dislocation displacement y as a function of the frequency ω of the applied stress $\sigma(\omega)$. Using this, the dislocation strain and the effective modulus (or ultrasonic wave velocity) and decrement (or ultrasonic wave attenuation) are easily found.

At low frequencies, where the dislocation motion is almost in phase with the applied stress, the displacement is given by

$$y = \frac{4b\sigma}{\pi A\omega_0^2} \sin \frac{\pi x}{l} \tag{2}$$

where ω_0, the resonant frequency, is given by

$$\omega_0 = \frac{\pi}{L} \left(\frac{C}{A}\right)^{1/2} \tag{3}$$

Using $C = Gb^2/2$ as a crude estimate for the tension, the maximum displacement is then approximately given by

$$\frac{\bar{y}}{b} = \epsilon_0 \left(\frac{l}{b}\right)^2 \tag{4}$$

With typical ultrasonic strain amplitudes ϵ_0 in the range 10^{-7} to 10^{-6}, and with $l = 10^4 b$, we obtain maximum dislocation displacements in the range of 10–$100\ b$. The modulus defect $\phi = \Delta E/E$, where E is the elastic modulus of a crystal containing no defects, and ΔE is the reduction produced by the dislocations, is given at low frequencies by

$$\phi = \frac{\Delta E}{E} = \frac{\epsilon_{dis}}{\epsilon_{el}} \tag{5}$$

where ϵ_{el} is the elastic strain and ϵ_{dis} is the dislocation strain. The latter is given by the Burgers vector times the area swept out by the dislocations or

$$\epsilon_{dis} = \Lambda b\bar{y} \tag{6}$$

where Λ is the total length of dislocation line per unit volume and $\bar{y}$ is the average dislocation displacement. The decrement Δ defined as the energy lost per cycle divided by twice the maximum energy stored per cycle, is given by

$$\Delta = \oint \frac{F\,dy}{(\sigma_0^2/G)} \tag{7}$$

where F is the dissipative force acting on the dislocation line, σ_0 is the amplitude of the oscillating stress acting, and G is the shear modulus.

If the drag force is taken to have the form

$$F = Bv^n \tag{8}$$

where v is the dislocation velocity, then the decrement can be evaluated from Eq. (7).

In Eq. (1), it has been assumed that $n = 1$ in Eq. (8), and the resulting expressions for the decrement and modulus change are

$$\Delta = \left(120\,\Omega\Delta_0\,\Lambda L^2\right)\left(\frac{\omega B L^2}{\pi^2 C}\right) \tag{9}$$

and

$$\phi = \frac{6\,\Omega\Delta_0\,\Lambda L^2}{\pi} \tag{10}$$

In Eqs. (9) and (10), an orientation factor Ω has been included to take account of the fact that only the resolved shear stress in the slip system produces dislocation motion. Also numerical factors are contained which provide for the fact that L is now to be interpreted as the average loop length of a random distribution of loop lengths l. The constant $\Delta_0 = (8 G b^2)/(\pi^3 C)$.

The most conspicuous and characteristic feature of Eqs. (9) and (10) is the fourth and second power dependence of the decrement and modulus defect on the average loop length which is predicted. We note further that if $n \neq 1$ in Eq. (8), then not only are different loop-length laws predicted, but also a different frequency dependence is predicted for the decrement. The decrement then becomes stress or strain-amplitude dependent. Thus the decrement depends on ωL^4 if and only if the drag force is linear in the dislocation velocity. As we shall see, the predictions of Eqs. (9) and (10) are fully borne out by the experiments, and this fact has important consequences for the discussion of high-speed dislocation drag, which we discuss later.

At high speeds, or frequencies, the dislocation drag becomes strong enough to prevent the dislocation from achieving the displacement permitted by the tension forces, and the dislocation shape changes as indicated in Fig. 1. In this case the dislocation moves more like a rigid rod over most of its length, coming down to zero displacement only near the pinning points. Thus we expect the effect of pinning points to be large at low frequencies and negligible at high frequencies. At the highest frequencies the motion is limited by inertial effects and the overall behavior (for a group of dislocations, all of the same length) is given in Fig. 2. The resulting decrement has a typical resonance-type frequency behavior, which depends, however, on the magnitude of the viscous damping constant B. The quantity D in the figure is defined by $D = \omega_0 A/B$. Theoretical estimates of the damping constant to be expected for dislocation interactions with phonons and electrons have been made by Leibfried,[7] Eshelby,[3] Mason,[8] and Holstein.[50] In the absence of both phonons and electrons, there is still a radiation damping. This effect is presently being studied at the University of Illinois by R. Schwenker and J. Garber. From these one expects: (1) that the phonon interaction is usually much larger than the electron interaction except at low temperatures in metals; and (2) that the phonon interaction is usually so large that the dislocation resonance is overdamped. This has the effect of broadening out the resonance and moving it to lower frequencies. Also when it is considered that not all dislocations have the same length, the expected maximum is broadened out further. For dislocation segments of length of order $1\,\mu$, the expected resonance

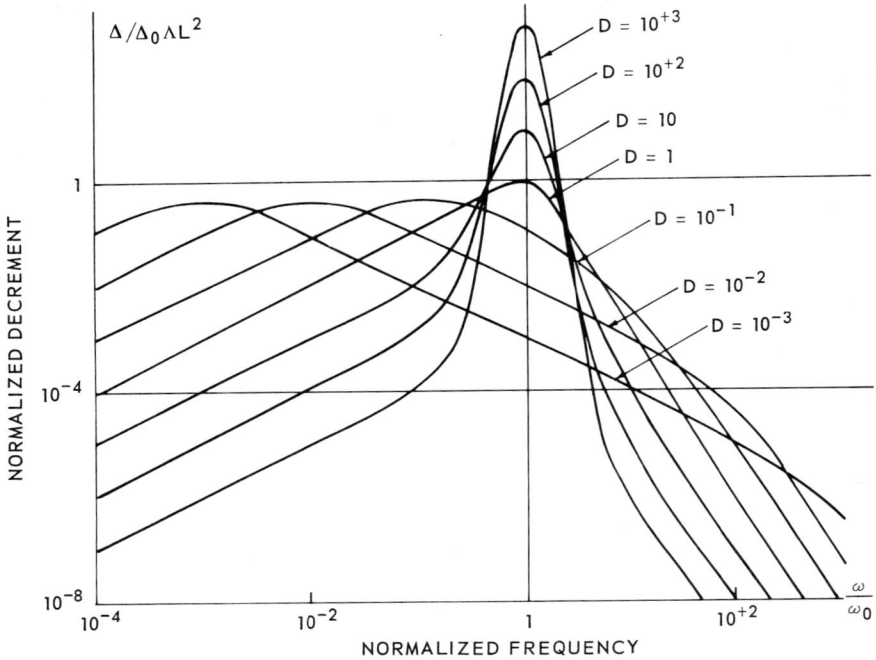

Fig. 2 The various curves show the frequency dependence of the decrement for various values of the damping constant for a delta-function distribution of loop lengths. The frequency has been normalized to the resonant frequency and the decrement has been normalized by the factor $\Delta_0 \Lambda L^2$, where Δ_0 is a constant of order one, Λ is the dislocation density, and L is the loop length. (After Granato and Lücke.)

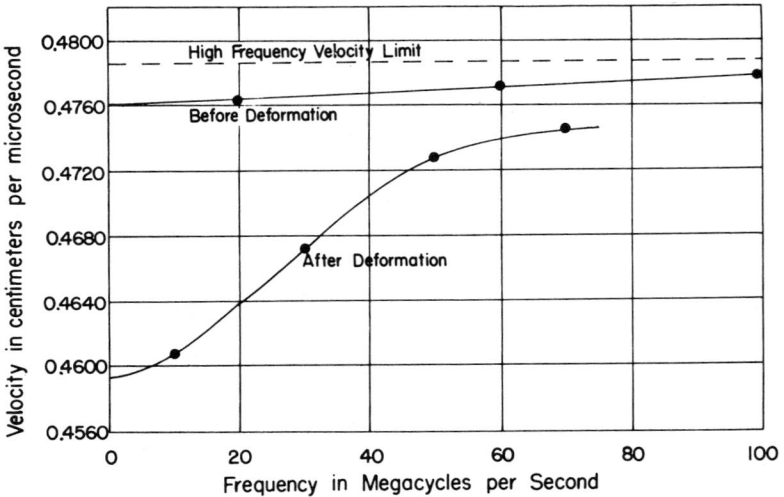

Fig. 3 Velocity dispersion for compressional elastic waves propagating in the $\langle 100 \rangle$ direction in NaCl. Deformation increases the magnitude of the dispersion from 0.5 to 4 per cent and moves it to lower frequencies. (After Granato, de Klerk, and Truell.)

frequency is about 10^9 cps. However, because of the large damping (we may picture the dislocation as a string moving in heavy molasses), the maximum may be brought down to frequencies as low as 100 kcps. At the same frequencies where the decrement goes through a maximum, a dispersion in the elastic constant is expected.

This dispersion effect is illustrated in Fig. 3 where the velocity of compressional waves was measured by Granato, de Klerk, and Truell[9] in a sodium chloride crystal as a function of frequency before and after a slight deformation. Before deformation there is only a small dispersion of 0.5% centered at about 75 Mcps. After the deformation, the magnitude of the dispersion has increased to about 4% and has moved to a lower frequency (35 Mcps). At room temperature, the dispersion was observed to gradually recover towards the initial condition, presumably as a result of dislocation pinning by deformation-induced defects. The interpretation of the effect according to the vibrating string model is as follows. At low frequencies, the dislocations are in phase with ultrasonic stress. When the stress is applied, the apparent elastic constant (and therefore the ultrasonic velocity) is reduced because the dislocation motion reduces the rigidity of the specimen. However, at high frequencies the dislocations can no longer follow the rapidly changing stress, and thus the modulus approaches the true elastic value.

A somewhat similar study was conducted by Merkulov and Yakolev,[10] who measured the attenuation as a function of frequency in NaCl deformed 1%. Their results are shown in Fig. 4. The

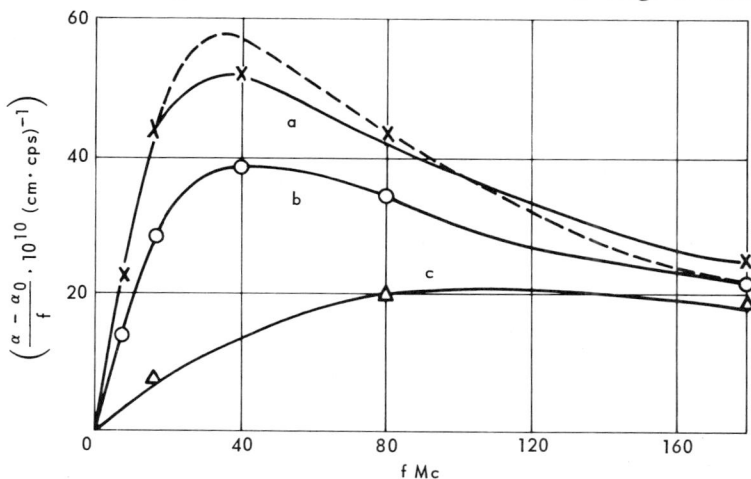

Fig. 4 Damping in NaCl as a function of frequency and time of recovery after deformation. (After Merkulov and Yakovlev.)

ordinate is proportional to the decrement, and Curves a, b, and c correspond to recovery times after deformation of 20 min, 2 hrs, and 60 hrs, respectively. The dashed line is a theoretical curve which deviates slightly from the measured curve, indicating that the maximum is somewhat broader than would be expected if all loop lengths were exactly the same. The overall agreement with the theory is good; the maximum moves to higher frequencies as the average loop length decreases during recovery; the attenuation is very different at low frequencies where it should be proportional to L^4, but the curves approach the same value at high frequencies, where the attenuation should be independent of loop length according to the string model. In addition, Merkulov and Yakovlev studied the dependence of the attenuation and velocity on plastic deformation, orientation, and aging time, and found further evidence for the string model. Also, they found that the attenuation was amplitude-dependent giving rise to curves similar to those found at kilocycle frequencies.

Irradiated copper has been studied by Alers and Thompson[11] and Stern and Granato.[12] Alers and Thompson made detailed studies of the attenuation and velocity as a function of temperature and orientation before and after irradiation. In addition, some measurements as a function of frequency were obtained. Stern and Granato studied in detail the change of the attenuation with time of cobalt gamma irradiation as a function of frequency. Moreover, some measurements as a function of orientation also were made. These studies complement each other. Both were able to give a quantitative account of their results in terms of the string model. For the measurements at 10 Mcps as a function of temperature, a dispersion was found, and it could be inferred that the damping measurements were on the low side of the maximum. The temperature-dependence of the damping constant could be deduced and was found to be linear, as would be expected if the damping was caused by the drag on dislocations from thermal phonons, according to Leibfried.[7] Measurements of the decrement on a different specimen as function of frequency were found to be on the high side of a maximum.

The frequency-dependence of the damping arising from dislocation motion is illustrated in Fig. 5. These are measurements by Stern and Granato[12] showing the effect of cobalt gamma irradiation on the damping of high-purity copper. Before irradiation, the damping has a maximum at a few megacycles. The gamma rays produce electrons which are energetic enough to displace lattice atoms, giving interstitials which can be effective as pinning points.

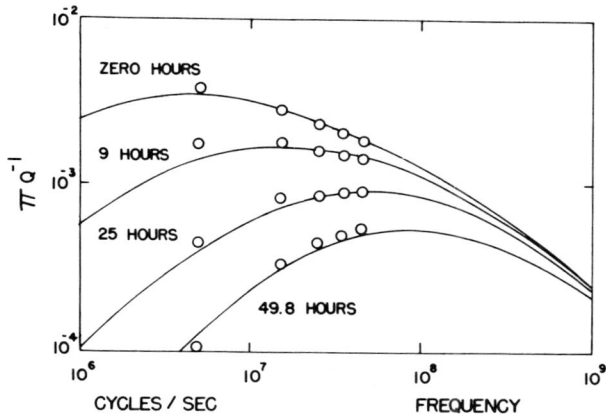

Fig. 5 Decrement as a function of frequency for several times during cobalt gamma irradiation in a 6000 curie source. The solid curves are theoretical. (After Stern and Granato.)

After 50 hr of irradiation in a 600°C cobalt source, the height of the maximum decreased, and the location increased to about 100 Mcps.   The damping at low frequencies is much more sensitive to the increased number of pinning points than that at high frequencies, as expected from our previous discussion of Fig. 1. According to the theory, the maximum decrement Δ_m is given by

$$\Delta_m = 2.2 \Omega \Delta_0 \Lambda L^2 \tag{11}$$

at the frequency ω_m, given by

$$\omega_m = \frac{0.084 \pi^2 C}{B L^2} \tag{12}$$

By comparing Eqs. (11) and (12) with (9) and (10), it is seen that measurements of the height and location of the maximum in the decrement are equivalent to measurements of the decrement and the modulus at low frequencies. Actually, the two quantities which can be determined from the measures are[12] the ratios Λ/B and L^2/C. Before dislocation densities and loop lengths can be determined, the damping constant B and tension C must be known. Later, we shall discuss means by which the damping constant can be determined.

The first significant evidence for the string model from low-amplitude measurements was that of Thompson and Holmes,[13] who irradiated pure single crystals of copper with neutrons. At room temperature, the radiation process would be expected to produce interstitials and vacancies, with the interstitials able to diffuse immediately to dislocations where they can act as pinning points. It is reasonable to assume, as did Thompson and Holmes, that the dislocation density does not change in this process so that observed changes in the damping and modulus should be caused only by changes in the average loop length. Assuming that the number of pinning points increased linearly with the irradiation time, Thompson and Holmes showed that the damping and modulus change depended upon the fourth and second power of the average loop length, respectively. In addition, they found that the size of the effect varied considerably from specimen to specimen, all of nominally the same purity. The specimens could be restored to their original condition by a suitable high-temperature annealing treatment. Also, it was found that there was a considerable background damping which was not removed by the radiation. The importance of these results for the further development of the field cannot be over-emphasized. They demonstrated that the low-amplitude damping effects follow definite characteristic laws, but that it is practically impossible to prepare two different specimens in exactly the same state. Radiation techniques provide a means for continuously varying the "purity" and studying the dependence on loop length in detail. Also the results showed that in order to attempt quantitative analysis of damping measurements, it is necessary to have some means of separating the effects from any unknown background effects. Once these conditions are realized, the method proves to be a means of extraordinary sensitivity in studying defect interactions.

However, no evidence was found for a linear frequency-dependence in the kilocycle measurements, because frequency dependence is difficult to determine in this range. On the other hand, as shown by Granato and Stern,[14] the megacycle and kilocycle measurements are compatible with each other. When the kilocycle measurements of the decrement and modulus are extrapolated into the megacycle range and vice versa using Eqs. (9), (10), (11), and (12), it was found that the kilocycle measurements are contained within the range of observations made in the megacycle region (see Fig. 6).

2.2 Amplitude Dependent Effects

A characteristic feature of dislocation damping effects is the pronounced dependence of the damping (and modulus) on the

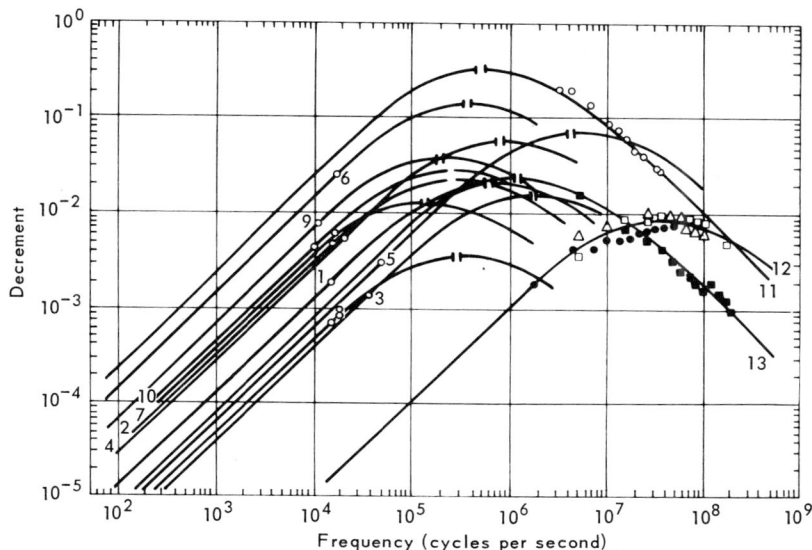

Fig. 6 Decrement of high-purity copper single crystals as a function of frequency. The data points shown are measured values. The solid lines are theoretical curves for a random distribution of pinning points. Curves 1-10 are curves fit to the decrement and modulus data of Thompson and Holmes on 10 different specimens. Curves 1-10 correspond, in order, to their specimens 2A, 2B, 2C, 4A, 83A, 1A, 3A, 3B, 5A, and 5B. The megacycle measurements have been normalized to the orientation and temperature of the kilocycle measurements, according to the procedure given in the text. Curve 11 is the data of Alers and Thompson for longitudinal wave propagation in the [110] direction at 198°K. Curve 12 is the data of Stern and Granato on a gamma-irradiated specimen for (a) [110] longitudinal waves at room temperature (Δ); (b) longitudinal [100] waves at room temperature (□); and (c) longitudinal [100] waves at 78°K (○). Curve 13 is the room temperature longitudinal [100] wave data on an irradiated specimen. The maxima for the unirradiated specimens lie between about 100 kc and 5 Mc. (After Granato and Stern.)

amplitude of the external stress. A typical example is shown in Fig. 7 for measurements in the kilocycle range. These are measurements by Read[15] of the damping of a 99.998% pure copper single crystal after compressive loads of 0, 60, 120, and 150 psi. The lowest curve is that for no applied stress, while the higher curves correspond to successively higher loads. Both the strain amplitude-dependent part of the damping and the residual damping at low strain amplitude increase with small plastic deformation. The damping is much more sensitive to small amounts of plastic deformation than is the stress-strain curve. Large changes in damping are obtained even though a stress-strain curve may not show any deviations from the elastic straight line. This is understandable since, in the stress-strain curve, the dislocation

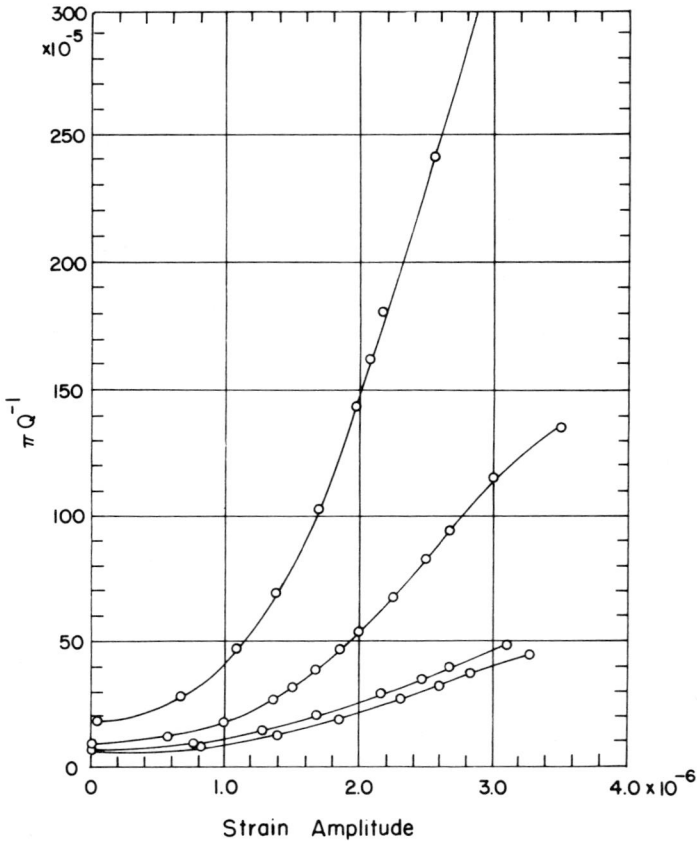

Fig. 7 Δ as a function of strain amplitude and applied compressive stress. (After Read.)

contribution to the total strain can only be detected if it is comparable to the elastic deformation, whereas with damping experiments the effects of the dislocations are observed directly.

Figure 8 shows the results of Read's study of the effect of orientation on the amplitude–dependent decrement in zinc. The orientation of the slip planes of four different specimens with respect to the longitudinal axis of the specimen are given on the figure. The resolved shear stress on the slip planes should be a maximum for 45° and zero for 0 and 90°. This general behavior is observed and provided the earliest and most convincing qualitative evidence that the loss is caused by dislocation motion.

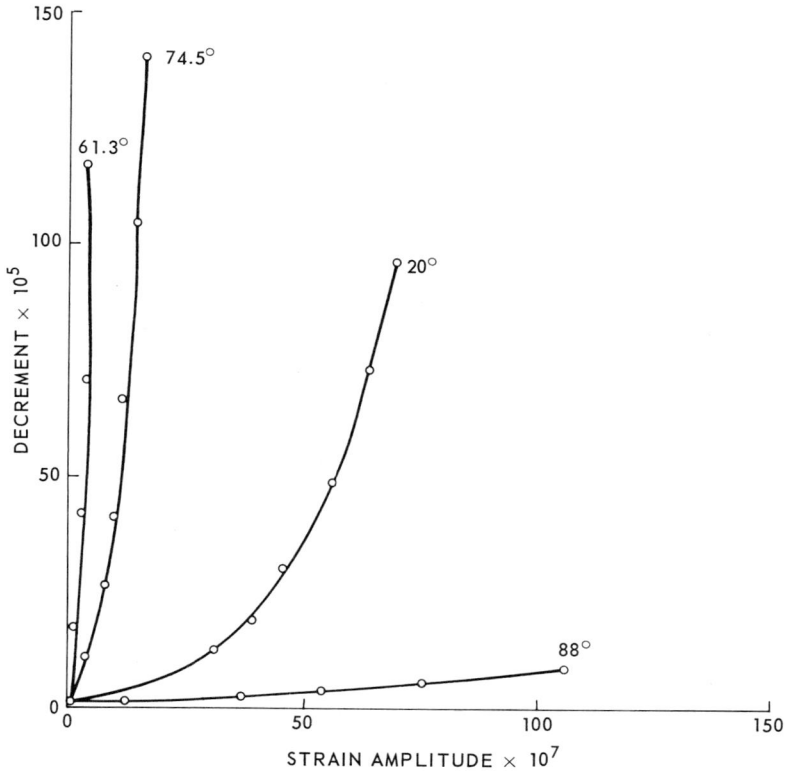

Fig. 8 Variation of decrement with strain amplitude in zinc crystals whose cylinder axes make various angles θ with the hexagonal axis. (After T. A. Read.)

The effect of annealing aluminum for various times at an annealing temperature of 84°C is shown by measurements by Baker in Fig. 9. As the annealing time increases, the stress necessary for breakaway increases. This indicates that point defects are diffusing to the dislocation lines, making breakaway more difficult.

The amplitude dependence of the damping is interpreted on the vibrating string model as resulting from a breakaway of the dislocation from weak pinning points as the stress amplitude is increased. This process is easily understood by means of Fig. 10. Here it is supposed that dislocations are pinned by the nodes of the dislocation network and also by impurities. The average length of a dislocation loop between two network points is called L_N and that between two impurity points L_c. The stress–strain curve corresponding to the dislocation displacements of Fig. 10 is shown in Fig. 11. For small applied stresses, the loops L_C bow out until a breakaway stress is reached. There is then a

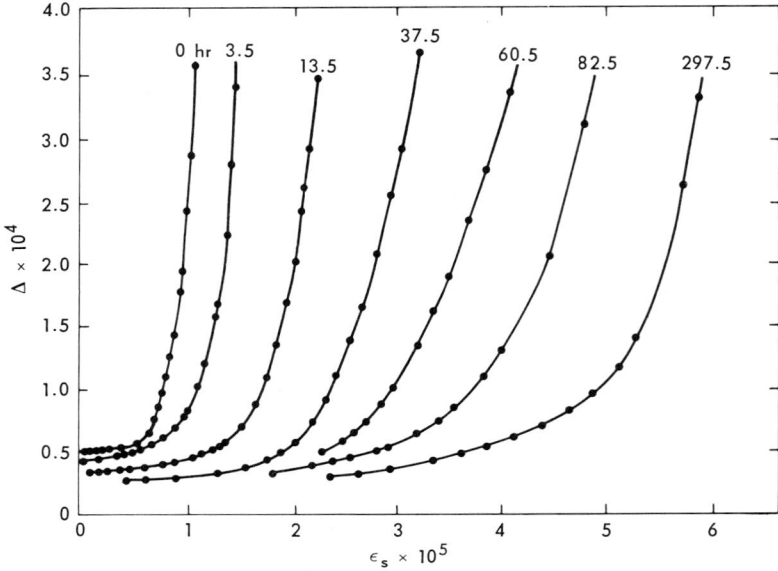

Fig. 9 Decrement as a function of strain amplitude and recovery time in aluminum annealed at 84°C. (After Baker)

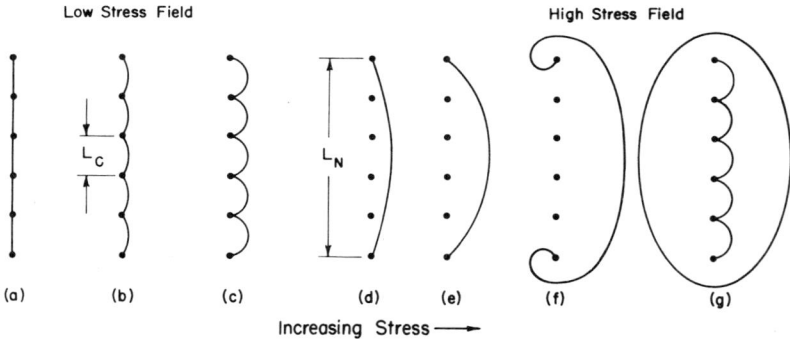

Fig. 10 The successive drawings indicate schematically the bowing out of a pinned dislocation line by an increasing applied stress. The length of loop determined by an impurity pinning is denoted by L_c, and that determined by the network by L_N. As the stress increases, the loops L_c bow out until break-away occurs. For very large stresses, the dislocations multiply according to the Frank-Read mechanism. (After Granato and Lücke.)

large increase in the dislocation strain (Fig. 11) for no increase in the applied stress. When the stress is reversed, the loops L_N contract so that a different path is followed in the stress-strain diagram for increasing and decreasing applied stress. This gives a hysteresis loss which is measured by the area in the stress-strain

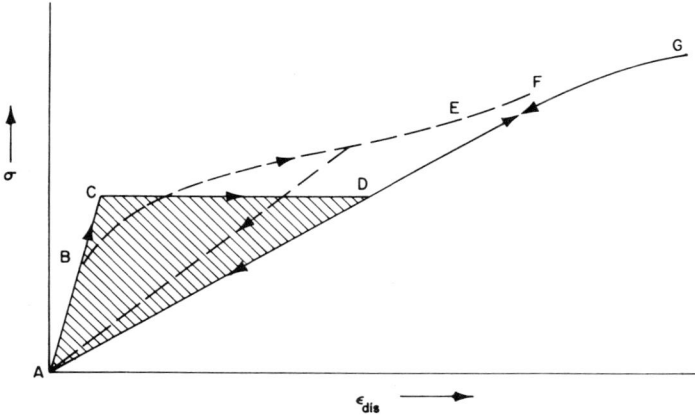

Fig. 11 The solid line shows the stress-strain law that results for the model shown in Fig. 10. The elastic strain has been subtracted out so that only the dislocation strain is shown. The path ABCDEF is followed for increasing stress; the path FA is followed for decreasing stress. The dashed line is that which would result if not all of the loops had the same length, but there was a distribution of lengths, L_c. (After Granato and Lücke.)

diagram (Fig. 11). This loss is frequency-independent for frequencies much less than that for which the low-amplitude decrement is a maximum. If there is a distribution of loop lengths L_C, then the longest loops break away first, giving a smoothly increasing damping with increasing strain amplitude. (A corresponding hysteresis loop is shown by the dashed line in Fig. 11.) The breakaway event is catastrophic in that whenever a loop breaks away within a network length L_N, the entire network breaks away. Thus the model provides two types of loss, one strain amplitude-dependent and the other not.

A calculation by Granato and Lucke[5] was based on the following assumptions: (1) all the network lengths L_N are the same size; (2) $L_N \gg L_c$; and (3) at zero stress, the loop lengths l are distributed randomly. These assumptions result in the following relation for the amplitude dependence of the decrement:

$$\Delta_H = \frac{c_1}{\epsilon_0} \exp\left(-\frac{c_2}{\epsilon_0}\right)$$ (13)

where

$$c_1 = \frac{\Omega \Delta_0 \Lambda L_N}{\pi L_c} c_2$$

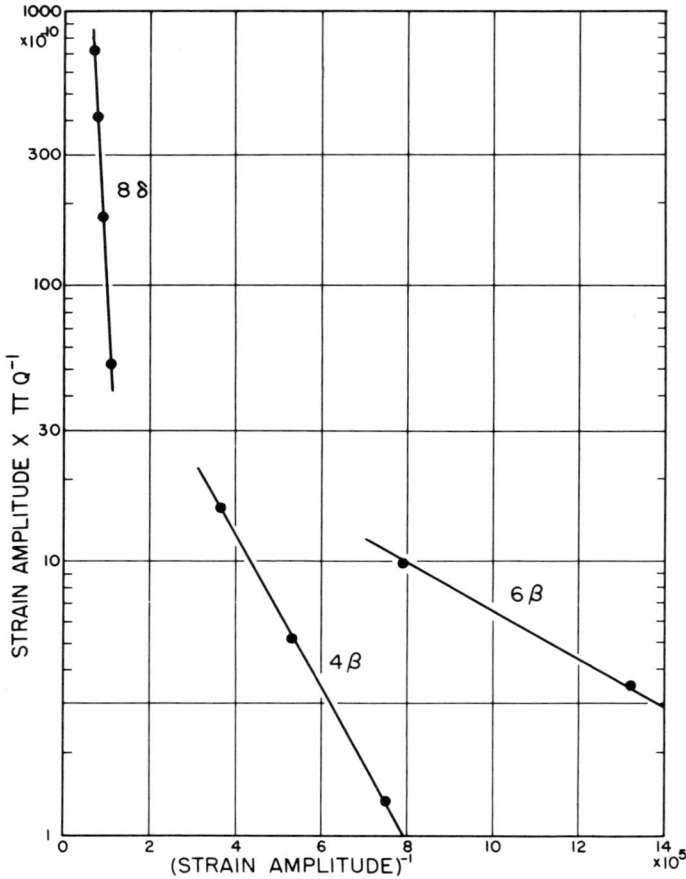

Fig. 12 The data of Weertman and Salkovitz are here plotted for lead with three different concentrations of added bismuth impurity. According to the theory, the points should lie on straight lines whose slopes are proportional to the concentration of impurity on the dislocation line. The deliberately added concentrations of bismuth in the specimens as given by Weertman and Salkovitz are 0.035, 0.053, and 0.65 at. per cent for samples 6β, 4β, and 8δ respectively. (After Granato and Lücke.)

and

$$c_2 = \frac{K\eta b}{L_c}$$

In the above equation $K = G/(4RE)$, where E is Young's modulus, R is the resolved shear stress factor, and η is Cottrell's misfit parameter. The subscript H is used to denote the hysteresis loss.

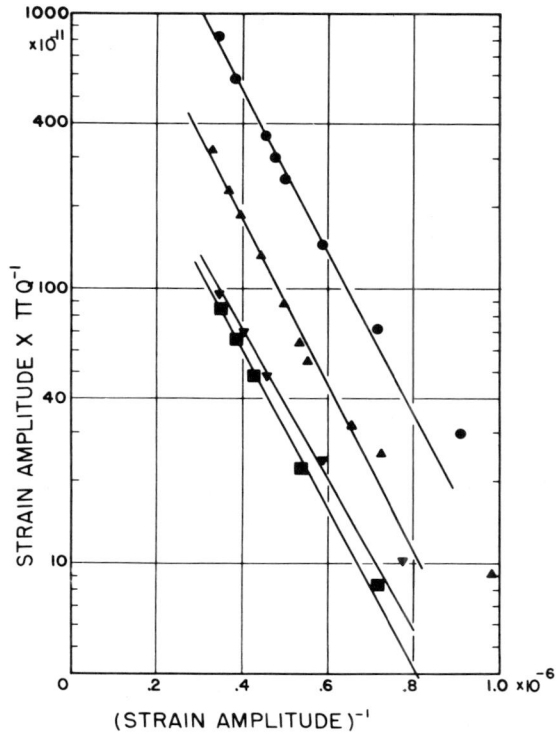

Fig. 13 The data of Fig. 7 replotted. (After Granato and Lücke.)

A modulus change $(\Delta E/E)_H$ of the same magnitude as Δ_H is also found from this model.

According to Eq. (13), $\log \Delta_H \epsilon_0$ should be a linear function of $1/\epsilon_0$, with slope proportional to the pinning point density and intercept proportional to the dislocation density. An example of such a test is shown in Fig. 12, in which data obtained by Weertman and Salkovitz[17] for lead of three different concentrations of added bismuth impurity is plotted. In Fig. 13, the data shown in Fig. 7 are replotted, showing that the expected increase of intercepts for increasing dislocation densities is obtained.

Generally speaking, fair agreement with the theory is usually found for not too pure materials, for which case the assumptions of the theory break down. The discrepancies are usually such as to indicate that the theory overestimates the rate of increase of the damping with strain amplitude. In fact, it is surprising that this simple model accounts for the results as well as it does since the theory is a zero-point theory and neglects the effect of thermal fluctuations on the breakaway process. Recent work, still not completed, by Teutonico, Granato, and Lücke[18] indicates that when thermal fluctuations are taken into account, the form of the results remains the same, with only a reinterpretation of the coefficient necessary in the damping vs. strain-amplitude relation.

A number of modifications and extensions of the amplitude-dependent theory have been proposed. Beshers[19] has pointed out that since the damping depends exponentially on the resolved shear stress amplitude on the various slip systems, then each slip system will contribute an exponentially increasing component, but with different slopes to the straight line plots. This would produce deviations from straight lines curving upward, as is often observed. Gelli[20] has proposed a qualitative model for amplitude-dependent dislocation damping in which two coaxial regions in the Cottrell impurity atmosphere are distinguished. In the internal region, a vibrating dislocation dissipates its energy by a hysteresis mechanism, both dynamically and statically, according to the mechanisms of Weertman and Salkovitz,[17] and an external region in which the damping follows the model described here. In the absence of detailed quantitative predictions, it is difficult to determine the merits of such a proposal. A development along somewhat similar lines emphasizing the effect of thermal fluctuations has been given by Friedel.[21] A modification taking into account the fact that some impurities may not interact with screw dislocations although they can pin dislocations with an edge component has been given by Swartz and Weertman.[22] Rogers[23] has made modifications intended to give a quantitative account of the maximum in the decrement expected at very high strain amplitudes. The effect of motion of pinning points along the dislocation line has been considered by Yamafuji and Bauer,[24] who found expressions for the resulting expected time dependence of the amplitude-dependent damping, and related the effect to pipe-diffusion measurements.

In a more general discussion of the model, the fact that the simple string analogy for dislocation motion is highly idealized, should be considered. The energy of a dislocation is distributed throughout a volume, whereas it is assumed that in the model there exists a "line energy," i.e., an energy per unit length of dislocation line. The validity of this assumption has been investigated by Leibfried,[25] who found that the line-energy concept is a useful approximation, except for large displacements near pinning points where interactions between neighboring loops become significant. Assuming the existence of a line energy, a "tension" can be defined

(deWit and Koehler,[26] Stern and Granato[12]), but the tension depends
upon the type of dislocation, being smaller for edge dislocations
than for screw dislocations. Thus a (continuous) distribution of
effective tensions should be taken into account in the theory. For
mixed dislocations, the displacement of a dislocation under an external stress will not be symmetric, as is assumed in the simple
theory. There is a further influence of anisotropy on the value of
the tension (deWit and Koehler[26]). Also, internal stresses in crystals will influence the detailed shape of the dislocation displacements. Some of these effects cannot be taken into account explicitly.
For example, the distribution of the types of dislocations and the
internal stress distributions usually are unknown. Nevertheless,
it appears to be a useful approximation to use the simple string
model where it must be realized that the tension represents some
kind of average or effective value as a result of these effects.

Because dislocation interactions are ignored, it should be expected that the string model would not hold at large dislocation
densities. In fact, it is found (Weertman and Koehler;[27] Hikata et
al.[28]) that the damping goes through a maximum after a few percent of plastic deformation. This indicates that dislocation interactions eventually are sufficiently effective in inhibiting dislocation
mobility to overcompensate for increasing dislocation densities.
Nevertheless, the predictions of the string model are often found
to be satisfied even after as much as 4% plastic deformation
(Granato et al.[29]).

The lattice interaction (Peierls' stress) is also neglected in the
string model. There is evidence that this is a good approximation
in f.c.c. and h.c.p. metals and alkali halides, less so in b.c.c.
metals (Chambers[30]), and probably a bad approximation in materials with large Peierls' stress, e.g., germanium or silicon.
Even in f.c.c. metals, however, one would expect a strong influence of the Peierls stress for dislocations oriented approximately
along crystallographic directions. Such dislocations should have
"kinks" where they cross over from one Peierls valley to the
next. The influence of such kinks on dislocation motion and damping is discussed in detail by Seeger and Schiller.[31] These dislocations are thought to give rise to the Bordoni peak at low temperatures. The number of dislocations with the required orientation
is estimated to be a small fraction of a percent of the total number
(Seeger[32]). Thus the mechanisms mentioned here and the Bordoni
peak can be discussed separately. Even when the dislocations do
not make small angles with the crystallographic directions, kinks
will occur on the dislocations. However, the kinks are now closely

spaced and merely put small wiggles on the dislocation line. During a dislocation displacement, the kinks move laterally but the overall motion is much like that of a string. It is possible to reinterpret the dislocation line tension in terms of kink interactions (Seeger and Schiller[31]), thus enhancing our understanding of the dislocation tension and also of the damping constant.

3   APPLICATIONS TO THE STUDY OF DEFECT INTERACTIONS

Since the internal friction effects depend entirely on the interactions of dislocations with other defects, we shall find it convenient to classify the effects in terms of defect interactions using the "particle" viewpoint introduced by Seitz.[34] The defects to be considered are dislocations, point defects, electrons and phonons. (Foreign atoms have been lumped into the point-defect category together with vacancies and interstitials.) First, we consider double interactions of dislocations with other defects, and select a few measurements to illustrate how these may be studied by internal friction methods.

3.1   Dislocation-Phonon Interactions

First, we note that phonons interact with dislocations in essentially two ways. Similar to the case of Brownian motion of a particle in an external field, a dislocation moving through a phonon gas under the action of an external applied stress is subject to both a viscous drag and to fluctuations in displacement.

The properties of the internal friction of copper at megacycle frequencies studied by Alers and Thompson[11] and by Stern and Granato[17] can be used to determine the magnitude of the dislocation-phonon drag. As already noted earlier, and as is easily seen from Fig. 5, at high frequencies, the attenuation depends only on the dislocation density and the damping constant and not on the loop length or dislocation line tension. By making an independent count of the dislocation density, the magnitude of the damping constant could be determined. Furthermore, the damping constant was found to be linear in temperature. This latter fact agrees with what is to be expected if scattering by phonons is the source of the damping, since the damping constant should then be proportional to the phonon density, according to Leibfried, or linear in temperature (at not too low temperatures). The magnitude of the damping constant found was somewhat in excess (about a factor of four) of that given by Leibfried's estimate. Both Alers and Thompson and Stern and Granato concluded that the physical source of the damping at megacycle frequencies was the scattering of phonons by the moving dislocations and that the vibrating string model is applicable. In addition, by extrapolation of megacycle results into the kilocycle range and vice versa, Granato and Stern[14] were able to show that the same mechanism accounts for the part of the damping observed in the kilocycle range that can be removed by irradiation pinning. Thus it seems safe to say that the question concerning the source of the damping is now understood.

An interesting side result here is that a question that arose early in dislocation theory—"can relativistic velocities of dislocations be achieved at stresses near the yield stress of soft crystals at room temperature?"—can now be answered from these results. The answer is no. From the value of the magnitude of the dislocation-phonon interaction strength deduced ultrasonically, it is found that in copper at room temperature, the relation between velocity and stress[12] is

$$\frac{v}{c} = \frac{75\sigma}{G} \tag{14}$$

where c is the shear velocity, and G is the shear modulus. Since the theoretical yield stress of a perfect crystal is of order of $G/30$, stresses of the order of the yield stress of perfect crystals would be required for relativistic velocities.

The remaining question is: "What is the mechanism of the scattering of phonons by dislocations?" It has not yet been determined whether this is due to scattering by the strain field (which changes the elastic constants in the vicinity of the dislocation) or whether the scattering is due to a reradiation of sound waves by the dislocation under the influence of the incident phonons.

As pointed out by Mason,[35] the same mechanism probably limits the velocity of dislocations at high stresses as found by Johnston and Gilman[36] in direct observations of the motion of etch pits. From these measurements, the damping constant for LiF is found from the relation

$$b\sigma = Bv \tag{15}$$

to be 7.0×10^{-4} dynes-sec/cm^2.

Another experiment by Baker[37] establishes the result that the velocities observed by Johnston and Gilman are not the instantaneous velocities of the dislocations at a given stress level, but only the velocities of the pinning points. Baker observes periodic dislocation motion of amplitude 1000 b at stress levels an order of magnitude below the macroscopic yield stress. These results are shown in Fig. 14. The velocity of the dislocations is then of order of 1 cm/sec ($v \sim d\omega$, where d is the amplitude of motion and ω is the frequency of oscillation). This result demonstrates that the Peierls force is ineffective in limiting dislocation motion, at least at room temperatures. The dislocations oscillate at high speeds between pinning points. The overall motion of the dislocations is limited by the speed of the pinning points. This result is a good example of ways by which internal friction measurements can help to distinguish between various postulated deformation mechanisms.

Evidence that the dislocations in NaCl are not Peierls stress limited even at low temperature can be obtained from the

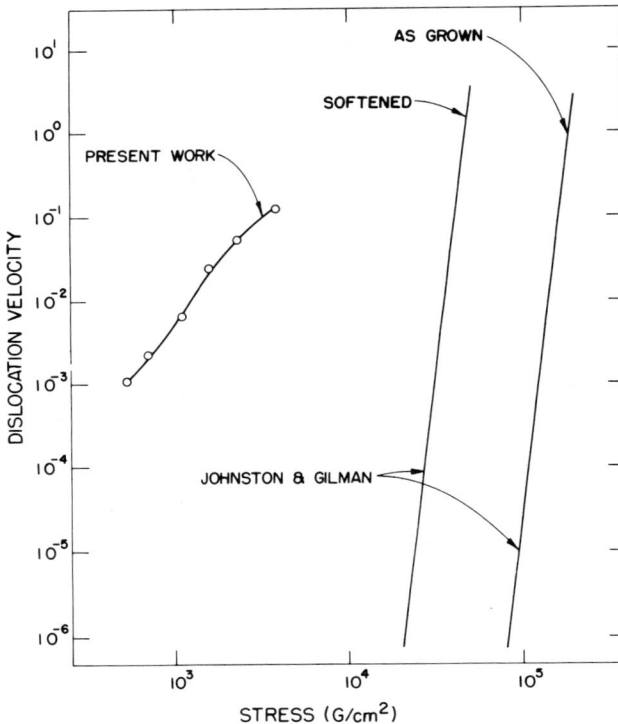

Fig. 14 Ultrasonically determined velocities in LiF compared with directly observed etch pit velocities. (After Baker.)

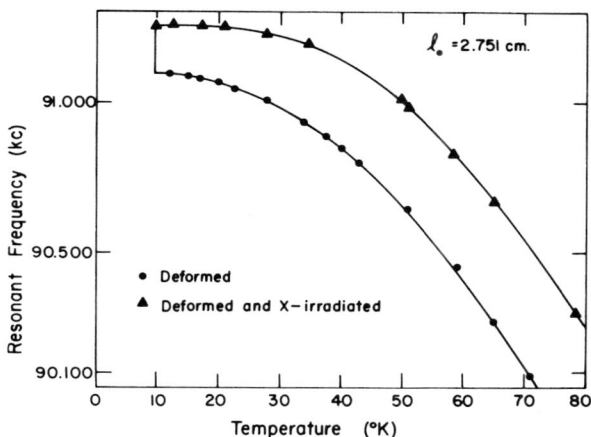

Fig. 15 Dependence of the resonant frequency of a deformed NaCl crystal, before and after x-irradiation, on temperatures between liquid-helium and liquid nitrogen temperature. l_o represents the room-temperature specimen length. (After Bauer and Gordon.)

low-temperature elastic modulus measurements of Bauer and Gordon[38] shown in Fig. 15. The fact that the modulus of unirradiated NaCl is lower than that in the irradiated state shows that the dislocations are mobile even at helium temperatures, and that no sudden increase in mobility occurs in the temperature range shown. This behavior is in contrast to that found in body-centered cubic materials by Chambers.[30] In that case the modulus defect freezes out at low temperatures, giving evidence of a Peierls stress.

Following the suggestion of Mason, a number of measurements of the damping constant B from ultrasonic measurements at high frequencies have been made for LiF and NaCl. For LiF the value 1.6×10^{-3} was obtained by Suzuki, Ikushima and Aoki[39] and the value 2.5×10^{-4} was given later by Mitchell.[40] In measurements on seven NaCl specimens, Moog[41] obtained values ranging from 2.5 to 10.5×10^{-4} with an average value of 4.5×10^{-4}. Moog accounts for the range of values by uncertainties in the dislocation density, which he finds to be very nonuniform in a given specimen.

There are also other limitations to the accuracy obtainable in ultrasonic measurements, particularly with respect to the problem of obtaining accurate values of the attenuation at high frequencies. In view of the significance of this measurement, we (Fanti et al.[42]) thought it important to make an effort to improve the accuracy of the measurement. A novel technique was employed for this purpose and many specimens were used to check the reproducibility of the

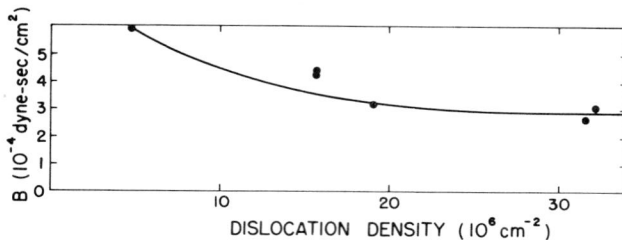

Fig. 16 Ultrasonically determined dislocation drag constant B as a function of total dislocation density. (After Fanti, et al.)

results. Measurements were made of both the attenuation and velocity changes at 10 and 30 Mcps resulting from pinning by cobalt 60 γ irradiation. The results are shown in Fig. 16 for LiF. It may be seen that reproducible results of improved accuracy were obtained. There is, in fact, a systematic dependence of the derived drag constant B on the dislocation density as determined by etch-pit measurements. Further experiments gave evidence that the large Λ values should be used since a fraction of the dislocations do not contribute to the damping at low dislocation densities. The values of the drag constants obtained (2.4×10^{-4} for LiF and 1.6×10^{-4} for NaCl are intermediate between the calculated Leibfried and Mason values, which themselves differ only by about a factor of two for LiF and NaCl at room temperature.

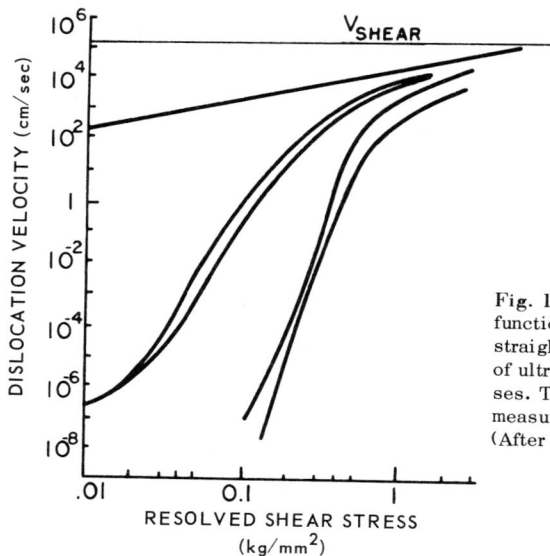

Fig. 17 Dislocation velocities as a function of applied stress in NaCl. The straight line is a linear extrapolation of ultrasonic velocities to higher stresses. The curved lines are the etch pit measurements of Gutmanas, et al. (After Fanti, et al.)

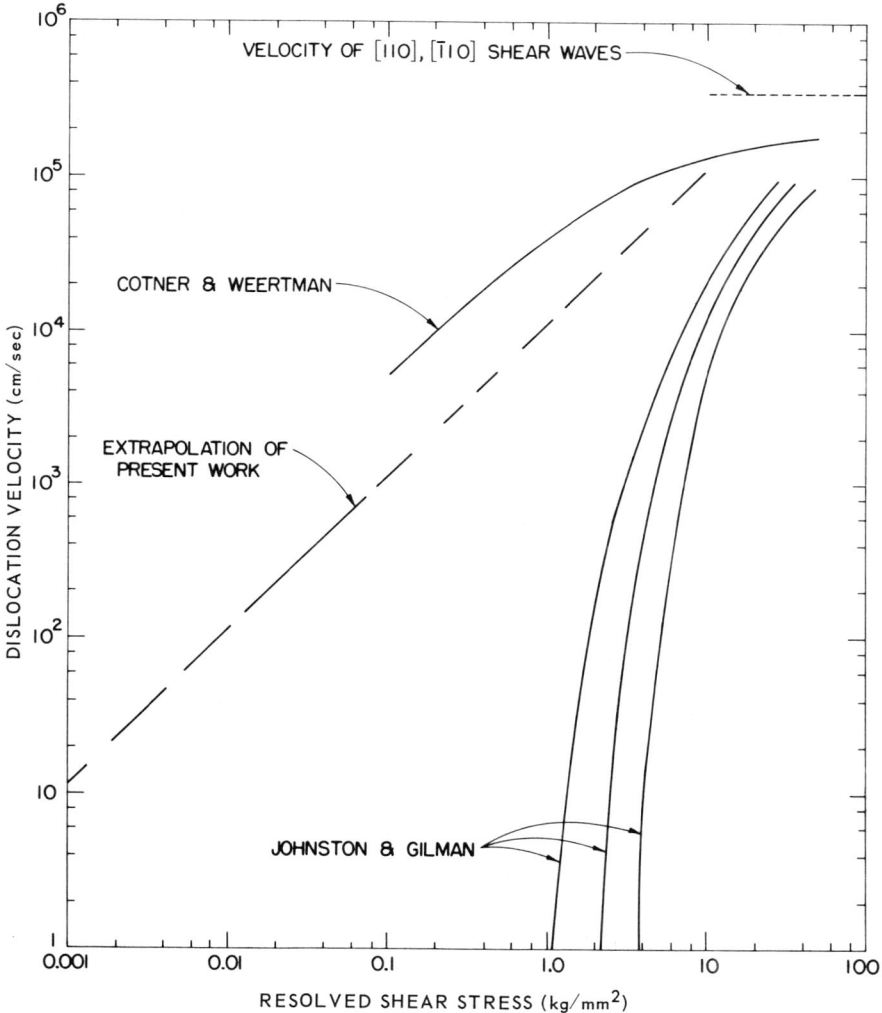

Fig. 18 Dislocation velocities as a function of applied stress in LiF. The straight line is a linear extrapolation of ultrasonic velocities to higher stresses. Also shown are etch pit measurements by Cotner and Weertman and Johnston and Gilman. (After Fanti et al.)

When these values are used to calculate the limiting dislocation velocities at high stresses, the results shown in Figs. 17 and 18 are obtained for NaCl and LiF. Figure 17 also shows the directly observed velocities at high stresses by the etch-pit technique of Gutmanas et al.[43] It may be seen that the directly measured velocities do not exceed, and appear to be limited by the velocities predicted by extrapolation of the ultrasonic results. This is strong evidence that the same drag mechanism is operative in both stress ranges on the intantaneous velocities. In Fig. 18 the corresponding comparison is made for LiF with the measurements of Johnston and Gilman[36] and of Cotner and Weertman.[44] The Johnston and Gilman velocities do not exceed the ultrasonic extrapolation line; in fact the measurements are at such high stresses that the sound-speed limitation should be as important here. However, the results of Cotner and Weertman do disagree with the behavior expected from the ultrasonic measurements. It should be noted that the ultrasonic measurements show that the instantaneous velocity is linear in stress up to velocities exceeding 10^2 cm/sec; thus while the stress ranges do not overlap, the extrapolation is, nevertheless, relatively narrow in range. Also the discrepancy is outside the limits of error of the ultrasonic determination, which we estimate to be of order 50%. It is our opinion that the Cotner and Weertman values may have been overestimated. At least in NaCl, it appears that the ultrasonically determined dislocation velocities relate well to those determined by direct observation at higher stresses. Also, when the damping constant is known, values for the dislocation densities in specimens can be determined from nondestructive ultrasonic measurements. This becomes a valuable technique for counting dislocations, particularly since the ultrasonic method works best in the region of densities between 10^6 and 10^8 cm^{-2}, a region of much practical interest.

3.2   Dislocation-Point Defect Interaction

The best known interaction of this type is that discussed first by Cottrell.[45] Because an oversized (or undersized) impurity can relieve the strain energy of the lattice by moving to the dilated (or compressed) region near a dislocation, the dislocation will be bound to the impurity. The binding strength will be small, at most of the order a few tenths of an electron volt, so that a sufficiently large stress can pull the dislocation away from immobile point defects. Interstitial atoms and vacancies can also act as pinning points. Other possibilities are jogs, dislocation nodes, intersections, and places where a dislocation may move out of the slip plane.

A significant advance in this area was made by Bauer and Gordon,[38] who showed that by combining ultrasonic and optical measurements, it is possible to identify the atomic configuration which is effective in pinning a dislocation in NaCl. They found that dislocation pinning in x-irradiated NaCl proceeded at the same rate at low temperatures as at room temperature. From this it was concluded that diffusion of point defects is not involved in the pinning process and that the pinning points must be produced at the dislocation core. Further, they discovered that dislocations pinned by irradiation at low temperature can be unpinned by light. This effect is shown in Fig. 19. Bauer and Gordon note that a model used to explain dislocation pinning must satisfy many conditions to be in agreement with their observations. First, defects must be created at, or in the immediate vicinity of, free dislocation segments; in addition, these defects must act as strong pinning points. Furthermore, the pinning defect must possess a characteristic optical absorption band and must be simple enough to be "dissolved" when it is ionized or excited. In the case of rock salt, it is found, for example, that unpinning is produced only by light with wavelengths within a fairly narrow band centered about 6300 A. Finally, the defect must occur generally in the alkali halides (with the possible exception of LiF). The model must also be capable of explaining how dislocation pinning can be reversed at low temperatures, but converted to a permanent type of pinning if the crystal is warmed to room temperature, and, moreover, how unpinning illumination is capable of unpinning dislocations at low temperatures while F illumination can cause additional pinning at all temperatures. The model suggested by Bauer and Gordon which fits all these experimental facts is one in which the pinning point is identified as a complex consisting of a jog on a dislocation formed by a Cl-ion and a F-center located one atom distance away and below the slip plane. Because this F-center is in a region where the crystal structure is dilated, its absorption band is shifted toward the red, that is,

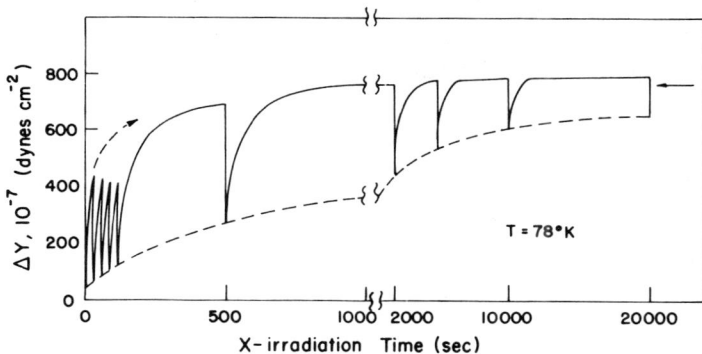

Fig. 19 Behavior of the modulus change of a deformed NaCl crystal, ΔY, during successive x-irradiations and exposure to visible illumination. (After Bauer and Gordon.)

from 4500 to 6300 A, an amount which is in agreement with the shift calculated from the strain field about a dislocation in rock salt. Ionization of the F-center by 6300 A light results in electro-static attraction between the negative-ion vacancy thus formed and the Cl-ion forming the jog on the dislocation; recombination of this Cl-ion and its neighboring vacancy causes the pinning point to "dissolve." The model predicts that if a crystal irradiated at low temperature is warmed up in the dark to a temperature where F-centers can diffuse, and then cooled down again, it should no longer be possible to remove the pinning points by illumination, as is observed.

As a second example of the study of dislocation-pinning point interactions, we may note the interesting observation made by Hikata and Tutumi[46] shown in Fig. 20. They find a striking simi-larity in the curves of creep rate and ultrasonic attenuation in aluminum at room temperature. In this case the pinning points must be carried along with the dislocations as they move through the lattice. This suggests that jogs may be the effective pinning points in these measurements. The similarity of the shape of the curves may be understood if both depend primarily on the distance between pinning points, and this distance is assumed to decrease

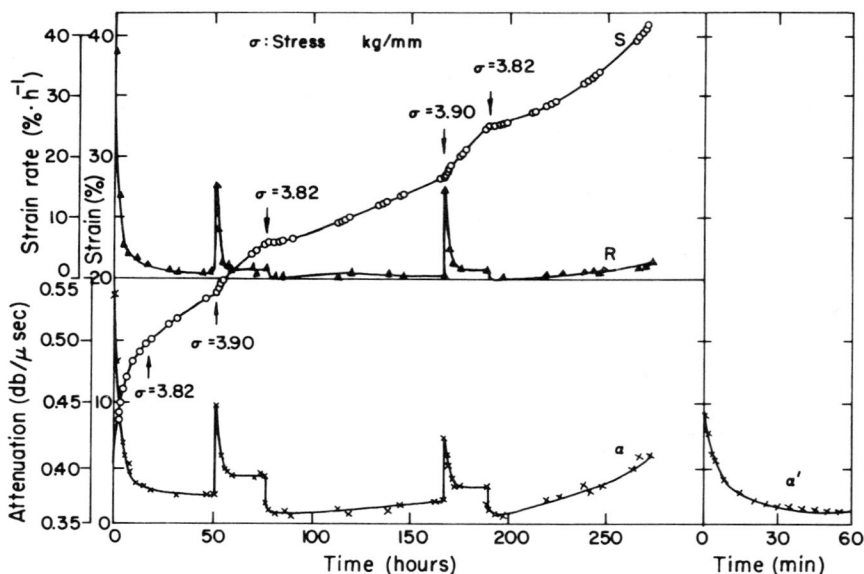

Fig. 20   Curve S: creep strain-time relation. Curve α: attenuation-time relation. Curve R: creep rate-time relation. Curve α': recovery of attenuation. (After Hikata and Tutumi.)

with time at constant load. This is because both attenuation and creep rate depend sensitively upon the distance L between pinning points. The former depends upon the fourth power of L, whereas the latter should depend exponentially upon L. Presumably L decreases because of jog formation in dislocation intersections. Similar measurements in NaCl by de Rosset are in progress at the University of Illinois in an attempt to relate the creep and ultrasonic data quantitatively.

By using dislocations as an intermediary, double interactions between points defects and phonons, as they give rise to diffusion, may be studied.

An example of the way in which point-defect, diffusion-migration activation energies can be determined is given by the analysis of Granato, Hikata, and Lücke[29] on the recovery data of Smith.[47] Smith deformed copper specimens by 1% and measured the recovery of the modulus as a function of temperature and recovery time. By assuming that the recovery mechanism is the pinning of dislocations by deformation-induced defects, an activation energy of 1 eV which was assumed to be that for vacancy migration energy was found. At the time of this assignment, all the others for this quantity were either near 0.8 eV or less, or 1.2 eV or more. The migration energy has been determined by Simmons and Balluffi,[48] and the ultrasonic value is the only previous assignment that does not disagree with their value. This suggests that the ultrasonic method may be useful for such studies.

A second example of the way in which ultrasonic effects can be used in the study of point-defect motion is provided by the measurements of Thompson, Blewitt, and Holmes,[49] shown in Fig. 21. In this experiment the modulus (or frequency) of a copper specimen was measured as a function of temperature after a neutron bombardment. At first, the modulus decreases with temperature in the normal way, but at about 40 °K this process is interrupted. Interstitials are known to move at this temperature, and presumably the normal decrease of modulus is being compensated by dislocation pinning by interstitials. After the pinning is complete, the modulus continues to decrease in the normal fashion. When the specimen is recooled, no anomalous effects occur. This indicates that the dislocations are now fully pinned and no longer contribute to the modulus. This measurement provides a good example of the selectivity property of ultrasonic measurements. Only the interstitials which migrate to dislocations are detected. Electrical resistance measurements show that annealing occurs at temperatures below 40 °K, but these defects do not travel to dislocations.

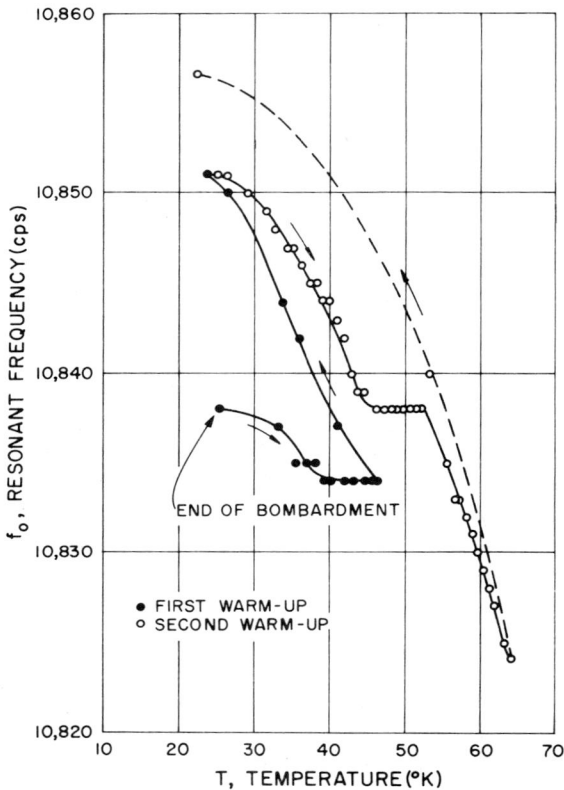

Fig. 21 The change in resonant frequency with temperature upon warming up from fast neutron irradiation at 20°K. (After Thompson, Blewitt, and Holmes.)

3.3   Dislocation-Electron Interactions

A new and interesting application of this model was recently introduced by Tittmann and Bommel,[50] who measured ultrasonic attenuation in superconducting lead. While studying the phonon-electron interaction in pure lead at 51 Mcps, an amplitude dependence, first found by Love and Shaw,[51] was observed. Some of Tittmann and Bommel's results are shown in Fig. 22. The upper curve shows the typical attenuation vs. temperature curve in the normal state, while the lower (dashed) curve shows that expected in the superconducting state. The measured values are intermediate and depend strongly on the amplitude of the ultrasonic wave as temperature is decreased. Tittmann and Bommel suggest a plausible

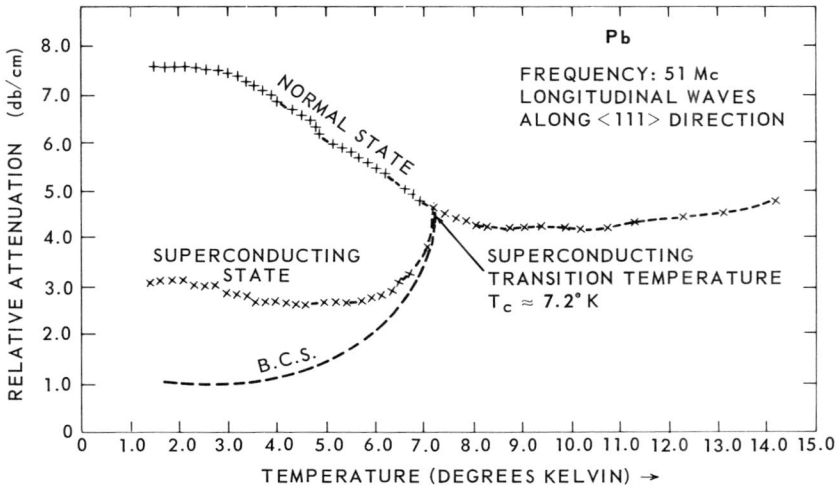

Fig. 22 Temperature dependence of ultrasonic attenuation at medium amplitude in 99.999 % pure single-crystal lead. (After Tittman and Bommel.)

model for the absence of dislocation damping in the normal state and its presence in the superconducting state. According to their model, the conduction electrons behave as a viscous gas which damps the motion of the dislocations. As the temperature is lowered below the critical temperature, the density of electrons able to damp the dislocation motion decreases, and the breakaway mechanism is able to operate.

A number of experiments were conducted to test these ideas. The amplitude-dependent damping was found to follow the expected strain amplitude-dependence (Fig. 23). Measurements of 99.9% pure lead crystals doped with tin as impurity showed a strong reduction in amplitude-dependence. The frequency dependence of the effect was found to be approximately constant, which is in agreement with the theory. The ranges studied were 30 Mcps to 1.0 Gcps for the impure lead and 30 to 150 Mcps for the pure lead. From these measurements it can be inferred that the (undamped) resonance frequencies of the dislocations in their pure lead are higher than 150 Mcps. The measurements demonstrate that the electron gas is an effective source of damping at low temperatures and present new possibilities for the study of dislocation-electron interactions. A quantitative account of these effects has recently been given by Mason[52] and by Holstein.[50]

3.4   Dislocation-Dislocation Interactions

It may be expected that with increasing deformation, dislocation-dislocation interactions will become important. A maximum in the

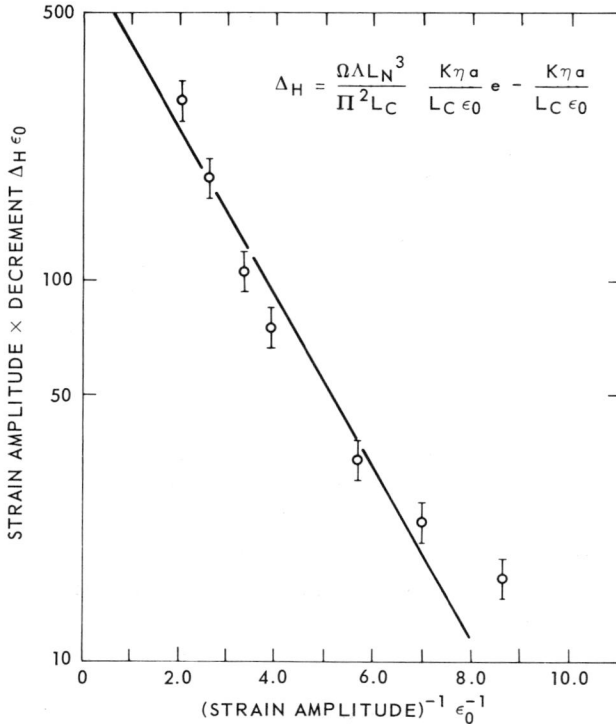

$$\Delta_H = \frac{\Omega \Lambda L_N^3}{\Pi^2 L_C} \frac{K \eta a}{L_C \epsilon_0} e^{-\frac{K \eta a}{L_C \epsilon_0}}$$

Fig. 23 Plot of the amplitude-dependent ultrasonic attenuation of 99.999 % pure single crystal of lead at 4.2°K. (After Tittman and Bommel.)

damping as a function of deformation is often observed [27, 28] indicating that these interactions become strong enough to immobilize the dislocations. However, for deformations of less than a few percent, the observed results seem to be understandable on the basis of the vibrating string model, neglecting dislocation interactions. When the interactions are strong, some cooperative effects might be expected, but as yet there is no proven theory for such effects. Some of the models already given [27, 53] may be useful when more detailed experimental studies are available.

3.5   Dislocation-Lattice Interactions

In the "string" model, we have abstracted away the lattice. If it is needed, we could formally bring it back as a "defect" (Peierls stress) in the "perfect" solid. This interaction plays an important part in the interpretation of the Bordoni peak. The literature abounds with data on this subject, and hence we will not discuss it here. Perhaps even more spectacular as a lattice interaction is the fact that dislocation motion normally occurs only on "slip planes" (where the Peierls stress is low). Some interesting results concerning the distribution of dislocations on various slip systems have recently been obtained by Hikata, Chick, Elbaum, and Truell.[54] In these experiments, ultrasonic attenuation and velocity as well as stress were measured continuously as a function of strain. The aluminum specimen was oriented for single slip, and ultrasonic waves of two different orientations were used. The results are shown in Figs. 24 and 25. In Fig. 24, results are given for the case where the ultrasonic wave had no shear-stress component in the primary glide system. After the initial rise, the

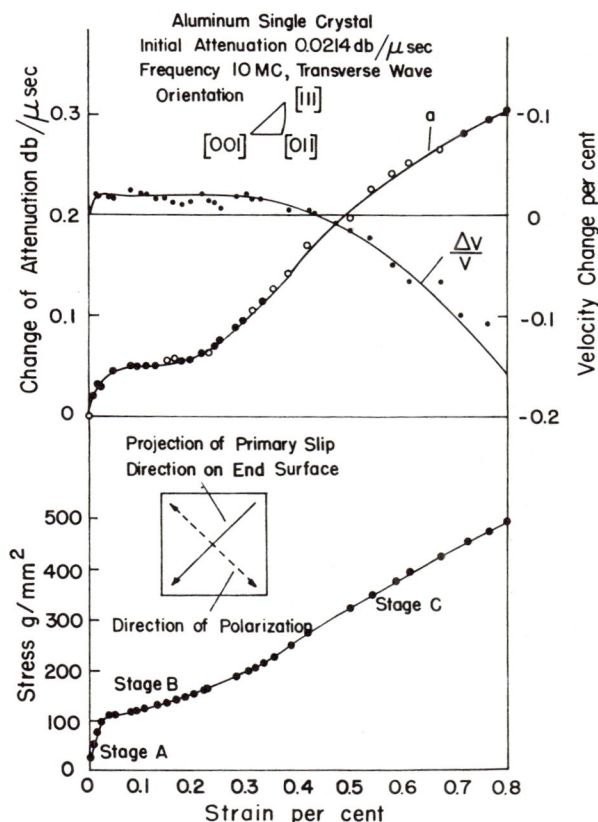

Fig. 24 Stress, shear wave attenuation, and velocity change as a function of total strain for <0.5> orientation. The polarization direction of the shear wave is perpendicular to the projection of primary slip direction on end surface. (After Hikata, Chick, Elbaum, and Truell.)

stress-strain curve clearly indicates the existence of easy glide
for tensile strain of approximately 0.2 percent. The corresponding
attenuation change is very similar to the stress-strain curve, with
little increase of attenuation during the easy glide region even
though the dislocation density is increasing greatly as can be seen
from the results of Fig. 24. The behavior of longitudinal waves,
which have shear-stress components in the primary glide system,
is quite different. As can be seen in Fig. 25, the attenuation increases quite rapidly from the beginning of the deformation with
increasing strain, and does not exhibit any special characteristics
associated with easy glide. These results show clearly the sensi-

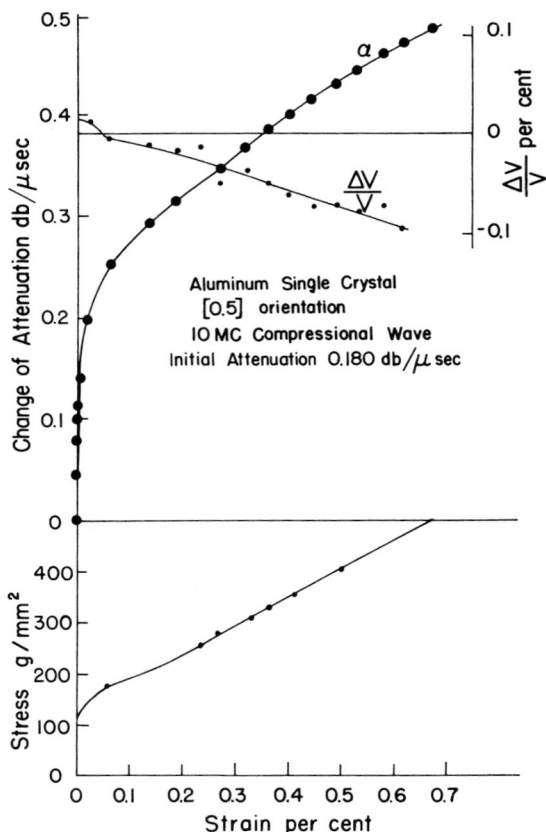

Fig. 25 Stress, longitudinal wave attenuation, and
velocity change as a function of total strain for <0.5>
orientation. (After Hikata, Chick, Elbaum, and Truell.)

tivity of the ultrasonic measurements to the distribution of dislocations. With further development of the technique, we may look forward to the possibility of having continuous plots of dislocation density and loop lengths as a function of strain in various slip systems.

3.6 Triple Interactions

Here, we will discuss only the dislocation-phonon-point defect interaction. A very strong effect is the temperature dependence of the breakaway stress. Thermal fluctuations, even below room temperature, are evidently very effective in drastically reducing the stress amplitude at which breakaway occurs, even when the loop lengths do not change with temperature, as seen by the amplitude-dependent damping.

At temperatures sufficiently high for diffusion to take place, the concentration of pinning points can attain an equilibrium value according to

$$\frac{a}{L_c} = C = C_0 \exp \frac{Q}{kT} \qquad (16)$$

where Q is Cottrell's[45] interaction energy between a dislocation and impurity atom, and a is the atomic spacing along the dislocation line.

Figure 26 shows the decrement as a function of strain amplitude and temperature measured by Fiore and Bauer[55] for a specimen of copper containing 0.0057 atomic percent germanium. The same data are replotted in Fig. 27. As the temperature increases, the slopes decrease. This is interpreted to mean that at high temperature, according to Eq. (16), the equilibrium density of pinning points is decreased. From this data, the value of the binding energy Q of the germanium to the dislocation is 0.3 eV, which is in excellent agreement with independent estimates. Fiore and Bauer made a systematic effort to check the theory using alloys ranging from 0.005 to 0.5 atomic percent germanium. They found that the theory adequately accounts for the results in the amplitude-dependent region, but that the values of the loop length L_c derived from the amplitude-dependent data do not agree with those needed for the amplitude-independent range.

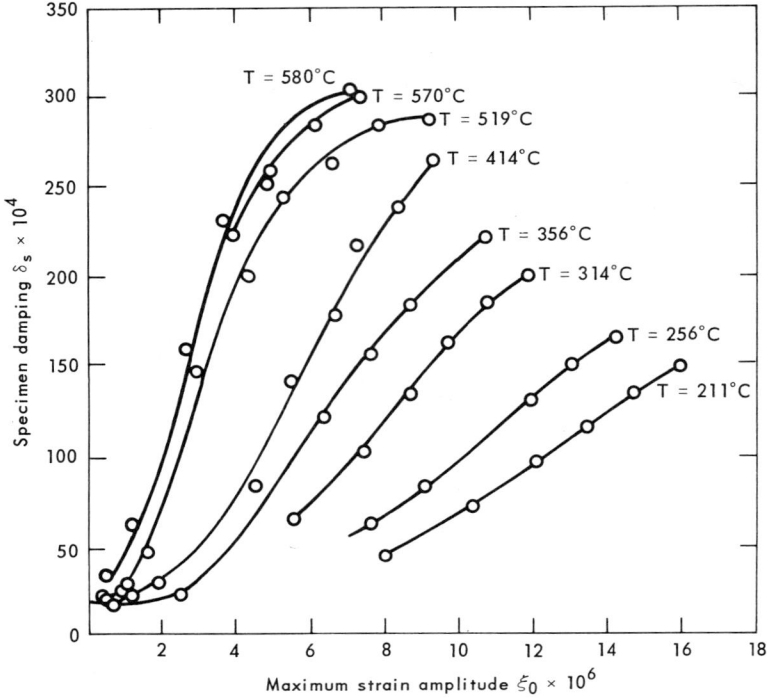

Fig. 26 Specimen damping versus maximum strain amplitude for a copper-germanium alloy (alloy E). (After Fiore and Bauer.)

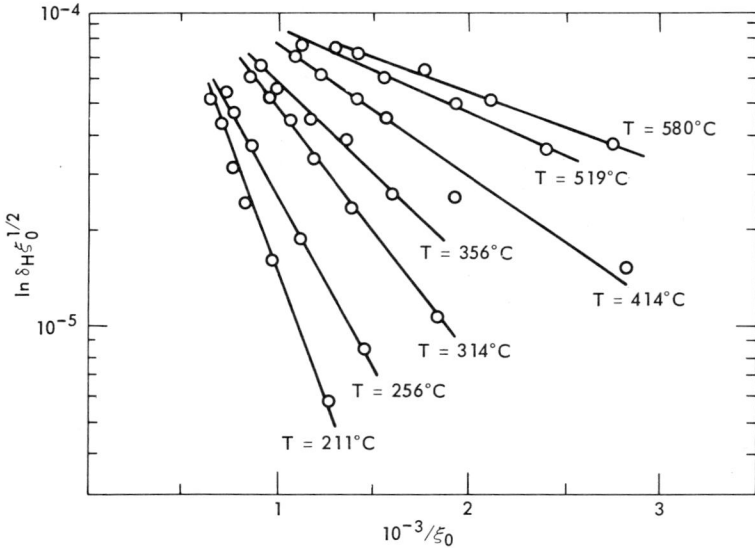

Fig. 27 Data of Fig. 26 replotted. (After Fiore and Bauer.)

4 CONCLUSION

It is clear that ultrasonic measurements provide us with a considerable amount of useful information concerning important defect interactions for understanding the mechanical behavior of crystals. In this report we discussed the qualitative features of a few selected results, but it should be emphasized that results in quantitative form also are obtained. Some of the conclusions obtained are as follows:

1. The vibrating string model is confirmed by the measurements. This model may be applicable even for moderate (up to a few percent) deformation. The theory is quantitatively reliable for low-amplitude effects, but less firmly based for amplitude-dependent effects.

2. The physical source of the sound damping is the dislocation-phonon or electron interaction. Relativistic dislocation velocities cannot be achieved in copper, LiF and NaCl at room temperature at the yield stress.

3. The dislocation drag constant in NaCl has a value which compares favorably with the theory and with values determined from direct measurements by etch-pit techniques of dislocation velocities at higher stresses. The results found demonstrate: (a) that relatively accurate values for the dislocation drag can be obtained ultrasonically; (b) that values for the dislocation densities in specimens can be determined from nondestructive ultrasonic measurements; and (c) that the ultrasonically determined dislocation velocities can be related to those determined by direct observation at higher stresses.

4. An atomic configuration making up a pinning point has been identified in NaCl.

5. Deformation in common alkali halides is determined by the motion of pinning points. The dislocation is free to move between such points. Similar considerations seem to apply for creep in aluminum.

6. The migration activation energies of point defects in small concentrations can be determined ultrasonically.

7. A detailed description of dislocation densities and loop lengths in various slip systems may be obtained ultrasonically.

8. Work currently in progress on the effect of thermal fluctuations on dislocation unpinning promises to provide us with useful information concerning the temperature dependence of the yield stress. However, there is a need for further theory and experiment describing the internal friction of heavily deformed materials.

ACKNOWLEDGMENT

The author wishes to acknowledge the support of the National Science Foundation in this work.

REFERENCES

1. Mott, N. F.: *Phil. Mag.* 7 43: 1151 (1952).
2. Friedel, J.: *Phil. Mag.* 7 44: 444 (1953).
3. Eshelby, J. D.: *Proc. Roy. Soc.* A197: 396 (1949).
4. Koehler, J. S.: "Imperfections in Nearly Perfect Crystals," p. 197, W. Shockley et al. (eds.), John Wiley & Sons, Inc., New York, 1952.
5. Granato, A. V . and K. Lucke: *J. Appl. Phys.* 27: 583 (1956).
6. Granato, A. V., and K. Lucke: *J. Appl. Phys.* 27: 789 (1956).
7. Liebfried, G.: *Z. Physik* 127: 344 (1950).
8. Mason, W. P.: *J. Appl. Phys.* 35: 2779 (1964).
9. Granato, A. V., J. de Klerk, and R. Truell: *Phys. Rev.* 108: 895 (1957).
10. Merkulov, L. G., and L. A. Yakovlev: *Akust. Zh.* 6: 244 (1960); *Soviet Phys.-Acoust.* (English Trans.) 6: 239 (1960).
11. Alers, G. A., and D. O. Thompson: *J. Appl. Phys.* 32: 283 (1961).
12. Stern, R. M., and A. V. Granato: *Acta Met.* 10: 358 (1962).
13. Thompson, D. O., and D. K. Holmes: *J. Appl. Phys.* 27: 713 (1956).
14. Granato, A. V., and R. M. Stern: *J. Appl. Phys.* 33: 2880 (1962).
15. Read, T. A.: *Phys. Rev.* 58: 371 (1940); see also *Trans. AIME* 143: 30 (1941).
16. Baker, G. S.: Ph.D. Thesis, University of Illinois, 1956.
17. Weertman, J., and E. I. Salkovitz: *Acta Met.* 3: 1 (1955).
18. Teutonico, L. J., A. V. Granato, and K. Lucke: *J. Appl. Phys.* 35: 220 (1964).
19. Beshers, D. N.: Ph.D. Thesis, University of Illinois, 1955.
20. Gelli, D.: *J. Appl. Phys.* 33: 1547 (1962).
21. Friedel, J.: *Natl. Phys. Lab., Gt. Brit., Proc. Symp.* 15: 409 (1963).
22. Swartz, J. C., and J. Weertman: *J. Appl. Phys.* 32: 1960 (1961).
23. Rogers, D. H.: *J. Appl. Phys.* 33: 781 (1962).
24. Yamafuji, K., and C. L. Bauer: *J. Appl. Phys.* 36: 3288 (1965).
25. Leibfried, G.: *Oak Ridge Natl. Progr. Rept.* ORNL 2829 (1959).
26. de Wit, G., and J. S. Koehler: *Phys. Rev.* 116: 1113 (1959).
27. Weertman, J., and J. S. Koehler: *J. Appl. Phys.* 24: 625 (1953).
28. Hikata, A., R. Truell, A. V. Granato, and K. Lucke: *J. Appl. Phys.* 27: 396 (1956).
29. Granato, A. V., A. Hikata, and K. Lucke: *Acta Met.* 6: 470 (1958).
30. Chambers, R. H.: *Appl. Phys. Letters* 2: 165 (1963).
31. Seeger, A., and P. Schiller: in "Physical Acoustics," W. P. Mason (ed.), vol. IIIA, Academic Press, New York, 1966.
32. Seeger, A.: *See discussion after paper by Lücke and Granato,* 1962 (1957).
33. Chambers, R. H., G. Alefeld, and J. Shultz: *Bull. Am. Phys. Soc.,* 2 10: 322 (1965).
34. Seitz, F.: "Imperfections in Nearly Perfect Crystals," chap. 1, John Wiley & Sons, Inc., New York, 1952.
35. Mason, W. P.: *J. Acoust. Soc. Am.* 32: 458 (1960).
36. Johnston, W. G., and J. J. Gilman: *J. Appl. Phys.* 30: 129 (1959).
37. Baker, G. S.: *J. Appl. Phys.* 33: 1730 (1962).
38. Bauer, C. L., and R. B. Gordon: *J. Appl. Phys.* 33: 672 (1962).
39. Suzuki, T., A. Ikushima, and M. Aoki: *Acta Met.* 12: 1231 (1964).
40. Mitchell, O. M. M.: *J. Appl. Phys.* 36: 2083 (1965).
41. Moog, R. A.: Ph.D. Thesis, Cornell University, 1965.
42. Fanti, F., J. Holder, and A. Granato: To be published.
43. Gutmanas, E. Y., Nadgornyi, and A. V. Stepanov: Fiz. Tverd. Tela 5: 1021 (1963); English Trans.: Soviet Physics–Solid State 5: 743 (1963).
44. Cotner, J., and J. Weertman: *Disc. Faraday Soc.* (G.B.) 38: 225 (1964).
45. Cottrell, A. H.: *Rept. Conf. Strength Solids,* p. 30, 1948; *Phys. Soc., London* (1948).
46. Hikata, A., and M. Tutumi: *J. Phys. Soc. Japan* 14: 687 (1959).
47. Smith, A. D. N.: *Phil. Mag.* 7 44: 453 (1953).

48. Simmons, R. O., and R. W. Balluffi: *Phys. Rev.* 129:1533 (1963).
49. Thompson, D. O., T. H. Blewitt, and D. K. Holmes: *J. Appl. Phys.* 28:742 (1957).
50. Tittman, B. R., and H. E. Bommel: *Phys. Rev.* 151:178 (1966). The calculation of the electronic drag constant by T. Holstein is given in an appendix to this paper.
51. Love, R. E and R. W. Shaw: *Rev. Mod. Phys.* 136:260 (1964).
52. Mason, W. P.: *Appl. Phys. Letters* 6:111 (1965).
53. Granato, A. V.: *Phys. Rev.* 111:740 (1958).
54. Hikata, A., B. Chick, C. Elbaum, and R. Truell: *Acta Met.* 10:423 (1962).

Part Two

PRELIMINARY CONSIDERATIONS

PEIERLS
BARRIER ANALYSIS

R. Hobart

Battelle Memorial Institute,
Columbus Laboratories,
Columbus, Ohio

ABSTRACT

The Peierls barrier in the Frenkel-Kontorova dislocation model is calculated by summing a series solution to the nonlinear difference equation of the model by complex variable residue techniques. The method relates the barrier to poles on an axis of imaginary lattice spacing.

1 INTRODUCTION

The Frenkel-Kontorova[1] model is perhaps the simplest dislocation model for study of the Peierls barrier. Indenbom[2] calculated the barrier in this model by an analytical approximation analogous to that used by Peierls[3] and Nabarro[4] in their original work.

As in that work, the barrier given by Indenbom decreases exponentially with dislocation width, a result contested by Lifshitz[5] who claimed an inverse fifth-power decrease.

The present author has introduced two techniques for calculating the Peierls barrier in the Frenkel-Kontorova dislocation model. The first[6] is a numerical technique which is applicable to the model with a general substrate function. Numerical calculation supports Indenbom over Lifshitz, but shows the barrier given by the former to be low by a factor of about a half. The second[7] is an analytical approximation technique applied to the Frenkel-Kontorova model with a third-degree polynomial replacing the usual sine function as substrate. This technique relates the barrier to a pole on an axis of imaginary lattice spacing. The barrier given by the analytical technique was compared with that given by the numerical technique and shown to be high by a factor decreasing from 4.33 to 1.92 over a 20 order of magnitude decrease in the barrier, resulting as the dislocation width increases from 2 to 12. The extremely small barriers for wide dislocations are not, of course, physically significant. However, the predominance of exponential dependence of the barrier on width within the model is a mathematical question and its numerical confirmation serves as a check on the assumptions made in the analytical approximation technique. In the following, the analytical technique is extended to apply to the sine substrate function of the original Frenkel-Kontorova model, and the barrier obtained is checked numerically.

2 FRENKEL-KONTOROVA MODEL

The Frenkel-Kontorova model represents an edge dislocation by replacing the atoms above the slip plane with a chain of mass points connected by identical springs, and replacing the atoms below the slip plane by a sinusoidal potential substrate upon which the chain rests. The equation for equilibrium of the nth mass point is

$$\alpha(\Psi_{n+1} - 2\Psi_n + \Psi_{n-1}) + \frac{2\pi A}{b} \sin \frac{2\pi \Psi_n}{b} = 0 \tag{1}$$

The value of A is expressed in terms of the shear modulus G by Frenkel's classic argument predicting the strength of a perfect crystal $A = Gb^3c/4\pi^2a$; the value of α is given in terms of Young's modulus E and Poisson's ratio ν by Indenbom as $\alpha = acE/b(1 - \nu^2)$, where b is the magnitude of the Burgers vector, a is the spacing of the planes parallel to the slip plane, and abc is the atomic volume.[6] It is assumed that the crystal is isotropic and that the dislocation is edge-type.

Let the position of the zeroth mass point be near zero in terms of the dimensionless displacements $Y_n \equiv 2\pi\Psi_n/b$, and assume[6] that the position of the first mass point can be adjusted so these two positions and the equilibrium equation give a strictly increasing sequence $Y_0, Y_1, Y_2, \ldots$ which approaches π. If the proper position Y_1 for a given Y_0 is unique, then this defines a function g by $Y_1 = g(Y_0)$. Similarly, the position Y_{-1} is adjusted so that $Y_0, Y_{-1}, Y_{-2}, \ldots$ is a strictly decreasing sequence approaching $-\pi$. This is essentially the same problem, and its solution is $Y_{-1} = -g(-Y_0)$. The procedure which is illustrated in Fig. 1, leads to equilibrium configurations in two symmetric cases: the I-configuration with $Y_0{}^I = 0$, and the II-configuration with $Y_1{}^{II} = -Y_0{}^{II}$. For intermediate configurations, $Y \equiv Y_0$ and $Y_0{}^{II} < Y < Y_0{}^I$, an external force $-(\alpha b)R(Y)$ on the zeroth mass point is generally required to hold the configuration, where

$$g(Y) - 2Y - g(-Y) + \frac{4\pi^2 A}{\alpha b^2} \sin Y = 2\pi R(Y) \tag{2}$$

Since the inverse of g satisfies

$$g(Y) - 2Y + g^{-1}(Y) + \frac{4\pi^2 A}{\alpha b^2} \sin Y = 0 \tag{3}$$

the equation for R simplifies to

$$-2\pi R(Y) = g(-Y) + g^{-1}(Y) \tag{4}$$

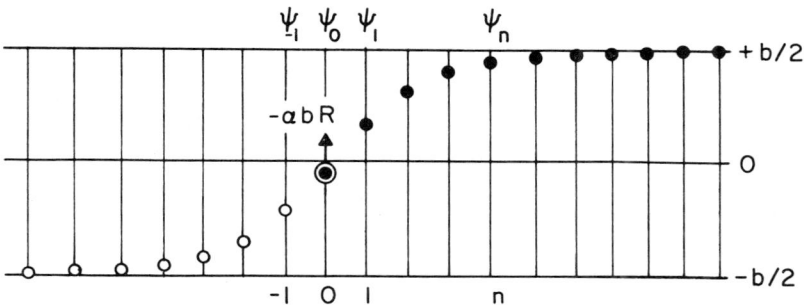

Fig. 1 Force on mass point $n = 0$ holding nonequilibrium Frenkel-Kontorova dislocation configuration.

The integral of the only force on the dislocation $-(ab)R(Y_0)$ over the displacement $(b/2\pi)dY$ between Y_0^{II} and Y_0^{I} is the energy of the I-configuration less the energy of the II-configuration $(ab^2)E_P$. This is the barrier provided $R(Y)$ is negative on $Y_0^{II} < Y < Y_0^{I}$, as is indicated by both numerical work and the following analysis. By symmetry $R(Y)$ is odd, thus

$$E_P = \frac{1}{2\pi} \int_0^{Y_1^{II}} R(Y)\, dY \tag{5}$$

It is convenient to consider A/ab^2 as not yet specified and to introduce $h \equiv 4\pi(A/ab^2)^{1/2}$ as a positive parameter. We substitute $\phi(x + nh) = (2\pi\Psi_n/b) + \pi$ into Eq. (1) and, without loss of generality, set $n = 0$ to obtain

$$\frac{\phi(x + h) - 2\phi(x) + \phi(x - h)}{h^2} = \frac{\sin\phi(x)}{4} \tag{6}$$

An appropriate analytical solution[8] for describing a single dislocation can be expressed as a power series in exponentials

$$\phi(x) = \sum_{\nu\ \text{odd}} (-1)^{(\nu-1)/2}\, b_\nu(h)\, \exp[\nu\sigma(h)x] \tag{7}$$

The coefficients $b_\nu(h)$ are obtained by inserting the series (7) in Eq. (6) and matching corresponding powers of the exponential as shown by Eq. (A3) in the Appendix. In the continuum limit of h vanishing (6) becomes a differential equation which is solved by

$$\phi(x) = \sum_{\nu\ \text{odd}} (-1)^{(\nu-1)/2}\, \frac{4}{\nu}\, \exp\frac{\nu x}{2} \tag{8}$$

3   RESIDUE TECHNIQUE

The scale of h was chosen so that the slope of the dislocation at the center in the continuum limit is unity. The value of ϕ at the center is π. This is the basis of the useful approximation

$$\phi(x) \simeq \pi + x + \Phi(x) \tag{9}$$

where it is assumed that $|\Phi(x)| \ll h$ for $|x| < h$. From $\phi(x + nh) = Y_n + \pi$, $Y_{-1} = -g(-Y_0)$ and $Y \equiv Y_0$, the equations $Y = \phi(x) - \pi$ and

$-g(-Y) = \phi(x - h) - \pi$ follow. Making the approximation (9), we obtain

$$g(Y) \simeq Y + h + \Phi(-Y) - \Phi(-Y - h) \tag{10}$$

and

$$g^{-1}(Y) \simeq Y - h + \Phi(-Y) - \Phi(h - Y) \tag{11}$$

Thus Eq. (4) becomes

$$-2\pi R(Y) \simeq [\Phi(Y) + \Phi(-Y)] - [\Phi(Y - h) + \Phi(h - Y)] \tag{12}$$

It is important to note that the approximate R depends only on the even part of Φ, and thus of ϕ which is

$$E\phi = \sum_{\nu \, \text{odd}} (-1)^{(\nu - 1)/2} b_\nu(h) \cosh[\nu\sigma(h) x] \tag{13}$$

In Eq. (A19) [see Appendix], an approximation $b_\nu(h) \simeq b(h\nu)/\nu$ for small positive h is introduced, where $b(z)$ is an even function analytic on the real axis. With $\sigma \simeq 1/2$

$$E\phi \simeq \frac{1}{2} \sum_{\pm\nu \, \text{odd}} (-1)^{(\nu - 1)/2} \frac{b(h\nu)}{\nu} \cosh\frac{\nu x}{2} \tag{14}$$

results. The sum (14) can be expressed in terms of the residues of

$$F(z) \equiv \frac{\pi b(z) \cosh(zx/2h)}{2z \cos(\pi z/2h)} \tag{15}$$

by

$$E\phi \simeq -\frac{1}{2} \sum_{\pm\nu \, \text{odd}} S(h\nu) \tag{16}$$

where $S(z_0)$ is the residue of the pole in F at z_0. It is assumed that $b(z)$ has poles at integer multiples of $\pm 4\pi i$, that otherwise $b(z)$ is analytic, and that the integral of $F(z)$ on a circular contour missing poles approaches zero for large radii provided $|x| < h$. Some support for these assumptions is given following Eq. (A21). According to Eq. (A21), the first pole on the imaginary axis is third-order

$$b(z) \simeq \frac{i\kappa}{(z - 4\pi i)^3} \tag{17}$$

for $z \simeq 4\pi i$ and κ real. If terms which go to zero with h faster than those shown are omitted, then the residue is given by

$$S(4\pi i) \simeq -\frac{\kappa\pi}{16\,h^2} \exp\frac{-2\pi^2}{h} \left[\left(\frac{\pi}{2} + \frac{h}{2\pi}\right)\cos\frac{2\pi x}{h} + x\sin\frac{2\pi x}{h}\right] \tag{18}$$

Since $F(z)$ is odd, the residues satisfy $S(-z_0) = S(z_0)$. The poles further out on the imaginary axis have negligible residues as the exponential factors of these are raised to higher powers than for Eq. (18). The only residue left to calculate is at zero. The circular contour integral is zero so the sum of all the residues is zero. Since $E\phi(0) = \pi$ and $S(0) = \pi b(0)/2$, cancellation of the residues in the limit $h \to 0$ gives

$$b(0) = 4 \tag{19}$$

But for small positive h cancellation gives

$$E\phi(x) \simeq \pi + S(4\pi i) \tag{20}$$

From Eq. (12) we find

$$R(Y) \simeq \frac{\kappa}{16\,h} \exp\frac{-2\pi^2}{h} \sin\frac{2\pi Y}{h} \tag{21}$$

which yields $Y_1^{II} \simeq h/2$ as the first zero. From Eq. (5) the final result is therefore

$$E_P \simeq \frac{\kappa}{32\,\pi^2} \exp\frac{-2\pi^2}{h} \tag{22}$$

4 DISCUSSION

The scale of the parameter h was chosen so $g(0)/h \simeq 1$ for small h. Therefore a suitable definition for the dislocation width is $l \equiv 2\pi/h$, the number of units of h horizontally between the intercepts at $\phi = \pm\pi$ of a line tangent to the dislocation at its center.

The assumptions of this analytical calculation are tested by comparing the barrier E_P given by Eq. (22) with the barrier E_P' given by the numerical method.[6] These results are shown in Table 1 for the value of κ given by Eq. (A24). The values of E_P' are accurate to the decimals shown for the model, therefore the ratios in the last column support the result (22) to the extent that they approach unity

TABLE 1 Test of Analytical Approximation

l	E'_P	E_P/E'_P
2	1.89×10^{-2}	1.87
4	4.66×10^{-5}	1.41
6	9.55×10^{-8}	1.29
8	1.87×10^{-10}	1.22
10	3.61×10^{-13}	1.19
12	6.88×10^{-16}	1.16
14	1.30×10^{-18}	1.15
16	2.46×10^{-21}	1.13

with increasing dislocation width. The agreement for the sine substrate is better than that for the cubic substrate reported[7] earlier (ratios from 4.33 to 1.92) although the calculation with the sine substrate involves all the assumptions of the cubic substrate calculation plus the questionable procedures associated with $N \to \infty$. The approximation here gives 113% of the correct barrier within the Frenkel-Kontorova model for a dislocation of width 16. Correspondingly, the approximation of Indenbom[2,6] gives 39%, and the approximation of Seeger and Schiller[10] gives 36%.

ACKNOWLEDGMENT

This study was supported by a grant from the U. S. Air Force Office of Scientific Research.

APPENDIX

The nonlinear integral equation (A9) and the consistency equation (A20) are obtained for a polynomial substrate function [right-hand side of (A1)] of odd degree N. They apply to the sine substrate function of the Text only in the limiting sense of Eqs. (A18) and (A21), in context.

Consider the difference equation

$$\frac{\phi(x + h) - 2\phi(x) + \phi(x - h)}{h^2} = \sum_{\mu \text{ odd}}^{N} (-1)^{(\mu - 1)/2} f_\mu \phi^\mu(x) \tag{A1}$$

where f_μ and h are positive. An appropriate analytical solution[8] for describing a single dislocation can be expressed as a series in the increasing exponential solution to the linear equation obtained from Eq. (A1) if all the f_μ except f_1 were zero

$$\phi(x) = \sum_{\nu \text{ odd}} (-1)^{(\nu - 1)/2} b_\nu(h) \exp[\nu\sigma(h)x] \tag{A2}$$

The coefficients $b_\nu(h)$ are obtained by inserting the series (A2) in Eq. (A1) and matching corresponding powers of the exponential to give

$$b_\nu(h) = \frac{\displaystyle\sum_{\mu \text{ odd}}^{\nu}{}' f_\mu \sum_{\Omega(\nu,\mu)} b_{\omega_1}(h)\, b_{\omega_2}(h) \ldots b_{\omega_\mu}(h)}{\left\{ \dfrac{4\, \sinh^2(\sigma h\nu/2)}{h^2} - f_1 \right\}} \tag{A3}$$

where $\Omega(\nu,\mu)$ is the set of μ-tuples $\omega_1, \omega_2, \ldots, \omega_\mu$ of odd integers with sum ν, and the prime indicates $\mu = 1$ is omitted. Each coefficient is given in terms of all prior coefficients and thus in terms of b_1.

In principle, Eq. (A3) gives the coefficients, but, in practice, it is far too difficult to use, except for small ν. An approximation to the coefficients can be obtained, however, on the assumption that for $b_1 = 4$

$$b(\tau) = \lim_{\nu \to \infty} \nu^{(N-3)/(N-1)} b_\nu(\tau/\nu) \tag{A4}$$

exists and is analytic for real τ. From Eq. (A3) it follows that $b(\tau)$ is even.

Parenthetically, by comparision with Eq. (8), the exponent in Eq. (A4) was first expected to be unity. However, unity does not work, but an unknown exponent γ is fixed at the value shown by the condition that all terms in Eq. (A3) be finite and not all be zero when the substitution

$$\left(\frac{h}{t}\right)^\gamma b(t) \simeq b_\omega(h) \tag{A5}$$

is made and h allowed to vanish with $t \equiv h\omega$ held positive. The sums for various μ over $\Omega(\nu,\mu)$ all become integrals, and all but the last $\mu = N$ vanish with h. Therefore, Eq. (A3) becomes

$$b(\tau) = \frac{f_N \tau^{\frac{N-3}{N-1}} \overset{\infty\,\infty}{\underset{0\ 0}{\int\int}} \cdots \overset{\infty}{\underset{0}{\int}} \dfrac{b(t_1)b(t_2)\ldots b(t_N)}{(t_1 t_2 \ldots t_N)^{\frac{N-3}{N-1}}} \delta(t_1 + t_2 + \cdots + t_N - \tau)\,dt_1 dt_2 \ldots dt_N}{2^{N+1}\sinh^2(\tau/4)}$$

(A6)

using $\sigma(0) = 1/2$.

We are interested in the analytic continuation of $b(\tau)$. Because of the denominator in Eq. (A6), poles may appear on the imaginary axis. An equation for $a(\tau) \equiv b(i\tau)$ can be obtained from Eq. (A6) by replacing h with ih in Eq. (A3), and τ with $i\tau$ in Eq. (A4):

$$a(\tau) = \frac{f_N \tau^{\frac{N-3}{N-1}} \overset{\infty\,\infty}{\underset{0\ 0}{\int\int}} \cdots \overset{\infty}{\underset{0}{\int}} \dfrac{a(t_1)a(t_2)\ldots a(t_N)}{(t_1 t_2 \ldots t_N)^{\frac{N-3}{N-1}}} \delta(t_1 + t_2 + \cdots + t_N - \tau)\,dt_1 dt_2 \ldots dt_N}{2^{N+1}\sin^2(\tau/4)}$$

(A7)

and $a(\tau)$ is real for $0 \le \tau < 4\pi$ by the equation replacing Eq. (A4).

From Eq. (A7) in the limit of small τ, the expression

$$\frac{f_N a(0)^{N-1}}{2^{N-3}} = \frac{\Gamma\big(2N/(N-1)\big)}{\Gamma^N\big(2/(N-1)\big)}$$

(A8)

is obtained from the Dirichlet integral formula[9] (which strictly should not be applied here since the integrand is not a continuous function). Equation (A7) can be written in more elegant form by the substitutions $\eta \equiv \tau/4$, $\theta(\eta) \equiv a(\tau)/a(0)$, and Eq. (A8):

$$\frac{\sin^2 \eta}{\eta^2}\theta(\eta) = \frac{\Gamma\big(2N/(N-1)\big)}{\Gamma^N\big(2/(N-1)\big)} \overset{\infty\,\infty}{\underset{0\ 0}{\int\int}} \cdots \overset{\infty}{\underset{0}{\int}} \dfrac{\theta(\eta x_1)\theta(\eta x_2)\ldots\theta(\eta x_N)}{(x_1 x_2 \ldots x_N)^{\frac{N-3}{N-1}}}$$

$$\times\, \delta(x_1 + x_2 + \cdots + x_N - 1)\,dx_1 dx_2 \ldots dx_N$$

(A9)

By the assumptions for $b(\tau)$, $\theta(\eta)$ is even and analytic at zero

$$\theta(\eta) = \sum_{\nu=0}^{\infty} \eta^{2\nu}\theta_\nu$$

(A10)

where $\theta_0 = 1$. We introduce the notation

$$\frac{\sin^2 \eta}{\eta^2} = \sum_{\nu=0}^{\infty} \eta^{2\nu} S_\nu \tag{A11}$$

Now express the product of thetas under the integral in Eq. (A9) in terms of series

$$\left[\theta_0 + \eta^2 x_1^2 \theta_1 + \cdots + \eta^{2\alpha_1} x_1^{2\alpha_1} \theta_{\alpha_1} + \cdots\right] \cdots$$

$$\times \left[\theta_0 + \eta^2 x_i^2 \theta_1 + \cdots + \eta^{2\alpha_i} x_i^{2\alpha_i} \theta_{\alpha_i} + \cdots\right] \cdots$$

$$\times \left[\theta_0 + \eta^2 x_N^2 \theta_1 + \cdots + \eta^{2\alpha_N} x_N^{2\alpha_N} \theta_{\alpha_N} + \cdots\right] \tag{A12}$$

Define $\alpha(N)$ as the set of N-tuples $\alpha_1, \alpha_2, \ldots, \alpha_N$ of nonnegative integers with sum n. The coefficients of the terms in Eq. (A9) proportional to η^{2n} are

$$\sum_{m=0}^{n} \theta_{n-m} S_m = \frac{\Gamma(2N/(N-1))}{\Gamma^N(2/(N-1))} \sum_{\alpha(N)} \left(\prod_{i=1}^{N} \theta_{\alpha_i}\right) \int_0^{\infty}\int_0^{\infty} \cdots \int_0^{\infty}$$

$$\times \frac{\prod_{i=1}^{N} x^{2\alpha_i} \delta(x_1 + x_2 + \cdots x_N - 1)}{(x_1 x_2 \ldots x_N)^{\frac{N-3}{N-1}}} dx_1\, dx_2 \ldots dx_N \tag{A13}$$

Using the Dirichlet integral formula, this becomes

$$\sum_{m=0}^{n} \theta_{n-m} S_m = \frac{\Gamma(2N/(N-1))}{\Gamma^N(2/(N-1))} \sum_{\alpha(N)} \left(\prod_{i=1}^{N} \theta_{\alpha_i}\right) \frac{\prod_{i=1}^{N} \Gamma\left(2\alpha_i + \frac{2}{N-1}\right)}{\Gamma\left(2n + \frac{2N}{N-1}\right)} \tag{A14}$$

Now collect together equal terms in the $\alpha(N)$ sum. To do this we define $\kappa(n)$ as the set of n-tuples of nonnegative integers $\kappa_1, \kappa_2, \ldots, \kappa_n$ which satisfy

$$\sum_{m=1}^{n} m\kappa_m = n, \quad \text{with} \quad \kappa_0 \equiv N - \sum_{m=1}^{n} \kappa_m \qquad (A15)$$

Collecting terms, we find

$$\sum_{m=0}^{n} \theta_{n-m} S_m =$$

$$\frac{\Gamma\left[2N/(N-1)\right]}{\Gamma^N\left[2/(N-1)\right]} \sum_{\kappa(n)} \frac{N! \, \theta_0^{\kappa_0} \Gamma^{\kappa_0}\left[\frac{2}{N-1}\right]}{\kappa_0! \, \Gamma\left[2n + \frac{2N}{N-1}\right]} \prod_{m=1}^{n} \frac{\left\{\theta_m \Gamma\left(2m + \frac{2}{N-1}\right)\right\}^{\kappa_m}}{\kappa_m!} \qquad (A16)$$

In the limit $N \to \infty$, Eq. (A16) becomes

$$\sum_{m=0}^{n} \theta_{n-m} S_m = \frac{1}{(2n+1)!} \sum_{\kappa(n)} 2^{\sum_{l=1}^{n} \kappa_l} \prod_{m=1}^{n} \frac{\left\{(2m-1)! \, \theta_m\right\}^{\kappa_m}}{\kappa_m!} \qquad (A17)$$

which recursively gives θ_1 in terms of θ_0 for $n = 1$, θ_2 in terms of θ_1, θ_0 for $n = 2$, etc. This gives $\lim\limits_{N \to \infty} \theta(\eta)$ as a series. The function $b(z)$ is thus

$$b(z) = 4 \lim_{N \to \infty} \theta\left(\frac{z}{4i}\right) \qquad (A18)$$

since $b(0) = 4$ by Eq. (19) and $a(0) = b(0)$.

Equation (A5) with γ unity in the limit N infinite gives

$$b_\nu(h) \simeq \frac{b(h\nu)}{\nu} \qquad (A19)$$

This approximation for small h and large ν also holds for small ν by Eqs. (8) and (19).

Consider now $\eta \lesssim \pi$ in Eq. (A9). Assume $\theta(\zeta) = \Lambda/(\pi - \zeta)^\rho$ with $\rho > 2$ for $\zeta \lesssim \pi$. The singular behavior of the integral results from the region $\eta x_1 \lesssim \eta \lesssim \pi$, $x_2 \gtrsim 0$, $x_3 \gtrsim 0$, ..., $x_N \gtrsim 0$ and the N similar regions obtained by cyclic reordering of the indices. Keeping only the dominant terms and integrating with the Dirichlet integral formula, we obtain the following consistency equation:

$$\rho = \frac{3N - 1}{N - 1} \tag{A20}$$

A third-order pole in $b(z)$ is thus indicated

$$\rho = \lim_{N \to \infty} \frac{3N - 1}{N - 1} \tag{A21}$$

the coefficient of which $i\kappa$ is imaginary since $a(\tau)$ is real for $0 \le \tau < 4\pi$.

In the text it is assumed that the integral $F(z)$ on a circular contour missing poles approaches zero for large radii provided $|x| < h$. Partial support for this assumption is as follows: From either Eqs. (A3) and (A4) or (A6) it is expected that $b(z)$ approaches zero faster than $1/\sinh^2(z/4)$ when the real part of z is large. This should dominate the other factors of $F(z)$ for large real part of z provided $|x| < h$ and $z \ne h\nu$. The cosine denominator is expected to ensure that $F(z)$ also vanishes exponentially for large imaginary part of z except for integer multiples of $\pm 4\pi i$.

As an approximation, take

$$\lim_{N \to \infty} \theta(\eta) \simeq \sum_{\mu=1}^{3} \gamma_\mu \left\{ \frac{1}{(\pi - \eta)^\mu} + \frac{1}{(\pi + \eta)^\mu} \right\} \tag{A22}$$

and expand as a power series in η^2. The coefficients of $\eta^{2\nu}$ yield

$$\pi^{2\nu} \lim_{N \to \infty} \theta_\nu \simeq [1] \frac{2\gamma_1}{\pi} + [2\nu + 1] \frac{2\gamma_2}{\pi^2} + [2\nu^2 + 3\nu + 1] \frac{2\gamma_3}{\pi^3} \tag{A23}$$

With the values of $\lim_{N \to \infty} \theta_\nu$ given by Eq. (A17) used in Eq. (A23) for three successive integers, we have three linear equations in three unknowns which numerically converge rapidly to $\gamma_1 = -1.$, $\gamma_2 = -7.5$, and $\gamma_3 = +23.31$. From the last

$$\kappa = 5967 \tag{A24}$$

follows by Eqs. (17), (A18), and (A22).

Erratum. Equation (21) does not follow from Eq. (20) in the earlier paper[7] dealing with the cubic substrate mode. The values $\beta_4 = +95.07$, $\beta_3 = -60.5i$, and $\beta_2 = +37.$ with β_1 masked by error given there should be corrected to the new values $\beta_4 = +95.07$, $\beta_3 = 0$,

$\beta_2 = +42.36$, and $\beta_1 = 0$. The constant $\beta \equiv \beta_4$ of the cubic model is analogous to the constant $\kappa = 256\gamma_3$ in the sine case, Eq. (18) in that paper being analogous to Eq. (22) here.

REFERENCES

1. Frenkel, J., and T. Kontorova: *Phys. Z. Sowjet.* 13:1 (1938).
2. Indenbom, V. L.: *Soviet Phys. - Cryst.* 3: 193 (1958).
3. Peierls, R.: *Proc. Phys. Soc.* london, 52:34 (1940).
4. Nabarro, F. R. N.: *Proc. Phys. Soc.*, London, 59: 256 (1947).
5. Lifshitz, I. M.: Unpublished [see V. L. Indebom and A. N. Orlov, *Soviet Phys. - Usp.* 5: 272 (1962)].
6. Hobart, R.: *J. Appl. Phys.* 36:1944 (1965).
7. Hobart, R.: *J. Appl. Phys.* 37:3573 (1966) [see erratum at end of present paper].
8. Hobart, R.: *J. Soc. Indust. Appl. Math.* 13:639 (1965).
9. Whittaker, E. T., and G. N. Watson, "A Course of Modern Analysis," p. 258, 4th ed., Cambridge University Press, Cambridge, England, 1958.
10. Seeger, A., and P. Schiller, "Physical Acoustics III A," p. 394, W. P. Mason (ed.), Academic Press, New York, 1966.

DISCUSSION *on paper by R. Hobart*

R. BULLOUGH: Does not the fact that the substrate is rigid invalidate this model for the purpose of calculating the Peierls stress?

R. HOBART: A rigid substrate is part of the price paid for the simplicity of the model. The Peierls stress to shear modulus ratios predicted by the Frenkel-Kontorova model are not, however, unreasonable (6×10^{-5} for Cu, Reference [6]).

F. R. N. NABARRO: Does your method not lose the spirit of the calculation by considering the force acting on just one atom?

R. HOBART: Only the energy difference of the two equilibrium configurations is of physical interest. The force R is merely a mathematical construction for calculating the difference.

A. W. SLEESWYK: Concerning the influence of the shape of the assumed potential, it might be asked whether it would not be possible to regard the sinusoidal Peierls potential assumed here as being the first term in a Fourier series. One would guess that already adding a second harmonic might give some relevant information on the size of the shape effect caused by various potentials.

R. HOBART: Inclusion of a second harmonic leads to some quite interesting features: for example, the Peierls-barrier minima versus stacking fault energy predicted in Hobart, R., *J. Appl. Phys.*, 36:1948 (1965). Seeger and Shiller describe an approximate analytical solution to this problem in Ref. [10].

ATOMIC CONFIGURATION AND NONLINEAR ASPECTS OF THE CORE OF DISLOCATION IN IRON

R. Bullough and R. C. Perrin

Theoretical Physics Division,
Atomic Energy Research Establishment,
Harwell, Didcot, Berks., England

ABSTRACT

An atomic model has been constructed to simulate body-centered-cubic iron using a nonequilibrium pair potential adjusted to satisfy the elastic and phonon scattering data. The model has been used to study the core configurations of certain dislocations in iron and to calculate the interaction between these dislocations and vacancies. The results indicate that the screw dislocation has reduced its core energy by releasing the shear misfit on two of the three {112} planes that intersect it; in contrast the cube edge dislocation has similarly reduced its core energy by microcracking on its tensile side. The vacancy interaction with the edge dislocation is rather complex but has an asymptotic form which agrees closely with the second-order inhomogeneity interactions. Thus, the behavior of the vacancy is similar to a soft spot with negligible associated dilation.

INTRODUCTION

The linear elastic continuum is frequently used as a model of the real crystalline solid for the calculation of various properties of defects. Thus, for example, elastic inclusions (usually spherical) embedded in an elastic continuum are used to represent point defects in a crystalline matrix, and Volterra dislocations are used to represent dislocations in crystals. Recent calculations have shown that this kind of simple elastic approximation not only leads to expected quantitative error in its representation of physical phenomena, but can lead to predictions that are qualitatively in error. For example, Bullough and Hardy[1] have shown that in a harmonic cubic lattice, purposely constructed to be exactly *elastically* isotropic, two centers of dilation has an asymptotic interaction energy that is proportional to the inverse fifth power of their separation and is both large and oscillatory when the two centers approach one another; in the similar isotropic *continuum* this interaction is identically zero. In addition to such effects, there are many defect properties that depend explicitly on the discrete nature of the crystal lattice; for example, the core configuration and mobility of a dislocation. The purpose of this report is to describe some of the results obtained via a discrete atomic model, for certain dislocations in iron. In particular, we have studied the core configuration of a pure screw dislocation with Burgers vector $b = a/2 <111>$ and a pure "cube" edge dislocation with $b = a <100>$. The screw dislocation was chosen because it is believed to have an extremely important influence on the plasticity of body-centered cubic metals. Our results which are given below, should contribute to the understanding of this influence. The cube-edge dislocation was chosen for two reasons: (1) crystallographically, it is a very simple defect, thereby providing an easily understood dislocation configuration; and (2) it is believed to play an important role in crack nucleation in the body-centered cubic metals. Again, our results on the core structure of this dislocation should assist in our understanding of the fracture process. In addition, we have calculated the interaction energy between a single vacancy and the cube-edge dislocation as a function of position around the dislocation. The results of this last calculation certainly emphasize the danger in assuming the validity of the analogous results which have been obtained using the simple elastic continuum model.

1   THE DISCRETE MODEL FOR IRON AND THE NUMERICAL METHOD

The atomic model consists of a large parallelepiped of discrete interacting atoms. For the studies of isolated straight dislocations, the parallelepiped is extensive in directions normal to the dislocation line and only the crystallographic repeat distance thick parallel to the dislocation. When point defects are introduced into the neighborhood of the dislocation the parallelepiped must be made extensive in all directions. The perfect body-centered cubic iron lattice is first achieved by allowing a nonequilibrium central force potential to act between the first and second neighboring atoms with appropriate pressure and periodic boundary constraints applied to the surfaces to maintain equilibrium.

The potential used in this work was that constructed by Johnson[4] and used by him to study various properties of elementary point defects in iron. It consists of a set of splines (cubic polynomials) interpolated to give a continuous and differentiable pair potential. Its precise form is deduced by fitting its slope and curvature at the first and second neighbor equilibrium positions to the linear elastic constants of iron, and at short range it is matched to the radiation damage potential of Erginsoy et al.[3] The potential is made to vanish, with zero slope, at a point half-way between the second and third neighbor positions. In addition, we find it to agree reasonably with the phonon dispersion data; that is, it leads to consistent frequencies at the center of certain selected zone faces. The potential used has the following analytic form:

$$
\begin{aligned}
\phi(r) &= -2.195976\,(r - 3.097910)^3 + 2.70406\,r - 7.436448 \\
&\qquad\qquad 1.9 < r \le 2.4 \\
&= -.639230\,(r - 3.115829)^3 + .477871r - 1.581570 \\
&\qquad\qquad 2.4 \le r \le 3.0 \\
&= -1.115035\,(r - 3.066403)^3 + .466892r - 1.547967 \\
&\qquad\qquad 3.0 \le r \le 3.44
\end{aligned}
\tag{1}
$$

where ϕ is in eV and r is in Angstroms.

The dislocation is introduced into the model by treating the discrete atoms as points in an elastic continuum and imposing the anistropic elastic displacements on the atoms appropriate to a perfect straight edge or screw dislocation. By ensuring that the "elastic" dislocation lies precisely midway between the atomic layers defining the particular slip plane, it is easy to avoid difficulty with the singularity in the elastic solution at the center of the dislocation. The periodic boundary conditions on the extensive faces (perpendicular to the dislocation line) are necessary since it is desired that the model be representative of the internal situation in a crystal, and also since the dislocation cannot terminate in the crystal. The atoms in the other faces of the parallelepiped are then held in the elastic dislocated positions—they are deemed to be sufficiently far from the nonlinear core region for the simple elastic solution to be appropriate. The pair potential above was then allowed to act between the atoms, and the general relaxation procedure entered with the boundary conditions above imposed on the assembly. The relaxation process involved the numerical integration of several thousand simultaneous classical equations of motion:

$$\ddot{x}_l^{ijk}(t) = m^{-1} \sum_{\substack{i'j'k' \\ \neq ijk}} F_l[x^{ijk}(t) - x^{i'j'k'}(t)] \tag{2}$$

where $x_l^{ijk}(t)$ ($l = 1, 2, 3$) are the Cartesian components of the position vector $x^{ijk}(t)$ of an atom at time t relative to a fixed laboratory system of axes; the triple integers ijk are convenient for identifying a particular atom, and F_l are the components of the force on the (ijk)th atom from the $(i'j'k')$th atom; in obvious notation:

$$F_l(\mathbf{r} - \mathbf{r'}) = \frac{\partial \phi(|\mathbf{r} - \mathbf{r'}|)}{\partial x_l} \tag{3}$$

To solve the set of differential equations (2) a method of discretization is used according to the following scheme:

$$\ddot{x}_l^{ijk}(t) = (\Delta t)^{-1} \left\{ V_l^{ijk}\left(t + \frac{\Delta t}{2}\right) - V_l^{ijk}\left(t - \frac{\Delta t}{2}\right) \right\}$$
$$= m^{-1} \sum_{\substack{i'j'k' \\ \neq ijk}} F_l \left\{ x^{ijk}(t) - x^{i'j'k'}(t) \right\} \tag{4}$$

and

$$V_l^{ijk}\left(t + \frac{\Delta t}{2}\right) = (\Delta t)^{-1}\left\{x_l^{ijk}(t + \Delta t) - x_l^{ijk}(t)\right\} \tag{5}$$

Thus from a configuration and velocity distribution at time t [the latter strictly at time $(t - \Delta t/2)$] the coordinates and velocity of each atom at a subsequent time are:

$$V_l^{ijk}\left(t + \frac{\Delta t}{2}\right) = V_l^{ijk}\left(t - \frac{\Delta t}{2}\right) + \Delta t(m)^{-1}\sum_{\substack{i'j'k' \\ \neq ijk}} F_l\left\{x^{ijk}(t) - x^{i'j'k'}(t)\right\} \tag{6}$$

$$x_l^{ijk}(t + \Delta t) = x_l^{ijk}(t) + \Delta t\, V_l^{ijk}\left(t + \frac{\Delta t}{2}\right) \tag{7}$$

Each free atom in the assembly is considered in turn and the force on it due to neighboring atoms is evaluated. From the difference equations (6) and (7) the mean velocity of the atom and, hence, its new position after the current time step, is deduced. When the position and velocity of each atom has been updated, a norm proportional to the total kinetic energy of the lattice is evaluated and the process is continued for the next time step. With a suitable choice of time step the kinetic energy passes through a maximum value after a few complete iterations. At this point the atoms are all brought to rest (the assembly is quenched) and the iterations restarted. This dynamic procedure ensures very rapid convergence to the absolute minimum in total potential energy, and appears to be more rapid than other "static" relaxation procedures that might be envisaged. In addition, it has the distinct advantage that metastable configuration can be avoided, if so desired, by adopting a suitably large time step.

2   RESULTS

2.1   The $a/2\langle 111\rangle$ Screw Dislocation

The assembly used for this dislocation was a rectangular parallelepiped with (111), $(\bar{1}01)$ and $(1\bar{2}1)$ faces; the screw dislocation was perpendicular to the extensive $\{111\}$ faces. Figure 1 shows a (111) projection of the fully relaxed atomic configuration around the screw dislocation. The initial position of the elastic screw dislocation is indicated by S at the center of the figure. The three

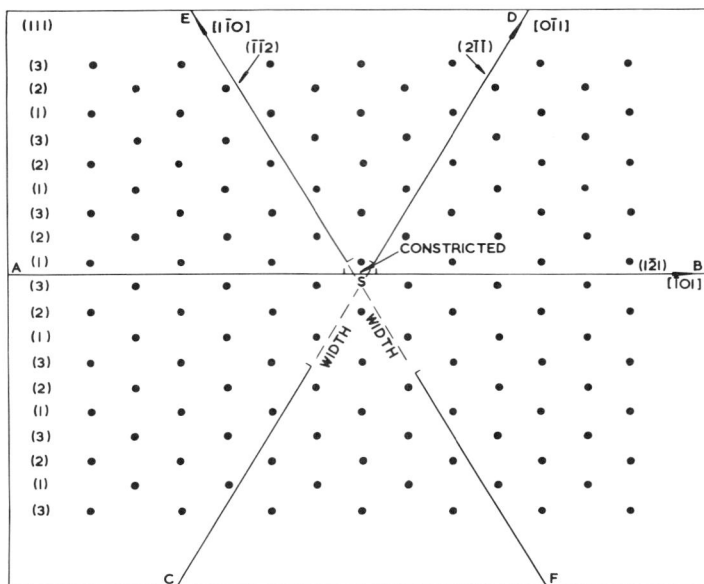

Fig. 1 A (111) projection of the atomic configuration around a screw dislocation with Burgers vector **b** = $(a/2)[111]$. The only significant displacements are in the [111] direction and are thus not visible in this projection. The numbers in parentheses indicate the particular (111) layer to which the atoms in the row belong. The core configuration shown in detail in the {112} projections (figures 3, 4 and 5) is schematically indicated.

(111) layers that make up the initial parallelepiped are distinguished by the numbers (1), (2) and (3) that appear on the left of the figure. It is important to note at the outset that during the relaxation (and the initial anistropic elastic dislocating process) the only significant atomic displacements are in the [111] directions, that is, parallel to the screw dislocation. Thus the projection in Fig. 1 is

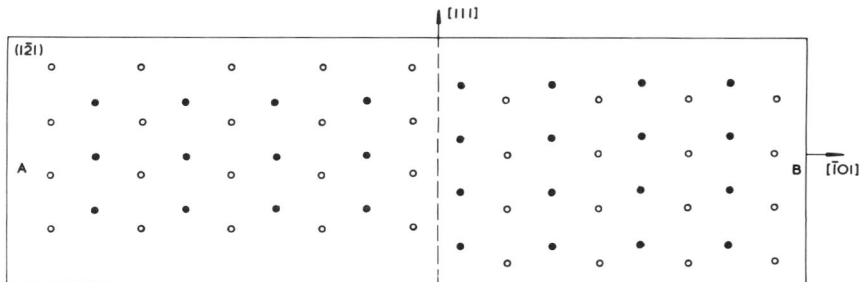

Fig. 2 A projection of the atom positions across the (1$\bar{2}$1) slip plane ●-(1) layer. ○- (3) layer. The screw is highly constricted on this plane; the width of the dislocation is almost b.

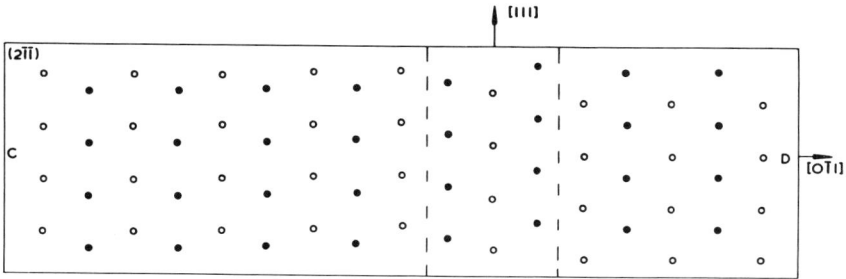

Fig. 3 A projection of the atom positions across the $(2\bar{1}\bar{1})$ slip plane. ●-(1) layer, o-(2) layer. The width of the dislocation is now about $3b$ with a formal $(1/6)$ $a[111]$ stacking fault present.

almost indistinguishable from the analogous projection of a perfect crystal.

To discover what has actually happened in the core region during the relaxation, the atomic configurations across the three {112} slip planes that intersect the initial screw axis must be examined. These planes are shown in Fig. 1 and distinguished by the letters AB on the $(1\bar{2}1)$ plane, CD on the $(2\bar{1}\bar{1})$ plane and EF on the $(\bar{1}\bar{1}2)$ plane. The projections across these three planes are shown in Figs. 2, 3, and 4, respectively.

It can be seen immediately that the spread of misfit in these slip planes is nonuniform. In the $(1\bar{2}1)$ plane (Fig. 2), the transition from A to B is completely abrupt and indeed no relaxation exists in this plane. However, in the other two {112} planes, the situation is quite different. In both cases (Figs. 3 and 4) detectable relaxation has occurred and the transition from C to D and from E to F is not abrupt. This transition is achieved over a distance of about $3b$ and is denoted by dashed lines in the two figures. Furthermore,

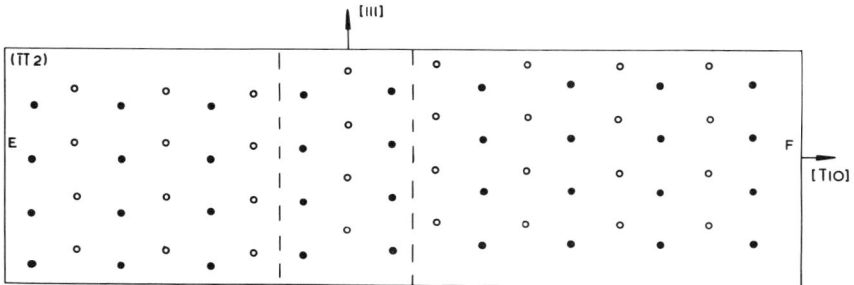

Fig. 4 A projection of the atom positions across the $(\bar{1}\bar{1}2)$ slip plane. ●-(1) layer, o-(2) layer. The width of the dislocation is now about $3b$ with a formal $(1/6)$ $a[111]$ stacking fault present.

by carefully comparing the particular atom rows in these figures with those in Fig. 1, we discover that this (formal) splitting of the screw dislocation on the $(1\bar{1}2)$ and $(2\bar{1}1)$ planes occurs as indicated in Fig. 1; that is, both split downwards with the "faults" forming an angle of 60° with each other. It should be emphasized that this situation arose entirely during the relaxation procedure; prior to that, the transition is equally abrupt in all these {112} planes. Therefore, it appears to be the genuine static core configuration associated with the screw dislocation. The particular choice of $(2\bar{1}1)$ and $(1\bar{1}2)$ planes to share the relaxation is due presumably to the specific initial position of the screw. We shall attempt to vary this initial position systematically and see how the choice of {112} planes is correlated with the screw position. Similar projections across the {110} planes that intersect the screw axis show a completely abrupt transition with no significant spreading of the misfit in these planes.

2.2   The a ⟨100⟩ Edge Dislocation

The assembly used for this dislocation is a particularly simple rectangular parallelepiped with all {100} faces; the edge dislocation is perpendicular to the two extensive faces. Figure 5 shows a (010) projection of the fully relaxed atomic configuration around the core of the edge dislocation. In this case, in contrast to the screw dislocation, the position of the dislocation is obvious.

The projection across the slip plane is shown in Fig. 6 where it is clear that the dislocation is extremely narrow (a very abrupt transition). In fact it is clear from Fig. 5 that in this case the relaxation in the core region has been achieved by a tensile relaxation below the slip plane rather than a spreading of shear misfit as we found in the screw dislocation core; that is, the edge dislocation has formed a microcrack immediately below the slip plane. This fact is, of course, particularly interesting if we wish to understand the part played by such dislocations in the fracture nucleation process in body-centered cubic metals.

The fact that the core is actually microcracked can also be demonstrated by introducing an interstitial atom into the core region. When the interstitial enters the core region, the large strains associated with it vanish almost entirely and, in addition, the binding energy of the interstitial to the dislocation is very large (~3.5 eV) and almost equal to the entire formation energy

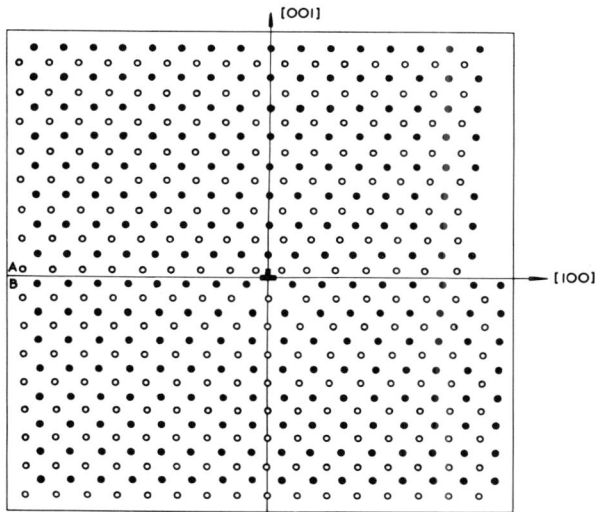

Fig. 5 A (010) projection of the cube edge dislocation with Burgers vector $a[100]$. There are no displacements in the [010] direction. The slip plane is indicated by the step on the right hand (100) surface of the assembly. The upper slip plane is denoted by A and the lower slip plane by B. The two (010) layers in the projection are distinguished by the symbols ● and ○.

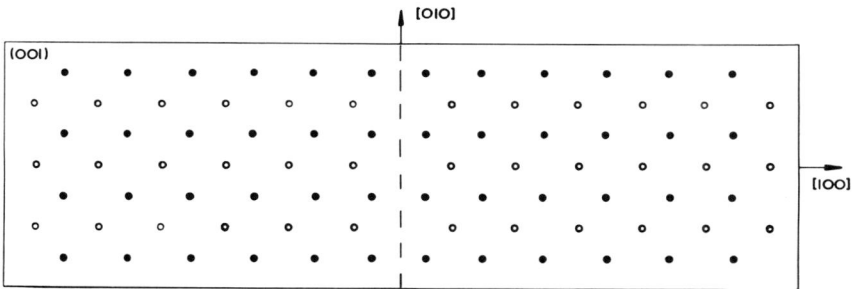

Fig. 6 A projection of the atom positions across the (001) slip plane for the cube edge dislocation with Burgers vector $a[100]$. ●- upper A layer, ○- lower B layer (see Fig. 5). The edge is highly constricted.

of the isolated interstitial. This means, of course, that the core is reacting to the interstitial like a local free surface and is thus, by definition, a microcracked region.

2.3   The Interaction between the Edge Dislocation and a Vacancy

To study the interaction between a vacancy and the cube dislocation, the parallelepiped mentioned earlier was considerably extended in the direction parallel to the dislocation line (from a for the isolated cube edge dislocation with only two (010) atom layers, to about 19 a with a total of twenty (010) atom layers). The atoms in this very large assembly (actually about 2000 atoms were involved) were then displaced to coincide with the fully relaxed configuration appropriate to the previous isolated cube edge dislocation; the total potential energy of this dislocation configuration E_D was then computed. A vacancy was then introduced at a variety of positions relative to the dislocation line. For each position the assembly of atoms was carefully relaxed; actual migration of the vacancy from its prescribed position was avoided by a judicious choice of time step (Δt) in the relaxation equations (6) and (7). When the total potential energy E_{D+V} was completely minimized for the particular vacancy position, its value was computed. A separate calculation was performed to obtain the potential energy of the assembly when it contained only a vacancy with no dislocation; if we denote this quantity by E_V then the required interaction energy E_I between the vacancy and the dislocation is given by:

$$E_I = E_{D+V} - E_D - E_V \qquad (8)$$

for each position of the vacancy.

The vacancy positions and interaction energies obtained are shown in Fig. 7 where the energies are given in eV. The maximum interaction energy or binding energy between the vacancy and the dislocation is large, about 0.7 eV, and occurs when the vacancy occupies the atomic site in the slip plane on the compressive side of the dislocation. However, the second largest interaction energy is again in the slip plane but on the *tensile* side of the dislocation. In fact, the interaction energy immediately *below* the dislocation (−0.2 eV) is only slightly smaller than the interaction energy immediately above the dislocation (−0.31 eV). This fact that the vacancy is almost equally attracted to the dislocation at all the first neighbor positions of the dislocation line strongly suggests that the interaction arises predominantly from the second-order inhomogeneity interaction; the isotropic continuum expression for this interaction is:[5]

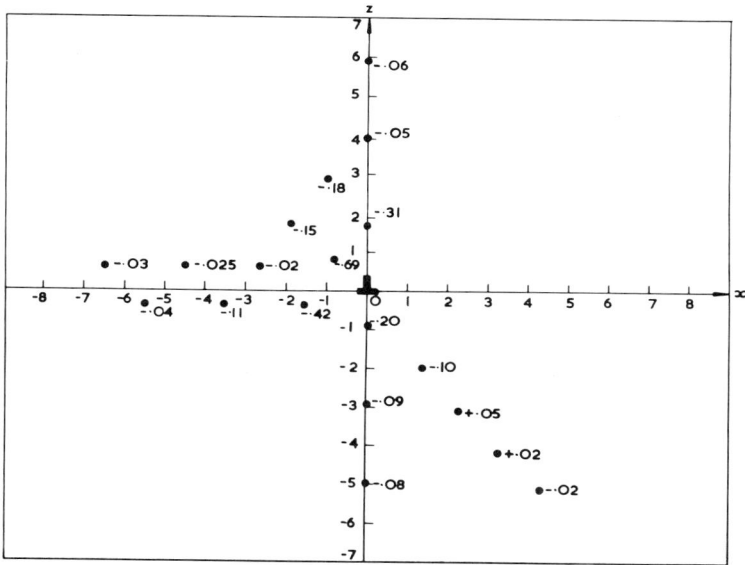

Fig. 7 The symbol ● indicates the vacancy positions around the cube edge dislocation. The interaction energies indicated are in electron volts and the distance is in units of half the cell size.

$$E_I = -\frac{K}{R^2} \{ 1 - \tfrac{1}{2} \sin^2 \theta \} \tag{9}$$

where (R, θ) defines the vacancy position relative to the dislocation axis which lies along the Z axis of the cylindrical polar coordinate system, and K is a constant involving the dimensions of the vacancy, the elastic shear modulus and Poisson's ratio. The factor $1/2$ in the bracket is obtained by assuming Poisson's ration to be approximately one-third in the general expression given in Ref. [5]. The expression (9) is always negative and has its maximum value when the vacancy is in the slip plane $(\theta \sim 0)$ where is is almost 50% greater than in the $\theta = \pm 90°$ position. All these features are satisfied qualitatively by our close neighbor results. However, in Fig. 8, where we plotted $R^2 E_I/\{1 - \tfrac{1}{2} \sin^2 \theta\}$ vs. R, it is seen that the results for larger R do not satisfy Eq. (9). Intuitively one might expect that the inhomogeneity interaction could only predominate when the vacancy is very close to the dislocation and that for greater separations, the first-order size effect should dominate. This is not the case; Fig. 9 is a plot of $R \, E_I/\sin \theta$ vs. R and certainly does not become constant at large R. Actually, most of the

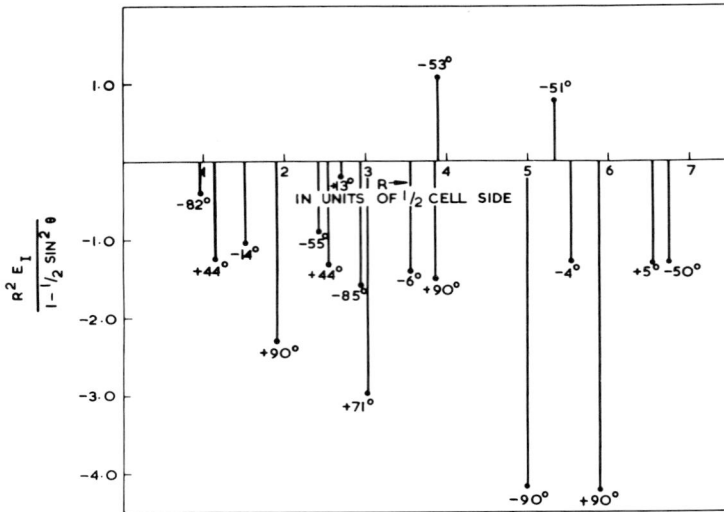

Fig. 8 The interaction energy E_I between a cube edge dislocation and a vacancy. The dislocation lies along the Z axis and the vacancy is located at (R, θ).

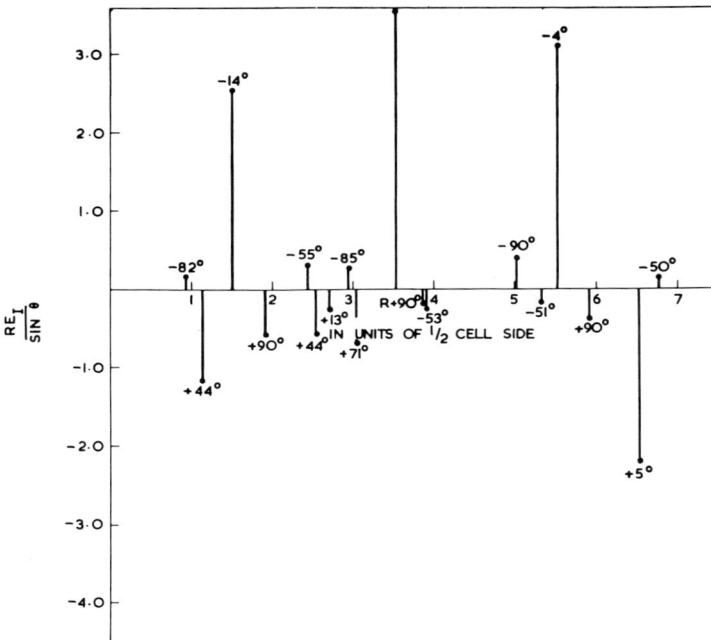

Fig. 9 The interaction energy E_I between a cube edge dislocation and a vacancy. The dislocation lies along the Z axis and the vacancy is located at (R, θ).

interaction energy values are reasonably satisfied by:

$$E_I \simeq \frac{-1.0}{R^2}$$

This empirical result suggests that the interaction between a vacancy and an edge dislocation in iron is due primarily to the vacancy being a soft spot in the matrix (the inhomogeneity effect) and not because the intrinsic strain field of the vacancy interacts with the dislocation stress field (the size effect). The angular dependence in the isotropic continuum result (9) can hardly be expected to be valid for the elastically anisotropic discrete iron lattice. However, the dependence of the induced or inhomogeneity interaction with defect separation R could well be the same in both models. A recent analytic calculation of the induced interaction between two vacancies in a face-centered harmonic lattice has shown[7] that the interaction varies with the separation of the two vacancies in precisely the same way as the continuum result.

3 CONCLUSIONS

1. The discrete iron lattice has been simulated by means of a large, fast digital computer. The discrete model was used to obtain the atomic configurations associated with a screw and edge dislocation.

2. The atomic configuration in the core of a screw dislocation is complex. The core energy is minimized by the spreading of the shear misfit in two of the {112} planes that intersect the screw. The misfit does not relax in the third {112} plane. Only a formal identification of the $\frac{1}{6} a <111>$ partial dislocations is possible; the core is simply wider $\sim 3b$ in two of the {112} planes compared with a completely abrupt transition $\sim b$ in the remaining {112} plane.

3. In contrast to the screw dislocation, the core energy of the cube edge dislocation is minimized by the lattice microcracking immediately beneath the dislocation.

4. The interaction energy between a vacancy and the cube edge dislocation has very little angular dependence. The form of the interaction suggests that it appears largely as a consequence of the second-order, induced (or inhomogeneity) interaction. There is no sign of the first-order size effect interaction. This, rather surprising result, lends some support to our previous hypothesis[5] that the second-order interaction is dominant in the vacancy-dislocation interaction and can, for example, thus provide the basis of an explanation on the aging kinetics in some quenched deformed crystals.[5]

REFERENCES

1. Bullough, R., and J. R. Hardy: *Phil. Mag.* 15: 237 (1967).
2. Johnson, R. A.: *Phys. Rev.* 134:1329 (1964).
3. Erginsoy, C., G. H. Vineyard, and A. Englert: *Phys. Rev.* 133:595 (1964).
4. Low, G. G. E.: *Proc. Phys. Soc.* 79:479 (1962).
5. Bullough, R., and R. C. Newman: *Phil. Mag.* 7: 529 (1963).
6. Bullough, R., and R. C. Newman: *Proc. Roy. Soc.* A266: 209 (1962).
7. Bullough, R., and J. R. Hrdy: *Phil. Mag.* 16: 405 (1967)

DISCUSSION *on Paper Presented by R. Bullough*

Z. S. BASINSKI: How sharply do the results depend on the potential chosen?

R. BULLOUGH: We have not systematically varied the potential but we believe the results will be fairly insensitive to minor changes in the potential. This is particularly true if we restrict the class of potentials to cover first and second neighbor interactions only. In this case the force constants (the slope and curvature of the pair potential) at the first and second neighbor equilibrium positions will be invariant, and thus the harmonic behavior of the lattice will be invariant. Since most of the interatomic strains are small, it follows that most of the atomic configuration will be independent of the form of the potential away from the equilibrium positions. Our results indicate a spatially continuous relaxation of shear misfit in the screw dislocation core instead of the formation of well-defined stacking faults; such gradual relaxations do not involve many large interatomic shear strains and thus will not tend to depend too sensitively on the form of the potential between the equilibrium positions.

N. BROWN: Is there a strong interaction between the interstitial atom and the [100] dislocation?

R. BULLOUGH: Yes, there is a very strong interaction. When the interstitial is in the core, the binding energy is almost exactly equal to minus the computed formation energy of the isolated vacancy, which is about 3.5 eV. This means the core of the cube edge dislocation is microcracked and presents a local free surface to an approaching interstitial. We have not yet calculated the interaction energy as a function of interstitial-dislocation separation; however, because of the dumbell configuration of the interstitial and its associated high strains, we would expect a dominant first-order, size effect interaction inversely proportional to the interstitial-dislocation separation with a rather complex angular dependence.

P. B. HIRSCH: You mention that your model gives a realistically large stacking fault energy for iron of about 3600 ergs/cm^2. What type of fault are you considering, and on what basis do you claim that the value obtained is realistic?

R. BULLOUGH: This value refers to the fault energy of two parallel {110} planes in the split dumbell configuration (i.e., split in the <110> direction). I would expect such a fault to have an energy comparable to the surface energy of iron; that is, several thousand ergs/cm^2. This is what I meant by realistic. This energy is the important driving force for the conversion of an initially faulted platelet of interstitials on a {110} plane to a perfect dislocation loop and, thence, by a glide to a dislocation loop on a {111} plane.

J. W. TAYLOR: In view of the fact that dilatational strains of the order of 10% are seen near the cores of edge dislocations, it would seem to be important that the interatomic potential should take explicit account of nonlinearities in the elastic interaction. Have you compared your potential functions against the compressibility of iron as measured at high pressure, or from shock wave data?

R. BULLOUGH: We have not fitted our potential to the high-pressure data you refer to. The potential is consistent with the linear elastic constants and the phonon dispersion data; this information defines the slope and curvature of the potential at the first and second neighbor equilibrium positions. At closer ion-ion separations, corresponding to very high compressive strains, the potential is fitted to the radiation damage potential developed by Erginsoy et al. In view of the fact that we are forced to make somewhat arbitrary allowance for the electronic part of the linear elastic constants, we do not think it worthwhile to include the third-order elastic data, since these would again have to be adjusted to extract the purely ion-ion parts.

A. L. BEMENT: Your model expresses a relatively weak angular variation in the interaction energy between a lattice vacancy and the core of an edge dislocation. To what extent should such an interaction be influenced by localized lattice relaxations, and to what extent does your simplified $\sin^2 \phi$ dependency adequately account for the asymmetry of the core? I would expect the angular variation in the interaction potential between such defects as interstitials and solute atoms and the dislocation core to be not only highly irregular but also to change sign at specific angles on either side of the slip plane.

R. BULLOUGH: We have calculated the interaction energy between a vacancy and the cube edge dislocation and it is true we find that the angular variation is not very pronounced. This is because the effective dilatation associated with the vacancy is, in our model, very small (less than −0.01); in units of half the cell size we find that the first neighbors of the vacancy are displaced inwards by 0.0509 along the $<111>$ directions and the second neighbors are displaced outwards by 0.0513 along the $<100>$ direction. From these displacements we can, using the fitted force constants, calculate the generalized forces that act on these neighbors in their displaced positions. It is then a simple matter to obtain the effective continuum dilatation. Under these circumstances, when the point defect has very low dilation, one would expect the second-order inhomogeneity interaction to become dominant. The isotropic elastic continuum form for this interaction is proportional to $(1 - \frac{1}{2} \sin^2 \theta)/R^2$ and thus at large distance from the dislocation one would expect our results to approach this variation. In fact, we find that the last two points shown in Fig. 8 do agree very well with the continuum inhomogeneity result when a vacancy radius of $\sqrt{3}/2$ a is assumed. Thus, the simple elastic inhomogeneity interaction does account for the long-range vacancy cube edge dislocation in our model of iron. The departure from this variation at small vacancy-dislocation separations are indicative of the effects of the discrete nature of the lattice on such an inhomogeneity interaction.

I agree that the point defects, such as interstitial atoms or solute atoms, with an appreciable dilatation (size effect) would have an interaction energy with strong angular dependence which was repulsive at some angles and attractive at others.

NONLINEAR INTERNAL FRICTION, MODULUS CHANGES, AND STRAIN RELAXATION CAUSED BY PEIERLS-NABARRO POTENTIALS

G. Alefeld

Kernforschung Jülich,
Institut für Festkörper- und Neutronenphysik,
517 Jülich, Germany

J. Filloux and H. Harper

John Jay Hopkins Laboratory for Pure and Applied Science,
General Atomic Division of General Dynamics Corporation,
San Diego, California

ABSTRACT

Starting from the assumption that the periodicity of a crystalline lattice causes periodic variations of the energy of a hypothetically straight moving dislocation, a set of nonlinear internal friction-modulus defect- and strain relaxation- phenomena can be predicted.

The nonlinear experiments are explained and typical results, mainly for aluminum and selected examples for tungsten are presented.

1 INTRODUCTION

Over the last decade, it has been demonstrated convincingly that dislocation motion can cause energy loss of vibrating crystals and that the elastic constants of crystals can be decreased significantly by dislocation contributions to the strain. In cold-worked materials one finds characteristic absorption spectra, part of which can be attributed to dislocations. Therefore, it is hoped that internal friction and modulus measurements will provide some insight into details concerning the elementary processes of plastic deformation.

The information that can be deduced from an amplitude-independent internal friction peak concerns the peak position (relaxation time), peak height (relaxation strength), the peak width (multiplicity of process), and the shift of the peak with frequency. Many completely different models, such as, point-defect relaxation, diffusion of geometric kinks in dislocations, interactions of point defects and dislocations, cause absorption peaks of the same general shape to appear in the amplitude-independent region. Thus the observation of a peak indicates only that a relaxation process is taking place and provides limited information about the atomistic process. However, the amount of information obtained can be increased considerably by extending the observation to nonlinear effects, that is, the dependence of modulus and absorbed energy on the amplitude of vibration or on a superimposed biasing static stress. The chance that different models will show the same behavior in the nonlinear region are not very likely. Over the last few years, much attention has been given to the question of the existence of finite periodic potentials for dislocations and the significance of these so-called Peierls-Nabarro potentials[1,2,3] for long-range dislocation motion. In this paper, only the first part of the question will be considered. We will examine the consequences of finite Peierls-Nabarro potentials (PNP) for internal friction and modulus changes, and show that PNP cause particular nonlinear effects which are alien to dislocations in materials without PNP. It is assumed that the reader is familar with the concepts of double-kinks and geometric kinks.[4,5,6]

2   RESTRICTED MOTION OF GEOMETRIC KINKS

The following argument has been used against the existence of PNP: If PNP exists, dislocation motion proceeds via double-kink-generation and therefore must be thermally activated. Consequently, at very low temperatures, no dislocation motion will be possible at stresses below the PN-stress. Nevertheless, the experimental observation was as follows:

The modulus of elasticity, as measured at He-temperature, increases in the process of annealing cold-worked materials. Thus it is seen that dislocation motion is possible at He-temperature. This observation holds for Cu, Al, etc.,[7] for W, Ta, Nb, etc.,[8] and for salts.[9] However, it has not yet been tested for Ge or Si.

The above-mentioned argument is not quite correct since it starts from an assumption which is highly improbable for most materials, that is, that all dislocations run parallel to close-packed directions at He-temperature. The existence of geometric kinks is ignored. However, the argument can be applied with the following modification:

At such temperatures and stresses where double-kink-generation is not possible, dislocation motion is restricted to that area over which geometric kinks can move (see Fig. 1). Once the geometric kinks are pressed against a new close-packed direction, further dislocation motion requires thermal activation. The experimental

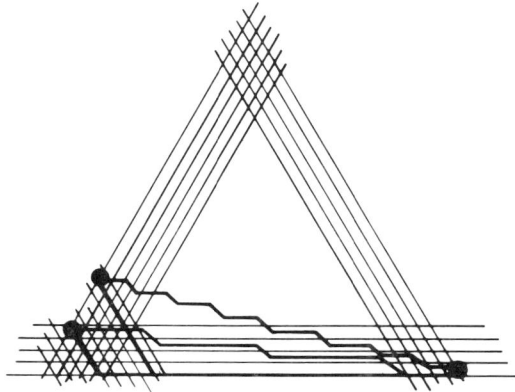

Fig. 1   Maximum area, over which kinked dislocations can move without double-kink-generation. The two dislocation segments correspond to different geometric kink density.

observation of mobile dislocations at He-temperature, therefore, shows only that in those materials the PNP for lateral motion of kinks is very small if not zero. In the following, three experimental observations which verify restricted dislocation motion at low temperatures will be presented. The experimental setup was a torsion pendulum.

2.1   High Amplitudes of Oscillation

At small amplitudes of oscillation, a chain of geometric kinks contributes the same amount of inelastic strain as a dislocation string. However, a slightly modified expression for the line tension[6,10,11] may have to be used. The period T of oscillation of a specimen is connected to the dislocation strain ϵ_a which is in phase with the stress according to the following formula:

$$ T \sim M^{-1/2} \sim (\epsilon_e + \epsilon_a)^{1/2} \sim \left(1 + \frac{1}{2}\frac{\epsilon_a}{\epsilon_e}\right) \tag{1} $$

where ϵ_e is the elastic strain, and M is the modulus of elasticity including contributions due to dislocations.

At high amplitudes, the dislocation strain ϵ_a of a kink chain will increase slower than proportional to the stress.[10,11] Thus the ratio ϵ_a/ϵ_e decreases for increasing amplitude of oscillation, and, consequently, the period of oscillation becomes shorter, and the material becomes stiffer. It should be pointed out that this postulate contradicts almost all of the experimental observations reported up to now. Up to amplitudes of $5 \cdot 10^{-5}$ where most measurements stop, one usually finds an increase of T. This softening of the material is attributed, generally, to the breakaway of dislocations from pinning points.[12]   Figure 2, in which the oscillation period for a tungsten specimen is plotted against amplitude, shows first an increase of the period. But at stress amplitudes of 1.5×10^{-3} G, a drastic decrease can be observed which continues on to stresses which are close to the flow stress of tungsten.[13] Since the decrease of T beyond $\epsilon_0 = 10^{-3}$ depends on the deformation state of the crystal and strongly on temperature, it is safe to conclude that we are not considering an intrinsic lattice nonlinearity but a dislocation effect.

It is plausible that at the beginning of increasing amplitudes the restricted kink motion will first be obscured by all those effects which correspond to an increase of the effective line length of dislocation segments (breakaway from pinning points is the most

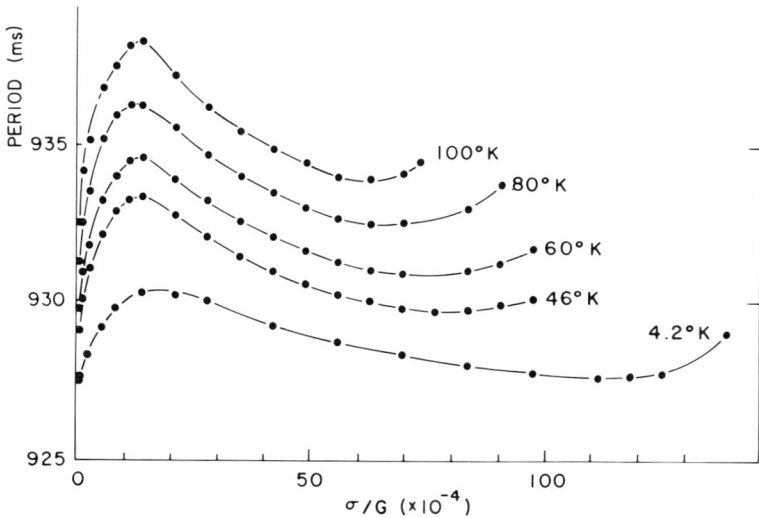

Fig. 2 Period of oscillation versus amplitude for high purity tungsten with 1.7 % deformation. (The figure is taken from ref. 13).

familiar one). But also the overcoming of internal stress hills can act similarly as if the line length is increased. Thus, three conditions must be fulfilled before the restricted motion of geometric kinks can be observed:

1) The oscillating stress σ_O must exceed the breakaway stress σ_B:

$$\sigma_O > \sigma_B \qquad (2)$$

2) The oscillating stress must exceed the internal stresses σ_i

$$\sigma_O > \sigma_i \qquad (3)$$

3) The oscillating stress must stay below the PNP stress σ_P

$$\sigma_O < \sigma_P \qquad (4)$$

Conditions 3 and 4 state that for observation of restricted kink motion, the Peierls stress must be large compared to the internal stresses. This condition implies that for most f.c.c. metals of standard perfection, the stiffening cannot be observed. Single crystals with very low dislocation density would be required.

2.2   Bias Stress Experiments

The restricted kink motion can also be verified as a reduction of the oscillation period via a biasing static stress on which the oscillating stress is superimposed. Figure 3 shows the result for tungsten. Here the abscissa denotes biasing stress. The parameter is the oscillation amplitude. For comparison, the curve T versus ϵ_O is included. The bias experiments do not show an increase of the period but only a decrease. This decrease becomes severe only when $\sigma_B > 10^{-3}$, a quantity which we are inclined to relate to the internal stress level. To make the influence of an external biasing stress observable, it must exceed the internal biasing stresses.

2.3   Stepwise Increase of Amplitude Dependence vs. Temperature

The Peierls potentials are definitely not the only cause of non-linear effects. Other sources are breakaway as well as the overcoming of internal stress hills (static hysteresis). Chambers[14] has pointed out that for dislocations which are parallel to close-packed direction, breakaway cannot occur at such temperatures where double-kink-generation is not possible. At the particular temperature

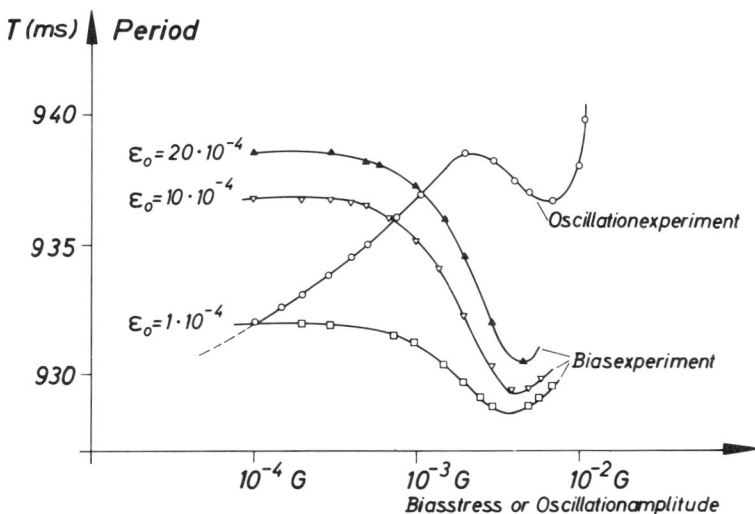

Fig. 3 Period of oscillation versus biasstress or oscillation amplitude at 78°K for tungsten with 1.6% deformation (ϵ_0 = oscillation amplitude). (The figure is taken from ref. 13.)

at which, for the given measuring frequency double-kinks can be generated, the process of breakaway will suddenly be initiated. This argument can easily be generalized to dislocations with geometric kinks. The stress σ on the first geometric kink in the kink pileup which is pushed against a pinning point, is equal to $N\sigma_a$ where N is the number of kinks in the pileup and σ_a is the resolved applied stress. For breakaway, $N\sigma_a$ must exceed σ_B. Clearly, a stepwise increase of the amplitude dependence is expected at the particular temperature for which $\omega\tau = 1$ (τ = relaxation time for double-kink-generation, $\omega = 2\pi f$, f = frequency of oscillation). The same argument applies for the mechanism of overcoming internal stress hills. The transition from restricted dislocation motion to unrestricted dislocation motion will be connected with a stepwise increase of the amplitude-dependent modulus defect and internal friction vs. temperature. To observe these phenomena, no such conditions like Eqs. (2), (3), (4) are required. Thus an observation on f.c.c. metals will be possible. Figure 4 gives an example for aluminum. For comparison, we have included the position of the Bordoni peak as observed after a 6% cold-working.[15] There is a remarkable coincidence between the Bordoni peak position and the step in Q^{-1} in Fig. 4. In the experiment of Fig. 4, the aluminum sample had only a very small deformation of 0.06%. The step in amplitude dependence becomes even more pronounced if the deformation is increased to 1%, which is still a small value. In Fig. 5 the internal friction for $\epsilon_0 = 3.4 \times 10^{-5}$ is temperature

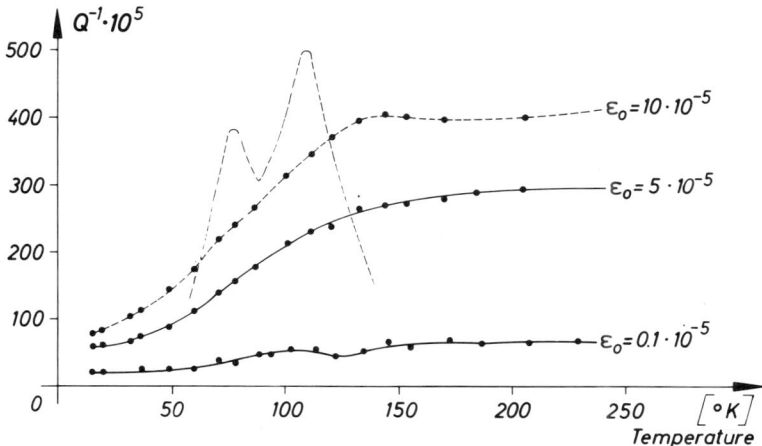

Fig. 4 Internal friction versus temperature for aluminum with 0.06% deformation at 16 cps (ϵ_0 = oscillation amplitude). (The dashed peak is the Bordoni peak after 6% deformation, taken from ref. 15).

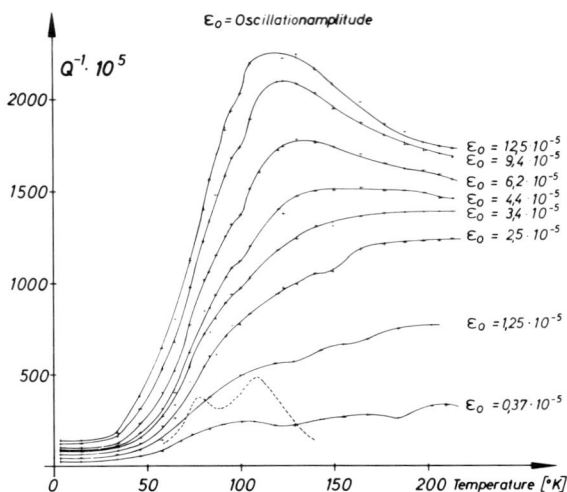

Fig. 5 Internal friction versus temperature for aluminum
with 1% deformation (ϵ_0 = oscillation amplitude, frequency
f = 16 cps)

independent below 50° and above 150°K, and increases by a factor
of 20 in the region of the Bordoni peak. The peak in Fig. 5 will be
discussed in the following section. A comparison of Figs. 4 and 5
shows that the plateau at 200°K is approximately a factor of 4 to 5
larger after cold-working. The deformation apparently introduces
long line segments, which are more effective in producing static
hysteresis by moving over internal stress hills. It should be mentioned that the curves in Fig. 5 have been measured for falling
temperature.

3   RELAXATION STRENGTH OF THE
DOUBLE-KINK-GENERATION PEAK

Seeger[16] pointed out that the process of double-kink-generation
should cause a peak of internal friction. The Bordoni peaks[17]
appeared to have the required properties in regard to the relaxation
time and the dependence of the relaxation strength on cold-working
and annealing. Nevertheless, some doubt concerning the interpretation of the Bordoni peak still remained. What criterion dis-
tinguishes between a double-kink-generation peak and, for example,
a point-defect dislocation interaction peak? Such a criterion appeared to be highly desirable, since, for example, in tungsten

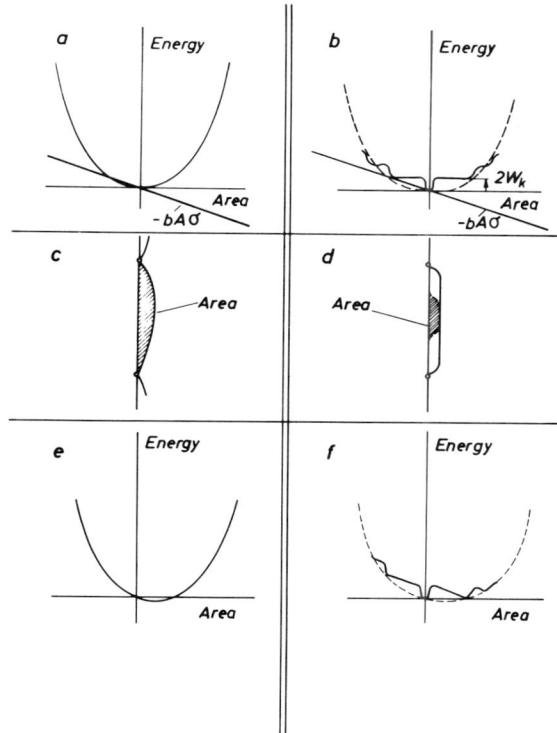

Fig. 6 Energy versus area swept over by a bowing dislocation string (Fig. a, c, e) and a kinked dislocation (Fig. b, d, f).

between He-temperature and 500°K, nine different peaks have been found after cold-working.[18]

We will now examine nonlinear properties of the double-kink-generation peak and compare a dislocation string in a material without PNP to a dislocation with kinks due to PNP. The diagrams in Fig. 6 demonstrate schematically the essential difference between a dislocation string and a kinked dislocation. Figure 6a shows the energy U of a dislocation string, plotted vs. the area A over which the bowing dislocation has moved. Figure 6b shows the same diagram for a dislocation segment which originally is parallel to the Peierls-Nabarro energy barrier. The abscissa "Area" in Fig. 6b is equivalent to the distance between the two kinks. The significant difference between Figs. 6a and b is a stepwise increase of the energy in Fig. 6b by $2W_k$ due to the Peierls potentials, before any bowing can occur.

For dislocation with geometric kinks, the same argument holds: The creation of a double-kink means a quantized increase of the energy, a concept which is alien to dislocation strings.

The double-kink-generation peak will be caused by a change in the thermodynamically determined population of the states without and with a new double-kink. Paré[19] has pointed out that the relaxation strength of this peak would be undetectly small because the state with kinks in Fig. 6b is hardly populated. Thus a small external stress cannot cause a significant change of population. It can be shown[20] that Paré's argument is correct for low-frequency measurements. Megacycle measurements will shift the peak to such high temperatures where, due to the high entropy of the state with double-kinks, this state will be equally populated. To make the observed peak height of the Bordoni peak consistent with the double-kink-generation model, Paré suggested that internal stresses will create a second, equally deep energy minimum which is required for the observation of the full strength of the relaxation process. The effect of a stress field on the energy diagram is shown in Figs. 6e and f. The energy in a stress field is proportional to the stress and the area, swept over, so that we do get a straight line as indicated in Figs. 6a and b. The energy diagram for kinked dislocation (Fig. 6f) is tilted by the stress energy and a second minimum appears. For the string the parabola is planely shifted. According to Paré, cold-working will create dislocation arrangements for which the mutual interaction is sufficiently large so that a peak will be observed, namely,

$$ablσ_i \geq 2W_k \tag{5}$$

($σ_i$ = internal stress, l = line length, a = lattice constant, and b = Burgers vector).

For well-annealed material, most dislocations are nearly straight and thus their contribution to the relaxation strength negligibly small. Paré was certainly aware of the fact that cold-working not only increases dislocation interaction, but also creates long line segments, another requirement for large relaxation strength.[21]

3.1 High Oscillation Amplitudes

From Figs. 6b and f it is immediately evident that instead of employing internal stresses, a second, equally deep or even deeper energy valley can be created by the oscillating measuring stress

itself. By sufficiently high amplitudes of oscillation, all dislocations will periodically reach a configuration in which the state
with a double-kink is energetically favored and, therefore, the dislocation bows out. We thus expect that the double-kink-generation
peak will increase by high-amplitude oscillations or rise out of
the background if it was unobservable at small oscillation amplitudes. This phenomena may be obscured from observation by the
stepwise increase of amplitude dependence (see Sec. 1.3) which
occurs at the double-kink-generation peak. Since the nonlinear
part of this peak is being added on a shoulder of amplitude dependence, it will only be resolved if it is larger than the step in amplitude dependence.

Figure 5 shows the result for aluminum, which had 1% deformation in torsion. The dashed line is the Bordoni peak, as it appears after 6% deformation.[15] In our experiment, the relaxation
strength of the Bordoni peak is below 1×10^{-3}. For rising amplitudes, the Bordoni peak first disappears, but rises at $\epsilon_o = 12.5 \times 10^{-5}$
to a height which has never been observed at low amplitudes after
cold-working. It is difficult to estimate the height of the amplitude-
dependent peak, but it is safe to say that it is an order of magnitude
larger than the value at low amplitudes. At 95°K, there seems to
be an inflection point signifying two separate peaks. This amplitude-
dependent peak is accompanied by an amplitude-dependent modulus
defect.[22]

3.2 Bias Stress Experiments

Instead of using internal stresses, it should be equally possible
to create in a reversible way the double-kink-generation peak by
static external stresses.[23] In Fig. 7 we have plotted the difference
in internal friction with and without a bias stress for a sample
which had only 0.06% deformation. Parameter is the oscillating
stress. The measuring sequence was as follows: For given temperatures, measurements were first made at different oscillation
amplitudes without bias stress. Then the bias stress was applied
and the measuring sequence was repeated. Again a measurement
without bias stress was made and finally a negative bias stress
was applied. We chose this sequence to ensure that the application of the bias stress did not change the state. We find the remarkable result that the sensitivity against external bias stress
is strongly peaked at that temperature where the Bordoni peak
appears after several percent deformation. The width of the nonlinear peaks in Fig. 7 is comparable to that of the Bordoni peak

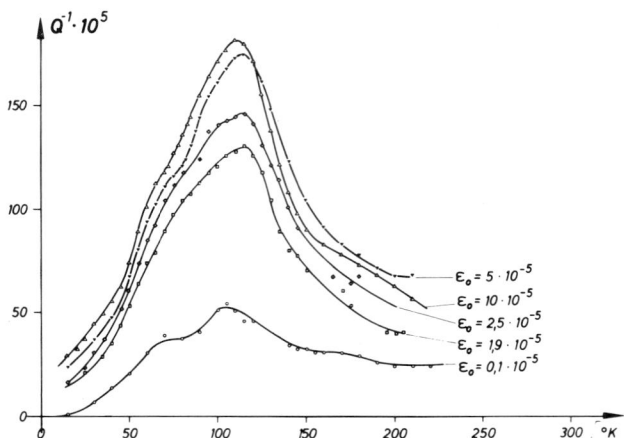

Fig. 7 Difference of internal friction with and without a bias stress $\sigma_B = 13.10^{-5}$, measured at different oscillation amplitudes (Al, 0.06 % deformation, $f = 16$ cps)

after cold work. It is also of interest that the nonlinear peak is accompanied by a nonlinear modulus change similar to the well-known one for linear peaks.[22] Figure 7 shows that the height of the nonlinear peak is enhanced by high amplitudes of oscillation, similar to Fig. 5. The interpretation of these results can again be found in Fig. 6. The bias stress is tilting the energy diagram so that at least one further energy minimum appears. For some of the dislocations the bias stress creates two equally deep energy minima, the well-known requirement for internal friction at low amplitudes. However, very special circumstances would be required if for the same bias stress, all dislocation segments would end up with two exactly equally deep, energy minima. We therefore need high-oscillation amplitudes in addition to the bias stress to cause an exchange of the population between energy valleys which are not equally deep. The latter explanation is the same as that for Fig. 5. Nevertheless, with the bias stress applied the internal friction is higher at all amplitudes than without bias stress.

4   THE ACTIVATION ENERGY AND THE ACTIVATION VOLUME FOR DOUBLE-KINK-GENERATION AND ANNIHILATION

Whereas in the preceding experiments the nonlinear behavior of the relaxation strength was considered, the experiments here

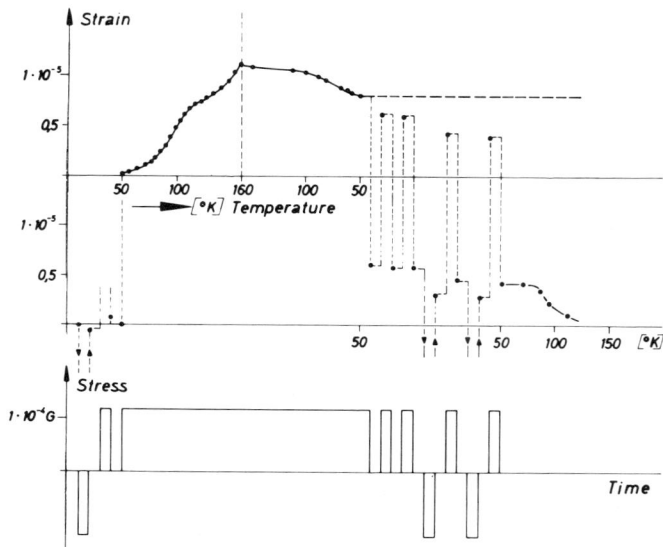

Fig. 8 Creep experiment (explanation in the text).

are concerned with the relaxation time. The following experiment was performed (Fig. 8). A static stress has been applied several times at 50°K in positive and negative direction, and the permanent strain after removal of the stress is recorded. It has been observed that a small but, nevertheless, permanent strain remains after removal of the stress. This observation belongs to the class of static hysteresis, probably caused by irreversibly shifting of dislocations over internal stress hills. Figure 8 shows that the static hysteresis is approximately 1×10^{-6} which is important for later observations. Now the stress is kept on and the temperature raised.

In the temperature region of the Bordoni peak, an extra amount of strain adds to the elastic strain. After having reached 160°K, the temperature is then reduced to 50°K still with a stress on. The remaining temperature-dependence of the strain is caused by that of the elastic strain. Now the amount of extra strain for which thermal activation has been required is recorded at 50° as 8×10^{-6}.

In the next step the stress is removed. If the Bordoni peak was due to, say, point-defect relaxation, one would expect that for the recovery of the thermally activated part of the strain, temperatures of 90 to 100°K are required as it was for the production of the strain. However, this was not the case in our experiment. We realized that 25% of the thermally activated strain recovers immediately at 50°K. The reading of the strain was at least a factor 10 more accurate than indicated by the size of the data points in Fig. 8. To ensure that there was no mistake in the reading of the strain, the positive stress is reapplied. One indeed verifies a second time that 25% strain is missing. This experiment has been reproduced from the very beginning several times at different stress levels and with different signs. Thus we have shown that the defect which is responsible for the Bordoni peak can move in one direction only with the aid of thermal fluctuations, whereas it can move back to the original position without the help of thermal activation, that is, as far as can be seen from the 50°K experiment. This result agrees with the energy diagram in Fig. 6 for dislocations with kinks.

In the untilted energy diagram, the activation energy for forward motion (double-kink-generation) is $2W_k$, whereas the activation for backward motion (double-kink-annihilation) is zero. Since the immediate recovery of the strain has not yet been observed at He-temperature, but only at 50°K, we can summarize the experimental observation as follows:

The activation energy for double-kink-annihilation is at least a factor two smaller than that for double-kink-generation.

The following continuation of the above-described experiment will provide information on the stress dependence of both activation energies. At 50°K a negative stress was applied and immediately removed again.

Figure 8 shows that even more strain has disappeared. Note that the reading of the strain must now be corrected by the amount of static hysteresis. Now a positive stress tries to recover the newly lost strain. In Fig. 8 it is seen that the strain does not come back at this temperature. Thus we have shown that thermally activated strain can be removed at 50°K by a stress, but not generated again by the opposite stress. This observation is also consistent with the double-kink and annihilation mechanism. From Fig. 8 we see that any significant lowering of the activation energy for double-kink-generation requires much higher stresses than a change of the activation energy for a double-kink-annihilation. The latter changes by a stress according to $U = abl\sigma$ (l = length). Therefore, we have verified experimentally that: *the activation volume for double-kink-annihilation is large compared to that for double-kink-generation.*

The experiments of Fig. 8 can be interpreted as follows: On the first removal of the positive stress, double-kink-annihilation of unbiased dislocation segments or segments originally biased in negative direction by internal stresses occur. On the application of a negative stress, double-kink-annihilation has been achieved for those dislocation segments which have originally been biased in positive direction by internal stresses. The experiment is completed by heating up and removing thermally activated dislocation strain.

Finally, it should be pointed out that the effects of Fig. 8 can only be observed in nonlinear creep experiments. It has been verified that a spontaneous recovery of strain at 50 °K cannot be observed for a creep experiment at a stress level of 10^{-6} G.

5   CONCLUSION

So far, we have listed seven different nonlinear elastic phenomena which we attributed to dislocations and, in particular, to the PNP for dislocations. This list is by no means complete. We have not considered nonlinear effects at high frequency, e.g., harmonic generation or change of velocity of sound by bias stress in that frequency region where the viscous damping and inertia effects of dislocations play a role. In addition, we have not considered the shift of the double-kink-generation peaks at high oscillation stresses. Finally, it should also be mentioned that oscillating a sample at high amplitudes below the double-kink-generation peak, increases the internal stress level without causing irreversible changes in the dislocation distribution. Dislocations, which are originally under internal stress and thus bowed by double-kink-generation, lose the double-kinks and become straighter by oscillation.[23] The dislocation network will be brought into a highe metastable state and will, after stopping the oscillations, approach a relative energy maximum with the relaxation time of the double-kink generation peak. In the experiments of Figs. 2 and 3, the dislocation network occurred after the first measurement in such a metastable state.

Summarizing the experimental results presented in this paper, we find that all observations are consistent with the concept of the Peierls-Nabarro potentials for dislocation. In particular, the reversible appearance of the Bordoni peak by either high-amplitude oscillation or by a biasing stress, strongly supports the interpretation by the double-kink-generation model. It is very unlikely that another dislocation process can be found which will manifest itself with the same linear and, especially, nonlinear properties as the kink and double-kink mechanisms.

One concluding remark: the arguments presented in this paper were based plainly on the assumption that periodic potentials exist. Consequently the experiments, if done on b.c.c. metals, will not be able to distinguish between "Peierls–Nabarro Potentials" and "Pseudo-Peierls–Nabarro Potentials"[24] as caused by splitting of dislocations into partials on different planes.

REFERENCES

1. Peierls, R. E.:*Proc. Phys. Soc.* 52:34 (1940).
2. Nabarro, F. R. N.:*Proc. Phys. Soc.* 59:256 (1947).
3. Leibfried, G., and H. D. Dietze: *Z. Phys.* 131:113 (1951). H. D. Dietze:*Z. Phys.* 131: 156 (1951); *Z. Phys.* 132: 107 (1952).
4. Shockley, W.:*Trans. Amer. Inst. Min. Met. Eng.* 194:829 (1952).
5. Mott, N. F., and F. R. N. Nabarro: "Report of a Conference on the Strength of Solids," p. 1, *Physical Society*, London, 1948.
6. Seeger, A., and P. Schiller:*Acta Met.* 10:348 (1962).
7. See, for example: Okuda, S.:*Sci. Papers of the Inst. of Phys. and Chem. Res.* 57:116 (1963).
8. Chambers, R. H.: in "Physical Acoustics," vol. III-A, W. P. Mason (ed.), Academic Press, New York, 1966.
9. See, for example: Gordon, R. B.:*Acta Met.* 10: 339 (1962); Fig. 4 (obtained by C. L. Bauer).
10. Suzuki, T., and C. Elbaum:*J. Appl. Phys.* 35: 1539 (1964).
11. Alefeld, G.:*J. Appl. Phys.* 36:2642 (1965); *J. Appl. Phys.* 36: 2633 (1965).
12. Granato, A., and K. Lücke:*J. Appl. Phys.* 27:583 (1956).
13. Chambers, R. H., T. E. Firle, and G. Alefeld: To be published.
14. Chambers, R. H.:*Appl. Phys. Letters* 2:165 (1963).
15. Völkl, J., W. Weinlander, and J. Carsten:*Phys. Stat. Sol.* 10:739 (1965).
16. Seeger, A.: "Theory of Lattice Defects," *Encyclopedia of Physics*, vol. VII, x 72, S. Flügge (ed.), Springer, Berlin-Göttingen-Heidelberg, 1955.
17. Bordoni, P. G.: *Ricera Sci.* 19: 851 (1949);*J. Acoust. Soc. Amer.* 26:495 (1954).
18. Filloux, J., H. Harper, and R. H. Chambers:*Bull. Am. Phys. Soc.* 9:239 (1964).
19. Pare, V. K.: *J. Appl. Phys.* 32:442 (1961); thesis, Cornell University, Ithaca, New York, 1958 (unpublished).
20. Alefeld, G.: in "Lattice Defects and their Interaction,"*Proceedings of the Symposium on Lattice Defects at Honolulu 1965* , R. R. Hasiguti (ed.) (in press).
21. For the line-length dependence of the double-kink-generation peak, see Ref. [20].
22. Alefeld, G., J. Filloux, and H. Harper: To be published.
23. Alefeld, G., R. H. Chambers, and T. E. Firle: *Phys. Rev.* 140:1771 (1965).
24. Nabarro, F. R. N., Z. S. Basinski, and D. B. Holt: *Adv. Phys.* 13:193 (1964).

DISCUSSION *on Paper Presented by G. Alefeld*

G. ALERS: You attribute the stiffening in the material at high oscillating stress or high bias stress to pushing the kinks to one end of the dislocation. One can estimate the stress needed to do this from a knowledge of the line tension and the loop length, and find that only 10 to 100 gm/mm^2 is needed. How do you explain the fact that experimentally you needed stresses of 1000 to 10,000 gm/mm^2?

G. ALEFELD: The reason is that the piling up of kinks cannot be observed as long as those processes dominate which cause the contrary effect, namely, an increase of the oscillation period. Those processes are, for example, breakaway and the motion of dislocations over internal stress hills. In our paper you find some criteria which must be fulfilled so that the piling-up effect can be observed. I did not present them here to save time.

P. HAASEN: You have given criteria for determining whether or not an observed relaxation peak corresponds to the formation of double kinks in the Peierls potential. In b.c.c. metals five or more peaks have been found experimentally. Does one know which one, if any, corresponds to kink formation? This would indicate which part of the T dependence of the yield stress of b.c.c. metals might possibly be due to such a process.

G. ALEFELD: To answer your question, I should refer you to Bob Chambers. The situation in body-centered materials is much more complicated than in aluminum. When I left General Atomic, we had not arrived at a firm conclusion as to which of the three main peaks in b.c.c. materials corresponds to double-kink generation. At that time, Bob Chambers tried to establish a connection between these peaks with long-range flow. His main point was that the flow stress of b.c.c. materials is not determined by just one process. In different temperature regions, a different process was rate-determining.

A. V. GRANATO: I have a comment and a question.

The comment is that a "hardening" in the modulus is not necessarily alien to the string model. The effective tension of an edge dislocation, increases as it is bowed out. Thus the existence of hardening, taken by itself without additional evidence, does not necessarily rule out a string model.

The question is: To what do you attribute the difference in behavior between your results in aluminum and those in copper as described by Caswell, who found no pronounced increase in the nonlinear damping above the Bordoni temperature?

G. ALEFELD: I agree with our comment. But if you try to think of dislocation strings at stresses of $10^{-2} G$, you are forced to assume such short segments that the modulus defect would not be observable at all.

I think that the pronounced increase in the nonlinear damping at the Bordoni peak is due to slightly cold-working the material, thereby creating long-line segments which will cause static

hysteresis by moving over internal stress hills. If you compare
Figs. 4 and 5, you will see a pronounced increase of this effect
after increasing the cold-working from 0.06 to 1%. I assume that
Caswell would have seen similar effects after introducing some
cold work.

G. T. HAHN: Could you elaborate on the internal stress? You argue
that you do not observe kink motion until the stress in tungsten
is G/1000 because the internal stress reaches high values.
However, the internal stress is periodic, so there must be a substantial volume of material experiencing stresses close to zero,
and some kink motion should be observed at low stresses.

G. ALEFELD: I do not argue that there is no kink motion at low
stresses. On the contrary, I mean that the reduction of kink
motion due to pileup on other Peierls potentials will be obscured as long as the applied stress stays below the internal
stress level. This is due to the motion of the dislocations over
internal stress hills, an effect which causes an increase of the
dislocation strain with stress. I agree with your second point.
Nevertheless, if you start to think about what the dislocation
arrangements are in a silightly deformed material, you will
have difficulty in finding many dislocations which are free of
internal stress.

Z. S. BASINSKI: How does your bias stress applied to aluminum
specimens compare with the flow stress of well annealed
pure single crystals?

G. ALEFELD: The largest bias stresses which we used were $1.3 \times 10^{-4} G$. A very well-annealed pure single crystal would have
flowed. Actually, we were also surprised that this high bias
stress could be used. Apparently, our specimen was not as
good as we thought it would be. I want to point out here that the
interpretations of our observations are not based on the assumption that we had high-quality single crystals. The amount of deformation introduced by the bias stresses was easily controlled
with an accuracy of 10^{-7} plastic strain. The accumulated deformation stayed below 10^{-5} plastic strain.

J. WEERTMAN: Can you give use numbers for the Peierls stress
of W and Al?

G. ALEFELD: If you use the stress value at 4.2 °K at which the modulus defect starts to rise, it would be $1.3 \times 10^{-2} G$ for tungsten.
For aluminum, I cannot give you an experimental value because
we did not observe the corresponding effect.

J. WEERTMAN: There always has been a conflict between the critical
resolved shear stress of f.c.c. metals measured at very low

temperatures, and the Peierls stress estimated from Bordoni experiments.

G. *ALEFELD*: This conflict still exists if Seeger's theory is applied in its present form. Nevertheless, it seems possible to me that the material can flow at Helium temperatures at stresses below the Peierls stress. Bob Chambers and I have entertained the following question: whether or not there are points along a dislocation line (for example, the corners at which the dislocation changes direction and goes from one Peierls valley into another) where the double-kink generation is easier than on a straight segment. This may be an answer to the apparent discrepancy.

A. R. *ROSENFIELD*: Have you examined your specimens by electron microscopy to determine whether your experiments avoid generation? It would appear that your bias-stresses are sufficiently high to cause such effects.

G. *ALEFELD*: We have not examined our specimens by electron microscopy, but we could easily control the amount of plastic deformation with an accuracy of 10^{-7} in plastic strain. Figure 8 shows you that a bias stress of 1×10^{-4} G did not introduce any plastic strain at 50°K. As I said already, our specimen was apparently not of extremely good quality, but the interpretation of the presented experiments does not depend on an assumption concerning the quality of the crystal.

INTERACTION OF DISLOCATIONS WITH HIGH-FREQUENCY SOUND WAVES

G. A. Alers and K. Salama

Scientific Laboratory,
Ford Motor Company,
Dearborn, Michigan

ABSTRACT

When a high-frequency sound wave passes through a solid containing mobile dislocations, the velocity of the wave is reduced and the energy in the wave is dissipated. Measurements of these changes in velocity and attenuation give information on the drag experienced by moving dislocations, as well as on the effective length of dislocation lines between pinning points. Experimental results obtained by using these techniques on copper at temperatures below 4°K show that some mechanism exists by which dislocation motion becomes "frozen out" as the absolute zero of temperature is approached. If this mechanism involves thermal activation, the effective activation energies must be distributed over a very broad spectrum. Experiments measuring the response of the dislocation to a small static stress were also performed to find evidence for a predicted anharmonic response of firmly pinned dislocations. Even though these measurements were carried out at low temperatures to suppress unpinning effects, they showed sound velocity changes which indicated that unpinning was still taking place, thus overshadowing any anharmonic effects.

INTRODUCTION

The use of internal friction and elastic modulus measurements in studies of dislocation motion has already been discussed by Granato.[1] His theory is based on the model that a dislocation behaves like a string stretched between fixed pinning points and capable of being bowed out under the action of an external stress. In the oscillating stress field of a high-frequency sound wave, the pinned dislocation vibrates and reduces the velocity of propagation by an amount given by

$$\frac{\Delta v}{v} = \frac{\rho v^2 \delta}{2} \frac{1}{1 + (\omega\tau)^2} \tag{1}$$

where v is the velocity of sound in the material when no dislocations are present, ρ is the material density, and ω is the angular frequency of the sound wave. The vibrating dislocation also dissipates energy and contributes an attenuation to the sound wave which is given by

$$\alpha = \frac{\text{log. dec.}}{\lambda} = \frac{\pi \rho v^2 \delta}{\lambda} \frac{\omega\tau}{1 + (\omega\tau)^2} \tag{2}$$

where λ is the sound wavelength. The dislocation parameters (distance between pinning points l, the total length of dislocation line per unit volume Λ, the intrinsic line tension or energy per unit length C and the parameter B which describes the damping force exerted by the material on the moving dislocation) are contained in the quantities δ and τ given by

$$\delta = \frac{8 b^2 \Lambda}{\pi^4} \frac{l^2}{C} \tag{3}$$

and
$$\tau = \frac{B}{\pi^2} \frac{l^2}{C} \tag{4}$$

where b is the Burgers vector. Thus, by measuring both α and v, it is possible to obtain information on two of the three unknowns

B, Λ and l^2/C. If some estimate or measurement of the dislocation density Λ is made, B and l^2/C can be obtained separately.

Previous experiments[2] have used this method to measure the temperature-dependence of the parameter B. The results showed that B was small at 4.2°K; therefore, the dislocations vibrated freely as if their microscopic motion did not involve any thermally activated process. However, more recent experiments,[3] extending down to 0.1°K, show that some mechanism comes into play which suppresses dislocation motion as the absolute zero of temperature is approached. These experiments will be discussed here as an example of the use of high-frequency sound waves to determine the properties of dislocations under conditions which make more direct measurements impossible.

As a second example of this use of sound waves, this paper also presents some new results wherein an attempt is made to measure the stress dependence of the parameter l^2/C at very low stresses. This is important because recently several proposals have been made[4-6] which postulate that the dislocation line tension parameter C should depend on the displacement of the dislocation line. This anharmonic behavior of the line could be measured by observing the changes in v and α produced by a static stress which is small enough to simply bow out the dislocations without producing any break away from the pinning points. Since the effects of temperature and stress are not related in an obvious way, the paper is divided into two sections: the first devoted to temperature-dependent effects; and the second, to stress-dependent effects.

1   EXPERIMENTAL TECHNIQUE

To use Eqs. (1) and (2), it is necessary to know the velocity and attenuation of the sound wave in the material when there is no dislocation contribution. This is because the equations describe the deviation of the velocity and attenuation from their values in the dislocation free state. Fortunately, Thompson and Holmes[7,8] have shown that in copper it is possible to irradiate the sample with a relatively small dose of neutrons or γ-rays and tie up the dislocations (i.e., make l very small); therefore, they no longer make a measurable contribution. Thus an absolute velocity and attenuation measurement must be made on an irradiated sample to define the value of v and α characteristic of the dislocation-free state.

Figure 1 shows the temperature variation of the elastic modulus C_{44} as deduced from the velocity of shear waves in a well-annealed,

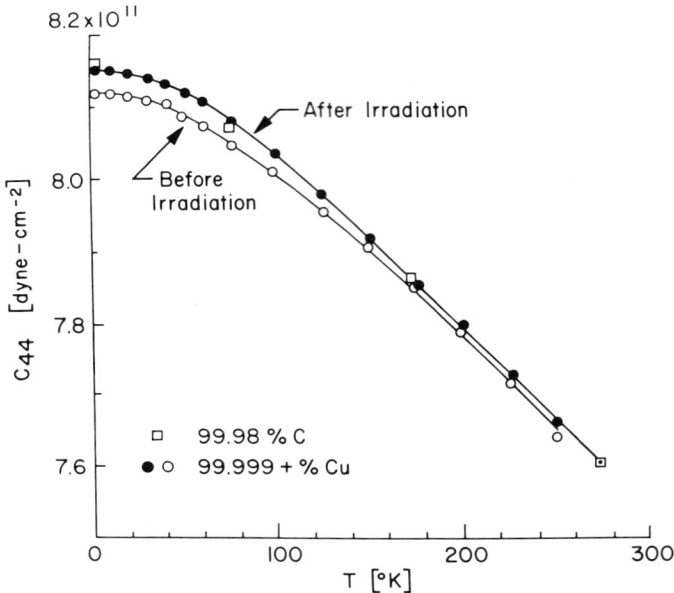

Fig. 1 The temperature variation of the elastic shear modulus C_{44} in annealed and neutron irradiated high purity copper. The difference between these two curves determines the temperature variation of the dislocation contribution to the modulus C_{44}.

single crystal of copper both before and after irradiation with neutrons. The difference between these two curves represents the temperature dependence of the dislocation contribution to the modulus C_{44} which is easily related to the velocity of sound called for in Eq. (1) [$C_{44} = \rho v^2$].

Figure 2 shows the temperature variation of the attenuation of this same shear wave expressed as the logarithmic decrement of the sound wave. Here, the difference between the two curves measures the dislocation contribution to the attenuation which is called for in Eq. (2). These data were obtained using a conventional ultrasonic pulse-echo technique[9] at a frequency of 10 Mcps, and have been described elsewhere.[2]

It is important to note that the dislocation contribution to the sound velocity (1/2 of $\Delta C_{44}/C_{44}$) is of the order of only a few tenths of a percent, while the effect on the attenuation is approximately a factor of five. This sensitivity of the attenuation to dislocations is the reason why internal friction has been used so often in studies of dislocation dynamics, even though it provides only half of the available information. To obtain additional information contained

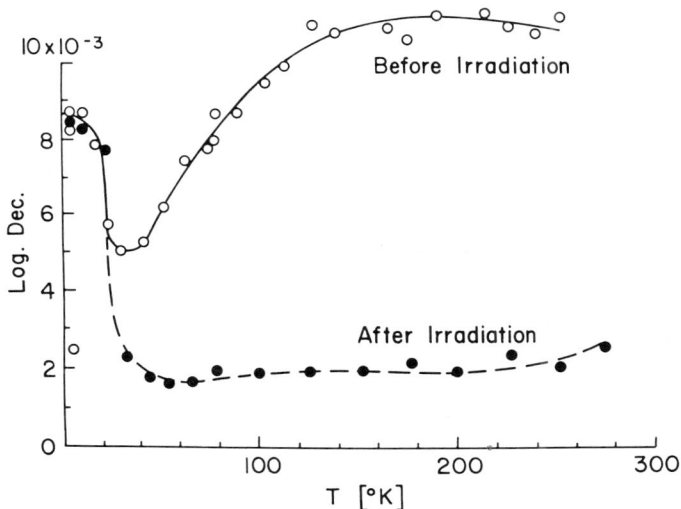

Fig. 2 The temperature variation of the attenuation of the ten mega-cycle sound wave used to measure C_{44} in Fig. 1.

in the sound velocity, very special techniques must be used which are particularly sensitive to small changes in the velocity of sound. The apparatus used in our experiments can detect velocity changes of less than one part in a million. It was designed and built by Forgacs[10] and is based on the ultrasonic "sing-around" method. In operation, it is actually a very stable oscillator whose oscillation period is determined by the time required for a pulse of 10-Mcps sound energy to travel the length of the specimen. A high-frequency counter is used to display this period of oscillation and thus presents the transit time of the sound wave to seven decimal places. Due to uncertain electronic and acoustic phase shifts, these seven decimal places cannot be used to define the transit time to seven significant figures, but they can be used to measure a change in transit time to one part in 10^7.

To convert the observed transit time changes to velocity of sound changes, some knowledge of possible changes in the path length must be introduced. In the case of measurements as a function of temperature below $10\,°K$, the thermal expansion data of Carr et al.[11] shows that the length-change corrections are completely negligible. For changes produced by the application of the static stress, a knowledge of the elastic compliances[12] permits a necessary correction to be calculated. However, these also turn out to be negligible.

2 LOW-TEMPERATURE EFFECTS

In Fig. 1, it can be seen that at low temperatures the dislocations can vibrate freely and produce a low value for the shear
modulus C_{44}. At the same time, the difference in the attenuation
between the irradiated and nonirradiated samples shows that the
vibrating dislocations no longer contribute significantly to the attenuation, i.e., the damping factor B is small at low temperatures.
(The increase of the attenuation observed below 20°K is due to the
interaction of the conduction electrons with the sound waves in the
high-purity copper specimen used in these experiments.[13] This
electron-produced attenuation can be suppressed by the application
of a magnetic field. The open-circle data point shown at a low
attenuation value at 4.2°K was observed when a magnetic field of
approximately 10 kilogauss was applied to the specimen.) It is
interesting to speculate about the possibility that at temperatures
below 4.2°K, the lower limit of the measurements shown in Fig. 1,
the dislocations could be "frozen out" and their vibrations stopped.
If this were the case, as would be expected from the usual theories
of dislocation motion which are based on thermally activated processes, then one might expect to observe the variation of the
modulus and attenuation shown by the dashed lines in Fig. 3 as the
temperature is lowered below 1°K. In an attempt to observe this
abrupt rise in modulus and the sharp peak in attenuation, the "sing-
around" system was used to monitor the velocity of sound and the
attenuation while the temperature of a well-annealed copper crystal was lowered from 4.2°K to below 0.2°K via adiabatic demag-
netization techniques to attain the necessary temperatures. The
results of these measurements[3] are shown as solid lines in Fig. 3.
(Since the "sing-around" system measures only relative changes
in the sound velocity, the position of the solid curve on the vertical
scale is somewhat arbitrary and was chosen to coincide with the
theoretical curve at about 6°K.) It is obvious that there is some
mechanism which impedes the motion of dislocations in copper at
very low temperatures, but this mechanism is not a thermally
activated process with one unique activation energy. To retain the
concept of thermally activated dislocation motion, it is necessary
to postulate an extremely broad distribution of relaxation processes
to account for the gradual rise in modulus and the nearly constant
attenuation that is observed. Such a distribution would have activation energies covering at least two orders of magnitude (0.1°K to
10°K) which raises rather serious questions concerning the validity

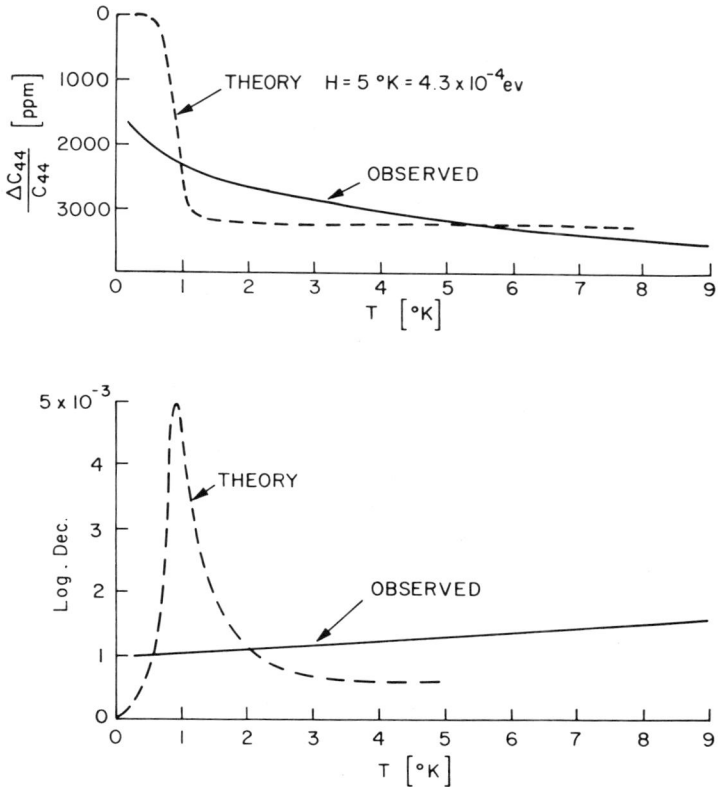

Fig. 3 The dashed lines show the temperature variation of the modulus C_{44} and its associated internal friction which would be expected if the motion of dislocations were thermally activated with a single activation energy of 5°K. The numerical magnitudes were chosen to fit the measurements given in reference 2. The solid line marked observed was taken from reference 3.

of even using thermally activated processes to describe dislocation motion at very low temperatures.

To show that the temperature dependence of C_{44} that was observed arises from dislocations, the sample was irradiated with γ-rays in a 4000 Curie Cobalt-60 γ-ray source for various lengths of time. Figure 4 shows that as the irradiation dose increased and more pinning was introduced, the variation of C_{44} with temperature decreased. In fact, a dose of 350 hours was sufficient to nearly remove the temperature dependence. A crystal containing 0.13 wt/% gold showed no temperature dependence[3] of C_{44} below about 10°K as would be expected from a specimen in which the dislocations were

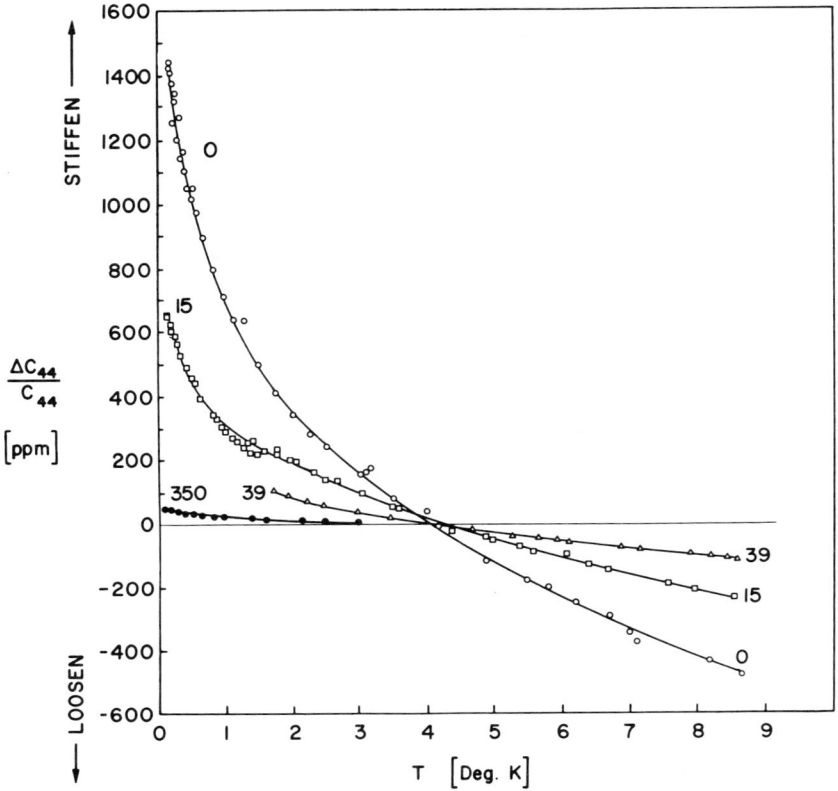

Fig. 4 Temperature dependence of the shear modulus C_{44} relative to its value at 4.2°K after various doses of γ-ray irradiation. The numbers on each curve specify the hours of exposure to a 4000 Ci γ-ray source.

pinned by an alloy element. A pure gold specimen which showed a small temperature dependence of its Young's modulus below 4.2 °K was cold-worked slightly, and subsequently showed an increased temperature dependence in the same temperature range. Thus, there seems to be no doubt that the effects arise from dislocations vibrating in the sound beam.

3   EFFECTS OF STRESS

The application of a small static stress would be expected to cause a dislocation loop to bow out between its pinning points. A superimposed oscillating stress will then produce vibrations about

this bowed-out position. If the effective line tension of the dislocation were to depend upon its displacement (i.e., a dislocation line
is an anharmonic oscillator), then the line tension parameter C in
Eqs. (3) and (4) would be modified by the external stress and one
could expect to observe a dependence of the sound velocity on the
static stress through Eq. (1). Since the dislocation displacement
must be symmetric for a change in sign of the stress, the line
tension can depend only on even powers of the stress which, in
turn, means that this dislocation contribution to the sound velocity
must vary as the square of the applied static stress. Elbaum[4] has
worked out a detailed model for this effect using the isotropic
continuum model of a dislocation. His results for the dependence
of sound velocity on the static stress σ is given as

$$\frac{\Delta v}{v} = \frac{A}{\rho^2 v_0}\sigma - \left(\frac{\Delta v}{v}\right)_0 - 3\pi C'(1 - \omega^2 r^2)\left(\frac{\Delta v}{v}\right)_0^2 \frac{\sigma^2}{\rho v_0^2 \Lambda C} \tag{5}$$

where C' is a geometrical quantity of order unity that describes
the edge or screw character of the dislocation. The stress-
independent term $(\Delta v/v)_0$ is the amount the velocity of sound is
lowered by the presence of dislocations, and is given by Eqs. (1),
(3), and (4) in terms of all the dislocation parameters. The quantity
A is a third-order elastic constant, therefore, the first term expresses the linear dependence of the sound velocity on stress to be
expected from the anharmonic character of the crystal lattice itself. For a copper crystal containing 4×10^6 dislocations/cm^2 and
in which $(\Delta v/v)_0 = 0.4\%$, the magnitude of the term which is quadratic
in stress is about 2 ppm at 10 gm/mm^2 and hence should be observable on the "sing-around" system. At 100 gm/mm^2, it would
contribute 200 ppm and would be clearly detectable.

On the other hand, Brailsford[5] has shown that under certain
conditions, the kink model of a dislocation will lead to a vanishing
of the anharmonic term in the line tension and, hence, there should
be no observable quadratic dependence of the sound velocity on
stress. It therefore appears that a measurement of this anharmonic
term could distinguish between the kink model and the elastic continuum model.

It is important to point out that these anharmonic effects are
based on the hypothesis that the pinning points at the ends of the
dislocation line are fixed and infinitely strong so that there is no
change in the loop-length parameter l when the stress is applied.
If the loop length l is not fixed but varies with stress, then the
$(\Delta v/v)_0$ term will be stress-dependent and could dominate the

observations. To estimate the effects on Eq. (5) of a stress-dependent $(\Delta v/v)_0$, we can differentiate Eqs. (1) and (2) with respect to stress assuming that l and Λ depend on stress. The results are

$$\frac{d\left(\frac{\Delta v}{v}\right)}{d\sigma} = \frac{1}{v}\frac{dv}{d\sigma} = -\left(\frac{\Delta v}{v}\right)_0 \left\{ 4L\left[\frac{1}{2} - \frac{(\omega\tau)^2}{1 + (\omega\tau)^2}\right] + D \right\} \qquad (6)$$

and

$$\frac{1}{\alpha}\frac{d\alpha}{d\sigma} = 4L\left[1 - \frac{(\omega\tau)^2}{1 + (\omega\tau)^2}\right] + D \qquad (7)$$

where $L = (1/l)(dl/d\sigma)$ and $D = (1/\Lambda)(d\Lambda/d\sigma)$. For the $(\Delta v/v)_0$ term in Eq. (5) to contribute less than 1 ppm at 10 gm/mm^2

$$\left(\frac{\Delta v}{v}\right)_0 [2L + D] \leq 10^{-6}$$

for a $d\sigma$ of 10 gm/mm^2. This means that for a $(\Delta v/v)_0$ of 4×10^{-3},

$$2\frac{dl}{l} + \frac{d\Lambda}{\Lambda} \leq 2.5 \times 10^{-4} \qquad \text{(for } \omega\tau < 1)$$

or that the average loop length and dislocation density can change by no more than 0.025% when 10 gm/mm^2 are applied. Since this is a very stringent criterion to meet, we can expect difficulty in separating the anharmonic effects from loop-length change effects.

Figure 5 shows the experimental apparatus for applying the static load. It consists of a 5 to 1 lever arm acting on a plunger which transmits the load to a hard, plastic hemisphere which in turn distributes the load over the top surface of the crystal. The two quartz transducers used in the sing-around circuit are on the sides of the cube-shaped crystal so that the sound is propagated at right angles to the static stress. The plunger also acts as the coaxial cables which transmit the 10-Mcps r.f. electrical signals to and from the transducers. Thus, it is a simple matter to perform the loading experiments at low temperatures because the specimen and the end of the apparatus can easily be immersed in a cryogenic fluid. The method of taking data was to note the change in transit time when a certain dead load was both applied

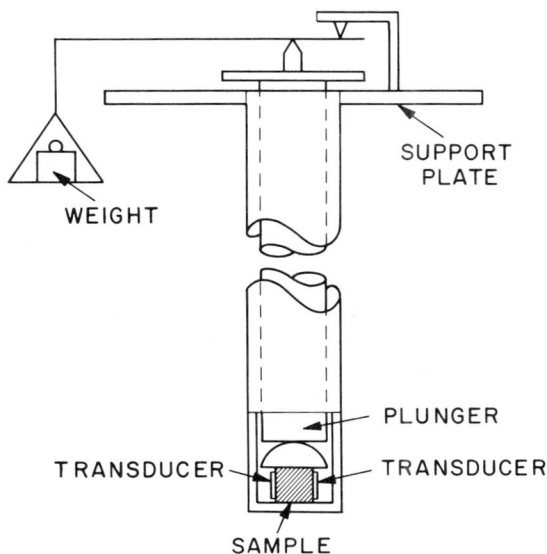

Fig. 5 Apparatus used to apply a static load to a copper sample in a low temperature bath. The two quartz transducers are used to measure the sound velocity at right angles to the static stress value.

and removed from the lever arm. Hence, irreversible effects could be detected immediately and the effects of temperature drift were minimized. The process was repeated several times with different loads to prepare the graph of velocity change vs. load. In all cases, no irreversible changes in sound velocity or attenuation were observed even for loads as high as 300 gm/mm^2. The specimen was an ultrahigh-purity copper single crystal in the shape of a cube approximately 5/8 inch on a side and grown from ASARCO material. Four faces were {110}-type crystal planes and two faces were {100}-type planes. After cutting the crystal specimen from the 3/4-in dia. ingot, it was annealed for 4 hrs at 450°C.

Figure 6 shows an example of the stress dependence of the sould velocity and the attenuation observed on this crystal at room temperature. No hysteresis was observed and the data fell on a straight line which extrapolated to zero stress at zero velocity change. Thus there is no indication of a critical stress above which dislocation effects suddenly appear. Although a velocity that changes linearly with stress is expected from the first term in Eq. (5), the slope of the line in Fig. 6 is much too large to be attributed to the natural anharmonicity of the lattice. Measurements[14]

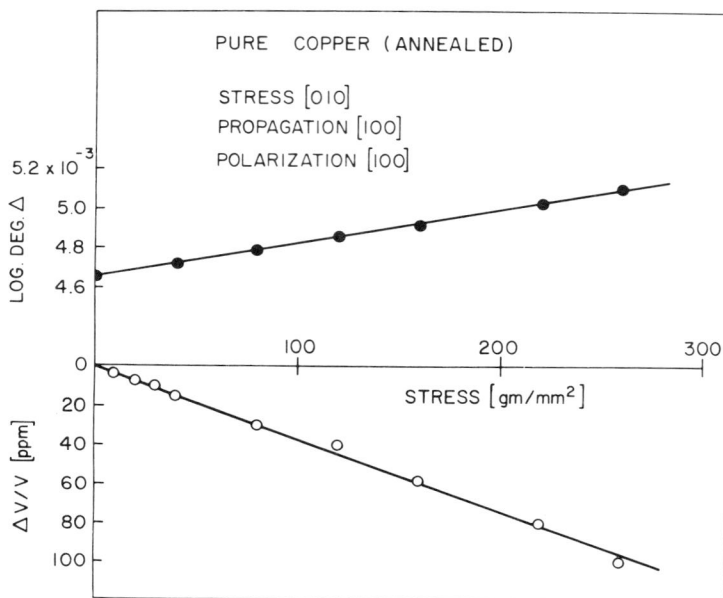

Fig. 6 Variation of the velocity of sound and the attenuation as a function of static stress on a copper single crystal at room temperature. Each data point represents the average of several measurements obtained by successively loading and unloading the sample at a particular stress value.

of the third-order elastic constants of copper lead to the conclusion that the lattice anharmonicity would predict a velocity change of only +1 ppm at 100 gm/mm^2. Furthermore, lattice anharmonicity would not be accompanied by the attenuation change which was observed here. Thus the entire effect of Fig. 6 must arise from dislocations. Since it is linear in stress, it cannot be explained by the quadratic stress-dependent term in Eq. (5). Therefore the effect must come from changes in the loop length and dislocation density as described by Eqs. (7) and (8).

The slopes of the lines shown in Fig. 6 define the left-hand members of Eqs. (6) and (7). Using γ-ray irradiation to pin the dislocations as described by Thompson and Holmes, it was possible to make an absolute determination of the velocity of sound in the crystal when the dislocations were free to vibrate and when they were pinned. From this data, $(\Delta v/v)_0$ was found to be 2.8×10^{-3} and a value of $\omega\tau$ was estimated from the ration of $(\Delta v/v)_0$ and α. Equations (6) and (7) could then be solved for L and D from which it could be concluded that the relative change in loop length and dislocation density produced by the application of 100 gm/mm^2 were

$$\frac{\Delta l}{l} \approx 1\%$$

and

$$\frac{\Delta \Lambda}{\Lambda} \approx 0$$

These results show that in a well-annealed crystal of copper, the application of a static stress produces an increase in the average effective loop length of only a few percent and a much smaller change in dislocation density. Furthermore, the effective loop length appears to be a linear function of the stress. These conclusions disagree with Young's results[15] which show dramatic increases in dislocation density when much smaller loads are applied. However, his crystals initially contain abnormally small dislocation densities while ours contain the usual 10^6 lines/cm^2.

To eliminate the change in loop length with stress and to make the $(\Delta v/v)_0$ term in Eq. (5) independent of stress, it is necessary to increase the strength of the pinning points. This might be accomplished by alloying, irradiation damage or by lowering the temperature to suppress the breakaway of thermally activated dislocations.

The results of a series of experiments using crystals which had been irradiated with γ-rays and neutrons, as well as having different impurity contents, are shown in Fig. 7. For the particular choice of sound wave and static stress used here, the lattice anharmonicity gives rise to a linear increase in sound velocity with stress which can be shown to agree with the measurements on the neutron-irradiated sample and the Cu-0.1% Au alloy sample. Thus these two samples have no dislocation effects and only the first term in Eq. (5) contributes to the measurements. The other two alloy crystals, the Cu-1.3% Mn and the OFHC copper sample (Cu-0.03% P), both show the existence of a small contribution from dislocation breakaway which is linear in stress. The crystal which had been irradiated with γ-rays shows the expected effects from a crystal where there is a critical stress below which the dislocations are effectively pinned, and above which there is an increase in effective loop length with stress. The ultrasonic attenuation observed on this γ-ray irradiated crystal is also consistent with this hypothesis. These results show that considerable effort is required to effectively pin the dislocations and make the loop length independent of stress so that the anharmonic terms can be observed separately. Unfortunately, these pinning techniques make $(\Delta v/v)_0$ very small, and since the term describing the

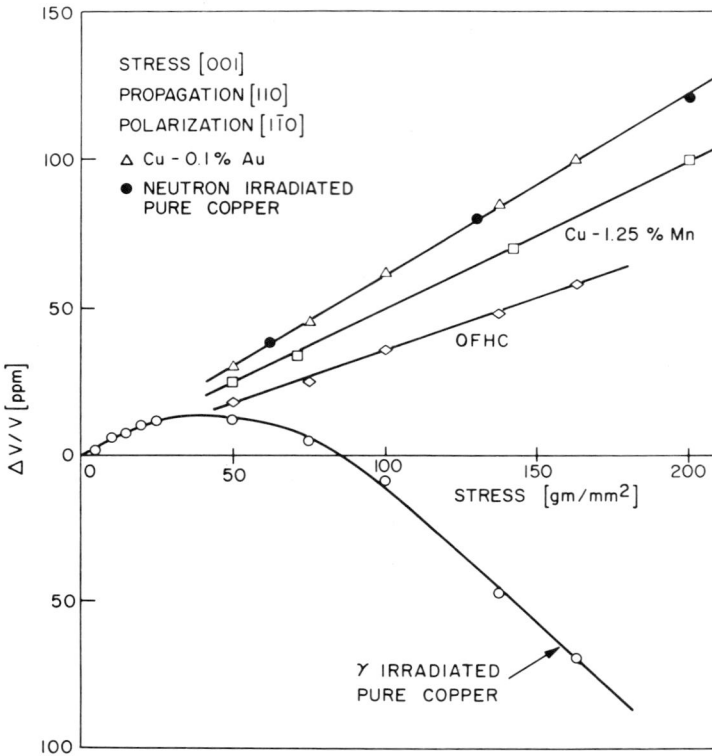

Fig. 7 Variation of the velocity of sound of a particular sound wave with static stress in samples of various purities and irradiation dosages. The true lattice anharmonicity is specified by the neutron irradiated sample. Hence all but the Cu-Au alloy exhibit some dislocation effect.

anharmonic properties of dislocations depends on $(\Delta v/v)_0^2$, it becomes impossible to observe any such effects in crystals hardened by alloying or irradiation.

To observe the quadratic stress dependence of the sound velocity arising from the anharmonic dislocation term, it is necessary to have a material in which $(\Delta v/v)_0$ is large but in which the dislocations are well pinned. This might be accomplished by performing the measurements on an annealed sample at 4.2°K where thermally activated unpinning would not be likely.

Figure 8 shows an example of the variation of the sound velocity with stress observed at 77 and 4.2°K. The dashed line represents the velocity variation contributed by the lattice anharmonicity. From the figure it is seen that at very low stresses there is still dislocation breakaway with the accompanying decrease in sound

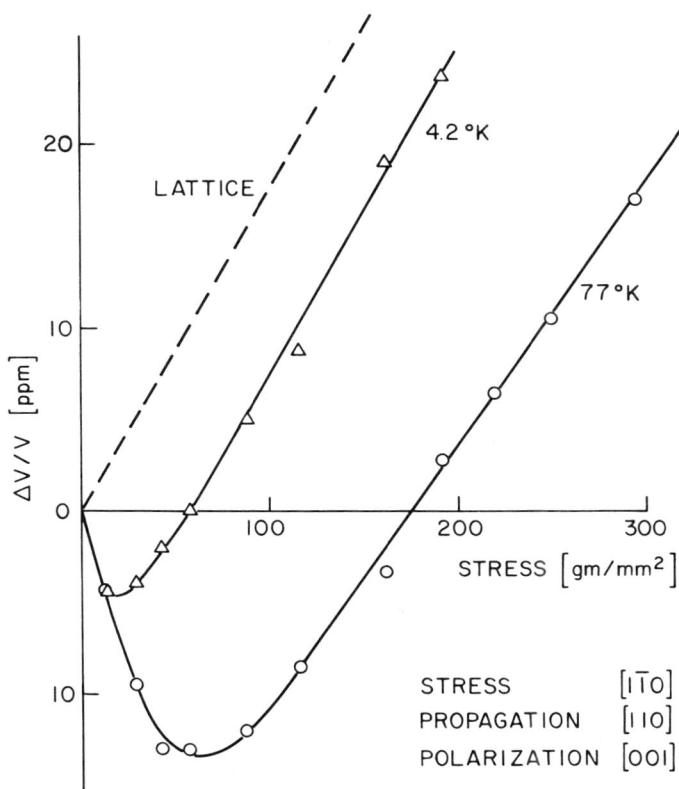

Fig. 8 Variation of the sound velocity with static stress observed on an annealed copper crystal at 4.2°K and 77°K. The dashed line represents the velocity variation contributed by the anharmonic lattice.

velocity expected from a small increase in the effective loop length. However, at higher stresses the loop length appears to become effectively independent of stress so that the velocity vs. stress curve becomes a straight line with a slope comparable to that which would be expected from the lattice term alone. This qualitative behavior was observed in all the other combinations of stress and sound-wave directions examined. However, it was not always true that the limiting slope at high stresses became comparable to the value contributed by the lattice anharmonicity term alone. (However, this might be due to not going to sufficiently high static stresses in these cases.) The attenuation changes accompanying these velocity changes were too small to be observed.

The shape of these curves cannot be described by the sum of a linear and a quadratic stress dependence as predicted by Eq. (5) over the entire stress range. Such an analysis might be applied to the low stress region but the behavior at high stresses cannot include a quadratic stress term. The fact that the high stress region is linear and very nearly equal to the lattice term implies that $(\Delta v/v)_0$ is stress-independent, and, hence, that no changes in effective loop length are taking place. At the stress levels being used in this region, it is anticipated that the measurements are characteristic of effects above the flow stress and therefore should be discussed in terms of the motion of dislocations through an array of pinning points rather than the more microscopic processes of bowing and breakaway. In the low stress region, the lowering of the velocity of sound indicates that breakaway is taking place which certainly makes it difficult to prove that a quadratic stress term exists. Therefore, we are forced to conclude that even though the quadratic term was not clearly observed, our experiments do not prove that it does not exist.

ACKNOWLEDGMENT

The authors would like to thank J. E. Zimmerman for his most valuable assistance with the very-low-temperature measurements. Also J. A. Karbon and R. C. Root deserve special thanks for their assistance with the experiments.

REFERENCES

1. Granato, A. V.: This colloquium, p. 117.
2. Alers, G. A., and D. O. Thompson: *J. Appl. Phys.* **32**:283 (1961).
3. Alers, G. A., and J. E. Zimmerman: *Phys. Rev.* **139**:A414 (1965).
4. Hikata, A., and C. Elbaum: *Phys. Rev.* **144**:469 (1966).
5. Brailsford, A. D.: *J. Appl. Phys.* **35**:2256 (1964); also *Ford Motor Co. Scientific Laboratory Report,* SL 64-40 (1964).
6. Alefeld, G.: This colloquium, p. 191.
7. Thompson, D. O., and D. K. Holmes: *J. Appl. Phys.* **27**:191 (1956).
8. Thompson, D. O., and D. K. Holmes: *J. Phys. Chem. Solids* **1**:275 (1957).
9. Forgacs, R. L.: *Proc. Natl. Electronics Conf.* **14**:528 (1958).
10. Forgacs, R. L.: *IRE Trans. Instr.* **9**:359 (1960).
11. Carr, R. H., R. C. McCammon, and G. K. White: *Proc. Roy. Soc.* London **A280**:72 (1964).
12. Waldorf, D. L.: *J. Phys. Chem. Solids* **16**:90 (1960).
13. Mason, W. P.: "Physical Acoustics and the Properties of Solids," p. 319, D. Van Nostrand Co., Inc., Princeton, New Jersey, 1958.
14. Hiki, Y., and A. V. Granato: *Phys. Rev.* **144**:411 (1966).
15. Crump, J. C., III, and F. W. Young, Jr.: *Bull. Am. Phys. Soc.* **12**:369 (1967).

DISCUSSION *on Paper Presented by G. A. Alers*

N. BROWN: In the curves of modulus defect vs. bias stress, are the curves independent of the order in which the stress bias is applied?

G. ALERS: Yes, the curves are independent of the order in which the stress was applied and were reversible and reproducible. We could repeat any modulus defect change at will or we could go back and fill in parts of the curve where we might like more data. The serious problem raised by this reversibility is that we observed it even at stresses which were high in comparison with the critical resolved shear stress of pure copper often mentioned at this conference (i.e., about 20 gm/mm^2).

G. ALEFELD: The possibility of explaining the temperature dependence of the modulus defect at very low temperature by geometric kink motion for which a very broad spectrum for the activation energies exists has been mentioned by you already. This explanation ought to be consistent with the observation that the friction constant B for copper decreases linearly with temperature down to 40°K. This implies that the activation energy for lateral motion should be below 0.003 eV so that your effect should be observable only below 3°K.

A second explanation of your results is possible in terms of higher-order Peierls potentials. For large kink density (kinks in regard to a <110>-direction) it appears energetically more favorable to straighten out the dislocation and thus it will experience higher-order Peierls potentials, for example, in the <112> direction. In aluminum, one finds indeed below the Bordoni peak another small peak at approximately 16°K (Lax and Tilson) and at lower temperatures a modulus defect change* similar to that which you found in copper.

Z. BASINSKI: Your crystal was prestressed to above 300 gm/mm^2 and therefore presumably had a dislocation density larger than 10^9; with strain amplitude used of 10^{-10}, the dislocations would move over an average distance of 10^{-11} to 10^{-12} mm. Can you envisage any pinning mechanism which could prevent such small movement?

G. ALERS: If the dislocation density was that high in my samples, it would have been obvious in the acoustic properties. Thus, I do not believe we had generated such a high density in our crystal. However, it might be worth checking. If the density does turn out

*Harper, H., J. Filloux, R. H. Chambers: *Bull. Amer. Phys. Soc.* 12: 369 (1967).

to be so high, and your estimated dislocation displacement correct, then I can envisage a pinning mechanism. The occurrence of such small displacements was the basis for postulating the kink model in the first place. A very small effective displacement of a dislocation loop can be achieved by a larger lateral motion of a kink. These motions could be pinned easily by an impurity.

J. HIRTH: With regard to the spectrum of apparent activation energies for the excitation of dislocation motion at $0.1\,°K$, I would like to comment on the possibility of a quantum effect explanation. Suppose that the motion corresponds to thermally activated motion of geometric kinks over a kink barrier. Applying standard rate theory, the probability of an activated jump is proportional to the ratio of the total partition function of the system in the activated state to that in the ground state. As a sample calculation to show the desired effect, let us suppose that the activated state can be described as having one vibrational degree of freedom, a local mode with vibrational frequency ν_1, while the ground state is a mode with frequency ν_2. The probability is then given by

$$P = \left[1 - \exp\left(-\frac{h\nu_2}{kT}\right)\right]\left[1 - \exp\left(-\frac{h\nu_1}{kT}\right)\right]^{-1} \exp\left(-\frac{E_0}{kT}\right) \tag{1}$$

where E_0 is the zero-point potential energy of activation. In terms of free energy of activation F,

$$P = \exp\left(-\frac{F}{kT}\right) \tag{2}$$

The activation entralpy H is given by:

$$H = k\,2\ln\frac{P}{2(1/T)} = E_0 - \frac{h\nu_1 \exp(-h\nu_1/kT)}{[1 - \exp(-h\nu_1/kT)]} + \frac{h\nu_2 \exp(-h\nu_2/kT)}{[1 - \exp(-h\nu_2/kT)]} \tag{3}$$

Now let ν_2 be a high-frequency mode (relative to $10\,°K$) so that the last term in Eq. (3) is zero. Also let ν_1 be the relaxed frequency with a value 2×10^9 per sec, with a corresponding Debye temperature of about $0.1\,°K$. Then at $T = 0\,°K$, $H = E_0$, while at $T \gg 0.1\,°K$, $H = E_0 - kT$. The drop in H will occur over a range in temperatures around $0.1\,°K$. It is evident that a number of modes could contribute to H in an extension of the above argument, giving a possible rationalization of Alers' data in terms of a process with a unique activation potential energy E_0.

In principle, another type of quantum effect is possible. If the applied stress raises the ground level near to the maximum energy configuration, quantum mechanical tunneling through the barrier is possible.

G. ALERS: Your proposal has a most interesting aspect. Presumably the frequency ν_2 would be the vibration of a kink within a potential well along the Peierls barrier while the frequency ν_1 could be the vibrational frequency of a kink moving along the Peierls barrier over the total length of the dislocation line. In this way the parameter ν_1 would be expected to depend upon the dislocation loop length and therefore we would have a wide distribution of values for the activation energy for the process.

INTERACTIONS BETWEEN
MOVING DISLOCATIONS

J. P. Hirth

Metallurgical Engineering Department,
Ohio State University

J. Lothe

Physics Institute,
Oslo University

ABSTRACT

Interactions between moving dislocations are treated in terms of a Lagrangian formulation. The resulting interaction forces are used in several applications to moving dislocations, including the extension of partials, the nucleation of dislocations, the formation of extrinsic fault configurations, the drag of sessile jogs, and the stability of a straight dislocation with respect to a zig-zag configuration.

1   INTRODUCTION

Interactions between dislocations at rest are well understood in the reasonable approximation that core-core interactions are negligible. The rest self-energy per unit length of a straight dislocation is

$$W_R^\circ = E^\circ \ln\left(\frac{R}{r_0}\right) \tag{1}$$

where R represents an appropriate outer cutoff in the strain energy integral, r_0 is the core cutoff radius and E° is the energy factor, $\mu b^2/4\pi$ for a screw, etc., with μ the shear modulus and b the modulus of the Burgers vector. Similarly, the rest energy of a system containing a pair of straight parallel dislocations separated by a distance r is

$$W^\circ = W_\infty^\circ - W_I^\circ = W_\infty^\circ - 2E^\circ \ln\left(\frac{r}{r_0}\right) \tag{2}$$

where W_∞° is the energy of the recombined dislocation $r = r_0$. Here, the entire r dependence is contained in the interaction energy W_I°.

The interaction force on a dislocation can be related directly to the change in strain energy of the system with virtual displacements of the dislocations,

$$F = -\frac{\partial W^\circ}{\partial r} = \frac{2E^\circ}{r} \tag{3}$$

A more direct definition involves the net stress σ existing at one dislocation core because of the presence of the other dislocation

$$F = \sigma b \tag{4}$$

This definition has the advantage of showing clearly that the dislocation will tend to move in the direction of the force, which does net work on the slip plane, which, in turn, is dissipated as heat. In the absence of friction, the core is unable to accelerate in the direction of the force.

In the case of a moving dislocation, the force is *defined* in terms of the stress acting over the core.[1] However, unlike the static case, the variation in total energy of the system with a virtual displacement does *not* yield the correct value for the force. However slow the change in configuration of the system, inductive energy terms must be taken into account.

In this report, interactions between moving dislocations are considered in terms of the concept of dislocation momentum, introduced by Nabarro[2] and Eshelby.[1] Analogous to classical mechanics, this concept is shown to lead to a description of interaction forces in terms of the Lagrangian for the system. This formulation is found to yield results consistent with those calculated directly by Weertman,[3] and it is also in agreement with the formal Lagrangian theory of Stroh.[4] The results then are applied to a number of problems in dislocation dynamics.

Since most problems in dislocation dynamics involve only glide, particularly at high velocities, the present development deals mainly with the glide case. Also, frictional forces, including radiation forces, are not considered; it is obvious how such forces would be included in a general force balance.

2   THE LAGRANGIAN FORMULATION

2.1   General Case

We hypothesize that forces on uniformly moving dislocations are given in terms of a Lagrangian (the possibility of such a development is suggested by the work of Eshelby[4]). Analogous to Eq. (1) the self-Lagrangian per unit length is

$$L = W_E - W_K = (E_E - E_K) \ln\left(\frac{R}{r_0}\right) = \Lambda \ln\left(\frac{R}{r_0}\right) \tag{5}$$

Here W_E and W_K represent the elastic strain energy and kinetic energy, respectively, in the moving system, the E values are the corresponding energy factors, and [cf. Eq. (1)] Λ is the Lagrangian factor. Also of interest is the total energy of the system

$$W = W_E + W_K \tag{6}$$

Proceeding as in Eq. (2), for a pair of uniformly moving, parallel, straight dislocations, we find

$$L = L_\infty - L_I = L_\infty - 2\Lambda \ln\left(\frac{r}{r_0}\right) \tag{7}$$

The interaction force per unit length then is given by

$$\sigma b \equiv F = -\frac{\partial L}{\partial r} = \frac{2\Lambda}{r} \tag{8}$$

The proof of Eq. (8) proceeds as follows. Consider a straight dislocation in an anisotropic medium, gliding with uniform velocity v in the x direction. The equilibrium equations are

$$\frac{\partial \sigma_{ij}}{\partial x_j} = \rho \frac{\partial^2 u_i}{\partial t^2} \tag{9}$$

where ρ is the density and the u_i are the displacements. At steady-state, uniform motion,

$$\frac{\partial}{\partial t} = -v \frac{\partial}{\partial x} \tag{10}$$

Hence

$$\frac{\partial \sigma_{ij}}{\partial x_j} = v^2 \rho \frac{\partial^2 u_i}{\partial x^2} \tag{11}$$

Multiplying Eq. (11) by u_i and integrating over a volume containing no singularity, we find

$$\int u_i \frac{\partial \sigma_{ij}}{\partial x_j} dA = v^2 \rho \int u_i \frac{\partial^2 u_i}{\partial x^2} dA \tag{12}$$

In Eq. (12), integration over unit length parallel to the dislocation line is inferred, so that only an area integral remains. By Stoke's theorem

$$\int_A u_i \frac{\partial \sigma_{ij}}{\partial x_j} dA = \int_l u_i \sigma_{ij} dl_j - \int_A \sigma_{ij} \frac{\partial u_i}{\partial x_j} dA \tag{13}$$

and

$$\int_A u_i \frac{\partial^2 u_i}{\partial x^2} dA = \int_l u_i \frac{\partial u_i}{\partial x} dl_x - \int_A \left(\frac{\partial u_i}{\partial x}\right)^2 dA \tag{14}$$

where l represents the line bounding the area A, $dl_x = -dy$ and $dl_y = dx$, a notation consistent with l considered as surface per unit depth. Substituting Eqs. (13) and (14) into (12) yields

$$\int_l u_i \left(\sigma_{ij} - v^2 \rho \, \frac{\partial u_i}{\partial x} \, \delta_{xj} \right) dl_j \;=\; \int_A \left[\sigma_{ij} \frac{\partial u_i}{\partial x_j} - \rho v^2 \left(\frac{\partial u_i}{\partial x} \right)^2 \right] dA \qquad (15)$$

where δ_{xj} is the Kronecker delta, i.e., $\delta_{xj} dl_j = dl_x$.
The strain energy density in the elastic field is

$$w_E \;=\; \frac{1}{2} \, \sigma_{ij} \frac{\partial u_i}{\partial x_j} \qquad (16)$$

while the kinetic energy density is

$$w_K \;=\; \frac{1}{2} \rho v^2 \left(\frac{\partial u_i}{\partial x} \right)^2 \;=\; \frac{1}{2} \rho \dot{u}_i^{\,2} \qquad (17)$$

using Eq. (10) for this steady-state result. Also, the quantity $-\rho v (\partial u_i / \partial x)$ is the momentum density p_i of the strain field. Thus Eq. (15) becomes

$$\frac{1}{2} \int_l u_i (\sigma_{ij} + v p_i \delta_{xj}) \, dl_j \;=\; \int_A (w_E - w_K) \, dA \qquad (18)$$

where the right-hand side is a Lagrangian.

Consider now the dislocation in Fig. 1, where l_2 and l_3 are two circular cylindrical bounding surfaces and l_1 is a cut in the x direction, the direction of motion. Consider the line integral in Eq. (18) over l_2. Only terms of logarithmic order are of interest, since the volume integral is evidently logarithmic. There are terms in u_i logarithmic in r for the edge dislocation solution but these terms are *constant* on the surface l_2. In addition, there are smaller varying displacements of order b, but these cannot contribute logarithmically (cf. core traction energies).[5] Thus, to logarithmic order, u_i is constant and

$$\int_{l_2} u_i (\sigma_{ij} + v p_i \delta_{xj}) \, dl_j \;\simeq\; u_i \int_{l_2} (\sigma_{ij} + v p_i \delta_{xj}) \, dl_j \;=\; 0 \qquad (19)$$

The integral vanishes at equilibrium as shown by the following arguments. The integral of $\sigma_{ij} dl_j$ is the net increase in momentum per unit time in the volume within l_2, fixed in space. In the same volume, moving in space, there is an additional change of momentum given by the integral of $v p_i \delta_{xj} dl_j$ because new portions of material with momentum p_i are enclosed in the moving volume at a rate proportional to v while an equivalent portion of material is removed behind the dislocation. Now, within the volume l_2 at steady state, the momentum is constant as expressed by Eq. (19).

A similar development holds for l_3 so only the surfaces l_1 need to be treated. Equation (18) becomes

$$\frac{1}{2} \int_{l_1} u_i \sigma_{ij} dl_j + \frac{1}{2} \int_{l_1} u_i v p_i dl_x = \int_A (w_E - w_K)\, dA \qquad (20)$$

With the cut *coinciding* with the direction of motion, $dl_x = 0$. Also, the discontinuity in u_i across the cut is b_i, so that Eq. (20) reduces to

$$\frac{1}{2} b_i \int_{l_1} \sigma_{ij} dl_j = \int_A (w_E - w_K)\, dA \qquad (21)$$

Introducing the resolved stress, $\sigma = B/x$ on the cut in the direction of b

$$\frac{1}{2} bB \ln \frac{R}{r_0} = \Lambda \ln \frac{R}{r_0}$$

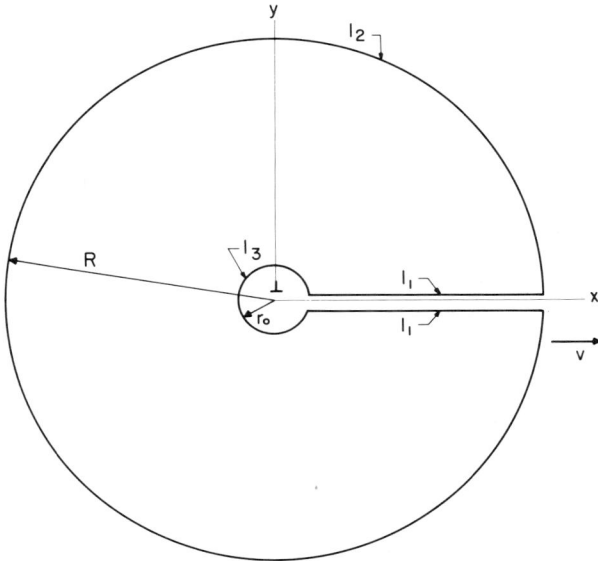

Fig. 1 Surfaces of integration, l_1, l_2, and l_3, for a dislocation moving with velocity v.

or
$$\sigma = \frac{2\Lambda}{bx} = \frac{2\Lambda}{br} \tag{22}$$

which verifies the original hypothesis, Eq. (8).

The relation (22) is valid only on the plane parallel to the direction of motion, where the integral dl_x in Eq. (20) vanishes, in contrast to Eq. (3) which holds on *any* cut.

Some other important quantities now are listed briefly and related to the terms in the above development. For a free dislocation such as that in Fig. 1, in rectilinear motion by glide in the x direction and with no net external forces present, the quantity

$$M = -\int \rho \dot{u}_i \frac{\partial u_i}{\partial x} dA \tag{23}$$

is converved during the motion.[4] The quantity M is the dislocation momentum per unit length;[2] its density is *distinct* from that of the momentum of the strain field p_i, as can be seen by a comparison with Eq. (18). The momentum is related to the kinetic energy, Eq. (17), by

$$M = \frac{1}{v} 2 W_K \tag{24}$$

in uniform motion. Similarly, the effective mass per unit length of the dislocation is given in terms of W_K by

$$m = \frac{2}{v^2} W_K = \frac{\rho}{v^2} \int \dot{u}_i^2 \, dA = \rho \int \left(\frac{\partial u_i}{\partial x}\right)^2 dA \tag{25}$$

As is illustrated below, it also follows for slow accelerations that

$$F = \sigma b = n \frac{dv}{dt} \qquad M = mv \tag{26}$$

Hence a general analogy with classical mechanics exists for dislocations moving in their glide planes. However, the analogy does not hold for interactions normal to the direction of motion as also discussed below.

2.2   Screw Dislocation Interactions

Consider the case of two like-sign* screw dislocations moving in the same glide plane (Fig. 2a). As shown in detail by Weertman,[3] the integration of Eqs. (16) and (17) for this case, and the determination of σ_{ij} from the displacements yields

$$E_E = \frac{\mu b^2}{4\pi} \frac{(1 + \beta^2)}{2\beta} \qquad E_K = \frac{\mu b^2}{4\pi} \frac{(1 - \beta^2)}{2\beta} \qquad E = \frac{\mu b^2}{4\pi} \frac{1}{\beta}$$

(27)

$$\sigma_{yz} = \frac{\mu b}{2\pi r} \beta \qquad \beta = \left(1 - \frac{v^2}{c^2}\right)^{1/2}$$

where c is the velocity of transverse sound waves. For the virtual displacement δr in Fig. 2a, the use of the preceding formulas shows that

$$\Lambda = \frac{\mu b^2}{4\pi} \beta$$

(28)

and that the repulsive force between the dislocation is indeed given by

$$\sigma_{yz} b = F = \frac{2\Lambda}{r}$$

(29)

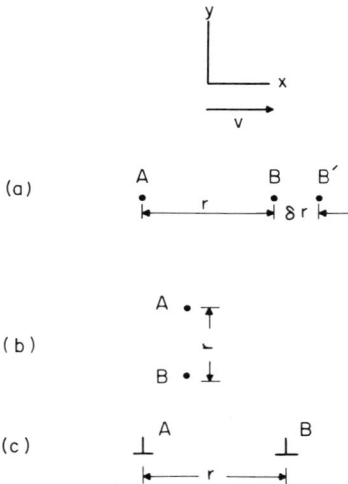

(a)

(b)

(c)

Fig. 2 Three cases of interaction of parallel, straight dislocations moving uniformly with velocity v in the x direction; (a) and (b) two like-sign screws, (c) two like-sign edges.

*It is clear that for the case of two opposite-sign dislocations not treated here, the the sign of the force would change, but all other conclusions would apply.

One might wonder why, on physical grounds, the interaction force is not given by the variation in energy of the system with the virtual displacement, i.e., whether energy is not conserved. It turns out that energy is conserved in the virtual displacement, but only when inductive terms are included in the energy balance.* When the external forces are relaxed to permit the small increase δr in the constant velocity case, the strain energy and kinetic energy decrease, respectively, by amounts δW_E and δW_K, corresponding to a decrease in dislocation mass. Hence, in the spontaneous process occurring without external constraints, to conserve dislocation momentum, a small increase in overall velocity will occur, which is smaller the greater the outer cutoff R, i.e., the heavier the entire configuration [Eqs. (24)–(26)].

An increase of W_K at *constant* velocity implies a momentum increase

$$M = \frac{2}{v} \delta W_K \qquad (30)$$

Hence, if the displacement δr takes place slowly during a time interval t, then to keep the overall velocity strictly constant, a net average external force

$$\delta F = \frac{\delta M}{t} = \frac{2}{vt} \delta W_K \qquad (31)$$

must be applied to balance out the inductive terms. The work extracted by relaxing the forces balancing the interaction forces is δL. Thus, in all, the mechanisms balancing the interaction forces extract an energy

$$\delta L + vt \delta F = \delta W_E + \delta W_K = \delta W \qquad (32)$$

in the process, and the energy account checks.

In the spontaneous process, i.e., with no external mechanism to keep v constant, the overall velocity will tend to increase. However, this is not general; later discussion shows that in some cases, the velocity decreases in the spontaneous process.

No matter how useful it is, the principle of conservation of dislocation momentum does not apply without limitation. Consider

---

* As the present work neared completion, Weertman[6] presented a discussion of the interaction of moving edge dislocations which clearly showed the importance of inductive energy terms.

the example of Fig. 2*b*. The appropriate stress component and energy factor are

$$\sigma_{xz} = \frac{\mu b}{2\pi r} \frac{1}{\beta} \qquad E = \frac{\mu b^2}{4\pi} \frac{1}{\beta} \tag{33}$$

Hence, the repulsive force between the screws is

$$F = \sigma_{xz} b = \frac{2E}{r} = -\frac{\partial W}{\partial r} \tag{34}$$

The force is simply related to the total energy change. Nevertheless, according to the definition of M [Eq. (23)], there is for fixed v a change of momentum in the direction of motion, namely, Eq. (30) again. In the present case the momentum change is *not* associated with a force in the direction of motion; momentum is not conserved in the spontaneous process; the spontaneous process proceeds *without* a change in velocity. Only when the screws are constrained to rectilinear motion, as in the example of Fig. 2*a*, does the principle of conservation of dislocation momentum hold. This limitation upon the principle of conservation of dislocation momentum is associated with the fact that there is no term analogous to the Lorentz force in the dynamic theory of dislocations.[7]

2.3 Edge Dislocation Interactions

For the edge dislocations shown in Fig. 2*c*, the appropriate values of the stress and of the energy factors, as determined[3] directly from the displacement and from Eqs. (16) and (17), and of the Lagrangian factor, are

$$\sigma_{xy} = \frac{\mu b}{2\pi r} \frac{4c^2}{v^2} \left[\frac{(1 + \beta^2)^2}{4} - \alpha \right] = \frac{2\Lambda}{br}$$

$$E = \frac{\mu b^2}{4\pi} \frac{4c^2}{v^2} \left[\frac{1}{\alpha} + 2\alpha - \frac{1}{4}\left(\frac{1}{\beta^3} - \frac{6}{\beta} - 7\beta\right) \right] \tag{35}$$

$$\alpha = \left(1 - \frac{v^2}{c_e^2}\right)^{1/2}$$

where c_e is the velocity of longitudinal sound waves. Hence, the interaction force is

$$F = \frac{\mu b^2}{2\pi r} \frac{4c^2}{v^2} \left[\frac{(1 + \beta^2)^2}{4} - \alpha \right] \tag{36}$$

For small v, the bracketed factor in Eq. (36) is positive, corresponding to a repulsive force; at the Rayleigh velocity v_R ($<c$) it is zero,[3] and for $v > v_R$ it is negative, corresponding to an attractive force.* When $v < v_R$, the behavior is qualitatively similar to that of two parallel screws in the same glide plane; the velocity tends to increase in the spontaneous process. However, when $v > v_R$, the energy would increase in the process taking place at constant velocity; thus, by a development similar to that presented for the screw case, the velocity tends to decrease in the spontaneous, attraction process. Hence the Rayleigh velocity is, in a sense, a *stable* velocity for a like-sign pair; the spontaneous process occurring either above *or* below v_R tends to change v towards v_R.

3   APPLICATIONS AT LOW VELOCITIES

3.1   Accelerations

Consider the applicability of the previous results when acceleration is present. With an acceleration a, Eq. (9) becomes, instead of Eq. (11), the expression

$$\frac{\partial \sigma_{ij}}{\partial x_j} = \rho v^2 \frac{\partial^2 u_i}{\partial x^2} - \rho a \frac{\partial u_i}{\partial x} \tag{37}$$

As an example, we consider the vibrating string model and classify terms as functions of frequency (ω) and amplitude X. In order of magnitude

$$v \sim X\omega \qquad a \sim \omega^2 X \tag{38}$$

With $(\partial u/\partial x) \propto r^{-1}$ and $(\partial^2 u/\partial x^2) \propto r^{-2}$, the acceleration term is seen to be the dominating perturbance in Eq. (37) in the region

$$r > X \tag{39}$$

Thus, in the major part of the moving stress field, the steady-state

---

*In the anisotropic elasticity description, like-sign *screw* dislocations can also attract and coalesce above a critical velocity less than c, whenever the screw has three displacement components.[8]

approximation is inadequate for determining the leading terms in the changes of strains. In the region $X < r \le R$, where

$$R \ll \frac{c}{\omega} \tag{40}$$

and, thus, where the strains follow the dislocation motion, the perturbing term in Eq. (37) is roughly $(\rho \omega^2 Xb/r)$, corresponding to changes

$$\Delta \frac{\partial u}{\partial x} \sim \frac{\omega^2 Xb}{c^2} \qquad \Delta W_E \sim \frac{\mu \omega^4 X^2 b^2 R^2}{c^4} \tag{41}$$

to be compared with a kinetic energy

$$W_K \sim \frac{\mu b^2 \omega^2 X^2}{c^2} \tag{42}$$

Thus

$$\frac{\Delta W_E}{W_K} \sim \left(\frac{R\omega}{c} \right)^2 \tag{43}$$

which shows that ΔW_E is negligible when condition (40) is fulfilled. In the same region, W_K can be calculated to the order $X^2 \omega^2$ from the static displacements as a function of time.

In particular, for the problem of a vibrating dislocation segment of length λ, a model prominent in internal friction theories,[9] one can determine the potential and kinetic energy of the vibrating system from "quasi-static" displacements to the leading terms for small ω, provided that condition (40), with $\lambda \sim R$ is fulfilled. The region $r \gg \lambda$ can be neglected in such a case because the displacements in that region will be much smaller than in the case of an infinite straight dislocation.

The above analysis suffices for a first-order discussion of the potential energy (restoring force) and the kinetic energy (inertia) of a vibrating segment pinned at its ends. A more detailed analysis of the effect of acceleration would include radiation effects and the associated damping of the system, but these effects are higher order in ω and can be ignored compared to the restoring force and inertial term at low frequencies. Radiation forces, important at higher frequencies, are discussed elsewhere for the cases of a vibrating screw segment[10] and a vibrating kink.[11, 12]

3.2   Line Tension

Having discussed the validity of the low-velocity approximation for the vibrating dislocation segment, let us now examine the applicability of the vibrating string model (involving explicitly such concepts as local mass per unit length and line tension) to such a problem. Lothe[13] found recently that energy is transported along a dislocation line while it is in motion. In principle, this phenomenon could lead to changes in the parameters of the vibrating string model, but, as shown in the following, no such changes are necessary.

Lothe shows that to third order in v, there is an energy flow

$$J = -v \frac{\partial W^\circ}{\partial \phi} + Cv^3 \tag{44}$$

along a moving dislocation line, where ϕ is the inclination of the line with respect to a base line in the glide plane. A term in v^2 in J is not possible; J is antisymmetric in v (recall that dynamic changes in strain are second order in v).

Consider the element δl in a dislocation with radius of curvature R and moving with velocity v (Fig. 3). The minimum radius of curvature is related to the amplitude of vibration X, in the small amplitude limit, by

$$R = \frac{\lambda^2}{8X} \tag{45}$$

As the segment dl advances, the following terms enter the energy

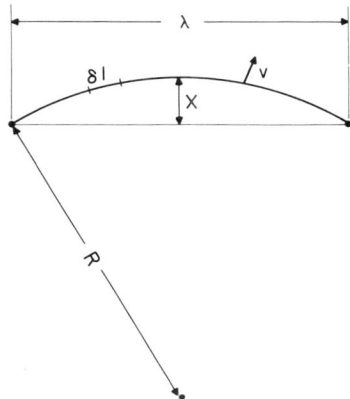

Fig. 3 Oscillating dislocation segment of length λ, pinned at both ends. Segment is shown bowed to a radius curvature R, with amplitude X, and moving with velocity v.

balance: Energy is supplied by the applied stress at a rate

$$\sigma b v \delta l \sim \sigma b X \omega \delta l$$

The strain energy increases at a rate

$$W^\circ \frac{d}{dt}(\delta l) = W^\circ \frac{v}{R} \delta l \sim \frac{W^\circ X^2 \omega}{\lambda^2} \delta l$$

This is a line-tension term; the approximations involved are the same as those in static problems.[14] The kinetic energy increases at a rate

$$\frac{d}{dt}\left(\frac{1}{2} m v^2\right) \delta l = m v \frac{dv}{dt} \delta l \sim m X^2 \omega^3 \delta l$$

The change in W_K with change in length is

$$\frac{1}{2} m v^2 \frac{d}{dt}(\delta l) \sim \frac{m X^4 \omega^3}{\lambda^2} \delta l$$

The energy flow into the element, to first order, is

$$-\frac{\partial J}{\partial l} \delta l = \frac{1}{R} \frac{\partial J}{\partial \phi} \delta l = -\frac{v}{R} \frac{\partial^2 W^\circ}{\partial \phi^2} \delta l \sim \frac{X^2 \omega}{\lambda^2} \frac{\partial^2 W^\circ}{\partial \phi^2} \delta l$$

The third-order term in Eq. (44) would contribute a term $(X^4 \omega^3 \delta l/\lambda^2)(\partial C/\partial \phi)$, which completes the energy balance. Retaining terms up to the order $X^2 \omega^3$, one finds that the energy sum corresponds to an equation of motion,

$$\sigma b - \frac{S}{R} = m \frac{dv}{dt} \tag{46}$$

where S is given by the usual deWit-Koehler expression[15] for the line tension

$$S = W^\circ + \frac{\partial^2 W^\circ}{\partial \phi^2} \tag{47}$$

Equation (46) is the precise mathematical statement of the vibrating string model. The approximations are the usual line-tension ones; the derivation of a local inertia from W_K is consistent with the line-tension approximation of the restoring force. Radiation damping is a higher-order effect in ω. The energy flow is taken into account in S; no new dynamic effects appear in the low-velocity approximation

4 HIGH-VELOCITY APPLICATIONS

4.1 Dissociation of Partials

At high velocities, it is anticipated that the Weertman effect, i.e., that like-sign edges attract above v_R, will have interesting consequences with respect to the extension of dislocations and cross slip. Previous work[16] on the moving screw dislocation in the Peierls-Nabarro model has shown that the core contracts with increasing velocity. Hence, in this model, profuse cross-slip is expected as v approaches c. However, when the screw dissociates into partials bounding a stacking fault, edge components appear, in general, in the partials and the behavior is more complex. As examples, we consider the partial extension of screws and edges in b.c.c. and f.c.c. crystals as a function of v. Only the dynamic effects of Sec. 2 are included. The effects of friction forces are neglected, and, consistently, so are forces produced by the applied stress. Both terms could significantly alter the results. Unfortunately, little is known about the friction forces in the interesting range of velocities. As a first approximation, the behavior in the absence of friction is of interest.

The dissociation for the f.c.c. lattice is that of a $\frac{1}{2} <110>$ dislocation into $\frac{1}{6} <112>$ partials bounding an intrinsic stacking fault. That for the b.c.c. lattice is a $\frac{1}{2} <111>$ dislocation dissociating into $\frac{1}{6} <111>$ and $\frac{1}{3} <111>$ partials bounding an intrinsic fault. At the equilibrium separation d, the interaction force balances the force per unit length γ associated with the stacking fault; γ is the stacking fault energy. The results in terms of the Lagrangians Λ_s for the screw [Eq. (28)], and Λ_e for the edge [Eq. (35)], are for a b.c.c. screw dislocation

$$d = \frac{4}{9\gamma} \Lambda_s \tag{48}$$

for a b.c.c. edge dislocation

$$d = \frac{4}{9\gamma} \Lambda_e \tag{49}$$

for a f.c.c. screw dislocation

$$d = \frac{1}{2\gamma} \left(\Lambda_s - \frac{1}{3} \Lambda_e \right) = \frac{1}{2\gamma} \Lambda_A \tag{50}$$

and for a f.c.c. edge dislocation

$$d = \frac{1}{2\gamma} \left(\Lambda_e - \frac{1}{3} \Lambda_s \right) = \frac{1}{2\gamma} \Lambda_B \tag{51}$$

The values of the various Λ factors are given in Table 1 as a function of v for the case where Poisson's ration $\nu = 0.30$, corresponding to $v_R = 0.925\,c$.

TABLE 1   The Λ Factors for Eqs. (48) to (51) as a Function of (v/c) and of $(v^2/c^2) = 1 - \beta^2$. The Factors Are Given in Units of $\mu b^2/4\pi$ for the Case $\nu = 0.30$. Here, b is the Modulus of the Burgers Vector of the Perfect Dislocation.

(v/c)	(v^2/c^2)	Λ_e	Λ_s	Λ_A	Λ_B
0	0	1.4286	1	0.52380	1.0953
0.010	10^{-4}	1.4285	0.99996	0.52379	1.0952
0.100	0.01	1.405	0.995	0.522	1.088
0.316	0.1	1.371	0.949	0.492	1.055
0.447	0.2	1.324	0.894	0.453	1.026
0.548	0.3	1.235	0.837	0.425	0.956
0.632	0.4	1.149	0.775	0.392	0.891
0.707	0.5	1.042	0.707	0.360	0.807
0.775	0.6	0.868	0.633	0.343	0.657
0.837	0.7	0.703	0.548	0.313	0.521
0.894	0.8	0.367	0.447	0.325	0.218
0.916	0.841	0.142	0.399	0.351	0.000
0.925	0.856	0	0.380	0.380	-0.127
0.949	0.9	-0.140	0.316	0.456	-0.245
0.975	0.95	-0.153	0.224	1.376	-1.227

The results show that the edge dislocation constricts completely in both lattices, while the screw does not completely constrict in either case. The value of d for the screw decreases monotonically with v for the b.c.c. case but goes through a minimum at $v = 0.8\,c$ for the f.c.c. case. The minimum value of d in the interesting f.c.c. case is 59% of the rest value. Hence, dynamic effects should not markedly increase the tendency for cross-slip in low stacking fault energy crystals in our approximation. We emphasize that frictional forces, neglected here, could lead to a change in the above conclusion.

Above v_R, other effects are possible; these are discussed in the next section.

4.2 Dislocation Nucleation

We suggest that above v_R edge dislocations become unstable and that profuse dislocation multiplication sets in by a mechanism related to the Weertman mechanism for coalescence. As indicated by Eq. (36), and as first noted by Weertman,[3] two like-sign edge dislocations moving at a velocity $v > v_R$ tend to coalesce to form a superdislocation, which, in turn, can attract a third dislocation, etc. (Fig. 4a). Now consider a single dislocation made up of two dislocations of opposite sign—one a superdislocation of Burgers vector 2b and one of Burgers vector b (Fig. 4b). At $v > v_R$, the dislocations *repel*, which indicates that a single dislocation is unstable with respect to spontaneous dissociation. As explained in Sec. 2, the velocity tends to decrease as the reaction proceeds. However, an applied stress could keep the superdislocation at a velocity above v_R; also such a stress would aid in tearing the dislocations apart. The superdislocation could then dissociate into a pair 3b and – b, etc. In principle, a sufficiently large superdislocation to comprise an incipient crack could form in this manner.

The above possibility is not limited to edge dislocations. Consider the f.c.c. 60° dislocation in Fig. 5, represented in

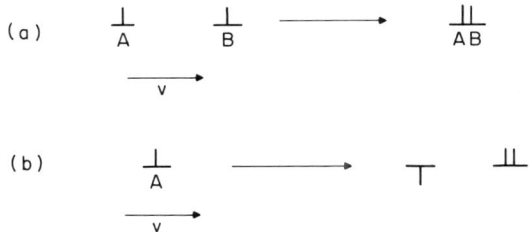

(a) ⊥   ⊥   ──────→   ⫠
     A   B   AB

     ──→
      v

(b) ⊥   ──────→   ⊤   ⫠
     A

     ──→
      v

Fig. 4 (a) Coalescence of two like-sign edges moving with velocity $v > v_R$. (b) Reaction of an edge dislocation with Burgers vector b to form a pair – b and 2b moving at velocity $v > v_R$

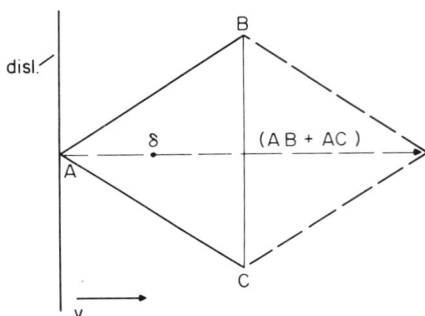

Fig. 5 An fcc dislocation moving at velocity v on glide plane (d).[17] The orientation of several Burgers vectors on this glide plane is also depicted.

Thompson's notation.[17] Dislocations with Burgers vectors AC or AB are completely contracted at v_R because the partial Aδ is pure edge and interacts only with the edge component of the other partial. A possible reaction at $v > v_R$ is

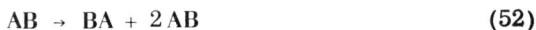

$$AB \rightarrow BA + 2AB \tag{52}$$

Perhaps a more likely dissociation is

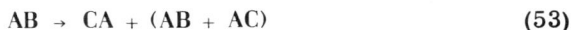

$$AB \rightarrow CA + (AB + AC) \tag{53}$$

because the superdislocation (AB + AC) is pure edge and interacts only with the edge component of CA. Also, the screw dislocation BC might dissociate by the reaction

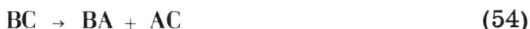

$$BC \rightarrow BA + AC \tag{54}$$

Above v_R both the edge and screw components in BA and AC repel; hence this reaction could even occur slightly below v_R. However, since the screw partials do not contract completely, the activation barrier for reaction (54) could be appreciable.

It may be asked why, by the same reasoning, an opposite sign pair of edge dislocations moving at $v > v_R$ cannot nucleate spontaneously within a region of perfect crystal. The answer is that the latter configuration does not have enough inertia to prevent the velocity from dropping immediately to a low value $v < v_R$, whereupon the pair would annihilate. Only when the initial configuration has a considerable amount of energy and momentum stored in the long-range strain field, such as for an edge dislocation moving at $v > v_R$, can the required energy and momentum be transferred in the nucleation process without severe retardation, leading to annihilation.

When the dislocations slow to below v_R, as they eventually must, the reverse reactions of the above will occur. However, similar to the case of work hardening, complete reversibility is not expected. Dislocation reactions and forest dislocations in secondary slip systems would cause the retention of high dislocation densities. Also a high stacking fault density could be retained if many faults were produced by screw dissociation. Finally, some of the superdislocations could undergo partial reactions which could stabilize them. For example, the superdislocation (AB + AC) can dissociate into three partials Aδ bounding adjacent ribbons of intrinsic and extrinsic stacking fault (Fig. 6a). This configuration, which has been observed experimentally and discussed by Gallagher,[18] is stable.[19]

As a final example of such effects consider a partial δC bounding an intrinsic fault and moving rapidly under the force γ (Fig. 6b). If it attains a velocity $v > v_R$, it can dissociate by the reaction

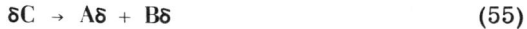

$$\delta C \rightarrow A\delta + B\delta \tag{55}$$

and form extrinsic fault at the edge of the receding fault configuration.

In all of the above, friction is neglected. Apart from the problem of whether near-sonic velocities are attainable in the presence of friction forces, the friction forces, including, generally, both Peierls forces and radiation forces, can depend on the direction of motion and on the character of the dislocation. One partial in an

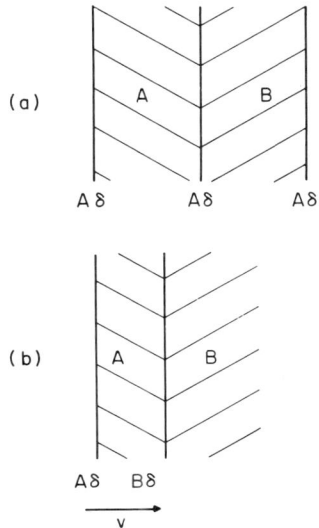

Fig. 6 Extended configurations on glide plane (d) in a fcc crystal. B and A represent, respectively, intrinsic and extrinsic stacking faults. (a) Rest equilibrium configuration.[18] (b) Configuration at the edge of a contracting fault.

extended pair could experience a different frictional force than the other, for example, producing a different dynamic extension than that predicted in Table 1. At high velocities and high stresses, such as those experienced in shock loading, such effects could be important. In b.c.c. crystals, an asymmetry in yield stress is observed for slip on $(11\bar{2})$ planes between slip in the [111] and [$\bar{1}11$] directions;[20] also an asymmetry exists in the direction of cross-slip from {110} toward {112}.[21,22] Such asymmetries are expected to become more pronounced at high velocities. Experimental results on asymmetries at high velocity would be valuable because the asymmetric higher-order terms of the applied stress in the dislocation velocity, σ^2, σ^4, etc., would be more pronounced (odd powers of stress are asymmetric terms for v as a function of σ). At low stresses, where v is proportional to σ, asymmetries do not appear.

4.3   Jogged Dislocations

An additional effect which can influence the attainment of sonic velocities is drag by sessile jogs, which can move forward only by point-defect production. For dissociated dislocations not only are screw jogs sessile, as for undissociated dislocations, but edge jogs are sessile also.[23] The extrinsic frictional effects are better understood than the intrinsic effects discussed above.

As a dislocation acquires sessile jogs, the stress must be high enough that the defects form without thermal activation if the velocity is to be maintained. If the energy per unit length of the defect trails is W_f, the applied stress must satisfy the inequality

$$\sigma > \frac{W_f}{b\lambda} \tag{56}$$

to maintain fast motion. If there is a barrier $W_f^* > W_f$ which must be surmounted in defect formation, and if the excess energy is not transferred back to the dislocation as it decends from the top of the barrier but is radiated away, condition (56) must be replaced by the more restrictive condition

$$\sigma > \frac{W_f^*}{b\lambda} \tag{57}$$

Typically, $W_f \sim \mu b^2/3$ and $(W_f^* - W_f)$ is of the order of the migration energy of a unit line of vacancy $\sim \mu b^2/5$. Thus Eq. (57) gives

$$\sigma \gtrsim \frac{\mu b}{2\lambda} \tag{58}$$

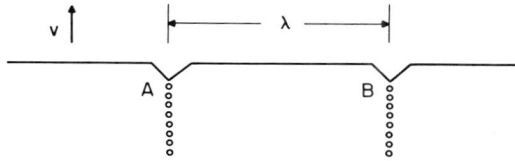

Fig. 7 A dislocation moving with velocity v and dragging sessile jogs A and B, separated by a distance λ, with the accompanying formation of point defect trails.

When the stress satisfies condition (58), the dislocation can accelerate from rest. As the velocity becomes high, the dislocation configuration is expected to be that of Fig. 7. The cusps at the jogs are expected to decrease in width and depth with increasing velocity, but to maintain an angle at the cusp so that the drag force is balanced by a line tension force.

Suppose that a dislocation moves a distance $x = 10^4 b$ before it acquires jogs at a spacing $\lambda \sim 10^3 b$. Then with an instantaneous stress $\sigma \sim 10^{-3}\mu$ applied, an energy $W \sim 10\mu b^2$ is supplied before the dislocation begins to retard. This is large compared to either the rest energy of a screw, $W^\circ \sim \mu b^2$, or its energy at v_R, $W \sim 2.5\mu b^2$. Thus, it appears that under special conditions of shock loading, dislocations could accelerate to near-sonic velocities before retarding effects are dominant.

4.4 Stability of a Straight Dislocation

A final question of fundamental importance is whether a straight dislocation moving at high velocity is unstable with respect to a zig-zag configuration. In other words, can the dynamic line tension become negative? In the static case the stability condition for a straight dislocation, analogous to similar criteria for the stability of flat surfaces, is that the line tension, Eq. (47), be positive [24]

$$\left(W^\circ + \frac{\partial^2 W^\circ}{\partial \phi^2}\right)_{\phi=0} > 0 \tag{59}$$

where ϕ is now measured from the initial straight configuration.

We postulate that the criterion for stability of a straight moving dislocation is that

$$\left(L + \frac{\partial^2 L}{\partial \phi^2}\right)_{\phi=0} > 0 \tag{60}$$

Below v_R, Λ increases with decreasing v (Table 1), contrary to the behavior of E. Thus, when wiggling up occurs in this velocity range, the Lagrangian per unit length increases. The dislocation would then be stable against wiggling, the dynamic effects actually tending to stabilize the dislocation. For $v > v_R$, in this model, the situation differs. The pure edge would still be stable. However, a pure screw, wiggling so that edge components arise, would lower its Lagrangian because of the large negative Lagrangian for the edge component; the screw would be unstable.

At present, criterion (60) is only an interesting suggestion. Further theoretical work is underway in an attempt to verify the hypothesis.[25] If criterion (60) is valid, important consequences then follow. Uniform motion in the range of v_R would then be impossible, and none of the particular effects possible in that region, as discussed above, could occur.

The most direct approach to the problem is to calculate directly the stress on the branches in a uniformly moving bend. For this, methods similar to those used in the calculations for static bends[26,27] are used. Nabarro[28] has developed general formulas for the displacements and stresses of arbitrarily moving dislocations; starting with his formulas and with the explicitly known Green's functions for isotropic solids, a solution should be possible.

5 CONCLUSION

1. A general, simple proof is presented, relating forces on moving dislocations to a Lagrangian function.

2. Dislocation velocity is shown to increase in spontaneous processes except for attracting edge or mixed dislocations moving with velocity greater than the Rayleigh velocity.

3. The line tension term in the equation of motion of a dislocation line is adequately given by the deWit–Koehler expression at low velocities.

4. Moving screw dislocations do not constrict completely at high velocities; thus the constriction energy must still be supplied before cross-slip can occur.

5. Dislocations nucleate spontaneously at moving edge dislocations when their velocity exceeds the Rayleigh velocity.

6. A criterion is postulated for the stability of moving straight dislocations; it resembles the deWit–Koehler expression, but with the Lagrangian L replacing the rest energy W°.

ACKNOWLEDGMENT

This research was supported in part by the Directorate of Materials and Processes of the U. S. Air Force Systems Command.

REFERENCES

1. Eshelby, J. D.: *Phys. Rev.* **90**:248 (1953).
2. Nabarro, F. R. N.: *Proc. Roy. Soc.* A209:278 (1951).
3. Weertman, J.: in "Response of Metals to High Velocity Deformation," p. 205, Interscience Publishers, New York, 1961.
4. Eshelby, J. D.: *Solid State Phys.* 3:105 (1956); see also the general Lagrangian formulation of A. N. Stroh: *J. Math. Phys.* 41:77 (1962).
5. Bullough, R., and A. J. E. Foreman: *Phil. Mag.* 9:315 (1964).
6. Weertman, J.: *J. Appl. Phys.* 37:4925 (1966).
7. Lothe, J.: *Phys. Rev.* **122**:78 (1961).
8. Teutonico, L. J.: *Phys. Rev.* **127**:413 (1962).
9. Granato, A., and K. Lücke: *J. Appl. Phys.* **27**:583, 789 (1956); see also the reviews in *Acta Met.* No. 4, 10 (1962).
10. Eshelby, J. D.: *Proc. Roy. Soc.* A197:396 (1949).
11. Eshelby, J. D.: *Proc. Roy. Soc.* A266:222 (1962).
12. Lothe, J.: *J. Appl. Phys.* **33**:2116 (1962).
13. Lothe, J.: *Phil. Mag.* 15:9 (1967).
14. Hirth, J. P., T. Jøssang, and J. Lothe: *J. Appl. Phys.* 37:110 (1966).
15. deWit, G., and J. S. Koehler: *Phys. Rev.* 116:1113 (1959).
16. Leibfried, G., and H. D. Dietze: *Z. Phys.* 126:790 (1949).
17. Thompson, N.: *Proc. Phys. Soc.* London B66:481 (1953).
18. Gallagher, P. C. J.: *Phys. Stat. Sol.* 16:95 (1966).
19. Hirth, J. P.: *J. Appl. Phys.* 32:700 (1961).
20. Taoka, T., S. Takeuchi, and E. Furubayashi: *J. Phys. Soc.* Japan 19:701 (1964).
21. Hook, R. E., and J. P. Hirth: *Acta Met.* 15:535 (1967).
22. Hirth, J. P., and J. Lothe: *Phys. Stat. Sol.* 15:487 (1966).
23. Hirth, J. P., and J. Lothe: *Can. J. Phys.* 45:809 (1967).
24. Mullins, W. W.: ASM Seminar "Metal Surfaces," p. 17, ASM, Cleveland, 1963.
25. Lothe, J.: To be published.
26. Lothe, J.: *Phil. Mag.* 15:353 (1967).
27. Brown, L. M.: *Phil. Mag.* 15:363 (1967).
28. Nabarro, F. R. N.: *Phil. Mag.* 42:1224 (1951).

DISCUSSION *on Paper Presented by J. Lothe*

J. WEERTMAN: Do you feel your theory disproves the existence of a Lorentz force?

J. LOTHE: The proof that the Lorentz force is nonexistent for dislocations in the sense that it cannot influence dislocation motion in the same manner as forces produced by applied stresses is given in papers by Lothe and by Stroh cited in our paper. Realizing the above point, one can handle the problem of the Lorentz force in two ways. First, one can always set it equal to zero. This has the advantage of leaving only forces which can cause dislocations to move in the force balance, but has the inconvenience that dislocation quasi-momentum is not converved in processes such as that in Fig. 2b of our paper. Second, one can include a Lorentz force with the consequence that quasi-momentum is conserved, but this has the disadvantage that one must exclude the Lorentz force from the forces producing dislocation motion. The first method is adopted in our paper.

In the above context, our results are consistent with the absence of a Lorentz force; no forces other than those we include are required in the force balance for Fig. 2b.

R. BULLOUGH: Is there any evidence that it is possible for a dislocation to reach the Rayleigh velocity? Does the fact that the Peierls model has zero width at such a velocity preclude higher velocities?

J. LOTHE: In answer to the first question, we can only say that present experimental evidence is ambiguous. There does not seem to be any evidence that would preclude sonic velocities at low temperatures. A crucial point in deciding this question involves the magnitude of the damping constant in the Rayleigh velocity region (see Agenda Discussion—High Speed Dislocations).

The zero width of the Peierls model at the Rayleigh velocity indicates that the Peierls model breaks down there. The continuum model also indicates that the region of compression, associated with the extra plane of atoms, contracts to zero width at the Rayleigh velocity and becomes a region of dilatation at higher velocities. The rate of change of displacements with velocity is continuous through the Rayleigh velocity region.

LINEAR ARRAYS OF MOVING DISLOCATIONS EMITTED BY A SOURCE

A. R. Rosenfield and G. T. Hahn

Metal Science Group,
Battelle Memorial Institute,
Columbus Laboratories,
Columbus, Ohio

ABSTRACT

The positions of straight coplanar dislocations emitted by a single source were calculated as a function of time. This was accomplished by combining expressions for the stress sensitivity of dislocation velocity with expressions for the stress exerted on an individual dislocation by all the other dislocations emitted and with an approximate source-activation criterion. The resulting equations were evaluated on a computer for ten to twenty dislocations, and the computer solutions were generalized to large numbers of dislocations by applying the continuous approximation.

These calculations indicate that the dislocations behave cooperatively, arraying themselves in a manner resembling a pile-up against the source. The lead dislocation quickly attains a steady-state velocity and the actual stress on it is found to be up to twice as large as the applied stress, depending on the source-activation stress and whether the dislocations produced are screw or edge. As a result, the steady-state velocity of the lead dislocation can be several orders of magnitude greater than that of an isolated dislocation under the same applied stress.

These results are shown to be consistent with data in the literature. In addition, the problems of evaluation of activation volume and of the ratio of edge to screw velocities are considered. In both cases the behavior of dislocations in arrays is shown to be markedly different from that of isolated dislocations. It is suggested that neglecting the interactions which occur in arrays can result in overestimates by a factor of two in activation volume and by one or two orders of magnitude in velocity ratio.

1   INTRODUCTION

A linear (or coplanar) array of dislocations is an appealing and useful idealization of a slip band. The linear dislocation pile-up has almost become synonymous with the blocked slip band even though real bands usually involve dislocations on a number of parallel planes. Although this complication has yet to be dealt with, it does not invalidate linear models. On the contrary, the success of the Bilby–Cottrell–Swinden linear dislocation model[1] in predicting real crack-tip displacement[2] and plastic zone-size[3] values is evidence of a close analogy between linear arrays and slip bands.

This paper considers arrays of long, straight, parallel coplanar dislocations emitted from a source. Two types of dislocation source are generally recognized: dislocation mills, and structural heterogeneities. The prototype of the dislocation mill is the Frank–Read source[4]—a mechanism by which a single dislocation loop can multiply itself into many loops by its own movement. Somewhat more obscure is the mechanism of generation at heterogeneities, although dislocations have been observed to extend from particles,[5] grain boundaries,[6] cracks,[2] scratches,[7] and hardness impressions.[8] The total number of dislocations produced by a Frank–Read source has been calculated by Fisher *et al.*[9] and Vreeland *et al.*,[10] while Worthington and Smith[6] and Weertman[11] have independently proposed that the BCS model provides a good analogy to a dislocation source at a heterogeneity. However, all four of these are essentially static calculations. We[12] have made a dynamic, nondislocation calculation of plastic relaxation in front of a moving crack, but it is not in a form to be a useful source calculation. In particular, the relation between crack length and source length is not at all clear.

The calculations to be described deal with the positions of moving glide dislocations emitted by a source along a single slip plane. This problem was first formulated by Weertman[13] who solved it, calculating the spacing of a limited number of dislocations as a function of time. Mendelson[14] simplified the problem by assuming that mobile dislocations were evenly spaced; in this way he estimated the stress on the leading dislocation. Later, Weertman[15] modified the BCS model so that the friction stress opposing dislocation motion decreases with a local increase of the product of dislocation density and velocity. By doing so, he was able to describe the Portevin-Le Chatelier effect. Mura[16, 17] has derived an expression for the stress at a point resulting from an arbitrary distribution of dislocations moving with arbitrary velocities. However, his general form is too complex to allow for quantitative predictions. Certain deductions (for example, the correctness of the Von Mises yield criterion) can be made by assuming that the velocity of each dislocation is directly proportional to the force on it.[18] Unfortunately, Mura's treatment does not offer simple numerical results, and the linear velocity-stress assumption is valid only at velocities on the order of 0.01-0.1 of the speed of sound.[19] The present calculations employ an exponential velocity-stress relation which approximates the behavior of dislocations under ordinary testing conditions.

2   PROCEDURE

Each individual dislocation in a linear array is assumed to obey an exponential velocity (v) — stress (τ) relation, which, for convenience, is taken to be:

$$v_i = A \exp(B\tau_{\text{eff}}) \tag{1}$$

where A and B are constants characteristic of the material, i is the index number of the dislocation, and τ_{eff} is the effective stress on it. It should be mentioned that some question exists concerning the proper functional relation between stress and dislocation velocity.[20] It will be shown that the calculations outlined here are valid provided the actual τ-v relation can be fitted by a semi-logarithmic relation over a two-order-of-magnitude range in velocity. Now

$$\tau_{\text{eff}} = \tau_a - \tau_f + \tau_{dj} \tag{2}$$

where τ_a is the applied stress, τ_f represents any long-range back stress, and τ_{dj} is the stress on a glide dislocation due to other dislocations in the array and due to the source. Calculations were carried out for low-speed screw dislocations so that

$$\tau_{dj} = \frac{Gb}{2\pi} \sum_{\substack{j=0 \\ i \neq j}}^{N} \left(\frac{1}{y_j - y_i} \right) = \frac{Gb}{2\pi} \eta_i \tag{3}$$

where G is the shear modulus, b is the Burgers vector, and the index $j = 0$ represents the source. The extension of the calculation to edge dislocations is discussed below. Combining Eqs. (1)–(3) and neglecting a back stress contribution, that is, $\tau_f = 0$:

$$v_i = A \exp(B\tau_a) \exp\left(\frac{BGb\eta_i}{2\pi} \right) \tag{4}$$

or
$$v_i = \frac{dy_i}{dt} = v_{\tau a} \exp(\beta \eta_i) \tag{5}$$

where $v_{\tau a}$ is the velocity of an isolated dislocation under applied stress, τ_a, and $\beta = BGb/2\pi$.

Equation (5) is actually a set of simultaneous differential equations which has been programmed for the CDC 3400 computer at Battelle-Columbus. Given the positions of all dislocations at some arbitrary time t, a time increment δt was selected. Values of δy for each dislocation were then calculated by the Runge-Kutta method.[21] If crossover occurred (that is, if a trailing dislocation was found to have passed one or more ahead of it), the value of δt was automatically decreased.

Some problem was encountered in simulating a source. Initially, the stress at $y = 0$ was monitored, and whenever, at the end of an iteration, $\tau_{\text{eff}}(y = 0) \geq \tau_s$, the postulated source activation stress, a new mobile dislocation was introduced at $y = 0$. This was found to be too sudden an introduction of a new dislocation, too close to the other mobile dislocations. The result was that unrealistic velocities and much crossover were encountered. In addition, this procedure was physically unreasonable. Consider, for example, the operation of a Frank-Read source. The source dislocation bows out under the influence of the applied stress, the distance it bows out being limited by its line tension and the repulsive fields of dislocations previously emitted. In turn, the source dislocation repels these dislocations, forcing them to expand. This allows the source dislocation to bow out further. Eventually, it frees itself from the source and simultaneously a new source dislocation is formed. To approximate this situation, the program described above was modified so that the existence of a "source dislocation" was postulated. At the end of each iteration, the position (y_s) where $\tau_{\text{eff}} = \tau_s$ was determined. Whenever $y_s \geq 0$ the source dislocation was allowed to become a mobile dislocation, the new position y_s was calculated, and a new "source dislocation" was introduced. Hence the number of dislocations could be continually increased. However, even in the modified program, the first dislocation was set in motion at $t = 0$, in accordance with the calculations of Campbell et al.[22] that the time required to initiate source operation $\lesssim 10^{-9}$ sec.

Values of B were taken which were representative of body-centered cubic metals below 0.25 of the melting point $(BG \sim 10^3\text{-}10^4)$,[7,23,24,25] and semiconductors $(B \sim 10^2\text{-}10^3)$.[26,27,28]. Higher values of $BG \sim 10^5$ [19,20] are found for alkali halides. A constant applied stress was employed and $v_{\tau a}$ was set equal to 10^{-4} cm/sec for all cases.

3 RESULTS

Calculations were carried out for a variety of values of stress sensitivity of dislocation velocity, source activation stress, and (constant) applied stress. A typical result is shown in Fig. 1, which plots the distance moved by some of the individual dislocations as a function of time. The most important characteristic of these curves is the constant velocity (v_1) attained by the lead dislocation shortly after the source began to operate. This velocity was 8.1 times the velocity of an isolated dislocation under the same applied stress $(v_{\tau a} = 1 \times 10^{-4}$ cm/sec). However, with other combinations of the chosen parameters, values of $v_1/v_{\tau a}$ in excess of 200 were calculated. The curves display a transient period during which the lead dislocation has a somewhat enhanced velocity. This may be a spurious feature of the calculations and assumptions, but it is interesting to note that both Stein and Low[7] and Schadler[23] noted a

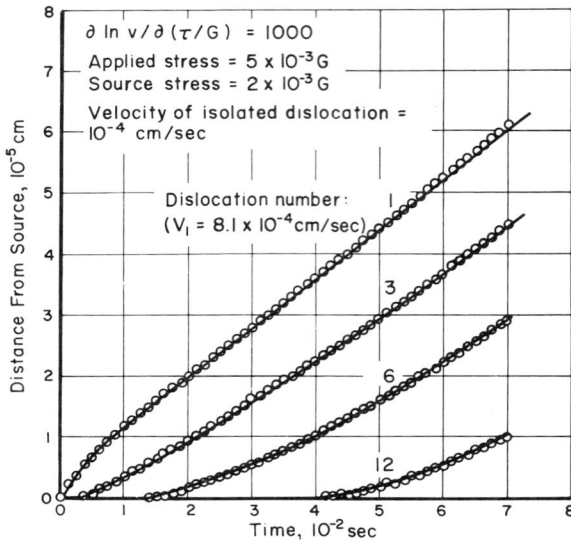

Fig. 1 Movement of some individual dislocations emitted by a source. A total of 20 dislocations was generated during the calculation.

similar effect experimentally. All subsequent dislocations in the array began with zero velocity and accelerated to a value approaching that of the lead dislocation. This difference in behavior can be ascribed to the back-stress on the trailing dislocations which diminishes as they move out from the source.

The variation of the velocities of individual dislocations with time for another set of conditions is shown in Fig. 2a. Here, the differences in velocity are emphasized, and it is seen that there is actually somewhat of a velocity gradient. However, two-thirds of the dislocations move at speeds within a factor of two of that of the lead dislocation. If the effective stress (τ_{eff}) on each individual dislocation is calculated from its velocity $(v_i = A \exp[B\tau_{eff}])$, the stress profile along the slip line can be evaluated and this is given in Fig. 2b, where it is found that τ_{eff} is relatively constant. The positions of the individual dislocations are given by

$$\tau_a + \frac{Gb}{2\pi} \sum_{\substack{j=0 \\ i \neq j}}^{N} \left(\frac{1}{y_i - y_j} \right) = \tau_{eff} \tag{6}$$

Alternatively, if the continuous approximation is used

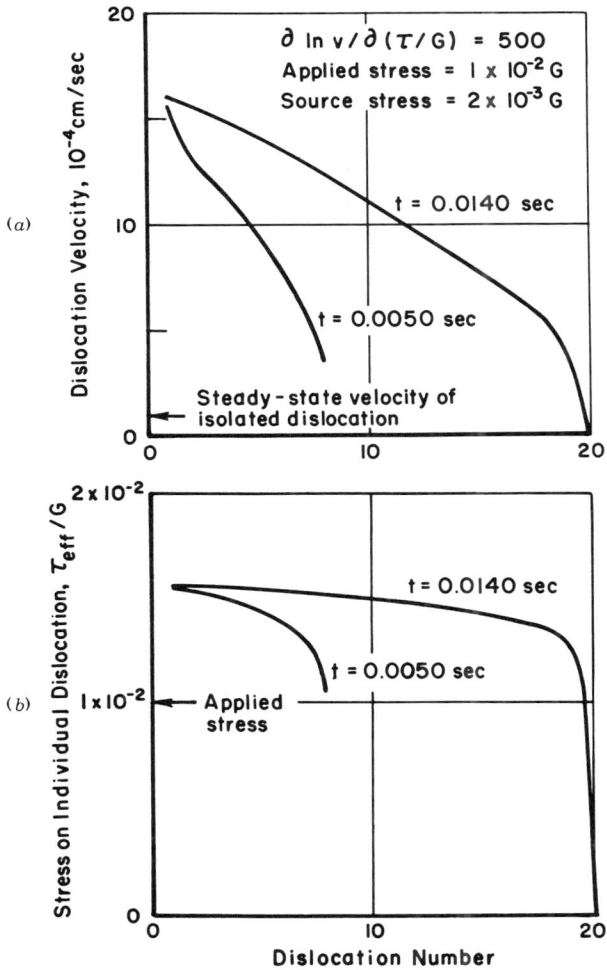

Fig. 2 Velocity of (top) and effective stress on (bottom) individual dislocations.

$$-\tau_{\text{eff}} + \tau_a + \frac{G}{2\pi} \int_0^{L(t)} \frac{f(y')\,dy'}{y - y'} = 0 \qquad (7)$$

where the distribution function $f(y)$ is defined in the usual way.[29] Since τ_{eff} has been shown to be roughly constant and greater than τ_a, our results suggest that the dislocation configuration in a slip band is approximately given by that of a pile-up against the source under constant stress. The upper limit of the integral was denoted by $L(t)$ to emphasize its time dependence.

The quantity $(\tau_{eff} - \tau_a) = \Delta\tau$ can be evaluated from the number of dislocations produced by the source, as well as from the velocity of the lead dislocation (v_1). To do so the number of dislocations emitted was plotted against the length of the slip line. The result was in all cases, a straight line (Fig. 3) whose slope is

$$\frac{dN}{dL(t)} = \frac{\pi\,\Delta\tau}{Gb} \tag{8}$$

The two estimates of $\Delta\tau/G$ generally differ somewhat (usually $\sim 10\%$); that obtained from Eq. (8) is always higher, as is demonstrated in Table 1. Coupled with the small variation of τ_{eff} along the slip line, this will introduce a slight error in the final result. Another small source of error arises from assuming that the velocity of the lead dislocation is equal to its steady-state value from the onset of source operation.

Taking the above results into consideration, the configuration of dislocations emitted by a source can be given approximately as

$$f(y, t) = \frac{2\,\Delta\tau}{G} \cot\theta \tag{9}$$

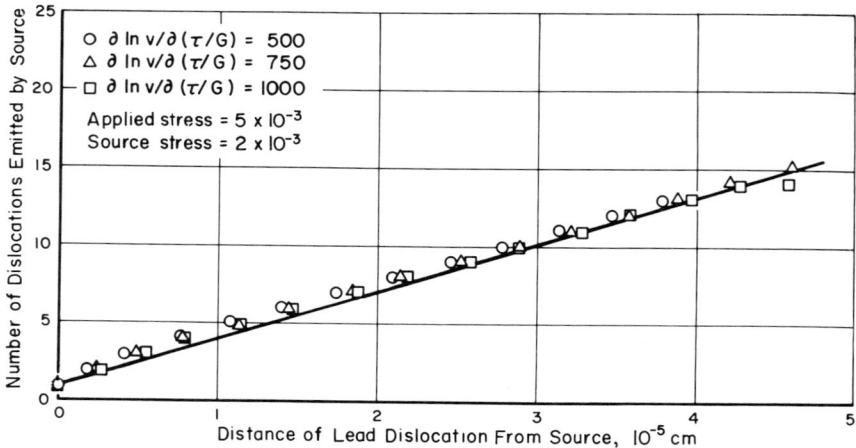

Fig. 3 Variation of the number of dislocations emitted by a source with distance traveled by the lead dislocation.

TABLE 1 Effective Stresses for Representation of Dynamic Dislocation Sources

$\dfrac{\tau_a}{G}$	$\dfrac{\tau_s}{G}$	$\Delta\tau/G$	
		From velocity of lead dislocation, v_1	From number of dislocations emitted
1×10^{-2}	0	6.1×10^{-3}	6.8×10^{-3}
	2×10^{-3}	5.2×10^{-3}	6.1×10^{-3}
	4×10^{-3}	4.0×10^{-3}	4.5×10^{-3}
5×10^{-3}	0	2.9×10^{-3}	3.9×10^{-3}
	2×10^{-3}	2.2×10^{-3}	2.4×10^{-3}
1×10^{-3}	0	7.2×10^{-4}	9.5×10^{-4}

where

or

$$\sin\theta = \left[\frac{y}{tv_{(\tau_a + \Delta\tau)}}\right]^{1/2} \tag{10a}$$

$$\sin\theta = \left[\frac{y}{tv_{\tau_a}\exp(B\Delta\tau)}\right]^{1/2} \tag{10b}$$

Using the average of the two values of $\Delta\tau$ found for a particular set of inputs, Eq. (10b) was integrated and evaluated for three different times as shown by the solid lines of Fig. 4. For comparison, the positions of the individual dislocations were plotted as points and the agreement was found to be very good. Since the first dislocation attains its steady-state velocity after about six dislocations have been emitted (Fig. 1), and since the rate of production is fixed after two dislocations have been emitted (Fig. 3), it appears that the formulation of Eq. (9) can be extended to any arbitrarily large number of dislocations and can be used as a general description of source operation under constant stress.

The effects of the variables τ_a (applied stress), τ_s (source activation stress), and B (stress sensitivity of dislocation velocity) remain to be evaluated. Figure 3, which depends on the number of dislocations emitted, indicates that $\Delta\tau$ is independent of B and this point can be confirmed by examining the velocity of the lead dislocation as a function of B, when τ_a and τ_s are constant. It is easily shown that

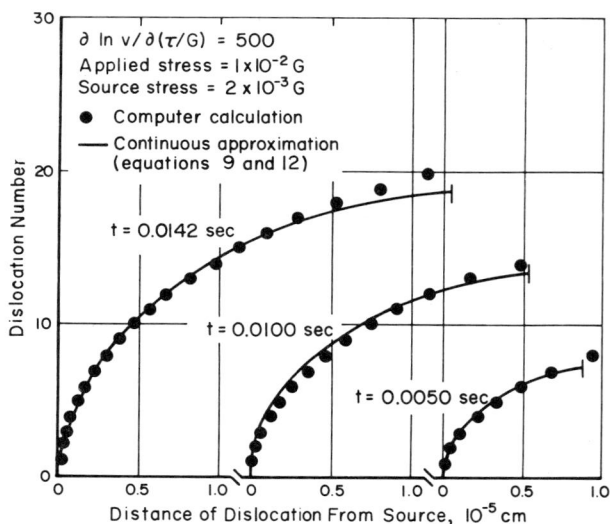

Fig. 4 Comparison of positions of individual dislocations as calculated by the computer with analytical expressions derived from the continuous approximation.

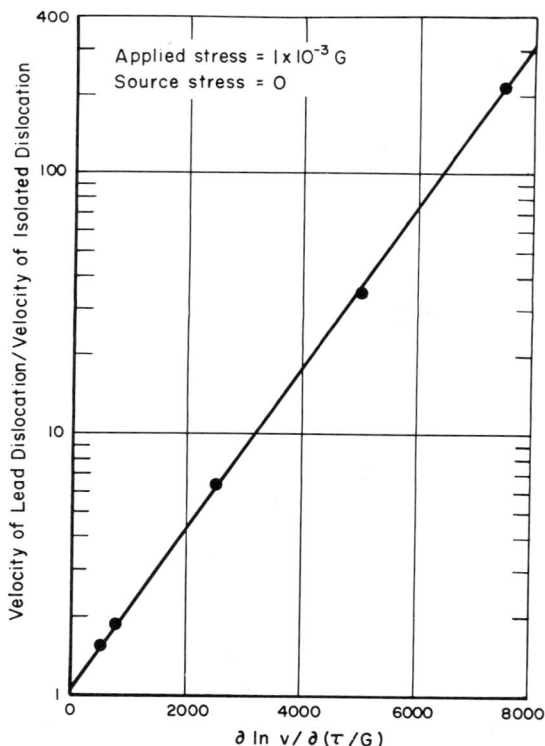

Fig. 5 Effect of stress sensitivity of dislocation velocity on the velocity of the lead dislocation.

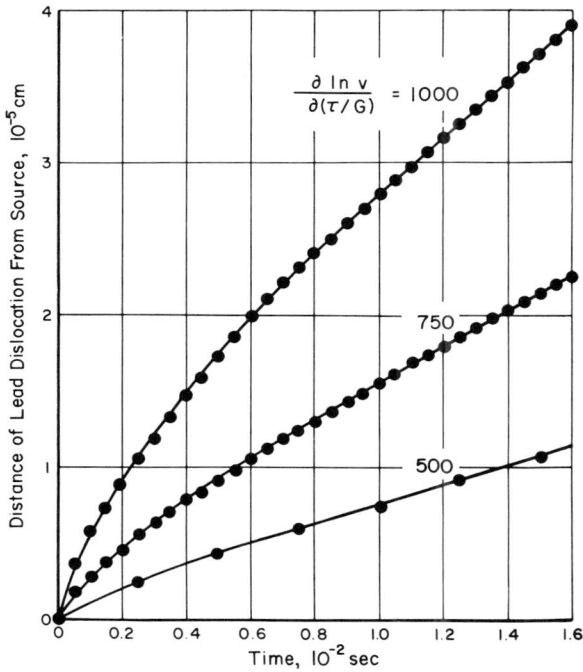

Fig. 6 Effect of stress sensitivity of dislocation velocity on the movement of the lead dislocation.

$$\frac{v_1}{v_{\tau_a}} = \exp(B\Delta\tau) \tag{11}$$

and that a plot of $\ln v_1$ vs. B will yield a straight line of slope $\Delta\tau$ and intercept $\Delta\tau = 0$ when $v_1 = v_{\tau_a}$. Figure 5 shows that this is the case. The dependence of $\Delta\tau$ with respect to B does not mean that rate sensitivity can be neglected in describing a source. Fixing all other variables, sources in the material with higher values of B will operate more rapidly as shown by the differences in positions of the lead dislocations plotted in Fig. 6. However, if each material were to be examined after the slip line emanating from the source had reached a given length, the configurations would be the same. In other words, the shape of the dislocation configuration is independent of B, while the rate of formation of the configuration is very sensitive to changes in B.

The variation of $\Delta\tau$ with overstress $(\tau_a - \tau_s)$ is illustrated in Fig. 7. Although the slope of the best dotted line through the points is not quite linear, the solid line of Fig. 7 represents a simple equation that should be sufficiently accurate for most purposes:

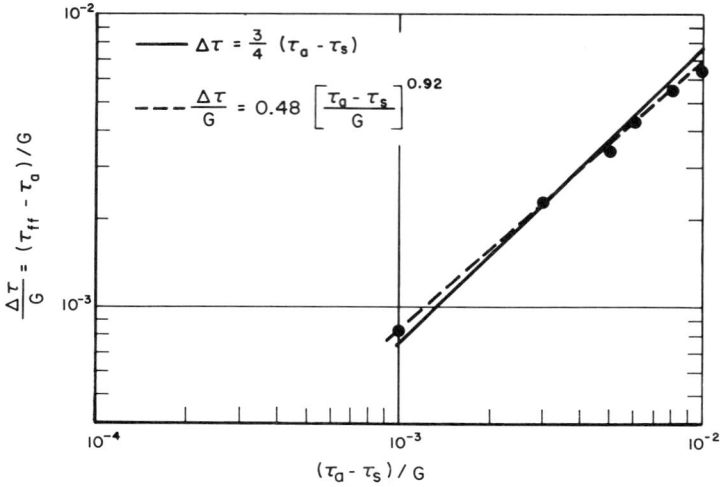

Fig. 7 Relation between the effective stress on an array emitted by a source, the applied stress, and the source activation stress.

$$\Delta\tau = \frac{3}{4}(\tau_a - \tau_s) \tag{12}$$

Equation (12) in addition to Eqs. (9) and (10) is a complete description of a dislocation source provided that the dislocations emitted are long, straight, and screw in nature, and provided that the back stress is zero.*

If the dislocations emitted by the source are not pure screw, but are still long and straight, $\Delta\tau$ is given simply by

$$\Delta\tau_{\text{mixed}} = \Delta\tau_{\text{screw}}\left(\cos\alpha + \frac{\sin\alpha}{1-\nu}\right) \tag{13}$$

where α is the angle between the dislocation line and its Burgers vector. In particular:

$$\frac{v_1\,(\text{edge})}{v_1\,(\text{screw})} = \frac{v_{\tau_a}\,(\text{edge})}{v_{\tau_a}\,(\text{screw})}\ \exp(B_{\text{edge}}\,\Delta\tau_{\text{edge}} - B_{\text{screw}}\,\Delta\tau_{\text{screw}}) \tag{14}$$

For the special case where the preexponential equals unity and $B_{\text{edge}} = B_{\text{screw}} = B$ (that is, both the absolute value of the velocity of an isolated dislocation under a given applied stress and its stress

---

*If back stress is not zero, τ_a in Eq. (12) is replaced by $(\tau_a - \tau_f)$.

sensitivity are independent of its orientation relative to its Burgers vector):

$$\frac{v_1 \text{ (edge)}}{v_1 \text{ (screw)}} = \exp\left(\frac{1}{2} B \Delta \tau_{\text{screw}}\right) \tag{15}$$

Thus, the source producing edge dislocations will operate more rapidly than the source producing screws, the ratio of the two rates of operation increasing with increasing stress and increasing stress sensitivity of dislocation velocity.

4 DISCUSSION

The principal results of these calculations are that glide dislocations move cooperatively along a slip plane and that the velocity of propagation of the array is much larger than the velocity of an isolated dislocation under the same applied stress. This suggestion has also been advanced by Fisher and Lally[30] to explain "acoustic emission" in a variety of metals during deformation, and by Mendelson[13] to account for the observed rate of formation of glide bands in NaCl. Mendelson[13] has estimated that $\Delta \tau \sim 2\tau_a$ for parallel slip planes which is somewhat higher than our value for single slip planes. The discrepancy is probably in the right direction. In addition, Kabler has reported[28] that the lead dislocation in a slip band in Ge moves several times as fast as an isolated dislocation under the same stress, as predicted by Fig. 1.

The constancy of the velocity of the lead dislocation provides the reason for the success of the Stein-Low scratch method[7] for measuring dislocation mobility. In fact, if the lead dislocation velocity were not constant, it would be very difficult to obtain any useful information from this technique. However, our results suggest that care must be exercised before applying information obtained from experiments involving arrays to describing mobility properties of isolated dislocations. An example of this is given in Fig. 8. These experiments[25] were made by the scratch technique and thus edge dislocation velocities were measured. For this case α [Eq. (13)] = 90° and $\Delta \tau = 3(\tau_a - \tau_s)/4(1 - \nu)$. The isolated dislocation line was drawn by assuming that $\tau_s = 0$, and there are marked corrections to both the absolute velocity and its stress dependence. For example, the stress dependence for an isolated dislocation $\sim \frac{1}{2}$ that of the lead dislocation. If the raw data are analyzed according to the theory of thermally activated yielding,[24] the resulting activation Volume (V^*) will be too large by a factor of two. Of course,

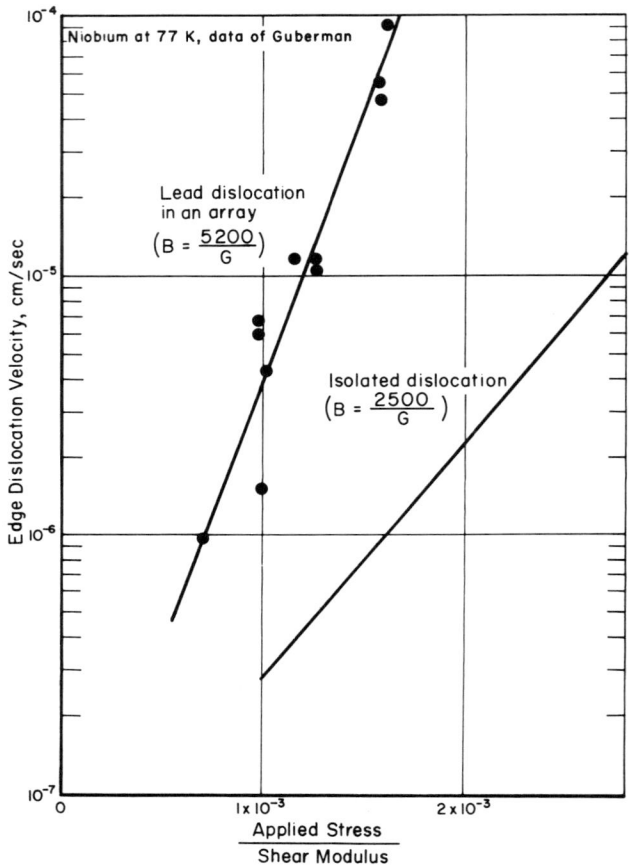

Fig. 8 Dislocation velocity as measured in an array by the scratch technique in Nb[25] corrected to illustrate the behavior of isolated dislocations.

for b.c.c. metals such as Nb, the Peierls process is usually considered as the rate-controlling mechanism, and any decrease in B would cause a corresponding decrease in V^* and provide stronger support for the Peierls mechanism. On the other hand, Suzuki and Kojima[27] have pointed out that activation volumes, in silicon, at least, are already too small to be described adequately by current formulations of kink energy.

The error in absolute value of dislocation velocity is much larger; for example, a factor of 70 at $\tau_a/G = 1.6 \times 10^{-3}$.* However,

*If $\tau_s \neq 0$, then the isolated dislocation curve of Fig. 8 will undergo a parallel shift to the left. The maximum value of τ_s is the lowest stress at which dislocation motion was actually observed and the velocity discrepancy will decrease from ~ 70 to ~ 10 at $\tau_a/G = 1.6 \times 10^{-3}$.

at present, this is not too serious, since predictions of dislocation velocity from theory are not well-developed. Furthermore, when dislocation dynamics calculations are made, the proper values of velocity to use are those associated with an array. Thus, Chaudhuri et al.[26] found that the configurations developed at their scratch were similar to those formed during tensile testing of unscratched semiconductors. Quite properly, they used the same velocity in calculations involving both experiments.

Equation (15), relating velocities of edge and screw dislocations, although it is strictly valid only for long, straight, isolated dislocation segments, can give a qualitative picture of the shapes of concentric loops. The equation predicts that, other things being equal, the ratio of the length of screw component to the length of edge component will increase with increasing stress and with increasing stress sensitivity of dislocation velocity. At present, there are few observations which give unambiguous information on the validity of these predictions. In particular: (a) the screw components in Fe-3Si have been observed to cross-slip extensively, causing the edge components to become heavily jogged;[31] (b) multiplication of individual dislocations in arrays has been observed;[32] and (c) glide on several parallel planes is common.[13] Qualitatively, it can be seen that loops are more elongated in materials with high B values (compare highly elongated loops observed by Low and Guard[31] for Fe-3Si where $B \approx 10^4/G$, and roughly equiaxed loops observed by Dash[33] for Si* where $B \approx 3 \times 10^2/G$), suggesting that at least part of the reported discrepancy between velocities of edge and screw components may be due to factors other than the intrinsic dislocation behavior. For example, evaluating Eq. (15) for a typical metal: $GB \sim 5 \times 10^3$, $\tau_a/G = 1\text{-}2 \times 10^{-3}$, and v_1 (edge)$/v_1$ (screw) $\sim 10^1\text{-}10^2$, that is, the screw component will be one or two orders of magnitude longer than the edge component. Ratios of this order have been observed in LiF[19] and inferred in Fe-3Si,[31] but there is still no adequate evidence to support their predicted stress dependence. Difficulties arise with small stress ranges, experimental scatter, and failure to distinguish experiments involving isolated dislocations from arrays.

Finally, it is possible to adjust these calculations for those materials where an exponential velocity-stress relation is not observed. The above calculation will always be appropriate for those cases where $v \approx \exp(B\tau)$ over the stress range between v_{τ_a}

---

* These observations were actually for α[Eq. (13)] $= 0°$ and $60°$, which should overestimate the asymmetry over pure screws and edges ($\iota = 0°$ and $90°$).

and $v_{\tau_a + \Delta \tau}$ (for example, about two orders of magnitude. Deviations from this relation at stresses below τ_a are inimportant since the change in dislocation position between successive iterations will be too small to influence the overall result. If the semilogarithmic form fails badly between τ_a and $(\tau_a + \Delta \tau)$, the proper value of $v_{\tau + \Delta \tau}$ from the actual $\tau-v$ relation will have to be inserted into Eq. (13a) for the length of slip line emitted.

ACKNOWLEDGMENTS

The authors are grateful to Mr. John Broehl for developing the computer program and making the numerical calculations. Discussions with A. H. Clauer and R. I. Jaffee were particularly useful. We also wish to thank R. M. Fisher of U.S. Steel Corporation for a helpful discussion. The research was supported by the Columbus Laboratories of Battelle Memorial Institute.

REFERENCES

1. Bilby, B. A., A. H. Cottrell, and K. H. Swinden: *Proc. Roy. Soc.* A272:304 (1963).
2. Hahn, G. T., and A. R. Rosenfield: ASTM *Symposium on Applications Related Phenomena in Titanium and its Alloys* (1967).
3. Cottrell, A. H.: *Proc. Roy. Soc.* A285:10 (1965).
4. Frank, F. C., and W. T. Read: *Phys. Rev.* 79:722 (1950).
5. Barnes, R. S., and D. J. Mazey:*Acta Met.* 11:281 (1963).
6. Worthington, P. J., and E. Smith:*Acta Met.* 12:1277 (1964).
7. Stein, D. F., and J. R. Low, Jr.:*J. Appl. Phys.* 31: 362 (1960).
8. Petroff, P., and J. Washburn: *J. Appl. Phys.* 37:4987 (1966).
9. Fisher, J. C., E. W. Hart, and R. H. Pry: *Phys. Rev.* 87:958 (1952).
10. Vreeland, T., Jr., D. S. Wood, and D. S. Clark: *Acta Met.* 1:414 (1953).
11. Weertman, J.:*Bull. Seismological Soc. Amer.* 54:1035 (1964).
12. Rosenfield, A. R., P. K. Dai, and G. T. Hahn: "Proceedings of the First International Conference on Fracture," p. 223, T. Yokobori *et al.* (eds.), Japanese Society for Strength and Fracture of Metals (Sendai), 1966.
13. Weertman, J.: *J. Appl. Phys.* 28:1185 (1957).
14. Mendelson, S.: *Phil. Mag.* 8:1633 (1963).
15. Weertman, J.: *Can. J. Phys.* 45:797 (1967).
16. Mura, T.:*Phil. Mag* 8:843 (1963).
17. Mura, T.:*Int. J. Engng. Sci.* 1:371 (1963).
18. Mura, T.:*Phy. Stat. Sol.* 10: 447 (1965); *ibid.*,11:683 (1965).
19. Johnston, W. G., and J. J. Gilman: *J. Appl. Phys.* 30:129 (1959).
20. Nadgornyi, E. M., and E. Yu. Gutmanas: "Physical Basis of Yield and Fracture Conference Proceedings," p. 266, A. C. Strickland (ed.), Institute of Physics and Physical Society (London), 1966.
21. Ralston, A., and H. S. Wilt: "Mathematical Methods for Digital Computers," p. 110, John Wiley & Sons, Inc., New York, 1965.
22. Campbell, J. D., J. A. Simmons, and J. E. Dorn: *J. Appl. Mech.* 28:447 (1961).
23. Schadler, H. W.:*Acta Met.* 12:861 (1964).
24. Prekel, H. L., and H. Conrad: This colloquium, p. 431; *Acta Met.* 15: 955 (1967).
25. Guberman, H. D.: *Report ORNL-4020*, Oak Ridge National Laboratory (1966), p. 41; see also: Abstract Bulletin of the Institute of Metals Division, *AIME* 2: [1], 105 (1967).
26. Chaudhuri, A. R., J. R. Patel, and L. G. Rubin: *J. Appl. Phys.* 33:2736 (1962).
27. Suzuki, T., and H. Kojima:*Acta Met.* 14:913 (1966).
28. Kabler, M. N.: *Phys. Rev.* 131:54 (1963).

29. Weertman, J., and J. R. Weertman: "Elementary Dislocation Theory," p. 126, The Macmillan Co., New York, 1964.
30. Fisher, R. M., and J. S. Lally: *Can. J. Phys.* 45: 1147 (1967).
31. Low, J. R., Jr., and R. W. Guard: *J. Appl. Phys.* 7:171 (1959).
32. Gilman, J. J., and W. G. Johnston: "Dislocations and Mechanical Properties of Crystals," p. 116, J. C. Fisher *et al.* (eds.), John Wiley & Sons, Inc., New York, 1957.
33. Dash, W. C.: *J. Appl. Phys.* 27:1193 (1956).

DISCUSSION *on Paper Presented by* *A. R. Rosenfield*

D. F. STEIN: The calculations presented in the paper indicate that the dislocation velocity at a given stress may be seriously over-estimated if one measures the velocity of the lead dislocation in a microslip band. However, the calculation has made certain assumptions about the structure of slip bands which are seriously at variance with observed slip-band structure, at least for Si-Fe and LiF. There are at least three variations.

(a) The slip bands are made up of dislocations of opposite signs in approximately equal numbers. Therefore, the stress at the tip of the slip band is probably seriously overestimated by using a model of the slip band in which all the dislocations are of the same sign.

(b) When dislocations move in Si-Fe and LiF, debris is generated by the moving dislocations. This debris is denser near the origin of the microslip band and therefore the internal stress would be expected to be greater near the origin. Both of these considerations would tend to produce reverse pile-ups in the microslip bands observed in dislocation velocity experiments.

(c) In addition, the dislocations are not long and straight, but in the form of loops of variable size.

For the above reasons it is clear that the model presented in this paper overestimates the effect of the slip band on the movement of the lead dislocation. However, it is also clear that there will be some effect of the slip band because the structure of those observed is not balanced equally along its length with plus and minus sign dislocations, thus there will be some net effect on the lead dislocations. To determine if this effect is not negligible would require a more refined calculation of the type presented by Rosenfield and Hahn.

A. ROSENFIELD: Since the model represents a first approximation, none of these effects has yet been taken into account. The first two can easily be incorporated into the formulation, while the last is discussed by Haasen in this Colloquium.

A. SLEESWYK: Concerning the arrays of screw dislocations, I would like to ask whether the authors have any proof of the two-dimensional pile-up configuration being stable. One would think off-hand that any number of screws above two would tend to be not lined up on one glide plane. This may be illustrated as follows:

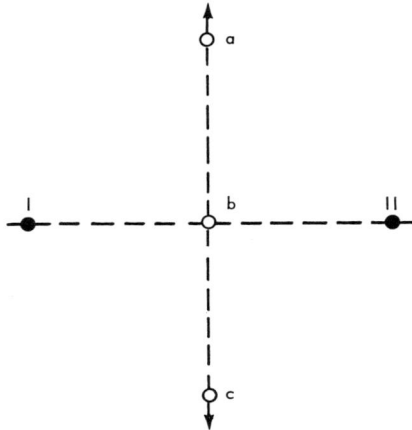

Consider the two screw dislocations I and II in this figure as being fixed in their lattice positions. A third screw may move on the plane halfway and perpendicular to the interconnecting I and II. No resultant force acts on the mobile screw at position *b*, but if the screw is at either *a* or *c*, the resulting interaction force on the mobile screw forces it away from the plane containing I and II. This implies that position b is not a stable one: if the mobile screw is perturbed slightly, it glides away from the plane containing the screws I and II under an increasing force.

This deduction would explain the observation that emissary dislocation arrays are broken up by cross-slip as soon as they attain the screw orientation.

A. ROSENFIELD: To the extent that the length of a slip band is much greater than its width, the model provides a reasonably realistic description of it. Indeed, in a crystal whose dimensions ~ 1 cm, the ratio of length to height can be $\sim 10^5$. A more serious problem is that glide often occurs simultaneously on a large number of parallel planes, and this requires more immediate attention.

P. HIRSCH: It seems to me that although the first few loops in a group moving away from a source should be elongated in the screw direction, the last few loops should be roughly circular if the velocity-stress dependence is the same for screws and edges. The pictures of Low and Guard for Fe-Si crystals seem to show that all the visible loops, even the innermost ones, are elongated in the screw direction.

A. ROSENFIELD: The micrographs of Low and Guard appear to show only the first few loops, thus, in fact, there is no evidence as to the configuration of the innermost loops.

J. HIRTH: Pertinent to the point raised by Sleeswyk, there are some factors which tend to stabilize screw pile-ups against cross-slip. Extension into partials on the glide plane is an obvious case. In addition, for fast-moving groups, the dislocation momentum is a stabilizing factor. Finally, in the anisotropic elastic calculation, pile-up shear stresses develop on a plane normal to the glide plane; in some cases, such as the alkali metals, these stresses act in the sense to stabilize the screw pile-up [see, J. P. Hirth, J. Lothe, *Phys. Stat. Sol.* 16:(1966)]. One should mention that for other metals the latter stresses can be unstabilizing.

A. ROSENFIELD: None of these effects have yet been considered. Extension into partial dislocations and anisotropy are within our current capability, but fast-moving dislocations represent a formidable problem.

T. SUZUKI: Although a question is now raised that Drs. Rosenfield and Hahn are correctly dealing with the arrangement of dislocation pits in a slip band in Fe-3Si, I have a good example to verify their theory: Many dislocations in Si generated by external stress from unidentified sources are arranged as their theory predicts as seen in Fig. 4 of our paper: *Acta Met.* 14:913 (1966).

GEOMETRICAL RELATIONS BETWEEN GRAIN BOUNDARIES AND DISLOCATIONS

W. Bollmann

Battelle Institute
Advanced Studies Center
Geneva, Switzerland

ABSTRACT

A method which allows the study of the pattern of lattice points in high-angle boundaries is described. The change in this pattern due to a dislocation ending at the boundary or crossing it is also investigated. It is shown further that discrete boundary dislocations may exist with Burgers vectors different from those in the crystal, and that the continuity (node condition) of the Burgers vectors is conserved even within the boundary.

1 INTRODUCTION

Among the lattice defects, dislocations and grain boundaries play a major part. Many interactions exist between both types of defects. For example, low-angle boundaries are formed by dislocation networks. Further, during recrystallization, moving grain boundaries annihilate dislocations, while, on the other hand, during plastic deformation grain boundaries can act as dislocation sources.

We shall discuss the geometry of some relations between grain boundaries and dislocations. First, an appropriate mathematical description of a grain boundary is developed. Then the meaning of a dislocation within a high-angle boundary is studied. Finally, the change in the grain-boundary structure due to a dislocation either passing through or ending at a grain boundary is investigated.

2   MATHEMATICAL DESCRIPTION OF A GRAIN BOUNDARY

We shall restrict ourselves to translation lattices which include the f.c.c. and the b.c.c. structures, but not the hexagonal close-packed ones. One approach for obtaining a description of a high-angle boundary is the "coincidence site lattice," the basis of which was essentially developed by Frank and others, and extended and reviewed by Ranganathan.[1] The idea is to consider lattice sites without space-filling atoms and let the lattices of the two crystals (lattices 1 and 2) "grow" through each other. Then a description of the common features of the two now interpenetrating lattices is given by all the sites where lattice points of both lattices coincide. These coinciding sites themselves form the "coincidence site lattices." A grain boundary will pass favorably through such coincidence sites since there the lattices match best. The lattice sites of the interpenetrating lattices can be interpreted as two sets of sites for one set of atoms. On one side of the boundary the atoms may lie in the positions of lattice 1 and on the other of lattice 2. The boundary itself with its near surrounding is essentially a remnant of the two interpenetrating lattices.

The disadvantage of the coincidence site lattice description is its "discontinuity": An infinitesimal change in the orientation of the two crystal lattices produces a complete change in the coincidence site lattice although it produces only an infinitesimal change in the physical situation. There are even an infinity of "irrational" orientations where no coincidence site lattice exists.

The concept of the coincidence site lattice can be generalized as follows: Let us start with a three-dimensional vector space of *real numbers* $x(x_1, x_2, x_3)$. The triples of *integer numbers* mark *lattice points*. Now we produce two lattices 1 and 2 by some tensor functions $F^{(1)}(x)$ and $F^{(2)}(x)$. The functions $F^{(1)}$ and $F^{(2)}$ must be such that three-dimensional vectors x furnish a three-dimensional lattice. Defined in this way, every lattice has its own coordinate system and the *coordinates* of an arbitrary point (not only a lattice point) are given by the real numbers (x_1, x_2, x_3). However, the *positions* of these points with respect to a "space coordinate system" are given by $F^{(1)}(x)$ and $F^{(2)}(x)$.

We may distinguish between "internal" and "external" coordinates. The internal coordinates are those within a unit cell $(0 \leq x_i < 1)$ and the external ones are the number of the unit cell. Internal coordinates remain unchanged by adding a vector with integer coordinates b (translation vector). In this perspective the coincidence site lattice is the set of points where internal coordinates of *zero* value coincide.

We now extend the condition such that we look for all the points where *internal coordinates* in both lattices *coincide regardless* of *their value*; expressed as an equation this means:

$$F^{(2)}(x) - F^{(1)}(x + b) = 0 \tag{1}$$

We call the points where the condition (1) is fulfilled the *O-points*. The *coordinates* of the *O*-points are given by the solutions x of Eq. (1) for all possible b-vectors. The *positions* of these *O*-points (with respect to our "space coordinate system") are given by $F^{(1)}(x + b)$ or by $F^{(2)}(x)$. It is not assumed that $F^{(1)}$ and $F^{(2)}$ are linear, they can be nonlinear, but first we shall consider the linear case.

3   THE LINEAR O-LATTICE

The general theory of the linear *O*-lattice is given in Ref. [2]. Here, we restrict ourselves to translation *lattices* produced by *homogeneous linear transformations*, i.e., transformations which leave the origin of the coordinate system unchanged. We consider crystals of the same nature, i.e., we study grain boundaries in contrast to phase boundaries although the general treatment holds true for both. We start with an arbitrary translation lattice which is defined by

$$F^{(1)}(x) = S x$$

where S is the matrix describing the transformation from an

orthogonal to the arbitrary lattice. The matrix by which lattice 2 may be transformed with respect to lattice 1 is denoted as R. Although all the relative orientations of crystals of the same phase can be obtained by orthogonal transformations, there are other transformations which may be better suited for our purposes; thus it does not have to be an orthogonal transformation. (This problem is discussed further in Ref. [2].) Equation (1) now becomes

$$RSx - S(x + b) = 0 \tag{2}$$

Multiplied from the left by S^{-1}

$$S^{-1}RSx - (x + b) = 0 \tag{3}$$

This means that we refer to lattice 1 as our basic coordinate system. We abbreviate

$$S^{-1}RS = A \tag{4}$$

Due to the condition that a three-dimensional lattice must be produced, A must have rank 3, i.e., the determinant

$$|A| \neq 0 \tag{5}$$

Equation (3) becomes

$$Ax - (x + b) = 0 \tag{6}$$

The solutions x of Eq. (6) are the *coordinates* of the O-point $x^{(OC)}$. The positions $x^{(OP)}$ are

$$x^{(OP)} = x^{(OC)} + b \tag{7}$$

which gives

$$A(x^{(OP)} - b) - x^{(OP)} = 0 \tag{8}$$

or

$$(A - I)x^{(OP)} = Ab \tag{9}$$

with I = unit transformation (identity). We abbreviate $x^{(OP)}$ to $x^{(O)}$ and understand $x^{(O)}$ the position of the O-point. Multiplying from the left by A^{-1} gives:

$$(I - A^{-1}) x^{(O)} = b \tag{10}$$

The b-vectors are all the possible translation vectors of lattice 1 (i.e., all the difference vectors between lattice points). If all these vectors are translated such that they start at a common origin, they again form a lattice, which as a whole is identical to lattice 1. We call this lattice of the difference vectors the *B-lattice* and imagine it as being placed separately from the interpenetrating lattices. The *O*-lattice then can be considered as an *image of the B-lattice onto the interpenetrating lattices 1 and 2* by means of Eq. (10). Instead of b we shall write $b^{(L)}$ to stress the fact that it is a lattice vector of the *B*-lattice. If the determinant

$$\left| I - A^{-1} \right| \neq 0 \tag{11}$$

we can solve Eqs. (10) directly and obtain

$$x^{(O)} = (I - A^{-1})^{-1} b^{(L)} \tag{12}$$

The general theory of the *O*-lattice and a few of its applications are are given in Ref. [2]. Here we summarize briefly a few of the essential points.

The *O*-lattice received its name from the following property: We define here lattice 2 out of lattice 1 by:

$$x^{(2L)} = A x^{(1L)} \tag{13}$$

(the index *L* indicates lattice vectors). Equation (13) means that every lattice point of lattice 1 has a "partner" in lattice 2 referred to the coordinate system of lattice 1 and its origin. We may now produce the *same lattice 2* out of the *same lattice 1* by the *same transformation* A but starting from a *different origin*. In that case points of lattice 1 and 2 are "paired" differently. The *O-lattice* is the *lattice of all the origins* fulfilling the above conditions. To change the origin to a specific *O*-point O'

$$x^{(O')} = (I - A^{-1})^{-1} b^{(L')} \tag{14}$$

we set

$$x^{(2L)} = x^{(2L')} + x^{(O')} \tag{15}$$

$$x^{(1L)} = x^{(1L')} + x^{(O')} \tag{16}$$

Introducing Eqs. (15) and (16) into Eq. (14) we obtain

$$x^{(2L')} = A\left(x^{(1L')} + b^{(L')}\right) \tag{17}$$

Here $x^{(2L')}$ is the same lattice point as $x^{(2L)}$ but referred to the new origin $x^{(1L')} + b^{(L')}$, it is again a lattice vector in lattice 1 (as $b^{(L')}$ is a translation vector) but different from $x^{(1L)}$. Equation (17) shows that if a point of lattice 2 has a partner $x^{(1L)}$ with respect to the origin O and $x^{(1L')}$ with respect to O'

$$x^{(1L')} - x^{(1L)} = b^{(L')} \tag{18}$$

where the image of $b^{(L')}$ by Eq. (14) is the new origin $x^{(O')}$.

An important problem is the behavior of the O-lattice on *translation* of lattice 2 with respect to lattice 1. We attribute its own coordinate system to every lattice. The coordinate system of lattice 1 is the reference system, the one of lattice 2 is given by Eq. (13) out of lattice 1, the one of the B-lattice is again the reference system and the system of the O-lattice is given by correspondence to the B-lattice. Then a shift of lattice 2 by $d^{(2)}$ with coordinates in lattice 2 (d_1, d_2, d_3) induces a shift of the B-lattice by $-d^{(1)}$, with numerically the same but negative coordinates $(-d_1, -d_2, -d_3)$, now in the system of the B-lattice. This causes a shift of the O-lattice by $-d^{(O)}$ also with the same coordinates $(-d_1, -d_2, -d_3)$ in the system of the O-lattice.

Shift of lattice 2	$d^{(2)}$	
Shift of the B-lattice	$d^{(1)} = -A^{-1}d^{(2)}$	(19)
Shift of the O-lattice	$d^{(0)} = (I - A^{-1})^{-1} d^{(1)}$	(20)

Usually, there are several different possible transformations $(AA' \dots)$ describing the relation between lattice 1 and 2. Hence different O-lattices can be constructed for two given interpenetrating lattices. The O-lattice with the largest unit cell is the one which relates effectively the nearest neighbors, and as such is of highest physical interest. Depending on the rank of the matrix $(I - A^{-1})$ the O-lattice can consist of the following kinds of elements:

Rank:	O-Elements:	Example of A:
3	points	expansion
2	parallel lines	roatation
1	parallel planes	shear

When the O-lattice is given, a *cell structure* can be introduced into the interpenetrating lattices with an O-element in the center of every cell. The *cell walls* are conveniently chosen as perpendicular bisectors on the connection between neighboring O-points (Wigner–Seitz walls).

Within an O-cell, the nearest neighbors in the two lattices are related by the transformation A, i.e., if the position of a lattice point of lattice 1 is given within an O-cell, its nearest neighbor in lattice 2 can be determined by the transformation, starting from the O-element of that cell.

A series of complications arising from the different ranks of the matrix are discussed in Ref. [2].

4 TWIST BOUNDARY

Here we shall restrict ourselves to rotation around an axis perpendicular to the plane of the boundary so that the rank of the matrix is 2. However, we shall formulate the problems two-dimensionally in which case the determinant $(I - A^{-1}) \neq 0$, i.e., we deal with O-points instead of vertical O-lines. We assume square lattices but the result can be interpreted with respect to a f.c.c. lattice with the rotation axis in [001]-direction and the boundary plane in the (001)-plane. Hence the general transformation is

$$A = \begin{pmatrix} \cos\theta & -\sin\theta \\ \sin\theta & \cos\theta \end{pmatrix}$$
(21)

Thus Eq. (12) becomes

$$\begin{pmatrix} x_1^{(O)} \\ x_2^{(O)} \end{pmatrix} = \begin{pmatrix} \dfrac{1}{2}, & \dfrac{1}{2}\cot\dfrac{\theta}{2} \\ -\dfrac{1}{2}\cot\dfrac{\theta}{2}, & \dfrac{1}{2} \end{pmatrix} \begin{pmatrix} b_1^{(L)} \\ b_2^{(L)} \end{pmatrix}$$
(22)

Hence the unit vectors of the O-lattice are

$$x_1^{(O)} = \left(\frac{1}{2}, \ -\frac{1}{2}\cot\frac{\theta}{2} \right)$$

$$x_2^{(O)} = \left(\frac{1}{2}\cot\frac{\theta}{2}, \ \frac{1}{2} \right)$$
(23)

5 PATTERN CONSERVATION ON TRANSLATION

We call the arrangement of lattice points of the interpenetrating lattice 1 and 2 a *"pattern,"* which can always be decomposed into *"pattern-elements."* A pattern-element is the pattern within one O-lattice cell. If a pattern is periodic it consists of a finite number of elements, and if it is not, then it consists of an infinite number of elements.

We may find the configuration of a pattern-element by tracing a line from the O-point to a point of lattice 1 within the O-cell, then to its partner in lattice 2 and back to the O-point (Fig. 1).

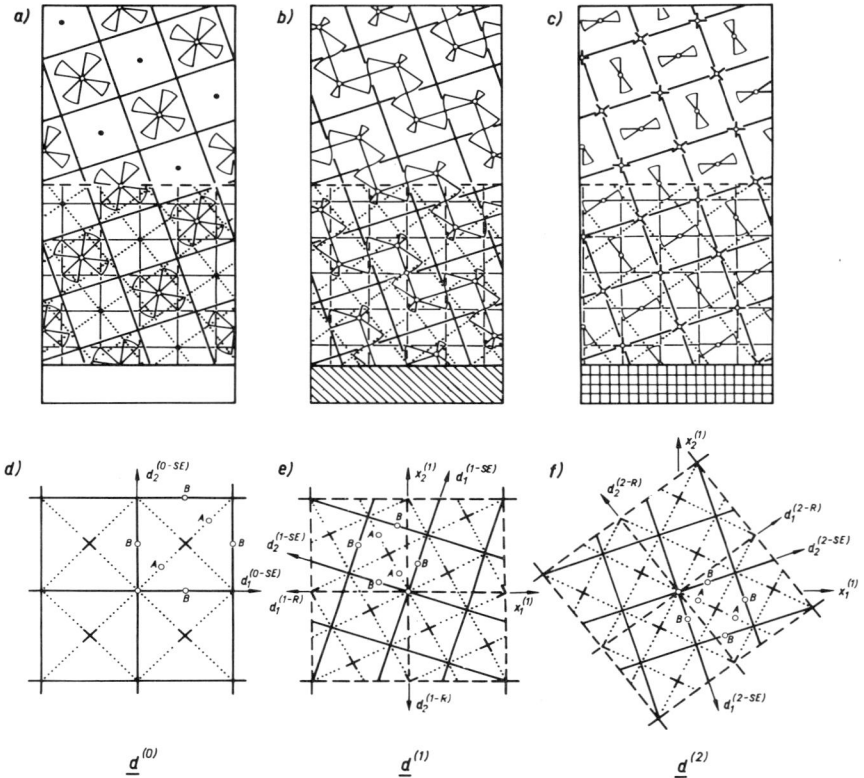

Fig. 1 (*a*) Pattern produced by a rotation of $\theta = 36°52.2'$ of a square lattice. The pattern consists of 2 elements with 0-points at internal coordinates $(0, 0)$ and $(1/2, 1/2)$. (*b*) Pattern produced from (*a*) by $d^{(2)}(1/2, 0)$ (in coord. of lattice 2) i. e. $d^{(0)}(-1/4, 3/4)$, (in coord. of lattice 1). Points A in (*d*)-(*f*). (*c*) Pattern produced from (*a*) by $d^{(2)}(1/2, 1/2)$ i. e. $d^{(0)}$ $(0, 1/2)$. Points B in (*d*)-(*f*). (*d*)-(*f*) $d^{(0)}$-, $d^{(1)}$- and $d^{(2)}$-lattice respectively. Enlarged 2.5 times.

The configuration of a pattern-element for a given transformation A *is fully determined by the position of the O-point within a unit cell of lattice 1* i.e., by the internal coordinates of the O-point in lattice 1. If the O-point falls on the corner of that unit cell of lattice 1 the pattern-element is a coincidence site. For a homogeneous transformation A the partner of a point at the origin is also at the origin, i.e., the two partners coincide.

The energy of a grain boundary will depend on the pattern of lattice points at the boundary surface. It is conceivable that there exist "minimum energy patterns" in the sense that any small change (translation, rotation) increases the energy of the boundary. It is expected that such a minimum energy pattern would show a high degree of symmetry and periodicity and thus would consist of only a few elements. The increase of boundary energy on a change might be such that the crystal prefers to conserve the minimum energy pattern and to introduce corrections in the form of occasional "dislocations." We shall assume that Fig. 1a is a "minimum energy pattern."

First, we must understand the meaning of a dislocation in a high-angle boundary. A dislocation represents a translation, i.e., if a dislocation moves through a crystal in a direction v, it virtually separates the crystal into two parts whereby the part marked by m

$$m = [l \times v] \tag{24}$$

is translated by the Burgers vector b with respect to the other part (l = line sense of the dislocation). This "movement rule" can be used as a definition of the Burgers vector in a high-angle boundary where a Burgers circuit might become questionable.

Thus, for a given transformation A, we must consider all the *translations* of lattice 2 with respect to lattice 1 which conserve the pattern as a whole. To fix the ideas let us assume that we have a coincidence site at the origin. Any other pattern-element could be present but a coincidence site is easily identified. We may plot these shifts d in a separate plot, the *D-plot*, which can be referred to $d^{(2)}$ the shift of lattice 2, $d^{(1)}$ the shift of the B-lattice or $d^{(0)}$ the shift of the O-lattice. Usually, $d^{(1)}$ will be traced but the plots of $d^{(2)}$ and $d^{(0)}$ might also be needed (Figs. 1d,e,f).

There are three kinds of translations which conserve the pattern as a whole.

1. The translation $d^{(2)}$ may be a lattice vector of lattice 2. Thus all the lattice positions which were occupied before the shift

are also occupied afterwards. This means that the total pattern is the same, and in the same position as before, i.e., the pattern is repeated. In the D-plot of $d^{(1)}$, all these shifts reform the B-lattice. We call this lattice the *"pattern-repeat-lattice"* or *DR-lattice* (R for repeat) [Figs. 1e,f, broken lines].

We may imagine lattice 2 slowly translated. Then also the B- and the O-lattice move continuously according to Eqs. (19) and (20). Thus all these points maintain their "individuality" although the pattern may change. Thus we can designate a B-lattice point and the corresponding O-point by the numbers (b_1, b_2, b_3) which correspond to the B-position before the movement. The actual positions of the B-points are given by $b + d^{(1)}$, and those of the O-point by $x^{(0)}(b) + d^{(0)}$ where $d^{(1)}$ and $d^{(0)}$ are related to $d^{(2)}$ by Eqs. (19) and (20). Thus during the movement an O-point, having originally been a coincidence site at the origin, is moved to another position, where it then represents another element of the pattern, while the configuration of the coincidence site has been taken over again by another O-point.

2.   The displacement of lattice 2 can be such that $d^{(0)}$ becomes a lattice vector of lattice 1. Then, a coincidence site having been at the origin is now moved to another lattice position in lattice 1, and with it the whole pattern is moved. All the movements of this kind form the *"elementary-pattern-shift-lattice,"* *DSE-lattice* (S for shift, E for elementary) [Figs. 1d,e,f, solid lines].

$$d^{(1-SE)} = (I - A^{-1})x^{(1L)} \tag{25}$$

3.   It was mentioned at the beginning of this section that a pattern consists of a (finite or infinite) number of elements and that the configuration of the element is determined by the position of the O-point within a unit cell of lattice 1. The translation might now be such as to be leading from the position corresponding to one element (e.g., the coincidence site) to a position corresponding to the configuration of another element within the same unit cell of lattice 1. Hence the coincidence site changes to another pattern-element but at the nearest possible position. All shifts of this kind form the *"complete-pattern-shift-lattice"* or *DSC-lattice* (C for complete) [Figs. 1d,e,f, dotted lines].

The displacements within the DSC-lattice are the smallest ones which conserve the pattern as a whole. Thus they are also the smallest possible Burgers vectors of dislocations within a high-angle boundary. These Burgers vectors have values different from those in the perfect crystal. In our example (Fig. 1), they are of the type [1/5, 2/5] instead of [1, 0].

In Ref. [2] it is shown how a dislocation network can be calculated as a result of a slight change in the transformation from the "minimum energy configuration." The Burgers vectors of these dislocations are lattice vectors of the *DSC*-lattice.

Shifts other than those of the *DSC*-lattice lead to new patterns (Figs. 1 *b,c*), but it can be shown that the number of different pattern-elements is invariant with respect to an arbitrary translation.

If we choose one of the patterns as the assumed "minimum energy pattern" (the existence of which would have to be shown by energy calculations) we may interpret other patterns as "stacking faults" limited by "partial boundary dislocations."

6 EFFECT OF DISLOCATION ON THE BOUNDARY PATTERN

To study the change in the grain boundary caused by a dislocation, we introduce the deformation due to the dislocation into one or both crystal lattices according to whether the dislocation either ends at or continues through the grain boundary. Then both the deformed crystal lattice and the distorted *O*-lattice are calculated and the complete pattern is plotted

We calculated the following two model cases:

A   An Edge Dislocation Passing Through a Twist Boundary

In this case the difference in orientation of the Burgers vectors on both sides of the boundary appears as a boundary dislocation and its effect can be seen in the change of the pattern.

B   An Edge Dislocation Ending at the Boundary

We do not consider a screw dislocation with a Burgers vector in the orientation of the axis of rotation between the two grains, because this would not leave a boundary dislocation but only a step in the boundary ending at the dislocation. Brandon[3] gives some considerations on energies of grain boundary dislocation reactions.

A   Edge Dislocation Passing Through a Grain Boundary

We consider our model grain boundary (Fig. 1a) which corresponds to a rotation by $\theta = 36°52,2'$ ($\tan(\theta/2) = 1/3$) and which leads to the transformation

$$A = (a_{ik}) = \begin{pmatrix} 4/5 & -3/5 \\ 3/5 & 4/5 \end{pmatrix} \tag{26}$$

and by use of Eq. (12) the O-lattice becomes

$$\begin{pmatrix} x_1^{(O)} \\ x_2^{(O)} \end{pmatrix} = \begin{pmatrix} 1/2 & 3/2 \\ -3/2 & 1/2 \end{pmatrix} \begin{pmatrix} b_1^{(L)} \\ b_2^{(L)} \end{pmatrix} \tag{27}$$

All the O-points with integer positions are coincidence sites; those with internal coordinates $(1/2, 1/2)$ are "crosses" (Fig. 1a). If an edge dislocation with a Burgers vector $b^{(1L)} = (1, 0)$ enters the boundary from lattice 1, it leaves it on the side of lattice 2 with the Burgers vector (referred to lattice 1) of

$$b^{(2L)} = A b^{(1L)} = (4/5, 3/5)$$

$$|b^{(2L)}| = 1 \tag{28}$$

The difference

$$b^{(2L)} - b^{(1L)} = (-1/5, 3/5) \tag{29}$$

is a lattice vector of the DSE-lattice. Applying

$$\begin{pmatrix} d_1^{(1-SE)} \\ d_2^{(1-SE)} \end{pmatrix} = \begin{pmatrix} 1/5 & -3/5 \\ 3/5 & 1/5 \end{pmatrix} \begin{pmatrix} b_1^{(1L)} \\ b_2^{(1L)} \end{pmatrix} \tag{30}$$

This is not the smallest possible Burgers vector (those are the ones of the DSC-lattice) but it could only be decomposed into two per-pendicular Burgers vectors so that in a crude energy considera-tion ($E \sim b^2$) nothing would be gained.

To calculate the configuration in the boundary with the disloca-tion crossing it, we formulate the problem in the form of Eq. (1) with a distortion of the lattices according to the solution for iso-tropic elasticity with the displacements[4]

$$u_1(x_1, x_2) = \frac{b}{2\pi}\left[\tan^{-1}\left(\frac{x_2}{x_1}\right) + \frac{1}{2(1-\nu)} \frac{x_1 x_2}{x_1^2 + x_2^2}\right]$$

$$u_2(x_1, x_2) = \frac{-b}{8\pi(1-\nu)}\left[(1 - 2\nu)\ln\left(x_1^2 + x_2^2\right) + \frac{x_1^2 - x_2^2}{x_1^2 + x_1^2}\right]$$

$$(31)$$

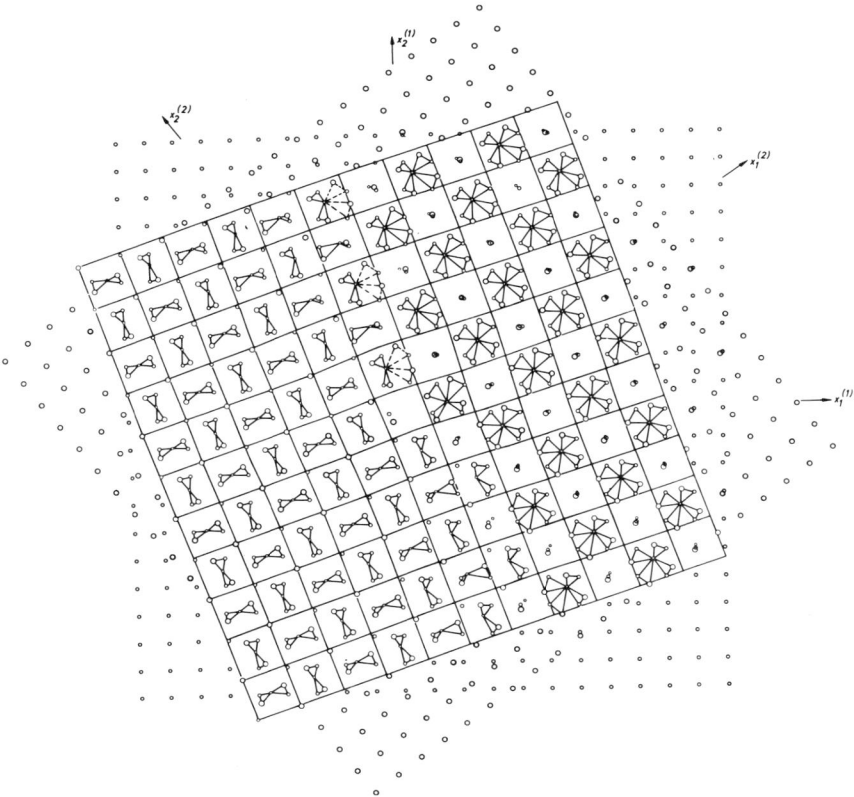

Fig. 2 Pattern produced by an edge dislocation passing through a boundary of the original pattern Fig. 1a.

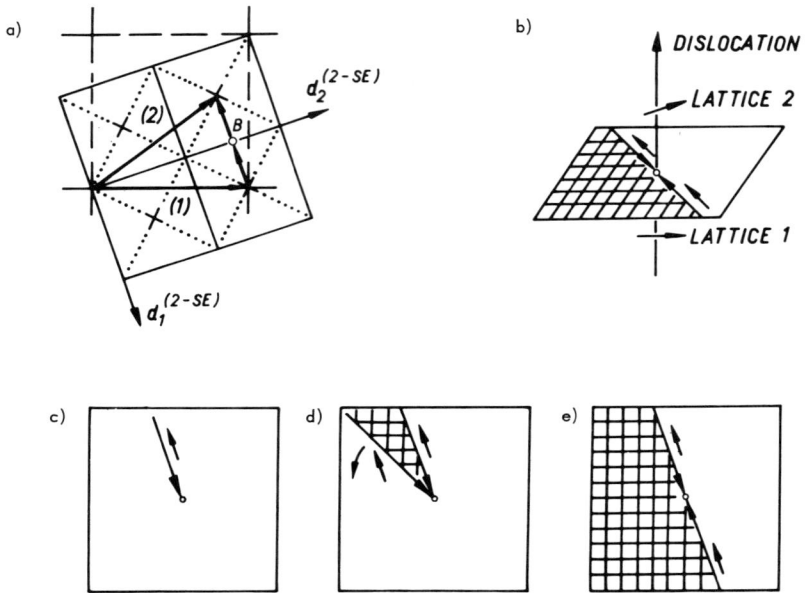

Fig. 3 Interpretation of the pattern Fig. 2. (a) Burgers vectors. (b) Perspective view. (c)-(e) Decomposition of the boundary dislocation (Hatching see Fig. 1, (a)-(c).

with $b = 1$ and $\nu = 1/3$. Thus the equations for the O-lattice become

$$A(x + u(x)) - (x + b + u(x + b)) = 0$$

or explicitely:

$$a_{11}\left[x_1 + u_1(x_1, x_2)\right] + a_{12}\left[x_2 + u_2(x_1, x_2)\right] - \left[x_1 + b_1 + u_1(x_1 + b_1, x_2 + b_2)\right] = 0$$

$$a_{21}\left[x_1 + u_1(x_1, x_2)\right] + a_{22}\left[x_2 + u_2(x_1, x_2)\right] - \left[x_2 + b_2 + u_2(x_1 + b_1, x_2 + b_2)\right] = 0$$

$$(32)$$

Where the solution x gives the coordinates $x^{(OC)}$ and $x + b + u(x + b)$ the positions $x^{(O)} = x^{(OP)}$ of the O-points.

The result of the computer calculation, programmed in Fortran for an IBM 1130, is shown in Fig. 2.

We see that the assumed "minimum energy pattern" now covers only half of the surface, while the other half is covered by a "stacking fault" which corresponds to the shift of exactly half the Burgers vector of the boundary dislocation. The picture can be interpreted such that the total boundary dislocation has split into two partials now pointing in opposite directions from the center and having

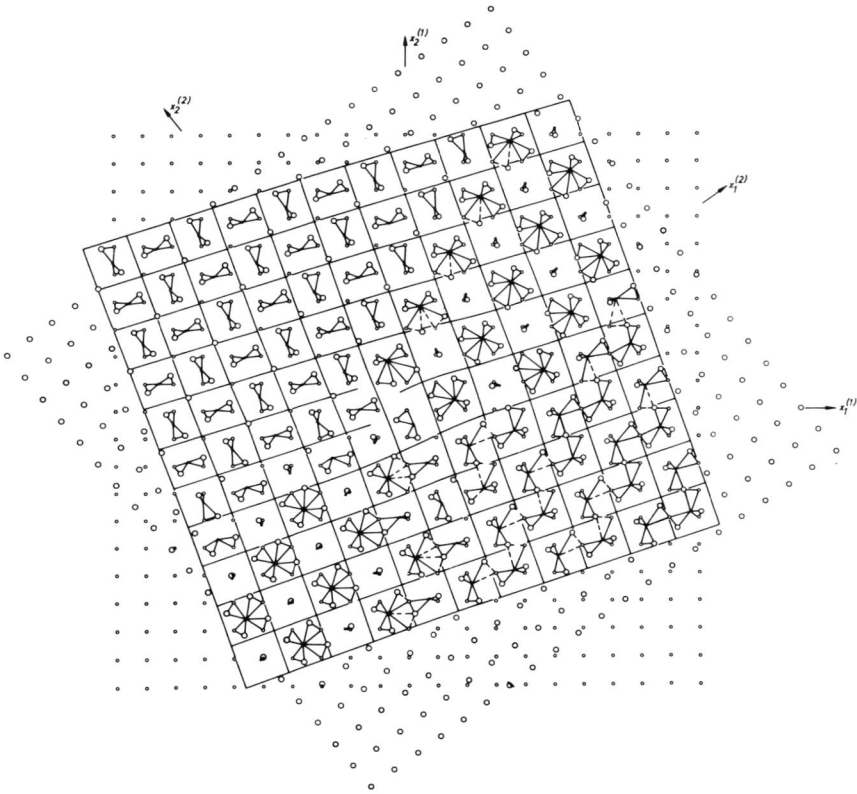

Fig. 4 Pattern produced by an edge dislocation ending at a boundary of the original pattern Fig. 1a.

produced the "stacking fault" (Fig. 3). This shows how the continuity condition for Burgers vectors in dislocation nodes holds within high-angle boundaries. It is to be remembered that all these considerations are "geometrical" and that the actual behavior of the boundary dislocation (regardless of whether it splits into partials) will depend on interaction energies. However, the fact that the Burgers vector remains conserved is independent of the energy.

B Edge Dislocation Ending at Grain Boundary

The calculation of the O-lattice for a dislocation ending at the grain boundary entering it from lattice 2 is carried out similar to

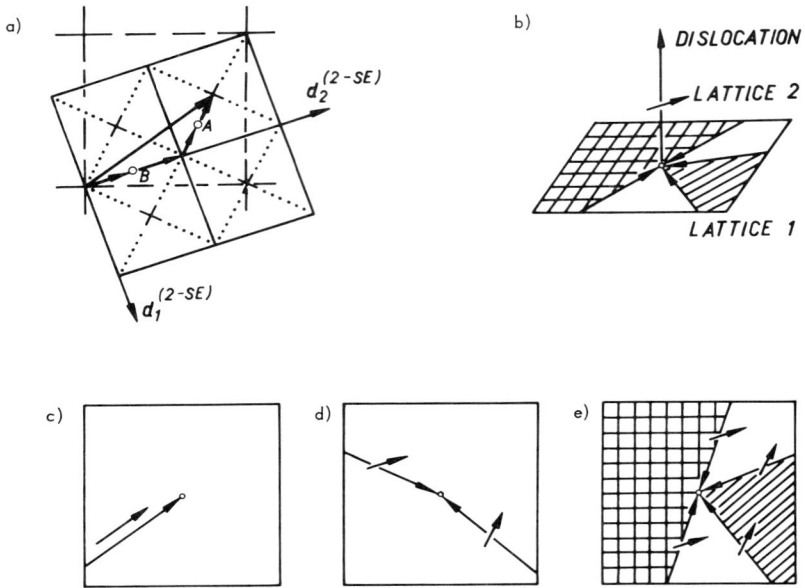

Fig. 5 Interpretation of the pattern Fig. 4 (analogous to Fig. 3).

that in the last section. The equations are:

$$a_{11}\left[x_1 + u_1(x_1, x_2)\right] + a_{12}\left[x_2 + u_2(x_1, x_2)\right] - \left[x_1 + b_1\right] = 0$$
$$a_{21}\left[x_1 + u_1(x_1, x_2)\right] + a_{22}\left[x_2 + u_2(x_1, x_2)\right] - \left[x_2 + b_2\right] = 0$$
$$\tag{33}$$

and the position of the O-point is given by $x^{(OP)} = x + b$.

The result (Fig. 4) shows that there is a "dislocation" in the O-lattice itself, i.e., in the configuration of the O-lattice points. The pattern can be interpreted first as a decomposition of the crystal dislocation into two boundary dislocations one of $(3/5, 1/5)$ and the other of $(1/5, 2/5)$ which together gives $(4/5, 3/5)$. This is the total Burgers vector in lattice 2 with the absolute value of 1 (Fig. 5). Both boundary dislocations appear split into equal partials, producing stacking faults of the corresponding patterns.

7   CONCLUSIONS

Using the O-lattice concept a description has been found which allows one to rationalize the geometrical problems of grain boundaries and, in a natural way, to relate them to the concept of dislocations. Thus the conservation of the Burgers vectors can be extended into the interior of the grain boundary. The boundary dislocations may have discrete and well-defined Burgers vectors but be completely different from those in the perfect crystal.

Although only a few simple examples have been shown here, the method can be applied in a general manner to all kinds of grain and phase boundaries.

ACKNOWLEDGMENTS

The author would like to thank Dr. A. J. Perry for his assistance in preparing the manuscript, Miss S. Koeppe for preparing the drawings and especially the BATTELLE INSTITUTE for financing this work.

REFERENCES

1. Ranganathan, S.: *Acta Cryst.* 21:197 (1966).
2. Bollmann, W.: *Phil. Mag.* 16:363 (1967); *Phil. Mag.* 16:383 (1967).
3. Brandon, D. G.: *Tokyo Conference on Strengthening of Metals and Alloys,* September 1967 (to be published).
4. Read, W. T.: "Dislocations in Crystals," p. 116, McGraw-Hill Book Company, New York, 1963.

DISCUSSION *on Paper by W. Bollmann*

R. ARMSTRONG: Would you discuss how one really knows where the boundary is on the basis of this analysis, i.e., I wish to see whether or how one may use this analysis to trace out the non-planar interface separating the two lattices.

W. BOLLMANN: The O-lattice is a means of determining where a boundary can pass. If the spacing of the O-points is marked by larger than the lattice spacing, the boundary will pass through O-points such that the misfit b per unit boundary surface becomes as small as possible. The path of curved or stepped boundaries can be investigated. If the spacing of O-points is comparable to the crystal lattice spacing, the energies of the patterns of lattice points must be considered.

R. BULLOUGH: Do you find it necessary to first reduce the A matrix, modulo the point group of the particular lattice for a grain boundary, or modulo the point group of the common holohedry of the two lattices for an interface between two different lattices.

W. BOLLMANN: For the time being, the formalism is restricted to translation lattices (these can always be treated as primitive lattices). However, it is so general that the special symmetries do not yet come into consideration.

GRAIN BOUNDARY STRENGTHENING AND THE POLYCRYSTAL DEFORMATION RATE

*R. W. Armstrong**

Brown University
Providence, Rhode Island

ABSTRACT

An analysis is developed to connect the study of low-tempera-ture, grain boundary strengthening with the study of the deformation rate dependence of the polycrystal flow strength. The analysis is interpreted to give additional support to the present understand-ing of polycrystal deformation gained by the largely independent development of these two fields of activity. A new indication is that variable deformation rate testing (e.g., creep testing) offers a sensitive technique for investigating the detailed nature of grain boundary strengthening.

1 INTRODUCTION

The mechanical strength of a polycrystalline aggregate at relatively low temperatures is generally greater, the smaller the

---

*Present address—University of Maryland, College Park, Md.

grain diameter. This strengthening effect due to grain-size refinement is usually demonstrated by performing unidirectional
tension (or compression) tests and showing that the yield or flow
stress σ, depends on the average grain diameter l, according to
the Hall-Petch relationship;

$$\sigma = \sigma_0 + kl^{-1/2} \tag{1}$$

where σ_0 and k are experimental constants.[1] The relation applies
quite widely for face-centered-cubic, body-centered-cubic and
hexagonal-close-packed metals and alloys.

The theory for understanding Eq. (1) rests on the notion that a
slip band is a stress concentrator and, also, that an exceptional
concentration of stress is generally required at the grain boundary
to propagate plastic flow between grains and, therefore, throughout the aggregate.[1,2] The stress-concentrating character of the
slip band may be modeled to varying degrees of sophistication, by
a viscous shear crack or in terms of a dislocation pile-up. However, the end result appears to be about the same, leading in either
case to the inverse size proportionality given by Eq. (1).

The strain rate is implicitly involved in Eq. (1) and this is the
main subject of this report. As will be seen, an understanding of
the strain-rate dependence in Eq. (1) is closely involved with another
aspect of the stress-grain size analysis—the relationship between
the deformation of the polycrystal and that of the single crystal.

2   THE STRESS-GRAIN SIZE ANALYSIS

The dependence of the yield strength on average grain diameter
for mild steel, 70-30 brass and nominally pure zinc is indicated in
Fig. 1.[2] The extent of the linear curves was made equal to the
actual range in $l^{-1/2}$ over which the data were obtained. The brass
data correspond to a sufficiently large range in $l^{-1/2}$ to indicate
that this plot is linear, while an l^{-1} or $l^{-1/3}$ dependence is not. A
recent study by Morrison[3] on various steels over a range of grain
size, $5 \le l^{-1/2} \le 78$ cm$^{-1/2}$, shows convincingly that his data are
only linear with $l^{-1/2}$. At this stage in the theoretical development
of the understanding for Eq. (1), the major reason for observing
other than an $l^{-1/2}$ relationship would appear to result from an
experimental variation of the condition of large grain vs. small
grain size specimens caused by possibly using different methods
to produce them.

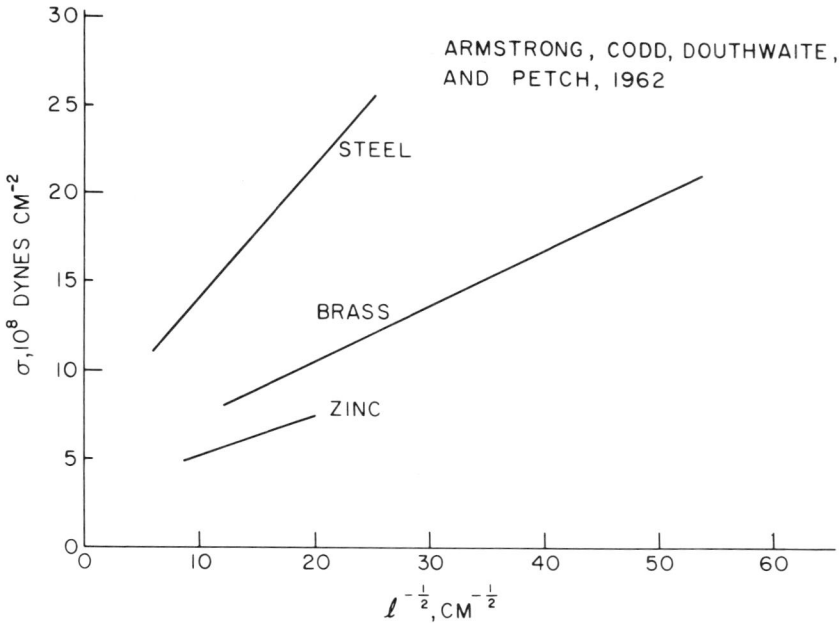

Fig. 1 Stress-grain size relation for steel, brass and zinc.

One theoretical explanation for Eq. (1) has been given based on connecting the properties of the polycrystal to the single crystal.[2] In the analysis, σ_0 is related by an orientation factor, known as the Taylor factor, to the single-crystal flow strength. Thus, the calculated single-crystal, critical resolved shear stress for the brass data shown in Fig. 1 is 1.5×10^8 dynes/cm^2 which compares favorably with the range of values 1.0-1.8×10^8 dynes/cm^2, measured, for example, by Mitchell and Thornton.[4] The value of k is determined largely by the product of two factors: one dependent on the available slip systems, and the other dependent on the obstacle properties of the boundary itself. The second factor is subject to the influence of impurities, solutes and precipitates in the vicinity of the grain boundaries. The more restricted the slip systems and the greater the segregation to the boundaries, the larger k is. The explanation for the large k value for mild steel and brass, as shown in Fig. 1, is due to the segregation effect. The k value for zinc is presumed to be due to the lack of available slip systems (a large orientation factor) and the need for activating additional slip and/or twinning systems in the vicinity of the grain boundary in order to maintain material continuity between the deforming grains. In principle, k may be decreased by establishing a preferred orientation for the grains of the polycrystal.

For different materials, k varies greatly. Some typical values which include the data of Fig. 1 are (in units of 10^7 dynes/$cm^{3/2}$): mild steel, 7.4; brass, 3.1; zinc, 2.2; and copper, 1.1. The temperature, strain rate, and strain, influence σ_0 and k in various ways for the different structures. For hexagonal-close-packed metals, changes in the temperature and strain rate appear to be reflected in changes in both σ_0 and k but, for face-centered-cubic and body-centered-cubic metals, the temperature and strain rate appear to have little influence on the determination of k by means of Eq. (1). The primary influence is on σ_0. The temperature dependence of σ_0 for yielding in α-iron follows the data for the temperature dependence of the single-crystal, critical resolved shear stress.[5] Table 1 shows that the influence of the strain rate is also to change σ_0 and not k for iron and steel.[6] From the flow-stress dependence on grain size at various values of strain by tensile testing, it appears that the major influence of increasing strain is to increase σ_0 and not change k appreciably, at least, until the onset of necking.

3   THE STRAIN RATE-GRAIN SIZE RELATION

The temperature and strain-rate dependence of the deformation of metal single crystals at relatively low temperatures may be described in many cases in terms of the thermally activated motion of crystal dislocations, as indicated by the extensive investigations

TABLE 1 Stress-Grain Size Parameters for Iron and Steel

Metal	T °K	$\dot{\epsilon}$ sec^{-1}	σ_0 dynes cm^{-2}	k dynes cm$^{-3/2}$
Iron*	300	1.0×10^{-3}	2.1×10^8	8.2×10^7
Iron*	300	9.6×10^2	34.5×10^8	8.2×10^7
Iron*	300	2.6×10^3	39.4×10^8	8.2×10^7
Steel†	194	1.0×10^{-6}	8.4×10^8	7.3×10^7
Steel†	194	2.0×10^{-4}	14.5×10^8	7.3×10^7
Steel†	194	2.0×10^{-3}	17.4×10^8	7.3×10^7

*Campell and Harding: "Response of Metals to High-velocity Deformation," p. 51, Interscience Publishers, N. Y., 1961.
†Heslop and Petch: *Phil. Mag.* **3**, 1128 (1958).

of Seeger[7] and Conrad.[8] One important feature of this approach is to separate into two parts on the basis of its temperature dependence, the applied resolved shear stress, τ: an athermal component, τ_G, and a thermal component, τ_{Th}. The magnitude of τ_{Th} determines the shear strain rate $\dot\gamma$, which may be written at large τ_{Th} as

$$\dot\gamma = \dot\gamma_0 \exp \left\{ -\frac{U_0}{RT} + \frac{v\tau_{Th}}{RT} \right\}$$

(2)

where T is the absolute temperature, R is Boltzmann's constant and $\dot\gamma_0$, U_0, and v are experimental parameters. In principle, the experimental parameters may be estimated for various dislocation models and, in fact, values of $\dot\gamma_0$, U_0 and v have been measured for single crystals of a number of the common metals. The quantity v, known as the activation volume for the thermally activated process, has been found to depend on τ_{Th}.[9] Because of this dependence, it is convenient to define a related quantity

$$v^* \equiv RT \frac{d\ln\dot\gamma}{d\tau_{Th}}$$

(3a)

where

$$v^* = v \left[1 + \frac{d\ln v}{d\ln\tau_{Th}} \right]$$

(3b)

From the strain-rate change tests made predominantly on single crystals, it is observed that v^* decreases as τ_{Th} increases until, at large τ_{Th}, v^* is approximately constant.

The thermal activation rate equation may be related to the stress–grain size equation through the value of σ_0, which applies to the single crystal strength, i.e.,

$$\sigma_0 = m \left[\tau_G + \tau_{Th} \right]$$

(4)

where m is the Taylor orientation factor mentioned earlier. Consequently, Eq. (2) may be expressed in terms of the tensile strain rate $\dot\epsilon$, as

$$\dot\epsilon = \dot\epsilon_0 \exp \left\{ -\frac{U_0}{RT} + \frac{v}{mRT} \left(\sigma - \sigma_G - kl^{-1/2} \right) \right\}$$

(5)

and at constant temperature and stress, Eq. (5) may be rewritten in the form

$$\ln\dot{\epsilon} = \ln\dot{\epsilon}_0^* - \frac{kvl^{-1/2}}{mRT} \tag{6}$$

If the dependence of v on τ_{Th} is taken into account, then

$$\frac{d \ln \dot{\epsilon}}{d\,l^{-1/2}} = - \frac{kv^*}{mRT} \tag{7}$$

The determination of kv^* from Eq. (7) may be compared with the value of k determined from Eq. (1) and the value of v^* determined from strain-rate change tests made on single crystals.

For α-iron and mild steel, prior experiments have been carried out over a large range of strain rate, temperature and grain size from which the comparison described above may be made. From the polycrystal data given in Table 1, a single value of stress

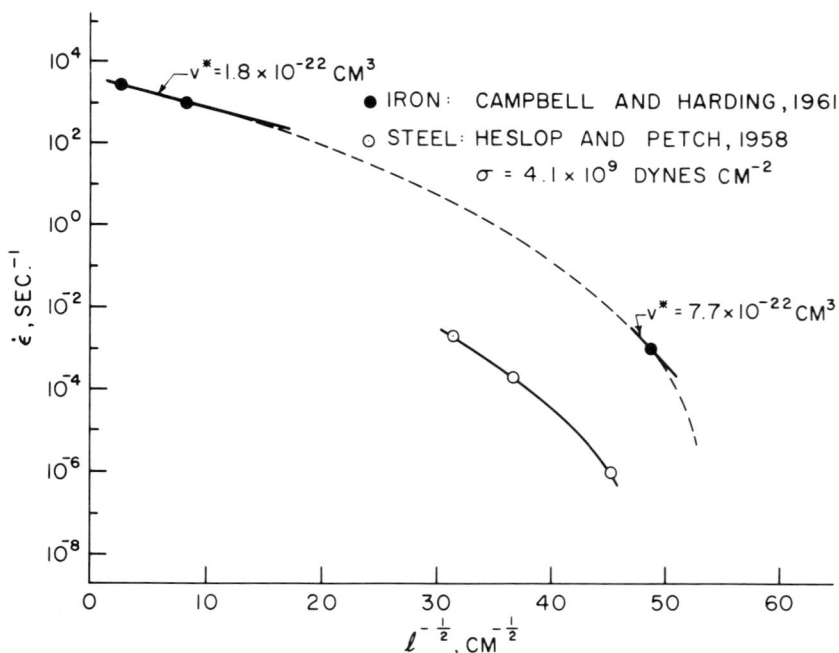

Fig. 2 Strain rate-grain size relation for iron and steel.

may be selected for which values of $l^{-1/2}$ may be calculated for each strain rate. The strain rates are shown vs. the corresponding values of $l^{-1/2}$ in Fig. 2. At the small values of $l^{-1/2}$ which apply for the data of Campbell and Harding[6a] the value of v^* is constant as is expected by virtue of the large values of τ_{Th} ($\sim 1.9 \times 10^9$ dynes/cm^2) which apply. Taking $m = 2$, $R = 1.38 \times 10^{-16}$ ergs/°K and $T = 298$°K, the value $v^* = 1.8 \times 10^{-22}$ cm^3 is determined from Eq. (7). This value of v^* corresponds to $11.3\,b^3$ where b is the dislocation Burgers vector. This value compares favorably with the limiting values $7.0 - 11.0\,b^3$ determined by Conrad[8] for values of $\tau_{Th} \gtrsim 1.9 \times 10^9$ dynes/cm^2. Conrad's values of v^* were determined at strain rates on the order of 10^{-4} sec^{-1} and temperatures nearer to 100°K. For the large values of $l^{-1/2}$ obtained from the data of Campbell and Harding, the slope of $\ln \dot{\epsilon}$ vs. $l^{-1/2}$ was calculated by taking a value of $v^* \simeq 7.7 \ 10^{-22}$ cm^3 from Conrad (for a value of $\tau_{Th} \simeq 10^8$ dynes/cm^2 and, in Fig. 2, the smooth dashed curve was drawn tangent to this slope. The form of this curve is similar to that drawn for the data of Heslop and Petch.[6b] In the latter case, the data occur at lower strain rates for the same grain size because of the lower temperature which applies. The correlation of all these data is taken as reasonable evidence for believing that Eq. (6) gives a useful description of the strain rate-grain size relationship.

The sign of the curvature of the strain rate-grain size relationship is also significant. The direct consequence of the variation of v^* with τ_{Th} in the temperature and strain-rate range of material behavior where the presence of grain boundaries strengthens the polycrystal, results in the inequality

$$\frac{d^2 (\ln \dot{\epsilon})}{d\,(l^{-1/2})^2} \leq 0 \tag{8}$$

When grain boundary strengthening occurs, at constant temperature and stress, the strain rate decreases at an ever increasing rate with decrease in grain size.

4 CREEP TESTING

The strain rate-grain size relationship suggests an alternative starting point from which to evaluate the strength of a polycrystalline material—consideration of the plastic deformation rate. From this viewpoint, creep testing is generally employed to determine the strain or strain-rate dependence of specimens subjected to constant stresses.[11] The technique is applied at both low and high temperatures, the latter being those for which diffusion processes appear to play an important role in reducing the strength of the material.

Low-temperature creep tests have been associated with a logarithmic dependence of the plastic strain on time. Conrad[12] has shown that if the value of τ_G implicitly involved in Eq. (2) is taken to contain a term dependent on the strain hardening of the material, then logarithmic creep gives a rate dependence in agreement with Eq. (2).* Also, Feltham and Meakin[13] have performed experiments showing that the steady-state creep rate of polycrystalline copper followed an equation directly analogous to Eq. (2). Furthermore, at high stresses, the creep rate was lower the smaller the grain size. These data were described by the authors in terms of a relation in which $\dot{\epsilon}$ was taken proportional to l^2. However, the qualitative reasoning given to explain the proportionality was that the grain boundaries serve to limit the distance which dislocations may travel within a slip band. This reasoning seems closely related to the theoretical basis given for Eq. (1).

On the basis of the preceding results, it is suggested that Eq. (6) may be applied to at least some of the results previously reported on the creep deformation of metals. In this case, $\dot{\epsilon}$ is taken as the steady-state creep rate. The proposal is that the analysis underlying Eq. (6) may be capable of explaining that portion of the observed creep rate which is due to the low-temperature strengthening associated with the thermally activated overcoming of obstacles within the grain volumes and the resistance of the grain boundaries to the motion of the dislocations out of the grains.

Figure 3 shows, following the prediction of Eq. (6), the dependence of the steady-state creep rate on grain size for copper, as determined by Feltham and Meakin,[13] and for 70-30 brass, as determined by Feltham and Copley.[14] A reasonably linear dependence is observed in both cases over the range of grain size studied. This grain-size dependence may be compared, for example, with that of the yield stress for copper, as determined by Eq. (1) and shown in Fig. 4, determined from the investigation of Carreker and Hibbard.[15] At 773°K, the value of k estimated from Fig. 4 is 0.31×10^7 dynes/cm$^{3/2}$, a very small quantity. A comparison of Figs. 3 and

---

*The relative insensitivity of k to strain, mentioned on p. 296, implies a similar insensitivity of the strain hardening to grain size.

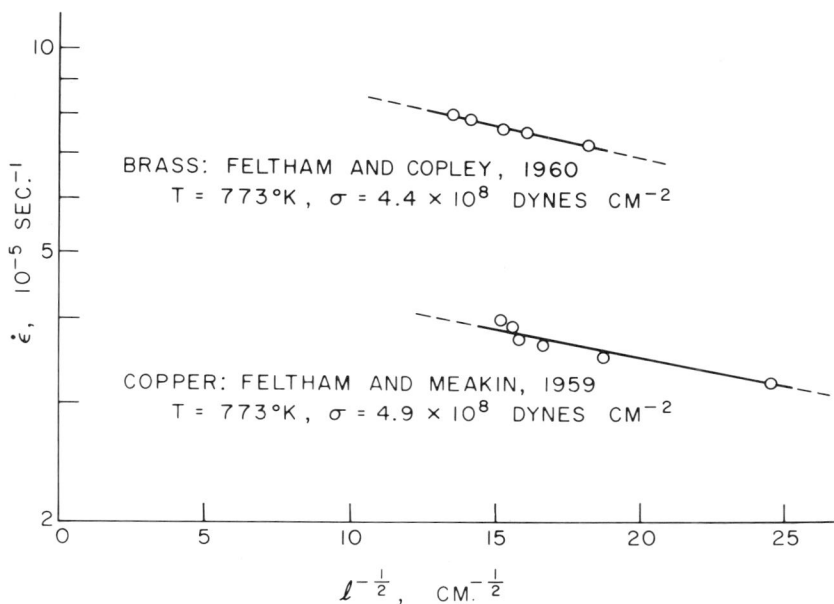

Fig. 3 Steady state creep rate-grain size relation for copper and brass.

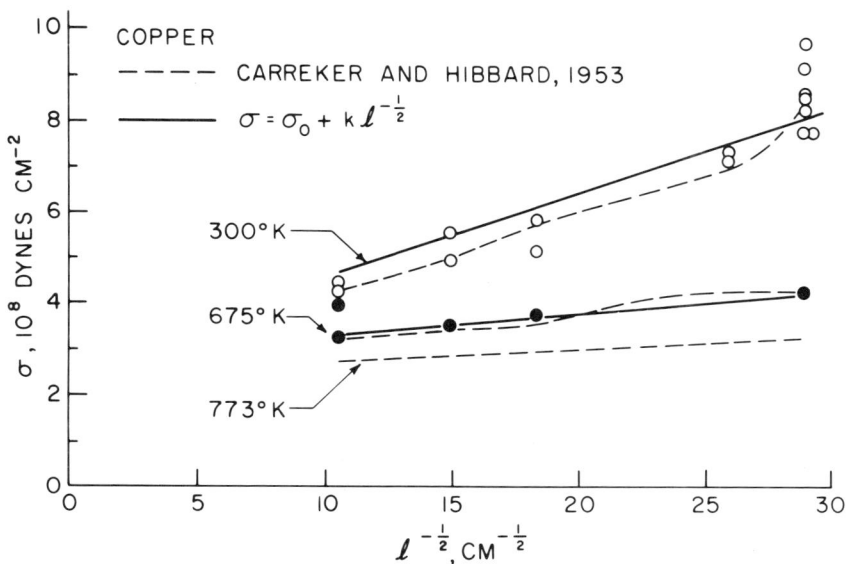

Fig. 4 Stress-grain size relation for copper.

4 indicates that the existence of grain boundary strengthening may be more reliably demonstrated from creep tests at constant (high) stress than from yield or flow-stress measurements at constant strain rate. However, additional measurements, say, on single crystals or polycrystals would be required to determine v^* and, therefore, allow a quantitative evaluation of k by means of Eq. (6). Another implication from the linear plots given in Fig. 3 is that an appreciable value of τ_{Th} applies and, thus, a reasonably strong temperature-dependence must obtain for the flow stress at a constant value of the strain rate. This agrees with the results given by Feltham and Meakin. It should be noted that the dependence of creep rate on grain size for the brass data would be expected on the basis of Fig. 1 to have been larger than is indicated in Fig. 4, but the experimental results of Feltham and Copley are

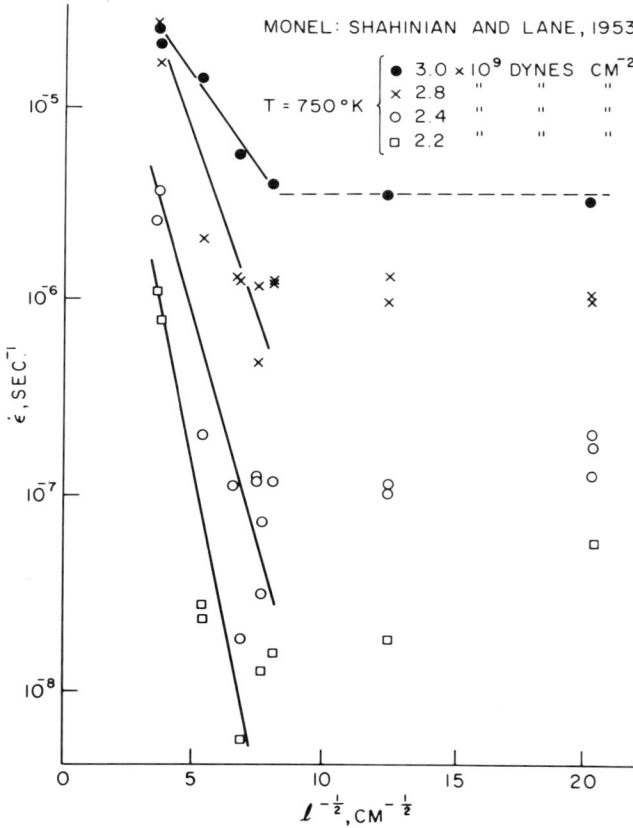

Fig. 5 Creep rate-grain size relation for monel.

consistent with their own separate observation[16] of a negligibly small k value determined according to Eq. (1).

Shahinian and Lane[17] have conducted creep tests on monel over a large range in stress and grain size at an intermediate temperature. They attempted to show that, as the grain size decreases, grain boundary strengthening first occurs and then grain boundary weakening occurs. Figure 5 shows the results for various applied stresses. At the large grain sizes, the strain rate-grain size dependence appears to follow that described by Eq. (6). The increase in slope with decrease in stress may be due to the anticipated increase in v^*. At a constant value of strain rate, the data have been used to determine a value of $k \approx 1.5 \times 10^8$ dynes/cm$^{3/2}$, according to Eq. (1). These data and the copper data have been used to estimate the values of v^* shown in Table 2. Although sufficient experiments have not been made to determine all of the parameters independently, these results are taken to indicate that the strain rate-grain size analysis may be applied successfully to creep data obtained at effective low temperatures.

The experiments of Shahinian and Lane do show that grain boundaries may contribute to material weakening. This weakening is envisioned to occur at very high temperatures and/or low stresses and/or very fine grain sizes. The weakening is associated with the enhancement of diffusion processes, grain boundary shearing and microstructural instability. With respect to these considerations, this analysis also leads to the suggestion that Eqs. (6)-(8) may be used for differentiating low-temperature creep and high-temperature creep. An example of the general case, e.g., is given in Fig. 6 which applies for tungsten.[18] Curves of similar shape have been reported for creep tests on lead,[19]

TABLE 2 Strain Rate-Grain Size Parameters for Copper and Monel

Metal	k dynes cm$^{-3/2}$	T °K	σ dynes cm^{-2}	kv^* dynes cm$^{3/2}$	v^* cm^3
Copper	‡ 3.1×10^6	773	4.9×10^8	§ 6.4×10^{-15}	2.1×10^{-21}
Monel▲	1.5×10^8	750	2.2×10^9	5.0×10^{-13}	3.3×10^{-21}
Monel▲	1.5×10^8	750	2.4×10^9	3.5×10^{-13}	2.3×10^{-21}
Monel▲	1.5×10^8	750	2.8×10^9	2.9×10^{-13}	1.9×10^{-21}

‡Carreker and Hibbard: *Acta Met.* 1, 654 (1953).
§Feltham and Meakin: *Acta Met.* 7, 614 (1959).
▲Shahinian and Lane: *Trans. Asm* 45, 177 (1953).

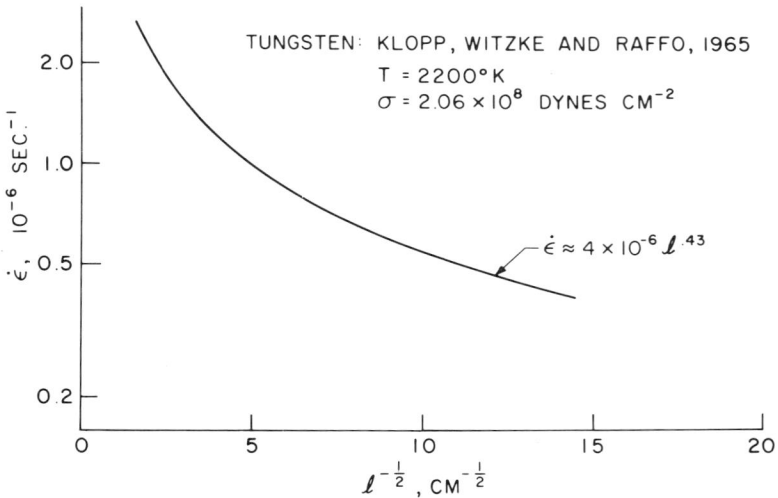

Fig. 6 Creep rate-grain size relation for tungsten.

copper[20] and nickel-cobalt alloys.[21] In these cases, even though the creep rate decreases as the grain size is decreased, the curvature is opposite to that given by Eq. (8) and it must be presumed that the deformation is complicated by the grain boundaries operating in a dual capacity or contributing to several plastic-flow mechanisms. The boundaries strengthen the material by blocking slip bands from spreading out of the grain volumes, while, at the same time, the boundaries contribute to weakening of the polycrystal, for example, through enhancement of diffusion processes. A quantitative assessment of the general case appears to be a much more difficult problem than the low-temperature situation described in this study.

5 SUMMARY

The (Hall-Petch) stress-grain size relation has been combined with the rate equation for the thermally activated motion of dislocations to quantitatively predict at constant stress and temperature, a logarithmic dependence of $\dot{\epsilon}$ on $l^{-1/2}$. For α-iron, prior experiments have been made over a range in strain rate, temperature and grain size from which the prediction appears to be substantiated. The results indicate that the dependence of strain rate on grain size is a very sensitive measure of grain boundary strengthening. On this basis, the analysis has also been applied with satisfactory results to experiments previously reported on the creep deformation of several face-centered-cubic metals at high stresses and reasonably high temperatures. It is suggested that the present analysis may be useful for differentiating between low temperature creep and high temperature creep, the latter case being one for which the presence of grain boundaries may reasonably lead to weakening of the polycrystal.

ACKNOWLEDGEMENT

This study was supported at Brown University by the U.S. Atomic Energy Commission, Contract Number AT(30-1)-2394.

REFERENCES

1. Hall, E. O.: Proc. Phys. Soc. London B64:747 (1951); N. J. Petch: J. Iron Steel Inst. 174:25 (1953).
2. Armstrong, R. W., I. Codd, R. M. Douthwaite, and N. J. Petch: Phil. Mag. 7:45 (1962).
3. Morrison, W. B.: Trans. Quart. ASM 59:824 (1966).
4. Mitchell, T. E., and P. R. Thornton: Phil. Mag. 9:1127 (1963).
5. Marcinkowski, M. I., and H. A. Lipsitt: Acta Met. 10:95 (1962).
6a. Campbell, J. D., and J. Harding: Response of Metals to High Velocity Deformation," p. 51, Interscience Publishers, New York, 1961.
6b. Heslop, J., and N. J. Petch: Phil. Mag. 3:1128 (1958).
7. Seeger, A.: "Dislocation and Mechanical Properties of Crystals," p. 243, John Wiley & Sons, New York, 1957.
8. Conrad, H.: J. Iron Steel Inst. 198:364 (1961); "Iron and Its Dilute Solid Solutions," p. 315, Interscience Publishers, New York, 1963.
9. Conrad, H., L. Hays, G. Schoeck, and H. Wiedersich: Acta Met. 9:367 (1961); H. Conrad, R. W. Armstrong, H. Wiedersich, and G. Schoeck: Phil. Mag. 6:177 (1961).
10. Armstrong, R. W.: J. Metals 14:718 (1962).
11. Garofalo, F.: "Fundamentals of Creep and Creep Rupture in Metals," The Macmillan Company, New York, 1965.
12. Conrad, H., and W. D. Robertson: Trans. Met. Soc. AIME 206:503 (1957); ibid., 212:536 (1958).
13. Feltham, P., and J. D. Meakin: Acta Met. 7:614 (1959).
14. Feltham, P., and G. J. Copley: Phil. Mag. 5:649 (1960).
15. Carreker, R. P., and W. R. Hibbard: Acta Met. 1:654 (1953).
16. Feltham, P., and G. J. Copley: Acta Met. 8:542 (1960).
17. Shahinian, P., and J. R. Lane: Trans. ASM 45:177 (1953).
18. Klopp, W. D., W. R. Witzke, and P. L. Raffo: Trans. Met. Soc. AIME 233:1860 (1965).
19. Feltham, P.: Proc. Phys. Soc. London B69:1173 (1956).
20. Parker, E. R.: Trans. ASM 50:52 (1958).
21. Davis, C. K. L., P. W. Davies, and B. Wilshire: Phil. Mag. 12:827 (1965).

DISCUSSION *on Paper by R. W. Armstrong*

A. BEMENT: In your discussion you stated that the parameter k, in the grain-size dependency term of the Hall-Petch relation appears to depend upon temperature and strain rate for hexagonal-close-packed metals. This is contrary to the experimental data for body-centered and face-centered cubic metals which show that state variables and structure effects appear to have little or no effect on the grain-size dependency of the lower yield strength. Such an apparent discrepancy in hcp metals might be accounted for both by preferred crystallographic orientations in highly textured material and by the high specificity of deformation systems in these metals. These effects can result in either smaller or larger effective grain sizes compared with the observed grain size. For example, in a highly textured sample, macroscopic plastic flow occurs by long-range cooperative strains across relatively small-angle grain boundaries. In this case, the effective grain size could be larger than that observed optically. Deformation twinning could cause the converse effect; that is, a smaller effective grain size. For example, if the applied stress is normal to the Burgers vector for the principal slip system, then deformation twinning will occur. However, the lattice rotation within the twin can be as high as 80-90°, which permits slip to occur within the twin.

Not only does plastic deformation occur in relatively restricted regions within the grains, but also slip (which accounts for the major amount of deformation) occurs in regions which have been slightly prestrained by the deformation twinning. Now, since it is not uncommon for the effective grain size as described above to vary with temperature and strain rate under conditions of heterogeneous deformation, the result would be an implied dependency of k on these variables.

R. ARMSTRONG: My statement concerning the existence of a temperature and strain rate dependence of k for hexagonal metals was not at all meant to imply that an incorrect grain size was being used in the experimental stress-grain size relation.

P. HIRSCH: Do you have a clear physical picture for the critical configuration corresponding to the activation energy in your expression for the strain-rate?

R. ARMSTRONG: No, I should just like to state my guess that the fact that this analysis works seems to imply that the constitutive equation for the thermal activation rate analysis is rather insensitive to the detailed process(es) which individual dislocations are really undergoing in the lattice. The situation is analogous to our present understanding of the Hall-Petch equation (at fixed strain rate): an $l^{-1/2}$ dependence follows for the yield or flow stress because a slip band is a stress concentrator and this dependence tells nothing of the actual mechanism by which yielding is propagated through the boundaries. Proceeding further, one must quantitatively predict the values measured for the constants. This should involve a proper statistical treatment of the problem.

J. D. CAMPBELL: Figure (i) shows values of $\sigma_0 = \sigma - kl^{-1/2}$, where σ is the lower yield stress plotted against the logarithms of the mean-plastic strain rate. The open circles denote values obtained from Ref. [6a], while the filled circle relates to recent tests by Harding on polycrystalline iron of similar purity to that used in the earlier work. The slope of the line at high stresses corresponds to an activation volume of 1.47×10^{-22} cm^3 or $9.7 b^3$, assuming a value of 2 for the orientation factor m. This may be a somewhat more accurate estimate than the value of 1.8×10^{-22} cm^3 or $11.3 b^3$ obtained from Fig. 7 of the paper, since the present method of analysis of the data is relatively unaffected by variations in the parameter k.

The use of the lower yield stress, however, raises the following difficulty. During the Luders elongation, the mean-strain rate $\dot{\epsilon}$ is governed by the number of Luders bands N, their speed of propagation V, and the rate of strain within the bands $\dot{\epsilon}_1$. Thus the rate of change of the lower yield stress (and

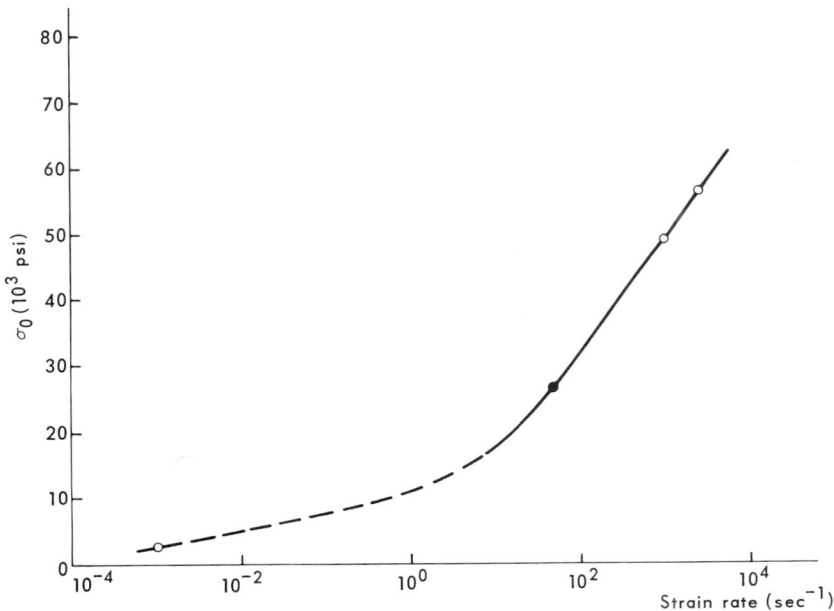

Fig. (i) Effect of strain rate on σ_0 for iron.

hence τ_0) with $\dot{\epsilon}$ will not be proportional to its rate of change with $\dot{\epsilon}_1$ unless N remains constant and V is proportional to $\dot{\epsilon}_1$. That V is not proportional to $\dot{\epsilon}_1$ is suggested by the increase in Luders strain with increase in lower yield stress;* some error is therefore to be expected in obtaining an activation volume from the rate of change of lower yield stress with mean-strain rate.

W. OWEN: I think it is well established that the Petch k for steels is not sensitive to temperature or strain rate. Only in very dilute noncarbon alloys is k a function of these variables. Thus, it is not surprising that the strain rate sensitivity is in the internal stress σ_i. It would be helpful if Armstrong could emphasize the new features of his theory.

R. W. ARMSTRONG: Steel was chosen as an example that everyone would probably be familiar with in respect to the qualitative features mentioned by Owen. My purpose on this point was to show the quantitative correlation of k and v^* values which were obtained in quite separate experiments. However, the main feature of this analysis, as is hopefully made clear in the paper, is that a quantitative creep rate (or strain rate) grain-size relationship is proposed which appears to have some special merit in quantiatively studying grain boundary strengthening.

H. CONRAD: An explanation for the $kl^{-1/2}$ in your thermally activated equation

$$\dot{\gamma} = \dot{\gamma}_0 \exp\left\{ - \frac{U_0 - v^*(\sigma - \sigma_G - kl^{-1/2})}{RT} \right\} \tag{1}$$

is that it reflects the effect of grain size on work-hardening and hence contributes to the σ_G term. This is supported by recent results that we have obtained on the effect of grain size on the flow stress and dislocation density in Nb[†‡] and T_i.[§] For a given grain size it was found that the dislocation density ρ as measured by electron transmission microscopy increased linearly with the strain ϵ so that

$$\rho = \rho_0 + A\epsilon \tag{2}$$

*Campbell, J. D., R. H. Cooper, "Proc. Conf. on the Physical Basis of Yield and Fracture," p. 77, Inst. of Physics Soc., London, 1966.

† Conrad, H., S. Fenerstein, L. Rice—submitted to Materials Science and Eng.

‡ Conrad, H., S. Fenerstein, L. Rice—submitted to Int. Conf. on Mechanical Properties of Metals, Tokyo 1967.

§ Conrad, H., R. Jones, F. Cooke, unpublished research.

Moreover, it was found that

$$A = \frac{\beta}{l} \qquad (3)$$

where β is a constant and l is the average grain diameter. Also, the results for all grain sizes yielded for the flow stress σ

$$\sigma = \sigma_i + \alpha \mu \, b \, \rho^{1/2} \qquad (4)$$

where $\alpha \simeq 0.5$, μ is the shear modulus and b is the Burgers vector and σ_i is the friction stress.

Substituting Eqs. (2) and (3) into Eq. (4) yields

$$\sigma = \sigma_i + \alpha \mu \, b \left(\rho_0 + \frac{\beta \epsilon}{l} \right)^{1/2} \qquad (5)$$

which gives for $(\beta \epsilon / l) \gg \rho_0$

$$\sigma = \sigma_i + \alpha \mu \, b (\beta \epsilon)^{1/2} \, l^{-1/2} \qquad (6)$$

in agreement with the Hall–Petch equation with

$$k = \alpha \mu \, b (\beta \epsilon)^{1/2} \qquad (7)$$

Reasonably good agreement was obtained between k calculated from Eq. (7) and that determined from a plot of σ versus $l^{-1/2}$. Also the σ_i from Eqs. (4) and (6) agreed with the σ_i from the Hall–Petch plot.

Hence in this work-hardening model, the $k l^{-1/2}$ term is associated with the long-range internal stress term σ_G.

R. W. ARMSTRONG: Conrad states on the basis of his model that k measures the strain hardening within the grain volume and k increases as the strain increases. The first point seems clearly to be a different physical description of why k exists at all as compared with the description I have given. The second point should rest on experimental confirmation: as far as the iron and brass data in Fig. 1 are concerned, reference to the original paper shows, for the brass, k is essentially constant for strains to 20% elongation and, for the steel, k decreases immediately after the Luders strain and then remains constant until 20% strain. The work-hardening is principally in σ_0.

Part Three

LOW-SPEED DISLOCATIONS

DISLOCATION CONFIGURATIONS IN COPPER IN THE FIRST STAGES OF PLASTIC FLOW

F. W. Young, Jr.

Solid State Division,
Oak Ridge National Laboratory,
Oak Ridge, Tennessee

ABSTRACT

Previous experiments on the motion and multiplication of dislocations in copper in the first stages of plastic flow have been reviewed and compared to simple dislocation theory. It is shown that the mechanism of dislocation multiplication and the factors which determine the yield stress cannot readily be deduced from these experiments. New observations via Borrmann x-ray topography of dislocation configurations after the application of small compression stresses corresponding to the first stages of flow are presented. These observations suggest that edge dislocations move large distances through the crystals at low stresses leaving long screw dislocations, and that the sources generally were accidentally introduced at the surface. It appears that dislocation multiplication and the yielding process in these crystals were not intrinsic processes dependent only on the dislocations initially in the crystal. Additional experiments on crystals deformed in bending and in tension show that edge dislocations move more easily and also leave long screw dislocations in these modes of deformation.

1   INTRODUCTION

For copper crystals of dislocation density $N \gtrsim 10^6/\text{cm}^2$, there is general agreement that the critical resolved shear stress (yield stress) is determined by dislocation interactions. However, for more perfect crystals, the factors governing dislocation mobility, dislocation multiplication, and a combination of these, which determine the yield stress, are not well-understood. As a result of a number of studies[1-6] on crystals of low N, it seems to be well established that: (a) fresh dislocations can move through the lattice under a stress (all stresses are resolved shear stresses) $\gtrsim 2 \text{ g/mm}^2$ and this lower limit is rather insensitive to N, for $N \lesssim 5 \times 10^4/\text{cm}^2$. There is some impurity pinning for aged dislocations, motion stress $\approx 4 \text{ g/mm}^2$. The number of grown-in dislocations which move under an applied stress increases with the stress. The distance moved is related to N and to the applied stress; (b) dislocation multiplication occurs under stresses as low as 7 g/mm^2;[12] (c) the yield stress, defined as that point of marked decrease in slope of a stress-strain curve, is larger than the multiplication stress, often 30-40 g/mm^2. It is relatively insensitive to strain rate, to initial dislocation density ($N \lesssim 10^5/\text{cm}^2$), and to initial dislocation configurations; (d) N increases with applied stress both prior to and after yielding—$N \sim 5 \times 10^5/\text{cm}^2$ at the yield stress for a group of crystals with initial $N \lesssim 10^5/\text{cm}^2$; (e) upon removal of the applied stress relaxation occurs, which is decreased in magnitude for stresses greater than the yield stress; and (f) these properties are not strongly temperature-dependent, although the experimental evidence is sparse.

These results are not readily explicable on the basis of existing theory or by analogy with investigations on other low N cubic crystals; for example, Ge, Si, LiF, MgO, and Fe-3%Si. The stress necessary to move dislocations through a dislocation forest of density N is $\sigma_f \approx Gb/2\pi r$; G = shear modulus = $4 \times 10^6 \text{ g/mm}^2$ for copper; b = Burgers vector, r = distance from any point in the lattice to the nearest dislocation $\approx N^{-1/2}$. Thus, for copper, $\sigma_f \approx 2 \text{ g/mm}^2$ for $N = 10^4/\text{cm}^2$, and this long-range stress field could not account for the motion stress for $N < 10^4/\text{cm}^2$. The insensitivity of the motion stress to temperature and its low value are difficult to reconcile with a thermally activated process, such as overcoming the Peierls stress as observed in Si and Ge. The observation of impurity pinning for aged dislocations raises the possibility of some sort of (presently undefined) impurity-dislocation interaction, but the observation of about the same value for the motion stress by several researchers on nominally 99.999% copper rather suggests an intrinsic property; this is assumed to be true in this paper. The fact that most of the dislocations do not move under the smallest stresses, and that the number which do move increases with the stress, can probably be related to the observation, via x-ray topography[7] that in well-annealed crystals the dislocations tend not to lie on slip planes; an alternate statement is that their jog density is very high. The observation that the distance moved is related to N suggests that this distance may be determined primarily by dislocation interactions.

The stress necessary for dislocation multiplication is $\sigma_x \sim \alpha Gb/l$, where l is the loop length, and α varies according to the specific model but can be assigned a value ≈ 0.3; if l is determined by the intersection with forest dislocations, $l \approx N^{-1/2}$. A somewhat lower multiplication stress than that observed would be predicted by this relation, and the indicated dependence of σ_x on N has not been found. The presence of small obstacles in the lattice could determine the multiplication stress. Following Kroupa and Hirsch,[8] $\sigma_x = \frac{1}{8} GbRn^{2/3}$ where R is the radius of the obstacles, and n is their density. It is possible for the vacancy clusters observed in crystals with $N \lesssim 5 \times 10^3/\text{cm}^2$ to act as such obstacles. Taking the reported values[9] $R = 1.2 \times 10^{-3}\text{cm}$, $n = 5 \times 10^5/\text{cm}^3$, then $\sigma_x \approx 8 \text{ g/mm}^2$. Hence it is possible for the multiplication stress to be determined by dislocation interactions for $N > 10^4/\text{cm}^2$ and by vacancy cluster obstacles for lower N, but there is no direct proof for this argument.

Since the yield stress is greater than the multiplication stress, it appears that ideas such as those above cannot account for yield. If the yield stress is considered to be the sum of thermal and athermal contributions $\sigma_y = \sigma_s + \sigma_a$, then the terms of thermal origin σ_s can be related to the dislocation velocity by $v = (\sigma/\sigma_0)^M$ and $\dot{\epsilon} = nbv$ ($\dot{\epsilon}$ = strain rate, n = density of gliding dislocations). Johnston[10] has shown that yielding in LiF can be accounted for on this basis. However, it is implicit in this approach that the yield stress be strain-rate dependent, and no appreciable strain-rate dependence of the yield stress has been reported. Also, Johnston showed that the lack of a yield drop in the stress-strain curve implied a large exponent M; a value of 200^{11} has been estimated for copper. More recently, the dislocation velocity has been measured directly,[5] and values $V \gtrsim 10^2$ cm/sec with $M \approx 2$ at stresses $\approx$ yield stress were reported. It is apparent that for $\dot{\epsilon} = 10^{-4}$/sec and $V \sim 10^2$ cm/sec the number of dislocations required to maintain the imposed strain rate is of order 10^2, whereas it was found that N always increased to $\sim 5 \times 10^5$/cm^2 at the yield stress.

Therefore, it seemed desirable to investigate further the pre-yield and yield phenomena to determine experimentally the sources of the new dislocations, the mechanisms of multiplication and the factors which determine the yield stress of a particular crystal. As part of this study, previously reported,[12] large $1 \times 1 \times 2$ cm parallelepipeds with $N < 10^3$/cm^2 were deformed, sectioned by acid sawing, and the lamellae examined by Borrmann x-ray topography. A novel feature of this study was that the dislocation configurations could be fixed in the crystal at any desired time by fast neutron irradiation. The idea that in crystals with such large dimensions, internal sources and evidences of multiplication mechanisms should be found, proved to be unrealized in experiment. The sources appeared to be at or near the original crystal surfaces and were unidentified. Neither were any dislocation multiplication mechanisms deduced. It was found that dislocation configurations varied markedly between stress-applied and stress-relaxed conditions. Also, it was deduced from these configurations that the velocity of edge dislocations was greater than screws. This paper is a report of continuing studies on this subject.

2   EXPERIMENTAL

Parallelepipeds $1 \times 1 \times 2$ cm with two 1×2 cm faces $(1\bar{1}1)$ and and $(\bar{1}11)$ and with long axis either near $[1\bar{1}0]$ or $[\bar{2}11]$ were acid sawed from 2.5 cm diameter 99.999% copper crystals grown by a Bridgman technique with $[111]$ growth direction. Either (111) or $(\bar{1}11)$ or both faces were acid polished, then the crystal was annealed at 1075°C in H_2 for two weeks.[7] After this treatment $50 < N < 5000$/cm^2 as determined by etch pits on $(1\bar{1}1)$ and $(\bar{1}1\bar{1})$, and there were no subboundaries. Such crystals also contain $\sim 5 \times 10^5$/cm^3 vacancy clusters.[9] The pre-yield deformation of these crystals was studied as follows:

1) Some crystals were deformed in compression either by dead-weight loading or by a spring to stresses less than the yield stress, then irradiated with $10^{17}/cm^2$ fast neutrons to pin the dislocations while the stress was applied. This dose of fast neutrons will immobilize dislocations for a stress < 750 g/mm^2, and the effects of the irradiation can be removed by annealing at $500\,°C$. Subsequently, the crystals were sectioned via acid sawing along the primary or other slip planes, and the dislocations in these lamellae were examined by Borrmann x-ray topography[13] and by etch pitting.

2) Some crystals were compressed, then irradiated after the stress was removed, sectioned and examined as above.

3) The stress necessary for dislocation multiplication was investigated in some nearly perfect crystals deformed by compression.

4) The yield stress was determined for some crystals by compression with an Instron machine.

5) The mode of dislocation motion in an annealed lamellae was investigated by Borrman topography.

6) Tensile specimens were prepared by acid sawing and stressed in a spring-loaded jig arranged so that Borrmann topographs could be taken of the unstressed specimen and after small stress increments were applied. Only stress was measured in this machine. The tensile specimen was cut from a larger crystal of low N which had been hardened by neutron irradiation, then annealed just prior to mounting in the machine. Great care was necessary.

3 RESULTS

A number of experiments will be described individually. Crystal I (see Fig. 1) was deformed (dead-weight loading) with a maximum stress of 24 g/mm^2 and irradiated with stress applied. Thus the resolved stress on the $[101]$-$(\bar{1}\bar{1}1)$ slip system was 9.7 g/mm^2. The crystal was sliced along $(\bar{1}\bar{1}1)$ [the primary slip plane is (111)], and the dislocation configurations in one of the lamellae were described previously in some detail.[12] A stereo pair of a $1\bar{1}\bar{1}$ topograph of a portion of another lamella is shown in Fig. 2. Of particular interest are the long screw dislocations and the much shorter edge segments. Also, note that in many instances the dislocations appear to be impeded in their motion by the vacancy clusters (black spots). As previously reported, the local curvature in the lines corresponds

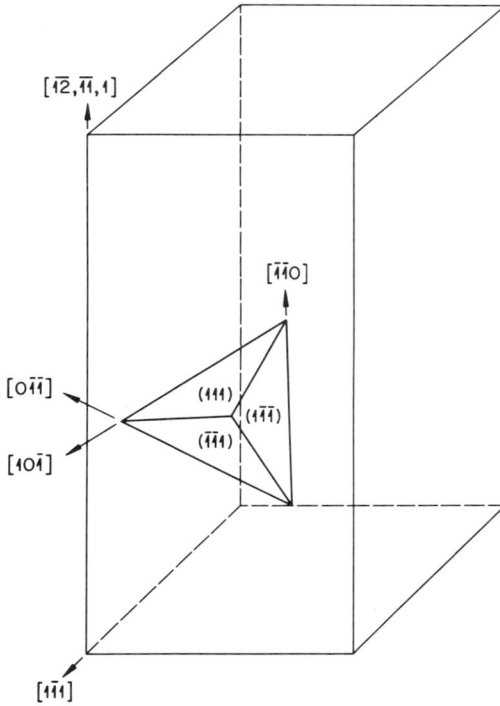

Fig. 1 Diagram of crystal I.

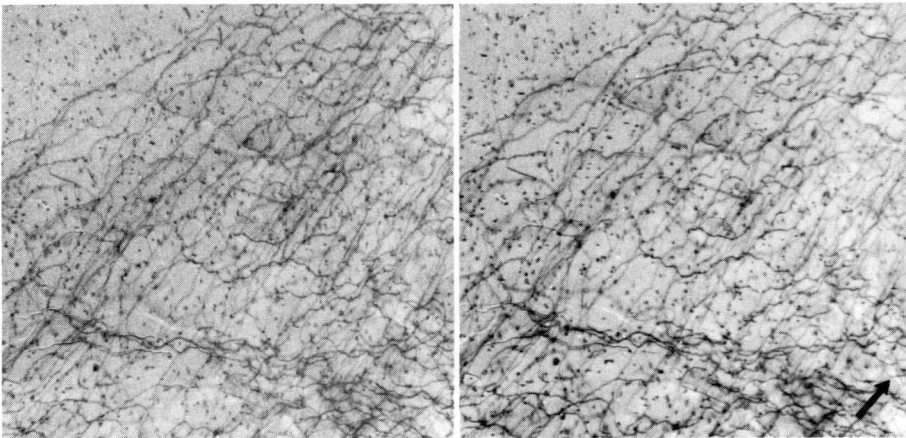

Fig. 2 Stereo pair of $\overline{1}1\overline{1}$ topographs of $1/3$ cm × $1/3$ cm of a lamella from crystal I. The arrow points in $[101] = b$. Thickness $t \approx 0.03$ cm.

to the applied stress. Although there were appreciable regions of no slip in the crystal, some dislocations were ~1 cm in length. The few as-grown dislocations were not affected by the applied stress. It was noted previously that the source(s) of the new dislocations appeared to be at or near the surfaces. This is shown more clearly in Fig. 3, which is a 111 topograph of a third lamella from this

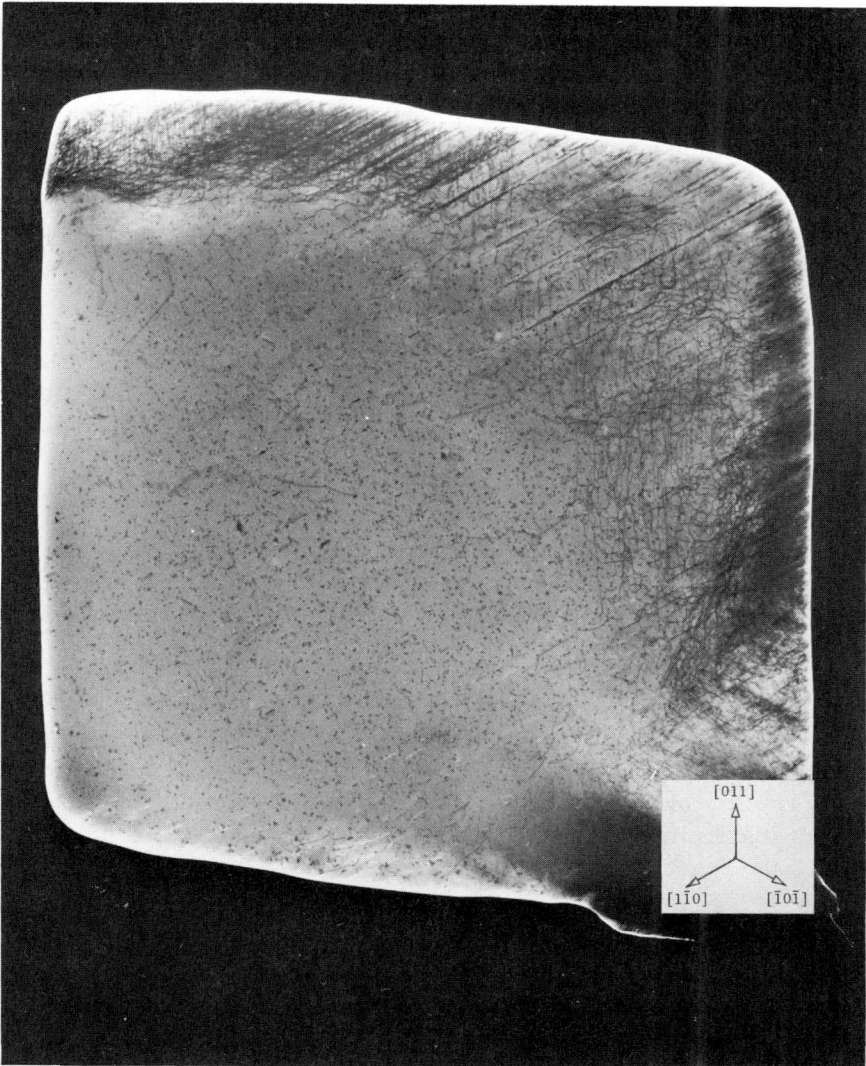

Fig. 3 A 111 topograph of a lamella from crystal I. Edge of lamella equals 1 cm. The long lines in the vertical direction are screw dislocations.

parallelepiped. The slip traces on (111) appear as ill-defined lines along [011], and it also appears that some dislocations in (1̄1̄1) have been generated from sources in these (111) traces. This experiment, along with several similar ones, suggested that the sources of new dislocations were associated with the surface. While the surfaces were not atomistically smooth (the crystal was electro-polished prior to deformation), it can easily be shown that the stress increment attainable from surface roughness is not nearly sufficient to promote dislocation generation from perfect crystal near the surface. However, minor mishandling could possibly introduce such sources which might then so act. Accordingly, a different compression jig, spring-loaded and better arranged to ensure crystal alignment, was made, and extra care was taken to avoid mishandling the crystal. Crystal II, compression axis $\sim[7\bar{6}1]$ and $N < 10^2/cm^2$, was compressed in this jig with a stress of 9 g/mm^2 on the primary system, and irradiated with the stress applied. It was sectioned along (111), and a topograph of a lamella ~ 0.1 cm thick is shown in Fig. 4. There was no indication of any deformation in any of the lamellae cut from this crystal. It is apparent from Fig. 4 that there were very few grown-in dislocations. A similar result was obtained on Crystal III, applied stress on primary system of 11 g/mm^2 but with $N \approx 500/cm^2$ prior to the attempted deformation. Although a complete analysis of all the grown-in dislocation in Crystal III would be too tedious, a partial analysis and close inspection using stereo pairs showed that few, if any, of the as-grown dislocations lay on the primary plane [i.e., the plane containing both b and a unit vector lying along the dislocation line was not (111)].

Crystal IV, compression axis [8̄5̄3] with initial $N = 1.2 \times 10^3/cm^2$, was deformed in this spring-loaded jig to a stress of 10 g/mm^2 on the primary system and irradiated with the stress applied. As determined by etch pits the dislocation density was increased to $2.1 \times 10^3/cm^2$ by the stress. It was then sectioned along (111); a topograph from a representative lammella is shown in Fig. 5. As would be expected from the etch-pit count, the number of new dislocations is small and their configuration is more-or-less in agreement with those reported previously for stress-applied crystals.[12] Again, the dislocations seem to have come from the surface of the parallelpiped in some way; their density was appreciably greater towards the end of the parallelepiped, suggesting an association of sources with the areas of contact with the load. Of particular interest in this and other lamellae from this crystal is the presence of some undefined strain centers (possibly small precipitates) dispersed throughout the volume, each with its own punched out dislocation loops. They appear as crosses or stars in the topographs. Such a phenomena was first reported by Jones and Mitchell in AgBr.[14] These centers have not acted as dislocation sources, although new dislocations have been moved large distances through this crystal by the applied stress.

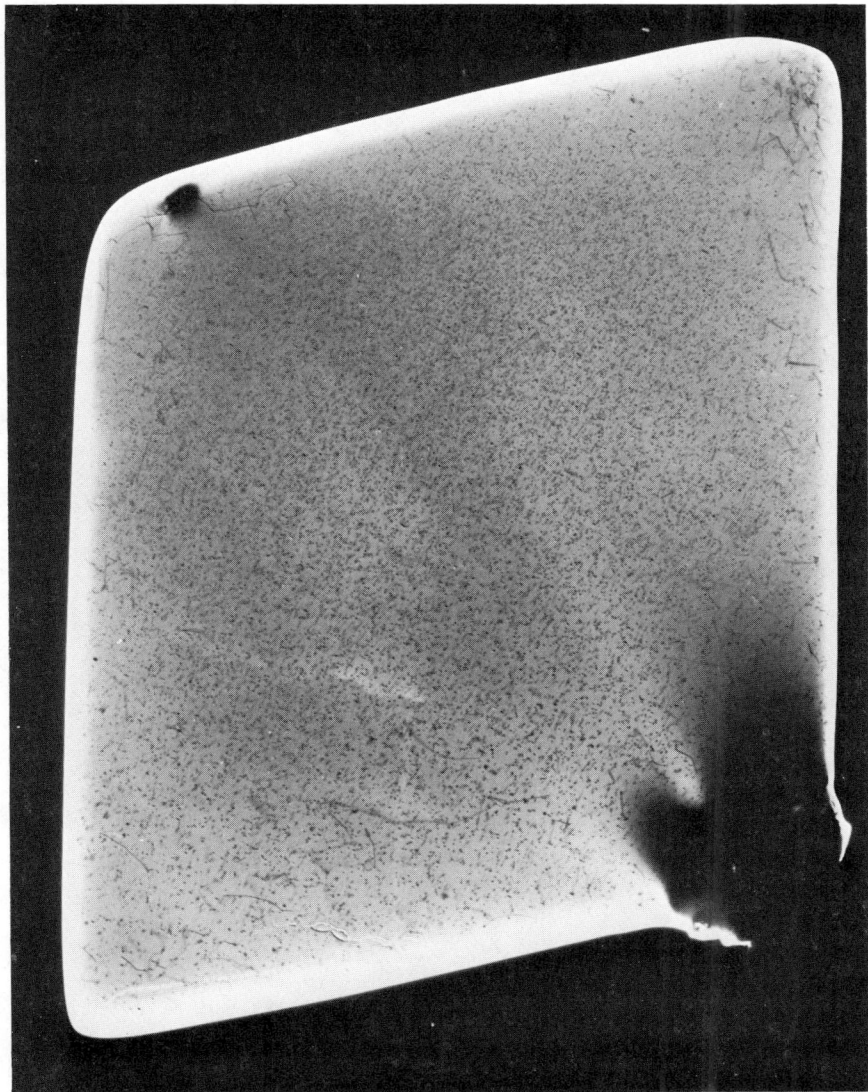

Fig. 4 Topograph of a lamella 1 × 1 × 0.08 cm from crystal II.

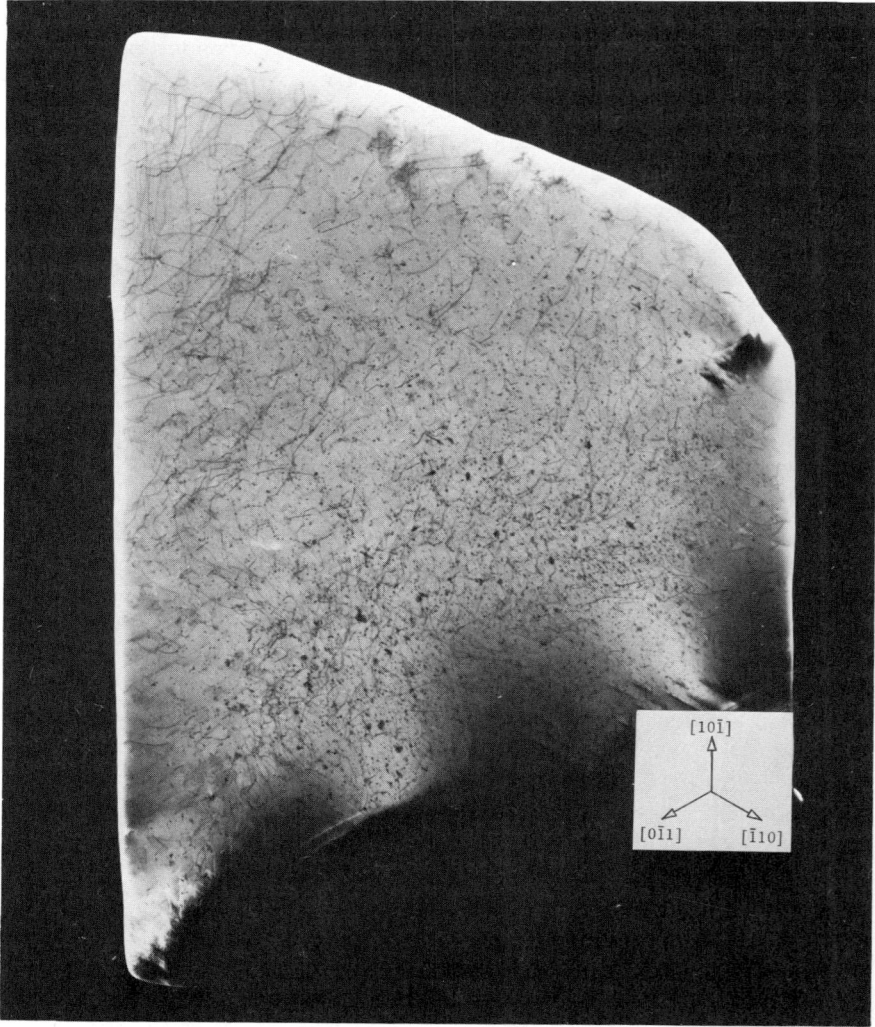

Fig. 5 A $1\bar{1}\bar{1}$ topograph of a lamella (about 1 cm width) from crystal IV. Black area at the bottom of the topograph is the result of accidental deformation.

Crystal V, compression axis $\sim[3\bar{2}\bar{1}]$ and $N = 540/\text{cm}^2$, was tested via etch pitting to determine the stress at which dislocation multiplication was initiated. This experiment consisted of a series of compressions in the spring-loaded jig using the multiple etching technique. There was no multiplication and no motion after stresses on the primary system of 7 g/mm², or of 12 g/mm². After a stress of 19 g/mm² N had increased to $\sim 10^5/\text{cm}^2$, and long slip traces were observed over all the crystal. The crystal was irradiated with the stress relaxed and sectioned along the primary plane, and a topograph of one of the lamellae is shown in Fig. 6. The configuration of the dislocations in this crystal was very similar to that reported previously for stress-relaxed crystals,[12] i.e., groups of dislocations resembling the bunches and tangles seen via electron

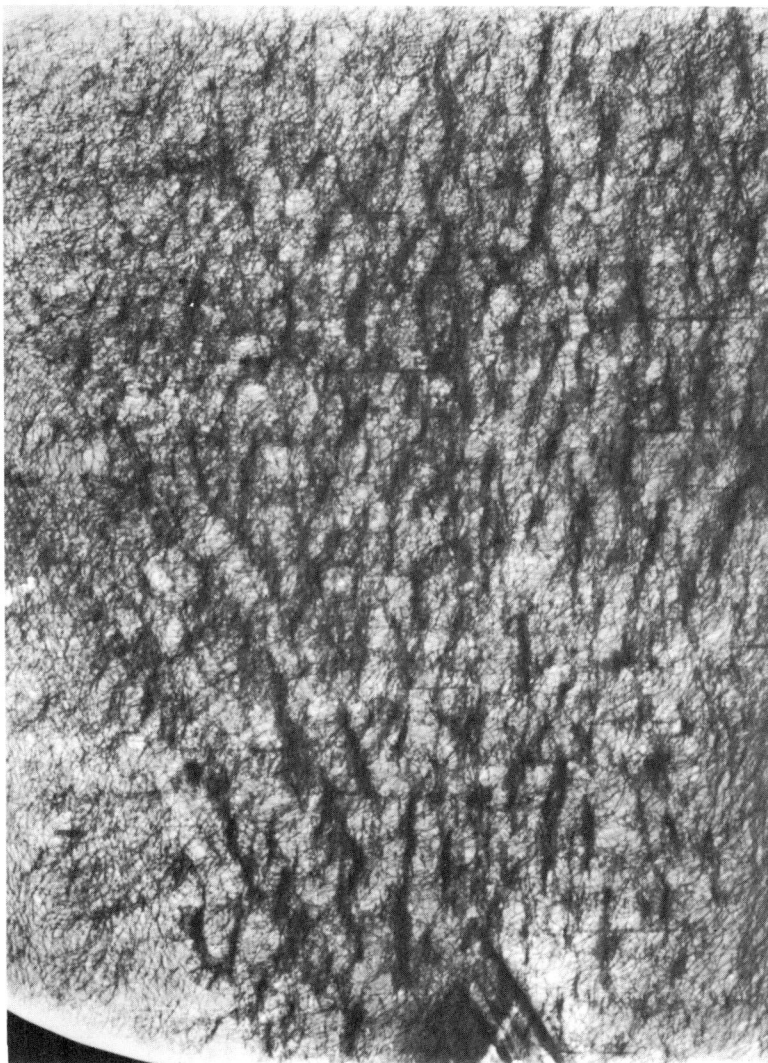

Fig. 6 A portion of a $\overline{1}\overline{1}1$ topograph of a lamella $t \approx 0.03$ cm from crystal V. Smaller dimension about 2/3 cm.

microscopy in more heavily deformed crystals. It is apparent that this dislocation density is near the upper limit for study with Borrmann topography.

Several parallelepipeds were tested in compression in an Instron machine ($\dot{\epsilon} = 4 \times 10^{-5}$) for yield stress and the slope of Stage I. It was reported previously[2] that a correlation seemed to exist in that all specimens cut from the same 2.5-cm diameter crystal exhibited approximately the same yield stress. This correlation was found again, although the number of tests was limited. Yield stresses from 14 to 19 g/mm^2 were measured; the yield stress could easily be determined to ± 1 g/mm^2. In particular, Crystal VI, compression axis a few degrees from [$2\bar{1}1$] and $N = 3 \times 10^3$/cm^2, which was cut from the same large crystal as Crystals IV and V, had a yield stress of 17 g/mm^2. After a stress of 20 g/mm^2 the crystal was etched and the dislocation density was $\approx 6 \times 10^4$/cm^2, averaged over the ($1\bar{1}1$) surface. There was a small band of glide polygonization at the two ends of the crystal in which $N \approx 2 \times 10^5$/cm^2. This crystal has not yet been studied via x-ray topography. The slope of Stage I (θ_1/G) in the crystals tested was $\approx 5 \times 10^{-4}$ which is approximately that previously reported for less-perfect crystals in tension.

The results of these experiments concerning dislocation motion and multiplication and yielding are consistent with the following account of the beginning of deformation in nearly perfect copper crystals. The dislocations remaining after the high-temperature anneal tend not to lie in slip planes, i.e., are heavily jogged. If accidental deformation prior to the deformation test is avoided and if N is $< 10^3$/cm^2, then it is possible to apply a load of 12 g/mm^2 with no plastic deformation. In other words, in previous experiments[12] as the copper crystals have been made more perfect the multiplication stress has decreased to an observed low of 7 g/mm^2, but for low N crystals carefully handled in the present experiments, the multiplication stress was somewhat higher; stresses as high as 12 g/mm^2 with no plastic deformation have been applied. Obviously, this stress is much less than the theoretical strength of copper, thus one must speculate as to the dislocation sources. It is possible that only accidentally introduced sources have been responsible for dislocation generation in all these experiments on low N crystals. It is probable that if all accidentally introduced sources could be avoided, short dislocation segments lying on or near slip planes would begin to act as generators at some higher stress.

The formulation of a slip band is not a simple process; the group of gliding dislocations are not on one glide plane, as determined by both etch pit and topographic observation.[1,2,12] However, this band does not broaden as the stress is increased. In addition, there is some evidence for generation of dislocations on other slip planes from these bands (Fig. 3).

If fresh dislocations have been introduced prior to the experiment, or possibly if the crystal has not been so well-annealed and thus has dislocation segments lying on slip planes (i.e., few jogs), then dislocation motion can occur at stresses less than the multiplication stress. The etch pit evidence previously reported for this conclusion is very convincing, although the present studies give little evidence of simple dislocation motion. However, the fact that dislocations are found in one region of a crystal with radii of curvature corresponding to a stress ≈ 10 g/mm^2 while large regions of the crystals have no dislocations (Crystal I), shows there is some impediment to motion. Further, the details of the configuration of such dislocations suggest that the edge dislocations move more easily than screws. Hence it is concluded that dislocation motion, in these crystals at least, is not continuous. The implication is that dislocation velocities reported may be low. There is no new evidence on the reason for the minimum stress (~ 2 g/mm^2) necessary for dislocation motion.

The combination of factors which determine the yield stress is still unknown. The yield stress is appreciably less in the crystals studied here than previously reported, but there is no apparent correlation between yield stress and either the initial dislocation density (it may be assumed that in these crystals, essentially all grown-in dislocations are fixed in position and may act as barriers) or the dislocation density at the yield stress. If it is assumed that the rate of multiplication is the limiting factor so that $\sigma_x = \alpha G b N^{1/2}$ (N now taken at the yield stress) should be related to the yield stress, a value too low for the yield stress would be predicted (experiments on Crystals V and VI). It is clear that dislocations moved large distances through the crystal under stresses less than the yield stress in some of these experiments. If it is assumed that the velocity of the dislocations is the critical parameter in determining the yield stress, then taking $V = 200$ cm/sec[5] and $\dot{\epsilon} = 4 \times 10^{-5}$ the calculated number of gliding dislocations is $\sim 10^2$/cm^2 compared to the experimental result of 6×10^4/cm^2 (Crystal VI). On the other hand, from the data of Crystal VI at the yield stress the computed value of v is $\sim 5 \times 10^{-2}$ cm/sec. Clearly, an explanation in terms of dislocation velocity is unsatisfactory. Another observation possibly of some significance is that there is some slip on all planes prior to yielding, after which slip occurs essentially only on the primary plane (for crystals oriented for single slip).

In spite of the great variation of pre-yield behavior and of the yield stress in copper crystals observed over a number of years by different researchers, the slope of Stage I is more nearly constant. The value $\theta_1/G \approx 5 \times 10^{-4}$ (measured in compression) is not very different from previous values reported by many other authors. The obvious implication is that the stress–strain relation in Stage I is determined by the interaction between the gliding dislocations in the copper lattice.

Up to this point, it had appeared that by investigating the deformation of more perfect crystals of large dimensions and making use of the techniques of etch pitting and Borrmann topography, definitive information about pre-yield and yielding could be obtained. However, it has been demonstrated that dislocation multiplication and the yielding process in copper are not intrinsic processes dependent only on the dislocations initially in the crystal. Consequently, some other experiments have been performed to investigate dislocation motion, sources and multiplication mechansims.

A topograph of a lamella cut from an undeformed radiation-hardened prallelepiped with $N \sim 500/cm^2$ is shown in Fig. 8a. The orientation of the lamella is shown in Fig. 7; the orientation in the topographs is a mirror image about [10$\bar{1}$] as a result of printing. After annealing for 1 hr at 500 °C to remove the irradiation damage, topographs were again taken of this crystal; it was noted that it had become deformed around the region of its handle, and single dislocations emanated from this region. A series of topographs was taken of this lamella over a period of a year. It was found that the very slight accidental stresses during the necessary handling caused some of the dislocations to move—some more than once. This result is shown in a topograph of the annealed crystal in Fig. 8b, a stereo pair in Fig. 9, and in Fig. 10 which shows a series of topographs for one region where the most dislocation activity occurred. The long lines lying in (11$\bar{1}$) along [1$\bar{1}$0] have $b = [\bar{1}10]$. Note that they curve up to the surface at one end (the other end was often ill-defined and connected to the deformation about the handle); hence they met the surface approximately in the edge orientation. The screw part stayed approximately fixed and the edge part moved along the direction of b, since the distance of the screw part from the surface changed little, if any. Although the stress on this system is unknown—the radius of curvature is difficult to measure accurately in the projection—it is possible that dislocation interactions were not insignificant in this motion. The smallest spacings in this group would correspond to a $\sigma_f \sim$ 1 g/mm^2. The distance to the surface was too great for image forces to affect the screw dislocations. Similar results were seen over this lamella on all three slip systems. For what was probably a bending deformation in this lamella, the mechanism of dislocation motion was clearly defined, and was consistent with the deduced mechanism of motion in Crystal I. In at least some instances the dislocations seem to be obstructed by vacancy clusters. If this mechanism of motion is general, then the implication is that dislocation velocities measured by Marukawai[5] are probably for edge dislocations.

Finally, some results have been obtained from the tensile-Borrmann topography experiments. The orientation and dimensions of the crystal are shown in Fig. 11. In this apparatus, $1\bar{1}\bar{1}$,

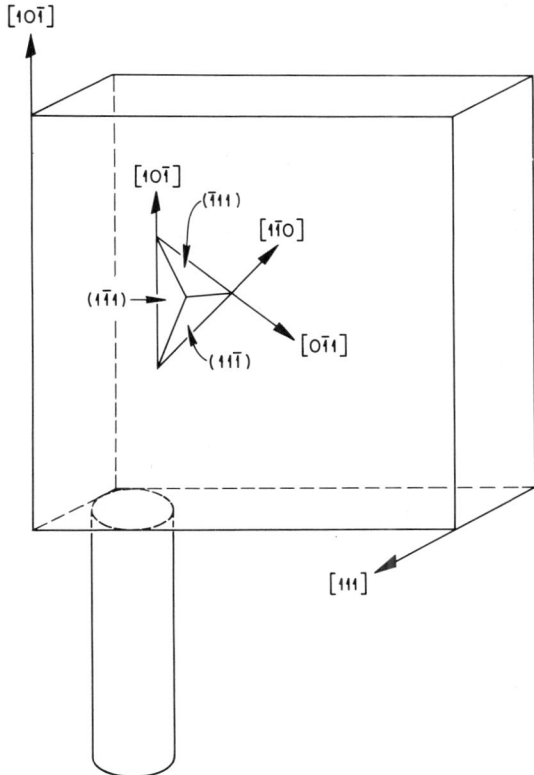

Fig. 7 Diagram of the orientation of the lamella seen in Figs. 8–10.

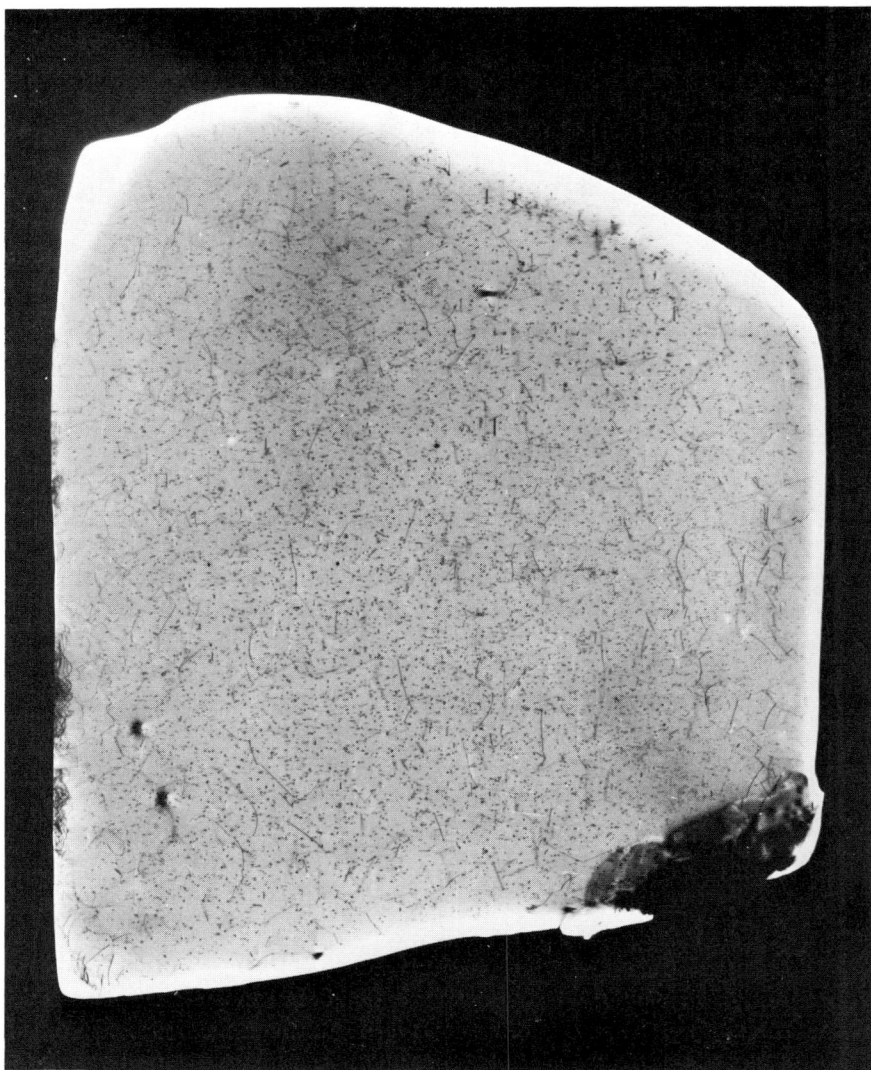

Fig. 8a $\overline{1}1\overline{1}$ topographs of a lamella of dimensions $1 \times 1 \times 0.03$ cm; orientation shown in
Fig. 7. Radiation hardened; many vacancy clusters and few dislocations.

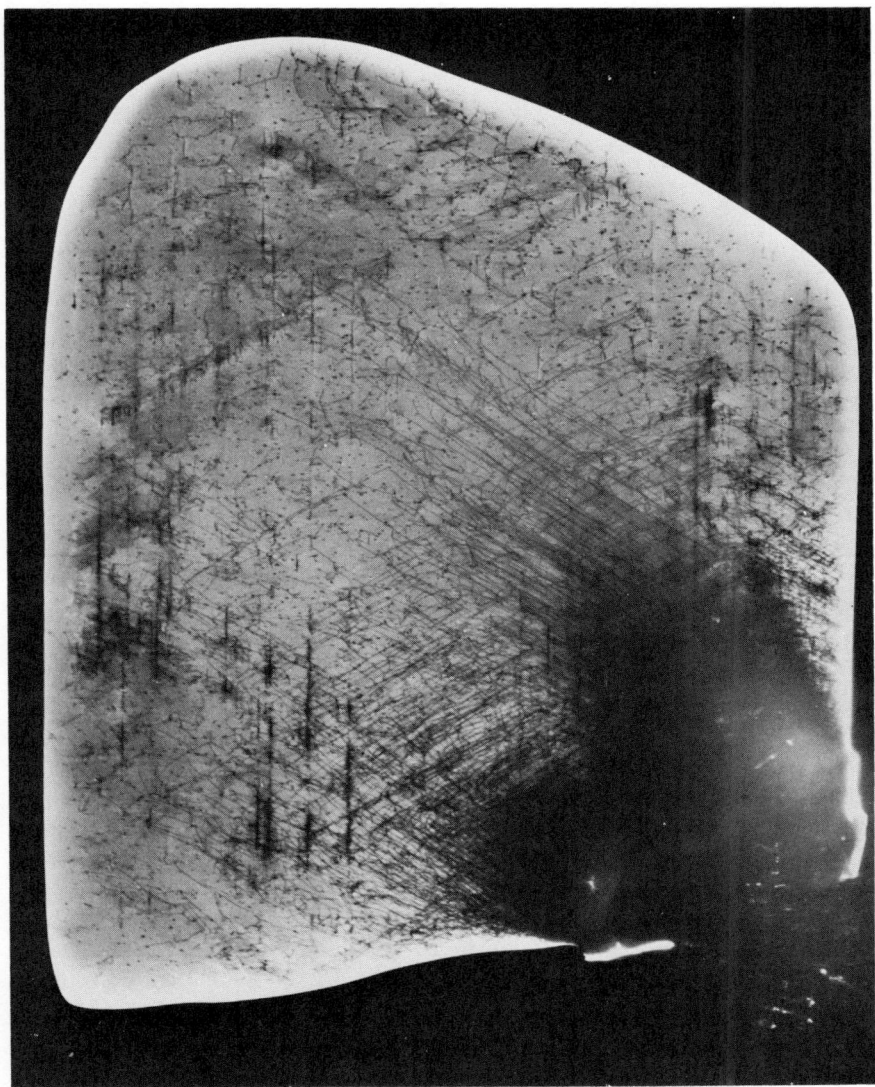

Fig. 8b $1\bar{1}\bar{1}$ topographs of a lamella of dimensions $1 \times 1 \times 0.03$ cm; orientation shown in Fig. 7. After 500°C anneal and accidental deformation.

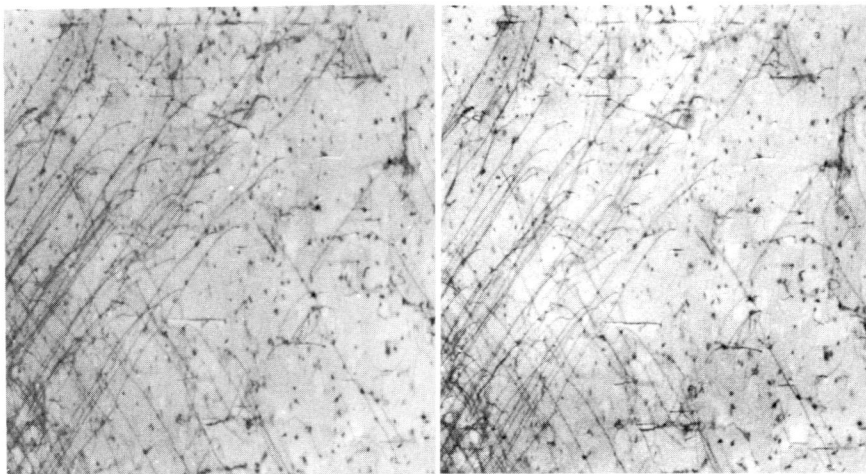

Fig. 9 Stereo pair from a region of the topograph in Fig. 8 b.

$\overline{1}\overline{1}1$, and $0\overline{2}0$ topographs can be taken, and without removing the crystal from the x-ray goniometer. Thus Burgers vectors can be determined. The first tensile specimen of thicknesses 0.075 cm was slightly mishandled so that topography at one end of the specimen was not useful. However, a series of stress increments were applied and topographs were taken of the good end. In Fig. 12 are $\overline{1}11$ topographs of a portion of this end after stresses of '0', 7.6, 9.1, and 9.1* g/mm^2 on the $(\overline{1}11)$-$[01\overline{1}]$ system. In fact, it was very difficult to know when zero stress was applied, but the incremental increases in stress were measured reasonably precisely. For the topograph indicated 9.1* there was some additional deformation which apparently occurred while rocking the crystal to a new position on the goniometer to take a $1\overline{1}1$ topograph (not shown here). These printes are mounted in Fig. 12 so that by using a stereo viewer the change in dislocation position after successive stress increments can be more easily recognized since they appear to be out of focus (these are not stereo views). The bowing-out of some lines can be seen. Other lines move to new positions and still others move but their new positions cannot be recognized; it is possible the latter move out of the crystal. The long lines which appear near the left edge of Figs. 12b, c, and d lie along $[1\overline{1}0]$ and are in $(11\overline{1})$, the slip system of maximum stress. In this region which originally had no dislocations, a slip band is formed with increasing stress which finally contains 7−8 dislocations. Their source is not obvious but appears to be connected

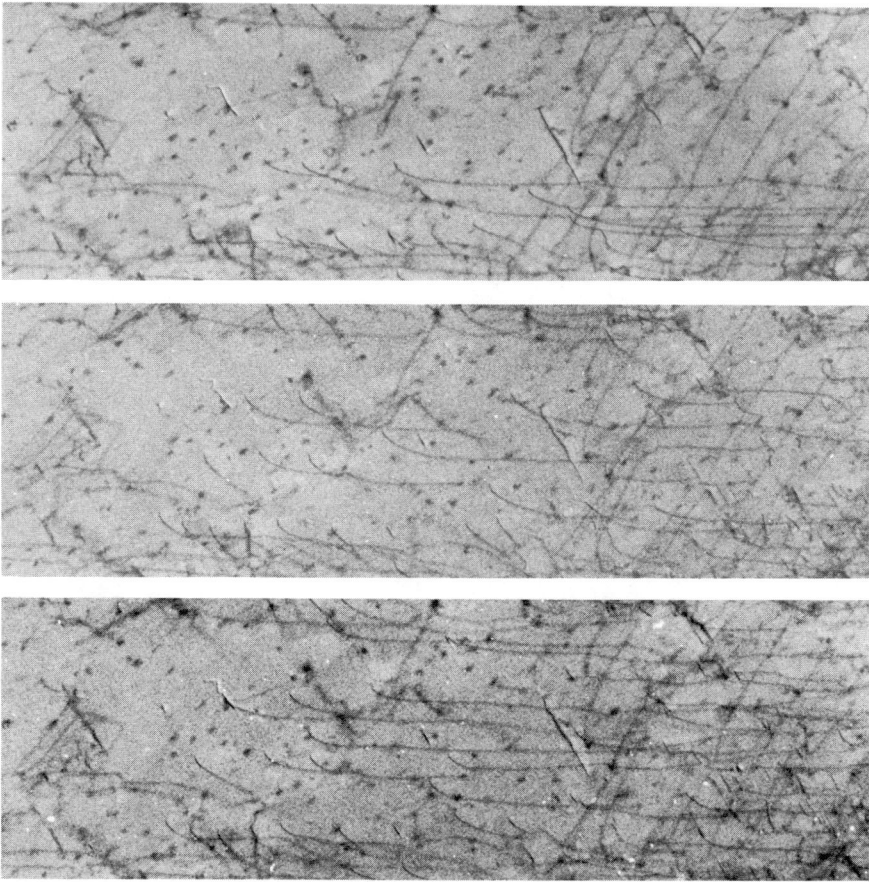

Fig. 10 Enlargements of portions of 1̄11̄ topographs from one region of the lamella of Figs. 8 and 9. The long dimension of the prints is along [1̄10] and the dislocations lying along this direction are screw in character. Small increments of stresses (top to bottom) have caused the dislocations to move from right to left.

with dislocations at the innermost end (Fig. 12*b*). In a 11̄1̄ topograph (not shown) other dislocations are visible in the region of this apparent source. Another group of dislocations with the same Burgers vector and parallel and to the right of this group, but somewhat shorter, can also be seen in Figs. 12*a*–*d*. Actually, if this latter group is viewed with a stereo viewer and one rapidly winks the appropriate eye, then the illusion of watching a source operate is possible. Slip and generation are also apparent in (1̄11) in Fig. 12; in 11̄1̄ topographs new dislocations could be seen in (11̄1)

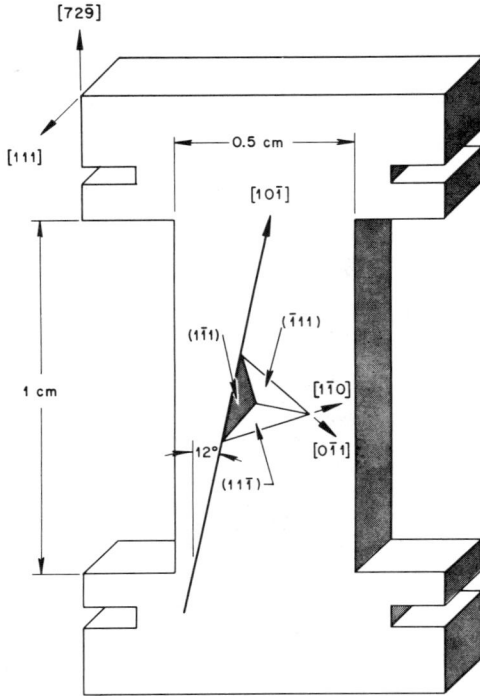

Fig. 11 Orientation and dimensions of the tensile specimen in Fig. 12.

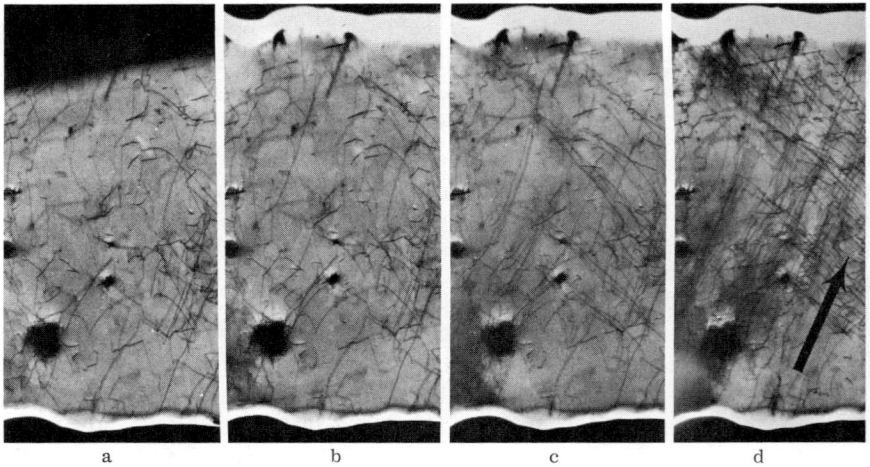

Fig. 12 $\bar{1}1\bar{1}$ Topographs from the same region of the tensile specimen after stresses of a) 0 g/mm², b) 7.6 g/mm². c) 9.1 g/mm² and 9.1* g/mm². The large black spots resulted from surface damage. The arrow indicates [1$\bar{1}$0].

even though the stress on that plane was less than half as great as on (111). Since the stress in the plane of the crystal (111) is approximately zero, the radii of curvature of the lines in Fig. 12 are not in (111) and thus cannot easily be converted into acting stresses. The long dislocation lines are all screw dislocations. Again, they appear to be left as the edge dislocations glide. In three types of deformation—compression, bending, and tension—at stresses less than the yield stress, screw dislocations have recently been observed in copper. In addition, the minimum dimension of these deformed crystals has varied from 0.04 to 1 cm and in all cases the edge dislocations have run from one side of the crystal to the other, leaving long screw dislocations behind. The fact that the dislocations have glided so far from their sources, making it generally very difficult to infer much about the generation mechanism, is probably a result of the high dislocation velocity at these applied stresses. This result might be contrasted to the observations in Si[15, 16] and Ge[17] of a family of new dislocations emanating from a generator. In this tensile experiment the new dislocations can be related to the old ones; surface sources apparently have been avoided, yet plastic deformation has occurred at low stress. The dislocation configuration at "zero stress" (Fig. 12a) certainly is not that of a well-annealed crystal, however.

It is somewhat disconcerting to have topographs which show the dislocation buildup, their Burgers vectors, etc., and still not be able to understand the yield and gilde mechanisms more completely. However, experiments are in progress (for example, the formation of slip bands as seen in Fig. 10 apparently is general) in a continuing effort to investigate these problems.

ACKNOWLEDGMENT

The expert help of F. A. Sherrill in taking the topographs is gratefully acknowledged.

This research was sponsored by the U. S. Atomic Energy Commission under contract with Union Carbide Corporation.

REFERENCES

1. Livingston, J. D.: *J. Appl. Phys.* 31:1071 (1960); *J. Aust. Inst. Met.* 8:15 (1963).
2. Young, F. W.: *J. Appl. Phys.* 32:1815 (1961); 33:963 (1962); 33:3553 (1963).
3. Hordon, M. J.: *Acta Met.* 10: 999 (1962).
4. Petroff, P., and J. Washburn: *J. Appl. Phys.* 37:4987 (1966).

5. Marukawa: *J. Phys. Soc.* Japan 22:499 (1967).
6. Brydges, W. T.: "The Early Stages of Plastic Flow in Crystals," Thesis, Mass. Inst. of Tech., 1966.
7. Young, F. W.: *J. Phys. Chem. Solids* 28:789 (1967).
8. Kroupa, F., and P. B. Hirsch: *Discussions of Faraday Soc.*, No. 38, p. 52 (1964).
9. Young, F. W., T. O. Baldwin, and F. A. Sherrill: "Lattice Defects and Their Interactions," R. R. Hasiguti (ed.), Gordon and Breach, in press.
10. Johnston, W. G.: *J. Appl. Phys.* 33:2716 (1962).
11. Nabarro, F. R. N., Z. S. Basinski, and D. B. Holt: *Advances in Physics* 13:193 (1964).
12. Young, F. W., and F. A. Sherrill: *Can. J. Phys.* 45:757 (1967).
13. Young, F. W., T. O. Baldwin, A. E. Merlini, and F. A. Sherrill: "Advances in X-Ray Analysis," vol. 1, p. 1, Plenum Press, 1966.
14. Jones, D. A., and J. W. Mitchell: *Phil. Mag.* 2:561 (1957).
15. Authier, A., and A. R. Lang: *J. Appl. Phys.* 35: 1956 (1964).
16. Dash, W. C.: *J. Appl. Phys.* 27:1193 (1956).
17. Gerold, V., and F. Meier: "Direct Observations of Imperfections in Crystals," p. 509, Interscience Publishers, New York, 1962.

DISCUSSION *on Paper by F. W. Young, Jr.*

P. HIRSCH: Is it possible to account for the broadening of the slip band by cross-slip at vacancy clusters, dislocation loops or small precipitates? Such sites are particularly favorable for initiating cross-slip for the following reasons: (1) The dislocation bowed out between two such centers exerts a force due to the line tension which is related to the stress τ by the expression τbl; the effect of the stress is therefore magnified by a factor l/D, where D is the effective diameter of the defect. (2) The strain field from small loops or precipitates falls off rapidly with distance from the defect. This has the effect of compressing the partials, since the repulsion by the defect of the second partial is less than that of the leading partial. Sometimes the situation is such that the first partial is repelled, and the second partial is attracted by the defect, thus considerably aiding constriction. (3) In the case of vacancy clusters, these can combine with the screws producing large jogs. On this mechanism it would be expected that the slip bands broaden further as they extend into the crystal. During their motion, if some of the defects, such as loops or vacancy clusters, are swept up by the leading dislocations, those following behind should deviate less, and might also be expected to move more rapidly.

It should be noted that the mechanism of nucleation of cross-slip at small defects is quite different from that usually considered to hold at the beginning of τ (Seeger's mechanism). In the latter case constriction is effected by the stress-concentration operative along the whole length of the dislocation. By contrast, at small defects, there is an effective stress concentration localized at the defect by the bowed-out dislocation, and, second, there is a stress gradient at the particle which also aids the construction.

It might be added that there is considerable evidence from electron-microscope observations of crystals containing small precipitates that cross-slip does occur at these sites (e.g., Hirsch and Lally (1965),* Humphreys (1967).[†]

F. YOUNG: Thank you for this comment. From the dislocations in Figs. 2 and 3, your $l \sim 100\mu$ and D is 10μ, if one takes the effective diameter as the size of vacancy cluster images. Then the magnification of the first effect would be about a factor of ten. Of course, I have no direct evidence on the other two effects.

Z. BASINSKI: Since the dislocation in your lamellae appeared to move in from some surfaces only, are the pictures necessarily an evidence for surface samples, or is it possible that nonaxial loading could be responsible?

F. YOUNG: While it is exceedingly difficult to rule out completely nonaxial loading, a number of self-consistency checks were made in these studies, the results of which were to convince me that nonaxial loading could not be responsible for the observed dislocation configurations. For example, the extent of activity of a given slip system did not change monotonically along a parallelepiped.

J. D. CAMPBELL: Can you estimate the plastic strain in the crystal with a dislocation density of 10^5 cm^{-2}?

F. YOUNG: The strain for crystals deformed to similar dislocation density has been measured by Hordoni[3] and Brydges;[6] I would estimate the strain in the indicated crystal to be about 10^{-4}.

R. ARMSTRONG: Would you say that one might be able to produce polycrystals having dislocation densities comparable to those obtained in your single crystals? If so, would you give some speculation on the stress-strain behavior you might expect for such polycrystals?

F. YOUNG: There is no experimental evidence on this question to my knowledge.

N. BROWN: Since you cannot detect dislocation motion of less than one micron, is 2 gm/mm^2 the stress to observe the movement of an etch pit, an intrinsic yield point?

F. YOUNG: I prefer to reserve the term-"yield point" for its usual definition in terms of the stress-strain curve; for example,

* Hirsch, P. B., J. S. Lally: *Phil. Mag.* **12**: 595 (1965).
† Humphrey, F. J.: Ph.D. thesis, University of Oxford (1967).

the yield point so defined in these crystals was 14-20 g/mm^2. As discussed in the paper, I do assume that the motion stress of $2\ g/mm^2$ is an intrinsic value.

G. *ALEFELD*: You gave a value for the motion stress of dislocations in Cu at room temperature, namely $2\ g/mm^2$ and mentioned a connection of this value to internal friction observations. For clarification, I want to point out that internal friction and modulus measurements (done on much less perfect crystals than yours) show that at stresses which are a factor 50 lower dislocation motion is still possible. Since the dislocation strain at these low stresses is proportional to stress, one can conclude that at room temperature there is really no minimum motion stress, at least for the less perfect crystals. The value of $2\ g/mm^2$ corresponds to a strain amplitude $0.5\ 10^{-6}$, a value, at which for relatively clean material, the anelastic strain starts to become non-linear with stress. It thus seems to me that the definition of a minimum motion stress as determined with x-ray topography may be coupled to that stress value at which the modulus defect becomes amplitude-dependent, although I must admit that we know nothing yet about modulus defect of crystals with your perfection.

Have you determined the motion stress of individual dislocations after you deformed the crystal slightly?

If you apply a stress close but below the motion stress, can you resolve a change in curvature of the dislocations, if you compare the unloaded and the loaded and pinned state? If yes, it would be interesting to do the loading at He temperature and unpin at this temperature. If there are Peierls potentials one may see a straightening out of the dislocation in the stressed state.

F. *YOUNG*: The observation of a motion stress of $2\ g/mm^2$ was made using etch-pit observation only, and this motion stress applies only to fresh dislocations. Studies of the bowing of dislocations under small stresses are in progress, but no definite results for stresses approximately $2\ g/mm^2$ have yet been obtained.

DISSIPATION OF ENERGY BY THE MOTION OF DISLOCATION THROUGH A FLUCTUATING INTERNAL STRESS FIELD

H. Kressel and N. Brown*

School of Metallurgy and Material Science,
Laboratory for the Research on the Structure of Matter,
University of Pennsylvania,
Philadelphia, Pa.

ABSTRACT

Dislocation damping during a slow microstrain experiment involves at least two mechanisms, one of which depends on prestrain. This mechanism has been attributed to the existence of ripples in the long-range internal stress field. A theory has been presented for the way in which ripples in a long-range internal stress field produce dislocation damping. Experimental observations on nickel have been interpreted in terms of the theory.

* Present address — RCA Laboratories, Princeton, New Jersey.

1 INTRODUCTION

Prestrained metals exhibit the stress–strain behavior shown in Fig. 1. A stress level exists, below which, the stress–strain curve is linear and retraceable and, therefore, elastic. Above a stress designated σ_E, purely elastic behavior ceases and closed hysteresis loops occur. Closed loops mean that all the strain is recovered, but energy is dissipated during the load–unloading cycle. At low temperatures, the closed hysteresis loops remain constant for a given amplitude of stress. This signifies that the state of the material remains constant in the range where closed hysteresis loops are observed. There is a stress amplitude σ_A, above which a permanent strain is observed after unloading. The existence of an open loop signifies that the state of the metal has changed in the sense that all the dislocations do not return to their initial position after being exposed to a stress σ_A. It must be emphasized that the values of σ_E and σ_A depend on the experimental strain sensitivity.

Prior experimental investigations[1,2,3] have shown that σ_E was much less sensitive to prestrain than σ_A, especially for large amounts of prestrain (i.e., greater than 1%). Also, for pure metal, σ_E was comparatively insensitive to temperature compared to σ_A.

The elastic limit σ_E, is closely connected to the frictional stress which is associated with the energy loss from a closed hysteresis

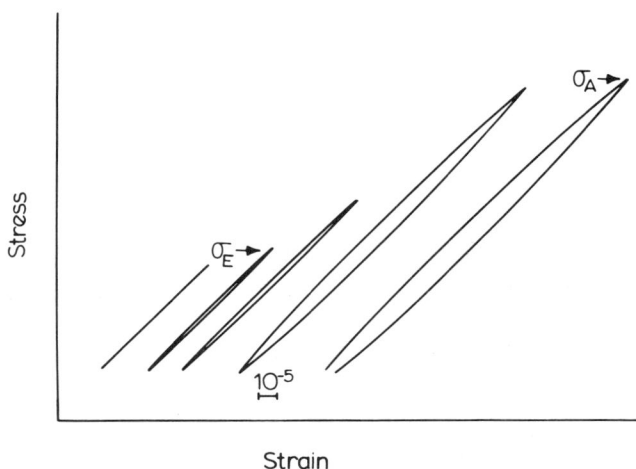

Fig. 1 Typical stress-strain loops in the microstrain region of a prestrained specimen.

loop. A simple model for the frictional[3,4] stress σ_f, shows that σ_E should be twice σ_f. Experiments show that σ_E is always greater than σ_f by a factor which is almost the order of two. Also, the temperature and prestrain dependence of σ_E and σ_f are generally the same.

In this paper, attempts are made to show the experimental aspects of the frictional stress and to give a theory which connects the frictional stress and the dynamic behavior of the dislocations under a fluctuating internal stress field.

The energy loss per cycle is obtained by measuring the area of the closed stress-strain loop. This energy has been[5] associated with a frictional stress through the equation

$$\Delta W = \oint \sigma_f d\epsilon_d \tag{1}$$

where ΔW = area of the loop or energy loss per cycle

σ_f = frictional stress, and

ϵ_d = strain produced by dislocations.

In general, σ_f is a function of ϵ_d. However, σ_f cannot be measured directly, but is obtained by averaging over a measured interval of strain. Thus, σ_f may also be defined as follows:

$$\Delta W = 2\bar{\sigma}_f \oint_0^{\epsilon_d(\max)} d\epsilon_d = 2\bar{\sigma}_f \epsilon_d(\max) \tag{2}$$

where $\bar{\sigma}_f$ is the frictional stress which has been averaged over the strain interval $\epsilon_d(\max)$, the amplitude of the dislocation strain. Equation (2) implies that the same frictional stress operates during loading and unloading. Thus, $\bar{\sigma}_f$ can be obtained from the plot of ΔW vs. $\epsilon_d(\max)$. The value of $\epsilon_d(\max)$ should be obtained by measuring the deviation of the stress-strain loop from the unrelaxed modulus. Since this measurement is not very accurate because the unrelaxed modulus is difficult to establish, the value of $\epsilon_d(\max)$ may be estimated from the maximum width of the stress-strain loop. In the remainder of this paper, the frictional stress will be designated simply as σ_f, but the average value of $\bar{\sigma}_f$ is implied. Also ϵ_d will be used instead of $\epsilon_d(\max)$.

Before presenting the main topic of this report, another definition of the frictional stress by Lukas and Klesnil[6] should be discussed, namely,

$$\left(\frac{dW}{d\epsilon_d}\right)_{\epsilon_d = 0} = 2\sigma_f \qquad (3)$$

where $(dW/d\epsilon d)_{\epsilon_d} = 0$ is the slope of the curve of loop area vs. dislocation strain evaluated at $\epsilon_d = 0$. One may generalize on this definition and obtain the frictional stress as a function of ϵ_d by means of the following type of experiment. If the frictional stress at a microstrain ϵ_d is desired, then the specimen is biased to that microstrain value and a series of closed loops is obtained at successively higher stresses above the bias stress. Then the curve of ΔW vs. $[\epsilon_d - \epsilon_d \text{(bias)}]$ may be obtained and the initial slope at the bias strain is related to the frictional stress at the bias strain as follow:

$$\left(\frac{dW}{d\epsilon_d}\right)_{\epsilon_d \text{(bias)}} = 2\sigma_f \qquad (4)$$

By varying the bias strain, $\bar{\sigma}_f$ may be obtained as a function of the strain. Equation (2) defines σ_f as a function of the strain amplitude at a fixed bias strain which is not necessarily zero. Equation (4) defines σ_f as a function of the bias strain at a fixed strain amplitude which is determined by the experimental strain sensitivity.

Extensive discussions[7,8,9,10,11] have been made concerning the manner in which frictional stress may be described. Some of the discussion has centered on the fact that energy is dissipated during only part of the dislocation motion. If the part of the dislocation motion during which energy is not dissipated cannot be measured, then only the average frictional stress can be recorded.

If a constant frictional stress exists (as was assumed in previous work[5]) then a plot of ΔW vs. ϵ_d should be linear. The experimental data for a number of metals, for example, copper,[12] alpha brass,[12] beryllium,[13] and iron,[6] indicate that such plots are usually linear only over limited values of ϵ_d and that particularly significant slope changes are observed for larger values of dislocation strain. The work of Bonfield and Li,[14] suggested that the frictional stress for large values of dislocation strains could be prestrain-dependent. The purpose of the work presented here was to investigate experimentally and theoretically the observed loss mechanism and, in particular, its dependence on the degree of specimen prestrain in polycrystalline nickel. Part of the theory is similar to that proposed by Weertman[15] to explain how dislocation damping occurs in dilute alloys. Before presenting our theory it may be of greater prior interest to present additional experimental observations.

2   EXPERIMENTAL METHOD

The specimens used were polycrystalline zone refined nickel rods provided by the Materials Research Corporation, Orangeburg, N.Y. The main impurities were (ppm): C, 20; O_2, 1; N,1; H, 0.07; Fe, 10 and Si, 10.

The specimens were annealed by the manufacturer at 800°C in a vacuum of 10^{-6} Torr, and their average grain diameter was 0.75 mm. The gauge section of the specimens was electrolytically polished in a solution of 40% H_2SO_4 in water after machining. This section was 1.25 cm long and had a diameter of 0.18 cm.

A brief description of the basic experimental technique will be given (greater detail is given in Ref. [3]). A ball–and–socket arrangement was used to connect the specimen to the Instron tensile testing machine. The extension of the specimen as a function of applied load was measured via a parallel plate capacitor secured to the shoulders of the specimen. The capacitor formed one arm of a bridge which was balanced at the beginning of each test. The out-of-balance signal from the bridge was fed to the x axis of an x-y recorder while the load was monitored on the y axis which was connected to the Instron load-cell circuit. The lower capacitor plate was mounted on a micrometer thread to permit ready changing of the plate separation and hence the sensitivity of the capacitor.

The strain sensitivity was maintained as constant as possible throughout the tests at a value of about 2×10^{-6}. The head speed of the machine was maintained at 0.04 inches per minute. The time to make a closed loop varied from about 10 to 30 seconds depending on its amplitude. The testing procedure consisted of successive cycles of applying a given load and immediately reducing it back to the minimum load required to keep the specimen assembly aligned.

To study the effect of prestrain on the microstrain parameters, the specimen was strained to the desired degree with the capacitor-plate spacing adjusted to a lower sensitivity value. The microstrain measurements were then performed after readjusting the plate separation back to its standard value.

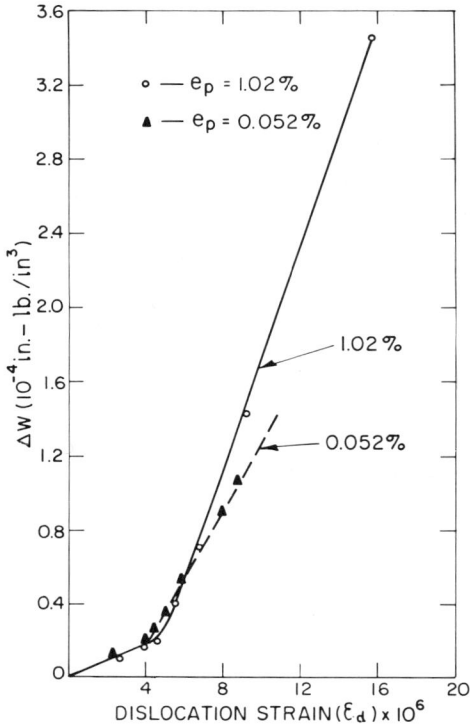

Fig. 2 Energy loss per cycle, ΔW, as a function of forward dislocation strain for two values of prestrain.

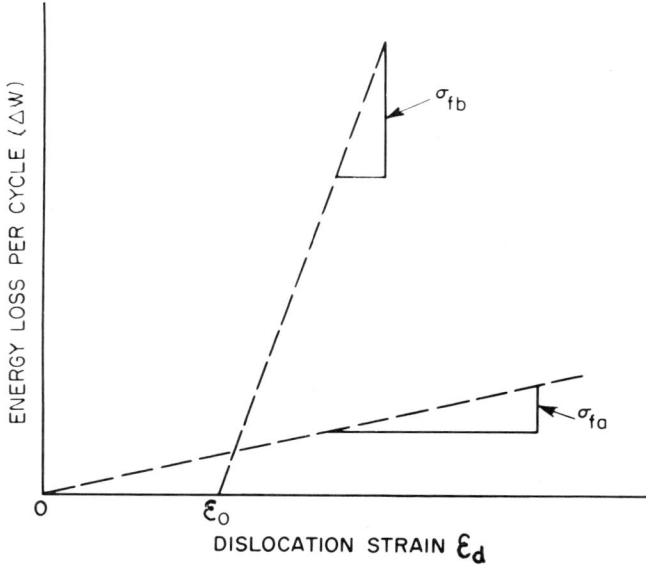

Fig. 3 Definition of the average frictional stresses σ_{fa} and σ_{fb}.

3   EXPERIMENTAL RESULTS AND ANALYSIS

A typical set of stress–strain curves is shown in Fig. 1. The forward dislocation strain ϵ_d was determined from a measurement of the maximum loop width. These measurements were made only on closed loops. The closed loops are reversible, so that within the sensitivity of the experimental technique, all the dislocations return to their original position after unloading. The open loop is produced at a stress which is high enough to produce a measurable permanent strain (irreversible dislocation motion). The energy loss per cycle, ΔW, was determined from the area of closed loops only.

Figure 2 shows a typical set of curves of ΔW vs. ϵ_d for two values of prestrain, $\epsilon_p = 0.052\%$ and $\epsilon_p = 1.02\%$ (note the curvature). This type of curve was observed in all the of the tests made. While the two parts of a curve frequently are not quite linear over an extended range of ϵ_d values, it is desirable for the purpose of analysis to separate the experimental curves into two linear components as shown in Fig. 3. The intercept of the larger slope curve on the ϵ_d axis is denoted by ϵ_0. From Fig. 3 two average frictional stresses are defined. If the dislocation strain does not exceed ϵ_0, then a loss mechanism described by σ_{fa} is operative, and the observed energy loss per cycle is then given by

$$\Delta W = 2\sigma_{fa}\epsilon_d \tag{5}$$

When $\epsilon_d > \epsilon_0$, a second loss mechanism, which is described by σ_{fb}, becomes operative and the energy loss is now given by

$$\Delta W = 2\{\sigma_{fa}\epsilon_0 + \sigma_{fb}(\epsilon_d - \epsilon_0)\} \tag{6}$$

From one-half of the slope of the ΔW vs. ϵ_d curve near the origin, one obtains σ_{fa}, and from the part of the same curve where $\epsilon_d > \epsilon_0$, one obtains σ_{fb}. The variation of σ_{fa} and σ_{fb} as a function of prestrain was determined from curves similar to those shown in Fig. 2. The resultant data are shown in Fig. 4, with the estimated error being largely determined from the assumption of linearity in the ΔW vs. ϵ_d curves.

The significant feature of these data is the contrast between the behavior of σ_{fa} and σ_{fb}. While σ_{fa} is constant within the experimental error, σ_{fb} increases by a factor of two in the first one-percent increase of prestrain. It is interesting to compare the variation of σ_{fb} with prestrain to that of flow stress σ_A, which is

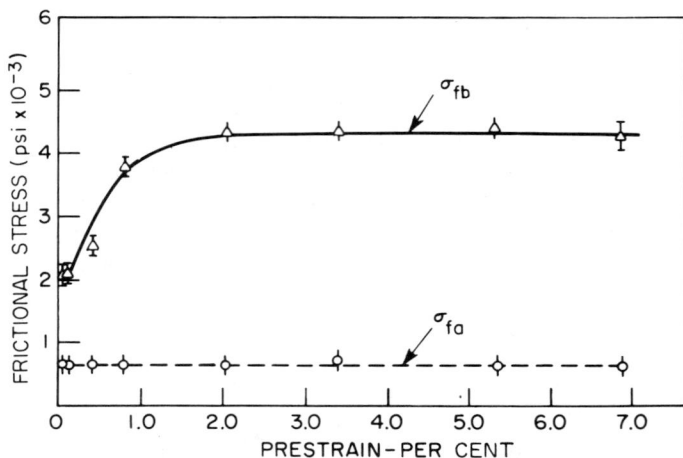

Fig. 4 Frictional stresses as a function of specimen prestrain.

now defined as the stress required to deform the specimen to a permanent plastic strain $\epsilon_p = 0.001\%$. A plot of σ_A vs. ϵ_p is shown in Fig. 5. A similarity between the behavior of σ_A and σ_{fb} exists in that both show the greatest rate of variation with prestrain for small values of prestrain. However, σ_{fb} is independent of prestrain beyond 2% deformation while σ_A is not. Also, σ_A the stress to

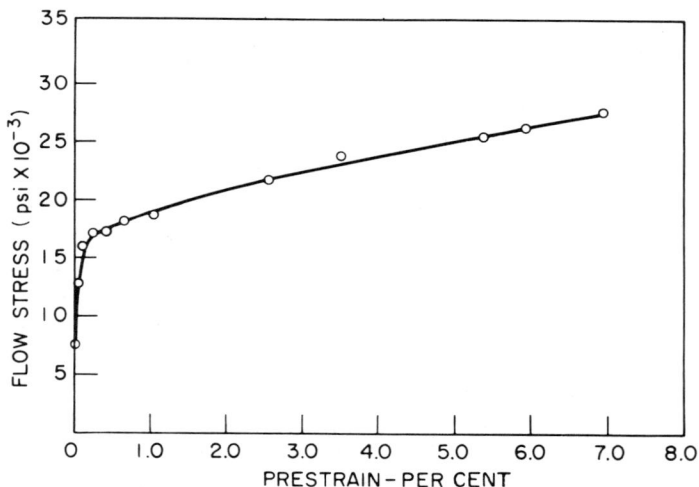

Fig. 5 Flow stress as a function of specimen prestrain.

produce a permanent strain of about (10^{-5}), is not very different from the usual macroscopic flow stress in magnitude and dependence on prestrain.

4 THEORETICAL ANALYSIS

The two energy-dissipation mechanisms which are described by σ_{fa} and σ_{fb} are believed to be quite different. However, σ_{fb}, but not σ_{fa}, shows a dependence on the dislocation density and morphology as evidenced by its variation with prestrain in the early stages of deformation. Lukas and Klesnil[6] have reported a similar result concerning σ_{fb} in pure iron. It was suggested by Bonfield and Li[14] that the energy loss in beryllium may be dependent on the amplitude of the internal stress field, which, in turn, depends on prestrain. It is not obvious, however, how elastic interactions among dislocations can give rise to an energy loss. The theory presented here explains how dislocations dissipate energy while moving through an internal stress field although the dislocation strain is completely recoverable upon unloading. The general concept is similar to that used by Weertman[15] to explain internal friction from impurities.

The existing data on dislocation morphology[16] show that the distribution of dislocations in a cold-worked metal is very complex and, consequently, the internal stress field produced by the dislocations should be equally complex. There should be long-range, high-amplitude stresses which affect the work-hardening in the region of macroscopic flow.[17] (See extensive papers from Stuttgart by Seeger.) There should also be smaller ripples in the internal stress field which are superposed on the long-range, high-amplitude internal stress field. The present theory is concerned with the motion of the dislocation through the small ripples of the internal stress field over sufficiently short distances so that the dislocation can still return to its original position under the combined restoring force produced by both the long-range internal stress field and its own line tension. Whether the long range restoring force is caused by the stress field of dislocations or by the line tension is not important in this analysis. Roberts and Brown[5] and Kuhlmann-Wilsdorf[18] have emphasized the effect of line tension. Both phenomena will be referred to as the long-range restoring force. The small ripples in the internal stress field are represented by an amplitude σ_0 and wavelength λ. Since we are interested in

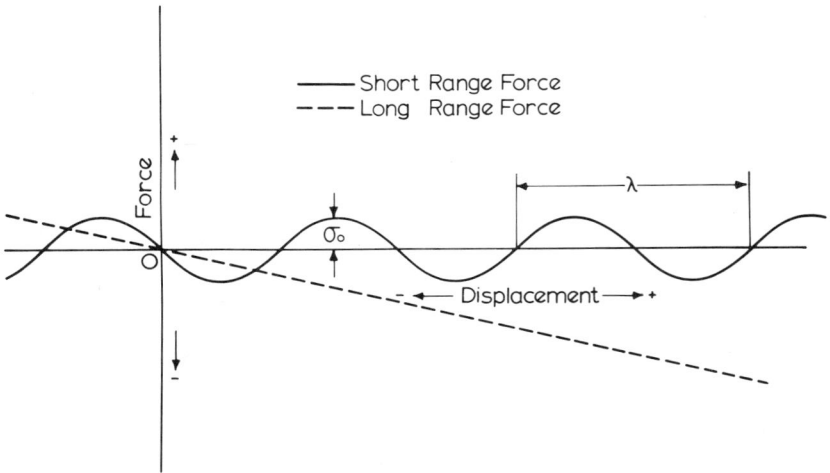

Fig. 6a Long range and short range internal stress fields in a crystal.

small displacements of the dislocation relative to the long-range stress field, we may linearize the long-range restoring force (Fig. 6a). In Fig. 6b the total internal elastic force on the dislocation is obtained by combining the linearized long-range restoring force with the ripples in the internal stress field. The ripples in the internal stress fields have been assumed to be sinusoidal only for the purpose of simplifying the analysis.

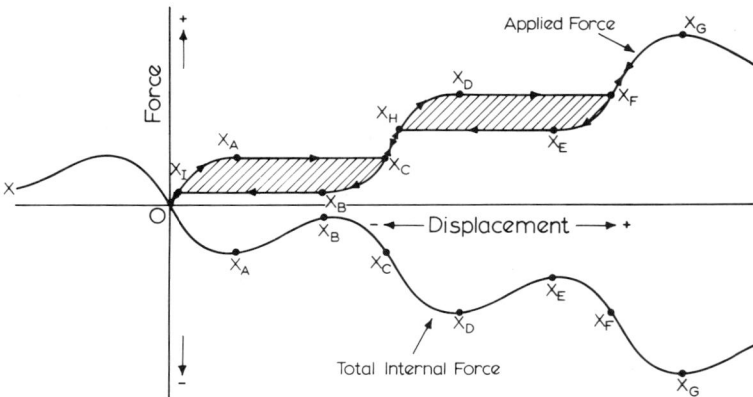

Fig. 6b Internal force and applied force as a function of the position of a dislocation.

Now the motion of an individual dislocation will be qualitatively described as it moves through the fixed internal force field (Fig. 6b) under the action of an externally applied stress. The externally applied stress increases slowly with time at a rate determined by the head speed of the testing machine. The net force on each dislocation is determined by the sum of the internal forces and the applied force. When the internal and external forces are in balance, then the dislocation moves at a rate determined by the rate of the testing machine. When the forces on a dislocation do not balance, the dislocation will accelerate very rapidly to a terminal velocity which is determined by the amount of unbalanced force and by a frictional stress that is assumed to depend linearly on velocity. When the dislocation moves to a position in the lattice where the unbalanced force is again zero, then the dislocation rapidly decelerates to a slow velocity determined by the testing machine. The rapid acceleration (and deceleration) of the dislocation is determined by the very small "mass" of the dislocation. The velocity-dependent friction is assumed to have the form

$$\sigma b = B \dot{X} \tag{7}$$

where $\dot{X}$ is the velocity and B has a value of about $10^{-4} \frac{dyne}{cm/sec}$ [19] and was determined by experiments involving forced vibrations at megacycle frequencies.

A detailed qualitative description of the behavior of an individual dislocation will be given with the help of Fig. 6b. The frictional stress σ_{fa}, is omitted from Fig. 6b to emphasize the velocity-dependent part of the energy loss. The upper curve gives the change in the applied force as the dislocation moves in the positive X direction. The internal force remains the same at each point during loading and unloading since it is assumed that the dislocation distribution is not modified appreciably during these small microstrains. The dislocation starts at point 0, and moves slowly from 0 to X_A at a rate determined by the testing machine. Beyond X_A, the internal force decreases, and, therefore, an unbalanced force exists which accelerates the dislocation to a high velocity which reaches a maximum value at point X_B. Beyond X_B, the dislocation decelerates because the unbalanced force is decreasing until it becomes zero again at point X_C where the dilocation velocity is again governed by the loading rate of the testing machine. From X_A to X_C, the dislocation dissipated energy because it was moving at a high speed, and the energy loss in accordance with Fig. 6b would be given by the area between the horizontal line from X_A and X_C and the curve for the total internal force.

It will be shown that only a short time is spent by the dislocation during an energy-dissipating interval such as from X_A to X_C. Since the dislocation has a low mass, it instantly reaches the velocity given by Eq. (7). This means that whatever the unbalanced stress $\Delta\sigma$ on the dislocation at a given position, its velocity at that position is practical given by $\Delta\sigma b/B$. The time constant for the dislocation to respond to changes in $\Delta\sigma$ is given by

$$\tau = \frac{m}{B} \approx \frac{\rho b^2}{2B} \tag{8}$$

where m is the mass of the dislocation, and ρ is the density of the metal. For nominal values of the parameters, $\tau \approx 10^{-12}$ sec. The time it takes the dislocation to maove from X_A to X_C is given by $t(AC) = (X_C - X_A)B/\Delta\bar{\sigma}b$ where $\Delta\bar{\sigma}$ is the average unbalanced stress on the dislocation between points X_C and X_A. If $(X_C - X_A) \simeq 1000\,b$ and $\Delta\bar{\sigma}$ is about one-tenth the applied stress or, in our case, about 100 psi, then $t(A_C) \simeq 10^{-8}$ sec. During this time, the testing machine would increase the applied load by a negligible amount. This fact is useful when solving the equation for motion in a later section of this paper.

Figure 6b indicates the change in applied stress as a function of the position of the dislocation. Starting from the origin, the path during loading is $0, X_A, X_C, X_D, X_F$, etc. The dislocation motion is in phase with the applied stress during intervals such as $0\,X_A$ and $X_C\,X_D$. Energy is dissipated during intervals such as $X_A\,X_C$ and $X_D\,X_F$. During unloading the path is $X_F, X_E\,X_H\,X_B, X_I, 0$. During intervals $X_F\,X_E$, $X_H\,X_B$, and $X_I, 0$ stress and dislocation motion are in phase, and during $X_E\,X_H$ and $X_B\,X_I$ energy is dissipated by the velocity mechanism. Thus, during a load-unload cycle between 0 and X_F the energy dissipated is given by the cross-hatched areas associated with the curve of applied force.

The differential equation that describes the dislocation motion is given by

$$m\ddot{X} + B\dot{X} + RX + \sigma_0 b \sin\left(\frac{2\pi X}{\lambda}\right) = [\sigma(t) - \sigma_{fa}]\,b \tag{9}$$

where R is the constant associated with the long-range restoring force, and $\sigma(t)$ is the applied stress which varies slowly with time. To calculate the energy loss associated with the damping during the acceleration of the dislocation, the above equation can be simplified as follows: (2) the term $m\ddot{X}$ can be neglected since m is very small; (2) $\sigma(t)$ can be taken as a constant over the region where the dislocation accelerates and decelerates. Therefore, during loading and over a region such as X_A to X_C, the above equation becomes:

$$B\dot{X} + RX + \sigma_0 b \sin\left(\frac{2\pi X}{\lambda}\right) = [\sigma(A) - \sigma_{fa}] b \tag{10}$$

where $\sigma(A)$ is the value of $\sigma(t)$ at $X = X_A$.

The energy dissipated when a unit length of dislocation moves over a distance $X = \lambda$ is given by

$$\Delta E = \int_{X_A}^{X_C} B\dot{X} \, dX + \int_0^\lambda \sigma_{fa} b \, dX \tag{11}$$

The first term represents the velocity–dependent energy loss while the second term is the velocity–independent loss. From Eqs. (10) and (11)

$$\Delta E = \int_{X_A}^{X_C} \left[(\sigma(A) - \sigma_{fa}) b - RX - \sigma_0 b \sin\left(\frac{2\pi X}{\lambda}\right)\right] dX + \int_0^\lambda \sigma_{fa} b \, dX \tag{12}$$

Integrating the above equation

$$\Delta E = (\sigma(A) - \sigma_{fa}) b (X_C - X_A) + \frac{\sigma_0 \lambda D}{2\pi}\left[\cos\left(\frac{2\pi X_C}{\lambda}\right) - \cos\left(\frac{2\pi X_A}{\lambda}\right)\right]$$
$$- \frac{1}{2} R\left(X_C^2 - X_A^2\right) + \sigma_{fa} b\lambda \tag{13}$$

The value of X_A may be found from the conditions that $X = X_A$ when $\dot{X} = 0$ and the slope of the internal force curve is also zero at that point. Hence, from Eq. (10):

$$R + \frac{2\pi\sigma_0 b}{\lambda} \cos\left(\frac{2\pi X_A}{\lambda}\right) = 0 \tag{14a}$$

or

$$X_A = \frac{\lambda}{2\pi} \cos^{-1}\left(\frac{-R\lambda}{2\pi\sigma_0 b}\right) \tag{14b}$$

Using the above value of X_A, the value of $\sigma_{(A)}$ is then found from Eq. (10):

$$\sigma(A) = \frac{R}{b} X_A + \sigma_0 \sin\left(\frac{2\pi X_A}{\lambda}\right) + \sigma_{fa} \qquad (15)$$

X_C is then found from a solution of Eq. (10)

$$RX_C + \sigma_0 b \sin\left(\frac{2\pi X_C}{\lambda}\right) = [\sigma(A) - \sigma_{fa}] b \qquad (16)$$

since $\sigma(A) = \sigma(C)$. Thus, ΔE can be calculated in terms of the physical parameters of the model, λ, σ_0, and R. Before making such a calculation the limitations on these physical parameters must be determined.

Conditions exist which must be satisfied by λ, R, and σ_0 for the dislocation to return to the origin when the applied stress is removed. The internal force (Fig. 6b) must be negative for all positive values of X. The critical condition corresponds to the situation when the first maximum is just tangent to the axis. This condition is determined by the simultaneous solution of the following equations:

$$RX + \sigma_0 b \sin\left(\frac{2\pi X}{\lambda}\right) \leq 0 \qquad (17a)$$

$$R + \frac{2\pi\sigma_0 b}{\lambda} \cos\left(\frac{2\pi X}{\lambda}\right) = 0 \qquad (17b)$$

The solutions of Eqs. (17a) and (17b) yields the desired condition

$$\frac{R\lambda}{\sigma_0 b} \geq 1.41 \qquad (18)$$

On the other hand, there will be no velocity-dependent energy loss by the proposed mechanism unless the internal force curve contains a region of positive slope (Fig. 6b). The critical condition for loss to occur in the range $0 \leq X \leq \lambda$ is determined by the simultaneous solution of the following equations:

$$R + \frac{2\pi\sigma_0 b}{\lambda} \cos\left(\frac{2\pi X}{\lambda}\right) \leq 0 \tag{19a}$$

$$\sin \frac{2\pi X}{\lambda} = 0 \tag{19b}$$

The first equation states that nowhere is the slope positive; the second equation states that the inflection point coincides with the point of zero slope. Equations (19a) and (19b) will be satisfied for $X = \lambda/2$ and

$$\frac{R\lambda}{\sigma_0 b} \leq 2\pi \tag{20}$$

Therefore, from Eqs. (18) and (20), we obtain the range of admissible values of the parameters which describe the internal stress field

$$1.41 \leq \frac{R\lambda}{\sigma_0 b} \leq 2\pi \tag{21}$$

These values hold only for the assumed sinusoidal variation in the internal stress field.

To illustrate the relation between the energy loss per cycle and the stress field parameters, we shall obtain a solution for ΔE with a value of $R\lambda/\sigma_0 b = \pi$. From Eq. (14b)

$$X_A \simeq \frac{\lambda}{3}$$

While from Eq. (15)

$$\sigma(A) = 1.91\sigma_0 + \sigma_{fa}$$

The value of X_C is found from Eq. (16)

$$X_C = 0.85\lambda$$

Substituting the above values into Eq. (13) we obtain

$$\Delta E = b\lambda(\beta\sigma_0 + \sigma_{fa}) \tag{22}$$

as the energy loss per unit length of dislocation when it moves a distance λ.

Here β is a constant equal to 0.21 for the particular set of conditions chosen. This constant will vary from zero when $R\lambda/\sigma_0 b = 2\pi$ to a maximum of about 0.45 when $R\lambda/\sigma_0 b = 1.41$.

The energy loss (for average dislocation displacements greater than λ) may be divided into two parts. One part $b\lambda\sigma_{fa}$ is associated with the constant frictional stress σ_{fa}. The other part is associated with the dynamic loss. The average frictional stress which is experimentally determined may be connected to the theory as follows. If the dislocations on the average move a distance $n\lambda$ (where n is an integer) and if the density of dislocations is N then:

$$\epsilon_d = Nbn\lambda \tag{23}$$

The observed energy loss per unit volume is

$$\Delta W_{\text{exp}} = 2Nbn\lambda\bar{\sigma}_f \tag{24}$$

The energy loss per unit volume according to the theory will be

$$\Delta W_{th} = 2nb\lambda(\beta\sigma_0 + \sigma_{fa})N \tag{25}$$

Therefore,

$$\bar{\sigma}_f = \beta\sigma_0 + \sigma_{fa} \tag{26}$$

Thus $\bar{\sigma}_f$ is directly related to the height of the ripples in the internal stress field for dislocation displacements *greater* than λ. If, however, the displacement of dislocations is *less* than $X_A = \lambda/3$, then the only loss mechanism present is due to σ_{fa} only and then $\bar{\sigma}_f = \sigma_{fA}$.

5   DISCUSSION

The essential results are that ripples in the internal stress field produce a frictional stress which is equal to $\beta\sigma_0$ where σ_0 is the amplitude of the ripples and β decreases as the parameter $R\lambda/\sigma_0 b$ increases. This parameter represents the ratio of the gradient of the long-range stress field to the average gradient of the short range stress field. When R is sufficiently large compared to $\sigma_0 b/\lambda$, no energy is dissipated because an unbalance in the internal stress field does not occur.

The theory shows that the energy loss is frequency-independent unless the applied strain rate was extremely high and on the order of the dislocation velocity during an energy dissipating interval such as $X_A X_C$ (Fig. 6b). It is expected that the above mechanism will be essentially independent of temperature as long as $\lambda \gg b$ because the activation volume should be on the order of $\lambda^2 b$. In three dimensions the internal stress field should be more like hills and valleys with λ^2 being the cross-sectional area of a hill.

One reason for the internal friction to be amplitude-dependent is that the frictional stress σ_{fa} would operate at low amplitudes and $\sigma_{fa} + \beta \sigma_0$ at higher amplitudes. Another reason is that at higher amplitudes of dislocation motion higher internal stress hills will be encountered. One should think of the internal stress field as consisting of a spectrum of hills of various heights.

The height of the internal stress field ripples σ_0 may be calculated from the theory and the experimental data. Using Eqs. (2), (6) and (26),

$$\sigma_0 = \frac{1}{\beta} (\sigma_{fb} - \sigma_{fa}) \left(1 - \frac{\epsilon_0}{\epsilon_d} \right) \tag{27}$$

The values of σ_{fb} and σ_{fa} in Fig. 4 may be obtained from the two slopes of curves such as those in Fig. 2. From Fig. 2 for the 1.02% prestrain, $\sigma_{fa} = 600$ psi, $\sigma_{fb} = 3600$ psi, and $\epsilon_0/\epsilon_d = 1/4$ and using an average value for $\beta = 1/4$, $\sigma_0 = 9000$ psi. For comparison σ_A at 1.02% prestrain is 18,000 psi and the macroscopic flow stress at 1% strain is about 25,000 psi.

In the case of nickel, σ_{fa} is less than $\beta \sigma_0$ and, in general, it is expected to be the case for a metal of high purity and low Peierls-Nabarro Force. All pure f.c.c. metals fall in this category. However, in concentrated solid solutions or metals with a Peierls-Nabarro Force, σ_{fa} may be almost as great or greater than $\beta \sigma_0$. Such is the case for alpha brass[12] and Fe - 3% Si alloys.

The frictional stress $\bar{\sigma}_f$ would only be exactly one-half of the elastic limit σ_E if a continuous friction such as σ_{fa} operated and the internal stress field were monotonic. In the case of the ripple mechanism $\bar{\sigma}_f$ would be less than σ_E, but the ratio would vary with the relative contributions of σ_{fa} and $\beta \sigma_0$.

In previous investigations[1,2] on iron at prestrains greater than 1%, the frictional stress was independent of prestrain, strain rate, and slightly dependent on temperature. It was then thought that frictional stress was due only to interstitial impurities and the Peierls stress. In the light of the present investigation, ripples in the internal stress field would also contribute because it is a general mechanism, independent of crystal structure.

Under conditions of forced vibrations where the applied stress is both positive and negative, condition (18) does not apply. When the applied stress reverses its direction, dislocations can be moved from much deeper valleys than when the applied stress oscillates without changing sign. Thus, greater values of the frictional stress may be expected when the applied stress changes sign. The last statement applies to the ripple mechanism and not to σ_{fa}.

During the complex motion of many dislocations, the occurrence of an unbalanced force should be very common. Therefore the acceleration of dislocations to high velocities by an unbalanced force should be an important mechanism for the conversion of the work of plastic deformation into heat.

ACKNOWLEDGEMENTS

This work was supported by the Laboratory for Research on the Structure of Matter of the University of Pennsylvania and the Office of Naval Research. R. Knox provided important technical assistance. We are grateful to the Radio Corporation of America for a David Sarnoff Fellowship.

REFERENCES

1. Ekvall, R. A., and N. Brown: *Acta Met.* 10:1101, (1962).
2. Kossowsky, R., and N. Brown: *Trans. AIME* 233:1389 (1965).
3. Brown, N.: Chapter on "Microplasticity," "Adv. in Materials Research," vol. II, John Wiley & Sons, New York, 1967, C. J. McMahon (ed.).
4. Cottrell, A. H.: "Dislocations and Plastic Flow in Crystals," p. 112, Oxford University Press, 1953.
5. Roberts, J. M., and N. Brown: *Trans. AIME* 218:454 (1960).
6. Lukas, P., and M. Klesnil: *Phys. Stat. Sol.* 11:127 (1965).
7. Galyon, G. T.: *Acta Met.* 14:1851 (1966).
8. Meakin, J. D., and A. Lawley: Ibid., 1854.
9. Galyon, G. T.: *Acta Met.* 15:410 (1967).
10. Roberts, J. M.: Ibid., p. 411.
11. Roberts, J. M.: Ibid., p. 569.
12. Rutherford, J.: Ph.D. *Thesis*, University of Pennsylvania, 1965.
13. Lawley, A., and J. Meakin: *Acta Met.* 14:236 (1966).
14. Bonfield, W., and C. H. Li: *Acta Met.* 13:317 (1965)
15. Weertman, J.: *J. Appl. Phys.* 26:202, (1955).
16. Nabarro, F. R. N., T. S. Basinski, and D. B. Holt: *Adv. in Physics* 13:193 (1964).
17. Seeger, A.: "The Relation between the Structure and Mechanical Properties of Metals." NPL Symposium no. 15, HMSO (1963) p. 3
18. Kuhlmann-Wilsdorf, D.: *Trans. AIME* 224:1047, (1962).
19. Stern, R. M., and A. V. Granato: *Acta Met.* 10:358, (1962).

DISCUSSION *on Paper Presented by N. Brown*

A. V. GRANATO: Are you able to apply both positive and negative stresses to determine whether the stress-strain diagrams close every half-cycle or every full cycle?

Can you determine the modulus from the slope of your stress-strain curves and separate the lattice strain from the dislocation strain?

N. BROWN: We have not applied both positive and negative stress. Since we start with a prestrained material and recognizing the existence of the Bauschinger Effect, I would suggest that the stress-strain curve would have the following form if we stress in both a positive and negative direction:

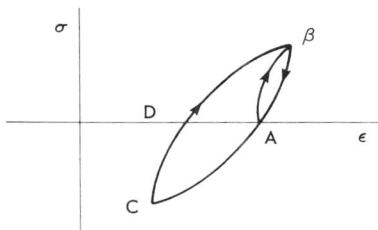

Stesss in one direction gives the closed loop AB. If the stress is reversed, the path would be $ABACDBA$. I do not believe the loop would go from C to A and form a butterfly loop.

R. BULLOUGH: Can your internal stress theory possible predict the eventually concave shape of the energy loss vs. dislocation strain curves?

N. BROWN: This point is not covered in the theory. I would say that, in general, as the dislocation moves further and further from its equilibrium position into the field of the other dislocations, it encounters more and stronger energy-dissipating mechanisms. The fluctuating internal stress field is only one of the many weak interactions that is capable of dissipating energy by the general mechanism proposed in this paper.

G. ALEFELD: Could you comment on why you need a second "friction stress," which acts similar to an applied stress. The string equation which you presented showed a term $B\dot{x}$, a friction term, which, according to you, is responsible for the energy loss in the periods of high speed of the dislocation. Therefore, why a second friction stress and of what nature is it? It seems highly unlikely that it could be a Peierls stress, as you mentioned.

At your frequency and at room temperature the Peierls stress will be insignificant in copper.

N. BROWN: The frictional stress, σ_{fa}, is not known in nickel except that it acts at the very beginning of the dislocation motion, and since our model suggests that the dislocation is initially moving very slowly in phase with the testing machine, it led to the assumption of σ_{fa} being independent of velocity. Possibly, very fine ripples in the internal stress field occur at strains less than ϵ_s; in that case σ_{fa} could be exactly like σ_{fb} except that very small fluctuation in the internal stress field are involved and such small fluctuations would have to be insensitive to prestrains beyond about 10^{-4}.

R. BULLOUGH: I should like to point out that the low initial friction stress is not always independent of prestrain. For example, Bonfield and Li found in Be that the initial friction stress was very sensitive to prestrain for prestrains less than 10^{-4}.

N. BROWN: Since beryllium does not exhibit a frictional stress which is independent of strain rate, then in the context of our model, it is suggested that very small ripples in the internal stress field are effective at strains of 10^{-6} which could act in addition to a frictional stress which is independent of prestrain and velocity. A frictional stress such as σ_{fa} would simply be additive.

DYNAMIC AGING EFFECTS IN IRON-NICKEL-CARBON MARTENSITE

W. S. Owen and M. J. Roberts

*Department of Materials Science and Engineering,
Cornell University,
Ithaca, New York*

ABSTRACT

The deformation kinetics and the effects produced by aging iron-nickel-carbon martensite during relaxation under stress (essentially a static stress) and during deformation at a constant applied strain-rate, are discussed in some detail. The experiments were carried out on two alloys—one with 0.003%, and the other with 0.12% carbon, quenched to cubic massive-martensite with no retained austenite. No strain-aging effects were produced in the martensite containing only 0.003% carbon. The higher-carbon martensite aged under stress in two stages: the first, detected at temperatures down to 213°K, is attributed to a Snoek interaction; the second, developed significantly only at temperatures above

273°K, is thought to be regular Cottrell segregation. The maximum increase in flow stress produced by static Snoek pinning is smaller then predicted theoretically. On aging a specimen strained continuously at a lower and a faster strain-rate than required to produce jerky flow, the moving dislocations are unaffected by the aging, the yield drop observed on increasing the strain-rate being the result of potential dislocation sources in the high density of static dislocations being pinned by carbon segregation.

At room temperature the velocity index m^* for dislocations moving in the low-carbon nonaging martensite is exactly the same as that for movement in the martensite with 0.12% carbon and is very close to the value of 5.0 found for annealed pure iron. The stress σ^* is also the same in both martensites. Thus, m^* is a property of the body-centered cubic lattice and is unaffected by the long-range internal stresses developed by the martensite substructure and the dissolved carbon. This confirms the general features of the model for the strength of martensite developed from earlier less sophisticated experiments.

1   INTRODUCTION

The primary differences between the mechanical behavior of martensite and of ferrite in iron-nickel-carbon alloys are a direct result of either the extremely high density of dislocations in the martensite substructure or the greater concentration of carbon in solution in the martensite lattice. The substructure results from the lattice-invariant deformation which is a necessary part of the transformation. This deformation always produces a high density of dislocations and, in some forms of martensite, many parallel twins about 200 Å thick. The twinned structure has some influence on the increase of flow stress with concentration of carbon, but the effect is small[1] and will not be considered here. Some martensites contain retained austenite which is transformed to martensite when the specimen undergoes small plastic strain. This effect, reinforced by the heterogeneous deformation in the early stage of straining due to internal stresses created by the transformation, produces preyield microstrain effects.[2,3] These have been avoided by studying the flow of martensite at strains of 0.006 and greater. None of the alloys under consideration here contained any austenite or significant volumes of twinned lattice when quenched to martensite.[4] Martensite specimens do not exhibit any evidence of a discontinuous yield (a yield drop) when first strained in a uniaxial tension test.

This may be because the density of mobile dislocations quenched-in is appreciable, but is more likely to be due to the influence of the heterogeneous quenched-in stresses.

All the results discussed here were obtained from alloys containing more than 18% nickel added to ensure that the M_s temperature was as low as possible consistent with the structure of rapidly quenched thin-sheet specimens (<0.010 inch thick) being massive-martensite. The nickel also improved the ductility of the specimens. The martensite was cubic. It has been shown previously[4] that a change from cubic to tetragonal martensite does not result in any significant change in plastic-flow behavior. Within the composition ranges considered, the variation in the solution-hardening effect of the nickel is negligible.

2   THE VARIATION OF FLOW STRESS WITH CONCENTRATION OF CARBON

The 0.6% flow stress increases with decreasing testing temperature. Addition of carbon raises the level of the flow stress at every temperature but does not change the rate of increase. Isothermally,

$$\sigma = \sigma_0 + \frac{\mu}{A} X_c^{1/2} \tag{1}$$

where σ is the 0.6% flow stress, σ_0 is the flow stress at zero concentration of carbon, X_c is the atom fraction of carbon, μ is the shear modulus. On testing at temperatures between 77% and 293°K at strain-rates of about 10^{-4} sec^{-1}, the constant A does not change with testing temperature (Fig. 1). Thus, carbon raises the flow stress by interacting with the glide dislocations by a mechanism which is not thermally activated to any significant extent. The effect is that of increasing the level σ_μ of the long-range stress field.

The nature of the plastic deformation of martensite was studied further by measuring the activation volume and energy by stress relaxation and strain-rate change experiments. The results are shown in Figs. 2 and 3. Below about 213°K, the activation energy decreases linearly with decreasing temperature suggesting that a single activated event is rate-controlling. The data from alloys within a range of carbon concentration fall on the same line. The activation volume V^* data also fail to reveal any effect of carbon.

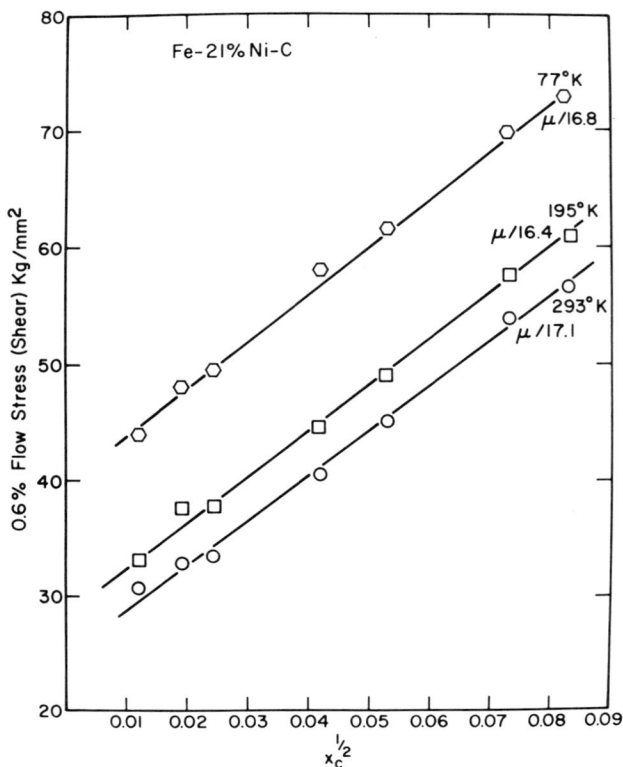

Fig. 1 0.6% flow stress of iron-21% nickel-carbon alloys as a function of the testing temperature and of the square root of the atom fraction of carbon.

Thus, the earlier conclusion that the thermally activated event does not involve the carbon is reinforced. Between 77° and 213°K, V^* is less than 30 b^3 and consequently the choice of rate-controlling mechanisms is very restricted. It is shown elsewhere[5] that the activation energy and volume data are in good agreement with the predictions of models which assume that the operative control is the overcoming of the Peierls barrier by a thermally activated double-kink mechanism either as formulated by Seeger[6] or by Dorn and Rajnak.[7]

Much larger activation energies and volumes are found by stress relaxation at a temperature above 213°K. An effect, related to carbon concentration, which caused an increase in flow stress with aging time at temperatures as low as 213°K was noted by Winchell

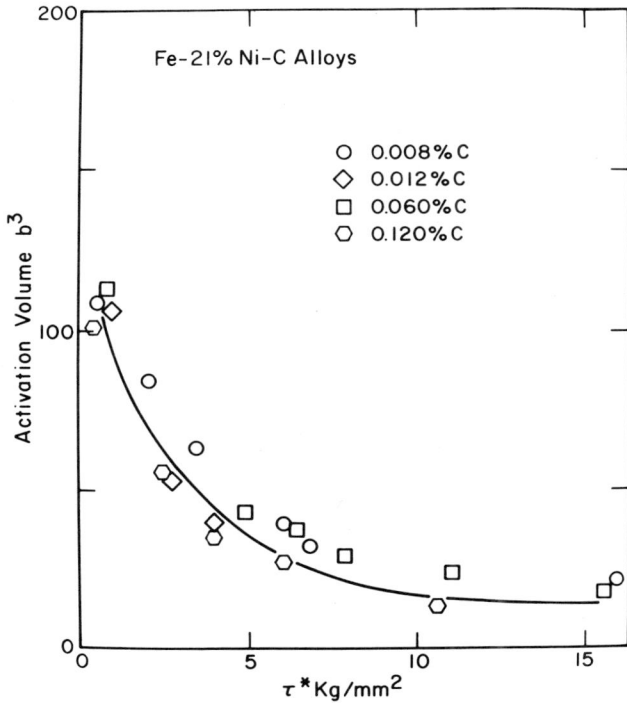

Fig. 2 Variation of the activation energy for deformation of iron-21% nickel-carbon alloys with temperature.

and Cohen in their studies of the strength of twinned martensite.[8] The effect revealed by stress relaxation of dislocated massive-martenite and that reported by Winchell and Cohen appear to be the same type of aging effect.

3 STRAIN AGING UNDER STRESS

In the aging experiments, and the strain-rate change experiments to be described in the next section, an iron-18% nickel alloy with a carbon concentration (0.12%) which is relatively high for the form of martensite was compared to an iron-21% nickel alloy which had been decarburized by wet and dry hydrogen treatments. This alloy contained less than 0.003% carbon. They will be referred to as the high- and low-carbon alloys. The difference in nickel content has no measurable effect on the flow stress or any of the

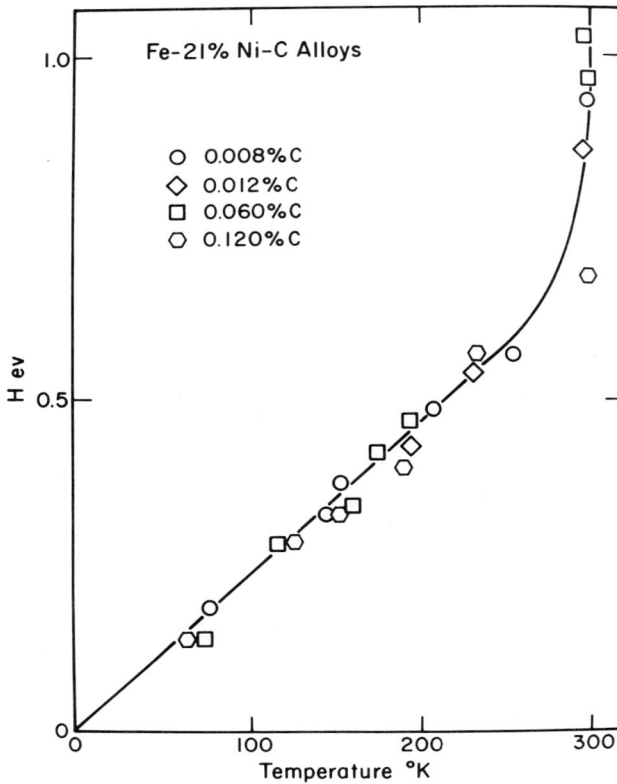

Fig. 3 The effect of the stress τ^* on the activation volume for the deformation of iron-21% nickel-carbon alloys.

parameters under discussion. The tensile specimens were made from 0.005 inch thick foil which had been austenitized at 925°C and quenched into iced lithium chloride solution.

Specimens were loaded at a constant strain-rate to a stress σ_f (Fig. 4) when the cross-head motion was stopped and the specimen relaxed for a measured time. On resuming the test at the original strain-rate, the variation of stress with strain was recorded. The stress increment $\Delta\sigma_y$ was measured in those cases where the stress reached an upper yield stress (σ_{uy}) before falling to the strain-hardening portion of the stress-strain curve. None of the relaxations or strain-aging treatments caused any change in the level of the strain-hardening curve. The upper yield stresses were reproducible and significant because the specimens were aged under stress so that no loading eccentricities were introduced.

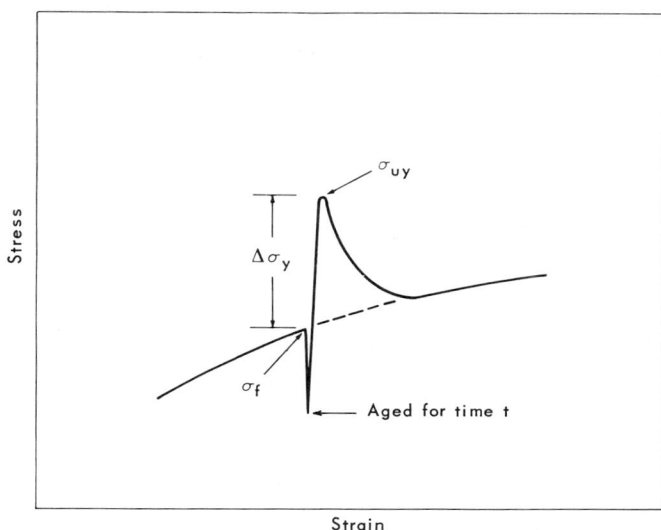

Fig. 4 Schematic representation of the stress-strain curve during a stress-relaxation experiment and on subsequent loading.

The increment $\Delta\sigma_y$ was measured as a function of the relaxation time. The low-carbon martensite did not show a yield drop at any temperature between 77° and 293°K after being relaxed for times of several hours or after repeated relaxations at progressively larger strains (Fig. 5). The high-carbon martensite showed no yield drops after relaxing below 213°K. However, measurable values of $\Delta\sigma_y$, which increased with relaxation time, were found at higher temperatures. The experimental data are plotted in Fig. 6. It is seen that $\Delta\sigma_y$ increases to a plateau value which at all temperatures lies close to 1.5 Kg mm^{-2}. At 273°K and 298°K, $\Delta\sigma_y$ continued to increase up to a maximum of about 14 Kg mm^{-2} with increasing relaxation time. At lower temperatures, no large second rise was found within the maximum time of the experiments (about 300 hours). The first rise to the plateau is thought to be due to a Snoek interaction, and, the second, to pinning by the formation of a Cottrell atmosphere.

Two facts support the proposition that the first increase in $\Delta\sigma$ is due to a Snoek effect. The time to reach the plateau value agrees reasonably well with the time for a single jump of a carbon atom in ferrite (Fig. 7). The agreement is not so good if the jump time is calculated from Hillert's diffusion coefficient for iron-carbon

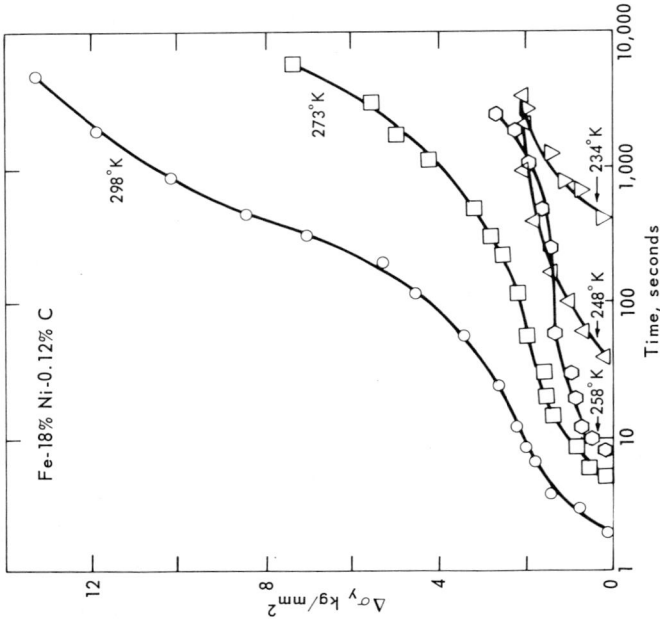

Fig. 6 The increase of σ_y with relaxation at different temperatures for the higher carbon martensite.

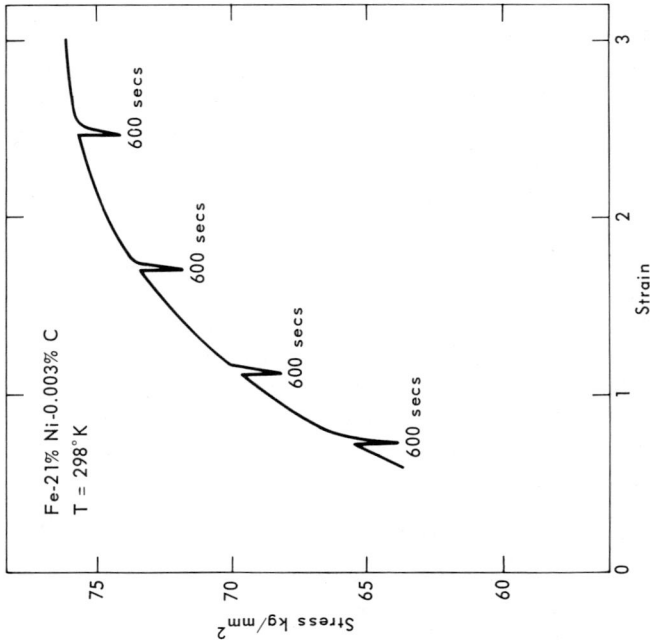

Fig. 5 Stress-strain curve for the low-carbon martensite specimen relaxed at 298°K at a constant crosshead separation for 10 minutes at each of four different strains. Note that no yield drop is developed on reloading.

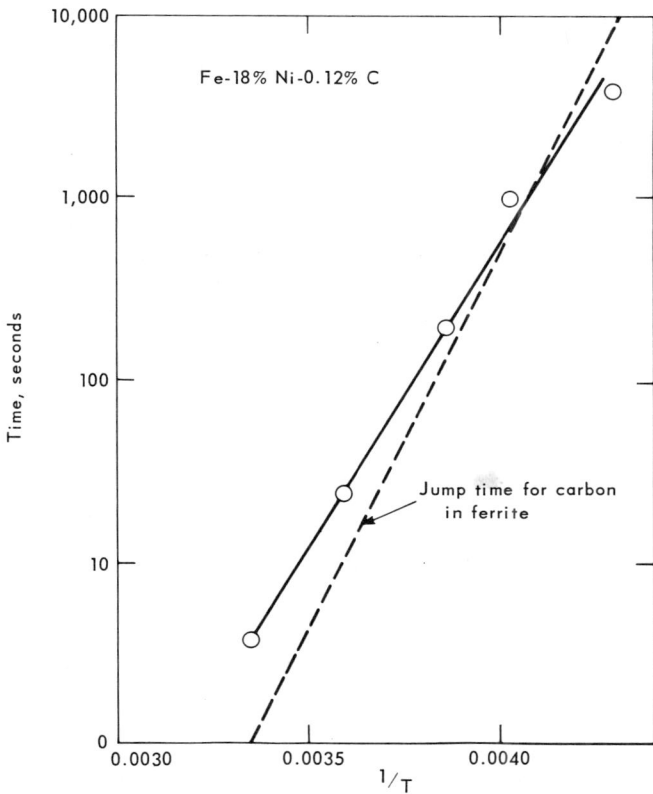

Fig. 7 The time to reach the Snoek plateau compared with the time for a single jump of carbon in ferrite at different temperatures.

martensite.[9] However, this is not surprising since the relaxation data were obtained using a cubic martensite. When the relaxation aging is repeated at successively larger strains, $\Delta\sigma_y$ is the same each time (Fig. 8). This shows that the pinning process does not remove carbon atoms from solution.

When aged for long periods at temperatures between 258° and 298°K so that $\Delta\sigma_y$ is greater than the plateau value (the second stage), at constant temperature, $\Delta\sigma_y$ increases as $t^{2/3}$ (where t is the time) in accordance with the prediction of Harper's modification[26] of the Cottrell-Bilby treatment[27] of the kinetics of strain-aging (Fig. 9). The activation energy for the second stage, obtained from a plot of the logarithm of the time to reach a selected value of $\Delta\sigma_y$ as a function of the reciprocal of the temperature (Fig. 10), is about 19.8 Kcal/mole^{-1} which is close to the activation energy

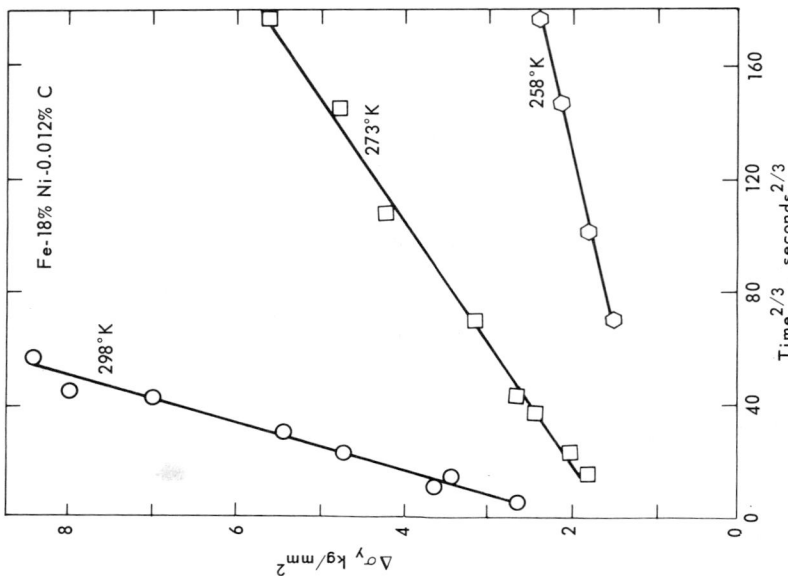

Fig. 9 The yield increment $\Delta\sigma_y$ on reloading as a function of the relaxation time.

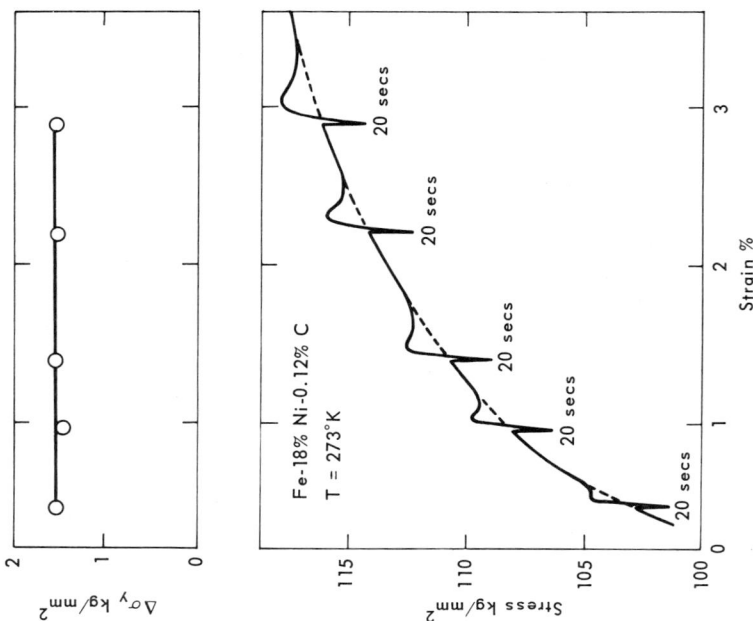

Fig. 8 The high-carbon martensite strained at $\dot{\epsilon} = 8.3 \times 10^{-4}$ sec^{-1} at 273°K and relaxed for 20 seconds at five successively larger strains. The yield increment $\Delta\sigma_y$ is shown as a function of the strain at which the relaxation occured.

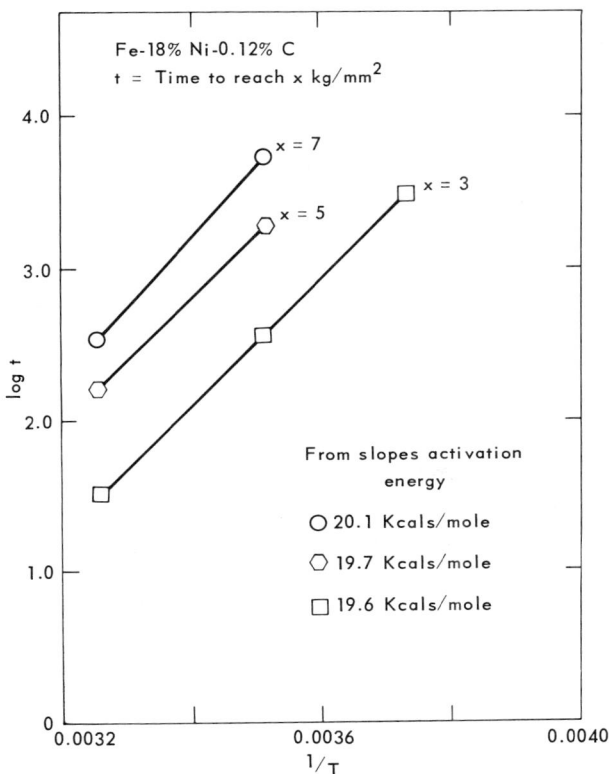

Fig. 10 Activation energy deduced from a plot of time to age to a selected value of $\Delta\sigma_y$ as a function of temperature.

for the diffusion of carbon in ferrite as required by the Harper theory. Thus, the second rise appears to be Cottrell strain-aging requiring long-range diffusion. On repeating the relaxation for the same time at a larger strain the yield drop on reloading was found to be smaller, presumably because the atmospheres formed during the first relaxation do not redissolve on further straining, that is, no unpinning occurs. It was not possible to repeat the relaxation more than two or three times with a specimen aged to give a large $\Delta\sigma$, because specimens relaxation-aged in this way had very little ductility.

Yield drops thought to be due to Snoek interaction have been observed in steel containing 0.039 w/o carbon and 0.0044 w/o nitrogen by Wilson and Russell.[10] The carbon and nitrogen in solution in the ferrite was changed by quenching after annealing at different temperatures in the ferrite range. Nakada and Keh[11] found a Snoek strain-aging effect in a dilute iron-nitrogen alloy (0.1 a/o nitrogen). The only report of a Snoek interaction in martensite is by Breyer[12,13] who studied the yield drops in a quenched and prestrained commercial 4340 steel which had been either aged under zero stress or relaxed at constant total strain while the specimen was under load. However, this evidence of a Snoek effect is not completely convincing. The primary yield drop increased with increasing prestrain, up to about 6% prestrain, and decreased at larger prestrains. The "activation energy" was stated to be 1600 to 1800 cal/mole which suggests that the rate-controlling process was not a simple effect requiring only the diffusion of carbon.

In all of the experiments discussed the specimens have been aged during stress-relaxation for times that are long in comparison with those required to reach nearly zero strain-rate. The relaxation experiments can be considered to be aging under static applied stress. It has been suggested that aging is accelerated when a stress is applied. Also, there is abundant evidence[10,14,15] that in ferrite the acceleration is measurable and that $\Delta\sigma_y$ developed under a static stress is slightly larger than when aged under the same conditions but with no stress applied. With martensite it was not possible to make the comparison between aging with stress and that without stress. This was because the aging occurred so rapidly under stress that the specimen could not be unloaded from the prestrain quickly enough to prevent it. Substantial aging, developing a $\Delta\sigma_y$ of about 2.0 Kg/mm^2, occurred in the first 10 seconds. In general, it seems that the more rapid aging in martensite can be accounted for by the greater carbon concentration. The martensite specimens aged under stress about ten times faster than the ferrite specimens studied by Almond and Hull[14] and the martensite specimens contained about ten times the concentration of carbon in the ferrite. The ferrite aged at a somewhat lower stress, but from the results of experiments on steels in the ferritic condition,[14,15] it is clear that the effect of stress on the kinetics is considerably smaller than could possibly account for the differences under consideration.

Neither the first nor the second stage of aging occurred in specimens from which the carbon had been removed by treatment with hydrogen. Thus, as in ferrite, both stages of aging of martensite require the presence of more than 0.003% carbon. In ferrite, $\Delta\sigma_y$ in the first stage increases with increasing carbon plus nitrogen in solution,[10] the maximum reported value obtained by strain-aging under stress being about 2.0 Kg/mm^{-2}, developed in a specimen with 0.11 a/o solute. In our investigation of martensite, the alloy with 0.12% (0.56 a/o) carbon in solution gave a value of $\Delta\sigma_y$ at 25°C of about 2.1 Kg/mm^{-2}. That is, about one-fifth the magnitude expected by extrapolation of the data for ferrite. $\Delta\sigma_y$ at the plateau at the end of the first stage does not vary appreciably with aging temperature. For martensite it was found to be between 1.5 and 2.9 Kg/mm^{-2} for aging temperatures between −39°C and 25°C. Schoeck and Seeger[16] estimate the maximum yield drop due to Snoek interaction in iron-carbon to be

$$\Delta\sigma_y = \frac{20.5\,A}{ba^3}\,X_c$$

where A is a constant in the Cochardt, Schoeck and Wiedersich[17] model for a carbon-dislocation interaction (for iron-carbon $A = 1.84 \times 10^{-20}$ dyn-cm^2 and X_c is the atom fraction. Taking the Burgers vector b as 2.48×10^{-8} cm and the lattice constant a as 2.8×10^{-8} cm gives $\Delta\sigma$ for the higher-carbon martensite as about 80 Kg/mm^2 which is many times larger than the measured value. On the same basis, the measured value for ferrite is closer to the calculated value. Calculations based upon a slightly more elaborate model of the Snoek interaction due to Eshelby[18] give almost the same results. The experiments show clearly that only a small fraction of the potential Snoek drag is developed in martensite. Thus either the models are not applicable to martensite or it must be assumed that only a small fraction of the carbon in solution is available to participate in a Snoek interaction.

Although $\Delta\sigma_y$ developed in the first stage of aging martensite under a static stress is smaller than might have been expected, the stress increment which can be produced by maximum aging in the second stage is as expected. The maximum value developed by relaxation-aging the higher-carbon martensite at 293°K is nearly 14 Kg/mm^{-2}, which is twice the value found for annealed ferrite containing 0.015% carbon and aged under comparable conditions.

4 AGING AT A CONSTANT STRAIN-RATE

Experiments were carried out on both the low- and the high-carbon martensite at the same temperatures as those used in the

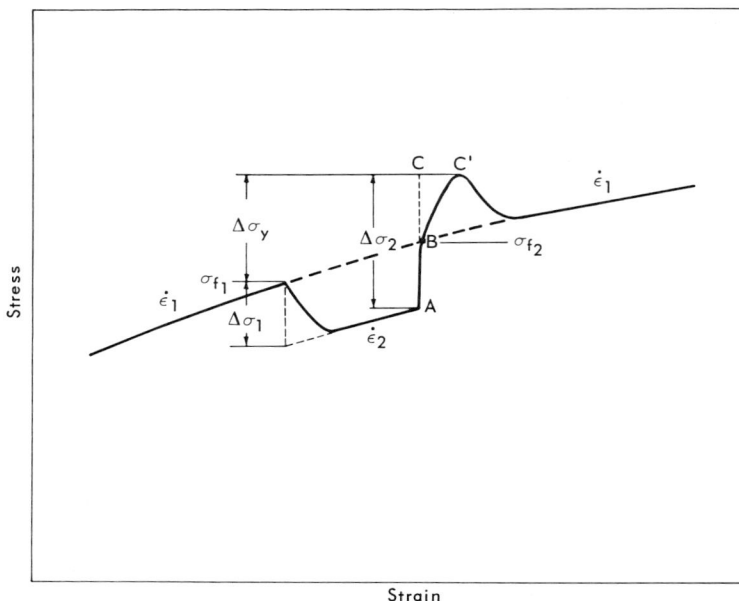

Fig. 11 Schematic representation of a stress-strain curve during one cycle of a strain-rate change experiment.

stress-relaxation tests. In these experiments, the incremental changes in stress ($\Delta\sigma_1$ on changing the strain-rate from $\dot\epsilon_1$ to $\dot\epsilon_2$ and and $\Delta\sigma_2$ on changing from $\dot\epsilon_2$ to $\dot\epsilon_1 (\dot\epsilon_1 > \dot\epsilon_2)$ were measured at progressively larger strains in an isothermal test. There was a little inertia in the Instron machine and the unloading was not quite instantaneous on changing from $\dot\epsilon_1$ to $\dot\epsilon_2$. The value of $\Delta\sigma_1$ was obtained by a short extrapolation of the stress-strain curve at $\dot\epsilon_2$ to the strain at which the change from $\dot\epsilon_1$ to $\dot\epsilon_2$ was imposed (Fig. 11). For the low-carbon martensite, $\Delta\sigma_1$ was exactly the same as $\Delta\sigma_2$ at *all* temperatures and the values of ΔH^* and V^* calculated from these data were in excellent agreement with the values deduced from the stress-relaxation experiments. This was true also when the two kinds of data for the high-carbon martensite at temperatures below −60°C were compared; at higher temperatures, $\Delta\sigma_2$ was always larger than $\Delta\sigma_1$ and on changing from the slower strain-rate $\dot\epsilon_2$, to the faster $\dot\epsilon_1$, a yield drop $\Delta\sigma_y$ occured. This $\Delta\sigma_y$ is included in $\Delta\sigma_2$.

The strain during the interval in which the specimen was extending at the slower strain-rate $\dot\epsilon_2$, was small so that σ_{f1} was nearly equal to σ_{f2}; there was no change in the rate of work-hardening at $\dot{\epsilon}_1$. It was found that in all cases ($\Delta\sigma_2 - \Delta\sigma_y$) was nearly the same as $\Delta\sigma_1$, the small difference being accounted for by work-hardening at the slower strain-rate $\dot{\epsilon}_2$. The increment $\Delta\sigma_1$ was the same as that measured in the low-carbon martensite tested under comparable conditions. This indicates that no carbon-segregation effects, dynamic or static, are involved in the $\dot{\epsilon}_1 \to \dot{\epsilon}_2$ change. There is no evidence of a yield discontinuity in this part of the stress-strain curve. In most of the experiments carried out at room temperature, the specimens were strained for about 15 seconds at the lower strain-rate. During this time in a static (relaxation) aging test at about the same stress level segregation of carbon occurs to an extent which produces a yield drop of about 2 Kg/mm^{-2} on reloading. If a dynamic strain-aging of similar magnitude occurred during straining at the slower strain-rate $\dot{\epsilon}_2$, by the end of the cycle the flow stress at $\dot{\epsilon}_2$ would increase by about 2 Kg/mm^{-2} above the level expected if only the usual work-hardening occurred. No such increase in stress occurs, but a yield drop $\Delta\sigma_y$ of about this magnitude is observed on increasing the strain-rate to $\dot{\epsilon}_1$. Thus, it is concluded that no dynamic aging occurs at either strain rate. The mobile dislocations are unaffected by the aging process. Evidently aging occurs by segregation of carbon to the high density of static dislocations which make up the martensite substructure. Thus it is the pinning of potential sources in this substructure that is responsible for the $\Delta\sigma_y$ which occurs on increasing the strain-rate. The behavior during stress relaxation is different. Interaction with carbon affects the stress relaxation kinetics as shown by the change in H and V with carbon concentration and temperature (Figs. 2 and 3). The aging during the relaxation reduces the density of moving dislocations and results in anomalously high values of H and V.

In these experiments the time of straining at $\dot{\epsilon}_2$ was between 10 and 20 seconds and, thus, by comparison with the results of the static aging tests, it is expected that the pinning at 293°K is by a Snoek interaction. The yield drop is of the expected magnitude (about 2 Kg/mm^{-2}). The data are shown in Fig. 12 in which the $\Delta\sigma_y$ developed after aging at different strain-rates $\dot{\epsilon}_2$ ($\dot{\epsilon}_1$ being the same in every case), are plotted as a function of the number of strain-rate cycles expressed as cumulative time at the lower strain-rate. There are two features of interest. There is evidence that in the case of aging during straining $\Delta\sigma_y$ decreases on repeated cycling even though other evidence suggests that the interaction is Snoek. Second, at the fastest strain-rate ($\dot{\epsilon}_2/\dot{\epsilon}_1 = 10^{-1}$) the specimen had to be taken through several cycles before the yield drop

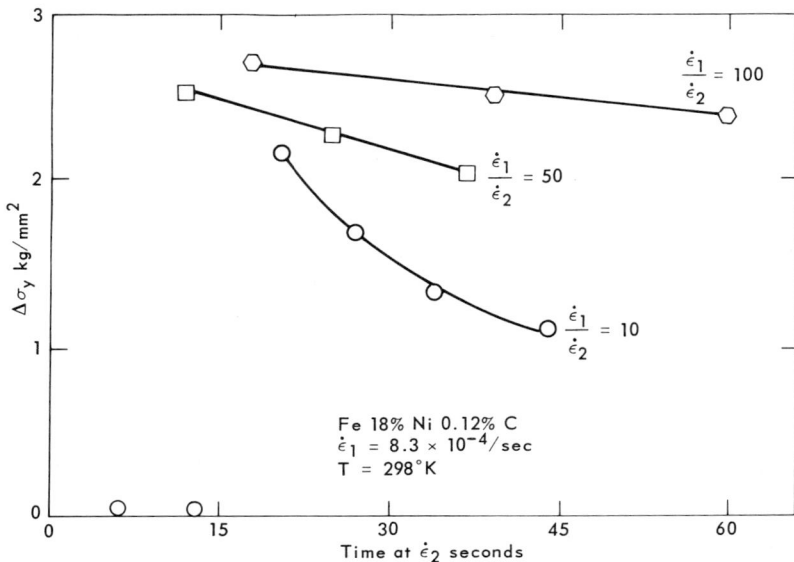

Fig. 12 The yield drop $\Delta\sigma_y$ developed on increasing the strain-rate from $\dot{\epsilon}_2$ to $\dot{\epsilon}_1$ as a function of the ratio $\dot{\epsilon}_1/\dot{\epsilon}_2$ (the faster rate $\dot{\epsilon}_1$ being the same in all experiments), the number of strain-rate cycles and the total accumulated time at the slower strain-rate, $\dot{\epsilon}_2$.

developed. Of course, a similar effect may have been found at slower strain rates if shorter initial time intervals had been used. It has been observed many times that usually some smooth plastic deformation precedes the onset of serrated yielding[19] and perhaps this is a related effect.

No jerky or serrated yielding was observed at either strain-rate or at any of the test temperatures employed. Again, compared with dilute solutions in ferrite, this is surprising. At $239°K$ and a strain rate of 10^{-4} sec^{-1}, a ferrite with 0.11 a/o carbon in solution exhibits jerky flow.[10] At the same temperature and nearly the same strain-rate the high-carbon martensite with more than five times the concentration of carbon in solution did not show any jerkiness in the stress-strain curves. This is further evidence that at the temperatures and strain-rates used in the strain-rate-change experiments no interaction between the moving dislocations and the segregating carbon atoms occurred. In additional experiments at higher temperatures jerky flow and serrated yielding in the high-carbon martensite have been observed at slow strain-rates.

The nature of the discontinuous yield on increasing the strain-rate from $\dot{\epsilon}_2$ to $\dot{\epsilon}_1$ should be considered in slightly more detail. The velocity $\bar{v}$ of the dislocations is considered to depend only on the stress. The strain-rate is determined by the product $\rho\bar{v}$. The density of moving dislocations at the end of the $\dot{\epsilon}_2$ strain ρ_A is assumed to remain unchanged upon instantaneously increasing the strain-rate to $\dot{\epsilon}_1$. At the stress σ_B (σ_{f2} in Fig. 11) the velocity will have increased to that corresponding to the strain-rate $\dot{\epsilon}_1$ and some plastic deformation occurs as is shown by the first deviation of the stress-strain curve from the vertical BC. However, because the sources have been locked by the aging, these dislocations cannot multiply sufficiently rapidly to produce a $\rho\bar{v}$ compatible with an imposed strain-rate $\dot{\epsilon}_1$ at a stress σ_B. Thus, the stress rises, along BC', until a stress level is reached at which many sources are activated and rapid multiplication occurs. The stress then drops to the level at which the condition $\dot{\epsilon} = b\rho\bar{v}$ is satisfied by the generation of dislocations from unpinned sources. Thus, the yield drop is produced, not by a change in the density of dislocations on increasing the strain-rate, but by the inability of the mobile dislocations inherited from $\dot{\epsilon}_2$ to multiply sufficiently rapidly at $\dot{\epsilon}_1$.

5 THE VELOCITY INDEX

Employing the usual terminology,[20] the velocity index $m = (\partial \ln v)/(\partial \ln \sigma)$, where v is the velocity of a dislocation, $m' = (\partial \ln \dot{\epsilon})/(\partial \ln \sigma)$ and $m^* = (\partial \ln v)/(\partial \ln \sigma^*)$. The effective stress σ^* is the difference between the applied stress and the internal long-range stress σ_μ. Here, m' is easily obtained from strain-rate-change experiments. The values for both carbon levels as a function of strain at 298°K and 77°K are given in Fig. 13. For the martensite with only 0.003% carbon, m' was obtained from the average of $\Delta\sigma_1$ and $\Delta\sigma_2$ but the values for the alloy with 0.12% carbon were deduced only from the stress increment $\Delta\sigma_1$ on decreasing the strain-rate. There does not appear to be any significant variation of m' with strain, but the limited ductility of the specimens restricted the measurements to a narrow range (0–5% strain). In annealed pure iron[21] and silicon-iron[22] m' increases markedly, and approximately linearly, with strain. The measured values of m' for the two martensites are much larger than those found for annealed ferrite at less than 5% strain, even discounting the low values during the Luders extension.[23] For polycrystalline ferrite, m' has been measured[21] at room temperature up to 0.392 plastic strain. Extrapolating

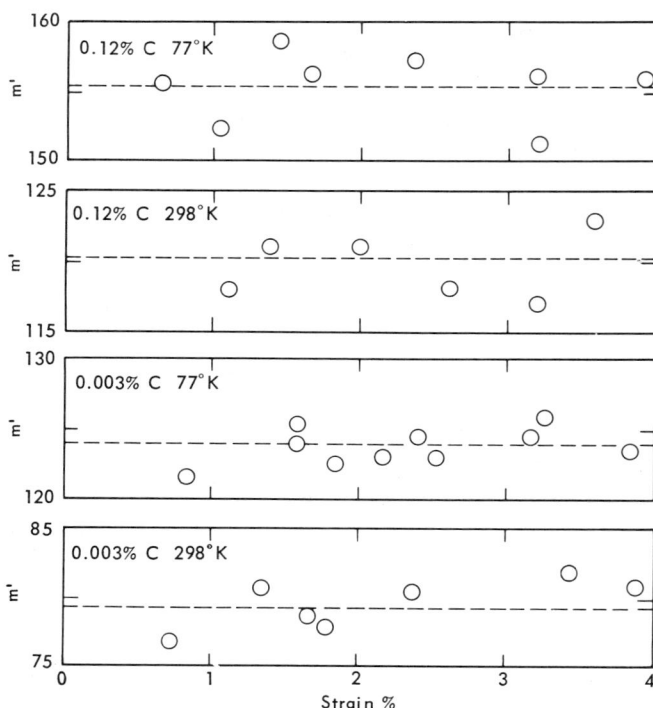

Fig. 13 The velocity index m' for the high and low-carbon martensite at 77° and 298°K as a function of strain.

these data suggests that the 0.003% martensite corresponds to
ferrite worked at 293°K by a plastic strain of 0.40 or at 77°K by a
strain of 0.90.

With the 0.12% carbon martensite the corresponding strains
are much larger, 0.70 at 293°K and 1.30 at 77°K. Evidently the
internal stress field builds up to a much higher value in the presence of carbon in solution. Of course, no quantitative significance
can be attached to the result of such an uncertain extrapolation,
and the values of m' determined at 77°K are usually suspect because it is probable that the density of mobile dislocations changes
on changing from $\dot{\epsilon}_1$ to $\dot{\epsilon}_2$. Nevertheless, the comparison does provide a dramatic illustration of the macroscopic effect of the very
high dislocation density in the martensitic structures.

The measured parameter m' is equal to the velocity index m
only if the density of mobile dislocations is unchanged on changing

from $\dot{\epsilon}_1$ to $\dot{\epsilon}_2$. It has been argued by Altschuler and Christian[21] and by Michalak[24] that at room temperature the increase in m' for ferrite with increasing strain is a result of a build-up of the long-range stress σ_μ and not, as had been supposed previously,[22] of a change in density of mobile dislocations. To examine this problem it is necessary to: (1) test the proposition that the density of mobile dislocations ρ_1 at $\dot{\epsilon}_1$ is the same as the density ρ_2 at $\dot{\epsilon}_2$; and (2) to determine the value of m^*. By definition

$$m^* = \left(\frac{\partial \ln v}{\partial \ln \sigma^*}\right)_T \tag{2}$$

As first pointed out by Li and Michalak[25]

$$\left(\frac{\dot{\epsilon}_2}{\dot{\epsilon}_1}\right)^{1/m^*} = \left(\frac{\rho_2}{\rho_1}\right)^{1/m^*} \left[1 - \frac{\Delta\sigma_\mu}{\sigma_1^*} + \frac{\Delta\sigma}{\sigma_1^*}\right] \tag{3}$$

and using $v = (\sigma^*/\sigma_0)^{m^*}$, where σ_0 is the stress to produce unit velocity, and $\dot{\epsilon} = b\rho\bar{v}$ gives

$$\Delta\sigma - \Delta\sigma_\mu = \sigma_0 \left[\left(\frac{\dot{\epsilon}_2}{\dot{\epsilon}_1}\frac{\rho_1}{\rho_2}\right)^{1/m^*} - 1\right]\left(\frac{\dot{\epsilon}_1}{b\rho_1}\right)^{1/m^*} \tag{4}$$

This equation can be used to test the proposition that $\rho_1 - \rho_2 = 0$ and $\Delta\sigma_\mu = 0$ by carrying out a sequence of experiments in which the ratio $\dot{\epsilon}_2/\dot{\epsilon}_1$ is held constant but $\dot{\epsilon}_1$ is varied over as wide a range as possible.[21] Figure 14 shows the results from measuring $\Delta\sigma = \Delta\sigma^*$. The plot is reasonably linear, indicating that the assumptions are correct and that the value of m^* is 5.0 for both the high- and the low-carbon martensite. Unfortunately, the alloy with 0.12% carbon could be tested only at two values of $\dot{\epsilon}_2$ because of the limited ductility. Michalak has used a slightly different procedure. Since $\dot{\epsilon} = \rho bv$

$$m^* = \left(\frac{\partial \ln \dot{\epsilon}}{\partial \ln \sigma^*}\right)_T \tag{5}$$

if $(\partial \ln\rho)/(\partial \ln\sigma^*)$ is zero. Then Eq. (5) can be written as

$$\frac{m^* \Delta\sigma}{\ln(\dot{\epsilon}_2/\dot{\epsilon}_1)} = \sigma^* + \frac{\Delta\sigma}{2} \tag{6}$$

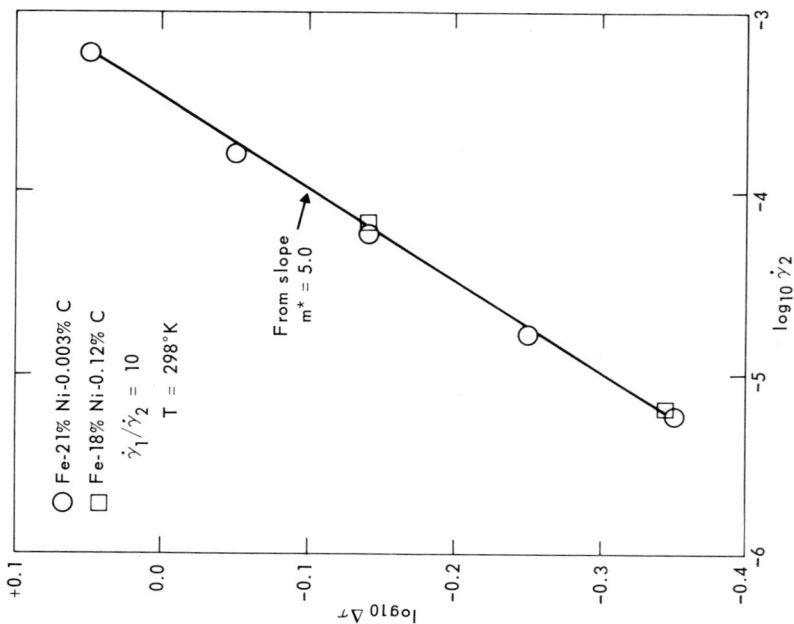

Fig. 15 The strain-rate change data plotted according to Eq. 6. $\Delta\tau$ is the change in shear stress.

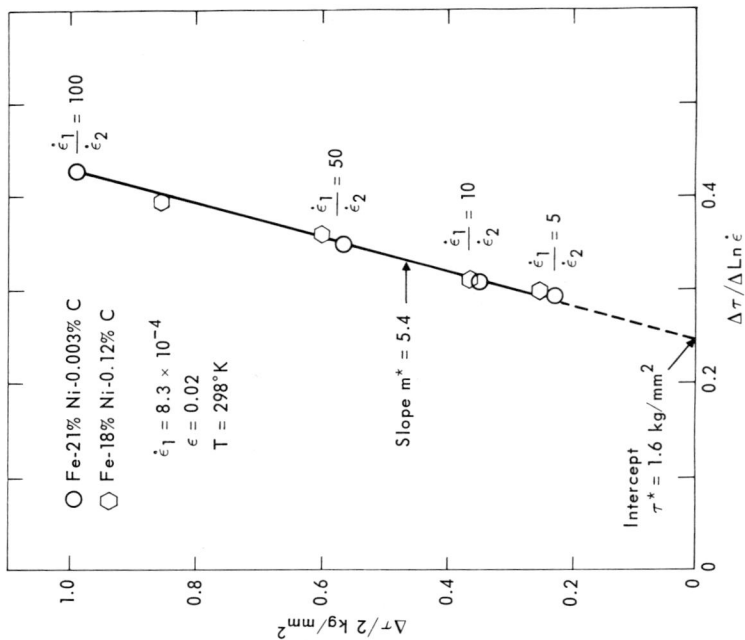

Fig. 14 The strain-rate change data plotted according to equation 4. $\Delta\tau$ is the change in shear stress and the strain-rates $\dot\gamma_1$ and $\dot\gamma_2$ are in terms of shear stress.

assuming that $\Delta\sigma_\mu = 0$ so that $\Delta\sigma = \Delta\sigma^*$. Thus, the assumptions can be tested by examining the linearity of a plot of $\Delta\sigma/2$ against $\Delta\sigma/[\ln(\dot{\epsilon}_2/\dot{\epsilon}_1)]$ when $\dot{\epsilon}_2$ is held constant and the ratio $(\dot{\epsilon}_2/\dot{\epsilon}_1)$ is changed over as wide a range as possible. Plots of this type, for both the martensites, are shown in Fig. 15. Again, the plots are linear and they coincide so that for both martensites the value of m^* is 5.4. Using this procedure it proved possible to test the 0.12% carbon martensite over a wider range than when the Altschuler and Christian method was used. The value of 5.4 should be compared with that of 5.5 found by Michalak[24] for pure zone-refined iron at room temperature. Thus, the value of m^* is virtually the same in pure annealed ferrite, low- and high-carbon martensite, indicating that m^* is a property of the body-centered cubic iron lattice and is unaffected, at least to a first approximation, by the density of dislocations or the carbon in solution. Experiments of this type did not provide useful information when carried out at temperatures below about 0°C because plots according to either Eq. (4) or (6) were too steep to be measured with significant accuracy. Neither test eliminates the possibility that, instead of $(\partial \ln\rho)/(\partial \ln\sigma)$ being zero, $(\partial \ln\rho)/(\partial \ln\dot{\epsilon})$ is a constant. However, a functional relationship between ρ and $\dot{\epsilon}$ is unlikely.[21]

The intercept (Fig. 15) gives $\sigma^* = 3.2 \, \text{Kg/mm}^2$ at 298°K and because the plots for the two martensites coincide the value is independent of carbon content. Thus, it is concluded that the effect of increasing the carbon content of massive-martensite is the same as that of increasing the work-hardening—the long-range internal stress σ_μ is increased but σ^* is unaffected. This is the same conclusion as that obtained by considering the data relating the flow stress, carbon concentration and temperature (Fig. 1) or the thermal activation parameters (Figs. 2 and 3).

ACKNOWLEDGMENTS

This work is part of a project supported by the Advanced Research Projects Agency, through the Materials Science Center at Cornell University.

REFERENCES

1. Roberts, M. J., and W. S. Owen: *Iron and Steel Institute Special Report* 93, 171 (1965).
2. Muir, H., B. L. Averbach, and M. Cohen: *Trans. ASM* 47:380 (1955).
3. McEvily, A. J., R. C. Ku, and T. L. Johnston: *Trans. AIME* 236:108 (1966).

4. Owen, W. S., E. A. Wilson, and T. Bell: "High Strength Materials," V. F. Zackay (ed.), p. 167, John Wiley & Sons, Inc., New York, 1964.
5. Roberts, M. J., and W. S. Owen: "Solid-Solution Hardening and Thermally Activated Deformation in Iron-Nickel-Carbon Martensites," submitted to *JISI*.
6. Seeger, A.: *Phil. Mag.* 1:651 (1956).
7. Dorn, J. E., and S. Rajnak: *Trans. AIME* 230:1052 (1964).
8. Winchell, P. G., and M. Cohen: *Trans. ASM* 55:347 (1962).
9. Hillert, M.: *JISI* 195:469 (1950); *Acta Met.* 7:653 (1959).
10. Wilson, D. V., and B. Russell: *Acta Met.* 7:628 (1959).
11. Nakada, Y., and A. S. Keh: *Acta Met.*, to be published.
12. Breyer, N. N., and N. H. Polakowski: *Trans. ASM* 55:667 (1962).
13. Breyer, N. N.: *Trans. Met. Soc. AIME* 236:1198 (1966).
14. Almond, E. A., and D. Hull: *Phil. Mag.* 14:515 (1966).
15. Brittain, J. O., and S. E. Bronisz: *Trans. Met. Soc.* 218:289 (1960).
16. Schoeck, G., and A. Seeger: *Acta Met.* 7:469 (1959).
17. Cochardt, A. W., G. Schoeck, and H. Wiedersich: *Acta Met.* 3:533 (1955).
18. Eshelby, J. D.: appendix to Ref. [10] and in other work unpublished.
19. Bailey, D. J., and W. F. Flanagan: Private communication.
20. Floreen, S., and T. E. Scott: *Acta Met.* 12:1459 (1964).
21. Altschuler, T. L., and J. W. Christian: Private communication.
22. Johnston, W. G., and D. F. Stein: *Acta Met.* 11:317 (1963).
23. Noble, F. W., and D. Hull: *Acta Met.* 12:1089 (1964).
24. Michalak, J. T.: *Acta Met.* 13:213 (1965).
25. Li, J. C. M., and J. T. Michalak: *Acta Met.* 12:1457 (1964).
26. Harper, S.: *Phys. Rev.* 83:709 (1951).
27. Cottrell, A. H., and B. A. Bilby: *Proc. Phys. Soc.* A62:49 (1949).

DISCUSSION *on Paper Presented by W. S. Owen*

A. S. KEH: Your data on the determination of the activation energy for Snoek ordering agrees with ours in that the slight disagreement between the jump time of interstitials determined from diffusional experiments and that from the Snoek ordering kinetics occurs at high temperatures. Is there any difference in the activation energy for diffusion due to the presence of a stress field?

W. S. OWEN: Of course, it is possible that the high internal stress field in martensite accelerates the diffusion of carbon but we have no direct evidence on this point. It seems more likely that the discrepancy at high temperatures is due to inaccuracies in the measurements in both of the investigations. The exact time at which the value $\Delta\sigma$ reaches the Snoek plateau is not very well defined and at high temperatures the time is very short so that it is easy to make an appreciable error in the measurement.

F. NABARRO: Why does the yield drop *decrease* with the time for which the specimen is held at a low strain rate? Can it be that the density of mobile dislocations is in fact constant, which is compatible with the low rate of hardening, while the density of locked dislocations gradually increases? On increasing the stress, the number of locked dislocations released would be an increasing function of the aging time, and the yield drop a decreasing function of the aging time.

W. S. OWEN: Professor Nabarro's suggestion is in line with our ideas about the events which occur during aging at the low strain-rate. The accumulation of locked sources is discussed in the paper. The only point on which we differ is that whereas Prof. Nabarro suggests that the density of mobile dislocations is constant at the constant strain-rate $\dot{\epsilon}_2$, we feel that we have to assume that the density decreases somewhat with increasing strain.

DYNAMIC STRAIN AGING IN IRON AND STEEL

A. S. Keh, Y. Nakada, and W. C. Leslie

Edgar C. Bain Laboratory for Fundamental Research,
U. S. Steel Corporation, Research Center,
Monroeville, Pennsylvania

ABSTRACT

This paper is a critical review of the literature and a presentation of new information on dynamic strain aging of iron and steel.

Dynamic strain aging is manifested in iron and steel by serrated stress–strain curves, high work-hardening rates, negative temperature, and strain-rate dependence of flow stress and reduction of ductility. The temperature and strain-rate dependence for the onset and disappearance of serrations can be represented by Arrheinus-type plots of different activation energies. The activation energy for the onset of serrations is comparable to that of interstitial diffusion in alpha iron (Q_D) [18,000–20,000 cal/mol]; that for the disappearance of serrations lies between 30,500 and 36,500 cal/mol, which can be interpreted as the sum of Q_D and ΔE, where ΔE is the binding energy between dislocations and interstitial solutes. The temperature at constant strain rate (or the strain rate at constant temperature) for the onset of serrations is very concentration-dependent; that for the disappearance of serrations is only weakly concentration-dependent.

The yield stress in the early stages of dynamic strain aging is nearly independent of temperature. The flow stress as a function of temperature at constant strain rate usually exhibits a maximum. The heights of these peaks are independent of strain rate, and they can be superimposed on a normalized $1/T$ scale. The increase of the work-hardening rate in the serrated region can be correlated with the increase in the rate of dislocation multiplication due to solute pinning during the test. This enhanced multiplication persists just beyond the disappearance of serrations, where the work-hardening rate remains very high.

1 INTRODUCTION

Dynamic strain aging denotes a series of aging phenomena which take place during the plastic deformation of a metal or alloy. Even before the turn of the century, it was noted that when a mild steel was heated to a blue color, it became brittle if worked.[1] "Blue brittleness" is thus one of the manifestations of dynamic strain aging. In 1909, LeChatelier[2] studied the tensile properties of mild steel, observed an increase in tensile strength associated with the blue brittle region, and noted that this increase was a function of deformation rate. Subsequently, several investigators studied in some detail the stress-strain curves of mild steels as a function of temperature and strain rate. Within certain combinations of temperature and strain rate, serrated stress-strain curves were observed. This phenomenon of serrated yielding and flow is generally referred to as the Portevin-LeChatelier effect, named after the original investigators of this phenomenon in aluminum alloys.[3] All the changes in mechanical properties in steel associated with dynamic strain aging were first related to static aging phenomena by Fettweiss in 1919;[4] however, it was not until the development of dislocation theory that an atomic model for dynamic strain aging was proposed. Nabarro[5] first suggested that serrated yielding in steel is observed when carbon or nitrogen atoms can diffuse fast enough to allow strain aging to take place during plastic deformation. Subsequently, Cottrell,[6] using the experimental data of Manjoine,[7] concluded that the temperature and strain rate for the start of serrated yielding in iron showed an Arrhenius relation yielding the same activation energy as that for the diffusion of nitrogen in alpha iron.

Various aspects of dynamic strain aging in steel have recently been summarized in two reviews.[8,9] The purpose of this paper is to update our knowledge of this topic. More specifically, we will demonstrate: (1) the temperature and strain-rate dependence of dynamic strain aging; (2) the concentration–dependence of dynamic strain aging; (3) the relation between serrations and concurrent changes in yield and flow stresses; and (4) a comparison of experimental results with existing theories.

2   MATERIALS AND PROCEDURES

In this investigation, a low-carbon rimmed steel and single crystals of several Fe-N alloys were used. The chemical analyses of these materials are shown in Table 1.

TABLE 1 Chemical Analyses of Iron and Steel (wt., %)

Low-C rimmed steel

C	Mn	P	S	Si	Cu	Ni	Cr	N
0.035	0.36	0.006	0.016	0.005	0.005	0.002	0.008	0.003

Fe alloys

	N	C
Fe-81	0.0009	0.0016
Fe-73-X	0.0202	0.0014
Fe-75-A	0.0156	0.0015
Fe-75-B	0.0049	0.0015

The "pure" iron single crystals were grown in strip form by a standard strain-anneal method. Nitrogen was introduced in some batches of crystals by equilibrating them in NH_3-H_2 atmospheres. All the crystals were quenched from 500°C to keep the nitrogen in solid solution.

For the major portion of this investigation, the low-carbon steel in the hot rolled form was annealed 72 hr at 250°C, followed by a brine quench. After this treatment, it was estimated that 0.003 wt/% of nitrogen was in solid solution, and substantially all the carbon was in the form of cementite particles. For a limited number of tests, this steel was quenched from 720°C after equilibration to study the effect of interstitial concentration in solid solution on dynamic strain aging. In this case about 0.022% C was put in solid solution in addition to the 0.003% N. The grain size was ASTM No. 5 after either treatment.

Tensile tests were performed at temperatures between $-10\,^\circ$C and 350 $^\circ$C, with the nominal strain rate varying from 10^{-5} to 10^{-2}/sec. Most of the steel specimens were tested on a Tinius-Olson tensile machine, and the single crystals and some steel specimens were tested on an Instron machine. No difference was observed in the nature of stress-strain curves of the steel specimens obtained from the two machines. All tensile specimens have reduced cross sections with 1 in. gage length. The tensile stress-strain curves for the polycrystalline specimens were converted into shear stress-strain curves using a Taylor factor of 2.78.

3   RESULTS

3.1   Dynamic Strain Aging in Fe and Fe-N Single Crystals

Single crystals of iron and Fe-N alloys of nearly identical orientation were strained at 25 and 250 $^\circ$C at a shear strain rate $\dot\gamma = 6 \times 10^{-4}$/sec. The stress-strain curves are shown in Fig. 1. At 250 $^\circ$C, the curves were serrated in all three interstitial concentrations. For the crystal containing 0.0025% interstitials serrations were absent in the first portion of the curve. With further deformation, serrations became more pronounced, but they gradually disappeared in the last stage of deformation. When the interstitial concentration was increased to 0.0064%, the serrations became much more pronounced at high strains, and the work-hardening rate was somewhat increased, especially in the early stage of deformation. With a further increase of interstitial content to 0.017%, the average amplitude of serrations remains the same, and the average frequency is slightly reduced. The dominant feature of this stress-strain curve is the high rate of work-hardening. These three stress-strain curves obtained at 250 $^\circ$C should be compared with the two obtained at room temperature. For the two alloys represented, the stress-strain curves are smooth. The rate of solid-solution hardening $(d\tau_i/dc)$ [where τ_i is the critical resolved shear stress (CRSS) and c is the interstitial concentration], at 25 $^\circ$C is about the same as at 250 $^\circ$C, but the large concentration-dependence of work-hardening observed at 250 $^\circ$C was absent at 25 $^\circ$C. For the crystal containing 0.0025% interstitials, the stress-strain curve at 25 $^\circ$C fails above that at 250 $^\circ$C, although the latter is serrated.

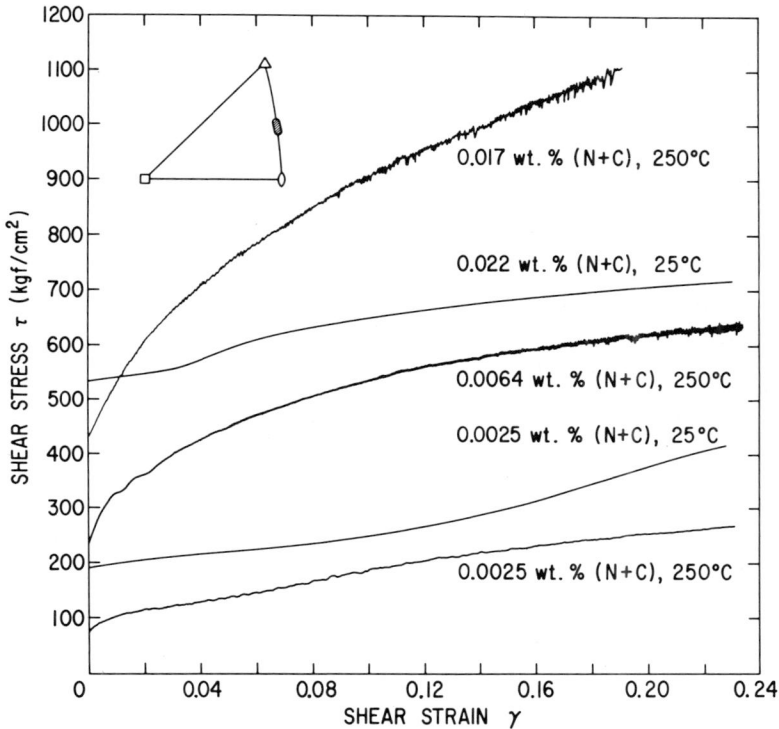

Fig. 1 Effect of nitrogen on the τ-γ curves of iron single crystals deformed in tension, $\dot{\gamma} = 6 \times 10^{-4}/\text{sec}$.

Figures 2 and 3 represent the temperature and strain-rate dependence of CRSS and flow stresses of the 0.02% N alloy near the onset of serrations. The CRSS is nearly independent of temperature and strain rate in the serrated region, but the flow stresses at constant strains increased with increasing temperature and decreasing strain rate.

Iron crystals containing 0.016% N were tested at 250°C, at which temperature all the nitrogen presumably was in solid solution. Serrated stress-strain curves were observed for all the strain rates employed, between 5×10^{-5} and $3 \times 10^{-2}/\text{sec}$. The strain-rate dependence of CRSS and flow stresses is shown in Fig. 4. The CRSS shows a strain-rate independent region, and a small increase at high strain rates. The work hardening rates for all strain rates are very high, but the flow stresses have peaks near $5.6 \times 10^{-4}/\text{sec}$ for all strains.

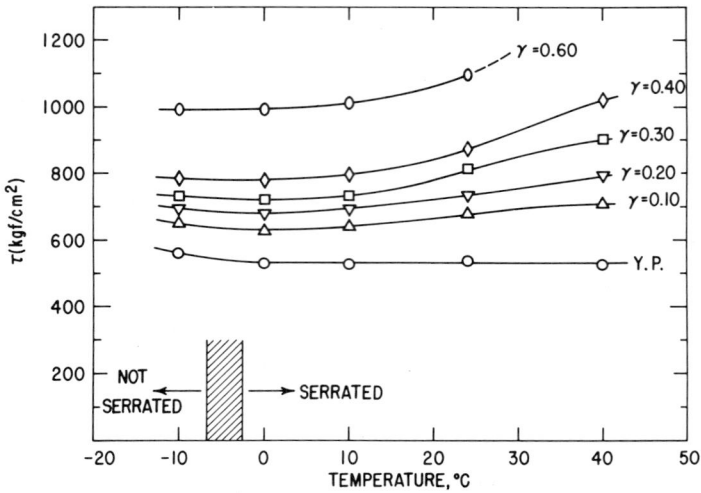

Fig. 2 Temperature dependence of yield and flow stresses of Fe-0.02% N single crystals strained in tension at constant crosshead speed of 0.01 in/min.

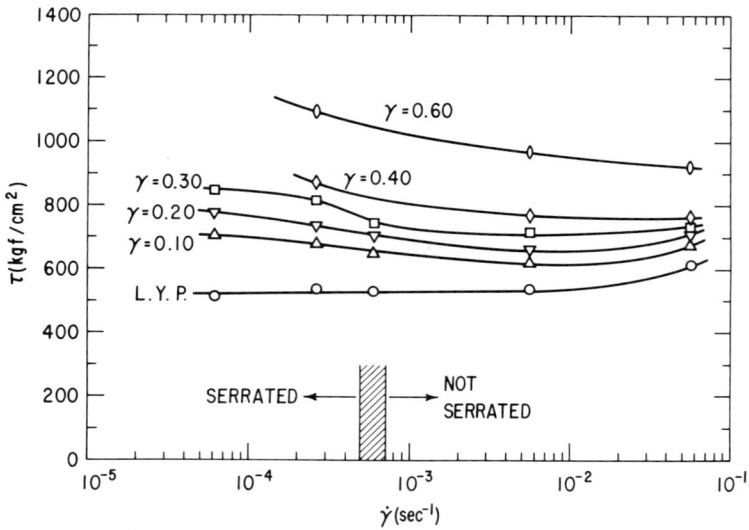

Fig. 3 Strain rate dependence of yield and flow stresses of Fe-0.02% N single crystals strained in tension at 25°C.

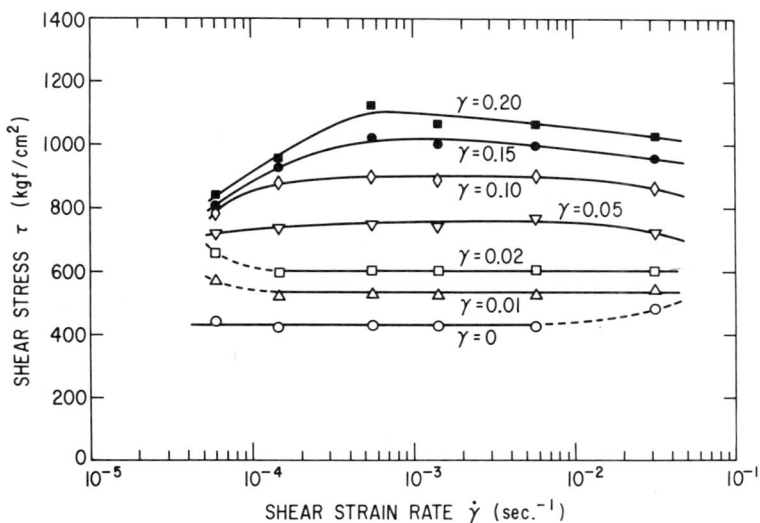

Fig. 4 Effect of strain rate on the yield and flow stress of Fe-0.016% N single crystals deformed in tension at 250°C.

3.1   Dynamic Strain Aging of Low-Carbon Steel

3.2.1   Temperature and Strain-Rate Dependence of Deformation Characteristics.

As shown by previous investigators, stress–strain curves of a mild steel undergo drastic changes in appearance in certain temperature regions in constant strain-rate tests. Figure 5 illustrates a series of such curves of the 0.035% C steel, quenched from 250°C and tested in tension between 50°C and 346°C at a constant cross-head speed of 0.0025 in./min. With an increase of testing from 50°C, the degree of serrations (their average amplitude and frequency) increased up to 162°C. At this temperature, the Luders band propagated discontinuously and contained regular serrations. At a higher temperature, serrations began to disappear from portions of the stress–strain curve until they completely disappeared at 250°C. With the appearance of serrations, the work-hardening rate began to increase with temperature to a maximum value and remained nearly constant until serrations completely disappeared; then it decreased with further increase of temperature. This change of work-hardening behavior with temperature is illustrated more explicitly

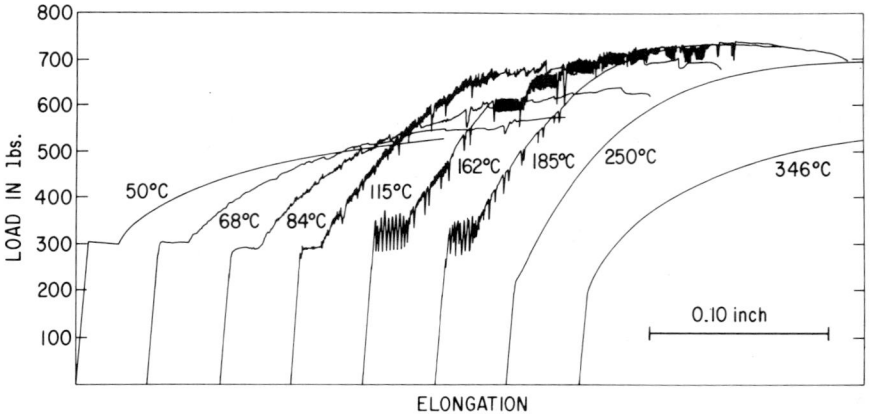

Fig. 5 Load-elongation curves of a 0.035% C steel strained in tension at crosshead speed of 0.0025 in/min. (Cross-sectional area $\sim$ 0.014 in^2, gage length = 1.00 in.)

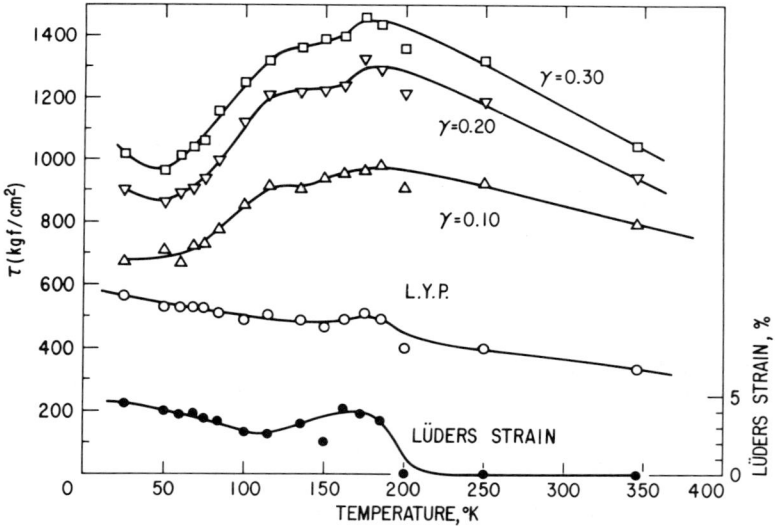

Fig. 6 Temperature dependence of Lüders strain, yield and flow stresses of a 0.035% C steel strained at constant crosshead speed of 0.0025 in/min.

in Fig. 6, where the yield and flow stresses are plotted as a function of temperature. At various shear strains, broad flow-stress peaks are observed, with peak height increasing with increasing strain. In contrast to the flow stresses, the lower yield stress first decreased slightly with increasing temperature and then developed a small peak to coincide with the flow stress maxima and the disappearance of serrations. The Luders strain changed in the same manner as the lower yield stress. In those cases where the Luders regions were serrated, the stress levels at the bottom of the serrations were taken as the lower yield stresses.

The temperature dependence of yield stress, flow stresses, and Luders strain was also obtained at three other strain rates. The heights of flow stress peaks at various strains were found to be independent of strain rate, but the positions of the peaks were shifted to higher temperatures with increasing strain rate. Manjoine[7] previously studied dynamic strain aging in a mild steel with strain rates varying from 10^{-6} to 10^3/sec. His data also suggested that the flow-stress maxima at constant strains and the ultimate tensile strength were independent of strain rate.

Depending upon the testing temperature, the yield and flow stresses undergo different changes with changing strain rate. At 25°C, after quenching from 250°C, serrated stress–strain curves were not observed for the range of strain rates used (10^{-5} to 10^{-2}/sec). Both the yield and flow stresses increased with increasing strain rate. However, at 100°C, when serrated stress–strain curves were observed at strain rates lower than 2.5×10^{-3}/sec, an inverse strain-rate dependence of the flow stresses was observed, as shown in Fig. 7. In this case, the yield stress still has the normal

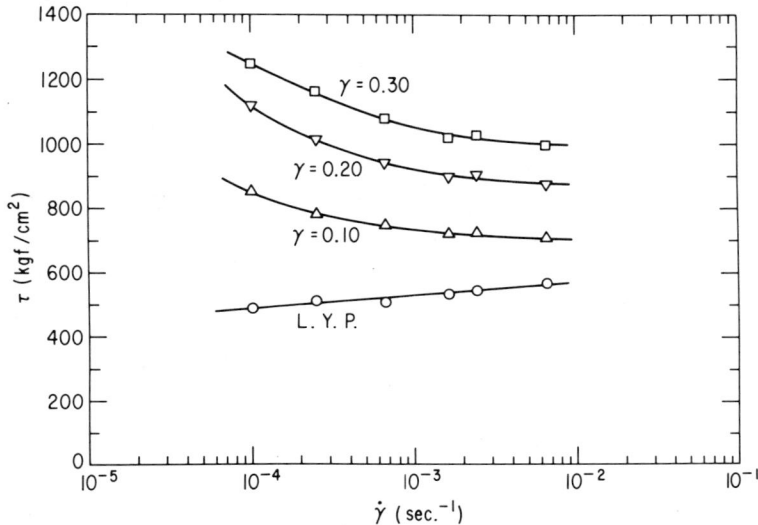

Fig. 7 Strain rate dependence of yield and flow stresses of a 0.035% C steel strained at 100°C.

strain-rate dependence. At higher temperature, flow-stress peaks were found (e.g., at $\dot{\gamma} = 1.5 \times 10^{-4}$ and $T = 150\,°C$), and·they shifted to higher strain rates at still higher temperatures. In general, the yield stresses either were strain rate-independent or had the normal strain rate-dependence at all temperatures in the dynamic strain-aging region. A small peak of yield stress was observed at $225\,°C$ at $\dot{\gamma} = 2.5 \times 10^{-3}/\mathrm{sec}$ in the stress vs. logarithm of strain-rate plot. At other temperatures, for the range of strain rate studied, similar peaks were not observed.

3.2.2 Apparent Activation Energies of Dynamic Strain Aging. As mentioned in the Introduction, Cottrell[6] first demonstrated that Majoine's data for the initial appearance of serrations in a mild steel can be represented by an Arrhenius-type relation between strain rate and temperature: $\dot{\epsilon} = \dot{\epsilon}_0 \exp(-Q/RT)$, with $Q = 18{,}200$ cal/mol. This value of Q, as pointed out by Cottrell, agrees closely with that for the diffusion of nitrogen in alpha iron. Sleeswyk[10] studied the onset of serrations of ingot iron by changing strain rate during deformation at various temperatures. Assuming that the critical strain rate $\dot{\epsilon}_{cr}$, at which serrations first appear, is proportional to the mean velocity of interstitial diffusion, a relation $T\dot{\epsilon} = K \exp(-Q/RT)$ was obtained. His experimental results were represented by a straight line on a $\log(T\dot{\epsilon})$ vs. $1/T$ plot, which yielded an activation energy of $20{,}400$ cal/mol. However, Sleeswyk's results can also be fitted on an Arrhenius plot, giving a value of $Q \simeq 19{,}000$ cal/mol.

Kinoshita et al.[11] studied both the onset and disappearance of serrations. The activation energies for both processes, obtained from an Arrhenius-type plot are shown in Table 2. The value they

TABLE 2 Activation Energies for the Onset and Disappearance of Serrations

	Onset of Serrations (Q)	Disappearance of Serrations (Q')
Manjoine[7]	18,200 cal/mole	34,000 cal/mole†
Sleeswyk[10]	20,400 cal/mole*	
Kinoshita et al[11]	19,800 cal/mole	36,500 cal/mole
Keh et al	19,000-20,000 cal/mole	30,500 cal/mole
Diffusion of N in α-iron	18,200 cal/mole	
Diffusion of C in α-iron	20,100 cal/mole	
Köster Peak		29,000-38,000 cal/mole

* Q obtained from $(T\dot{\epsilon}) = K \exp(-Q/kT)$.

† Q' obtained from the peaks of UTS vs. $1/T$ plot at various strain rates.

obtained for the disappearance of serrations was quite similar to
that for the cold-work internal friction peak (Koster peak) which
suggests that these two processes are probably related. We repeated Kinoshita's experiments and found similar activation energies for both processes (Table 2).

The disappearance of serrations in steel was accompanied by a
maximum on the flow stress vs. temperature plot, as observed by
Kinoshita et al.[11] and by ourselves. Because Manjoine[7] did not report the temperatures and strain rates at which the serrations disappeared, we assume that they disappeared when the flow stress
was maximum. By replotting the temperature and strain rate at
which the flow stress peaks appeared in Manjoine's investigation,[7]
a value of $Q = 34,000$ cal/mol was obtained. Therefore, as shown
in Table 2, if one assumes a simple Arrhenius-type relation between strain rate and temperature, good agreement exists between
the values for activation energies for the onset and disappearance
of serrations, as reported by various investigators. The value of
Q for the onset of serrations (Table 2) agrees very well with the
values of carbon or nitrogen diffusion in alpha iron and the second
value Q' can be considered to be the sum of the activation energy
for interstitial diffusion (Q_D) and the binding energy of interstitial
atoms to dislocations (ΔE), as in the Koster internal friction peak.

3.2.3 Correlation of Flow Behavior with Nature of Serrations.

To gain more insight into the dynamic strain aging
phenomena, the nature of serrations and the yielding and flow
parameters at various stages of aging was examined very closely.
In Fig. 8, the logarithm of the average shear strain rate is plotted
against the reciprocal of absolute temperature for various degrees
of serrations. The closed circles represent completely serrated
stress-strain curves; the open circles represent perfectly smooth
curves; and the partially filled circles represent various degrees
of serrations. The two heavy lines bound the region of serrated
stress-strain curves, for this 0.035% C steel, quenched from
250°C. The slopes of these lines gave the activation energies
shown in Table 2. The other straight lines drawn in between the
two heavy lines represent stress-strain curves of the same degree
of serrations and similar appearance. For example, one of the
curves in Fig. 5 (162°C) shows serrations in the Luders band region and several steps on the flow curves. The same type of stress-
strain curves were also found at other strain-rate and temperature
combinations. Thus a straight line was drawn through the points
(see Fig. 8). It is reasonable to assume that along each of these
straight lines the form of the stress-strain curve is controlled by

TEMPERATURE, °C

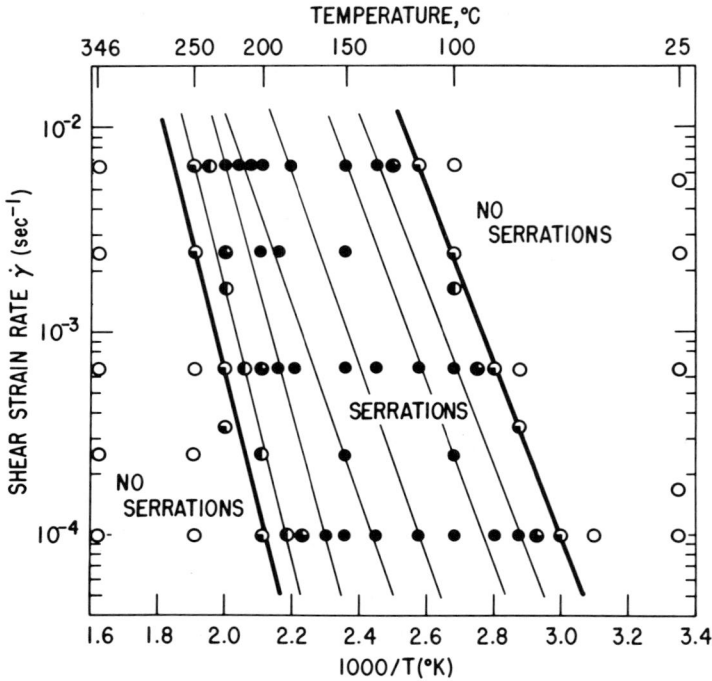

Fig. 8 Linear relation between log $\dot\gamma$ and $1/T$ for various degrees of serration on τ-γ curves of a 0.03% C steel, quenched from 250°C.

a single, thermally activated process, and the activation energy for the process can be determined from the slope of the line. Indeed, it was observed that the conditions which produced each straight line also produced nearly identical stress–strain curves. To show the validity of this observation throughout the serrated region, the flow stresses at various strains were plotted on a "normalized" $1/T$ scale for four different strain rates (see Fig. 9). For the lowest strain rate, a linear $1/T$ scale was used. For the three, higher strain rates, the $1/T$ scales were expanded to allow for the nonparallel ln $\dot\gamma$ vs. $1/T$ plots in Fig. 8. Indeed, for each plastic strain, all the data obtained at different strain rates can be fitted to a single curve, with very little scatter. Two peaks similar to those in Fig. 6 are again revealed. Although not shown in Fig. 9, there is a temperature-independent region at low temperature and a small peak at higher temperature for the yield stress and Luders strain. A close examination of the data from Manjoine[7] reveals

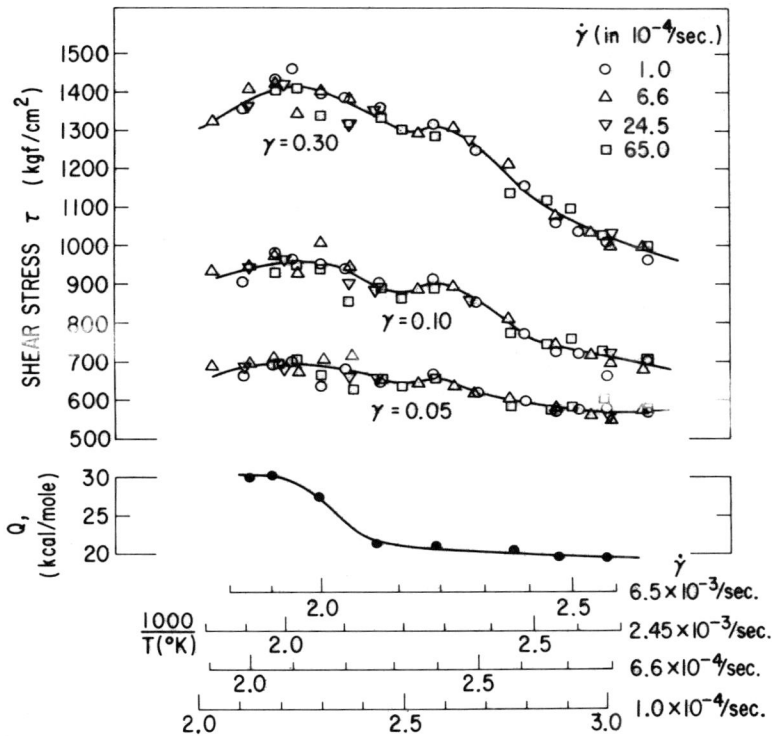

Fig. 9 Temperature and strain rate dependence of flow stresses, and activation energy for serrated flow of a 0.035% C steel strained in tension.

that: (1) flow-stress maxima at various strains between 2% and the ultimate tensile strength are strain-rate independent; and (2) the widths of the flow-stress peaks decreased with increasing strain rate. Both facts suggest that his data could be normalized in the same manner as our own data in Fig. 9.

The change in activation energy during dynamic strain aging, obtained from the slopes of the straight lines in Fig. 8, was replotted in Fig. 9. Recently, Blakemore and Hall[12] found the activation energy for the first appearance of serrations on the primary Luders plateau to be 20,100 cal/mol, which is close to the 21,300 cal/mol determined in this investigation for the same process. To justify the Arrhenius plots used by various investigators, a detailed atomic model for dynamic strain aging is needed. Some of the proposed models will be discussed later. For now, the Q values shown in Fig. 9 can be considered as apparent activation energies for dynamic strain aging.

3.2.4 Structural Changes during Dynamic Strain Aging.

The fact that stress–strain curves are frequently serrated in the dynamic strain-aging region suggests that deformation is nonuniform. As shown in Fig. 5, pronounced serrations were sometimes observed even within a Luders strain. Hall[13] first showed that the serrated Luders plateau corresponds to a discontinuous propagation of the Luders band. We performed a similar experiment with the results illustrated in Fig. 10. The specimen surface was prepolished. It was deformed at 200 °C at a nominal strain rate of 3×10^{-4}/sec. The number of bands corresponds to the number of serrations. Apparently, one band started from one end of the gage section, and as it propagated, the dislocations were pinned and the propagation was arrested. A new band started in the adjacent region, where the stress concentration was highest, and later stopped. Thus, the Luders band propagated discontinuously throughout the gage length before general work-hardening began. In the work-hardening region it was observed that serrations also correspond to slip bands appearing on the specimen surface. It appears that these band formations are localized events, superimposed on a more uniform work-hardening through the specimen.

T. A.

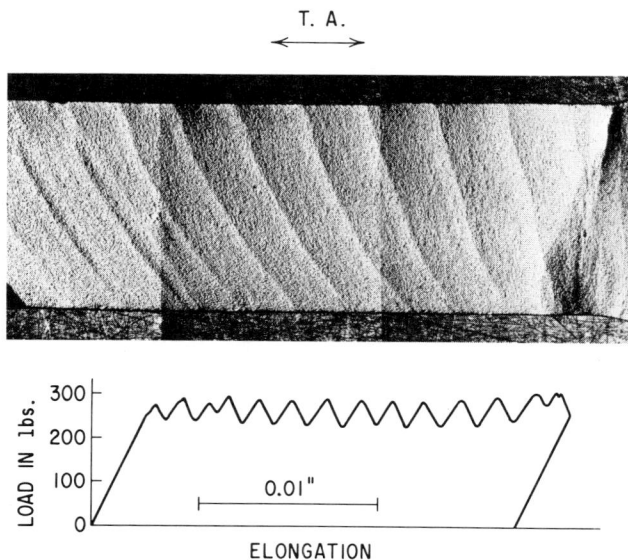

Fig. 10 Discontinuous propagation of Lüders bands in a 0.035% C steel strained in tension at 200°C at a crosshead speed of 0.02 in/min.

Although this correlation between surface structure and serrations is quite interesting, a quantitative study of the dislocation structure by transmission microscopy is more informative. In 1963, Keh and Leslie[14] first showed qualitatively that the high work-hardening rate of a mild steel during dynamic strain aging is associated with a high rate of dislocation multiplication. In a discussion of a paper by Baird and Jamieson,[15] Keh further showed that the rate of dislocation multiplication at 200°C under conditions of dynamic strain aging is two to three times greater than that at room temperature. These results, with a few new data points, are reproduced in Figs. 11 and 12. In Fig. 11, the stress-strain curves of the 0.035% C steel strained at 25°C and 200°C with a shear strain rate of $\dot{\gamma} = 6 \times 10^{-4} \text{sec}^{-1}$ are shown. The 200°C curve is serrated and shows a high work-hardening rate compared to the 25°C curve. Another specimen of the same steel was strained at 250°C at the same rate. The stress-strain curve was smooth, but the work-hardening rate is the same as in the 200° curve. The dislocation densities as a function of strain and deformation temperature are shown in Fig. 12. The rapid rate of work-hardening and the high flow stress of the specimens strained at 200° and 250° can be accounted for by the higher dislocation density observed. These results were confirmed qualitatively by the work of Baird

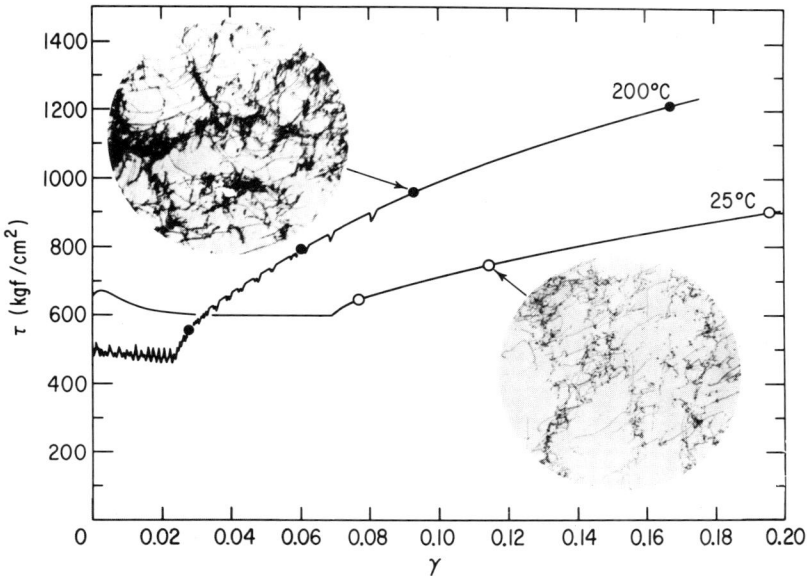

Fig. 11 Stress-strain curves and dislocation structures of a 0.035% C steel strained in tension at two temperatures with same strain rate $\dot{\gamma} = 6 \times 10^{-4}/\text{sec}$.

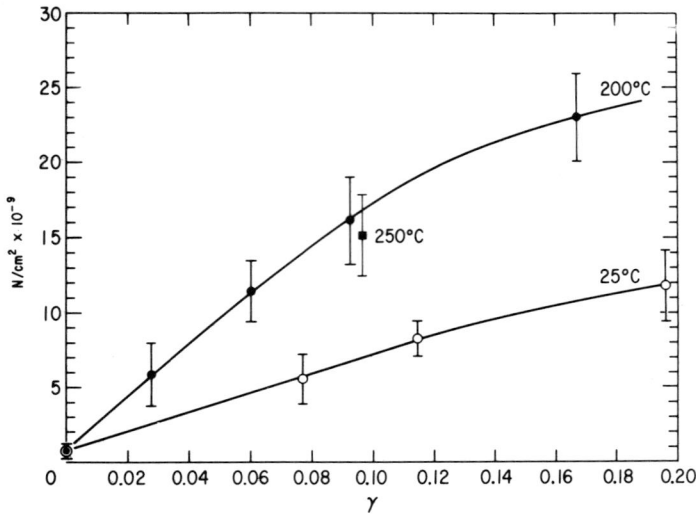

Fig. 12 Average dislocation density (N) vs. plastic shear strain (γ) for a 0.035% C steel deformed in tension at three temperatures, $\dot{\gamma} = 6 \times 10^{-4}/\text{sec}$.

and MacKenzie,[16] and Brindley and Barnby.[17] However, the latter authors showed that in one of their steels, strained at a temperature and strain rate just outside the serrated region, the dislocation density at a constant strain was reduced, although the flow stress remained high. This conclusion was not supported by the transmission electron micrographs in their paper, and is also in contradiction to our own findings.* In other alloy systems, Edington and Smallman[18] found a similar enhancement of dislocation multiplication during dynamic strain aging in vanadium, and Wilcox and Smith[19] found the same effect in Ni-H alloys.

3.2.5   Concentration Dependence of Dynamic Strain Aging. The results discussed in previous sections were obtained from the 0.035% C steel quenched from 250°C (condition A). To study the effect of the concentration of interstitials in solid solution on dynamic strain aging, the steel was equilibrated at and quenched from 720°C (condition B). After this treatment, dynamic strain aging was observed at much lower temperatures than after quenching from 250°C (Fig. 14). These results are comparable to

---

*Our results agree with those of Dingley and McLean.[39] The latter paper appeared after the completion of this manuscript.

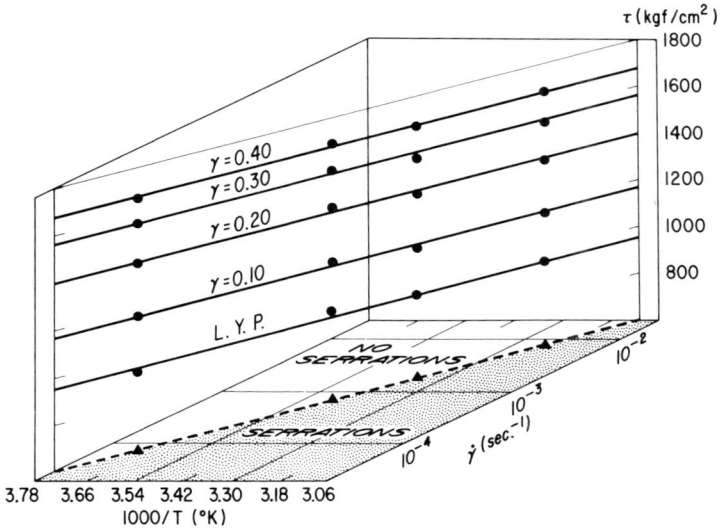

Fig. 13 Strain rate and temperature dependence of the onset of serrations, lower yield points and flow stresses of a 0.035% C steel quenched from 720°C.

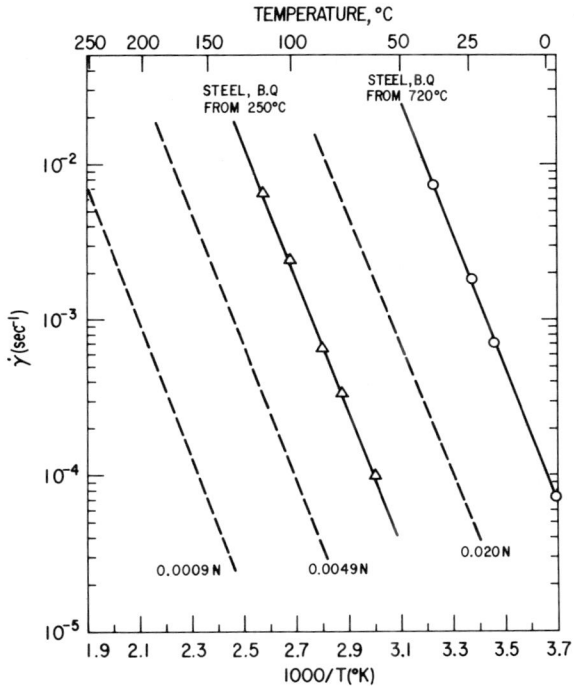

Fig. 14 Temperature and strain rate dependence of the onset of serrations in a 0.035% C steel and in three Fe-N alloy crystals.

those of previous investigators[20,21,25] although they did not discuss their results in terms of the concentration of interstitials. The temperature and strain-rate dependence of yield and flow stresses are similar to those shown in Figs. 2 and 3 for Fe-0.02% N single crystals. Again, the yield stress shows the normal temperature and strain-rate dependence, but the change of flow stress with strain rate and temperature again suggests dynamic aging behavior.

The onset of serrations can be represented by the dashed line on the ln $\dot{y}$ vs. $1/T$ plot of Fig. 13. For any combination of temperature and strain rate defined by the dashed line, the yield stress and flow stresses are constant, independent of either temperature or strain rate. Figure 14 illustrates the effect of the concentration of interstitials in solid solution on the onset of serrations. The two solid lines represent conditions A and B of the 0.035% C steel, quenched from 250°C and 720°C, respectively. The total concentration of interstitials in solution should be about 0.003% N in the former and 0.025% (C + N) in the latter condition. The dotted lines are similar results obtained from Fe-N single crystals with three concentrations of interstitials. Within experimental error, these lines are parallel, with an activation energy varying from 19,000 cal/mol to 20,000 cal/mol. It is interesting to note that dynamic strain aging can be observed in Fe single crystals containing less than 10 ppm interstitials in solid solution. However, in an Fe-0.15% Ti alloy, similar to the one used by Leslie and Sober[22] in their study of the temperature-dependence of yield and flow stress, dynamic strain aging was not observed. Previously, several investigators[15,23,24] showed that in steels containing enough aluminum to combine with nitrogen, yield stress decreased continuously with increasing temperature without showing a strain-aging peak.

Despite the observation of serrated flow, at temperatures as low as 0°C, in Fe-C alloys quenched from high temperatures in the γ region, no serrations were observed at 23°C on the flow curves of a series of carbon martensites, freshly quenched from the γ region.[25] These martensites, containing from 0.12 to 0.41% C, demonstrated other attributes of dynamic strain aging, such as negative strain-rate dependence and negative temperature-dependence of flow stress and anomalously high flow stress at 23°C, but no serrations occurred even at strains as great as 0.07. We must, therefore, consider the possibility that there is an upper limit of dislocation density for the appearance of serrations. At this upper limit, it would not be possible to produce serrations by an sudden avalanches of dislocations. It is possible that serrations may appear in the flow curves of martensites strained at elevated temperatures, because of carbide precipitation during the test.

4 DISCUSSION

A satisfactory model of dynamic strain aging should explain the following:

1) the temperature and strain-rate insensitivity of yield stress;
2) the strong temperature and strain-rate dependence of flow stresses;
3) the change of the nature of serrations and the concurrent change of activation energy in the strain rate and temperature region of dynamic strain aging; and
4) the strong concentration-dependence of the onset of serrations.

It is generally agreed that dynamic strain aging in steel is due to some form of interaction between dislocations and interstitial solute atoms. There are several mechanisms proposed based on either an impurity drag model or an impurity pinning model. The experimental observations just summarized will be discussed in light of these mechanisms.

4.1 Temperature and Strain-Rate Dependence of Yield Stress

Attempts to relate the measured lower yield stress of a solid solution to atomic theories of solid-solution strengthening are complicated by at least two factors: (1) the presence of residual stresses in quenched specimens; and (2) the difficulty of defining a plastic strain rate at the lower yield stress. Nevertheless, the lower yield stress, when it is well defined, is the best criterion for solid-solution strengthening compared to microyield stress or macroscopic flow stresses.

The fact that the lower yield stress in iron and steel is independent of temperature and strain rate in the range of dynamic strain aging suggests that it is athermal in nature. The origin of an athermal stress could be (1) long-range internal stress due to dislocations or (2) internal stress fields due to solute atoms. Several investigators have suggested that solid-solution hardening could be due to the presence of either a higher grown-in dislocation density in the alloys compared to the pure material, or a rapid increase of dislocation density in the microstrain region. In Fe-N single crystals, these suggestions are not correct.[29] The Mott-Nabarro theory [26,27] of solid-solution hardening could explain the temperature-independence of yield stress. However, the predicted hardening rate is one order of magnitude lower than the observed one. In the temperature region of dynamic strain aging, interstitial solutes are mobile. Schoeck and Seeger[28] proposed that the yield stress in iron in this temperature region is due to Snoek ordering of interstitial solute atoms around dislocations. Their calculated hardening rate due to Snoek drag can be expressed as

$$\frac{d\tau}{dc} \simeq \frac{10\,A}{b}$$

where τ = CRSS
c = concentration of interstitial solute
A = interaction parameter $\sim 1.84 \times 10^{-20}$ dynes/cm^2 for carbon in iron
b = Burgers vector

Our recent experiments in Fe-N single crystals[29] yield a value of $d\tau/dc$ at room temperature about 10% lower than the predicted value by this theory. Nevertheless, the conclusion of Schoeck and Seeger that the yield stress should be temperature-independent was apparently deduced from the fact that the calculated force to move a dislocation as a function of velocity has a maximum value, which is temperature independent. If this model were used, a peak in yield stress should be observed in a constant strain-rate test, assuming a constant mobile dislocation density. These peaks at various strain rates deduced from the force-velocity curve of the Schoeck-Seeger paper, assuming a mobile density of 10^8/cm^2, are shown in Fig. 15. Although the drag stress has the same maximum value at various strain rates, at a given strain rate the peak is very sharp as shown in Fig. 15. Schoeck and Seeger in their paper also compared their theoretical prediction with the results of Heslop and Petch[30] on polycrystalline iron. From the latter paper, the rate of solid-solution hardening due to interstitials is essentially constant at 18°C, −128°C, and −196°C, which agrees with our data from Fe-N single crystals.[29] It is likely that interstitial solutes form Snoek atmospheres around dislocations during the test at 18°C. However, at temperatures below −100°C, Snoek atmosphere cannot form during the test at usual strain rate. In view of the above objections, we conclude that the athermal stress is not due to Snoek ordering.

Cottrell and Jaswon[31] proposed that dynamic strain aging can be caused by the force exerted on dislocations due to a moving solute atmosphere trailing behind the dislocations. They showed that the critical (i.e., maximum) velocity at which a dislocation

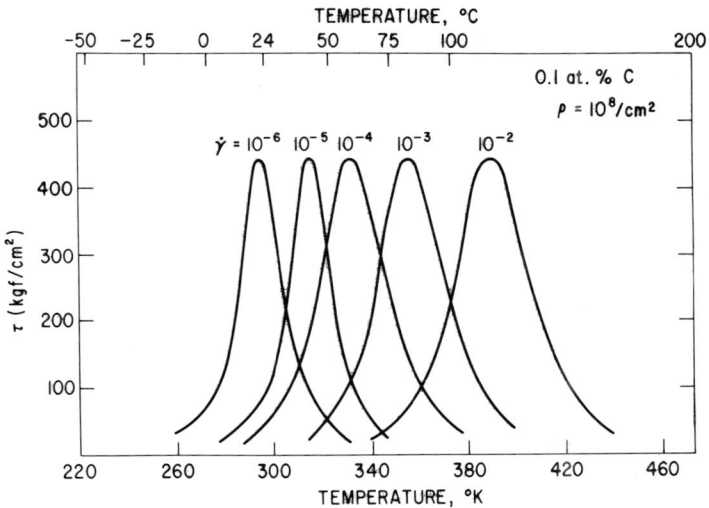

Fig. 15 Dragging stress from Snoek ordering of carbon atoms in the stress field of dislocations.

can still drag an atmosphere along can be expressed as:

$$V_c = \frac{4DkT}{A}$$

where D = diffusivity of solute atoms
       A = interaction parameter

and the maximum stress imposed by dragging an atmosphere has the form

$$\tau_c = 17 A C_0 \frac{N}{b}$$

where C_0 = solute concentration
       N = total number of atoms per unit volume
       b = Burgers vector

Using Cottrell and Jaswon's model, one would also expect a peak on the stress–temperature curve.

Hahn et al.,[32] using Manjoine's data from mild steel, noted that within a restricted range strain rate in the lower yield stress was independent of strain rate. They suggested that this was due to the combined effect of several peaks superimposed on the "true" strain-rate dependence of yielding in pure iron, which they assumed to be linear with the logarithmic of strain rate. The data at 200°C can be explained on this basis by adding together a calculated Snoek ordering peak at low strain rate, a Cottrell drag peak at high strain rate, and an enrichment peak in between. However, at lower temperatures, this explanation failed, and they had to invoke the possibility of hydrogen ordering, which is highly unlikely. The best explanation on the observed temperature and strain-rate independence of yield stress in iron-containing interstitials is given by Li[33] in these proceedings. He modified the Mott-Nabarro theory and showed that the hardening due to solute atoms is nearly athermal and its rate can be expressed as

$$\frac{d\tau}{dc} = 2\mu\epsilon$$

where μ = shear modulus
ϵ = solute misfit = $(1/a)(da/dc)$ [change of lattice parameter with concentration]

Inserting the known value of the various parameters, the predicted hardening rate agrees quite well with the observed value.[29]

In the 0.035% C steel, grain size also contributes to the athermal yield stress, in addition to the solid-solution effect. The small peak observed at high temperature near the end of serration (Fig. 6) could be due to the drag mechanism proposed by Cottrell and Jaswon.[31] However, the measurements of dislocation velocity as a function of temperature in Si-Fe by Stein and Low[34] gave neither an indication of Snoek ordering at 273°K nor of Cottrell drag at 373°K.

4.2   Temperature and Strain-Rate Dependence of Flow Stresses

The flow-stress peaks produced by either varying the temperature or strain rate are a major feature of dynamic strain aging. From the dislocation density measurements (Fig. 12), it is quite clear that these peaks are due to the enhanced rate of dislocation multiplication. This enhancement could be a result of dislocation drag of either the Cottrell or Snoek type which reduces the average velocity of dislocations, but a more plausible explanation of this effect is the pinning and regeneration of dislocations. Our previous work[35] clearly illustrated that dislocations can be pinned by interstitial solutes either by a Snoek or Cottrell atmosphere during low-temperature aging. It is reasonable to assume that at the temperatures and strain rates required for dynamic strain aging, dislocations can be pinned during the test. To maintain an imposed strain rate, more dislocations have to be generated. This immobilization and generation process is apparently one of the origins of serrations. The dual flow stress peaks in Figs. 6 and 9 could be due to solute pinning of dislocations by a Snoek and a Cottrell atmosphere, respectively. Quantitatively, the increase in flow stress can be accounted for by the increase in dislocation density assuming that the two quantities bear the relationship:

$$\tau = \tau_0 + \alpha \mu b \sqrt{N}$$

where α = const
μ = shear modulus
b = Burgers vector
N = total dislocation density

In this case $\tau_0 \simeq 170$ kgf/cm^2 at 25°C and $\tau_0 \simeq 0$ at 200° and 250°C, and $\alpha \simeq 0.43$. The exact meaning of τ_0 thus obtained is subject to question, as pointed out recently by Bailey and Flanagan;[36] however, it does not invalidate our argument.

4.3   Appearance and Disappearance of Serrations

As shown in Fig. 8, the $\ln \dot{\gamma} - (1/T)$ plot was divided into three regions. At low temperatures, interstitial solutes are immobile with respect to moving dislocations. Thus no serrations were observed on the stress-strain curve. At high temperatures (in the region at the extreme left of Fig. 8), solute atoms are so mobile that they can always follow dislocations (serrations again are absent). Between these two regions, the dislocation mobility is influenced by the mobile interstitial atoms. Therefore, it is not surprising that the activation energy for the onset of serrations should be the same as that for interstitial diffusion in iron. In this case the controlling process is apparently the locking of dislocations by solute atoms. To produce the same degree of locking of a dislocation, a higher temperature has to be used for the alloy with lower interstitial concentration. This positive concentration dependence is indeed observed as shown in Fig. 14. However, the magnitude of the concentration-dependence is much higher than one would expect from a simple diffusion model. As suggested by Hirth,[37] the apparent concentration-dependence could be much larger than that predicted from a simple diffusion model because of the presence of other dislocations or other competing sinks for interstitials.

The constancy of activation energy in the early stages of dynamic strain aging as illustrated in Fig. 9 suggests that the basic process at these stages is the same, but the degree of locking varies. By further increasing the temperature at a constant strain rate, dislocations can drag solute atoms with them. At some stage, it is conceivable that as a dislocation moves, solute atoms will flip between their favorable positions in the dislocation core and their next favorable positions adjacent to the core. This process should yield an activation energy equal to the sum of the activation energy for solute diffusion plus the binding energy between a solute and a dislocation. Indeed, this was observed as serrations began to disappear from the stress-strain curves.

Recently, Bailey and Flanagan[38] developed a model in terms of diffusion-controlled dislocation multiplication to explain the critical strain requirement for the onset of serrations. In our investigations, we found that the critical strain is not a well-defined quantity. In addition, the dislocation multiplication rates deduced from their theory are much larger than those observed in the serrated region. Therefore, this theory will not be discussed further here.

5 CONCLUSIONS

The following conclusions can be made with regard to dynamic strain aging in iron and steel.

1) In the serrated region, the yield stress of iron-containing interstitials is nearly independent of strain rate and temperature, but the flow stress is strongly dependent on these two parameters.

2) The increase of work-hardening rate in the serrated region is due to the enhancement of the dislocation multiplication rate as a result of dislocation pinning by solute.

3) The activation energy for dynamic strain aging changes from Q_D, that for interstitial diffusion in iron at the onset of serrations, to $(Q_D + \Delta E)$, where ΔE is binding energy of interstitial solutes to dislocations at the disappearance of serrations.

4) The temperature and strain rate required for the first appearance of serrations on the stress-strain curve is strongly concentration-dependent.

5) Stress-strain curves are identical along each line of constant activation energy on the $\ln \dot{\gamma}$ vs. $1/T$ plot.

ACKNOWLEDGMENTS

We wish to thank J. P. Hirth and J. C. M. Li for stimulating discussions and W. B. Seens, K. Phillips, and T. P. Churay for technical assistance.

REFERENCES

1. Stromeyer, C. E.: *Minutes of Proc. of Inst. of Civil Engr.* 80:114 (1885).
2. LeChatelier, F.: *Revue de Métallurgie* 6:915 (1909).
3. Portevin, A., and F. LeChatelier: *Compt. Rend. Acad. Sc.* 176:507 (1923).
4. Fettweiss, F.: *Stahl und Eisen* 39:34 (1919).
5. Nabarro, F. R. N.: "Report on Strength of Solids," p. 38, London: Physical Soc., 1948).
6. Cottrell, A. H.: *Phil. Mag.* 74:829 (1953).
7. Manjoine, M. J.:*Trans. ASME* 66:241A (1944).
8. Baird, J. E.: "Strain Aging of Steel—A Critical Review," Iliffe Publications Ltd., London, 1963.
9. Klein, M. J., and C. N. Reid: Metal Deformation Processing, VI, *DMIC Rept.* 208: 114 (1964).
10. Sleeswyk, A. W.:*Acta Met.* 8:130 (1960).
11. Kinoshita, S., P. J. Wray, and G. T. Horne: *Trans. AIME* 233: 1902 (1965).
12. Blakemore, J. S., and E. O. Hall:*JISI* 204:817 (1966).
13. Hall, E. O.:*JISI* 170:331 (1952).
14. Keh, A. S., and W. C. Leslie: in "Materials Science Research," VI, H. H. Stadelmaier and W. W. Austin (eds.), p. 208, Plenum Press, New York, 1963.
15. Baird, J. D., and A. Jamieson: "Proc. of NPL Symp. on The Relation between the Structure and the Mechanical Properties of Metals, p. 362, Her Majesty's Stationery Office, London, 1963.
16. Baird, J. D., and C. R. MacKenzie: *JISI* 202:427 (1964).
17. Brindley, B. J., and J. T. Barnby: *Acta Met.* 14:1765 (1966).
18. Edington, J. W., and R. E. Smallman: *Acta Met.* 11:1313 (1963).
19. Wilcox, B. A., and G. C. Smith: *Acta Met.* 12:371 (1964).
20. Sleeswyk, A.:*Acta Met.* 6:598 (1958).
21. Wilson, D. V., B. Russell, and J. D. Eshelby: *Acta Met.* 7:628 (1959).
22. Leslie, W. C., and R. J. Sober: *Trans. ASM* 60:79 (1967).
23. Glen, J., J. Lessells, and R. R. Barr: *International Conference on Creep*, 1963.
24. Wilson, D. V.: *Acta Met.* 9:618 (1961).
25. Leslie, W. C., and R. J. Sober: *Trans. ASM*, 60:459 (1967).
26. Mott, N. F., and F. R. N. Nabarro:*Proc. Phys. Soc.* 52:86 (1940); "Report on Strength of Solids," London: Phys. Soc., 1948.
27. Nabarro, F. R. N.:*Proc. Phys. Soc.* 58:669 (1946).
28. Schoeck, G., and A. Seeger:*Acta Met.* 7:469 (1959).
29. Nakada, Y., and A. S. Keh: To be published in *Acta Met.*
30. Heslop, J., and N. J. Petch:*Phil. Mag.* 1:866 (1956).
31. Cottrell, A. H., and M. A. Jaswon: *Proc. Roy. Soc.*, London, 199:104 (1949).
32. Hahn, G. T., C. N. Reid, and A. Gilbert: *Acta Met.* 10:747 (1962).
33. Li, J. C. M.: This colloquium, p. 87.

34. Stein, D. F., and J. R. Low: *J. Appl. Phys.* 31:362 (1960).
35. Nakada, Y., and A. S. Keh: *Acta Met.* 15:879 (1967).
36. Bailey, D. J., and W. F. Flanagan: *Phil. Mag.* 15:43 (1967).
37. Hirth, J. P.: Private communication.
38. Bailey, D. J., and W. F. Flanagan: Private communication.
39. Dingley, D. J., and D. McLean: *Acta Met.* 15:885 (1967).

DISCUSSION *on Paper Presented by A. S. Keh*

P. HAASEN: Concerning the concentration (*C*) dependence of the critical strain rate $\dot{\epsilon}$ for serrated yielding which you find empirically: I would like to point out that: (a) the simple dynamic model* of serrated flow which equates the dislocation velocity to the drift velocity of foreign atoms within an interaction radius $R = \mu T/A$ (*A* is the interaction strength between foreign atoms and dislocations) would predict an inverse relation between $\dot{\epsilon}$ and *c* (at constant stress) contrary to your experimental results. Since serrated flow basically represents an instability of slip, the overall strain rate is difficult to interpret, however, on an atomic scale.

A. KEH: The concentration dependence for the onset of serrations which I referred to is not at constant stress but at constant temperature. As the supersaturation of interstitials increases, the *CRSS* also increases. Prof. Hirth in his discussion of our paper, gave a plausible explanation of the concentration-dependence we observed.

R. ARMSTRONG: I should say that one suggestion has been put forward [†] that the flat portion of the stress–temperature curve for the polycrystal is due to the $Kl^{-1/2}$ term in the stress–grain size analysis; i.e., the level of this stress increases as the grain size decreases and the eventual fall-off of this stress at high temperatures is due to *K* decreasing to nearly zero (the pure iron value).

A. KEH: We have observed that a flat portion of the stress-temperature curve in both single and polycrystalline iron-containing interstitials. In the case of single crystals, we believe that this flat portion is due to the internal stress created by the interstitial solutes. The difference in the stress level between the single and polycrystals is due to the $K^{-1/2}$ term in the Petch analysis.

*For example, Haasen, P.: "Alloying Effects in Concentrated Solid Solutions," Massalski (ed.), Gordon and Breach, 1966.

[†]Armstrong, R. W., R. T. Begley, J. H. Bechtold: "Refractory Metals and Alloys II," p. 195 Interscience Publishers, New York, 1963.

J. P. HIRTH: Concerning the point raised by Keh and by Haasen, a positive concentration dependence for the *onset* of serrated flow appears reasonable. Below the critical temperature, at T_1, the velocity V_c of a dislocation with its atmosphere is too low to enable flow to occur according to the usual strain-rate equation

$$\dot{\epsilon} = \rho b V \qquad (1)$$

and hence breakaway occurs, fast dislocations with a velocity V_p, controlled, say, by phonon drag, form and control plastic flow. At the critical temperature, serrated flow begins when the fast dislocations can be *repinned*. The repinning should be diffusion-controlled, so that the diffusional activation energy is reasonable. Also the concentration-dependence of the diffusional process should be positive, corresponding to the appropriate solution to Fick's second law, giving a positive dependence of the critical rate for serrations. As discussed by Keh, the presence of competing carbon sinks, such as, other dislocations or precipitates, at low interstitial concentrations and high temperature, would give an apparent positive concentration-dependence with an exponent even larger than that for a simple diffusion solution, say, for a line sink in an infinite medium.

At a higher temperature T_2 the velocity of the pinned dislocation increases at a given stress until V_c is adequate to satisfy Eq. (1), whereupon serrations end. Thus the *ending* of serrations should correlated with Cottrell or Snoek drag, with inverse concentration-dependence, or core drag, with exponential concentration-dependence. A final point of interest is that if the competing sinks at the onset of serrations are efficient, then by the end of the serrations the carbon content should be that in equilibrium with these sinks and not the quenched-in concentration. Hence it is consistent with the above suggestion that the observed ending of serrations is relatively concentration-independent.

A. KEH: We agree with Prof. Hirth that the rate-controlling step for the onset of serrations is the pinning of fast-moving

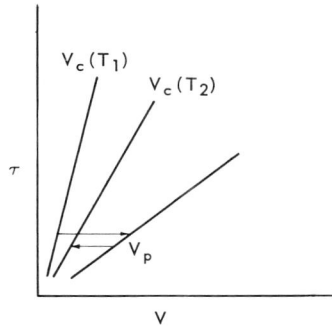

Fig. 1 "Stress (or Force)—Velocity Curve"

dislocations, which were either generated in the early stage of deformation or freed from pinned atmospheres. As I mentioned earlier, from a simple diffusion model of solute atoms to dislocation, the concentration-dependence for the onset of serrations should be positive with exponent of 1 or 2. It is quite reasonable that the very large apparent concentration-dependence observed in this investigation is due to the presence of competing sinks for interstitial atoms, as Prof. Hirth suggests. A bried discussion of this point was included in the text.

As to the disappearance of serrations, its concentration-dependence cannot be established quantitively due to the possible complication of precipitation. However, it appears that is is very small.

THERMALLY ACTIVATED DISLOCATION MOTION AND ITS APPLICATION TO THE STUDY OF RADIATION DAMAGE

F. A. Smidt, Jr. and A. L. Bement

Pacific Northwest Laboratories
Division of Battelle Memorial Institute
Richland, Washington

ABSTRACT

Radiation damage in alpha iron was studied by an analysis of thermally activated flow to determine the basic dislocation–defect interactions. A method which circumvents the difficulties encountered in separating the arthermal and thermal components of flow stress was used in the analysis and is briefly described. It consists basically of experimentally determining the stress dependence of the activation energy from an analysis of the temperature dependence of the flow stress.

The experiments on irradiated iron show that the defects present cause only slight modifications in the mechanism of thermal activation, but produce large increases in the athermal stress component which are dependent on fluence and purity of the sample. It is concluded that the Orowan mechanism of hardening is most consistent with the experimental observations.

1   INTRODUCTION

This investigation was undertaken to determine if the dislocation-defect interactions which give rise to the changes in mechanical properties in irradiated iron could be identified by an analysis of thermally activated flow. To achieve this objective several aspects of the deformation process in unirradiated iron samples were first investigated. New insights from this work as well as the results of the radiation damage studies are reported here.

Certain difficulties should be recognized in the use of the basic rate equation of thermally activated flow,

$$\dot{\gamma} = \dot{\gamma}_0 \exp\left(\frac{-G(\tau^*)}{kT}\right) \tag{1}$$

and the derived quantities which relate it to measurable parameters

$$V^* = -kT\left(\frac{\partial \ln(\dot{\gamma}/\dot{\gamma}_0)}{\partial \tau^*}\right)_T \tag{2}$$

and

$$G = -TV^*\left(\frac{\partial \tau^*}{\partial T}\right)_{\dot{\gamma}/\dot{\gamma}_0} \tag{3}^\dagger$$

Among these difficulties are the experimental determination of τ^*, the treatment of mobile dislocation density, and the identification of the rate-controlling process. It will be shown that a knowledge of $G(\tau^*)$ is useful in resolving these difficulties.

2   EXPERIMENTAL METHODS

In the experiment, Ferrovac E, a vacuum melted iron; a zone-refined iron obtained from Battelle-Columbus (BMI); and sintered

---

†Neglecting the temperature dependence of the shear modulus as a minor contribution in the b.c.c. metals.

molybdenum, each with the purities shown in Table 1 were used. The experiments for molybdenum were confined to a few activation volume measurements.

TABLE 1 Composition of Materials Studied
(Impurity concentrations in ppm)

	C	N	O	Si
Fe E	40	1	200	200
BMI	3.3	<1	2.6	10
Mo	10	8	11	10

A prestrain of 3.5% at 295°K was given to all iron specimens prior to any measurements. This treatment ensured a uniform substructure and dislocation density at the initiation of the tests and also reduced the tendency toward twinning at low temperatures.

Conventional strain-rate cycling tests between the base strain rates .002 min^{-1} and .02 min^{-1} were used to determine activation volumes. A temperature-change test designed to determine T_0, the temperature at which $\tau^* = 0$, and the temperature dependence of the flow stress supplemented the strain rate-cycling tests. In the latter, the stress increment accompanying a decrease in temperature of 10 to 15 °K was determined from the difference between the flow stress at T_1 and ϵ_1 and the yield stress at T_2 and ϵ_1, thus measuring $(\Delta\tau^*/\Delta T)_{\dot{\gamma}}$. The constancy of $\Delta\tau^*$ over a wide range of strain for a given temperature increment was experimentally verified by cycling between fixed temperature baths.

3 RESULTS AND DISCUSSION

3.1 Derivation of $G(\tau^*)$ Relation

Temperature-change tests on Ferrovac E (Fig. 1), showed no change in slope of $(\Delta\tau^*/\Delta T)_{\dot{\gamma}}$ vs. T curves indicative of a change in mechanism in this temperature range. As would be expected, T_0, the temperature at which τ^* becomes zero, was dependent on strain rate. The linear form of $(\Delta\tau^*/\Delta T)_{\dot{\gamma}}$ permitted integration to a form

$$\tau^* = AT^2 - BT + C \tag{4}$$

At $\tau^* = 0 \ (T_0)$, the experimental coefficients determined from the

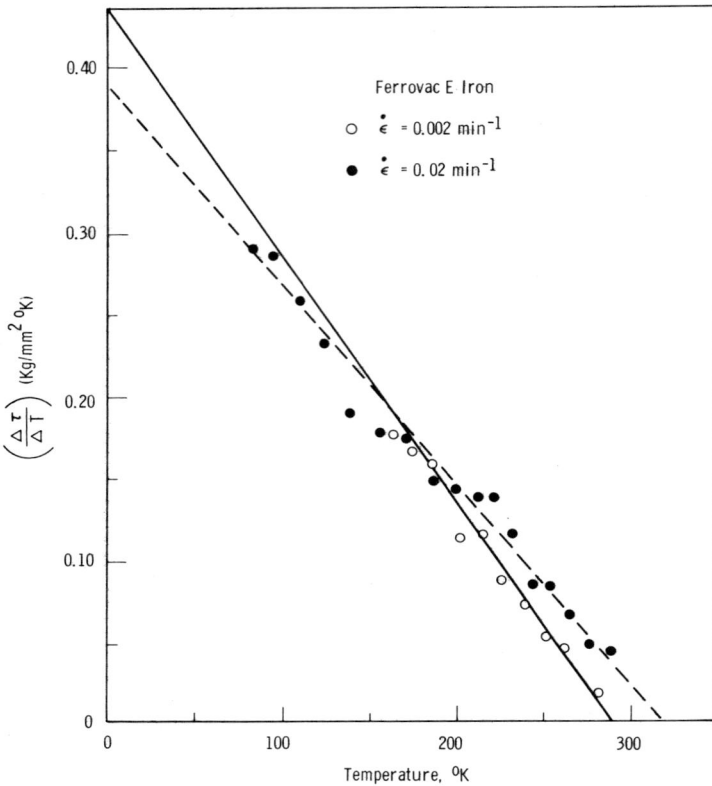

Fig. 1 $(\Delta\tau/\Delta T)_{\dot{\epsilon}}$ data from temperature change tests on Ferrovac E iron. Open circles (solid line) represent points from .002 min^{-1} strain rate tests.

slope and intercept are in the relation $B^2 = 4AC$; therefore the discriminant of this equation is zero and the roots are real and equal. If τ_p the effective stress at 0 °K is substituted for C and T_0 for $B/2A$, the equation becomes

$$\frac{\tau^*}{\tau_p} = \left(1 - \frac{T}{T_0}\right)^2 \quad \text{or} \quad \frac{T}{T_0} = 1 - \sqrt{\frac{\tau^*}{\tau_p}} \tag{5}$$

Since $G/G_0 = T/T_0$ at constant strain rate, an explicit relation between G and τ^* is obtained directly from the experimental data,

$$\frac{G}{G_0} = 1 - \sqrt{\frac{\tau^*}{\tau_p}} \tag{6}$$

The values of the coefficients A, B, C, and T_0 for the two strain rates studied are given in Table 2. It should be noted that the experimentally determined τ_p has the same value for both strain rates as required for consistency. Differentiating Eq. (6) with respect

TABLE 2 Coefficients in Eq. (4) for Ferrovac E

	A	B	$C(\tau_p)\,(\mathrm{kg/mm^2})$	$T_0\,(^\circ\mathrm{K})$
at $\dot{\epsilon} = .002$ min^{-1}	7.20×10^{-4}	0.425	62.8	295
at $\dot{\epsilon} = .02$ min^{-1}	6.12×10^{-4}	0.392	62.8	320

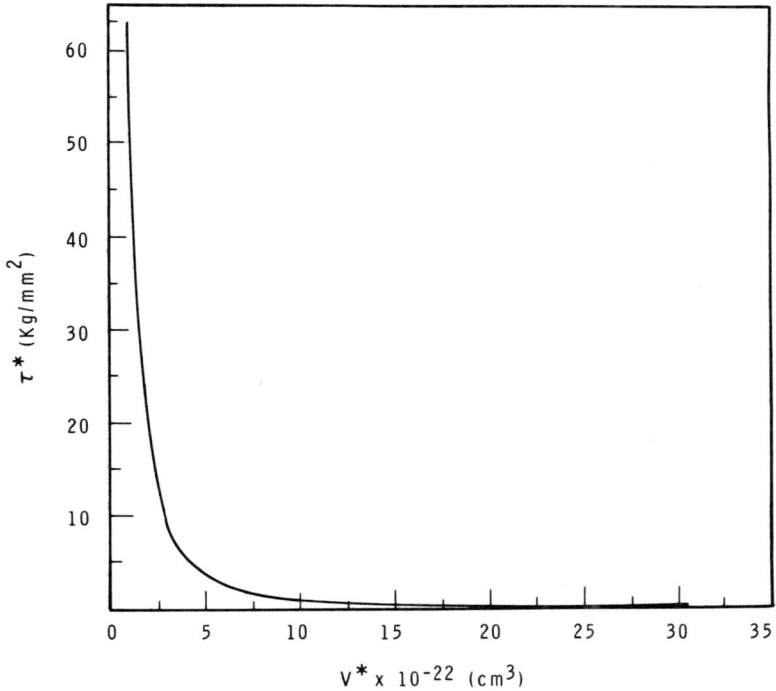

Fig. 2 V^* vs. τ^* curve for iron from the empirical relation $V^* = \frac{1}{2}\,G_0\,(\tau^*\,\tau)^{-1/2}$. The curve shown was drawn using the experimental parameters for Ferrovac E, $G_0 = .93$ eV and $\tau_p = 62.8$ Kg/mm^2, but within the limits of graphical error also represents the data for BMI, $G_0 = .99$ eV and $\tau_p = 63.0$ Kg/mm^2.

to τ^* gives an analytical expression for V^* (Fig. 2)

$$V^* = \frac{1}{2} G_0 (\tau_p \tau^*)^{-1/2} \tag{7}$$

This expression shows that V^* approaches a limiting value at 0 °K and at very small values of τ^* tends to ∞.

Several methods can be used to evaluate G_0 in these empirical equations, but none are entirely free of error. Simultaneous solution of the rate equation using the T_0 from the two strain-rates yields a value of .75 eV, but $\dot{\gamma}_0$ may not be the same for both tests. Substitution of the limiting value of V^* at 0 °K and the value of τ_p obtained from temperature-change tests into Eq. (7) yields a value of .93 eV for G_0 but is subject to errors in determining V^* at 0 °K and τ_p. The conventional method of measuring G [Eq. (3)], yields values from 0.8 to 0.9 eV, but is inaccurate near G_0 as will be shown below.

An error in the determination of activation volumes is introduced by the use of the approximation

$$\left(\frac{\Delta \ln(\dot{\gamma}/\dot{\gamma}_0)}{\Delta \tau^*} \right)_T \quad \text{for} \quad \left(\frac{\partial \ln(\dot{\gamma}/\dot{\gamma}_0)}{\partial \tau^*} \right)_T$$

which gives a secant to the curve rather than the desired tangent, and, as can be seen in Fig. 3, becomes poor in the low-stress region where the curvature is changing rapidly. In addition, the secant approximation applies to the midpoint of the measurement interval not the end points. A case is illustrated in Fig. 3 where the tangent varies from ∞ to 4.4 at the end points of a ten-fold increase in strain rate. Accurate calculations of G by Eq. (3) require that V^* and $(\Delta \tau / \Delta T)_{\dot{\gamma}}$ be determined for the same value of τ^*, i.e., identical temperatures and strain rates. As is shown, the difficulties in determining V^* accurately in the low-stress region lead to poor values of G and nonlinear G-T curves in the neighborhood of G_0.

Several other formulations for $G(\tau^*)$ of interest were considered in this investigation, among them an equation derived from the Johnston-Gilman velocity equation[1] and the Dorn-Rajnak equation[2] for the nucleation of a double-kink. The former is consistent with present knowledge of thermally activated flow in the form

$$v = v_0 \left(\frac{\tau^*}{\tau_p} \right)^n \tag{8}$$

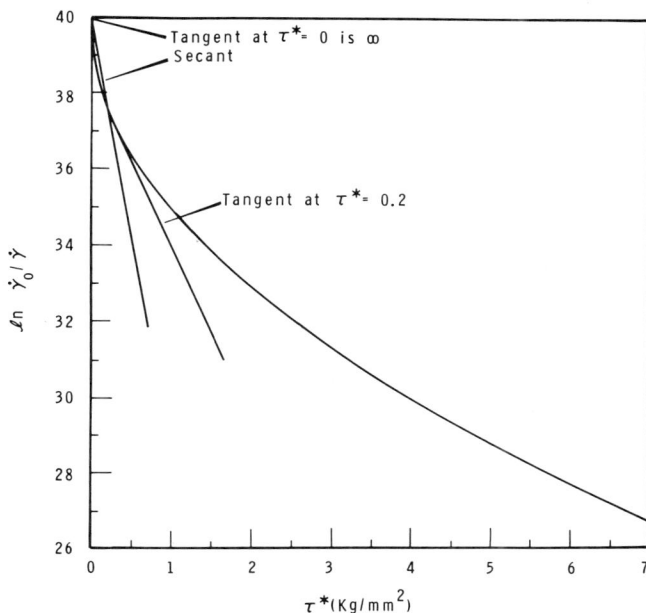

Fig. 3 $\ln \dot{\gamma}_0/\dot{\gamma}$ vs. τ^* relationship in low stress region as derived from empirical $G(\tau^*)$ relation using the following values: $G_0 = 1.00$ eV, $T = 295°K$, and $\tau_p = 63.0$ Kg/mm^2. The slope of the curve as determined at the initial and final strain-rates and from the secant is shown for a strain-rate change experiment in which the strain rate is increased by a factor of ten.

which gives for the stress-dependent activation energy

$$G = -nkT \ln \frac{\tau^*}{\tau_p} \qquad (9)$$

This equation was found to describe the data fairly well at a constant temperature but the complex temperature dependence of n limits its usefulness. In addition, G tends to ∞ at $\tau^* = 0$.

The Dorn-Rajnak relation fits the data quite well in the low-temperature region and has a theoretical basis; however, the range over which this mechanism is theoretically operable does not correspond with the entire range where thermally activated flow occurs in iron. Another mechanism may be operative in the low-stress range,[3] but a change in slope in $(d\tau/dT)_{\dot{\gamma}}$ vs. T or in the G vs. T curves indicative of a change in mechanism was not noted. A compariosn of these three forms of τ^*/τ_p vs. T/T_0 are shown in

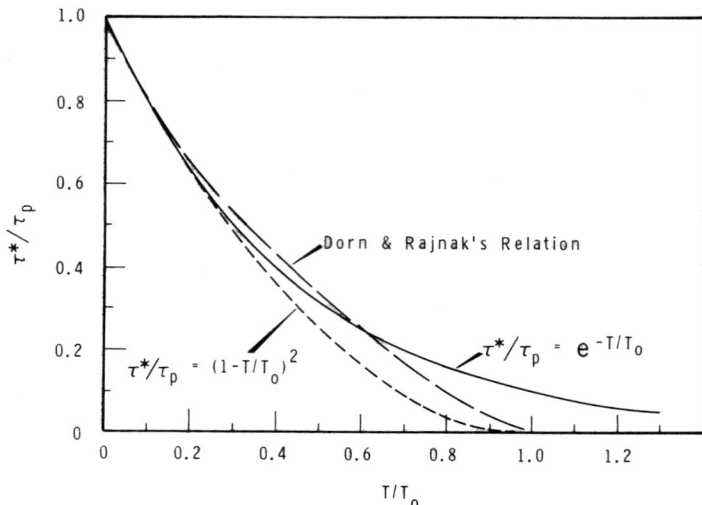

Fig. 4 A comparison of $\tau^* - T$ relationships for Dorn and Rajnak's theory, an expression derived from dislocation velocity measurements $(\tau^*/\tau = C^{-T/T_0})$, and the empirical relation found in this study $[(\tau^*/\tau_p = (1 - T/T_0)^2]$.

Fig. 4. The advantages of the empirical $G(\tau^*)$ relation derived in this paper are its good description of the data, correct behavior at the limits, and the determination of T_0 and τ_p directly from experimental data.

3.2   Application of $G(\tau^*)$ Relation

The empirical $G(\tau^*)$ relation [Eq. (6)] was used to deduce an expression for the dislocation velocity by substitution into Eq. (1) and $\dot{\gamma} = \rho b v$

$$v = v_0 \exp\left[-\frac{G_0}{kT}\left(1 - \sqrt{\frac{\tau^*}{\tau_p}}\right)\right] \tag{10}$$

This predicts that $\ln v$ vs. $\sqrt{\tau^*}$ should fall on a straight line for each temperature at which data are taken. The relation was tested using the data of Stein and Low on Si-Fe[4] and a value of 22 kg/mm^2 for the internal stress σ_i determined by a stress-relaxation test,[5] $(\sigma^* = \sigma - \sigma_i = 2\tau^*)$. The excellent agreement of the data with this expression is demonstrated in Fig. 5. As evident from Eq. (10),

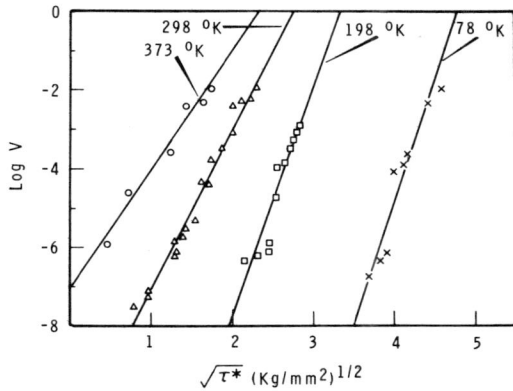

Fig. 5 A plot of Stein and Low's dislocation velocity for Si-Fe to test the fit for the log v vs. $\sqrt{\tau^*}$ relation predicted from the empirical relation found in this study. A σ_i value of 22 Kg/mm^2 was subtracted from each point of the original data, $\sigma^* = \sigma - \sigma_i = 2\tau^*$.

if the intercept of the lines at $\tau^* = 0$ in Fig. 5 is plotted against $1/T$ for each test temperature, G_0 can be obtained from the slope; and, similarly, if the slope of the lines in Fig. 5 is plotted against $1/T$, then τ_p can be obtained from the slope of the resulting plot. In both of the above procedures, the points for the 78 °K data did not fall on the same line as the other three points, and the estimated values of 1.07 eV for G_0 and 20 Kg/mm^2 for τ_p seem to be in the correct range, but not too accurate. In view of the experimental errors in dislocation velocity measurements and the extent of the extrapolations, this agreement is probably as good as can be expected.

The influence of changes in the density of mobile dislocations can also be examined by the use of these empirical equations. The density of mobile dislocations is generally assumed to remain constant during strain-rate cycling tests for lack of a better treatment. Calculations have shown that dislocation density influences the shape of the initial yield point,[6] and since density appears in the preexponential term ($\dot{\gamma}_0$), it clearly could influence the measurements. Changes in activation volume with strain, such as those seen in the BMI control sample in Fig. 8, may arise from changes in the density of mobile dislocations.[3] This possiblility can be examined by combining the rate equation (1), and the V^* equation (7), to give

$$\ln \frac{\dot{\gamma}_0}{\dot{\gamma}} = \frac{G_0}{kT} - \frac{1}{2}\left(\frac{(G_0)^2}{kT\tau_p}\right)\left(\frac{1}{V^*}\right) \tag{11}$$

Before the actual changes in $\dot{\gamma}_0$ can be estimated, however, a correction should be made for the decrease in true strain rate as the sample deforms. Such a correction, at a true strain of .30, would give a true strain rate .74 of the initial strain-rate and a true activation volume of 14.8×10^{-22} cm^3. Parameters used in Eq. (11) for this latter calculation were $G_0 = .99$ eV, $\tau_p = 63.0$ kg/mm^2, V^* at zero strain $= 13.7 \times 10^{-22}$, and $T = 295°$K. If the remaining discrepancy between the observed V^* at $\epsilon = .30$ and the calculated true V^* is ascribed to a change in $\dot{\gamma}_0$, it can be accounted for by a modest increase in $\dot{\gamma}_0$ from 5.2×10^{10} sec^{-1} to 13.7×10^{10} sec^{-1}. The behavior of Ferrovac E is just the reverse of the above case (see control sample in Fig. 7); after applying the correction for true strain-rate, $\dot{\gamma}_0$ is found to decrease from 9.4×10^{10} at zero strain to 7.0×10^{10} sec^{-1} at .30 strain. Such changes in $\dot{\gamma}_0$ are not inconsistent with changes which might be expected in the density of mobile dislocations during deformation. The relationship between a drop-in-load yield point and rapid multiplication could explain the difference in behavior of Ferrovac E, which has a yield drop and $\dot{\gamma}_0$ decreases, and BMI, which does not have a yield drop and $\dot{\gamma}_0$ increases with deformation.

The studies of BMI zone-refined iron uncovered another interesting phenomena which is illustrated in the temperature-change data for the control sample shown in Fig. 10. At temperatures from T_0 to 225°K, the behavior was very similar to Ferrovac E (Fig. 1) although T_0 was slightly higher. Below 225°K, however, the data points fell below the expected line and the scatter from sample to sample was quite large. In addition, it was noted that the flow stress subsequent to yielding at each lower temperature decreased over a range of about .005 strain, but then began to increase again. Data points taken after the sample began to work-harden fell much closer to the line extrapolated from above 225°K, however, irradiation removed this anomaly entirely. This will be discussed in a later section. Analysis of the data from the initial portion of the test ($T > 225°$K) yielded a τ_p of 63.0 Kg/mm^2 and G_0 of .99 eV, which are very close to the results on Ferrovac E. If a mechanism change were occurring, one would expect high values of $(\Delta\tau/\Delta T)_{\dot{\gamma}}$.

The lower values appear to result from a decrease in τ_i when the sample is deformed at temperatures below 225°K. It is probably another manifestation of the behavior of zone-refined iron reported by Michalak,[7] who found that samples deformed at 78, 125, and 140°K and then restrained at room temperature, had a flow stress 25% lower than that for an equivalent strain at room temperature only.

Dislocation substructures dependent on deformation temperature and sample purity have also been reported.[8] Therefore, it is concluded that the anomalous behavior observed in these temperature-change tests of BMI iron below 225°K have their origin in a rearrangement of substructure under increasing stress at low temperatures.

4 RADIATION DAMAGE-EFFECTS ON THERMALLY ACTIVATED FLOW

Irradiation of b.c.c. metals produces defect aggregates such as those seen in the transmission electron micrograph of molybdenum (Fig. 6). However, it is not apparent whether these visible defects

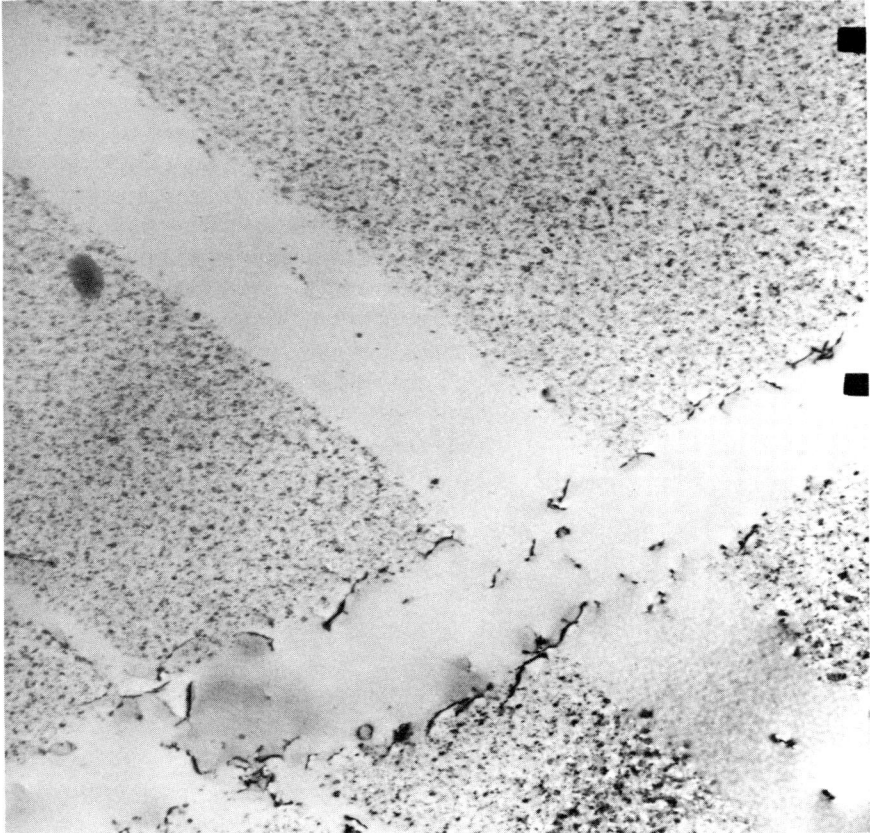

Fig. 6 Electron transmission micrograph of molybdenum irradiated to a fluence of 1×10^{19} n/cm² ($E > 1$ MeV) and subsequently deformed. Dark areas are high density of clustered defects, light areas free of defects correlate with the active slip systems in molybdenum. 75,000× magnification, neg. No. 2207B.

are the cause of the changes observed in the mechanical properties. Investigations of f.c.c. metals [9, 10] have shown that irradiation causes a change in the rate-controlling mechanism of deformation with resultant changes in the temperature dependence of the flow stress and activation volumes. Ferrovac E and BMI iron irradiated at 60°C to neutron fluences of 1.4×10^{18} and 8.3×10^{19} n/cm^2 $(E > 1 \text{ MeV})$ were studied by the methods described previously to determine if changes in mechanism similar to those observed in f.c.c. metals were occurring.

The results of activation volume measurements at 295°K on these samples in the irradiated condition and after a post-irradiation anneal at 350°C for two hours are compared with V^* measurements on control samples in Figs. 7 and 8. From the data on Ferrovac E (Fig. 7) it should be noted that the initial measurements on the irradiated samples (low strain) are significantly lower than V^* for the control sample. It should also be noted that annealing at 350°C results in complete recovery. The influence of carbon distribution in the furnace-cooled control samples was studied by quenching the control samples from 800°C and aging at 300°C for 24 hours. Quenching increased V^* by 1.4×10^{-22} cm^3

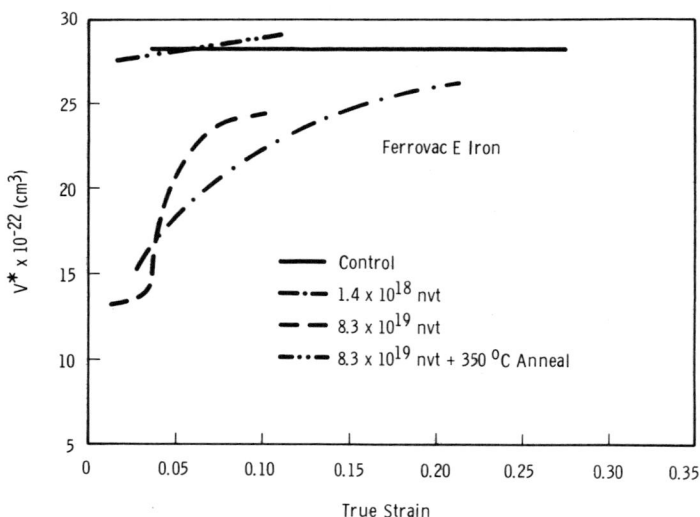

Fig. 7 Activation volume measurements at 295°K as a function of strain for Ferrovac E iron for the following conditions: control sample, irradiation to fluence of 1.4×10^{18} n/cm^2 $(E > 1 \text{ MeV})$, irradiation to fluence of 8.3×10^{19} n/cm^2, and irradiation to 8.3×10^{19} n/cm^2 followed by an anneal at 350°C for 2 hrs.

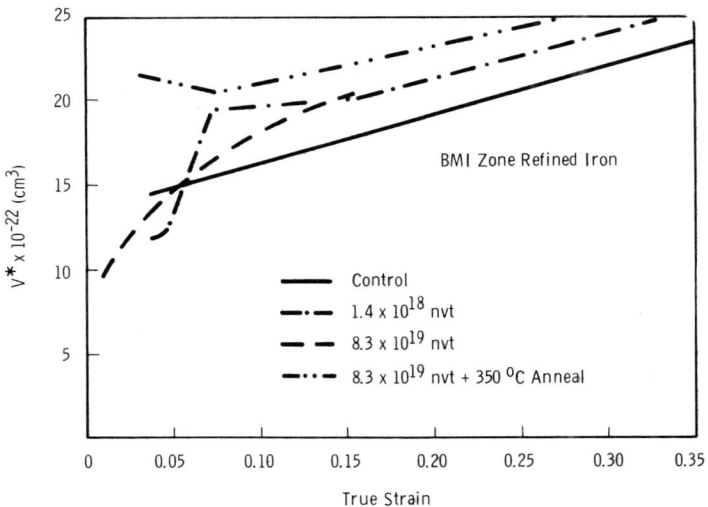

Fig. 8 Activation volume measurements at 295°K as a function of strain for BMI zone-refined iron for the following conditions: control sample, irradiation to a fluence of 1.4×10^{18} n/cm^2 ($E > 1$ MeV), irradiation to a fluence of 8.3×10^{19} n/cm^2 and irradiation to 8.3×10^{19} n/cm^2 followed by an anneal at 350°C for 2 hrs.

and aging for 24 hours decreases V^* by 7.2×10^{-22} cm^3. A two-hour anneal of the control at 350°C had no effect on V^*. The changes produced by irradiation are greater than those produced by a change in carbon distribution, but do not appear to be large enough to indicate a change in the mechanism of thermal activation. Data obtained at lower temperatures show that V^* at lower temperatures is affected even less by irradiation.

Another feature of these results which should be noted is the increase in activation volume toward the values of the control as the sample is deformed. In molybdenum, highly localized deformation was observed (Fig. 6).[11] The first dislocations moving through the lattice clear a defect-free "channel" which gradually widens as the deformation proceeds. If the initial decrease in activation volume is due to the defects introduced by irradiation, then their removal from the slip plane by interaction with moving dislocations should produce an increase in V^* toward the control as deformation proceeds, irrespective of the details of the mechanism.

The same general trends are seen in activation volume measurements on BMI iron (Fig. 8). One unique feature of the BMI results is that an anneal at 350°C produces a V^* higher than the unirradiated control sample; moreover, the V^* values after 7-10% strain also are higher than the control. Quenching BMI iron decreases V^* by only 0.8×10^{-22} cm^3 and aging does not affect V^*.

The temperature-change tests previously described should also provide some insight into the influence of irradiation on the thermally activated flow process. Results on Ferrovac E for a flux of 1.4×10^{18} n/cm^2 showed a decrease in the slope of $(\Delta\tau/\Delta T)_{\dot{\gamma}}$ vs. T after irradiation and an increase in T_0 of about 15°K. The data for BMI iron in Fig. 10 show the changes found to be typical of irradiated material, a decrease in slope of $(\Delta\tau/\Delta T)_{\dot{\gamma}}$ vs. T and an increase in T_0. In the BMI samples T_0 increased only 5°K. An analysis of the temperature-change data by the methods previously described yielded a value of 57 Kg/mm^2 for τ_p for both Ferrovac E and BMI iron compared to 63 Kg/mm^2 for the control samples. One notable feature of the BMI results was the inhibition of substructure rearrangement below 225°K in the specimens irradiated to the higher fluence.

Estimates of G_0 in the irradiated samples were made assuming the slope dG/dT, was the same as in the control samples, but that T_0 had changed as indicated. In Ferrovac E, G_0 increased from .93 to .97 eV and increased from .99 eV to 1.01 eV in BMI iron—very minor changes. The changes noted in activation volume are also quite minor as can be seen from the change they represent in τ^* (Fig. 2). Thus it is concluded that the changes in these parameters of thermally activated flow represent only a minor modification in the deformation process.

Activation volume measurements have also been performed on molybdenum irradiated to a fluence of 5×10^{18} n/cm^2 ($E > 1$ MeV) and after annealing at two temperatures.[12] The results of these measurements at several deformation temperatures are summarized in Fig. 9. As shown previously, the increase in V^* with strain in the unirradiated samples is probably due to an increase in the density of mobile dislocations. Unlike iron, irradiation produces an increase in V^* and as the damage anneals out, or is removed from the slip plane during deformation, V^* decreases toward the value for the control sample.

Significant recovery in V^* occurs after an anneal at 185°C ($.16T_m$), a temperature at which recovery in the resistivity and stored energy has also been noted.[13] No changes in microstructure have been observed at this temperature.[14] These facts indicate the recovery is associated with the movement of point defects but needs further investigation to resolve the detailed mechanism. Complete recovery of V^* occurs after an anneal at 600°C, a

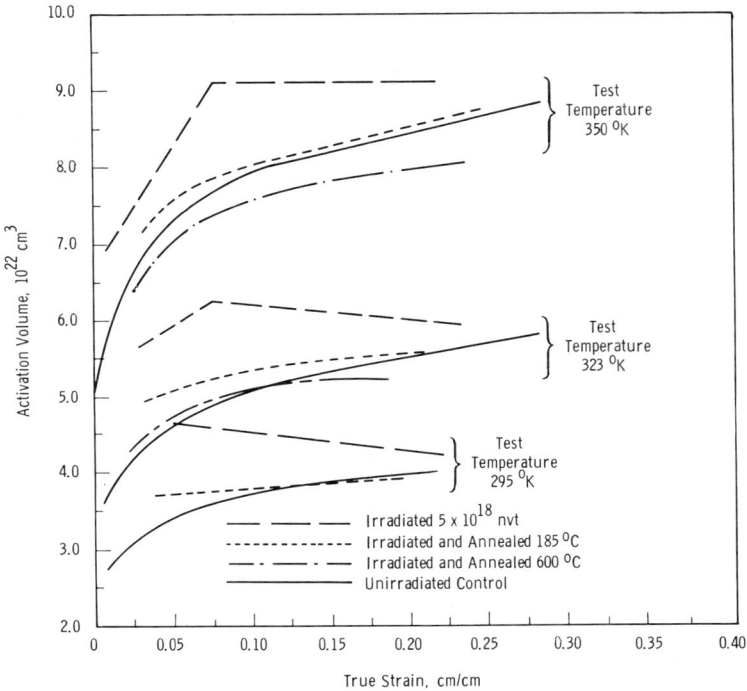

Fig. 9 Activation volume measurements at 295°K, 323°K and 350°K as a function of strain for molybdenum in the following conditions: control sample, irradiation to a fluence of 5×10^{18} n/cm^2 ($E > 1$ MeV), irradiation followed by an anneal at 185°C, and irradiation followed by an anneal at 600°C.

treatment which produced a slight decrease in the density of visible defect clusters.[14]

4.1 Radiation Damage—Effects on Athermal Flow Stress

In contrast to the minor changes in τ^* produced by irradiation, the increase in the athermal component of flow stress τ_i, is very pronounced. Heterogeneities in the lattice, such as, precipitate particles, solute atoms, and dislocation networks, which have long-range stress fields associated with them, contribute to τ_i. The increase in lower yield stress at 295°K was used as an indication of the change in τ_i in these experiments for fluences of 1.4×10^{18} and 8.3×10^{19} ($E > 1$ MeV). As seen in Fig. 11, the hardening is much greater in Ferrovac E than in BMI iron at equivalent fluences. This material has the higher carbon content and it would appear

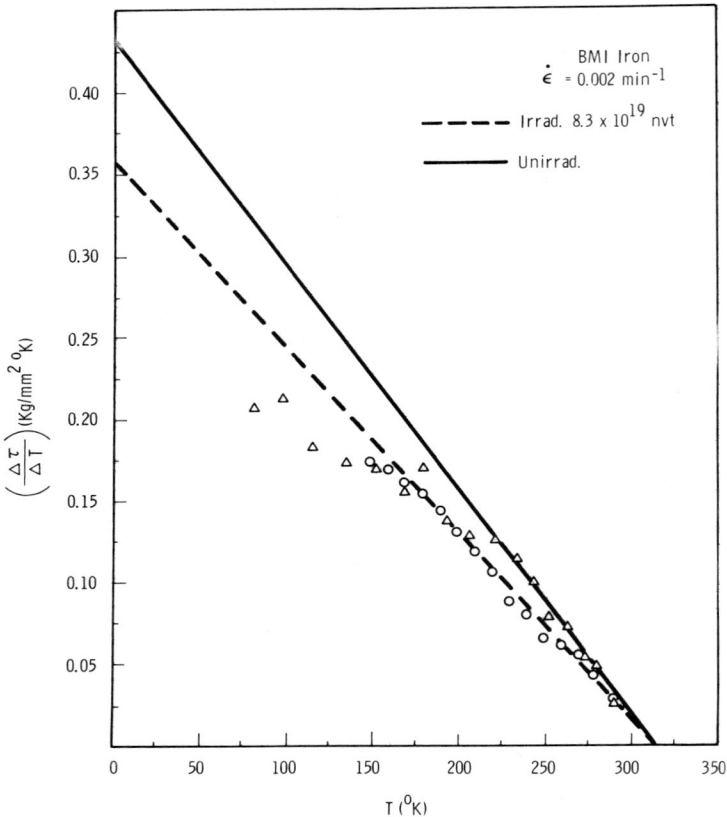

Fig. 10 $(\Delta\tau/\Delta T)_{\dot\epsilon}$ data for temperature change tests on BMI zone-refined iron. Triangles (solid line) represent points from the control sample and circles (dotted line) represent points from a sample irradiated to a fluence of 8.3×10^{19} n/cm^2 ($E > 1$ MeV).

that the carbon is playing a role in the strengthening. Grain size is not a factor because first, it is not very important at these exposures, and, second, the BMI iron which has the larger grain size (50 μ vs. 25 μ), whould harden more if grain size were a factor.[15, 16]

Studies of the recovery of mechanical properties after a two-hour anneal at 350°C also reveal differences in the behavior of these two materials. Table 3 summarizes these property changes for the lower yield stress $\Delta\tau_i$, and for the rate-sensitive stress $\Delta\tau^*$. The recovery in $\Delta\tau_i$ was found to be greater at the lower fluence for both materials, but at equivalent fluence for both

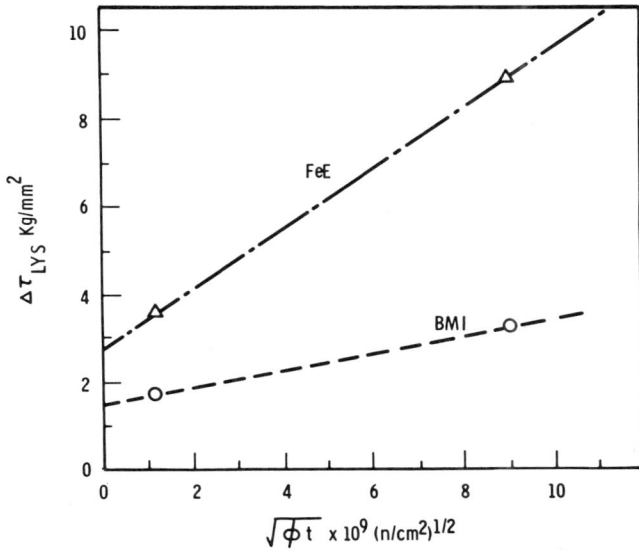

Fig. 11 The increase in lower yield stress as a function of square root of neutron fluence ($E > 1$ MeV) in Ferrovac E and BMI zone-refined irons.

materials more recovery occurred in BMI iron. It is difficult to generalize the $\Delta\tau^*$ recovery data because of the anomalous behavior of BMI iron (see discussion of activation volumes), but it is noteworthy that $\Delta\tau^*$ recovers more completely at the higher fluence for both materials which is just the reverse of $\Delta\tau_i$ recovery.

TABLE 3 Recovery in Mechanical Property Changes in Irradiated Iron Produced by a 2-hr Anneal at 350°C

Material	$\Delta\tau_i$ (LYS)	$\Delta\tau^*$ ($\dot{\epsilon}$ change)
Ferrovac E — 1.4×10^{18} n/cm^2	92 %	80 %
Zone-refined — 1.4×10^{19} n/cm^2	100 %	130 %
Ferrovac E — 8.3×10^{19} n/cm^2	18 %	100 %
Zone-refined — 8.3×10^{19} n/cm^2	47 %	175 %

4.2 Radiation Damage—Mechanisms

The identification of the defects or defect complexes causing hardening is essential to the formulation of a hardening mechanism.

However, a satisfactory solution to the problem of identification still has not been found. Pertinent to the identification of these defects are studies of the microhardness recovery in iron[17] which show that recovery occurs from 300-500°C. Also, microscope observations show that this hardness recovery correlates with growth of small spot defects 25-50 Å in diameter to form prismatic loops at about 350°C.[18] These loops which are stable to 500°C have been identified as interstitial loops.[17] Isolated vacancies are not likely to contribute to hardening.[19] Interstitials, although they cause hardening,[20] are mobile at the 60°C irradiation temperature ($\Gamma \sim 10^9$ sec^{-1}) and the absence of hardness recovery below 300°C indicates that contributions from free interstitials can be neglected. Similarly, the presence of large loops at 500°C, after recovery is essentially complete, eliminates this type of defect from consideration. Thus we are left with clusters or small loops as the most likely cause of the hardening. Furthermore, calculations of the migration distances of vacancies and interstitials prior to annihilation indicate that carbon atoms in solution could serve as nucleation sites for interstitial clusters, thus explaining the greater hardening observed in the higher carbon content Ferrovac E.

Kimura and Maddin[19] have reviewed possible mechanisms which could account for athermal hardening by loops or small clusters. The ones most likely to be applicable are the elastic interaction between prismatic loops and dislocations discussed by Kroupa and Hirsch,[21] Friedel's mechanism for pulling a dislocation away from a loop with which it has formed a low-energy configuration,[22] and Orowan's mechanism for bowing a loop between obstacles.[23] The first two predict stresses of ⅛ to ⅕ $\mu bdN^{2/3}$, where μ is the shear modulus, d is the diameter of the defect, and N is the volume density of defects: Orowan's mechanism predicts stresses of $\mu b(Nd)^{1/2}$. Assuming a defect size of 25 Å and densities of 10^{15}, only Orowan's mechanism can give increases in stress large enough to account for the observed hardening.

However, microscopy evidence indicates these obstacles are not impenentrable, and if they are clusters and sheared or loops in which a jog is formed, one would expect these processes to be aided by thermal activation. But as previously noted, there appears to be little change in the thermally activated process.

Arsenault[24] has considered the possibility of observing the Fleischer hardening mechanism in metals where the Peierls mechanism is operative and concluded that it would be observed only in the neighborhood of T_0 or above. The minimal increase in

T_0 observed in these experiments is not consistent with the increase predicted above for the Fleischer mechanism. It appears that in cases where the short-range stress field from the defect is less than that for the normal rate-controlling obstacles, no great change will be observed in the thermally activated flow. The question remaining then concerns the cause of small changes in the thermally activated process.

Guyot and Dorn[25] have considered several ways in which the basic Peierls mechanism could be slightly modified and hence result in small changes such as observed in these experiments. The most likely are a change in the Peierls stress τ_p, and a reduction of the free length of dislocation segment by pinning points closer than 20-30b. A high density of defect clusters could reduce the length of dislocation in which kinks could be nucleated, thereby decreasing V^* slightly. As the defects were cut by moving dislocations and destroyed, the thermal activation parameters would return to their original values. Since the areas in which defects have been removed from the slip plane are much weaker, the deformation will tend to be highly localized after the first dislocations to move have cleared a "channel."

The defects causing this effect may be different from those producing the athermal hardening, however, since densities around 10^{18} would be required to obtain enough pinning points.

It has been suggested that the observed changes in activation volume with strain[18] might be due to an increase in true strain rate caused by the localized deformation. However, if this were the case, then an increase in fluence or a decrease in deformation temperature should increase the tendency toward localized deformation and accentuate the change of V^* with strain; a behavior which has not been observed. Another difficulty with this suggestion is encountered in the deformation behavior of irradiated molybdenum which is the exact opposite of that predicted by this hypothesis. It is therefore concluded that the changes in V^* with strain, which are observed in irradiated iron, are attributable to the destruction of defects on the slip plane.

5 CONCLUSIONS

This study of deformation in iron has shown that the thermally activated component of the flow stress can be described by a relation of the form $G = G_0(1 - \sqrt{\tau^*/\tau_p})$ where $G_0 = .95$ eV and $\tau_p = 63$ Kg/mm^2. This relation has been used to suggest a new dislocation velocity-stress relationship, to calculate changes in dislocation density during deformation, and to investigate changes in the deformation process caused by neutron irradiation.

It has been concluded that the defects produced by irradiation do not alter the basic thermally activated flow mechanism in iron but do produce small changes in the flow parameters. The major effect of irradiation is an increase in the athermal stress, most probably caused by Orowan hardening by small defect clusters. The first dislocations to move interact with these clusters and remove them from the slip plane thus creating weak spots in the lattice in which deformation is localized.

ACKNOWLEDGMENTS

We wish to express our appreciation to Prof. J. E. Dorn for helpful discussions of the results and to H. A. Flaherty and T. J. Larson for assistance in performing the experiments. This work was supported by the U.S. Atomic Energy Commission under Contract No. AT(45-1)-1830.

REFERENCES

1. Johnston, W. G., and J. J. Gilman: *J. Appl. Phys.* **30**:129 (1959).
2. Dorn, J. E., and S. Rajnak: *Trans. AIME* **230**:1052 (1964).
3. Wynblatt, P., A. Rosen, and J. E. Dorn: *Trans. AIME* **233**:651 (1965).
4. Stein, D. F., and J. R. Low, Jr.: *J. Appl. Phys.* **31**:362 (1960).
5. Ohr, S. M.: "Radiation Metallurgy Section Solid State Division Progress Report Oak Ridge National Laboratory,;; p. 26 ORNL-3949, (1966).
6. Hahn, G. T.: *Acta Met.* **10**:727 (1962).
7. Michalak, J. T.: *Acta Met.* **13**:213 (1965).
8. Michalak, J. T., and L. J. Cuddy: "Role of Substructure in the Mechanical Behavior of Metals," p. 141, Report No. ASD-TDR-324, (1962).
9. Koppenaal, T. J., and R. J. Arsenault: *Phil. Mag.* **8**:951 (1965).
10. Diehl, J., G. P. Seidel, and L. Nieman: *Phys. Stat. Solid* **11**:339 (1965).
11. Brimhall, J. L.: *Trans. AIME* **233**:1737 (1965).
12. Laidler, J. J., and F. A. Smidt, Jr.: "Radiation Effects on Materials," W. F. Sheely (ed.), Gordon and Breach, N. Y. In Press.
13. Kinchin, G. H., and M. W. Thompson: *Journ. Nucl. Energy* **6**:4, 275 (1958).
14. Mastel, B., and J. L. Brimhall: *Acta Met.* **13**:1109 (1965).
15. Mogford, I. L., and D. Hull: *J. Iron and Steel Inst.* **201**:55 (1963).
16. Chow, J. G. Y., S. B. McRichard, and D. H. Gurinsky: "Radiation Damage in Solids," vol. 1, p. 277, IAEA, Vienna, 1962.
17. Eyre, B. L., and A. F. Bartlett: *Phil. Mag.* **12**:261 (1965).
18. Bryner, J. S.: *Acta Met.* **14**:323 (1966).
19. Kimura, H., and R. Maddin: "Lattice Defects in Quenched Metals," p. 319, R. M. J. Cotterill, M. Doyama, J. J. Jackson, and M. Meshii (eds.), Academic Press, N. Y., 1965.
20. Ono, K., and M. Meshii: *AERE-R-5269*, p. 539 (1966).
21. Kroupa, F., and P. B. Hirsch: *Disc. Faraday Soc.* **38**:49 (1964).
22. Friedel, J.: "Electron Microscopy and Strength of Crystals," p. 605, G. Thomas, and J. Washburn (eds.), Interscience Publishers, New York, 1963.

23. Orowan, E.: "Symposium on Internal Stresses in Metals and Alloys," p. 451, Institute of Metals, London, 1948.
24. Arsenault, R. J.: *ORNL-3993* (1966).
25. Guyot, P., and J. E. Dorn: *Can. J. Phys.* 45:983, 1967.

DISCUSSION *on Paper Presented by F. A. Smidt*

J. LI: I want to ask two questions: (1) I believe that there is no way to obtain a stress-temperature relation simply, because velocity is a power function of stress. In order to do that, you have to assume a relation between the velocity-stress exponent and temperature. What relation did you assume? (2) It seems difficult to understand why the activation volume becomes smaller after irradiation. I just wonder whether you compared the two activation volumes before and after irradiation at the same *effective* stress.

F. SMIDT: *Question (1)* Dr. Li's comment is correct, I assumed m, the velocity-stress exponent was inversely proportional to T. In fact, n has a more complex dependence upon temperature but this approximation is used to illustrate the very important point that any $G\text{-}\tau^*$ function with a $\ln\tau^*$ term in it behaves badly in the low-stress region and therefore cannot be a representation of the true state of affairs. This also shows that V^* cannot be represented by $1/\tau^*$ in this region.

Question (2) The test was intended to determine if a change in the mechanism of thermally activated flow had been modified by irradiation. At constant temperature a change in V^* of course indicates a change in τ^*. I did not measure V^* at the same τ^* as the control samples, but from the change in $(d\tau/dT)$ it can be seen that irradiation produces a very minor perturbation of the basic mechanism and if I had measured V^* at the same τ^* as the control sample, the value would have been essentially the same as for the unirradiated control.

DISLOCATION VELOCITY AND DEFORMATION DYNAMICS IN MOLYBDENUM

H. L. Prekel and H. Conrad***

*The Franklin Institute Research Laboratories,
Philadelphia, Pennsylvania*

ABSTRACT

Direct measurements of the velocity of dislocations in molybdenum are compared with results obtained from studies of micro $(10^{-6} < \epsilon < 10^{-3})$ and macro $(\epsilon > 10^{-3})$ deformation dynamics.

Dislocation velocities were measured as a function of stress at room temperature. These were found to obey the equation

$$v = v_0 \left(\frac{\tau}{\tau_0}\right)^m$$

*Now at the University of Pretoria, South Africa.

**Present address—University of Kentucky, Lexington, Ky.

with m = 6.38. Comparing dislocation velocity measurements with Seeger's model for thermally activated nucleation of kinks it is shown that

$$H_K = 4kTm = 4\tau^* v^* = 4.3 \times 10^{-2} \mu b^3$$

where H_K is the kink energy, k is Boltzmann's constant, T is the temperature, τ^* is the thermal component of the stress and v^* is the activation volume. Moreover, in accordance with the theory, v^* is proportional to the reciprocal of τ^* for dislocation velocity measurements and for strain rate-change tests over the entire range from microstrain to macrostrain. Also, the value of H_K derived from the micro and macro deformation dynamics is in good agreement with that derived from the dislocation velocity measurements.

Comparisons are also made between the value of the stress exponent m obtained from the dislocation velocity measurements with that obtained from strain rate-changes in the micro- and macrostrain regions. Taking the value of the athermal component τ_μ as the stress at the flat portion of the stress vs. temperature curves at high temperatures for macro deformation, or the value of $\tau^* = \tau - \tau_\mu$ as half the stress at which a loop first occurs in the micro deformation, the value of m derived from the relation

$$m = \frac{\partial \ln \dot{\epsilon}}{\partial \ln (\tau - \tau_\mu)}$$

is equal to that derived from dislocation velocity measurements, where τ is the applied stress and τ_μ is the long-range internal stress.

It is concluded that these results clearly establish that the deformation dynamics in both the micro- and macrostrain regions in molybdenum is directly related to the thermally activated motion of individual dislocations.

A discussion of the significance of the various stresses observed during microstraining of metals in terms of long- and short-range obstacles to dislocation motion is also presented.

1   INTRODUCTION

The plastic deformation dynamics of crystalline solids is related to the velocity of dislocations through the equation

$$\dot{\gamma} = \rho b v \tag{1}$$

where $\dot{\gamma}$ is the shear strain rate, ρ is the density of mobile dislocations and v is the average dislocation velocity. Two expressions have been used to describe the effects of shear stress τ and temperature T on the dislocation velocity:

  1. An empirical expression which is based on the experimentally observed linear relation between log v and log τ and gives

$$v = v_0 \left(\frac{\tau}{\tau_0}\right)^m \tag{2}$$

where v_0 is usually taken as 1 cm sec^{-1} and τ_0 as the stress required to give $v = 1$ cm sec^{-1}. The constants m and τ_0 may vary with temperature and structure in an undefined manner.

  2. A thermally activated expression where it is assumed that the motion of dislocations is thermally activated and can be described by an Arrhenius-type rate equation of the form

$$v = v_0 \exp\left\{-\frac{\Delta G(\tau, T)}{kT}\right\} \tag{3}$$

where $\Delta G = \Delta H - T\Delta S$ is the Gibbs free energy of activation and $v_0 = (A/l^*)v^*$ is termed the frequency factor. l^* is the length of the dislocation segment involved in the thermal activation, A is the area of the slip plane swept out per successful thermal fluctuation, and v^* is the frequency of vibration of the segment of length l^*. In this approach it is often assumed (and experimentally confirmed) that ΔH is primarily a function of the effective stress τ^* given by the difference between the applied stress τ and the long-range internal stress τ_μ and that $v = v_0 \exp\{\Delta S/k\}$ is relatively independent of stress and temperature. One then replaces $\Delta G(\tau, T)$ with $\Delta G(\tau^*, T)$ and assumes that $\Delta S \neq f(\tau^* \text{ or } T)$.

An advantage of the thermally activated approach is that it permits a more direct comparison of experimental results with theory.

In this report, results obtained at room temperature on the effect of strain rate on the flow stress in both the microstrain $(10^{-6} < \epsilon < 10^{-3})$ and macrostrain $(\epsilon > 10^{-3})$ regions in high-purity molybdenum are compared through Eqs. (1) to (3) with the effect of stress on the velocity of dislocations as determined by etch pits.

Moreover, values of ΔH and $v^* = -(\partial \Delta G / \partial \tau)_T$ derived from the present data are compared with those derived previously[1,2] for single and polycrystalline molybdenum of various impurity contents, and with those predicted for thermally activated nucleation of kinks over the Peierls hills as the rate-controlling mechanism.

2   EXPERIMENTAL PROCEDURE

2.1   Material

The high-purity molybdenum single crystals of low initial dislocation density ($\sim 2 \times 10^5$ cm^{-2}) used in this investigation were grown by the floating-zone, electron-beam technique and then annealed in hydrogen at 2000°C. A detailed description of the growing procedure, the annealing treatments, and the etch-pitting technique has been published previously.[3,4] The single crystals were oriented for single siip on a {110} <111> system.

2.2   Microstrain Testing

The microstrain testing ($\epsilon < 10^{-3}$) was performed in compression on a right cylindrical specimen 0.3 cm dia. × 0.6 cm high in the jig and in the manner described by Meakin.[5] The stress and strain sensitivity was of the order of 0.01 Kg/mm^2 and 10^{-6}, respectively. Instantaneous strain-rate changes between 1.5×10^{-6} sec^{-1} and 7.5×10^{-6} sec^{-1} were made during the microstraining to establish the microstrain deformation dynamics. Moreover, load-unload tests were conducted at a constant strain rate of 1.5×10^{-6} sec^{-1} and the values of σ_E (stress at which a closed loop first forms upon repeated loading and unloading), σ_A (stress at which an open loop first forms upon repeated loading and unloading), and W_{irr} (the energy associated with the closed loops generated upon loading and unloading to stresses between σ_E and σ_A) were determined on specimens which had been prestrained to strains of 10^{-4} to 10^{-2} at room temperature. There was no effect of prestrain on σ_E and W_{irr} over the range investigated.

2.3   Macrostrain Testing

The macrostrain tests ($\epsilon > 10^{-3}$) were conducted at a strain rate of 0.6×10^{-4} sec^{-1} using the same specimen geometry and the same compression jig as for the microstrain tests. Instantaneous strain rate changes were made between 0.6×10^{-4} sec^{-1} and 3.0×10^{-4} sec^{-1} to establish the macrostrain deformation dynamics.

2.4 Dislocation Velocity Measurements

The technique used to make the dislocation velocity measurements is presented in detail elsewhere[6] and will only be mentioned here briefly. Fresh dislocations were introduced into specimens of shape and orientation similar to those used by other workers[7,8] by indenting with a sapphire needle at intervals along a line in the tension surface. The etch-pitting technique used to reveal the dislocation sites was that described previously[4] where it was shown that a one-to-one correspondence exists between the etch pits and dislocations. The specimens containing the fresh dislocations were then loaded in three-point bending using a special jig which employed a dead weight to yield square stress pulses. Surprisingly, dislocations were only observed to move on {112} planes in the stress range examined, in spite of the fact that the resolved shear stress was higher on the {110} planes (Schmid factors were ~ 0.38 and ~ 0.48, respectively). This behavior is, however, in qualitative agreement with the results of Erickson[9] who showed that for Fe-3% Si above a certain temperature, the stress required to move dislocations at a particular velocity was less for {112} than for {110}. Dislocation velocities were calculated in the usual way by dividing the distance the dislocations farthest from the center of their introduction had moved by the time the load was applied to the specimen. Figure 1 shows the dislocation motion used to calculate the velocities.

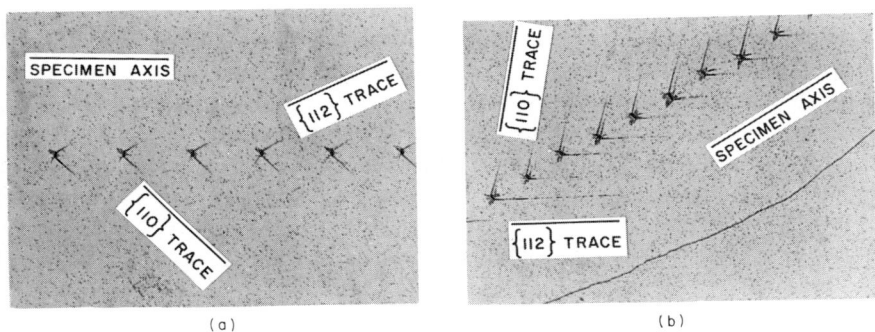

(a)   (b)

Fig. 1 An example of the dislocation motion in molybdenum used to determine the dislocation velocities. (a) As-indented; (b) After loading (125 ×).

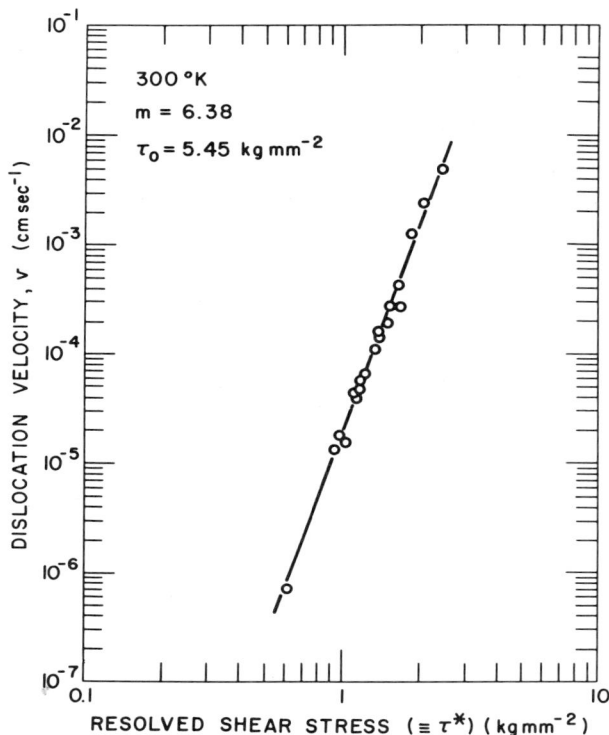

Fig. 2 Log-log plot of the effect of stress on the dislocation velocity in molybdenum at room temperature.

3   EXPERIMENTAL RESULTS

A log–log plot of dislocation velocity vs. resolved shear stress is given in Fig. 2. The data can be considered to lie on a straight line in accordance with Eq. (2) and yield $\tau_0 = 5.45$ Kg/mm^2 and $m = 6.38$. A semilog plot of the velocity vs. stress is presented in Fig. 3. Values of the activation volume v^* were derived from the slop at various points along the curve through the relation derived directly from Eq. (3), namely,

$$v^* = -\left(\frac{\partial \Delta G}{\partial \tau}\right)_T = kT\left(\frac{\partial \ln v}{\partial \tau}\right)_T \tag{4}$$

It is interesting to note a comparison of the microstrain and macrostrain deformation dynamics with the dislocation motion

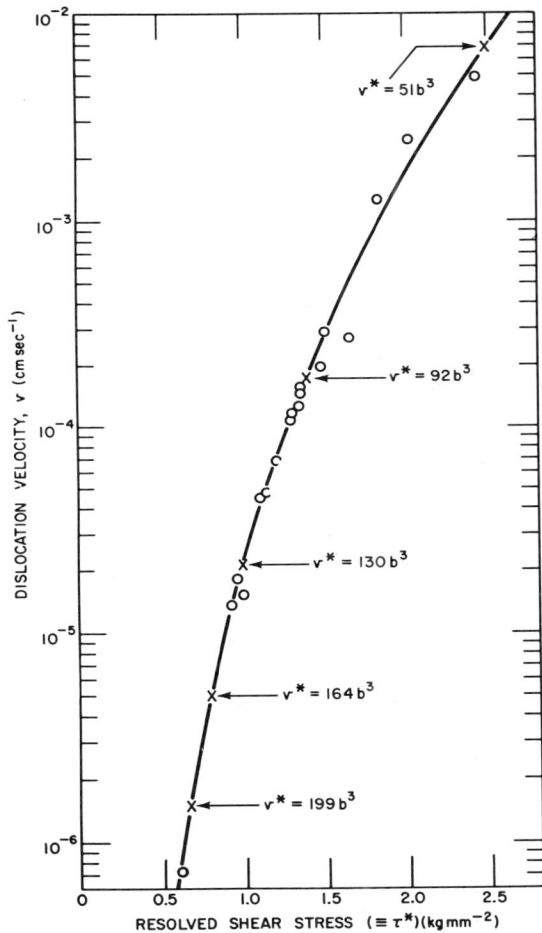

Fig. 3 Semi-log plot of the effect of stress on the dislocation velocity in molybdenum at room temperature.

dynamics, i.e., a comparison of the values of m and v^* obtained from the changes in strain rate with those derived directly from the dislocation velocity measurements. For such a comparision the value of $\tau^* = \tau - \tau_\mu$ at room temperature must be known. For the macrostrain tests the value of τ^* was determined in the usual manner,[10] i.e., by taking the relatively flat region of the flow stress vs. temperature curve at high temperatures as the value of τ_μ and subtracting this from the applied stress at a given temperature T. This is illustrated by the results of Figs. 4 and 5. Macro

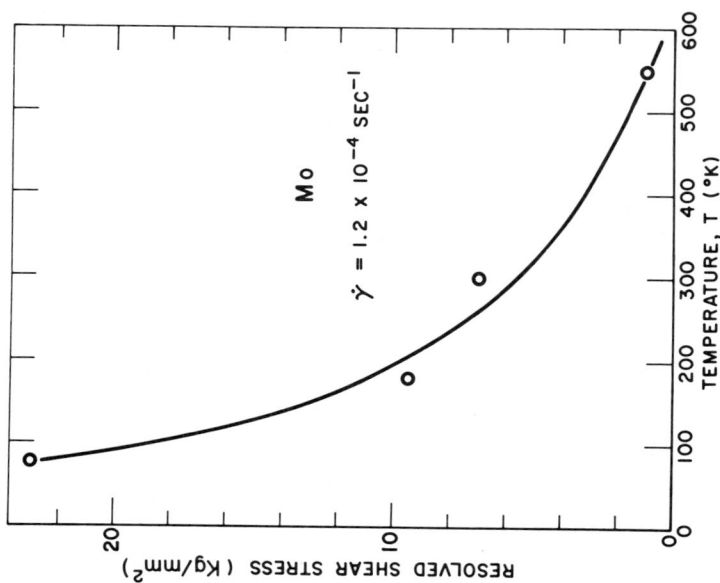

Fig. 5 Effect of temperature on the macrostrain ($\epsilon_p > 10^{-3}$) yield stress of single crystals of high purity molybdenum.

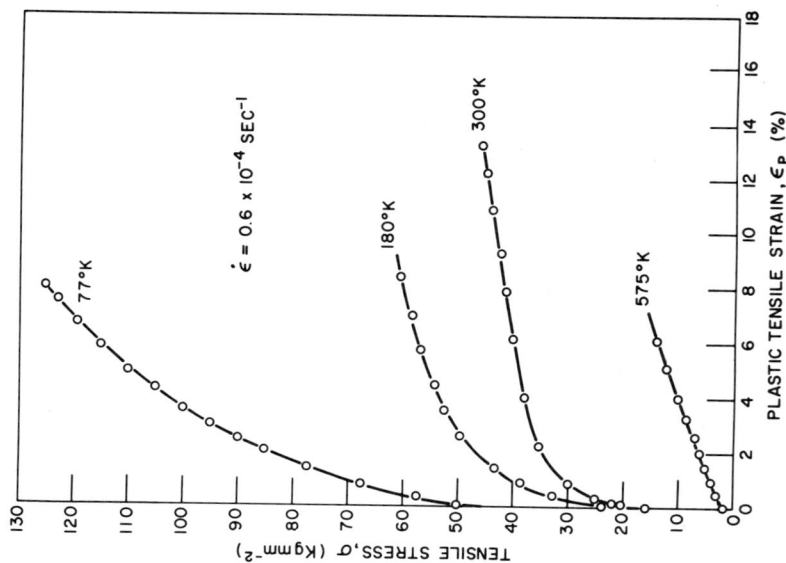

Fig. 4 Stress-strain curves as a function of temperature for high purity molybdenum single crystals.

stress-strain curves were determined at several temperatures (Fig. 4). A plot of the initial macro yield stress ($\epsilon_p \approx 10^{-3}$) vs. temperature taken from these curves is given in Fig. 5. It is seen here that $\tau_\mu < 0.5$ Kg/mm^2 for the present high-purity crystals and hence $\tau^* \simeq \tau$ for the macrostrain tests.

The value of τ^* for the microstrain tests was determined from the repeated loading tests through the relations applicable in such tests,[11-13] namely,

$$W_{irr} = 2\tau_f \gamma_{pm} \tag{5}$$

$$\tau_E = 2\tau_f \tag{5a}$$

where τ_E is the elastic limit, W_{irr} is the energy loss per loop, τ_f is the friction stress associated with the short-range obstacles, and γ_{pm} is the maximum nonelastic forward shear strain at the maximum stress. Thus, τ_f is obtained directly from the value of τ_E and indirectly from the initial slope of a plot of W_{irr} vs. γ_{pm} or from the intercept of a plot of W_{irr} vs. the maximum stress τ_m. That τ_f is in fact τ^* is discussed in the Appendix.

Plots of W_{irr} vs. the maximum forward plastic strain and of W_{irr} vs. the maximum stress amplitude are presented in Figs. 6 and 7. From the slope and intercept, respectively, of these curves we obtain $\tau^* = 0.053$ Kg/mm^2 and $\tau^* = 0.056$ Kg/mm^2, indicating good agreement. The constant of 1/4 rather than 1/2 is used for these determinations since compression stresses and strains are plotted rather than shear stresses and strains. A value of 0.12 Kg/mm^2 was obtained for τ_E from the microstrain stress cycling tests, giving a value of m in good agreement with that derived in Figs. 6 and 7.

The following parameters were calculated from the strain rate-change tests in the microstrain and macrostrain regions

$$n = \frac{\ln\left(\dot{\epsilon}_2/\dot{\epsilon}_1\right)}{\ln\left(\tau_2/\tau_1\right)} \equiv \frac{\partial \ln \dot{\gamma}_p}{\partial \ln \tau} \tag{6}$$

$$m = \frac{\ln\left(\dot{\epsilon}_2/\dot{\epsilon}_1\right)}{\ln\left\{(\tau_1^* + \Delta\tau)/\tau_1^*\right\}} \equiv \frac{\partial \ln \dot{\gamma}_p}{\partial \ln \tau^*} \tag{7}$$

$$v^* = kT \frac{\ln\left(\dot{\epsilon}_2/\dot{\epsilon}_1\right)}{\tau_2 - \tau_1} \equiv kT \frac{\partial \ln \dot{\gamma}_p}{\partial \tau^*} \tag{8}$$

where τ_1 is the shear stress immediately preceding a strain rate-change from $\dot{\epsilon}_1$ to $\dot{\epsilon}_2$, τ_2 is the shear stress immediately following

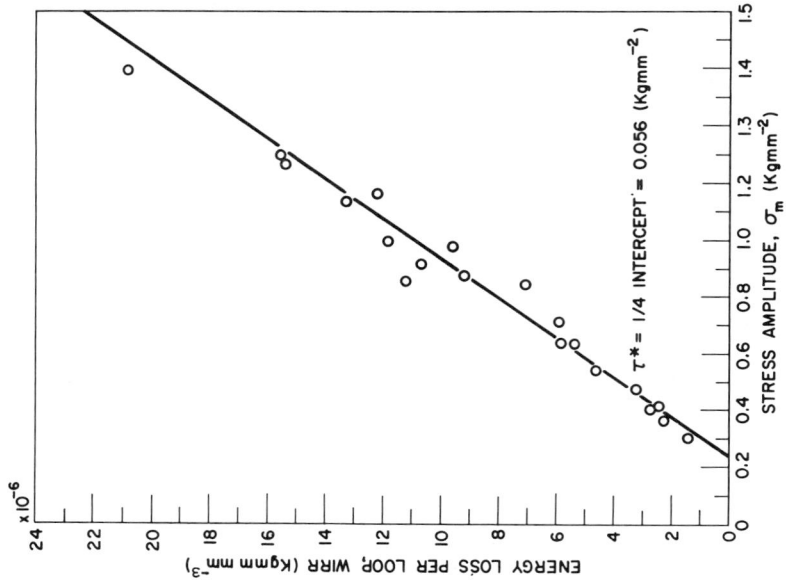

Fig. 7 Energy loss per loop versus maximum tensile stress amplitude for high purity molybdenum single crystals.

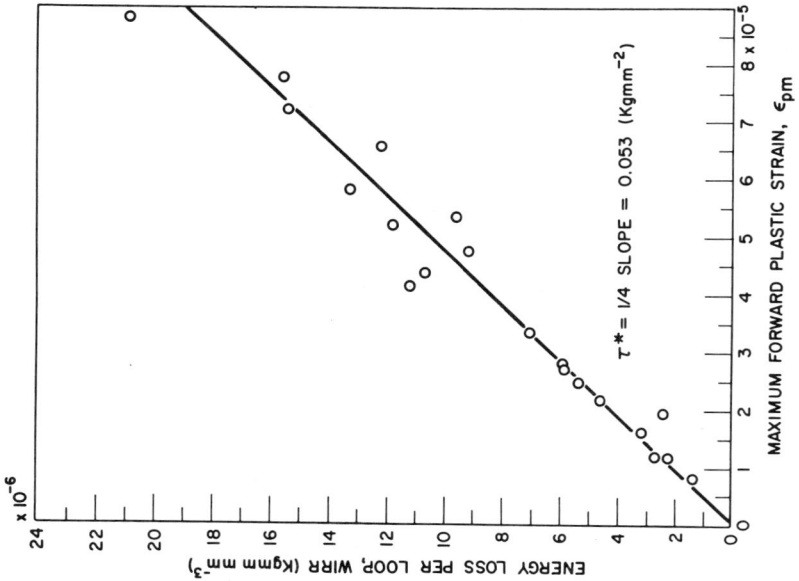

Fig. 6 Energy loss per loop versus maximum forward plastic strain for high purity molybdenum single crystals.

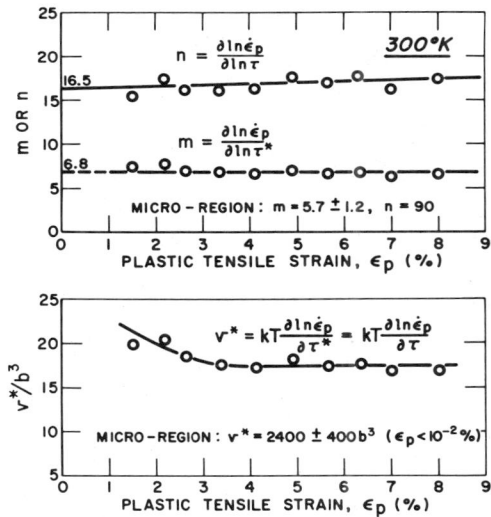

Fig. 8 Plot of the parameters n, m and v^* versus plastic strain for high purity molybdenum single crystals.

the strain rate change and $\Delta \tau = \tau_2 - \tau_1$. A plot of n, m, and v^* vs. strain is given in Fig. 8. Here it is seen that, for the macrostrain region, n and m are essentially independent of strain, whereas v^* initially decreases with strain to about 3% plastic strain and then remains essentially constant at 17.5 b^3. Both n and v^* are substantially higher in the microstrain region compared to the macrostrain region; however, m is essentially constant over the strain range 10^{-6} to 10^{-1}.

The value of m obtained from the strain rate–change tests (deformation dynamics) is in good agreement with that for the dislocation velocity measurements (Fig. 2). On the other hand, the extrapolation of n to zero strain, which is frequently used to obtain m, does not yield the same value as that for the dislocation velocity. Further agreement between the deformation dynamics and dislocation dynamics is indicated in Fig. 9 where $\log v^*$ is plotted vs. $\log \tau^*$. In the figure it is seen that the values derived from the strain rate–change tests lie on the same straight line (with a slope of −1) as do those obtained from the slopes of Fig. 3.

The results of Fig. 9 give

$$v^* = 130 \frac{b^3}{\tau^*} \qquad (9)$$

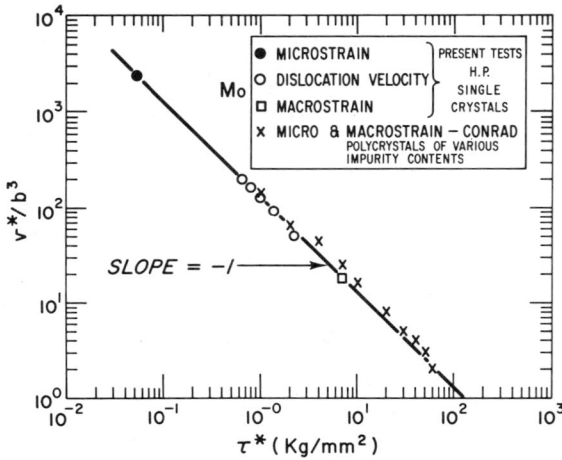

Fig. 9 Effect of stress on the activation volume for the plastic flow of high purity molybdenum single crystals.

for the deformation and dislocation dynamics of molybdenum. Moreover, comparing Eqs. (7) and (8) it is expected if m is a constant that

$$v^* = \frac{kTm}{\tau^*}$$ (10)

which agrees with the experimentally observed variation of v^* with τ^*. Combining Eqs. (9) and (10) then gives

$$m = \frac{130 \, b^3}{kT}$$ (11)

Taking $b = 2.73$ Å one obtains $m = 6.3$, in good agreement with the values derived from dislocation velocity measurements and those derived directly from the strain-rate change tests through Eq. (7).

4   DISCUSSION

The present experimental results indicate that the deformation dynamics of molybdenum single crystals can be expressed by a single set of equations over the entire strain range from the microstrain region ($10^{-6} < \epsilon < 10^{-3}$) to the macrostrain region ($\epsilon > 10^{-3}$).

The only parameter which varies over this strain range is the effective stress τ^*. Support for this conclusion is provided by the fact that $m = (\partial \ln \dot\varepsilon)/(\partial \ln \tau^*)$ is a constant and that v^* is a single function of τ^* over the entire strain range.

The present data further indicate that the deformation dynamics in both the microstrain and the macrostrain regions can be related directly to the dislocation dynamics if comparisons are made on the basis of the effective stress τ^* rather than the applied stress. This is seen by comparing the values of m and v^* for the dislocation velocity measurements with those from the effect of strain rate on the flow stress. The agreement between the values provides direct proof that the long-range internal stress (τ_μ) and the number of mobile dislocations (ρ) remain essentially constant during a strain rate-change, which is generally assumed for such tests, but has often been questioned.[14-16]

Finally, these results provide additional support for the earlier conclusion[1,2] that the deformation dynamics of the b.c.c. metals at temperatures below $0.2\ T_m$ is associated with the thermally activated overcoming of the Peierls-Nabarro stress by the nucleation of kinks according to the model proposed by Seeger.[17] The activation energy for this process at low stresses is

$$\Delta H = H_K\left(1 + \frac{1}{4}\ln\frac{16\,\tau_p^\circ}{\pi\,\tau^*}\right) \tag{12}$$

where H_K is the kink energy and τ_p° is the Peierls-Nabarro stress. Substituting Eq. (12) into Eq. (13) and rearranging we obtain

$$v = \left[\nu_0\exp\left(\frac{\Delta S}{k}\right)\right]\left[\exp\left(-\frac{H_K}{kT}\right)\right]\left[\left(\frac{\pi\,\tau^*}{16\,\tau_p^\circ}\right)^{H_K/4kT}\right] \tag{13}$$

$$v \equiv \mathrm{const}(\tau^*)^m \tag{13a}$$

with $m = H_K/4kT$. Equation (13a) has the same form as Eq. (2) except that τ is replaced by τ^*.

The fact that dislocations obey Eq. (2) has caused some controversy in the literature on the correctness of using τ^* rather than the total applied stress τ, whereas, in fact, no inconsistency exists. Dislocation velocity measurements are generally performed on specially treated crystals, i.e., crystals of high purity and low grown-in dislocation density. Since τ_μ is due to the presence of other dislocations, impurity atoms, precipitates, etc., it must be very small in such crystals and hence $\tau^* \approx \tau$, as was found for the present molybdenum crystals. Therefore, taking $\tau^* = \tau$ is a reasonable assumption in such crystals and it is immaterial whether Eq. (1) or Eq. (13a) is used.

Using the value of m from Fig. 2 and the relation $H_K = 4mkT$ gives $H_K = 4.3 \times 10^{-2} \mu b^3$ ($\mu = 12.7 \times 10^3$ Kg/mm^2) which compares well with the value of $\sim 4 \times 10^{-2} \mu b^3$ derived entirely from macro deformation dynamics.[1] Moreover, from Eq. (12) one obtains for the activation volume

$$v^* = \frac{1}{4} \frac{H_K}{\tau^*} \tag{14}$$

which agrees with the stress dependence of the activation volume indicated in Fig. 9. Taking $\frac{1}{4} H_K = 130 \, b^3$ Kg/mm^2 [Eq. (9)] gives $H_K = 4.1 \times 10^{-2} \mu b^3$, in good agreement with that derived from the value of m.

These results thus provide strong support to the earlier conclusion[1,2] that a single dislocation mechanism is rate-controlling during the plastic deformation of the b.c.c. metals at low temperatures and that this mechanism is the thermally activated nucleation of kinks according to the model of Seeger.[17] Especially significant is the good agreement between the values of v^* and H_K derived in the present tests from both deformation and dislocation dynamics on high-purity single crystals, and those derived earlier[1] for molybdenum with a wide range of impurity contents, grain sizes, and other microstructural variations.

Moreover, these results indicate that the same dislocation mechanism operative during microstraining is operative during macrostraining. This is contrary to the conclusion of Kossowsky and Brown[18] and Meakin[5] who postulated that different dislocation mechanisms operate, based on their observations that the temperature-dependence of flow stress and the activation volume differ in the two regions. These differences can now be explained as being due to a difference in the plastic strain rate (and hence a difference in dislocation velocity) in the two strain regions, which in turn produces a difference in τ^*, rather than that different mechanisms operate. In the microstrain region, the elastic strain provides a large contribution to the total strain and hence the plastic strain rate is relatively much smaller for the same overall specimen rate than in the macrostrain region. Thus, the smaller temperature-dependence of the flow stress in the microstrain region observed by Kossowsky and Brown[13] can be attributed to the much smaller plastic strain rate (smaller value of τ^* and of dislocation velocity) in this region compared to the macrostrain region. Similarly, the sudden rapid decrease in the activation volume observed by Meakin[5] upon increase in strain in the range of $\epsilon \simeq 10^{-3}$ occurs exactly in the strain range where the stress-strain curve is beginning to bend over sharply and hence where the plastic strain rate is rapidly increasing (see Fig. 4). With the rapid increase in strain rate there occurs a large change in τ^*, and a rapid change in v^* is expected. Analogously, the initial decrease in v^* with strain indicated in Fig. 8 can be attributed directly to a gradual increase in plastic strain rate associated with a decrease in the strain-hardening coefficient $d\tau/d\gamma$ in this strain range. A constant value of v^* with strain occurs in the region where the strain-hardening coefficient is relatively constant and hence the plastic strain rate is constant.

5   CONCLUSIONS

The results in this paper give strong support to the following conclusions regarding the plastic deformation of molybdenum at low temperatures ($T < 0.25\,T_m$):

1. The deformation dynamics can be expressed by a single set of equations over the entire strain range from the microstrain region ($10^{-6} < \epsilon < 10^{-3}$) to the macrostrain region ($\epsilon > 10^{-3}$).

2. Dislocation dynamics can be deduced directly from the deformation dynamics if comparisons are made at the same values of the effective stress τ^*.

3. The number of mobile dislocations and the long-range internal stress remain essentially constant during strain rate-change tests.

4. The rate-controlling mechanisms during microstrain and macrostrain plastic deformation at low temperatures is thermally activated nucleation of kinks over the Peierls hills.

It is expected that these same conclusions apply equally well to all of the Group V and Group VI b.c.c. refractory transition metals and iron.

ACKNOWLEDGMENT

We wish to acknowledge support of this research by the Office of Naval Research under contract No. Nonr-4434(00). We also wish to acknowledge helpful discussions with Dr. J. D. Meakin regarding microstrain testing.

APPENDIX

Determination of τ^* from Hysteresis Loops Obtained in Microstrain Tests

During a load-unload cycle in the microstrain region a loop such as that illustrated schematically in Fig. 10 was obtained for stresses less than the anelastic limit τ_A (which is the stress at which an open loop first occurs). The area of such a loop gives the energy per unit volume which is dissipated in going around the loop and is

$$A = \frac{\tau_E}{\cos\theta}\, \gamma_{pm}\, \cos\theta = \tau_E \gamma_{pm} \tag{A1}$$

τ_E is the elastic limit and γ_{pm} is the maximum forward plastic strain. Since the loop closes, γ_p is completely reversible. The dynamic modulus E_D and the relaxed modulus E_R are also indicated in Fig. 10.

The stress-strain behavior illustrated in Fig. 10 can be explained in the following manner. As a dislocation moves through a crystal it experiences long- and short-range stress fields (obstacles) which oppose its motion, similar to those in Fig. 11. Above absolute zero, anelastic nonreversible plastic flow occurs when the applied stress τ is equal to the sum of the long-range internal stress τ_μ and the thermal component of the stress τ^* associated with the short-range internal stress field near the top of the long-range stress hill. This value of the applied stress defined the anelastic limit τ_A, since once the dislocations have passed over the top of the hill they cannot return to their original position upon removal of the stress. For applied stresses less than τ_A, the dislocations can return to their original position and the strain is reversible.

The behavior for stresses less than τ_A can be better understood by referring to Fig. 12, which represents an enlarged section of Fig. 11 near the position where τ_μ is nearly zero. Here it is assumed that τ_μ is approximately constant between the short-range obstacles A-B and B-C. Also, the short-range stress fields are drawn rectangular in shape for simplification. A positive internal stress represents one which opposes the applied stress.

Referring to Fig. 12, when the applied stress is reduced to zero, the long-range internal stress presses the dislocation against the short-range obstacle A and motion in a direction opposite to that of the previously applied stress just ceases when $\tau_{\mu_0} = \tau^*$. To cause

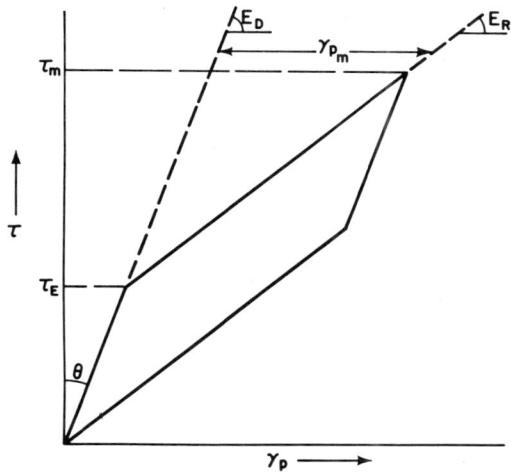

Fig. 10 Schematic of the hysteresis loop observed during microstrain tests.

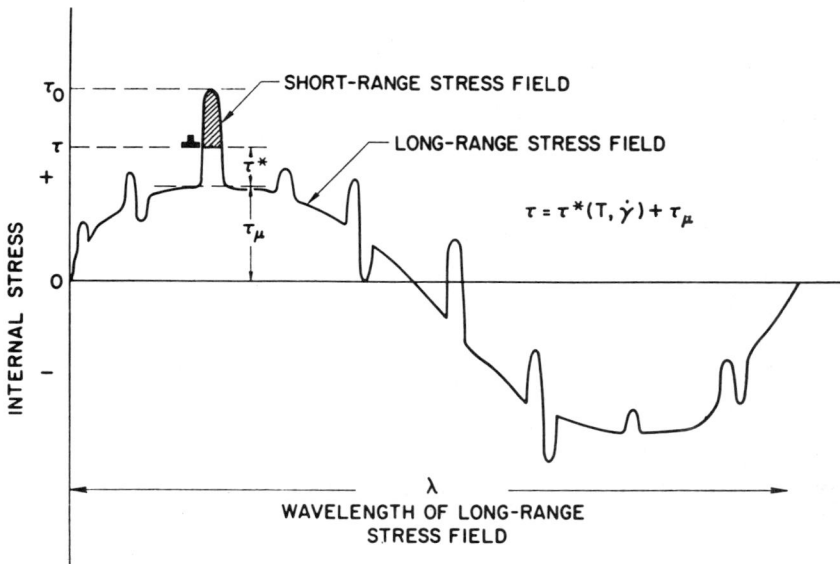

Fig. 11 Schematic of internal stress fields encountered by a dislocation moving through the crystal lattice.

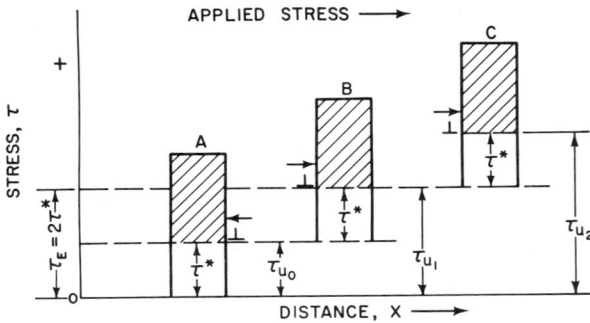

Fig. 12 Schematic of internal stress versus distance relating to microstrain hysteresis loops.

the dislocation to move in the forward direction, a stress must first be applied which overcomes the long-range internal stress τ_{μ_0}; this value of stress will take the dislocation to the base of the short-range obstacle at B. For thermally activated flow to occur (i.e., for an energy-dissipative process to occur) the stress must be increased further by an amount equal to τ^*. Thus, energy loss occurs only for stresses greater than τ_E, where

$$\tau_E = \tau_{\mu_0} + \tau^* = 2\tau^* \tag{A2}$$

Further increase of the applied stress to τ_m gives additional thermally activated motion as the short-range obstacles further up the long-range stress hill are overcome.

Now let the stress be decreased again. Similar reasoning indicates that the initial reverse motion will be "elastic" and that thermally activated motion in the reverse direction will not begin until τ decreases by an amount $\tau_E = 2\tau^*$. The stress-strain curve will then follow the "elastic line from τ_m to $\tau_m - \tau_E$ and then return to the original position forming the parallelogram shown in Fig. 10.

It should be mentioned at this point that since τ^* is associated with a thermally activated process, τ_E will be temperature and strain rate-dependent. It should also be noted that the foregoing explanation is valid only if the load reversals are sufficiently rapid to prevent relaxation. This condition was always met in this work since the Instron cross-head speed can be reversed almost instantly.

Now let us consider the energy loss when going around a loop. In general, this is given by

$$W_{irr} = \oint \tau \, d\gamma \tag{A3}$$

with $\tau = \tau^* + \tau_\mu$ and $\gamma = \gamma_E + \gamma_p$ (γ_E is the elastic component and γ_p the plastic component of the total strain γ). Since γ_E and γ_p are independent

$$dy = d\gamma_E + d\gamma_p \tag{A4}$$

and

$$W_{irr} = \oint (\tau^* + \tau_\mu)(d\gamma_E + d\gamma_p) \tag{A5}$$

$$= \oint \tau\, d\gamma_E + \oint \tau^* d\gamma_p + \oint \tau_\mu\, d\gamma_p$$

The first integral is zero since it represents the purely elastic energy which must disappear if the crystal returns to the starting point. Therefore,

$$W_{irr} = \oint \tau^* d\gamma_p + \oint \tau_\mu\, d\gamma_p \tag{A6}$$

$$= \int_0^{\gamma_p = \gamma_{pm}} \tau^* d\gamma_p + \int_{\gamma_p = \gamma_{pm}}^{\gamma_p = 0} (-\tau^*)\, d\gamma_p + \oint \tau_\mu d\gamma_p \tag{A6a}$$

$$= 2 \int_0^{\gamma_p = \gamma_{pm}} \tau^* d\gamma_p + \oint \tau_\mu d\gamma_p \tag{A6b}$$

Also since $\tau^* = \tau^*(T, \dot{\gamma}_p)$ but not a function of γ_p, it follows that

$$W_{irr} = 2\tau^* \gamma_{pm} + \oint \tau_\mu d\gamma_p \tag{A7}$$

Now, since from Eq. (A1), $W_{irr} = A = \tau_E \gamma_{pm}$ and $\tau_E = 2\tau^*$, it follows upon substituting into Eq. (A7) that $\oint \tau_\mu d\gamma_p = 0$, which is expected since τ_μ is essentially elastic in nature. Thus, it will be generally true for a closed loop that

$$W_{irr} = \tau_E \gamma_{pm} = 2\tau^* \gamma_{pm} \tag{A8}$$

The effective stress τ^* (thermal component of the flow stress) acting on a dislocation in the microstrain region can thus be determined from hysteresis loop experiments. The following three methods can be used to determine τ^* from such measurements:

1. from a measurement of τ_E which gives $2\tau^*$ directly;
2. from the slope of a plot of W_{irr} vs. γ_{pm} which gives $2\tau^*$; and
3. from the intercept of a plot of W_{irr} vs. the applied stress τ, which gives $2\tau^*$.

Since τ^* is a function of temperature and strain rate, these must be maintained constant during the determination of τ_E and W_{irr}.

REFERENCES

1. Conrad, H.: NPL Symp., "The Relation between Structure and Mechanical Properties of Metals," p. 475, Her Majesty's Stationery Office, 1963.
2. Conrad, H.: "High Strength Materials," p. 436, John Wiley & Sons, Inc., New York, 1965.
3. Prekel, H. L., and A. Lawley: "Electron and Ion Beam Science and Technology," Gordon and Breach, New York, to be published.
4. Prekel, H. L., and A. Lawley: *Phil. Mag.* 14:545 (1966).
5. Meakin, J.: *Can. J. Phys.* 45:1121 (1967).
6. Prekel, H. L., A. Lawley, and H. Conrad: to be submitted to *Acta Met.*
7. Stein, D. F., and J. R. Low: *J. Appl. Phys.* 31:362 (1960).
8. Schadler, H. W.: *Acta Met.* 12:861 (1964).
9. Erickson, J. S.: *J. Appl. Phys.* 33:2499 (1962).
10. Conrad, H., and H. Wiedersich: *Acta Met.* 8:128 (1960).
11. Cottrell, A. H.: "Dislocations and Plastic Flow in Crystals," p. 111, Clarendon Press, Oxford, 1953.
12. Roberts, J. M., and N. Brown: *Trans. AIME* 218:485 (1960).
13. Meakin, J. D., and A. Lawley: *Acta Met.* 14:1854 (1966).
14. Guard, R. W.: *Acta Met.* 9:163 (1961).
15. Johnston, W. G., and D. F. Stein: *Acta Met.* 11:317 (1963).
16. Floreen, S., and T. E. Scott: *Acta Met.* 12:758 (1964).
17. Seeger, A.: *Phil. Mag.* 1:651 (1956).
18. Kossowsky, K., and N. Brown: *Acta Met.* 14:131 (1966).

DISCUSSION *on Paper Presented by H. Conrad*

J. P. HIRTH: At low stresses and high temperatures the critical double-kink configuration is denoted by A in Fig. 1. This gives a $\sqrt{\sigma}$ dependence of the activation energy for plastic flow on

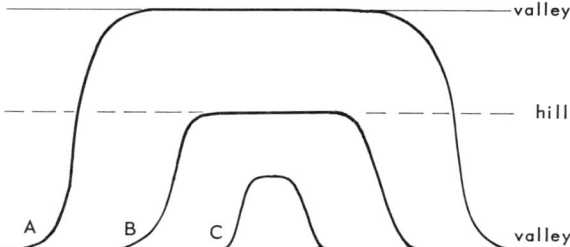

Fig. 1 Kink configurations at a Peierls energy ridge.

stress σ and a $1/\sqrt{\sigma}$ dependence of the activation volume. At higher stresses and lower temperatures, possible critical configurations of the double kinks are B and C in Fig. 1. These give $\ln\sigma$ and $1/\sigma$ dependences of the activation energy and $1/\sigma$ and $1/\sigma^2$ dependence of the activation volume. You correlated your work with a dependence of the

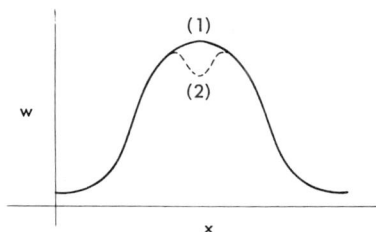

Fig. 2 Peierls energy models.

sort corresponding to configuration B. Did you try the other possibilities?

Configuration B is the original configuration suggested by Seeger (1957). As noted by Seeger in the discussion of Haasen's paper (p. 000), other models can also give the $\ln\sigma$ dependence. However, it is now accepted that configuration B is not a realistic critical configuration for a Peierls energy variation such as (1) in Fig. 2. A possibility which would restore configuration B as a feasible critical configuration is that the Peierls variation is that of (2) in Fig. 2. This would stabilize the segment of line lying along the top of the energy ridge and have the consequence of extending the stress and temperature range where $\ln\sigma$ activation energy dependence obtains.

H. CONRAD: Yes, we tried other correlations but for the range considered the fit was best with the activation energy with stress being of the form $H = H_0 - \alpha \ln\tau^*$, which yields $v^* = \alpha/\tau^*$.

P. HAASON: Is there experimental evidence in b.c.c. metals that the exponent m in the velocity–stress relation really increases in proportion to T^{-1} as you decrease the temperature? This is suggested by your derivation. I might add that in diamond-structure crystals which appear to deform by the Peierls mechanism, m is found to be independent of temperature over quite a temperature range. In this case the internal stress has been considered in the velocity–stress relation in the appropriate way.

H. CONRAD: Since the writing of the manuscript we have completed dislocation velocity measurements on molybdenum at 77° as well as 300°K. For the (112) dislocations the value of m at 78°K is 22.3 ± 1.5. Comparing this value with that at 300°K ($m = 6.4$) we note that $m_{300}/m_{78} \simeq 78/300$; that is, that m is proportional to $1/T$.

G. ALEFELD: (1) You established a formula which relates the m-factor in the velocity stress relation to a unique quantity, namely, the kink energy. On the other hand, you stated that m is dependent on structure. How do you explain this?

(2) What value do you get for the kink energy of molybdenum?

H. CONRAD: (1) The exponent m will depend on structure if it is determined for the total applied stress rather than the effective stress τ^*. It is only when it is determined for the effective stress that m becomes independent of structure and the method used to obtain it.

(2) The kink energy derived is $(4.1 \pm 0.2 \times 10^{-2}) b^3$ or approximately $0.60 - 0.65$ eV.

G. ALEFELD: You emphasized in your paper that one has to subtract very carefully from the applied stress the internal stress to get structure-independent results from the m-factor. Your model was based on double-kink generation. I want to bring to your attention a paper by G. Schoeck in *Phys. Stat. Sol.* 1964 or 1965, in which Schoeck points out that for the Peierls model the effective stress is not $\sigma_a - \sigma_i$ but $\sigma_a + \sigma_i$. Schoeck's conclusion is based on the argument that for double-kink generation the dislocation has a choice where along the line the double kinks are being generated and thus the generation will always occur at that spot, where the internal stress aids the external stress. I do not completely agree with Schoeck's argument, since the dislocation will not always have such a spot where the internal stress is positive. Nevertheless, Schoeck has a point if one assumes that the circles can be generated everywhere along the line. The effective stress is $\sigma^* = \sigma_a - \sigma_i$, but for σ_i one should not use the uniaxial internal stress, but only a fraction of this. Which fraction is certain a difficult question.

SOME EFFECTS OF CRYSTAL ORIENTATION AND PURITY ON THE DYNAMICAL BEHAVIOR OF DISLOCATIONS

D. F. Stein*

General Electric Research and Development Center, Schenectady, New York

ABSTRACT

Experimental results on the effect of crystal orientation and purity on the yield stress, temperature dependence of the yield stress, and strain-rate sensitivity of several b.c.c. metals will be presented. These results will be interpreted on the basis of their effect upon the parameters in the deformation equation based on dislocation dynamics. The approach is simply to relate the observed macroscopic behavior to the parameters in the deformation equation and to carefully examine what assumptions are necessary to relate dislocation dynamics to observed macroscopic yielding behavior.

*Present Address—University of Minnesota, Minneapolis, Minn.

1   INTRODUCTION

A consistent theory that readily accounts for many of the phenomena associated with the mechanical testing of materials has been developed on the basis of dislocation dynamics. This theory explains many of the macroscopic phenomena associated with yielding, such as yield points,[1,2] delay times,[1] Luders band velocity,[2] and the temperature dependence of the yield stress.[3,4] In addition, it has recently been shown[5] that many of the effects measured in microstrain experiments are compatible with the dislocation dynamics theory. Therefore, dislocation dynamics has proven to be a powerful concept in understanding the deformation behavior of materials.

In this paper it will be assumed that the stress–strain behavior of the materials to be considered (W, Mo, Fe) can be represented accurately by an equation based on the theory of dislocation dynamics. Experimental observations on the effect of orientation and purity on the stress–strain curves, the critical resolved shear stress and the temperature dependence of the critical resolved shear stress will be presented. These experimental observations will then be compared to the theoretical curves calculated using the theory of dislocation dynamics to determine what values of the various parameters such as the drag stress, initial dislocation density, dislocation multiplication parameter, dislocation velocity exponent, and work-hardening are needed to reproduce the experimentally observed results.

2   EXPERIMENTAL RESULTS

2.1   Polycrystals

At times it is tempting to measure the properties of polycrystals in an attempt to determine the dynamical properties of dislocations. While it has been demonstrated[2] that many of the concepts of dislocation dynamics are compatible with the behavior of polycrystals, here it will be shown that in some cases the interpretation of polycrystal data in terms of dislocation dynamics can lead to confusion.

The friction stress and temperature dependence of the friction stress have been assumed to be directly related to the yield stress of single crystals because such a correlation has been established for Si-Fe[3] and W.[4] However, it has been shown that, in general,[5] the temperature dependence of the yield stress for polycrystals is significantly greater than that for single crystals.

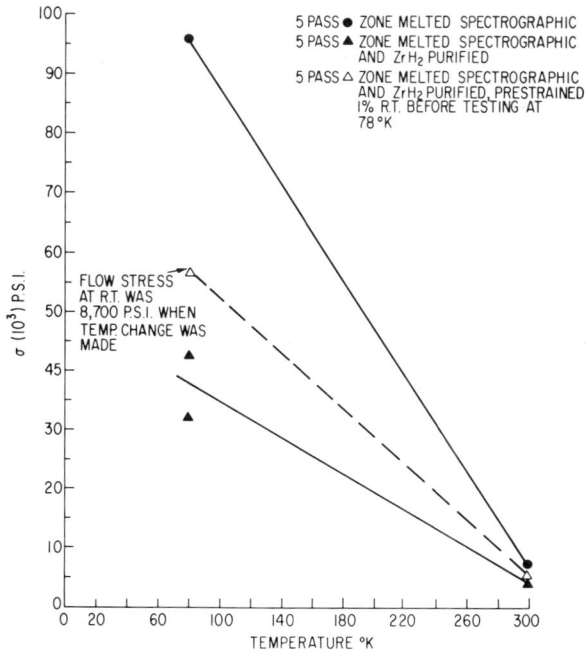

Fig. 1 Temperature dependence of flow stress in zone melted polycrystalline iron.

Figure 1 illustrates the results on zone-melted iron and zone-melted plus high-purity, hydrogen-purified iron. It is seen that the purified iron has a lower yield stress and temperature-dependence of the yield stress. It should be pointed out after ZrH_2 purification, as shown in Fig. 2, the grain size was considerably larger. Thus a direct comparison on yield stress cannot be made. Figure 3 presents results on a high-purity iron that was prepared by swaging to rod before being prepared into tensile specimens. In this case, the ZrH_2 purification results in a decreased yield stress, but an increasing temperature dependence of the yield stress. Therefore, if one were to assume that the temperature dependence of the yield stress correlated directly with the friction stress, one would conclude in the first case that the temperature dependence of the friction stress decreased with increasing purity and in the second that it increased.

The different behavior of the two irons was confusing at first, but a better understanding of the behavior is now available since we know the critical resolved shear stress law is not obeyed by

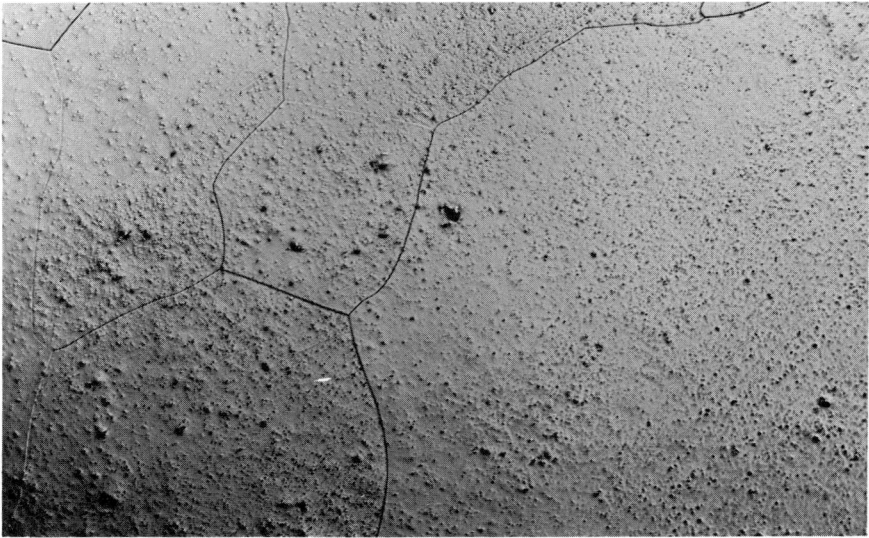

(a)

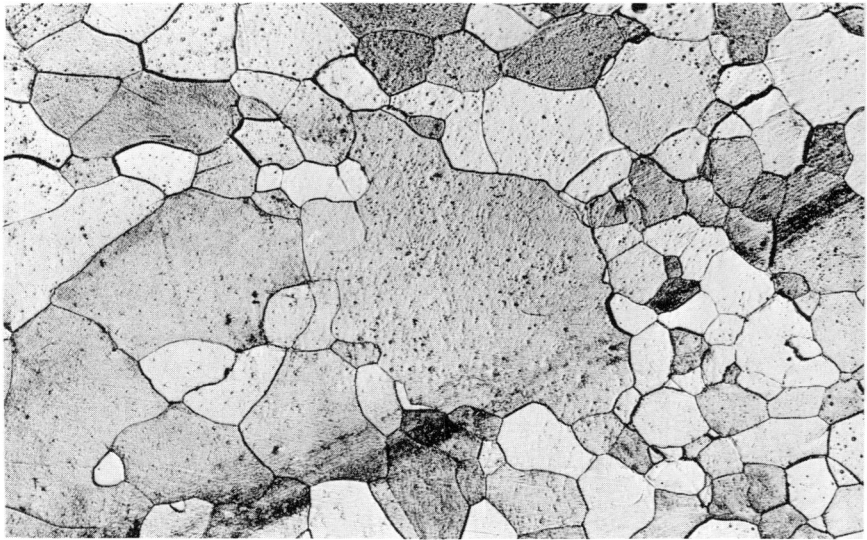

(b)

Fig. 2 Grain size of iron 100×. (a) Zone melted and ZrH$_2$ purified. (b) Zone melted only.

iron at low temperatures. In preparing the swaged iron, a texture was developed in the iron as shown in Fig. 4. It is observed that the texture is strongly <110> as expected, and in a later section, it

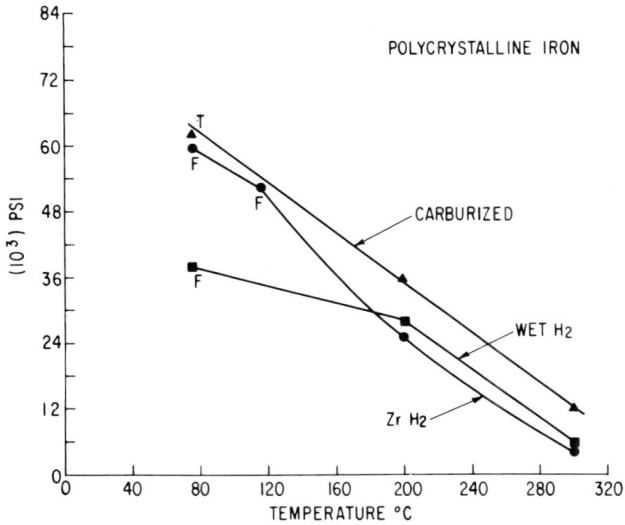

Fig. 3 Temperature dependence of flow stress for iron polycrystals prepared by swaging for different treatments.

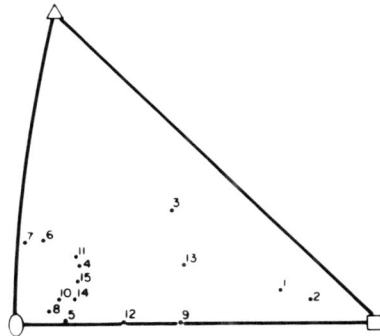

Fig. 4 Grain orientation by reflection of light for swaged iron.

will be shown that the temperature dependence of this orientation single crystal is much greater. In addition, there is indication that the temperature dependence of the yield stress for iron single crystals of that orientation increases with purification. Therefore, the single-crystal behavior is compatible with that of the poly-crystal. However, these results indicate that extreme caution should be used in relating dislocation dynamics to polycrystalline behavior. Therefore, the experimental results that shall be examined on the basis of dislocation dynamics will all be for single crystals.

2.2   Stress-Strain Behavior

Stress-strain curves for high-purity iron[6] and high-purity molybdenum[7] at liquid-nitrogen temperature are presented in Figs. 5 and 6. The separate curves in each figure represent the observed behavior for single crystals of the same purity, but different crystal orientation. For the case of iron it is observed that resolved shear stress values at the same strain are approximately twice as great for a <110> tensile axis specimen as for a <491> (easy slip) orientation crystal. Figure 6 compares stress-strain curves for molybdenum oriented with the compression axis <100> and <110> tested at 78°K. At the proportional limit yield stress ($\approx 5 \times 10^{-4}$ strain) the value of the resolved shear stress for the <110> oriented crystal is approximately three times that for the <100> crystal. The work-hardening rate of the <100> oriented crystal is greater and, therefore, the difference decreases with increasing strain, but even at six percent strain, the value of the flow stress is considerably lower than for the <110> oriented crystal.

The information presented represents the extreme effects of orientation, and other orientations behave in an intermediate way.[6,7,8,9] In general, those crystals tested with the tensile or compression axis along the trace connecting the (110)-(111) poles of the stereographic triangle show the highest values of the critical resolved shear stress and the lowest rate of work-hardening. The lowest values of the critical resolved shear stress are observed to be associated with the <110> tensile or compression axis in molybdenum[7,10] and tungsten,[11,12] but in iron[8,9] the easy slip orientation appears to have the lowest values for the critical resolved shear stress.

2.3   Effect of Orientation on CRSS

Figure 7 shows the effect of temperature on the values of the critical resolved shear stress for zone-melted molybdenum. It is observed that the temperature dependence of the critical resolved shear stress is approximately three times as great for crystals

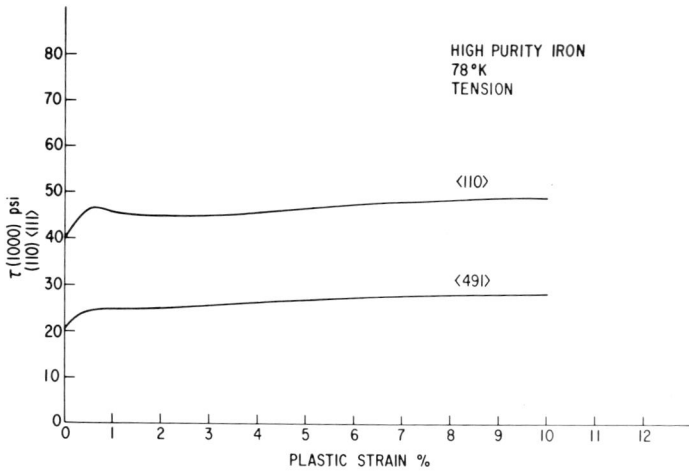

Fig. 5 Stress-strain behavior of high purity iron with different tensile oreintation, tested at 78°K.

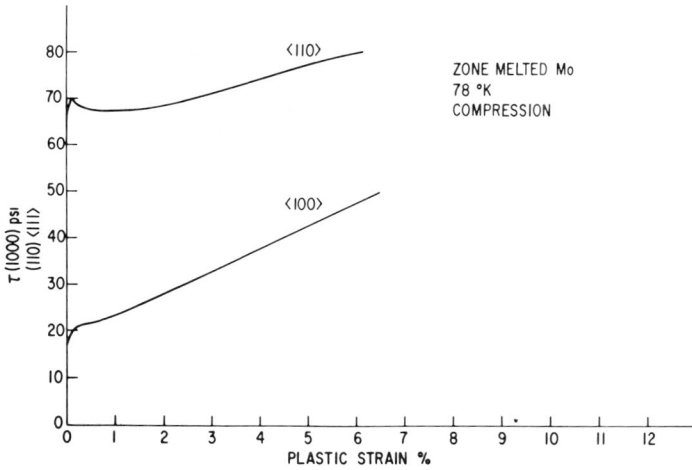

Fig. 6 Stress-strain curves of zone melted molybdenum with different compression orientation, tested at 78°K.

oriented with a $<110>$ compression axis as for a $<100>$ compression axis. It is also observed that the failure of the critical resolved shear-stress law persists to room temperature. This is in contrast to the results presented for iron in Fig. 8 showing a compilation of data for hydrogen-purified iron where it is observed that the critical resolved shear-stress law is nearly obeyed at temperatures above approximately 150°K. At 78°K, there is a discrepancy of more than a factor of two in the value of the critical resolved shear stress. There is also a large difference in the values of the critical resolved shear stress reported by different authors. Cox, Horne, Mehl[13] measured the critical resolved shear stress for slip in crystals of different orientation and concluded that the critical-resolved-stress law was obeyed with the values reported shown by the solid line. Allen, Hopkins, and McLennon[14] observed similar results. Recently, Tomalin and Stein,[6] Stein and Low,[8] and Keh and Nakada[9] all observed large deviations from the critical resolved shear-stress law. However, the carbon content for the iron crystals tested in these later experiments was considerably less than in the irons used in the earlier investigations. Keh and Nakada[9] observed that the higher-purity crystals had an increased value of critical resolved shear stress in all orientations except the easy slip orientation when compared to the early work, while Stein and Low[8] found decreased values for the critical resolved shear stress in all orientations except for the $<110>$. Tomalin and Stein[6] found decreased values for the $<110>$ tensile orientation. The interstitial impurity content reported by each of the investigators is shown in Table 1.

At present, it appears that the breakdown of the critical resolved shear-stress law is associated with very high-purity iron crystals and an investigation is needed to establish if this is true.

TABLE 1

	ppm		
	C	N	O
Tomalin and Stein[6]	<.005	<2	6 ± 3
Stein and Low[8]	<.005	<2	20 ± 3
Keh and Nakada[9]	11*	3	70

LECO Analysis - Strain-aging experiments indicate that the carbon may be lower than this value.

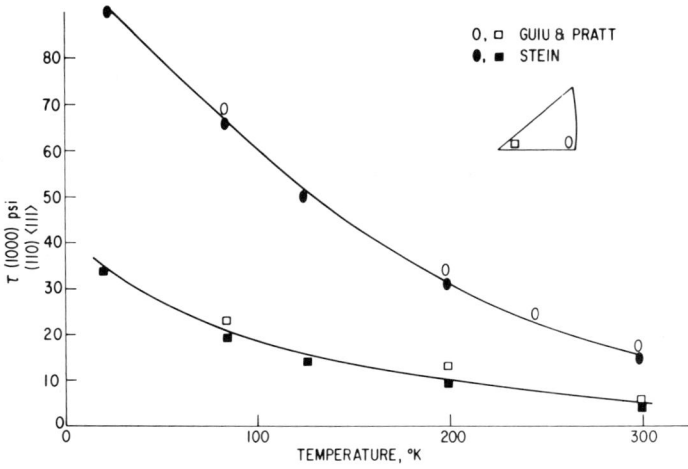

Fig. 7 Temperature dependence of the proportional limit critical resolved shear stress of molybdenum single crystals having $\langle 110 \rangle$ and
$\langle 100 \rangle$ compression axes.

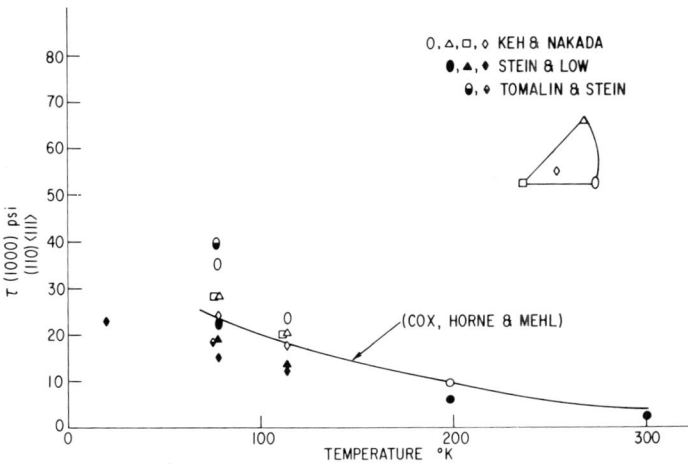

Fig. 8 Temperature dependence of the critical resolved shear stress of
iron single crystals of different orientations, as determined by different
investigators. Keh and Nakoda data was flow stress; other data are
proportional limit.

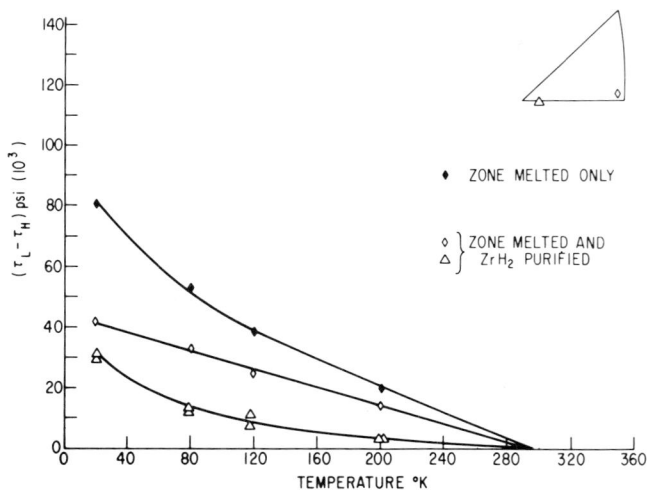

Fig. 9 Temperature dependence of the critical resolved shear stress in molybdenum single crystals showing the effects of purity and of orientation.

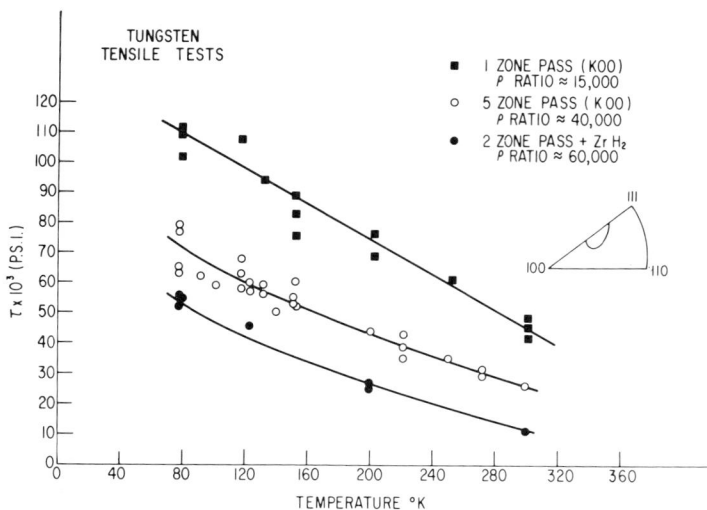

Fig. 10 Temperature dependence of the critical resolved shear stress in tungsten single crystals showing the effect of purity.

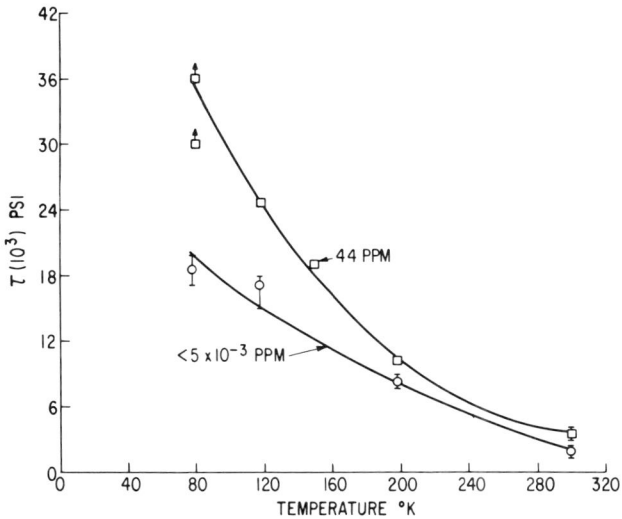

Fig. 11 Temperature dependence of the critical resolved shear stress in iron single crystals, showing the effect of purity.

2.4 Effect of Purity

Figures 9, 10, and 11 show the effect of increasing purity on the temperature dependence of the yield stress for molybdenum, tungsten, and iron. For each metal, the values reported are for the crystal orientation noted in the stereographic triangle. It is observed that increasing the purity of the material decreases the values of the critical resolved shear stress and the temperature dependence of the yield stress. The most marked effects occur in iron and molybdenum, with a discernible effect in tungsten.

3 DERIVATION OF DEFORMATION EQUATION

To compare the measured plastic behavior of these materials with the theory of dislocation dynamics, it is necessary to develop a deformation equation. The derivation of the deformation equation to be used to fit the experimental results is quite similar to that given by Johnston.[1] Following Johnston and Gilman,[1, 15] the cross-head motion of the tensile machine is related to the stress and ultimately to the elastic and plastic strain in the crystal. The tensile machine is considered to be infinitely hard, so that the displacement of the cross-head must be absorbed entirely by the crystal as elastic and plastic strain. The total strain ϵ_T in each crystal being compressed is made up of two components—the plastic strain ϵ_P and the elastic strain ϵ_E. If the rate of cross-head motion (of the tensile machine) is designated by S, where $S = dx/dt$, the total length change is then equal to St for constant rate of cross-head motion. The elastic strain is given by $\epsilon_E = \sigma/E$, where σ = stress, and E = Young's modulus. Thus the plastic strain is

$$\epsilon_P = \frac{St}{l_0} - \frac{\sigma}{E} \tag{1}$$

where l_0 = initial length of specimen and is set equal to 1 inch. Johnston[1] has derived an equation relating the plastic strain-rate to the dynamic properties of dislocation; the following equation is so similar to his that the derivation will not be repeated:

$$\dot{\epsilon}_P = 2b\,(\rho_0 + C\epsilon_P)\left(\frac{\tau}{D_{\text{eff}}}\right)^m \tag{2}$$

where ρ_0 = initial dislocation density
b = Burgers vector
C = dislocation multiplication parameter
m = dislocation velocity exponent
τ = resolved shear stress
D_{eff} = the effective stress

There are two differences between Eq. (2) and Johnston's equation (6):

1. The term $(\rho_0 + C\epsilon_P)$ replaces the expression $\alpha\epsilon_P$ to allow the initial dislocation density to be altered independently of the plastic strain.

2. D_{eff} replaces the parameter D, where $D_{\text{eff}} = D + H\epsilon_P$ and D is the stress to maintain a velocity of 1 cm/sec in a nonwork-hardened lattice and H is the work-hardening coefficient. The calculation is made much easier by introducing the work-hardening in this way, and it is consistent with the experiments of Gilman and Johnston[16] in which they show that the dislocation velocity-stress curve in LiF is simply shifted to a higher stress for increased dislocation density.

Differentiating Eq. (1) with respect to time and combining with Eq. (2) results in the following expressions:

$$\frac{d\tau}{E\,dt} = S - 2b(\rho_0 + C\epsilon_P)\left(\frac{\tau}{D + H\epsilon_P}\right)^m \tag{3}$$

Substituting $\epsilon_P = (St - \sigma/E)$ into Eq. (3) and using the following relation:

$$\sigma = \frac{P}{A} \qquad \frac{\tau}{R} = \sigma \qquad \tau = \frac{PR}{A} \qquad \frac{dx}{S} = dt$$

where P = load, A = area, and R = Schmid factor, gives

$$\frac{S}{RE}\frac{d\tau}{dx} = S - 2b\left\{\rho_0 + C\left(x - \frac{\tau}{RE}\right)\right\}\left[\frac{\tau}{D + H(x - \tau/RE)}\right]^m \tag{4}$$

Equation (4) is the expression that is to be integrated on the computer to determine the necessary parameters to fit the experimental data.

4 APPLICATION TO STRESS-STRAIN BEHAVIOR OF SINGLE CRYSTALS

4.1 Restrictions

In Eq. (4) there are five parameters (m, D, H, C, ρ_0) which can be considered variable. With this number of free parameters, it is possible to fit nearly any imaginable stress-strain behavior and, therefore, it is necessary that certain restrictions consistent with other experimental evidence on b.c.c. metals be placed on the calculation. Direct measurements of m have been reported on Si-Fe,[3] W,[4] Nb,[17] and Mo.[19] In addition, measurements of strain-rate sensitivity[18] have been used to measure the value of m in other b.c.c. metals. It was found that the lowest values of m reported are around 5 for high-purity iron[20,21] and the highest value is near 45 for SiFe.[3] Moreover, it has been found that the values of m decrease[20] with increasing purity and for most high-purity b.c.c. metals the range of the value of m is between 5 and 10.[4,17,19,20,21] Therefore, the values of m considered in these calculations will be limited to between 5 and 10.

The multiplication rate has been found to vary from approximately 1×10^{10} [22]/in^2/unit strain (for tungsten) to approximately 1×10^{13} [23]/in^2/unit strain (for columbium). In the calculation of the effect of the multiplication rate, a larger range of values will be used to demonstrate the rather insensitive dependence of the yield stress on this parameter. However, it will be seen that this parameter does affect the shape of the stress-strain curve.

The initial density of dislocations can be extended over a very wide range of values. Using very careful single-crystal growing techniques, it would be possible, in principle, to reduce the number of mobile dislocations in a crystal to near zero, and by cold-working it might be possible to attain dislocation densities near 10^{13} in^{-2}. However, for the cases we will consider, initial dislocation densities near 10^9 in^{-2} will be appropriate.

The values of D and H are left completely free and are allowed to vary in any way needed to fit the observed stress-strain behavior.

4.2   Calculation of Stress-Strain Curves

Figures 5 and 6 show experimentally determined stress-strain curves for iron and molybdenum, respectively. Using Eq. (4), an attempt was made to calculate the curves keeping the value of the drag stress constant for both orientations. This would be consistent with the position that dislocation velocity at a given temperature would be a function of shear stress only. The only remaining parameter that might be expected to give the large change in the stress-strain behavior would be the multiplication rate C. Therefore, choosing values of C equal to 2.7×10^6 and 2.7×10^{12} to cover the extremes which seemed possible, the stress-strain curves were calculated for two different work-hardening rates. The results of this calculation are shown in Fig. 12 and it is seen that the multiplication rate has little effect on the value of the 10^{-3} percent yield stress although it does alter the stress-strain curves considerably. Therefore, it appears that the drag stress is the only parameter that can be changed to obtain the observed behavior. Figures 13 and 14 show calculated stress-strain curves that accurately represent the experimental observation. The values of the parameter chosen to give a close fit with the experimental data are given in Table 2.

The values of m, E, C, S, and b were obtained either from other experiments or were a condition of the test. The values of ρ_0, H, and D

TABLE 2 Parameters of Eq. (4)

Molybdenum				
298°K		78°K		Units
<100>	<110>	<100>	<110>	
D	1.8×10^4	5.0×10^4	4.6×10^4	18.5×10^4	psi
H	50×10^5	30×10^5	19.0×10^5	17.0×10^5	psi/unit strain
m	6	6	10	10	—
ρ_0	1.3×10^9	1.3×10^9	1.3×10^9	1.3×10^9	$1/in^2$
E	51.8×10^6	43.5×10^6	51.8×10^6	43.5×10^6	psi
C	4.5×10^{11}	4.5×10^{11}	4.5×10^{11}	4.5×10^{11}	$1/in^2$ unit strain
S	.02	.02	.02	.02	in/min
b	1.1×10^{-8}	1.1×10^{-8}	1.1×10^{-8}	1.1×10^{-8}	in

Iron				
298°K		78°K		
		<491>	<110>	
D	—	—	5.5×10^4	12.0×10^4	psi
H	—	—	9.0×10^5	11.0×10^5	psi/unit strain
m	—	—	10	10	—
ρ_0	—	—	1.5×10^9	1.5×10^9	$1/in^2$
E	—	—	25.0×10^6	32.0×10^6	psi
C	—	—	5.0×10^{11}	5.0×10^{11}	$1/in^2$ unit strain
S	—	—	.02	.02	in/min
b	—	—	1×10^{-8}	1×10^{-8}	in

were obtained by the best fit with the experimental stress–strain behavior. The value for ρ_0 (initial mobile density) was determined primarily from the yield point obtained in the molybdenum specimen tested at 78°K having a <110> tensile axis. The value was then held constant for all temperatures and orientations. The high-value of this parameter was somewhat surprising but it probably represents the difficulty in accurately aligning a compression specimen.

The value of H (work-hardening) was found to be not very sensitive to orientation and to decrease with decreasing temperature.

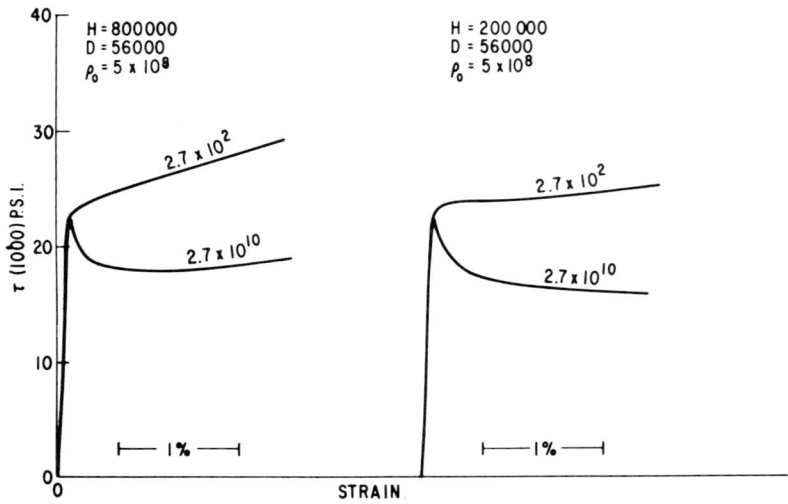

Fig. 12 Calculated stress strain curve showing the effect of multiplication rate at two different values of the work-hardening.

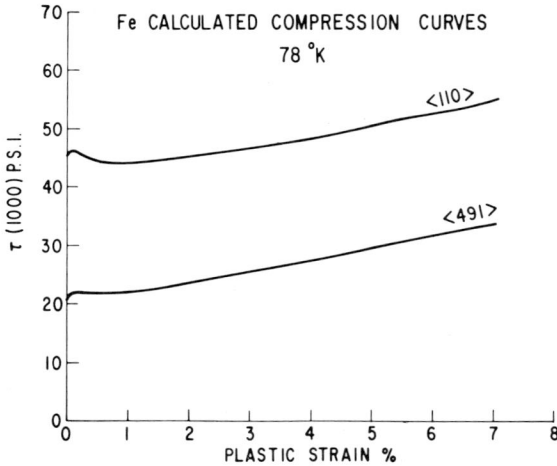

Fig. 13 Calculated compression stress strain curves for iron single crystals with (110) and (100) compression orientation. In comparing, remember that the data in Fig. 5 was measured in tension.

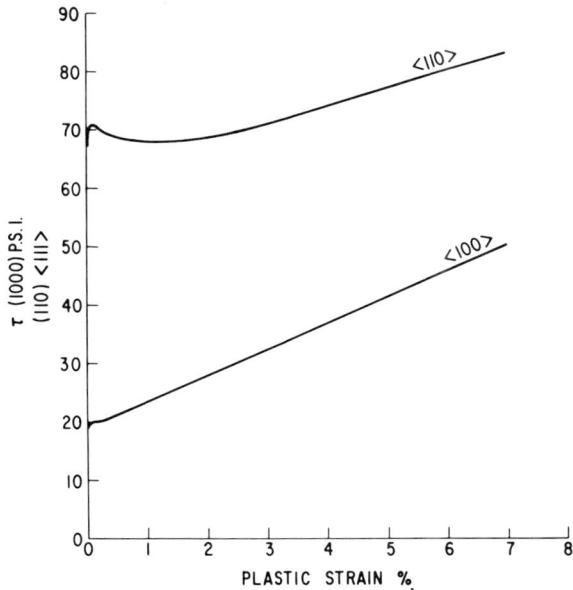

Fig. 14   Calculated compression-strain curve for molybdenum single crystals with (110) and (100) compression orientation.

The values of D (drag stress) turn out to be a very strong function of orientation with a difference of approximately four at 78°K, and three at room temperature in molybdenum. The difference is approximately a factor of two in the iron tested at 78°K.

Therefore, it is apparent that if the dislocation dynamics treatment is applicable to the deformation of iron and molybdenum single crystals, the values of the drag stress must be a strong function of orientation.

4.3   Temperature Dependence

Using the same data on molybdenum, the effect of temperature on the drag stress can also be estimated. It was found that the drag stress increases by about 28,000 psi in the <100> orientation and about 135,000 psi in the <110> orientation. If one considered the ratios of the drag stress for the two orientations, the ratios $D_{78°K}/D_{298°K}$ for the <100> specimens is 2.6 and for the <110> specimens 3.7. Therefore, even on a ratio basis, the drag stress increases much more rapidly for the <110> oriented molybdenum specimen.

TABLE 3 Parameters of Eq. (4)

		Molybdenum 78°K <110>	
	Zone-melted	Zone-melted plus ZrH$_2$ anneal	
D	18.5×10^4	15.0×10^4	psi
H	17.0×10^5	17.0×10^5	psi/unit strain
m	10	10	–
ρ_0	1.3×10^9	4.3×10^9	1/in^2
E	43.5×10^6	43.5×10^6	psi
C	4.5×10^{11}	4.5×10^{11}	1/in^2 unit strain
S	.02	.02	in/min
b	1.1×10^{-8}	1.1×10^8	in

4.4   Effect of Purity

Table 3 compares the values of the parameter used to calculate the stress strain curves for <110> oriented crystals of molybdenum, one of which was zone melted only and the other zone melted plus hydrogen annealed for 150 hours at 1200°C. As shown in Fig. 9, the proportional limit yield stress exhibits a lower temperature dependence for the hydrogen annealed specimens than for the zone-melted specimen.

Calculated curves using Eq. (4) were fit to the experimental stress-strain curves to determine the parameter that gave a best fit. The values determined are presented in Table 3. Although the work-hardening was allowed to remain a variable, the best fit was obtained in both materials using the same value for H. However, the drag stress was reduced about 20 percent by the annealing, and the initial free dislocation density was increased by a factor of three. As previously reported,[7] a variation in the chemical composition could not be detected as a result of hydrogen annealing, but the impurity levels were near the limits of detectability for the analytical techniques used and, therefore, they may not be significant.

The change in the drag stress as a result of hydrogen annealing (and assumed purification) is only about 20% in contrast to about a 400% change when the tensile axis of the specimen is changed. This demonstrates that, while purity does affect the drag stress,

by far the most important contribution to the drag stress in some
mechanism associated with the stress state.

5 DISCUSSION

The effect of orientation, temperature, and impurities can be
well explained if the assumption is made that the drag stress is
affected by these three variables. It is not surprising that temperature and impurities would affect the drag stress, but the effect
of orientation on the drag stress is much more puzzling. This difficulty is compounded by recent microstrain experiments by Garfinkle[24] on tungsten and Davies and Gilbert[25] on molybdenum.
Garfinkle found that the stress to produce a strain of 10^{-5} in
tungsten single crystals at room temperature was independent
of orientation. A very similar result was obtained by Davies and
Gilbert on molybdenum and, in addition, they found that the microstrain yield stress in molybdenum crystals was independent of
temperature. They have explained these results on the basis that
multiplication rate is stress and temperature sensitive. Therefore
they define a multiplication stress and a dislocation mobility
stress, but they felt that the yield stress was determined by the
multiplication stress and not the mobility stress.

This position is tenable for molybdenum because direct measurements of dislocation mobility have not been made. However,
in tungsten the direct measurements of mobility have been made,
and it is found that the dislocation mobility is a strong function of temperature. Since both molybdenum and tungsten exhibit
the same microstrain behavior at room temperature and there is a
great deal of similarity in their mechanical behavior, it seems
unlikely that the dislocation mobility will not be a function of temperature in molybdenum also. In addition, Schadler[22] did not find
the multiplication rate of dislocations to be a function of stress if
it is measured on a strain basis. Obviously, the dislocation density
will increase most rapidly near the yield stress because at this
point the plastic strain rate is greatly increased, but the increased
plastic strain rate would be a consequence of dislocation mobility.
In the presentation of the deformation equation it was assumed that
the density of dislocations increased linearly with plastic strain,
which means that the number of new dislocations per unit time is
proportional to the area swept out by existing dislocations per unit
time. Therefore, it is the rate of dislocation multiplication per
unit strain that should be examined to determine if the multiplication
is stress sensitive. Otherwise one may confuse dislocation multi-

plication rate with dislocation mobility. Since there are no direct experiments which indicate that the multiplication rate is a function of stress in Mo, W, or Fe, it cannot be assumed that this is the correct explanation of the orientation and temperature effects.

Another possibility would be that the value of m (dislocation velocity exponent) is orientation sensitive. The work of Guiu and Pratt[10] indicate that this may be true because they found that the yield stress was more sensitive to strain rate for <110> oriented crystals than for <100> crystals. If one calculates the value of m from their results, it is found that at room temperature m is equal to 10 in the <110> orientation and 13 in the <100> orientation. Determination of m from strain-rate sensitivity usually results in values that are higher than the true m, but this does indicate that a variation in m may be occurring in molybdenum as a function of crystal orientation. If this is true, the microstrain behavior can be understood because it has been shown[5] that it is very sensitive to the value of m. It was found that for small values of m, the microstrain yield stress was a smaller fraction of the flow stress; therefore it would be expected that the microstrain yield stress for a <110> oriented crystal would be a lower fraction of the flow stress than that for a <100> oriented crystal, resulting in an approach to the same value of a microstrain yield stress. Therefore, if m is allowed to vary, it would be possible to explain the microstrain experiments without recourse to a stress-sensitive multiplication rate.

6   CONCLUSION

Using the concepts of dislocation dynamics, experimental measurements on the stress-strain behavior of molybdenum, tungsten, and iron have been examined. If the assumptions made in the analysis are correct, the following conclusions can be made:

a. the breakdown of the critical resolved shear stress in the body-centered cubic metals is a consequence of an orientation dependence of the drag stress;

b. the temperature dependence of the critical resolved shear stress is a consequence of an increased drag stress at low temperature, but the drag stress does not have the same temperature dependence as does the yield stress;

c. the temperature dependence of the drag stress is decreased by increasing the purity, but the effect is much less than that of orientation (20% compared to 300% for molybdenum).

ACKNOWLEDGMENT

This research was sponsored in part by the Army Research Office at Durham, North Carolina.

REFERENCES

1. Johnston, W. G.: *J. Appl. Phys.* **33**: 2716 (1962).
2. Hahn, G. T.: *Acta Met.* **10**:727 (1962).
3. Stein, D. F., and J. R. Low, Jr.: *J. Appl. Phys.* **31**: 362 (1960).
4. Schadler, H. W.: *Acta Met.* **12**:861 (1964).
5. Stein, D. F.: "A Dislocation-Dynamics Treatment of Microstrain," in "Microplasticity," to be published.
6. Tomalin, D. S., and D. F. Stein: *Trans. AIME* **233**: 2056 (1966).
7. Stein, D. F.: *Can. J. of Phys.* **45**:1063 (1967).
8. Stein, D. F., and J. R. Low, Jr.: *Acta Met.* **14**:1183 (1966).
9. Keh, A. S., and Y. Nakada: *Can. J. of Phys.* **45**:1101 (1967).
10. Guiu, F., and P. L. Pratt: *Phys. Stat. Sol.* **15**:1539 (1966).
11. Rose, R. M., D. P. Ferriss, and J. Wulff: *Trans. AIME* **224**: 981 (1962).
12. Argon, A. S., and S. R. Maloof: *Acta Met.* **14**:1449 (1966).
13. Cox, J. J., G. T. Horne, and R. F. Mehl: *Trans. ASM* **49**:118 (1957).
14. Allen, N. P., B. E. Hopkins, and J. E. Malennan: *Proc. Roy. Soc.* London **234**:221 (1959).
15. Johnston, W. G., and J. J. Gilman: *J. Appl. Phys.* **30**:129 (1959).
16. Gilman, J. J., and W. G. Johnston: *J. Appl. Phys.* **31**:687 (1960).
17. Guberman, H. D.: Stress Dependence of Dislocation Velocity in Irradiated and Unirradiated Niobium, *AIME Annual Meeting*, p. 105, Feb., 1967, *Abstract Bulletin*.
18. Hahn, G. T., C. N. Reid, and A. Gilbert: *Tech. Doc. Report* No. ASD-TDR-63-324, 1963.
19. Prekel, H. L., and H. Conrad: This colloquium, p. 433.
20. Stein, D. F.: *Acta Met.* **14**:99 (1966).
21. Michalak, J. T.: *Acta Met.* **13**:213 (1965).
22. Schadler, H. W., and J. R. Low, Jr.: Final Report ONR Contract No. Nonr-2614(00), 1962.
23. Gregory, D. D.: ASD-TDR-62-354, 1962.
24. Garfinkle, M.: *Trans. AIME* **236**: 1372 (1966).
25. Davies, R. G., and A. Gilbert: *Acta Met.* **15**:665 (1967).

DISCUSSION *on Paper by D. F. Stein*

A. W. SLEESWYK: (1) From your discussions this morning, are you willing to say that impurities in b.c.c. metals have only minor effects on the temperature dependence of yield and flow stresses?

(2) Is it true that the dislocation velocity measurements in Si-Fe on (110) plane by you and Dr. Low, and on (112) plane by Erickson show the same change with temperature as the yield stress change on the two planes?

D. F. STEIN: (1) The experimental results on b.c.c. metals indicate that the temperature dependence of the proportional limits is considerably reduced, but that the flow stress is decreased considerably less. The work by Schadler on tungsten indicates that the temperature dependence of the dislocation mobility can best be calculated from the proportional limit. Therefore, I would take the position that the experiments I reported indicate that purity has a strong effect on the temperature dependence of the stress to maintain a constant dislocation velocity, especially in the low-velocity range.

(2) That is correct. However, the dislocation velocity exponent in Si-Fe is very high (35–40). Under these conditions I have shown* that one would expect the stress to maintain a consistent dislocation velocity to have the same temperature dependence as the yield stress. However, for material with a low value of m this need not be true* and the result of Haasen reported in the conference experimentally demonstrates that there is not a correspondence in Ge.

P. HAASEN: I would like to emphasize the warning that the velocity stress exponent m should not be extracted uncritically from yield stress-strain measurements. In our theory of the yield phenomenon in Ge, the lower yield stress τ_0 is related to strain rate $\dot{\epsilon}$ by $\tau_0^{2+m} = C(T)\dot{\epsilon}$. The strain-rate stress exponent is due to dislocation interaction loops larger than the velocity stress exponent. For other parts of the stress-strain curve the stress dependence of the dislocation multiplication process may also enter the flow stress (see paper VII-4).

---

*Stein, D. F.: "A Dislocation-dynamic Treatment of Microstrain," *Microplasticity*, to be published.

Agenda Discussion:
LOW-SPEED
DISLOCATIONS

F. R. N. Nabarro*

University of the Witwatersrand,
Johannesburg, South Africa

H. Conrad**

Materials Science and Engineering Division,
The Franklin Institute,
Philadelphia, Pennsylvania

The chairman first gave an opportunity for discussion of the introductory lectures of the first day. He then called for a general discussion of the factors controlling the flow stress of body-centered cubic metals at low temperatures.

There was extensive discussion of Gilman's lecture on the plastic response of solids. His interpretation of the Haasen-Kelly effect (the appearance of a transient increase in load when the straining of a specimen has been interrupted and the test is then started again at the initial cross-head speed) was questioned by Weertman, who thought that inertial effects would be unimportant in the motion of dislocations under these conditions. Gilman explained that he was not concerned with the inertia associated with

*Chairman
**Secretary; Present address—University of Kentucky, Lexington, Ky.

the effective masses of individual dislocations, but with collective interactions which can be thought of as scattering events. These introduce a transient drag as a set of dislocations starts changing its average velocity. This is analogous with the transient drag that appears in traffic flow theory. Weertman could not see why the dislocations should not all start moving together, and Gilman agreed that they would, providing they were all either independent or rigidly coupled. Weertman also thought that the neglect of the interaction of one member of the procession with all other members except that immediately in front of it was justified in the traffic flow problem, but not in the dislocation problem, and suggested an approximate method of solution in which the procession of dislocations was smeared out into a continuous distribution. Gilman was pressed by Seeger to specify the number of dislocations which might have to take part in a procession in order to produce a Haasen-Kelly effect, but Gilman replied that a highly nonlinear problem such as this one can only be solved by numerical methods beyond the scope of an oral discussion. Basinski asked if Gilman could explain the dependence of the Haasen-Kelly effect on the degree of unloading. Gilman replied that it does not always depend on the degree of unloading.

Haasen questioned Gilman's interpretation of the motion of dislocations in silicon and germanium in terms of electron tunneling because the electronic energy gap is a function of temperature [see Varshni, *Physica*, 34: 149 (1967)] and the activation energy for dislocation motion is found to be independent of temperature. Haasen (who presented a paper on the subject later in the Colloquium) claimed that the observations on silicon and germanium could be explained by a Peierls mechanism with due regard for kink and line energies, and no consideration of tunneling. Gilman replied that his remarks were restricted to the low-temperature regime where Haasen's work does not apply. Also, the temperature dependence of the gap is about the same as that of the shear modulus; and therefore of the same order as that expected for the Peierls barrier.

Basinski doubted if it was possible to find satisfactory definitions of a dislocation and its Burgers circuit in glass. Gilman, supported by Brown, said that one could relate the actual positions of the oxygen atoms to those they would occupy in a reference lattice having cells with dimensions equal to the mean dimensions of the distorted cells in the glass, the actual cell sizes being distributed fairly narrowly about this mean. Also, regardless of the microstructure, one can take the line integral of the elastic displacement field to find the Burgers displacement vector. Brown and Hobart said that one could take the Burgers vector to be the average displacement of the oxygen atom from its original unslipped position, and Nabarro thought to himself that this might require one to have made preliminary observations on the original grain of sand before it had been incorporated in the glass. Sleeswyk said that a structure must be singly connected before one could define a Burgers vector independent of the Burgers circuit, and Hirsch said that in a glass the expected value of the Burgers vector would increase systematically as the Burgers circuit was enlarged. Sleeswyk pointed out that one had only to insert a grain boundary, which would not be noticed in a glass, and a dislocation line could terminate in this boundary. Gilman suggested that the surface-step gold decoration technique could be used to look for surface steps on glass deformed at low temperatures. Thus the point might be settled experimentally. Seeger remarked that there are really two situations. In the first, we are presented with the configurations before and after deformation. Then an analysis of the deformation in terms of dislocations, defined as the boundaries between slipped and unslipped regions, will usually be possible. In the second, we are simply presented with the atomic configuration in a piece of glass. Then an analysis of its dislocation structure may well not be possible.

In discussion of Li's lecture, Campbell pointed out that the average dislocation velocity might depend critically on the form of the distribution of internal stresses. If these stresses varied randomly, and the applied stress only slightly exceeded their upper bound, as indicated in Fig. 1a, the dislocation velocity would be very small in isolated regions. These regions would therefore be the dominating influence in controlling the mean dislocation velocity, as given by the relation

$$\bar{v} = \frac{X}{\displaystyle\int_0^X \frac{dx}{v}}$$

Thus $\bar{v}$ would be approximately equal to $X/\sum_i t_i$, t_i being the time spent in the ith region of high internal stress (see Fig. 1b). Granato said that in activation energy problems of this kind the entropy factor $\exp(S/k)$ need not be taken to be unity since it can be calculated explicitly. In fact, where calculations have been made, this factor is greater than one, so that a dislocation line pinned at

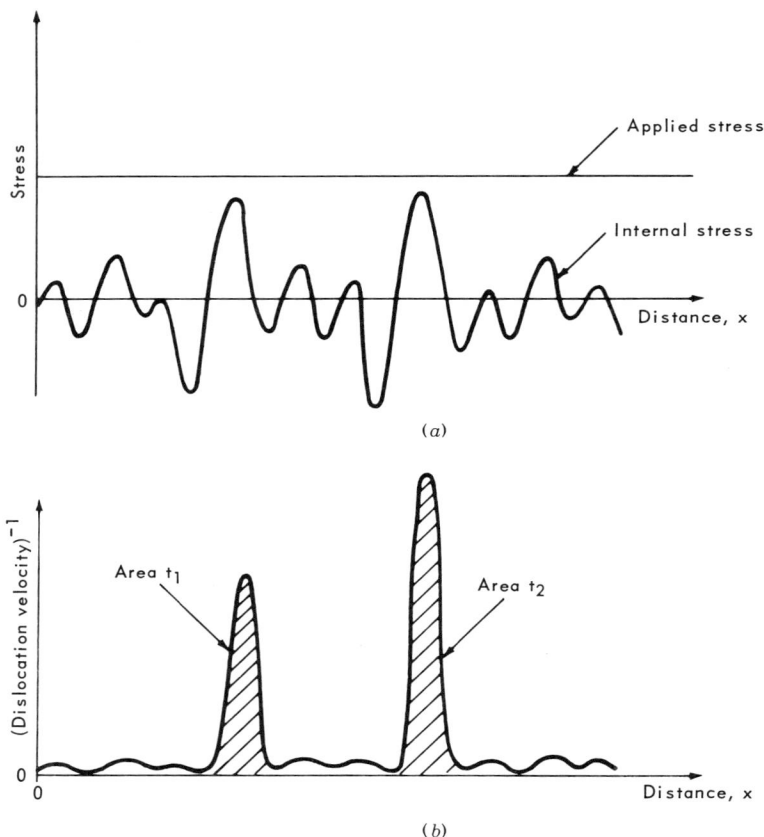

Fig. 1 Effect of internal stress distribution on dislocation velocity. (a) Stress distribution along the path of a dislocation. (b) Variation of reciprocal of velocity along the path of a dislocation.

its center with a binding energy of 0.1 eV has an effective attack frequency of about 10^{10} sec^{-1} independent of the length of the line. Li said that experimentally the entropy factor is usually close to unity. According to Seeger, studies in Stuttgart showed that the entropy was dependent on stress and that this led to a spectrum of activation energies, bu Li and Basinski did not think that the problem was impossibly complicated. Basinski called attention to Cottrell's analysis of the observations of Wyatt, using the model of an advancing energy front. Seeger agreed with Basinski that a distribution of activation energies was tractable; a distribution of activation volumes was more troublesome. The inverse problem of finding the nature and distribution of the obstacles from the macroscopic observations was more difficult. Arsenault and Li have recently written a paper [*Phil. Mag.* 16:1307 (1967)] on this subject. Conrad asked if the real difficulty lay in the incompleteness of the theory or in the fact that the experimental techniques did not yield the right type of information. Was there any case in which there was no doubt of the validity of the theory, and the experimental techniques failed to disclose the nature of the obstacles to dislocation motion? Seeger explained his view by giving two practical examples. When analyzing the temperature and strain-rate dependence of the flow stress of plastically deformed copper crystals in the usual manner in terms of a single thermally activated process, it is found (R. Zeyfang and O. Buck, Stuttgart, unpublished results; J. M. Galligan and J. L. Davidson, Columbia University, *J. Appl. Phys.*, to be published) that the preexponential factor (essentially the density of mobile dislocations involved) depends strongly on temperature, stress and strain rate. This result contradicts the basic assumption of the analysis. [For the additional complications connected with the softness of the testing machine and the problems of practically determining small changes in flow stress, see, e.g., H. Mecking and K. Lücke, *Mater. Sci. Eng.* 1:349 (1967).] The second example concerns the plastic deformation of neutron irradiated copper single crystals, where, according to the work by Diehl and his associates [see J. Diehl and G. P. Seidel, *Phys. Stat. Sol.* 17:43 (1966); J. Diehl, G. P. Seidel, and L. Niemann, *Phys. Stat. Sol.* 11:339 (1965); 12:405 (1965)] and by M. Ruhle at Stuttgart (to be published), these difficulties are aggravated. Furthermore, as shown by the detailed work of Schwink and associates in Munich on the cinematography of slip lines [C. Schwink and G. Grieshammer, *Phys. Stat. Sol.* 6:665 (1964); C. Schwink and H. Neuhauser, *Phys. Stat. Sol.* 6:679 (1964); 17:35 (1966); C. Schwink, *Phys. Stat. Sol.* 7:481 (1964)], the volume of the crystals participating in the slip process is dependent on the strain rate and presumably also on the deformation temperature. Seeger said that in such cases the usual experiments for determining the thermally activated processes cannot be analyzed in a meaningful way without additional information that may either come from other experimental techniques or from theory. Such a complete analysis is being undertaken in the case of neutron irradiated copper single crystals by M. Ruhle, W. Frank, and M. Saxlova. Conrad then asked whether such complications also occurred when the velocity of individual dislocations was considered as, for example, in etch pit studies, rather than the behavior of a plastically deformed crystal where considerable interactions between dislocations could occur. Seeger said that evidence on dislocation velocity in copper as determined by etch pits also appears to indicate certain complications, related to the question of whether, during the observation interval, the dislocations had spent part of their time lying in front of obstacles, waiting for thermal activation.

The rapporteurs note that in informal discussions throughout the Colloquium and in the formal discussions of Haasen's paper, it seemed generally agreed that the distinctions brought forward in this exchange may prove to have been among the most important results of the Colloquium, making explicit an issue which had previously been only vaguely formulated. Internal friction measurements determine the drag on an isolated dislocation moving through distances small compared with the distance between obstacles. Etch pit studies of isolated dislocations may, and apparently often do, determine the same quantity. Studies of the flow stress, or etch pit studies in which the dislocations move large distances in comparison with the spacings between obstacles, determine a different mean velocity more analogous to that discussed by Campbell. When dislocations move together in a slip band, there seems to be a moderate increase in their mean velocity as a result of their cooperation in overcoming obstacles. Finally, in studies of the macroscopic strain rate one is concerned very deeply with the stress dependence of dislocation multiplication.

On specific materials, Armstrong reported preliminary results on the deformation of niobium single crystals that he and R. E. Reed obtained at the Oak Ridge National Laboratory this past summer. The results appear relevant to those reported by Guberman (also at Oak Ridge), as quoted by Li. In Armstrong's experiments two niobium crystals of the same crystal orientation were produced from the same starting material but each was electron beam "grown" at a different rate. The two crystals differed in the measured value of critical resolved shear stress by just under a factor of two and gave m values near to the two values obtained by Guberman. The faster growth rate gave a larger shear stress and m value. However, the value of $B \equiv v^*/RT$ determined from the strain-rate change tests as a function of strain at room temperature was reasonably the same for both crystals. The value agreed with that reported by other investigators on other starting material. Stein doubted the analysis of the observations on potassium in terms of a Peierls stress. He said that LiF gave similar results for stress as a function of temperature at various impurity levels, but that a Peierls mechanism was not involved in that case. Li argued that a stress–temperature relation alone was not sufficient to pin down a mechanism. Other factors such as activation enthalpy and activation area should all be taken into consideration. However, it was generally agreed that there was no complete analysis on these lines of the behavior of the b.c.c. transitional metals.

In discussion of Granato's lecture, Weertman said that the decrement did not fall off with decreasing frequency in the neighborhood of 1 cps as fast as the theory indicated. Granato said that no one has ever obtained experimentally the theoretical frequency or loop length dependence in this range, and that presumably other sources of damping predominated.

There followed a long discussion on the factors controlling the flow stress of body-centered cubic metals at low temperatures. Nabarro's original agenda read:

"Three factors are considered, (a) dissociation of screw dislocations, (b) Peierls force, and (c) impurity locking. Is (b) distinct from (a), or is (a) simply the mechanism of (b) which is most important in this lattice? Is it true that if (b) and (c) are both present, (c) will reinforce (b), so that a complete distinction between them is hardly possible?"'

It seems convenient to take the discussion of (a) and (b) first, and then to turn to (c) and its possible relation to the complex represented by (a) and (b).

The discussion of the (a)—(b) complex was started by Dorn, and continued by Stein and Hirsch. A considerable proliferation of terminology occurred, with the experimentalists appearing to favor what was called the "modified pseudo-Peierls mechanism." The following analysis is based on, rather than an account of, the discussion. We are concerned with the situation in which the energy of a straight dislocation lying along a low-index direction in the crystal varies with the period of the lattice as the dislocation moves through the crystal. The corresponding potential is called the Peierls potential. While it is different for edge dislocations and screw dislocations, the difference is not fundamental, and dislocations of either type might turn out to be more mobile than those of the other type. In face-centered cubic metals, or in the basal planes of hexagonal close-packed metals, dislocations can dissociate *on the glide plane*. This dissociation occurs in essentially the same way for edge and for screw dislocations, and reduces the Peierls potential. A more interesting situation arises if one or more partial dislocations can move *off the glide plane*. Then glide can occur only if the partial dislocations are pulled onto the glide plane again. The stress required to cause this is referred to as the "modified-," "pseudo-," or "modified pseudo-" Peierls stress. It may happen that the process occurs along the whole length of the dislocation without the dislocation passing through an activated configuration of maximum energy, or it may happen that an activated region has to be produced over a short length of dislocation, in which it is constricted. For the mechanisms of dissociation which have been considered in the b.c.c. lattice, only screw dislocations can dissociate. It seemed to Vreeland, Hirsch and Nabarro in private discussion that the converse situation might be present in the case of slip on second-order pyramidal planes in zinc, as described in Vreeland's paper. Here the *edge* components lie simultaneously in a prismatic and a basal plane. The Burgers vector is a + c, and it is only the edge component which can dissociate. Two dissociations are possible, and it is not clear which is relevant. Either a can move in the basal plane and c in the prismatic plane, in which case there is little or no long-range force between the two dislocations and no stacking faults are involved, or a partial of a can move in the basal plane, generating a stacking fault under the repulsion of the sessile residual dislocation. There is a clear possibility that thermally activated processes may lead to the formation of sessile configurations from edge dislocations, so that there is some possibility of explaining the observed low mobility of edge dislocations and the negative temperature coefficient of dislocation mobility.

The calculations of Bullough and Perrin, and of H. Suzuki, show that the actual separation of the partial dislocations in iron is likely to be very small, so that descriptions of "dissociation" are picturesque rather than precise.

In deciding which mechanism in the (a)–(b) complex is effective in b.c.c. metals, one must consider:

(1) the anisotropy in push-pull experiments;
(2) the failure of the Schmid law of resolved shear stress;
(3) the possibility that the mechanism of deformation in microstrain is the same as that in macrostrain; and
(4) the observed results on the mobility of edge dislocations.

Dissociation on the glide plane cannot explain (1) and (2), although it is compatible with (3). Dissociation off the glide plane can hope to explain (1) and (2), but, since only screw dislocations are assumed to dissociate, it cannot explain (4). There is thus still no wholly acceptable mechanism.

In discussing the Peierls mechanism, Alefeld suggested that the double kink in the Peierls mechanism might form in a corner region where the orientation of a dislocation passes from one close-packed direction, rather than along a straight section. It is expected that the energy required to form a kink in the corner would be lower than that required along the straight region, so that kinks would form at singular positions.

Escaig presented some comments on his calculations of the activation energy associated with the splitting model in b.c.c. metals. This was of the form [*Journ. de Phys.* 28:171 (1967)]

$$U = U_c + U_R$$

where U_c is $f(d)$, and $U_R \propto l_c \propto (1/\sigma)$. Here U_c = constriction energy, d = width of the splitting, U_R = recombination energy, l_c = critical length. At sufficiently high temperatures and sufficiently low stresses, $l_c \gg d$. The constriction term thus becomes negligible, and the use of elasticity to describe the dissociated core region is limited to the recombination energy per unit length. Escaig suggested that as long as this parameter is weakly stress dependent, the main result of his calculation should be valid, i.e., $U \propto \sigma^{-1}$. He reported good agreement with the experimental dependence of the flow stress on temperature, mainly in the high-temperature range (above 150°K). However, Hirsch disagreed with Escaig's method of calculation, which is based entirely on elastic considerations. Seeger drew attention to a paper on the thermally activated cross-slip in hexagonal metals [G. Schottky, A. Seeger, and V. Speidel, *Phys. Stat. Sol.* 9:231 (1965)] in which theoretical methods were developed which should also be suitable for other calculations involving dislocations with small extensions in their glide planes.

Regarding the actual geometrical behavior of dislocations in iron single crystals, Keh reported the following observations upon straining at 78°K after prestraining about 5% at room temperature:

(a) tangled dislocations at room temperature;

(b) upon restraining at 78°K in the "elastic" region there is a tendency for the dislocations to align in the screw orientation;

(c) upon continued straining at 78°K, the screws move.

There was no agreement on the influence of impurities on the low-temperature flow stress of b.c.c. crystals. Dorn claimed that the bulk of the evidence was that they were not important and that the work of Conrad, Christian, Dorn, and Owen all supported this view. Stein was less certain. The orientation effect is strongly influenced by the presence of impurities and the temperature dependence of the proportional limit is also affected. Both Dorn and Keh emphasized the influence of interstitial impurities on the athermal component of the flow stress, but Keh said that in a number of systems, such as Fe-N, alloying slightly decreases the temperature dependence of the thermal component of the flow stress. This was possibly because alloying facilitates the nucleation of kinks.

On the general question of the effect on the mechanical properties of point defects which produced tetragonal strains, Seeger called attention to the work of W. Frank, *Z. Naturforsch.* 27a: 365, 377 (1967); *Phys. Stat. Sol.* 18:459 (1966); *Phys. Stat. Sol.* 19:234 (1967). The principal results were:

(a) In regard to impurities one must consider those at a distance from the glide plane as well as those on the plane, and must remember that they can rotate under the stress of an approaching dislocation.

(b) The theory has been applied to irradiation hardening of Cu and Al, to divalent impurities associated with vacancies in alkali halides, and to α-Fe.

Seeger reported that according to Frank the available experimental results on the remaining temperature dependence of the flow stress of the iron with extremely small carbon contents are such that they could still be attributed to the presence of oxygen rather than to an intrinsic lattice effect. In this regard Stein pointed out that the temperature dependence in iron increased as the oxygen was reduced from 20 ppm to 3 ppm. Hirsch was not sure whether one can subtract the impurity effect from the intrinsic effect. It may be that the two energies must be combined to give the total energy, for example, in flow controlled by the formation of kinks, one might expect to find a strain rate given by $\dot\epsilon \sim \exp[-(H_k + H_i)/2kT]$, where H_i is the impurity interaction associated with the lateral motion of a kink. One can only subtract the impurity contribution to the stress if its contribution is solely to the athermal component.

The discussion as a whole left the impression that the dislocation theory of the flow stress was emerging from the doldrums in which it lay for some time. Major progress may well occur in the next few years in the confrontation and reconciliation of microscopic and macroscopic data on dislocation mobilities, and in the theory of the plastic properties of pure and slightly impure single crystals of body-centered cubic metals.

Part Four

HIGH-SPEED DISLOCATIONS

DISLOCATION DRAG MECHANISMS AND THEIR EFFECTS ON DISLOCATION VELOCITIES

Warren P. Mason

Department of Engineering and Applied Sciences,
Columbia, University

ABSTRACT

This paper will cover the recent phonon viscosity drag coefficient for which considerable experimental confirmation has been found. This mechanism tends to lose effectiveness at high velocities and would allow dislocation velocities approaching the speed of sound.

1 INTRODUCTION

There is considerable experimental evidence that, for the twinning process in tin and zinc, dislocations can travel at nearly

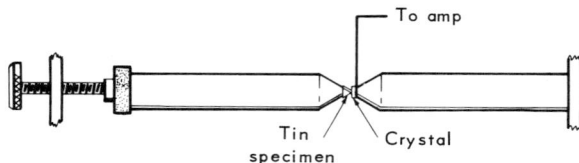

Fig. 1 Method for applying static stress to a tin crystal and crystal pick up device for measuring displacement.

the speed of sound. One of the first experiments indicating this result was that performed by Mason, McSkimin and Shockley[1] using the experimental apparatus of Fig. 1. Here, two glass rods, each 3 feet long, were tapered on the adjacent ends to a circle 1/8 inch in diameter. On the end of one, a tin specimen 1/8 inch in diameter on one end and tapered to a circle 1/16 inch in diameter on the other was soldered to a silver paste baked on the glass rod. The tin specimen was prepared from a 99.9 percent pure tin having a grain size in excess of 1/16 inch. Hence the specimen approaches a single crystal.

A quartz crystal 0.5 mm in thickness and 1/8 inch in diameter was soldered to the other glass rod. The sensitivity of this crystal was uniform from a few kilohertz to 5 MHz and by calibration it was found that a total force of 1000 dynes would produce a displacement of about 3/8 inches on the oscilloscope. Pressure was exerted on these two rods, lined up in a V block, by turning a screw on one end which puts pressure on the combination through a rubber pad. With this arrangement, about 15 photographs were taken of the voltage generated by the twinning process. Figure 2 shows two traces copied from the photographs by a tracing process.

During the time of observation (approx. 30 μs), the ultrasonic distrubance does not have sufficient time to reach the ends of the glass rods, and hence the rods act as very long transmission lines which have characteristic impedances which match the impedances of the tin crystal and the quartz crystal quite accurately. Hence no appreciable reflections occur at the interface. The transient voltage was always of a sign to indicate a relief of pressure on the crystal face and since it is small compared to the constant load, the deformation takes place at essentially constant load.

It can be shown that the velocity of yield $\dot{\xi}$ is equal to

$$\dot{\xi} = \frac{F}{dvs} = 3 \times 10^{-2} \text{ cm/sec} \tag{1}$$

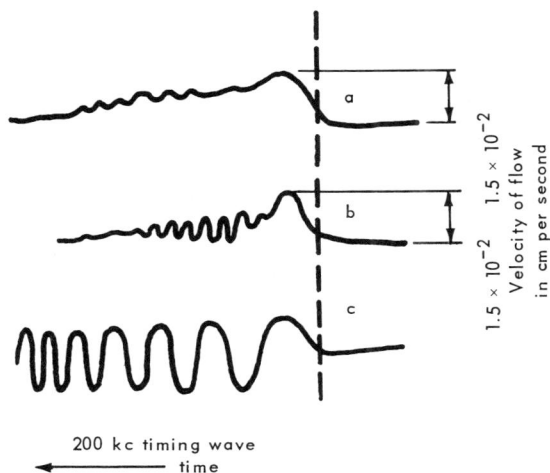

Fig. 2 Oscillograph traces showing voltages measured in the twinning process. Scale on right shows velocity of yield. Bottom curve shows timing wave.

where F is the total force on the crystal, d is the density of the tin specimen (7.1 g/cc), v is the longitudinal velocity (3.1 $\times 10^5$ cm/sec) and s is the cross-sectional area of 0.02 sq. cms. The equation has been evaluated for $F/s = 1000$ dyne/cm^2 which appears to be the average value measured. The velocity scale is shown by the abscissa of Fig. 2. The time scale is determined by the 200-kilocycle trace at the bottom of the curve. The 15 curves were divided into (a) type curves shown at the top and (b) type curves shown at the bottom of Fig. 2. The (a)-type curve rises to a velocity of 2×10^2 cms/sec in about 2 μs and dies off exponentially to a value $1/e$ in about 30 μs. The total volume change amounts to about 1.2×10^{-8} cc. For the (b)-type curve the area under the curve is about half this amount.

The fine structure is not the same for any two traces, which suggests that it is a characteristic of the twinning process itself. The distance moved in one cycle of the fine structure is about 2.5 Å for (a) and about 1/2 Å for (b). The period of the fine structure corresponds to 2.5 μs for (a) and 1 μs for (b).

The results are in general agreement with passage near the speed of sound (shear velocity 1.92 to 1.97 $\times 10^5$ cm/sec) of twinning dislocations from side to side of the specimen. The most likely path for the twinning dislocations is at 45° to the applied stress, since this is the direction of the greatest shearing stress. The distance across the specimen from one edge along a 45° path to the length is about 0.24 cms. Hence the velocity for a $2.5\text{-}\mu$s spacing between pulses is in the order of 1.9×10^5 cms/sec or under the slowest shear speed. The shorter time of (b) may be connected with a path less than the complete width of the specimen which can occur if the crystal grain does not extend through the complete specimen.

This procedure was applied to aluminum single crystals with negative results. Apparently, a twinning process is required before dislocations of this velocity can be produced. Koehler[2] found similar results in tin and zinc but no results for aluminum. A number of other experiments are reported by Friedel[3,4] who concludes "the creation of a twin neucleus requires much larger stresses than their subsequent growth. At the end of the nucleation, the twinning dislocations are then under stresses large enough to accelerate them to large speeds near the speed of sound." According to Cahn[4] the start of the twinning process in zinc requires stresses in the order of $\mu/10$ while the stress value to propagate the dislocation drops down to about $10^{-4}\mu$ after the dislocation has been accelerated to high speeds, where μ is the shearing modulus in the glide plane.

The question arises as to how the dislocation can move with velocities approaching the speed of sound for stresses of $10^{-4}\mu$ with the drag coefficients B that have been measured for low velocities, i.e., $B = 7 \times 10^{-4}$ dyne-sec/cm^2 for zinc.[5] It appears that the damping must decrease at the high velocities to permit the dislocation to continue with a speed near that of sound for applied stress levels of $10^{-4}\mu$. Two damping mechanisms that have this property are electron and phonon damping of dislocations proposed in earlier reports.[6,7] The purpose of this paper is to discuss the application of this type of damping to low- and high-velocity dislocation motions.

2 MEASUREMENTS OF DISLOCATION DAMPING MECHANISMS AT LOW VELOCITIES

Two methods have been used to measure the drag coefficient of dislocations in insulating and metallic crystals. One method makes use of the fact that dislocations in crystals are overdamped and at high frequencies produce an attenuation which becomes independent of the frequency. If this dislocation loss is separate from other sources of attenuation, the drag coefficient is given by the equation[8]

$$B = \frac{\overline{N} R \mu b^2}{2\alpha_L v} \tag{2}$$

where $\bar{N}$ is the number of dislocations per square cm, R an orientation factor which relates the strain in the acoustic wave to the average strain in the glide plane, μ is the shearing modulus in the glide plane which is $(c_{11} - c_{12} + c_{44})/3$ for face-centered-cubic crystals, b is the Burgers distance, α_L is the limiting attenuation at high frequencies in nepers per cm—1 neper is equal to 8.68 dB—and v is the sound velocity of the acoustic wave used to make the measurement.

The second method employs a pulse of high stress level and known duration to produce a movement of the dislocation. In one system employed by Johnston and Gilman[9] and by Gutmanas, Nadgorni and Stepanov,[10] the initial and final positions of the dislocation are determined by etch-pit techniques and the velocity is determined from the displacement distance divided by the duration of the stress pulse. In another system used by Dorn,[11] and his co-workers a strain-rate method is used in which an explosive charge actuates the crystal in a known direction and the time rate of displacement is measured. This method requires approximating the number of dislocations at each stage and is somewhat qualitative as is also the internal friction method.

Recently, the internal friction method has been used to evaluate the drag coefficient for single crystals of lead,[12] aluminum[8] and copper. The method involves measuring the attenuation of waves in known directions in the crystal and separating the dislocation component from the square-law component by combining two standard cannonical forms. The details of the separation are discussed in two earlier papers[8,12] and will not be repeated here. The resulting drag coefficients for lead, aluminum and copper are shown by Figs. 3, 4 and 5. These figures show also that the measurements are in fair agreement with the sum of three dislocation mechanisms that are discussed in Sec. 3.

Aluminum has been measured by the pulse-displacement rate method[11] and the result is shown by the dot-dash line in Fig. 4. While the drag coefficient measured by the pulse method may include some damping due to interaction with impurity atoms and other obstructions, the values are sufficiently similar to indicate damping by both phonons and electrons.

3 MECHANISMS FOR DISLOCATION DAMPING

A number of mechanisms have been proposed for the interactions of dislocations with lattice waves. One of the first was the

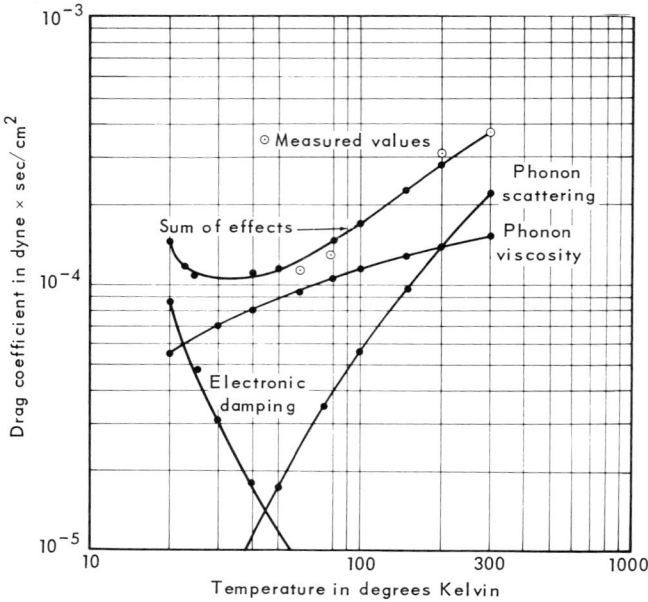

Fig. 3 Measured drag coefficient for lead (open circles) and the three components of dislocation drag.

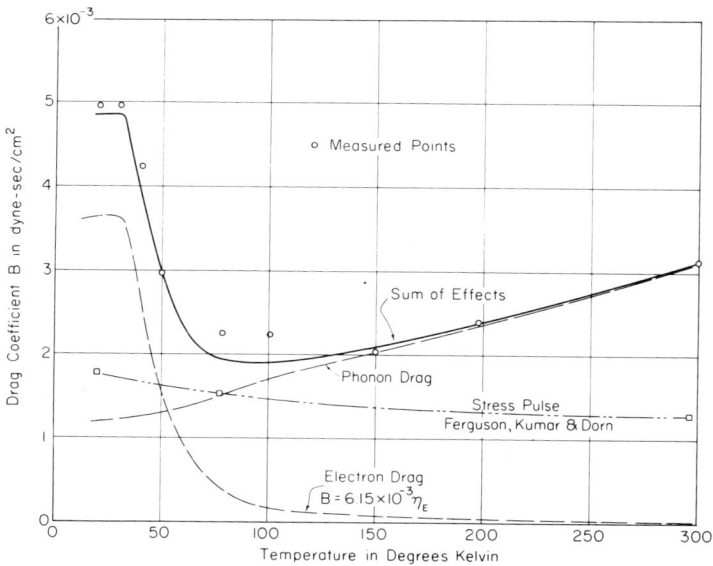

Fig. 4 Dislocation drag coefficient for aluminum. Dashed curves show two components of damping. Dot dash curve shows measurements of Ferguson, Kumar and Dorn by stress pulse method.

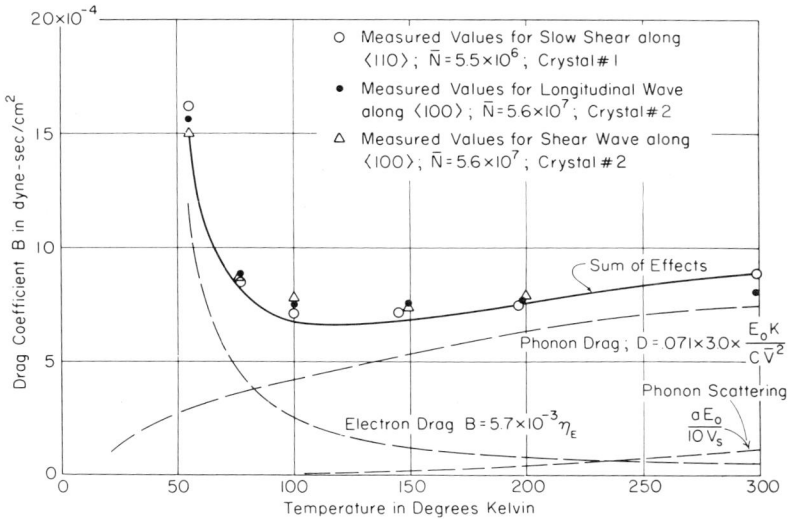

Fig. 5 Dislocation drag coefficients for copper and the three components of dislocation drag.

mechanism of Leibfried[13] in which the drag results from the difference in the phonon scattering from the front and back of the moving dislocation. This results in a drag coefficient B equal to

$$B = \frac{aE_0}{10\,v_s} \tag{3}$$

where a is the lattice constant, E_0 is the thermal energy density, and v_s is the shear velocity. A somewhat similar drag coefficient for electrons has been proposed by Holstein.[14] This is similar in form to Leibfried's calculation if we replace E_0 by the Fermi energy and v_s by the Fermi velocity of the electrons. This source gives a drag coefficient independent of the temperature.

In Refs. [8] and [12] it was pointed out that since a dislocation is surrounded by a strain field which changes at a given point as the dislocation moves along, one should expect a conversion of dislocation energy to thermal energy by the same mechanisms that damp moving sound waves, namely, electron and phonon viscosity. By determining the energy loss at each point and integrating over the volume surrounding the dislocation, it is shown that the drag on the dislocation was proportional to the velocity and equal to

$$B = \frac{b^2 \eta}{8\pi a_0^2} \tag{4}$$

for a screw dislocation. Similar results hold for an edge dislocation with the addition of an effect for a compressional viscosity. This, however, is small and can usually be neglected.[12]

This drag coefficient depends critically on the radius a_0, on which there are two limitations. The first of these has to do with the size of the region around the dislocation which can exhange energy with the surrounding phonon field. It has been suggested that material inside a radius equal to the mean-free phonon path should be excluded from the calculation. It was shown first by Suzuki, Ikushima and Aoki[15] that this suggestion holds only when the dislocation is moving with the speed of sound. For slower speeds there is more time to interchange energy between a suddenly stressed region and the phonons. Suzuki et al suggest that the radius should be

$$r = \frac{\bar{l}u}{\bar{v}} \tag{5}$$

where $\bar{l}$ is the mean-free phonon path, u is the velocity of the dislocation and $\bar{v}$ is the Debye average sound velocity. Hence it is only if u is equal to the sound velocity, that $\bar{l}$ determines the excluded radius.

This result can be seen from the fact that in the neighborhood of the dislocation, the strain changes discontinuously when the extra plane moves over by the Burgers vector b. To determine whether the material at the edge of the dislocation should be included in the calculation, the criterion is whether the phonon modes can be equilibrated in a time less than that between jumps. If they cannot, then energy is not lost to the phonons but is returned to the dislocation. This criterion results in the inequality

$$\frac{r}{t} = \frac{r}{b/u} \leq 1 \tag{6}$$

where t is the time the strain remains constant. If all the material is to be included up to the dislocation edge

$$u\tau = \frac{u\bar{l}}{\bar{v}} \leq b \tag{7}$$

For all the measurements obtained by internal friction methods this criterion is satisfied.

The other limitation is that the concept of the phonon as an acoustic wave transmitted through the medium may breakdown sufficiently close to the dislocation because of the nonlinear terms in the elastic energy. An estimate of the relative amounts of energy stored in the various terms can be obtained from the expression

$$ w = \tfrac{1}{2}! \, c_{ijkl} S_{ij} S_{kl} + \tfrac{1}{3}! \, C_{ijklmn} S_{ij} S_{kl} S_{mn} + \cdots \tag{8} $$

For measured values of third-order moduli of metals,[16] the second-order terms have energies about 50 percent of the first-order terms for strains of about 20 percent. This corresponds to a value of $a_0 \doteq (3/4 \, b)$ for a screw dislocation. Using this value of a_0, the drag coefficient is

$$ B = 0.0706 \, \eta_s \tag{9} $$

It is well known that acoustic waves are damped principally at low temperatures by the presence of free electrons in a metal. The action is equivalent to a viscosity, as was first pointed out in Ref. [17]. For a free-electron model the viscosity was shown to be

$$ \eta_e = \frac{9 \times 10^{11} \hbar^2 (3\pi^2 N)^{2/3}}{5 e^2 \rho} \tag{10} $$

where $\hbar$ is Planck's constant h divided by 2π, N is the number of free electrons per cc, e is the charge 4.8×10^{-10} esu, and ρ is the electrical resistivity in ohm-cms. This value gives good agreement with experiment for those materials for which the Fermi surface approximates a sphere, notably copper, gold, silver, sodium and potassium. If, however, the Fermi surface differs substantially from a spherical surface, the amount of damping becomes anistropic and may differ from the free-electron value.

For aluminum, for which data are given by Figs. 6 and 7, the measured values agree with Eq. (10) provided that we assume 1.43 electrons per atom. This results in the equation

$$ \eta_e = \frac{9 \times 10^{11} \hbar^2 (4.3 \pi^2 N)^{2/3}}{5 e^2 \rho} = \frac{1.64 \times 10^{-8}}{\rho} \quad \text{(aluminum)} \tag{11} $$

since $N = 6.11 \times 10^{22}$ atoms per cc for aluminum. For copper the data of Fig. 8 shows that Eq. (10) is in good agreement with the square-law attenuation measured at 150 MHz.

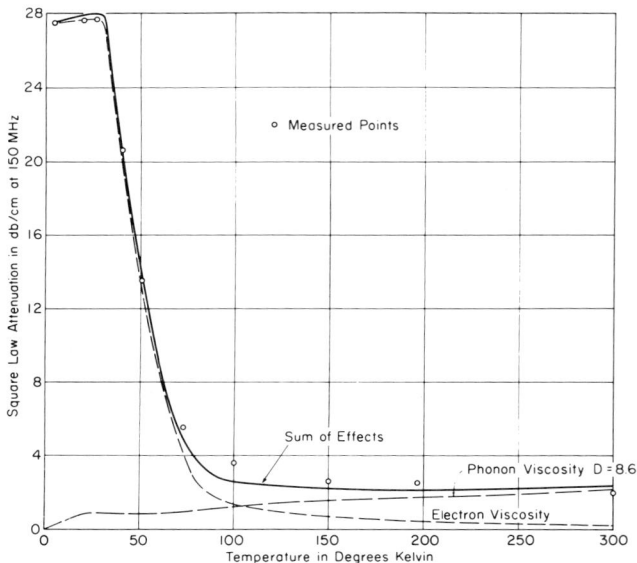

Fig. 6 Square law attenuation for slow shear wave ($v = 3.11 \times 10^5$ cms/sec) in single crystal aluminum. Dashed curves show electronic and phonon damping terms. D is evaluated as 8.6.

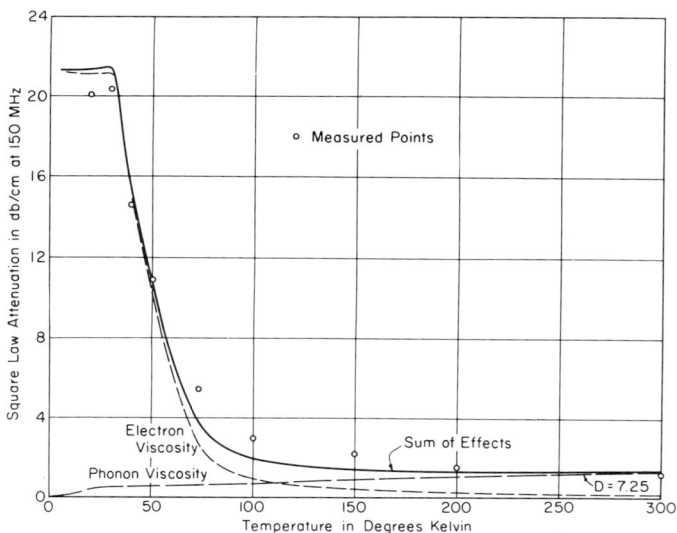

Fig. 7 Square law attenuation for fast shear wave ($v = 3.41 \times 10^5$ cms/sec) in single crystal aluminum. Dashed curves show electronic and phonon damping terms. D is evaluated as 7.25.

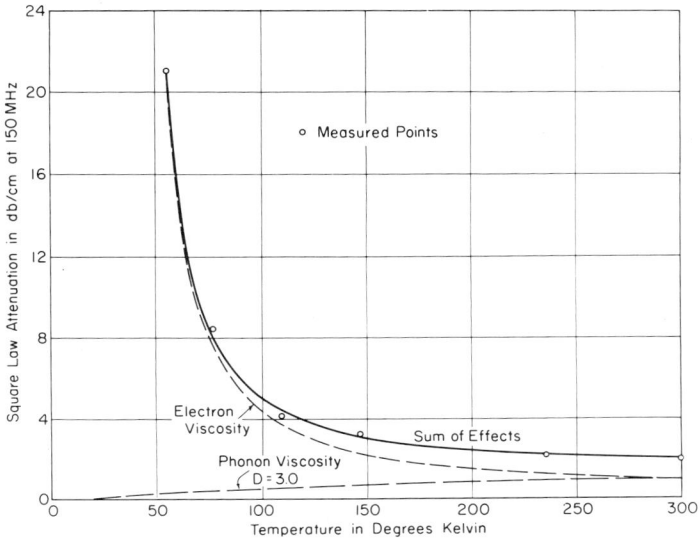

Fig. 8 Square law attenuation for slow shear wave in single crystal copper plotted as a function of the temperature. Dashed curves show electronic and phonon damping terms. D is evaluated as 3.0.

The attenuation measurements of Figs. 6, 7, and 8, dashed lines, give the attenuation associated with the electron viscosities of Eqs. (10) and (11). The difference between the solid and dashed lines must represent the effect of phonon damping on the shear waves. The results are in good agreement with the theory of phonon viscosity discussed previously. This indicates that the value of the phonon viscosity η_p should be equal to

$$\eta_p = D\left(\frac{E_0}{C} \frac{K_p}{\bar{v}^2}\right) \tag{12}$$

where E_0 is the thermal energy density, C is the specific heat per unit volume, K_p is the thermal conductivity due to phonons and $\bar{v}$ is the Debye average velocity. The thermal conductivity due to phonons can be separated from the total thermal conductivity—most of which is due to electrons—by measurements at two different impurity levels[17] or by calculations such as those due to Leibfried and Schlomann.[18] The nonlinearity constant D can be calculated from third-order moduli[19] if these are known. The measurements of Figs. 6, 7, and 8 give these values directly for shear waves in aluminum and copper.

TABLE 1 Quantities Necessary to Calculate the Phonon Viscosity and
Phonon Damping of Aluminum

Temp. °K	$\dfrac{E_0}{\rho C_v}$ in °K	K_p (watts/cm^2)	$\dfrac{E_0 K}{\rho C_v \bar{v}^2} \times 10^3$	$B \times 10^4$
300	187	0.356	5.44	31.5
200	104	0.535	4.55	26.2
150	66.5	0.715	3.88	22.2
100	34.4	1.07	3.01	17.3
77	23.2	1.39	2.63	15.2
66	18.4	1.62	2.43	14.0

$\theta = 425°K$; $\bar{v} = 3.5 \times 10^5$; $B = (b^2 D/8\pi a_0^2)(E_0 K_p/C\bar{v}^2)$; $D = 8.15$; $a_0 = 3/4\ b$

TABLE 2 Quantities Necessary to Calculate the Phonon Viscosity and
Phonon Damping of Copper

Temp. °K	$\dfrac{E_0}{\rho C_v}$ in °K	K_p (watts/cm^2)	$\dfrac{E_0 K}{\rho C_v \bar{v}^2} \times 10^3$	$B \times 10^4$
300	200	0.125	3.7	7.85
200	114.8	0.187	3.17	6.7
150	74.5	0.25	2.74	5.8
100	39.6	0.375	2.2	4.65
77	26.2	0.486	1.88	3.98
58	17.1	0.64	1.64	3.48

$\theta = 343°K$; $\bar{v} = 2.6 \times 10^5$; $B = (b^2 D/8\pi a_0^2)(E_0 K_p/C\bar{v}^2)$; $D = 3.0$; $a_0 = 3/4\ b$

With these values determined for aluminum and copper, the phonon damping coefficients are shown by Tables 1 and 2. These are shown plotted in Figs. 4 and 5. The electron viscosity also produces a damping of dislocations by the same mechanism. For electrons there is nothing corresponding to a compressional viscosity; thus Eq. (4) holds for both screw and edge dislocations. The nonlinearity radius a_0, however, is determined by the non-linearity effects of the Fermi surface rather than the elastic constants. Actual determinations for the ratio of the nonlinearity radii for phonons and electrons can be obtained for copper and aluminum by comparing the values of η_p to η_e at the temperature for which they are equal. For this case

$$\frac{a_{oe}}{a_{op}} = \sqrt{\frac{\eta_e}{\eta_p}} \tag{13}$$

For copper this ratio appears to be about 3.5 while for aluminum it is 3.4. The dashed curves of Figs. 4 and 5 are plotted on the assumption that $a_{op} = (3/4)\,b$ while a_{oe} is determined by the above ratios. The experimental points are obtained by assigning values $\overline{N}$ of dislocations per sq/cm to match the theoretical value at room temperature. For lead the number is 10^7 for aluminum $\overline{N} = 1.6 \times 10^7$/cc. The values for copper are shown by the figure. While these values are arbitrary, they are close to those found by etch-pit methods.

4 INSTABILITY OF DRAG COEFFICIENTS

The pulse method for measuring velocities produces much higher values than the internal friction method. Figure 9 shows the measurements of Johnston and Gilman[9] for lithium fluoride plotted on a linear scale. After subtracting the force required to break the dislocation away from obstructions, the indicated drag coefficient is 7×10^{-4} dyne-sec/cm^2. The highest velocity attained is 4.5×10^4 cm/sec. Since the thermal relaxation time

$$\tau = \frac{3K}{C\bar{v}^2} = 4.7 \times 10^{-13} \text{ sec} \tag{14}$$

for lithium fluoride, this velocity gives the product

$$4.5 \times 10^4 \times 4.7 \times 10^{-13} = 2.11 \times 10^{-8} < b = 2.86 \times 10^{-8} \tag{15}$$

which is less than the Burgers distance b for lithium fluoride. Hence the velocity is not large enough to produce instability according to the criterion of Eq. (6). The velocities attained for aluminum were less. Hence, so far, at least, the velocities attained are not high enough to test the criterion for instability of the drag coefficient.

For a twinning process, a very high stress is required to start the process and this can accelerate the dislocation beyond the instability point. After this, a much smaller stress is required

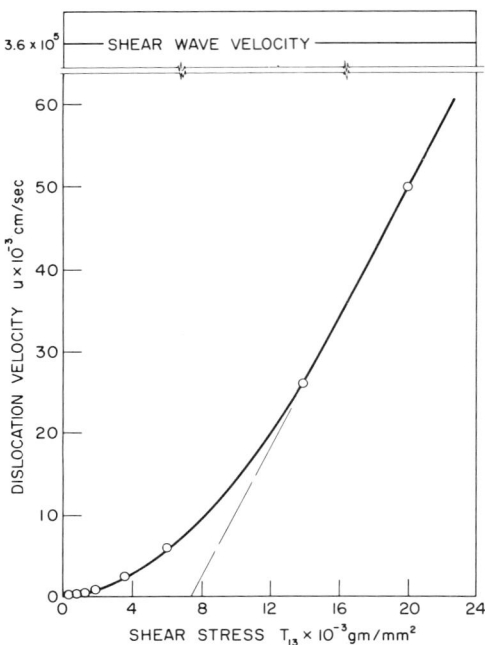

Fig. 9 Velocity of edge dislocations plotted as a function of the applied shear stress (after Johnston and Gilman).

to keep the dislocation at a speed close to the speed of sound. Since some numerical values [3,4] have been given for zinc, this crystal will be considered. Experimentally,[5,18] the drag coefficient B has been found to be about 7×10^{-4} dyne-sec/cm^2 at room temperature. Since the shearing modulus in the glide plane—the plane perpendicular to the hexagonal axis—is about 6×10^{11} dyne/cm^2, a stress of $(1/10)\mu = 6 \times 10^{10}$ dynes/cm^2 will give a velocity of

$$\frac{T_{13}\,b}{B} = \frac{6 \times 10^{10} \times 2.66 \times 10^{-8}}{7 \times 10^{-4}} \doteq 2 \times 10^6 \text{ cm/sec} \qquad (16)$$

and hence the drag coefficient is not the limiting value in attaining the sound velocity which is 2.9×10^5 cm/sec for this mode.

However, if the shearing stress T_{13} drops to $10^{-4}\,\mu = 6 \times 10^7$ dyne/cm^2, this is not large enough to maintain the velocity of sound with the drag coefficient of 7×10^{-4}. In fact, the velocity would be limited to 2.3×10^3 cms/sec. If now we assume that all of the drag is due to phonon and electron viscosity, as discussed in the previous section, we can show from the resistivity—6.6×10^{-6} ohm-cms at $300°K$ and $N = 6.56 \times 10^{22}$ atoms per cc—that the electronic viscosity is in the order of 2.1×10^{-3} *poises* at room temperature. From the ratio of drag coefficient to electron viscosity of about 6×10^{-3} found for aluminum and copper, the electron drag coefficient should be about $B_e = 1.2 \times 10^{-5}$ dyne sec/cm^2. This term will not decrease at the sound velocity since the product of the velocity by the electronic relaxation time

$$\tau = \frac{9 \times 10^{11} m}{Ne^2 \rho} = 8.5 \times 10^{-15} \text{ sec} \tag{17}$$

is still less than the Burgers distance 2.66×10^{-8} cms for zinc.

If the remainder of the measured drag is due to phonon viscosity, it is readily shown that this part will relax down to a small value at the speed of sound. This follows since the phonon relaxation time

$$\tau = \frac{3K_p}{C\bar{v}^2} = \frac{3 \times 0.1 \times 10^7}{2.76 \times 10^7 (2.7 \times 10^5)^2} = 1.5 \times 10^{-12} \text{ sec} \tag{18}$$

The product of this term by the sound velocity 2.5×10^5 cm/sec is 3.75×10^{-7} which is greater than the Burgers distance 2.66×10^{-8} cms. Under these conditions the excluded radius increases to the phonon mean-free path—Eq. (5)—which is

$$\lambda = \bar{v}\tau = 2.7 \times 10^5 \times 1.5 \times 10^{-12} \doteq 4 \times 10^{-7} \text{ cms} \tag{19}$$

Hence the phonon drag coefficient decreases to

$$B = 7 \times 10^{-4} \left[\frac{(3/4) b}{\lambda} \right]^2 = 1.75 \times 10^{-6} \text{ dyne sec/cm}^2 \tag{20}$$

The sum of the two effects will limit the velocity to 2.5×10^5 cms if the driving force is 1.3×10^8 dyne/cm^2 which is $2 \times 10^{-4}\mu$. This is within the experimental error of that measured for the value found after the dislocation is accelerated to high speeds.

On the other hand, if the two scattering coefficients discussed in Eq. (3) also exist then they will have values of 9×10^{-5} (phonon), 1.4×10^{-4} (electron). It requires a force about 40 times the observed value to produce a velocity close to that of sound. It is the author's opinion that there is no reason for the electron type of drag coefficient to decrease with sound velocity, since this is very small compared to the Fermi electron velocity. Hence, the possibility exists that this value has been overestimated. However, for the phonon drag coefficient, discussions in this meeting indicate that it also vanishes for dislocation speeds equal to the speed of sound.

ACKNOWLEDGMENT

This research was supported by Contract 266(91) with the Institute of Fatigue and Reliability of Columbia University.

REFERENCES

1. Mason, W. P., H. J. McSkimin, and W. Shockley: *Phys. Rev.* 23:1213 (1948).
2. Koehler, J. S.: *J. Phys. Soc. Japan* 10: no. 8, 669 (1955).
3. Friedel, J.: "Dislocations," p. 182, Pergammon Press, 1964.
4. Cahn, R. W.: *Advances in Physics* 3:363 (1954).
5. Ferguson, W. G., A. Kumar, and J. E. Dorn: *J. Appl. Phys.* 38:1863 (1967).
6. Mason, W. P.: *J. Acous. Soc. Amer.* 32:458 (1960); W. P. Mason: *J. Appl. Phys.* 35: 2779 (1964).
7. Mason, W. P.: *Applied Physics Letters* 6:111, (1965).
8. Mason, W. P., and A. Rosenberg: *Phys. Rev.* 151:434 (1966).
9. Johnston, W. G., and J. J. Gilman: *J. Applied Phys.* 30:129 (1959).
10. Gutmanas, E. Y., E. M. Nadgornyi, and A. V. Stepanov: *Sov. Phys. Solid State* 5:743 (1963).
11. Dorn, J. E., J. Mitchell, and F. Hauser: *Exp. Mech.* 5:353 (1965).
12. Mason, W. P., and A. Rosenberg: *J. Appl. Phys.* 38:1929 (1967).
13. Leibfried, G.: *Z. Physik* 127:344 (1950).
14. Bommell, H., and B. R. Tittman: *Phys. Rev.* 151:178 (1966).
15. Suzuki, T., A. Ikushima, and M. Aoki: *Acta Met.* 12:1231 (1964).
16 Hiki, Y., and A. V. Granato: *Phys. Rev.* 144:411 (1966).
17. Mason, W. P.: *Phys. Rev.* 97:557 (1955).
18. See also T. Vreeland, Jr.: This colloquium, p. 529 (values of $B = 8 \times 10^{-4}$ for copper and 7×10^{-4} for zinc at room temperature are given).

DISCUSSION *on Paper by W. P. Mason*

A. ROSENFIELD: Is there evidence that viscous damping effect in b.c.c. metals is of less relative importance than in f.c.c. metals? It it possible to calculate this?

W. MASON: I do not know of any experiments. It should be possible to calculate it if the third-order moduli and the thermal conductivity are known. The electron damping should be inversely proportional to the resistivity. The constant will depend on the shape of the Fermi surface and on the effect of the deformation potentials. However, it is easily measured for a given crystal.

G. ALERS: Your graphs showed that phonon viscosity was the cominant contribution to the attenuation at room temperature. You also showed that the phonon viscosity could be calculated from the third-order elastic constants. There are now complete-sets of third-order constants for Si, Ge, Cu, Au, Ag, and NaCl in the literature. How well do your calculated attenuations agree with the measured attenuations for these materials?

W. MASON: A recent paper, *J. Acous. Soc. Amer.*, compares the results for the six crystals, Si, Ge, NaCl, KCl, MgO, and Yttrium iron with measured third-order moduli. The results are always within a factor of 2 and for the well-measured crystals, usually within 50 percent. For copper the added attenuation for longitudinal waves is close to the calculated one. For shear waves, the attenuation is not so close. For Granato-Hicki measurements, the calculated value is about three times the measured. For the Breazeale-Gauster constants the agreement is closer. It is believed that these constants are not accurate enough to give a very good result. The silver and gold attenuation measurements are not known.

A. SEEGER: Mason says B at low temperature is very large due to electron damping. If it is that big, dislocations would not move and no modulus defect should be observed.

G. ALERS: A modulus defect is observed so the damping cannot be as large as Mason predicts. As a matter of fact, the measurements by myself and Thompson give both the damping and the modulus defect from which B can be uniquely determined. We deduced B as a function of temperature from our data and showed that B did not rise at low-low temperatures.

W. MASON: The results of Alers and Thompson that the attenuation difference between pinned and unpinned dislocations gets less at lower temperatures while the dislocations remain mobile—all while the electronic component is increasing—can be accounted for provided that the loop length is decreasing at the low temperatures. This effect has been observed at low frequencies by the measurements of Thompson and Holmes and is discussed in detail by the author in an invited speech at the 5th International Congress—Minneapolis 1966 (See book on that conference).

The attenuation varies as the fourth power of the loop length whereas the modulus varies only as the second power. Hence the attenuation becomes smaller at low temperatures in the face of an increase in B at low temperatures. If the product $(\mu l_0^2 B/6\mu b^2)$ is less than unity, then the modulus will not change appreciably. For a ratio of 400 to 1 between the resistivity at room temperature and the resistivity at 4.2°K, the drag coefficient will be around 2×10^{-2}. From the above reference $l_0 = 5 \times 10^{-5}$ cms;

$6\mu b^2 = 1.8 \times 10^{-3}$. Hence $(wl_0^2 B/6\mu b^2) = 0.8$ and the modulus defect will not change much up to the measuring frequency.

The model has been criticized since the radius of action found at low velocities is less than the mean-free path. This criticism can be avoided by going to a kink model with the kink larger than the mean-free path. Then, as shown by Seeger in his discussion of the kink model in the Prandtl, Dehlinger, Frenkel, Kontorova approximation, the phonons can approach the dislocation and exchange energy over the whole kink length up to a radius determined by the nonlinearity radius. For a fast moving dislocation, the interaction is limited to the mean-free path.

T. SUZUKI: I wonder if you have any idea to explain a rather large effect of solute atoms in copper on B. Dr. Ikushima and Dr. Kaneda,* and myself[†] studied the alloying effect on B, as shown in Figs. 1 and 2. It will be important to note that Pt and Pd gave almost equal effect on B, because these solutes give a nearly equal amount of misfit to the matrix, but they have a large difference in mass, i.e., the ratio of mass is about 2:1. This experiment* seems to give a negative answer to theoretical predictions made by Takamura[‡] and Lyubov,[§] in which the importance of local phonons with solute atoms will be expected to give the lattice friction to moving dislocations.

W. MASON: Impurities could affect the damping of a dislocation because they can add their strain field to that of the dislocation. Since the energy loss is proportional to the square of the strain rate, an integration across a given region containing impurities or other dislocations would be larger than for that over a pure material region.

A. GRANATO: (1) It is my understanding, that in the case of phonon damping at least, your formula does use the low-temperature side of the thermal conductivity. Wasn't this the basis of the explanation by Mitchell for the maximum of the drag constant she observed at low temperature in LiF? This kind of explanation puzzles me since the low-temperature side of the thermal conductivity maximum is determined by phonon interaction with defects at large distances, such as the surface of the specimen, for example, and not with interactions near the core of the dislocation.

---

*Ikushima, A., T. Kaneda: unpublished.

[†]Suzuki, T., A. Ikushima, M. Aoki: *Acta Met.* **12**: 1231 (1964).

[‡]Takamura, J., T. Morimoto: *J. Phys. Soc. Japan* **18**: Suppl. I, 28 (1963).

[§]Lyubov, B. Ya., G. M. Chernizer: *Soviet Phys. Doklady*, **10**, 373 (1965).

(2) It has never been clear to me that your "phonon viscosity" mechanism and the "phonon scattering" mechanism were distinct mechanisms. My impression is rather that your formula and Leibfried's formula are too different temperature dependences predicted constitutes a discrepancy between them. Do you have any comments on this?

W. MASON: (1) For all the metals only the slope on the right-hand side of the temperature peak is of interest since at lower temperatures, electrons will be the dominant damping mechanism. For lithium fluoride measurements were carried down to 20°K which is at the height of the temperature thermal conductivity peak. The theoretical curve should stop at this point.

(2) I agree that the phonon viscosity mechanism and the phonon scattering mechanism should be different aspects of the same phonon reaction picture. One pertains to the dissipation of energy in the medium while the other has to do with the radiation pressure from the phonons in the medium. The two, however, have different temperature variations since the radiation pressure falls off as the number of phonons in the medium decreases, whereas the energy loss in the medium varies approximately as FKT where F is a factor which goes from 1/4 to unity as the temperature increases. Since K is usually inversely proportional to the temperature, then the energy loss decreases much less rapidly than the absolute temperature. A difference also occurs in the velocity response for dislocations since the radiation pressure difference should be directly proportional to the velocity, whereas the energy loss may decrease due to the effective increase in radius of the excluded region about the dislocation.

CROSS-SLIP AND TWINNING IN B.C.C. METALS

A. W. Sleeswyk

University of Groningen

ABSTRACT

It is shown that the $1/2 <111>$ dislocation will decompose according to the equation $1/2 <111> \rightarrow 1/4 <111> + 1/4 <111>$. The $1/2 <111>$ edge or screw dislocation on $\{112\}$ will have its two $1/4 <111>$ partials interconnected by an extrinsic fault, while the fault on $\{110\}$ will be intrinsic. The reasoning leading to this model is based on the hard-sphere model for $\{112\}$, and on the stability of the b.c.c. lattice relative to close-packed structures for $\{110\}$. Extrinsic faults of this type may occur on planes other than $\{112\}$, of which the indices are given by: $\{hkl\} = \{011 + n(112)\}$. A distinction between "easy" and "difficult" cross-slip is introduced on the basis of the relative energy of the screw dislocation. Following Hirsch and Kroupa and Vitek, it is assumed that the rate-determining process is "difficult" cross-slip; it is demonstrated that the activation energy as a function of stress, determined experimentally by Sleeswyk and Helle, is in good agreement with the type of expression obtained by Vitek and Kroupa. Twinning will occur if the Peierls force exceeds a critical value. A discussion of the various stages of twinning in connection with their nucleation, and accomodation by emissaries, is given.

1   INTRODUCTION

In this paper concepts will be used such as the hard-sphere model or the extended dislocation in the b.c.c. lattice, of which it is well known that at best they possess only partial validity. Nevertheless, the fact that they are used can be explained only by their being so simple that they can be handled with confidence, which leads to certain conclusions that can be compared with experimental findings. These condluctions might never have been reached if the initial concepts had been more sophisticated. The resulting ideas on dislocation movements in the b.c.c. lattice are used as a guide for some of the experimental work that is starting now in the physical laboratory at the University of Groningen.

2   THE DISSOCIATION OF A PERFECT
DISLOCATION ON PARALLEL PLANES

Recently, it was shown by Ogawa[1] and Ogawa and Maddin,[2] that if the assumption is made that a $1/2 <111> \{112\}$ edge dislocation can dissociate spontaneously on three adjacent $\{112\}$ planes, then the resulting configurations of edge partials are of the same type as those derived earlier[3] for the $1/2 <111>$ screw dislocations decomposed on $\{112\}$ planes. The suggestion seems a valuable one, and it is, therefore, unfortunate that no explanation was given as to why, or under what conditions, this unorthodox process would take place. In addition, it would seem, at least at first sight, that the claim by Ogawa and Maddin that they observed emissary dislocations—i.e., $1/2 <111>$ dislocations on every third $\{112\}$ plane—produced by a deformation twin is difficult to reconcile with the idea that $1/2 <111>$ edge dislocations are decomposed in three $1/6 <111>$ partials lying on three adjacent $\{112\}$ planes. It is not clear how such dislocations could ever be emitted by a twin, or what would prevent them from coagulating into a single large twin once they would have been created. We shall explore somewhat further the question concerning the conditions under which the spontaneous dissociation of a perfect dislocation on parallel stacking fault planes is to be expected.

First, we consider the simple case of a two-dimensional lattice in which the only interatomic interaction is that between nearest-neighbors. In the example given in Fig. 1, the stacking fault vector f is equal to 1/3 of the Burgers vector b of the perfect dislocation. The example was chosen to resemble as closely as possible the situation of the dissociating $1/2 <111> \{112\}$ dislocation in the three-dimensional b.c.c. lattice. If we assume that the dissociation of the dislocation first produces a stacking fault on the glide plane by:

$$1/2 <11> \rightarrow 1/3 <11> + 1/6 <11> \cdots (1/2 <111> \rightarrow 1/3 <111> + 1/6 <111>)$$

$$(1)$$

it would seem that there are two obvious conditions for further accomodation of the extra half-plane in the lattice by the spontaneous creation of a stacking fault with f-vector f´ on the plane adjacent to the glide plane:

1. The remaining thickness of the partly accomodated half-plane should exceed the f-vector f´, or:

$$f´ < b - f \qquad (2)$$

otherwise it would require extra energy to create the partial on the second stacking fault plane.

2. An alternate means of creating a stacking fault on the glide plane would be by a displacement over b − f; not only must the stress required for this alternate process exceed that for creating the f stacking fault—this condition was clearly defined first by Barrett[4]—but it must also exceed the stress needed for creating the stacking fault f´ adjacent to the glide plane.

The first condition is obviously fulfilled in the example; in addition, it will be assumed that the second condition is also fulfilled. This allows a second $1/6 <11>$ partial to glide away from the half-plane rim.

The remaining configuration possesses a Burgers vector of $1/6 <11>$; if a third partial were to glide away, creating a third stacking fault, the accomodation of the original perfect dislocation in the lattice would be complete—the situation is shown in Fig. 1— as the Burgers vector of the configuration then remaining would be equal to zero. This half-plane configuration would, therefore, be insensitive to shear stresses; only the twin boundaries would exert

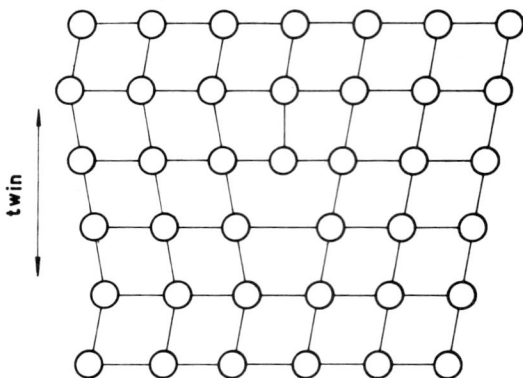

Fig. 1 Complete accomodation of a perfect $\frac{1}{2} < 11 >$ dislocation by a twin. The two-dimensional lattice is an analogue of the $\{110\}$ plane in the three-dimensional B.C.C. lattice.

a force F_T on it. The completely accomodated dislocation would still require a Peierls force F_P to move it through the lattice. One may distinguish two possibilities:

1. $F_T < F_P$. The configuration cannot move, the three partials may glide away freely under the influence of the applied shear stress, and a three-layer twin of infinite length may be created.

2. $F_T \geq F_P$. It is not possible that a third partial can detach itself in this case; if it somehow would do so, the mobile configuration would immediately catch up with it. The half-plane configuration with Burgers vector $\frac{1}{6} < 11 >$ may move under the influence of the shear stress, and the force F_T exerted by the twin boundaries; in the dislocation formalism the configuration may be rendered as a $\frac{1}{3} < 11 >$ partial combined with a $-\frac{1}{6} < 11 >$ partial on an adjacent plane. The original perfect dislocation has dissociated on two adjacent stacking faults, according to the dislocation reaction equation:

$$\frac{1}{2} < 11 > \;\rightarrow\; \frac{1}{6} < 11 > + \; \frac{1}{6} < 11 > + (\frac{1}{3} < 11 > - \; \frac{1}{6} < 11 >) \tag{3}$$

and the extended $\frac{1}{2} < 11 >$ dislocation is still completely mobile.

In conclusion, it may be stated that a spontaneous dissociation of a perfect edge dislocation in a two-dimensional analog of a $\{110\}$ b.c.c. plane on two adjacent stacking faults may indeed be possible. Depending on whether F_T or F_P is larger, a shear stress may either cause the extended dislocation to glide through the lattice, or cause it to dissociate further on a third stacking fault, resulting in the creation of a three-layer twin.

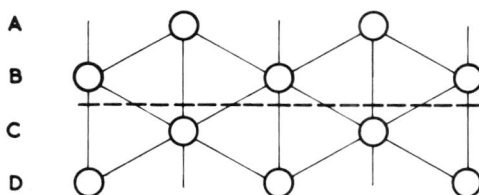

Fig. 2 Projection of the b.c.c. lattice on {111}, showing positions of interatomic bonds relative to the {112} plane.

3 THE DECOMPOSITION OF THE ½<111> DISLOCATION ON {112}

The structure presented in Fig. 1 is not really representative of the b.c.c. lattice. Real {112} b.c.c. planes are interconnected by atomic bonds as shown in Fig. 2; interatomic bonds not only extend from one {112} to the nearest {112} planes, but also to the next-nearest {112} planes. Thus, every atom in the {112} plane is linked to atoms on four different other {112} planes.

In metals, stacking faults arise because of the multiple bonding; directional bonds (if they exist) such as presupposed in Fig. 1, do not determine the structure of a metal. This implies that the discussion of stacking fault stability in metals must take into account the multiple nearest-neighbor relationships. For example, if the upper half of the crystal in Fig. 2 glides relative to the lower half, as indicated by the glide plane, the nearest-neighbor relationships between the atoms in the layers B and D, B and C, A and C, must be considered in a hard-sphere model. In the projection given in Fig. 3, this is a realized by ascribing *two* atomic radii to each atom; for the sake of clarity, only half of the atoms are depicted in this manner. In analogy with the preceding section, faults on one, two and three adjacent {112} planes in a hard-sphere model of the b.c.c. lattice will be examined. In the right-hand side of Fig. 3 the stacking possibilities of a single layer of atoms are illustrated by means of the curve described by the center of a sphere as it rolls over two lower {112} layers of spheres. The position of the top layer is stable when the atoms are at the deepest cusps in the rolling trajectory, whereas the stacking fault positions are given by the shallow cusps, in accordance with the well-known model first presented by Barrett (*loc. cit.*). The stacking fault vector $f = \frac{1}{6} <111>$ generally accepted for intrinsic faults on {112} would be exact if the fault had been created by a pure twinning shear on a single {112} plane; it is a fair approximation in the hard-sphere model.

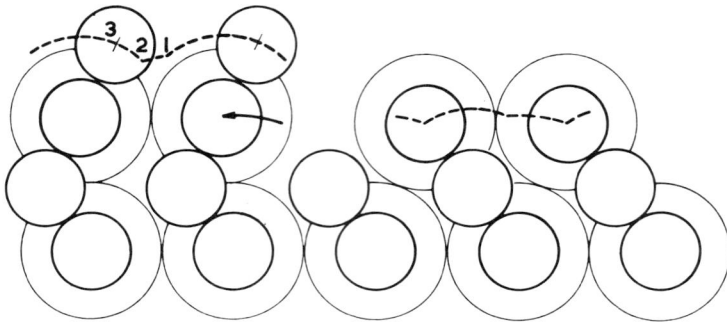

Fig. 3 Projection of a hard-sphere model of the b.c.c. lattice on {110}, giving rolling trajectories of single atoms on the {112} plane (Barrett's[4] procedure), for: (a) the undisturbed lattice, at right, (b) for lattice containing an intrinsic fault, at left. The enumerated positions are occupied by atoms if: (pos. 1) the original lattice is continued, i.e. if no more than a single stacking fault will be present, (pos. 2) an extrinsic fault is formed in the lattice, as explained in the text, (pos. 3) a multi-layer twin is created.

However, for an extrinsic or two-layer fault it is *not* obvious that the stacking fault vector $f = \frac{1}{3} <111>$, based on pure twinning shear on two {112} planes, may be used as an approximation in the hard-sphere model. In the latter model, the cusps in trajectories produced by spheres rolling over a faulted layer will be displaced relative to those obtained on the virgin material. This is not accounted for in a "two-layer twin" model. The point is illustrated in the left-hand side of Fig. 3 where the various positions of interest are enumerated. Position 1 is the cusp corresponding with the approximate position of the atom if the original lattice was to be continued, i.e., if there was only to be a single stacking fault in the lattice, position 3 is the one predicted by the "two-layer twin" model, and position 2 is the deepest cusp, located half-way between positions 1 and 3. The f-vectors corresponding to positions 1 and 3 would be $f_1 = \frac{1}{6} <111>$ and $f_3 = \frac{1}{3} <111>$. The f-vector connected with position 2, which is not predicted by the twinning analogy, can be approximated accurately by the mean of the other two: $f_2 = \frac{1}{4} <111>$.

The question concerning which of the three positions will be preferred, and what condition determines the preference, exposes the weak point in the procedure of determining cusps in rolling trajectories; the method takes into account only the structure of the material in the lower half of the crystal, neglecting that in the upper half. Fortunately, for positions 1 and 3 the situation is quite obvious; position 1 will be preferred if the upper half of the crystal possesses the same structure as the lower half, i.e., if only an intrinsic fault on a single {112} plane is present; position 3 will be preferred—in spite of its apparent instability in Fig. 3—if the structure in the upper half of the crystal continues as the *twin* of that in the lower half. The structure of the upper half of the crystal then ensures the stability of position 3.

The only remaining possibility for stability of position 2 is the case when the original structure is continued after the second fault plane, i.e., if an extrinsic fault is created. The converse is not true; position 3 is not *a priori* excluded as a possible stable position for the extrinsic fault. It should be added that position 2 as given in Fig. 3 does not satisfy the symmetry requirement that can be formulated for the {112} plane. It may be observed in Fig. 2 that the projection of the {112} plane is a line around which the symmetry operation of 180° rotation and glide can be performed, and this should still be possible if an extrinsic fault on {112} is present. Position 3 satisfies this requirement, position 2 does not, but it is a simple matter to slightly rearrange the stacking of the planes in such a manner that position 2 also fulfills the requirement.

The rearrangement can be performed most easily by shifting slightly, of the three atomic layers between which the two faults are present, the middle layer relative to the top and bottom layers until the configuration is symmetrical. The resulting shift of position 2 is negligible. In Fig. 4 the extrinsic fault thus obtained, which satisfies the symmetry requirement, is shown in a model giving seven {112} layers. The rolling trajectory for the atomic layer "C" on the upper boundary is shown; it may be observed that although the two cusps are further apart than in Fig. 3, the deepest cusp, corresponding to position 2, is still the one occupied. In addition, it may be noticed that the lateral position of the atoms in layer "A" on top is half-way between that of the atoms in the bottom layer "A"; in the undisturbed lattice the two layers would have been perpendicularly above each other.

Almost certainly the two-layer fault structure as presented in Fig. 4 will require correction in detail. However, it does not seem likely that the f-vector of the extrinsic fault will be much changed by it, or that the relative order of the depth of the cusps will be reversed. It would seem, therefore, that an extrinsic stacking fault based on position 2 in the rolling trajectory is more acceptable than one based on position 3. Its f-vector will be equal to $1/4 <111>$; since this value is different from that of a true two-layer twin, a distinction should be made between "extrinsic" and "two-layer twin" stacking faults.

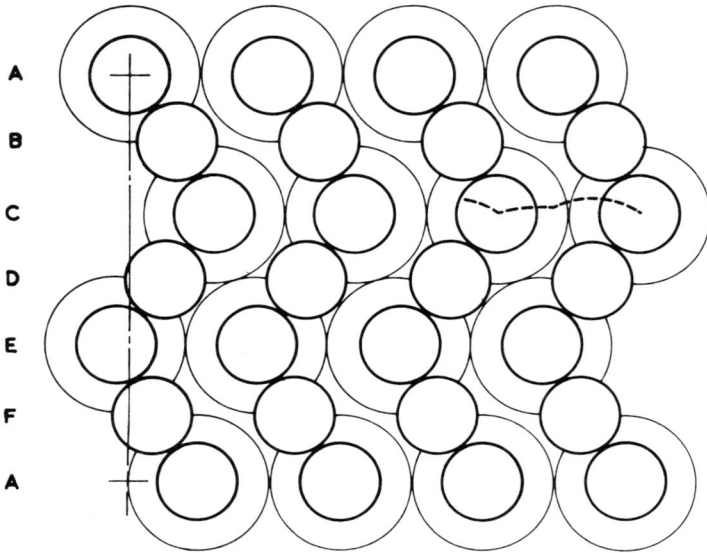

Fig. 4 Extrinsic fault, based on pos. 2 in Fig. 3, between atomic layers C-D and D-E. Stability of this fault is illustrated by the position of the atoms in layer C at the deepest cusp in the rolling trajectory. The f-vector is $1/4 <111>$.

Considering the perfect $1/2 <111>$ $\{112\}$ edge dislocation, it would seem that there is no obstacle to its decomposition on two stacking fault planes. The process might be given by the equation:

$$1/2 <111> \rightarrow 1/4 <111> + 1/4 <111> \tag{4}$$

in which the two partials, interconnected by an extrinsic fault, are not equivalent or symmetrical, although possessing the same Burgers vector. In comparison with the "classical" decomposition given by Eq. (1), in which an intrinsic fault is produced, it may be noted that the partials in Eq. (4) repel each other more strongly than those in Eq. (1), the interaction force being proportional to the product of the Burgers vectors of the interacting dislocations. This would favor the decomposition as given by Eq. (4).

Earlier, the author considered the decomposition of $1/2 <111>$ screw dislocations on $\{112\}$ planes,[3] using the twin layer fault model. It was found that of the two stable configurations possible, the one possessing lowest energy consists of a two-layer twin, with two $1/6 <111>$ partials in its advancing edge. The extrinsic fault created by the spontaneous dissociation of an edge dislocation will be stabilized by the screw components of the dislocation line. It is possible in the dislocation formalism to consider the process as hypothetically occurring in two stages, the first stage being the creation of an intrinsic fault, to which an extra layer is added by a further dislocation dissociation:

$$\tfrac{1}{3} <111> \rightarrow \tfrac{1}{4} <111> + \tfrac{1}{12} <111> \tag{5}$$

We shall call the $\tfrac{1}{4} <111>$ partial produced by this dissociation the "tail" $\tfrac{1}{4} <111>$ partial, and the $\tfrac{1}{4} <111>$ partial resulting from joining the $\tfrac{1}{12} <111>$ partial to the $\tfrac{1}{6} <111>$ partial on the adjacent plane the "head" $\tfrac{1}{4} <111>$ partial. The head $\tfrac{1}{4} <111>$ partial will, of course, have a tendency to remain dissociated, the tail partial will not dissociate as long as the edge dislocation is mobile.

If the Peierls force F_P and the total force exerted by twin boundaries F_T on the perfectly accomodated half-plane configuration are defined in the same manner for three dimensions as in the preceding section for two, the tail $\tfrac{1}{4} <111>$ partial will dissociate if a three-layer twin is formed. An extra layer is added to the extrinsic fault by the passage of a $\tfrac{1}{6} <111>$ partial on an adjacent $\{112\}$ plane, e.g., between the atomic layers E and F in Fig. 4, while

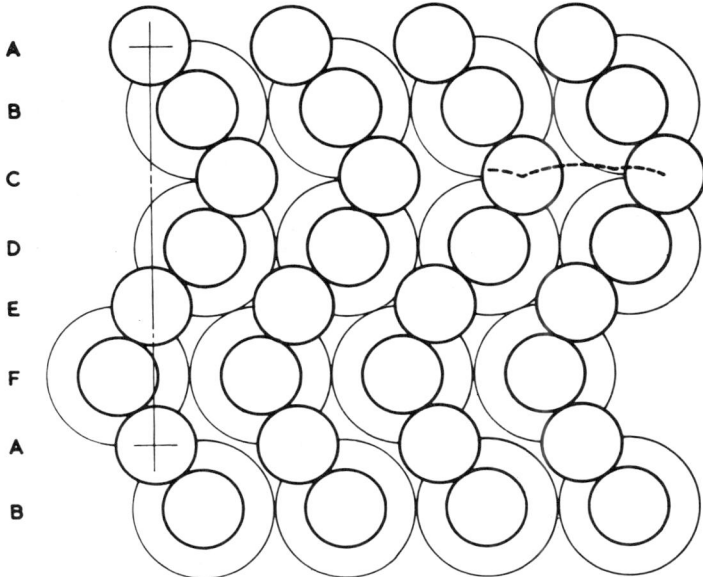

Fig. 5 Three-layer twin, based on pos. 3 in Fig. 3, which causes a shear displacement of $\tfrac{1}{2} <111>$.

the extrinsic fault is converted into a true two-layer twin by the passage of two $1/24$ <111> partials between the atomic layers C and D, and D and E in Fig. 4. The result is a true three-layer twin, which may be verified by the position of the upper layer "A" being perpendicular above the bottom layer "A" in Fig. 5.

4   STACKING FAULTS ON {110}

The stacking fault positions on the {110} b.c.c. plane that are proposed in the following are based on an attempt to describe the geometrical f.c.c → b.c.c.-transition model given by Bain[5] in terms of a change in the pattern of maximum and minimum potential positions for the stacking of atomic planes of one type in the f.c.c. lattice.

For such a description of the micromechanism governing this phase-transition not to be ambiguous it must be homogeneous, i.e., the proposed change must apply to all planes of the same type in the f.c.c. lattice. As will be shown in the following, the {111} f.c.c. plane satisfies this condition. If we consider the matrix expression for the correspondence relations between crystallographic directions in the Bain transformation given by Jaswon and Wheeler:[6]

$$(fDa) = \begin{pmatrix} 1 & 0 & \bar{1} \\ 0 & 1 & 0 \\ 1 & 0 & 1 \end{pmatrix} \tag{6}$$

We may observe that the [010] f.c.c. direction is transformed in the [010] b.c.c. direction. In other words, not only each of the two lattices but the correspondence relation itself possesses a maximum fourfold symmetry. This implies that the type of plane for which we try to define a change in the "micromechanical condition" should not occur more than four times in the f.c.c. lattice.

As the f.c.c. lattice belong to the $m3m$ crystal class only the {100} f.c.c. and the {111} f.c.c. planes fulfill the latter condition (the unit cell is centro-symmetrical). On the {111} f.c.c. plane transforms homogeneously; this may be verified in the matrix for the transformation of indices of planes:

$$(fPa) = \frac{1}{2} \begin{pmatrix} 1 & 0 & \bar{1} \\ 0 & 2 & 0 \\ 1 & 0 & 1 \end{pmatrix} \tag{7}$$

by filling in the indices of the three {100} f.c.c. planes, and of the four {111} f.c.c. planes which are transformed in {110} b.c.c. planes.

If we now consider the {111} f.c.c. plane and the corresponding {110} b.c.c. plane given in Fig. 6, two major differences may be noted:

1. One out of every three nearest–neighbor bonds in f.c.c. has become a next–nearest–neighbor bond in b.c.c.
2. Whereas the stacking positions of the atoms of the next layer in f.c.c. are Δ or ∇ sites, in b.c.c. the stacking positions are the n-n-n (next–nearest–neighbor) saddle points.

In Fig. 7 three stages of the Bain transformation are illustrated both by projection of a hard–sphere model on the {111} f.c.c. plane, and by corresponding changes in the Thompson tetrahedron. Stage 1 gives the f.c.c. lattice; in stage 2 the length of the vertex AD of the Thompson tetrahedron has been increased by 15.4% to the distance between next–nearest–neighbors in b.c.c., with the result that the positions of the atoms in the next layer come very close to n-n saddle points; this model is due to Bogers and Burgers.[7]

The change in the micromechanical condition corresponding to the step between stages 1 and 2 is that the original stable Δ sites have become unstable as stacking positions and that now the saddle point sites are preferred instead.

If we consider the Thompson tetrahedron in stage 2, it may be remarked that the first step in the transition has converted the

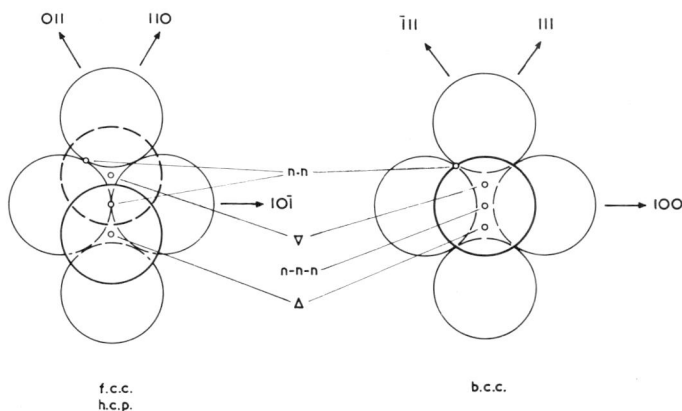

Fig. 6 Corresponding stacking sites on the close-packed plane and on the {110} plane in b.c.c. that results from such a plane by a phase transition. The stable stacking position in b.c.c. is halfway the $\Delta \rightarrow \nabla$ twinning movement in f.c.c.

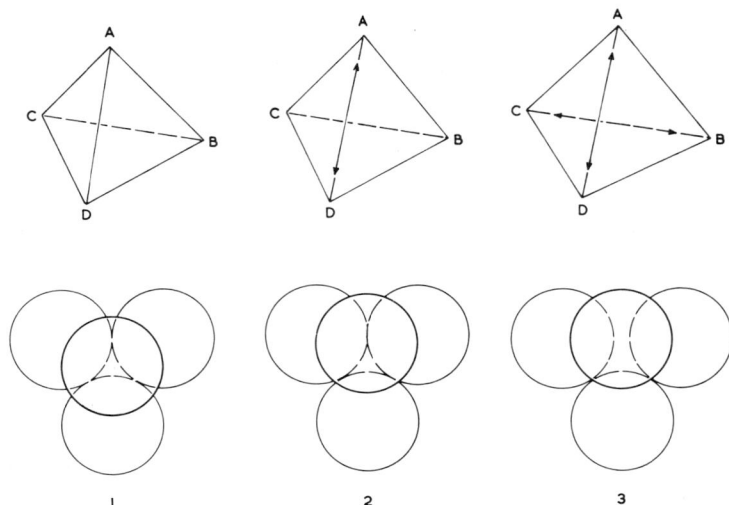

Fig. 7 The Bogers-Burgers model of the Bain transition from f.c.c. to b.c.c., illustrated by changes in the Thompson tetrahedron and the corresponding stacking model of hard spheres.

ABD and *ACD* planes from {111} f.c.c. type planes into {110} b.c.c. type planes; the *ABC* and *BCD* planes have remained unaltered. The structure may be described alternatively as a stacking of {111} f.c.c. planes on *n-n* saddle points or as a stacking of {110} b.c.c. planes on Δ or ∇ sites. These are not stable positions for either lattice, and, consequently, in the second step from stage 2 to sage 3 in Fig. 7 the stacking of {110} b.c.c. planes is changed from one on ∇ sites to one on saddle point sites. The process amounts to elongating the vertex *BC* by 15.4%, and thus simulataneously converting the two remaining {111} f.c.c. type planes into {110} b.c.c type planes. All four {111} f.c.c. planes of the Thompson tetrahedron have now become {110} b.c.c. planes; the final structure is b.c.c.

As far is known, in reality the transformation takes place in one single step, instead of in two. As the stacking of the four {111} f.c.c. layers changes simultaneously from stacking on Δ or ∇ sites to one on saddle points—as the symmetry of the crystal requires—the saddle points themselves are transformed from *n-n* sites to *n-n-n* sites. The change in the micromechanical condition, viz., that *n-n* and *n-n-n* saddle points have become more stable than Δ or ∇ positions, must apply to all intermediate types of saddle point positions as well so that the continuous phase transition may be carried to its final stage.

From these considerations it would follow that in the b.c.c. lattice, the order of stability of stacking positions on the {110} plane are: (1) the *n-n-n* sites; (2) the *n-n* sites; and (3) the Δ or ∇ sites. Therefore, stacking faults on this plane will occur preferentially by occupation of the *n-n* sites, and not of the Δ or ∇ sites as has been proposed by Crussard[8] and Cohen et al.[9] on the basis of the geometry of the hard-sphere model of the {110} b.c.c. plane. Thus, it is found that Eq. (4) not only gives the decomposition of the ½ <111> dislocations on {112}, but also that on {110} planes; the difference is that the partials on {112} planes are separated by intrinsic faults, whereas in the latter planes these faults are extrinsic.

It may be remarked that the reverse of this transition or a transition to h.c.p. would occur if the Δ or ∇ sites in b.c.c. on the {110} plane would be stable stacking positions. Although many b.c.c. metals do not exhibit the b.c.c.⇌f.c.c. or b.c.c.⇌h.c.p. transition, the above considerations on the instability of Δ or ∇ stacking sites would seem to apply to all b.c.c. metals; if not, an explanation why these metals would not spontaneously be transformed into f.c.c. or h.c.p. would appear to be necessary. The explanation of the stability of the b.c.c. lattice in terms of stacking preferences as given in the above may be regarded as a refinement of the model proposed by Leibfried,[10] in which an ''internal pressure'' is in equilibrium with the directed bonding forces between the atoms.

5 EXTRINSIC FAULTS ON PLANES NEAR {112}

The {112} plane in an ''S-plane'' (Hartman and Perdok[11]) or corrugated plane composed of alternating narrow strips of different {110} planes, each defined by two rows of atoms in the close-packed <111> direction. However, in the foregoing it was established that the stacking faults on {112} are different from those on {110}. It seems that the smallest units into which {112} stacking faults may be subdivided lie on pairs of different {110} planes. In Fig. 8, the stacking fault on (112) is shown as being subdivided, alternatively, in two different extrinsic stacking fault (esf) units.

Now the extrinsic fault on (112) can be considered as being composed entirely of either of the two different esf units. However, if two different esf units are joined together, a stacking fault lying on a double strip of {110} plane defined by three <111> rows

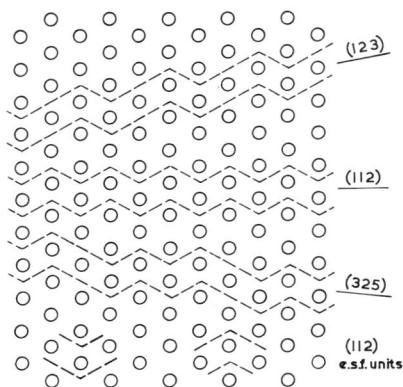

Fig. 8 Projection on {111} of extrinsic stacking faults. The {112} stacking fault can be broken up in extrinsic stacking fault units that may be recombined to give stacking faults of the same type on {123}, {325}, etc.

results; this gives the maximum width for a strip of {110} plane derivable from the {112} extrinsic fault. Evidently, the extrinsic stacking fault plane deviating most from {112} is composed of single strips of {101} alternating with double strips of {011}. These planes are of the {123} type; in general, the indices of the planes on which extrinsic {112} type stacking faults occur are given by:

$$\{hkl\} = \{011 + n(112)\} \tag{8}$$

in which $n = 1, 2, 3, \ldots$, etc.

In Figure 8 the projections of a {123} and a {235} plane ($n = 1$ and $n = 2$, respectively) are shown. The orientations of the planes for $n > 1$ are all confined between those of the {112} and {123} planes.

It seems likely, but at present it is not known with certainty, that if the {110} steps in these composite planes are longer than four <111> rows the {110} character of the plane predominates, and that the faults on these planes will be intrinsic. The same will be true for "mean" {123} planes, such as discussed by Chen and Maddin,[12] composed of two different {110} planes in a 2:1 ratio, but with a minimum length of each {110} step that is a multiple of that of the genuine {123} plane.

6 CROSS-SLIP

It is useful to distinguish two categories of cross-slip processes:

1. "easy" cross slip, during which the enrgy of the screw dislocation does not change, and which process inherently requires no extra energy; and

2. "difficult" cross-slip, in which the energy of the cross-slipping dislocation changes. In this an additional stress is always required to accomplish cross-slip, either to bring the dislocation line to the screw orientation—if the energy of the dislocation then increases—or to tear the dislocation away from the screw orientation if it possesses less energy in that position.

The general distinction between these two types of cross-slip may be applied to the case of slip in b.c.c. on the basis of the model developed above. The extended screw dislocations that are discussed here are, of course, idealized models: it is well known that the stacking fault energies are relatively high, which results in "stacking fault ribbons" that are a few atomic distances wide in most metals. However, these models will at least give an indication of the anisotropy of the dislocation cores.

In Fig. 9 the two basic forms of extended screw dislocations— or anisotropic dislocation cores—that are possible are depicted: the two $1/4 <111>$ partials are either interconnected by an extrinsic fault (Fig. 9b), or by an intrinsic fault (Fig. 9c). The $1/4 <111>$ partials may dissociate further on $\{112\}$ planes:

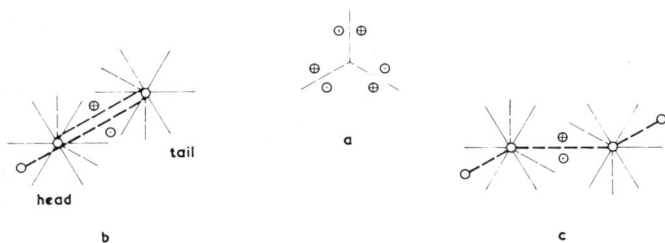

$$1/4 <111> \to 1/6 <111> + 1/12 <111> \tag{9}$$

Fig. 9 Dissociated $1/2 <111>$ screw dislocations projected on $\{111\}$. Figure (a) gives the shear on $\{112\}$ caused by an intrinsic fault, while Figs. (b) and (c) show dislocations on $\{112\}$ and $\{110\}$, respectively, in the same orientation.

in which an intrinsic fault is created. It seems unlikely that dissociations of this type will occur on {112} planes in which the $1/6$ <111> partial approaches the other $1/4$ <111> partial more closely; for this reason the tail $1/4$ <111> partial in Fig. 9b is given as undissociated. The extended dislocation in Fig. 9b possesses the same configuration as the edge dislocation; it would appear that the screw dislocation extended on {112} may proceed by an "easy" cross-slip process to another plane of {112} type (between {123} and {112}).

The screw dislocation extended on {110} may have both its $1/4$ <111> partials dissociated according to Eq. (9), as shown in Fig. 9c. The configuration may be produced by a dislocation line on {110} that attains the screw orientation; only then is the dissociation of the $1/4$ <111> partials possible. For further glide on {112} one of the dissociations must be reversed, for further glide on {110} both $1/4$ <111> partials must be contracted; most evidently, this type of extended dislocation occurs during "difficult" cross-slip.

In comparing these models with experimental findings in the literature, it must first be stated that this paper is not intended to be a review and cannot contain more than a choice of the extensive literature. The wavy slip lines commonly observed on the surfaces of plastically deformed specimens of b.c.c. metals are currently explained by "noncrystallographic" or "banal" slip—the term is due to Elam[13]—i.e., slip on continuously orientable surfaces belonging to the <111> zone, or by slip on the {110} planes belonging to this zone (Chen and Maddin, loc. cit.). The recent experiments of Taoka et al.[14] indicate that the orientation dependence of the critical shear stress is not compatible with banal slip, nor with the model of Chen and Maddin. On the other hand, the electron micrographs published by Duesbery et al.[15] strongly suggest the possibility of a continuous change of slip plane during cross-slip. It would seem that the concept of "easy" cross-slip, i.e., noncrystallographic glide on planes confined within narrow regions around {112}, would reconcile both experimental findings.

The idea that slip on {112} and {123} is different from that on {110} because the stacking faults are of different types is well supported by experiment. Already Barrett, Ansell, and Mehl[16] concluded from their classic study of slip in Fe and Fe-Si that the critical shear stress for slip showed the same temperature dependence for all planes for Si contents less than app. 3 1/2%, but that for higher Si contents the slip on {110} takes place at lower stresses than that on {112} and {123}, especially at temperatures below 120°K. This conclusion was fully corroborated by the work of Taoka et al.[14]

on single crystals of the same material. The experimental results of Kossowsky and Brown[17] on microyielding of iron were recently analyzed by Escaig[18] in terms of intrinsic and extrinsic stacking fault energies, on the basis of the earlier work of Sleeswyk (*loc. cit.*). These energies were determined as: $\gamma_{\{110\}} \approx 200$ ergs/cm^2 and $\gamma_{\{112\}} \approx 260$ ergs/cm^2 for the intrinsic and extrinsic faults, respectively. In 4% Si-Fe deformed in liq. N_2 the slip is predominantly of the crystallographic type on $\{110\}$, which is readily explained by the relatively low intrinsic stacking fault energy, causing the slip on $\{112\}$ and $\{123\}$ to be suppressed.

It may be remarked in Fig. 9b that it will make a difference whether the screw dislocation moves "head first" or "tail first" on the $\{112\}$ plane. If the former obtains, easy cross-slip to, e.g., a $\{123\}$ plane, may occur, if the latter, this is not possible, but difficult cross-slip to $\{110\}$ is more likely. The marked asymmetry of the critical shear stress for slip on $\{112\}$ reported by Steijn and Brick,[19] Sestak et al.,[20,21] Taoka et al.,[14] Reid et al.[22] may well be due to this effect. This would imply that the cross-slip mechanism determines the stress level, which is in accordance with the hypothesis proposed by Hirsch[23] and Mitchell, Foxall, and Hirsch,[24] viz., that the strain rate-determining process is the difficult cross-slip caused by the relatively low energy of the extended screw dislocation, rather than overcoming the Peierls barrier during dislocation movement. The idea was worked out by Víték[25] and Víték and Kroupa,[26] using as a model the high-energy configuration of the extended screw dislocation on $\{112\}$ given by Sleeswyk (*loc. cit.*), and the dissociation proposed by Crussard and Cohen et al. (*loc. cit.*) for $\frac{1}{2}<111>$ dislocations on $\{110\}$. Although it is clear from the treatment given in the preceding sections that this author does not consider this model to be correct in detail, the main conclusions reached by Víték and Kroupa, viz., that the rate-controlling mechanism is the thermal activation of cross-slip—"difficult" cross-slip in the present context—and that the energy for cross-slip H_c may be given by an expression of the type:

$$H_c = A \frac{B - \sigma}{\sigma} \tag{10}$$

remain valid. An expression of this type has previously been derived by Friedel.[27] The values of A and B are dependent on the model, and at present it would seem better to use these values as constants to be adapted to the experimental findings.

The activation energy for plastic deformation may be derived in a variety of ways. Assuming an Arrhenius-type equation:

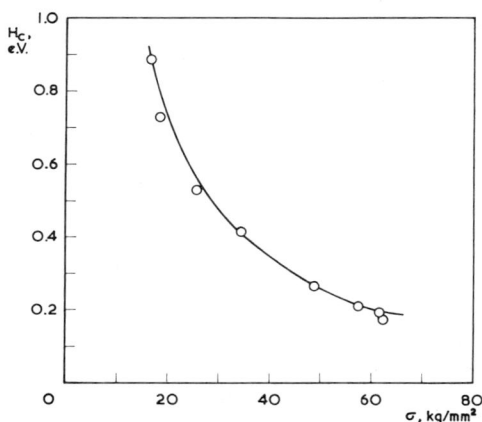

Fig. 10 The experimentally determined activation energy for plastic deformation determined by Sleeswyk and Helle[27] are compared with a curve obtained from Eq. (10), which is based on the work of Viték and Kroupa[26].

$$\dot{\epsilon} = C \exp\left(-\frac{H_c}{kT}\right)$$

in which the symbols have the usual meaning, differentiation yields the expression:

$$H_c = -k\left(\frac{\partial \ln \dot{\epsilon}}{\partial\ 1/T}\right)_\sigma \qquad (11)$$

which is convenient to use in the derivation of H_c from experiment.

Sleeswyk and Helle,[28] by determining the quotient $\left(\dfrac{\Delta \ln \dot{\epsilon}}{\Delta\ 1/T}\right)_\sigma$ from tensile tests performed alternatingly at two deformation velocities on Fe in the temperature range from 78° to 300°K, obtained the values for H_c as a function of the applied stress σ plotted in Fig. 10. The curve gives H_c as the function of σ given in Eq. (10), with $A = 0.06$ eV and $B = 264$ kg/mm^2; it would seem that the curve is in reasonable agreement with the experimental points, which appears to support the notion that difficult cross-slip is the rate-determining process in the plastic deformation of b.c.c. metals.

7   TWINNING AND SLIP

From earlier work it is evident that the most difficult step in the nucleation of a twin is the creation of a stacking fault, or

elementary twin, of sufficient length that it can meet a screw dislocation. In this way a source mechanism may allow the fault to
grow laterally. It would seem justified to regard the formation of
such a fault as a prerequisite for twin formation, i.e., $F_P > F_T$.
However, the criticism given in Sec. 2 on the postulate of Ogawa
is still valid; it is highly unlikely that emissary dislocations can
be formed as three-layer twins; in other words, a condition for the
formation of emissary dislocations would be $F_P < F_T$.

These seemingly contradictory requirements do not really exclude each other. First, it must be stated that the Peierls force
as used in the present context should incorporate the effect of thermal fluctuations. This *effective* Peierls force F_P' is strongly dependent on temperature and strain rate, while F_T changes only
slightly with temperature because of the changes in elastic constants. Consequently, at temperatures where $F_P \approx F_T$, rapid
straining may give: $F_P' > F_T$, slow straining: $F_P' < F_T$. Second,
the drops in load that occur in b.c.c. metals during twinning suggest that the shear stress on the twin plane during nucleation of a
twin is higher than that during growth and accomodation. It is
possible to give a consistent model of the various stages of twin
formation if the stress and strain rate–fluctuations are taken into
account. These stages are:

1. formation of three-layer twins from dissociated $1/2 <111> \{112\}$
dislocations, for which the necessary stress is given by:

$$b \cdot \tau = F_T < F_P' \tag{12}$$

in which b is the Burgers vector of the $1/2 <111>$ dislocation and τ
is the shear stress on the $\{112\}$ plane.

The effective Peierls force F_P' is a function of dislocation velocity,
and there exists a critical dislocation velocity V_C determined by
the equality $F_P' = F_T$. At the condition given by Eq. (12), this is the
velocity of the tail partial, while that of the head partial exceeds it,
or

$$V_H > V_T = V_C \tag{13a}$$

The tail and head partials are not separated if their velocity is
below V_C or

$$V_H = V_T < V_C \tag{13b}$$

2. rapid lateral grown of the twin by a source mechanism during which the stress τ decreased due to stress relaxation;

3. accomodation of the twin by emissary dislocations if the force at the twin tip necessary for creating an emissary dislocation F_E, has been reached. The force at the twin tip is approximated by:[28] $(n - 1) \cdot b' \cdot \tau$, in which n is the number of twinning partials in the noncoherent twin boundary and b' the Burgers vector of the twinning partial (or: $b' = \frac{1}{3} b$). For the emissaries to glide on the {112} plane, these extended dislocations must be composed of partials interconnected by two-layer faults: the stress must fulfill the inequality $b \cdot \tau < F_T$. The condition for the formation of emissary dislocations may be summarized in the following expression:

$$F_P < \frac{3}{n - 1} \cdot F_E < b \cdot \tau < F_T \tag{14}$$

The strain rate in this stage would be slow in comparison to that in stage 1;

4. coagulation of piled-up emissaries into single twins if the stress τ increases to $\tau \geq F_T/b$. It is this effect which may explain the observations by Priestner and Leslie,[29] which appeared to indicate that twins are created from arrays of emissaries. Formation of new twins by this process and by processes as described under 1, 2, and 3.

In this model most of the emissaries have only an ephemeral existence. Soon after their creation they are either annihilated at the free surface or, as the stress increases, formed into new twins. This may account for the fact that emissary dislocations are observed relatively rarely.

Pinning of the tail partial by interstitial impurities will locally low V_C, which would favor twinning if the first step in twin formation, viz., the creation of a three-layer twin, would be the most difficult one. The experiments of Churchman and Cottrell[30] and of Rosenfield et al.[31] on aging effects in Fe-C alloys would support this view. It might seem that as accomodation of the twin in the matrix probably absorbs most of the energy for the formation of a twin lamella, the dislocation movement by which this occurs would determine the twinning characteristics. This may well be so, but these characteristics would not be those of single dislocations in the case of accomodation by emissaries. The regular three-dimensional pileups of emissaries would be thermally activated much more easily. The phonon scattering cross section of a dislocation is proportional to b^2, and, consequently, as was pointed out by Nabarro,[32] the dislocations in a pileup have a larger cross section.

In addition, the thermally activated jump of one dislocation in a pileup gives rise to stresses on the other dislocations which favor similar movements.

Presently, the influence of twinning on the thermal activation of plastic flow is being investigated experimentally at Groningen. Preliminary results would indicate that twinning lowers the incremental stress $\Delta\sigma$ at liq. N_2 temperature by some 30%.

REFERENCES

1. Ogawa, K.: *Phil. Mag.* 11:217 (1965).
2. Ogawa, K., and R. Maddin: *Acta Met.* 12:713 (1964).
3. Sleeswyk: *Phil. Mag.* 8:1467 (1963).
4. Barrett, C. S.: "Cold Working of Metals," p. 65, Am. Soc. of Metals, 1948.
5. Bain, E. C.: *Trans. AIME* 70:25 (1924).
6. Jaswon, M. A., and J. A. Wheeler: *Acta Cryst.* 1:216 (1948).
7. Bogers, A. J., and W. G. Burgers: *Acta Met.* 12:233 (1964)
8. Crussard, C.: *Compt. Rend. Acad. Sci.* Paris, 252:373 (1961).
9. Cohen, J. B., R. Hinton, K. Lay, and S. Sass: *Acta Met.* 10:894 (1962).
10. Leibfried, G.: *Z. Physik* 129:307 (1951).
11. Hartman, P., and W. G. Perdok: *Acta Cryst.* 8: 49 (1955).
12. Chen, N. K., and R. Maddin: *Trans. AIME* 191:461 (1951).
13. Elam, C. F.: "The Distortion of Metal Crystals," Oxford University Press, Oxford, 1935.
14. Tacka, T., S. Takeuchi, and E. Furabayashi: *J. Phys. Soc. Japan* 19:701 (1964).
15. Duesbery, M. S., R. A. Foxall, and P. B. Hirsch: *J. Physique* 27:C3-193 (1966).
16. Barrett, C. S., G. Ansel, and R. F. Mehl: *Trans. ASM* 29:702 (1937).
17. Kossovsky, R., and N. Brown: *Acta Met.* 14:131 (1966).
18. Escaig, B.: *J. Physique* 27:C3-205 (1966).
19. Steijn, R. P., and R. M. Brick: *Trans. ASM* 46:1406 (1954).
20. Sestak, B., and S. Libovicky: *Acta Met.* 11:1190 (1963).
21. Sestak, B., and N. Zarubova: *Phys. Stat. Solidi* 10:239 (1965).
22. Reid, C. N., A. N. Gilbert, and G. T. Hahn: *Acta Met.* 14:975 (1966).
23. Hirsch, P. B.: 5th Int. Congress Cryst. (1960).
24. Mitchell, T. E., R. A. Foxall, and P. B. Hirsch: *Phil. Mag.* 8:1895 (1963).
25. Vitek, V.: *Phys. Stat. Solidi* 18:687 (1966).
26. Vitek, V., and F. Kroupa: *Phys. Stat. Solidi* 18:703 (1966).
27. Friedel, J.: "Dislocations," p. 71, Pergamon Press, 1964.
28. Sleeswyk, A. W., and J. N. Helle: *Acta Met.* 11:187 (1963).
29. Preistner, R., and W. C. Leslie: *Phil. Mag.* 11:895 (1965).
30. Churchman, T., and A. H. Cottrell: *Nature* 167:943 (1951).
31. Rosenfield, A. R., B. L. Averbach, and M. Cohen: *Acta Met.* 11:1100 (1963).
32. Nabarro, F. R. N.: *Proc. Roy. Soc. Lond.* Ser. A209:278 (1951).

DISCUSSION *on Paper by A. W. Sleeswyk*

J. HIRTH: Some of the early configurations that you showed resemble zonal dislocations in h.c.p. crystals. I suppose, however, that the motion of all of the partial dislocations on {112} planes involve no shuffles. Are shuffles involved in your latter examples, such as the {123} glide?

A. SLEESWYK: This depends on the way the partials are assigned to the head and tail configurations (the way this can be done is not unique). It is possible, however, to describe all shuffles that are involved, for instance, if the two-layer {112} fault is changed to a three-layer twin, in terms of the movement of partials, as was done in the paper.

DISLOCATION VELOCITY
IN COPPER AND ZINC

T. Vreeland, Jr.

W. M. Keck Laboratory of Engineering Materials,
California Institute of Technology

ABSTRACT

The stress dependence of dislocation velocity in copper and in the basal and second-order pyramidal slip systems of zinc has been measured. Dislocations were observed by the etch-pit and Berg-Barrett x-ray techniques before and after application of short-duration stress pulses. The velocity of dislocations of mixed edge-screw orientation in copper and of basal edge dislocations in zinc was found to be a linear function of stress at room temperature. This data gives a value of the dislocation damping constant of 7×10^{-4} dyne-sec-cm^{-2}, in good agreement with the value deduced from internal friction measurement in copper. Evidence of an unusual stress and temperature dependence for the velocity of dislocations in slip bands on the second-order pyramidal slip system of zinc is reported.

1 INTRODUCTION

The stress-strain behavior of the f.c.c. metals is relatively insensitive to strain rate compared, for example, to that of the b.c.c. metals. Thus, the inverse strain-rate sensitivity of the flow stress $\partial \ln \dot{\epsilon} / \partial \ln \tau$ is relatively large in the f.c.c. metals. This has been taken as an indication that the dislocation velocity is a sensitive function of stress in these metals, i.e., that the mobility exponent $\partial \ln v / \partial \ln \tau$ is relatively large.[1]

Adams et al.[2] measured the inverse strain-rate sensitivity of the flow stress in the basal system of zinc and found it was also relatively large. Measurements of the velocity of basal slip band growth in zinc[2] indicated that $\partial \ln v / \partial \ln \tau$ was significantly smaller than $\partial \ln \dot{\epsilon} / \partial \ln \tau$. However, an abrupt increase in velocity was found at about 0.5 Mdyne/cm^2, the velocity increasing from 0 to about 10 cm/sec. It was proposed that the abrupt increase in dislocation velocity occurs when attractive junctions between the basal and nonbasal dislocations are broken. This, in effect gives rise to a very large mobility exponent at the breakaway stress, with a significantly smaller mobility exponent at higher stresses. The strain-rate sensitivity of the flow stress appears to be influenced more by the large mobility exponent associated with the breakaway process than with the small mobility exponent found at higher stresses. This phenomenon can be described in terms of a change in the density of mobile dislocations with stress. The mechanism which limits the dislocation velocity in zinc at stresses in excess of 0.5 M dyne/cm^2 was not established, but phonon drag was considered likely because of the relatively high velocities and low stresses.

Dislocations in the f.c.c. metals might behave in a similar manner. A discontinuous velocity vs. stress relationship could exist for dislocations which have to cut through the forest, the velocity increasing from zero to a high value at the breakaway stress. This could account for the very small strain-rate sensitivity. Further, the stress sensitivity of dislocation velocity at stresses above the yield stress could be relatively small. The investigations reported here were undertaken to explore this possibility, and to gain further information on the stress dependence of dislocation velocity in zinc.

Direct measurements of dislocation displacements in copper and in the basal system of zinc were made. The displacements were produced by short duration, essentially, square-pulse loading, and dislocation velocities were calculated from the observed displacement and pulse length. Dislocation-dislocation interactions were minimized in these experiments, and the drag force on moving edge-screw dislocations in copper and basal edge dislocations in zinc was determined.

The torsional loading system devised to limit the displacement of high-velocity dislocations to distances that are small compared to the initial dislocation spacing (in crystals of good quality), is first described. Torsional loading of copper and etch-pit observations on (100) surfaces, and torsional loading on zinc with Berg-Barrett x-ray observations of (0001) surfaces is then described. These observations are discussed and compared to the behavior of dislocations in other materials.

2 VELOCITY MEASUREMENTS

2.1 Torsional Loading System

Single torsional stress pulses of microsecond duration were applied to the specimens by means of the machine described by Pope et al.[3] This machine utilizes zero-order torsional waves in cylindrical bars which are generated in the following manner. An initial static torque is applied to a section of a cylindrical rod which is a part of the torsion rod system shown schematically in Fig. 1. The torque is applied to the top of the section by dead-weight loading of a cranking disk attached to the rod through a rubber sleeve. A 0.015-cm thick glass disk cemented to the bottom of the section transmits this torque to a bakelite fixture which is attached to a fixed bearing tube surrounding the rod. A 0.005-cm thick aluminum foil is cemented between the glass disk and bakelite with Eastman 910 adhesive. The lower section of the torsion rod is attached to the opposite side of the glass disk by means of a butt adhesive bond. This section of the torsion rod is coaxial with the upper section but does not carry any static torque. The specimen assembly is attached to the bottom end of this lower section of the torsion rod, and the bottom end of the specimen assembly is the free end of the torsion rod system. The section of the rod above the cranking disk (damping section) is coated with a viscoelastic material in order to attenuate the waves propagating away from the specimen.

The application of the stress pulse to the specimen is initiated by a high-voltage capacitor discharge through the aluminum foil. Explosion of the foil releases the static torque and results in elastic waves which propagate away from the glass disk interface

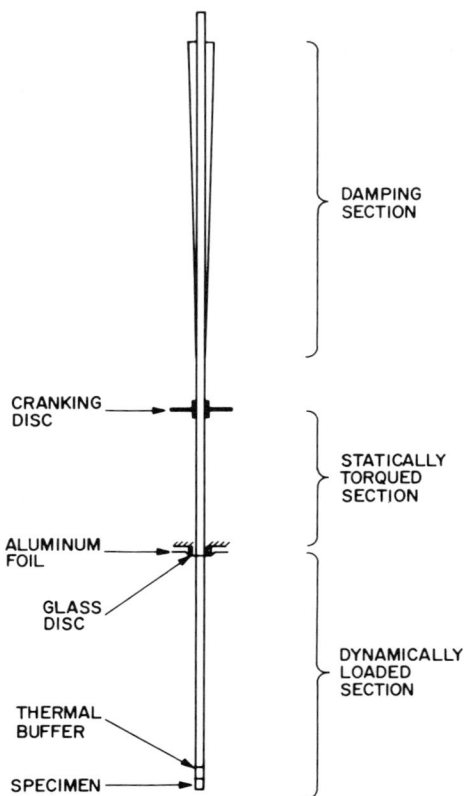

CRANKING DISC

ALUMINUM FOIL

GLASS DISC

THERMAL BUFFER

SPECIMEN

DAMPING SECTION

STATICALLY TORQUED SECTION

DYNAMICALLY LOADED SECTION

Fig. 1 Schematic of torsion pulse system.

of the torsion rod. The amplitude of the dynamic torque in the lower section of the torsion rod is one-half the initial static torque applied to the machine, and, therefore, proportional to the weight hung on the cranking disk. The duration of the stress pulse at any point in the specimen is the time required for the wave to propagate from that point to the free end and return.

The stress generated by the release of the static torque propagates without dispersion through the isotropic elastic rod attached to the glass disk because the elastic zero-order mode torsional waves are nondispersive. The propagation of zero-order torsional waves in $<100>$ axis copper specimens and in [0001] axis zinc specimens is also nondispersive and the distribution of shear stress over a cross section of a specimen is the same as that in an isotropic material.

The materials and dimensions of the rod system of the torsion machine were chosen so as to make the acoustic impedance of all the rod sections nearly equal to that of the specimen. Torsional wave reflections at the interfaces between different materials in the load train were thereby minimized. Steel rods of 1.27 cm diameter were used in the tests on copper and titanium rods of the same diameter were used in the tests on zinc. A thermal buffer of polycrystalline copper was cemented between the steel rod and the copper specimen. A monel thermal buffer was used in the zinc tests. The buffers prevented spurious thermal stresses in the specimens because the differential expansion of the specimen and buffer was negligible. All sections of the specimen-torsion rod system were joined together by means of Eastman 910 adhesive for the copper tests. A special bonding technique was devised to join the monel to the zinc so that dislocations were not displaced when the bond was made or when it was removed. This technique is described in a later section.

The amplitude and duration of the stress pulse were measured by means of silicon strain-gages, cemented to the torsion rod. The strain-gage output was displayed on a dual beam oscilloscope. The upper beam was triggered by the capacitor discharge employed to initiate the torsion wave and had a sweep rate of 200 μs/cm. This provided information of any reflected stress pulses applied to the specimen at relatively long times after the primary stress pulse. The initiation of the lower trace was appropriately delayed and its sweep rate was 20 μs/cm. The lower trace provided detailed information regarding the primary stress pulse. A record of strain gage output vs. time with no specimen attached to the titanium elastic rod system is shown in Fig. 2. The gages were located

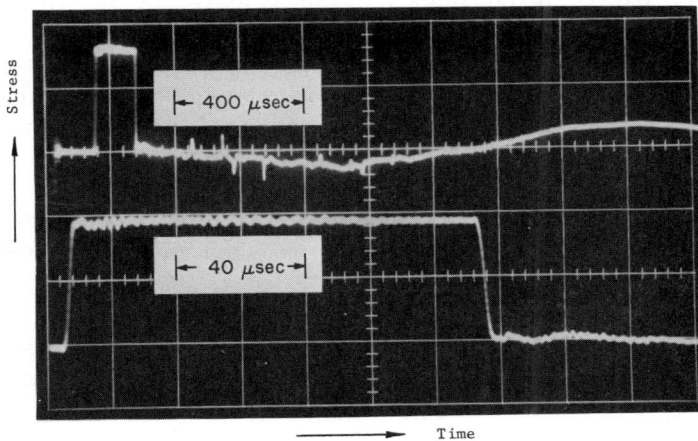

Fig. 2 A stress pulse produced by the torsion loading system.

19.5 cm from the free end of the rod system for this record, and the pulse length is just the time required from elastic shear wave in titanium to travel from the gage to the free end and return.

The dislocation displacements in the specimen crystals which are induced by the stress pulse must give rise to plastic strains which are small compared to the elastic strains associated with the pulse if dispersion is to be avoided. Thus, only a low density of moving dislocations can be tolerated when the dislocation velocity is large.

2.2   Testing of Copper Crystals

Experiments were performed on single crystal copper specimens in the form of right circular cylinders 1.25 cm in diameter. The cylindrical axis was parallel to the [100] crystal axis to within $\pm \frac{1}{2}°$. The cylindrical surface was modified by four, flat {100} observation surfaces, each about 0.3 cm wide, spaced at 90° intervals around the circumference and extending the full length of the specimen.

These test specimens were machined from larger single crystals. These crystals were grown in graphite crucibles from a charge of 99.999% copper by the modified Bridgman technique employed by Young and Savage.[4] Rough machining of specimens was performed by trepanning and wire slicing using spark-erosion machines. Finish machining was accomplished by chemical lapping using a saturated solution of cupric chloride in concentrated hydrochloric acid and a rotating cloth-covered Lucite wheel. Material of at least 0.04 cm thickness was removed by chemical lapping, thus removing the mechanical surface damage due to spark-erosion machining. The test specimens were annealed for about 100 hr at 1030 ± 10°C in a hydrogen atmosphere after finish machining.

Fresh dislocations were introduced into specimens by scratching the {100} observation surfaces in a controlled manner with a diamond phonograph stylus or an alumina whisker by means of a special scratching apparatus. Scratching was performed after the surface had been chemically polished in the manner employed by Livingston.[5]

Specimens were etched prior to stress application either by the solution used by Livingston[5] or by that used by Young.[6] The etch revealed grown-in dislocations on the {100} observation surfaces and the fresh dislocations produced by scratching. A permanent record of this dislocation configuration was obtained by making a replica of the specimen surface on a cellulose acetate film.

Four tests were performed with nominal resolved shear stress at the outer radius of the specimen ranging from 2.5 to 25 M dyne/cm^2. A different specimen was used in each test.

Analysis of the stress pulse records shows that the major portion of the torsional wave in the specimen propagated in an elastic manner. Thus, to a first approximation, the magnitude of the stress is independent of position along the specimen and the duration of stress is linearly proportional to distance from the free end of the specimen.

The stress vs. time relation at the loaded end of the specimen may be determined from the test record even when plastic wave propagation takes place. This is because both the incident and reflected waves which propagate between the strain gages and the loaded end of the specimen are purely elastic and propagate at known wave velocities. The actual stress pulse shape at the loaded end for each test was determined from the four records. The pulses were trapezoidal, rather than square. The dislocation displacement corresponding to the loaded end of the specimen was found by extrapolation of measurements of displacements vs. distance from the free end as explained below.

The specimen was reetched and the observation surfaces were replicated again immediately after testing. Photomicrographs of the replicas of specimen number 3-3-1 taken before (a) and after (b) the application of the stress pulse are shown in Fig. 3. The scratch (horizontal in the photomicrographs) is paralleled to the cylindrical axis of the specimen. The replicas were carefully examined to find the original and final positions of individual dislocations. The distance between the original and final positions of these dislocations was measured to within ± 4 μ.

The dislocation lines extending into the interior of the test specimen from the observed etch pits are probably oriented nearly at right angles to the trace of the {111} slip planes on the {100} observation surface. A pure edge dislocation so oriented has zero resolved shear stress applied to it. The dislocations which moved were therefore presumed to be of mixed edge-screw orientation.

Mean values of measured dislocation displacement are plotted as a function of distance from the free end of the specimen in Fig. 4 for the four tests conducted. The number of individual dislocation displacements measured is indicated near each plotted point. The vertical lines through the plotted points represent the standard deviation of the measurements. The maximum standard deviation is 13% of the mean dislocation displacement. The straight lines in Fig. 4 were drawn to represent the data. This shows that the

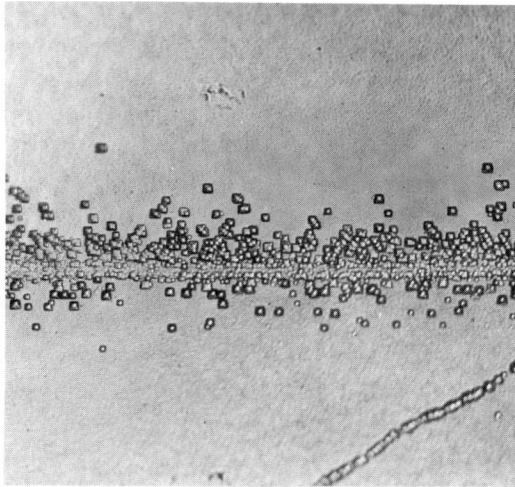

[101]

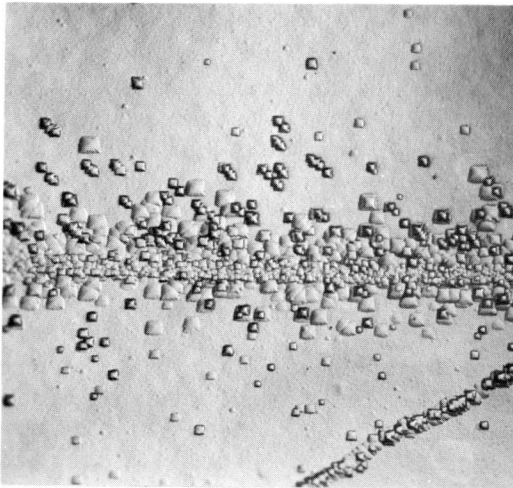

100 μ

Fig. 3 Etch pits near a scratch on a copper specimen
before and after a stress pulse was applied.

dislocation displacement is linearly proportional to distance from
the free end of the specimen. The points marked "x" at the end

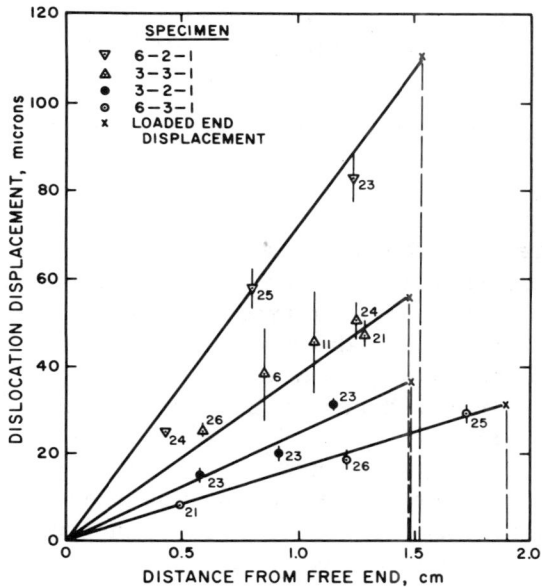

Fig. 4 Dislocation displacement vs. distance from the free end of copper specimens.

of each line in Fig. 4 represent the extrapolated value of dislocation displacement at the loaded end of each specimen.

The observed linear relationship between dislocation displacement and distance from the free end of the specimen is consistent with the assumptions that the stress wave propagates through the specimen in an elastic manner and that the acceleration time for the dislocations is negligible. If we assume a square, rather than a trapezoidal pulse, the duration of stress Δt is related to distance from the free end of the specimen x by

$$\Delta t = 2 \frac{x}{c} \tag{1}$$

where $c = 0.290 \times 10^6$ cm/sec is the velocity of propagation of an elastic torsional wave along the $<100>$ specimen axis. Thus, if Eq. (1) is employed to convert the abscissa scale of Fig. 4 to a scale of time duration of stress, the slopes of the lines are equal to the dislocation velocity. First approximation values of dislocation velocity v, determined in this manner, are fround to be approximately proportional to the $m = 0.7$ power of the resolved shear τ, when these data are fitted to an equation of the form:

$$v = v_0 \left(\frac{\tau}{\tau_0}\right)^m \tag{2}$$

where m and τ_0 are material constants and v_0 is unit velocity. This result indicates, however, that a significant portion of the total dislocation displacement occurred during the rising and falling portions of the trapezoidal stress pulses. Hence a more refined method of analysis of the experimental measurements was employed which used the actual stress-pulses shape to determine corrected values of dislocation velocity as a function of resolved shear stress.

When the corrected values of dislocation velocity vs. stress are fit to a power-law functional relationship, the "best fit" is still obtained by employing a stress component $m = 0.7$. The "fit" is not quite as good if it is assumed that the exponent is $m = 1$. However, the scatter and uncertainty in the experimental measurements are such that the value $m = 1$ may be the the true one. Thus, the resistance to the motion of individual dislocations in 99.999% pure copper may be described approximately as a simple linear viscosity, at least in the dislocation velocity range of 100 to 1000 cm/sec. A plot of the corrected dislocation velocity vs. the resolved shear stress is given in Fig. 5. The solid line is given

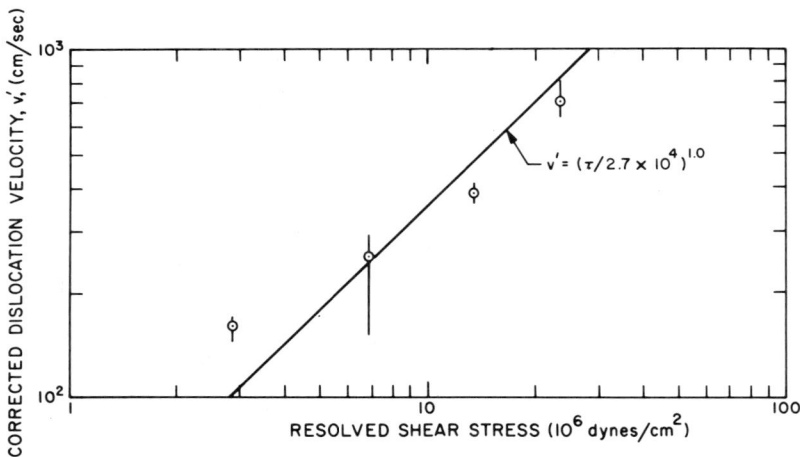

Fig. 5 Corrected dislocation velocity vs. resolved shear stress in copper.

by the equation

$$v = v_0\left(\frac{\tau}{2.7 \times 10^4}\right) \qquad (3)$$

with $v_0 = 1$ cm/sec and τ in dynes/cm^2.

2.3 Testing of Zinc Crystals

Torsion pulse tests were made on three [0001] axis zinc crystals. Figure 6 shows the specimen assembly used. Strain gages were placed 0.15 cm from the monel-specimen interface, and titanium rods were bonded to the other end of the specimen to obtain the desired stress pulse duration.

Prior to torsion pulse loading, edge dislocations were generated near the (0001) surface which was to be bonded to the monel thermal buffer. The dislocations were generated by scratching the surface using an Al$_2$O$_3$ whisker and a 50-mg load. When the scratch direction is perpendicular to a $<11\bar{2}0>$ slip vector, parallel edge dislocations of the same sign move out from the scratch and are revealed in a Berg-Barrett photograph. The technique employed by Armstrong and Schultz[7] was used to obtain the Berg-Barrett photograph shown in Fig. 7. Basal dislocations which lie within 5 μ of the surface are revealed in these photographs. The dislocations parallel to the three scratch segments shown in the $(01\bar{1}3)$ reflection are extinct in the (0004) and $(10\bar{1}3)$ reflections so they must have a Burgers vector in the $[1\bar{2}10]$ direction. Specimens were scratched along three diameters of an (0001) end surface, the scratches were made in short segments and were perpendicular to a $<1\bar{2}10>$ direction. The scratched surface was then bonded to a monel thermal buffer, and a stress pulse was applied.

Monel   Zinc   Titanium

Fig. 6 A zinc specimen bonded to the monel strain gage rod and the titanium extender.

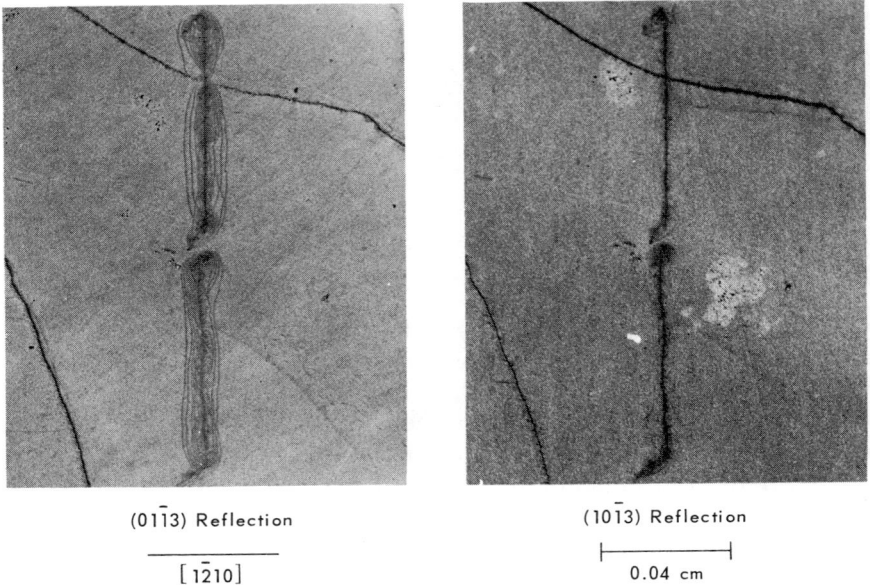

(01$\bar{1}$3) Reflection

[1$\bar{2}$10]

(10$\bar{1}$3) Reflection

|————————|
0.04 cm

Fig. 7 Berg-Barrett photographs of a scratched (0001) surface of zinc using two different reflecting planes. Scratch is along [10$\bar{1}$0], Co Kα radiation.

The bonding was found to be a very critical operation because conventional techniques produced sufficient shear stress to move the fresh dislocations generated by the scratch. Phenyl salicilate (salol), an organic crystalline material, was found to be a satisfactory bonding agent. It readily supercools to room temperature but undergoes a large volume change when it solidifies. The large volume change does not cause difficulties if the specimen surface is coated with a seed layer of salol (by spraying it with an artists air brush). The seeded surface is then brought into contact with the monel rod which has a layer of supercooled salol on its surface The salol bond is readily dissolved in acetone, and a comparison of Berg-Barrett photographs taken before bonding and after dissolving the bond showed only minor changes in dislocation positions (<0.002 cm).

Specimens were bonded and tested immediately after scratching. The bond was then dissolved, and Berg-Barrett photographs were taken. A Berg-Barrett photograph was not taken after scratching since it was found that the fresh dislocations became relatively immobile with aging.

Berg-Barrett photographs of the dislocation arrangement around three scratch segments on an untested and on a tested specimen

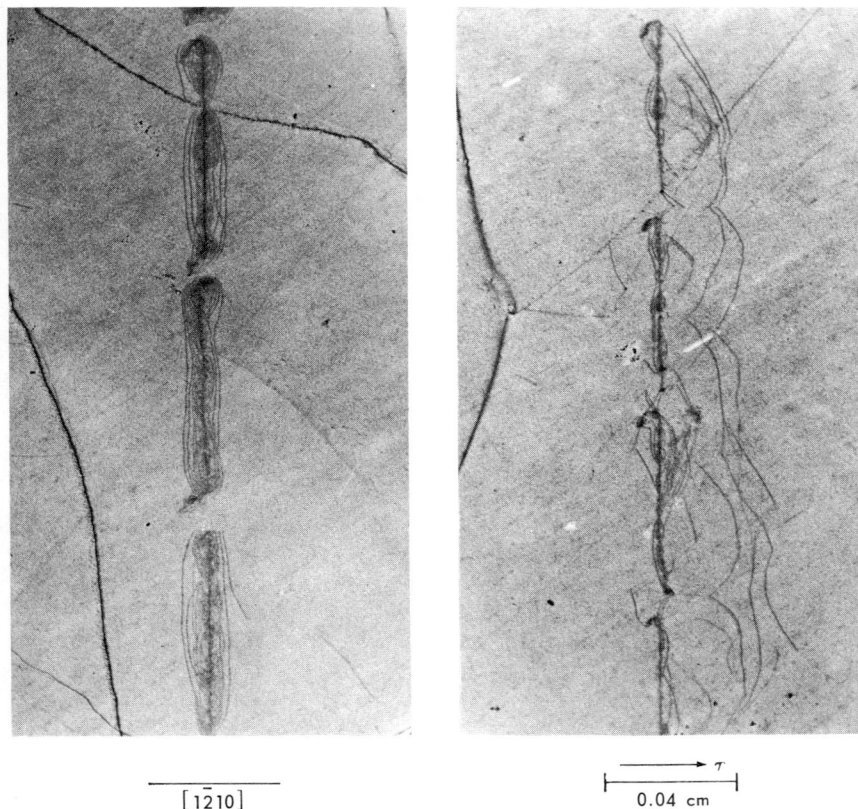

[1̄210]

0.04 cm

Fig. 8 Berg-Barrett photographs of dislocations near a scratch on an untested and on a tested zinc specimen (different specimens). Note dislocation displacements are in the direction of the applied stress.

are shown in Fig. 8. Dislocation motion took place in the direction of the applied shear stress, and varied with radial position on the (0001) surface.

Measurements of the maximum dislocation displacement from each scratch segment were made, and are plotted in Fig. 9 as a function of radial position. These displacements are the sum of the displacement due to scratching and that due to the stress pulse. The displacement due to scratching was taken as the intercept of the line (least-squares fit) thourgh the experimental points. This displacement was subtracted from the measured displacement to obtain the displacement which was produced by the stress pulse.

The plastic strain in the specimens was small compared to the elastic strain produced by the stress pulse. This can be shown by

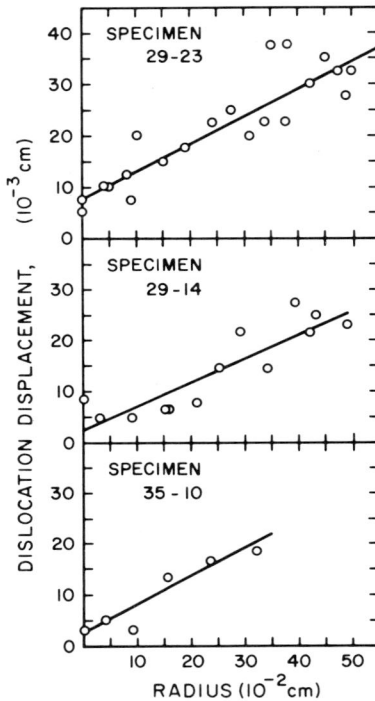

Fig. 9 Dislocation displacement plotted against radial position on the zinc specimens.

considering the dislocation displacements observed and the known density of basal dislocations in the bulk of the crystals (approximately 10^4 cm^{-2}). Therefore, the stress distribution was that associated with the elastic waves, namely, a stress which varies linearly with radius. The maximum stress occurs at the outer radius of the specimen and this stress was deduced from the strain-gage signal. The stress-pulse shape was essentially square, and its duration was independent of radial position at the monel buffer-specimen interface.

Since both stress and dislocation displacement were linear functions of radial position, dislocation velocity is a linear function of applied shear stress. The velocity corresponding to each displacement measurement was calculated (taking into account the small deviations from a square stress pulse) and the results are given in Fig. 10 as a function of stress. The data in Fig. 10 are fitted with the straight line

$$v = v_0 \left(\frac{\tau}{2.94 \times 10^4} \right) \tag{4}$$

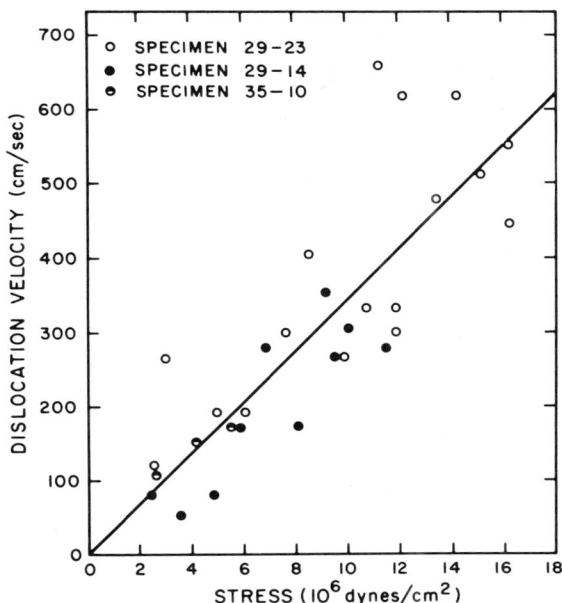

Fig. 10 Edge dislocation velocity in zinc as a function of applied resolved stress.

with $v_0 = 1$ cm/sec and τ in dynes/cm^2. The scatter in the velocity vs. stress data is greater than experimental uncertainties in the measurements of stress ($\pm 10\%$), dislocation displacement ($\pm 2 \times 10^{-3}$ cm), and time ($\pm 5\%$). Sources of this scatter are considered below.

The line tension of a curved dislocation could modify the net force on the dislocation. A typical radius of curvature of the dislocations on a tested specimen is 0.018 cm. The stress required to maintain this radius (0.5 M dynes/cm^2) is small compared to the stresses applied to the dislocations whose displacements were measured.

The final spacing of basal dislocations which moved from the scratch was sufficiently large that their interaction stresses were negligible compared to the applied stress. It was concluded that they were spaced widely enough during most of their motion so that interaction forces between them could be neglected. Interaction of the basal dislocations with forest dislocations was probable. The forest density ranges between 10^2 and 10^3 cm^{-2} in the zinc crystals tested, thus moving basal dislocations intersected forest dislocations every 0.05 cm on the average. Since basal dislocation displacements measured in this work varied from about 0.01 to 0.04 cm, interaction with forest dislocations could be responsible for much of the observed scatter.

The effect of drag due to point defects cannot be determined. Every attempt was made to minimize the number of point defects at the dislocations. High-purity material was used (99.999+% zinc), and the aging time after scratching and before testing was held to a minimum.

3 DISCUSSION

The mixed screw-edge dislocations in copper and the basal edge dislocations in zinc have almost the same linear velocity vs. stress relationship. This velocity vs. stress behavior is compared in Fig. 11 to that in some other materials on which direct measurements have been made. The copper and zinc measurements are unique in that they exhibit the highest velocities in the low-stress range. This indicates relatively weak interactions between the moving dislocations and other lattice defects. In fact, the majority of dislocation drag in the tests on copper and zinc may have been due to phonon interaction.

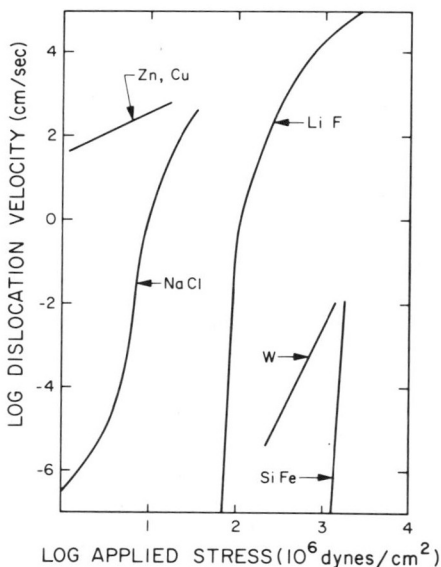

Fig. 11 A comparison of the velocity vs. stress relationship in Cu, Zn, LiF[8], SiFe[9], NaCl[10], and W[11].

A comparison of the direct dislocation drag measurements reported here, and the drag deduced from internal friction measurements in copper is given in Table 1. The damping constant derived from the results of this study, $B = b\tau_0/v_0$ in the notation used here, is in good agreement with the room-temperature measurements of Alers and Thompson[12] and Stern and Granato.[13] Tests are in progress to determine if B is reduced at lower temperatures as predicted by the phonon interaction theories.[15,16] It is is interest to note that the dislocation displacements in the tests reported here were several orders of magnitude greater than those in the internal friction experiments.

TABLE 1 Dislocation Damping Constant in Copper at Room Temperature, 10^{-4} dyne-sec/cm^2	
This study	7
Alers and Thompson[12]	8
Stern and Granato[13]	6.5
Suzuki, et al[14]	0.79

The small rate sensitivity of the flow stress in copper and in basal slip in zinc may be attributed to a strong stress dependence of the density of moving dislocations. Adams et al.[2] suggested that a strong stress dependence of the moving dislocation density occurs when the internal stress amplitude is comparable in magnitude to the applied stress. When there is a small but finite breakaway stress and the internal stress amplitude is comparable in magnitude to the applied stress, dislocations will be trapped whenever the total stress drops to the breakaway stress, and very small changes in the applied stress can change the velocity of a significant number of dislocations from zero to a relatively high value corresponding to the breakaway stress. The rate sensitivity of the flow stress is thereby very small.

The velocity of dislocations on the second-order pyramidal slip system on zinc have a very unusual stress and temperature dependence. The experiments will be reported in detail elsewhere. Briefly, the following observations have been made on dislocations in slip bands on the $\{1\bar{2}12\}$ $<1\bar{2}13>$ slip system in high-purity, acid-machined crystals.

1. At temperatures from 77° to 110°K thermally activated motion seems probable with a stress independent activation energy of about 1/4 eV. Screw dislocations in slip bands exhibit velocities approximately 7× *higher* than edge dislocations.

2. A more complex mechanism of dislocation motion is indicated at temperatures above 110°K. The mobility exponent increases rapidly with temperature above 250°K and the ratio of screw to edge velocity is unity above 250°K.

3. The stress required to maintain a given dislocation velocity goes through a minimum as the temperature is varied between 77°K and 320°K.

ACKNOWLEDGMENTS

The experiments on copper reported here were conducted by my former student W. F. Greenman. The experiments on basal slip in zinc were conducted by my former student D. P. Pope, now a research fellow at the W. M. Keck Laboratory of Engineering Materials. The interesting results on second-order pyramidal slip were obtained by my student Mr. R. C. Blish as part of his thesis research. Stimulating discussions with these students and with Mr. A. P. L. Turner who contributed greatly to our ability to effectively use the Berg-Barrett technique are gratefully acknowledged. The patient counsel of my colleague, Professor D. S. Wood, throughout this work is sincerely appreciated. Mr. G. R. May gave valuable assistance in the growth of crystals and the preparation of test specimens.

This work was sponsored by the U. S. Atomic Energy Commission.

REFERENCES

1. Cottrell, A. H.: "The Relation between the Structure and Mechanical Properties of Metals," Vol. II, p. 456, Her Majesty's Stationery Office, London, 1963.
2. Adams, K.H., T. Vreeland, Jr., and D.S. Wood: Mater. Sci. Eng. 2:37 (1967).
3. Pope, D. P., T. Vreeland, Jr., and D. S. Wood: Rev. Sci. Instr. 35:1351 (1964).
4. Young, F. W., Jr., and J. R. Savage: J. Appl. Phys. 35:1917 (1964).
5. Livingston, J. D.: J. Aust. Inst. Metals 8:15 (1963).
6. Young, F. W., Jr.: J. Appl. Phys. 32:192 (1961).
7. Schultz, J. M., and R. W. Armstrong: Phil. Mag. 10:497 (1964).
8. Johnston, W. G., and J. J. Gilman: J. Appl. Phys. 30: 129 (1959).
9. Stein, D. F., and J. R. Low, Jr.: J. Appl. Phys. 31:362 (1960).
10. Gutmanas, E. Yu., E. M. Nadgornyi, and A. V. Stepanov: Soviet Phys. Solid State 5: 743 (1963).
11. Schadler, H. W.: Acta Met. 12:861 (1964).
12. Alers, G. A., and D. O. Thompson: J. Appl. Phys. 32: 283 (1961).
13. Stern, R. M., and A. V. Granato: Acta Met. 10:358 (1962).
14. Suzuki, T., A. Ikushima, and M. Aoki: Acta Met. 12:1231 (1964).
15. Liebfried, G.: Z. Physik 127:344 (1950).
16. Mason, W. P.: J. Acoust. Soc. Am. 32:458 (1960).

DISCUSSION on Paper by T. Vreeland, Jr.

J. D. CAMPBELL: In your specimens the total strain is presumably proportional to radius, and any local plastic strain must be accompanied by a relaxation of stress. Is it possible that such a stress relaxation is sufficient to affect the dislocation velocity significantly?

T. VREELAND: When the plastic strain is small compared to the elastic strain, the stress relaxation is small compared to the elastic stress. This condition was met in our experiments; thus stress relaxation was small and did not appreciably affect the dislocation velocity.

P. HAASEN: To what extent are the dislocation velocities measured for copper influenced by dislocation interactions (multiple slip, inversed pileups, line tension, presence of the scratch)?

T. VREELAND: Attempts were made to minimize the influence of dislocation interactions in the experiments on Cu, but the success of these attempts is not as easy to estimate as in the measurements on Zn. We measured the displacement of etch pits in Cu which:

(a) started at positions furthest removed from the scratch or at relatively isolated positions;

(b) moved away from scratch damage or substructure;

(c) did not appear to interact with other dislocations along an intersecting glide path.

I estimate interaction stresses equal to a *maximum* of ±20% of the applied stress in the low-stress test, and less for the other tests.

J. WEERTMAN: A student of mine (S. Parameswaran) has been measuring dislocation velocities in Zn using the Gilman-Johnston technique. His preliminary results are:

Stress kg/mm^2	Dislocation velocity cm/sec
5.3	1.3×10^4
7.5	2.4×10^4
16.0	2.8×10^4
22.0	4.0×10^4
49.0	6.0×10^4

The velocity dependence is less than linear. The velocities are comparable, but somewhat slower, with those calculated with your value of $B = 7 \times 10^{-4}$ cgs.

T. VREELAND: The agreement is really quite good for the first two data points (at the low stresses). This gives additional support for a linear velocity dependence up to stresses of at least 10 kg/mm^2.

A. V. GRANATO: (1) Have you been able to measure velocity of edge and screw dislocations separately in the same materials?

(2) Do you have any information on the temperature dependence of dislocation velocities?

T. VREELAND: We have measured edge and screw velocities as a function of stress and temperature in zinc on the second-order pyramidal slip system and the results are briefly summarized in my paper. We have only observed the room-temperature stress dependence of basal edge dislocation velocity in zinc and mixed edge-screw dislocation velocity in copper.

G. HAHN: Could you speculate why you did not observe dislocation arrays in the zinc.

T. VREELAND: The x-ray observations reveal that basal dislocations are only in subboundaries and around the scratches (in the 5 μ layer which is observed). Some of the subboundaries emit dislocations during the stress pulse, but their number is small. This could indicate the necessity for thermal activation for extensive source activity at subboundaries. The edge dislocations which move from the scratches do not cross slip so no loops or debris are left in their wake. We might see a different picture when we displace screw dislocations over comparable distances.

G. ALEFELD: You presented a linear relation between velocity and stress. Do we agree on the following? This relation cannot hold at small stresses, as far as your experiment is concerned. But it will hold at small stresses as far as internal friction measurements are concerned. Both experimental methods are separated by an intermediate stress region, where this linear relation cannot hold. If it could hold at small stresses for your experiments, the dislocation would not be stabilized at a certain position. You need some barriers, many impurities to stabilize the dislocations. The stresses which you apply are sufficient to move dislocations over these obstacles without thermal activation, whereas in internal friction measurements the dislocations oscillate between these obstacles. There is no continuous transition between these two methods, although both methods study the same mechanism.

T. VREELAND: We estimated that our measurements are made at stress levels where internal obstacles such as impurities and forest dislocations are passed without thermal activation. At lower stress-levels dislocation motion must be thermally assisted and dislocation velocity will be very discontinuous when viewed on a scale where the obstacles can be seen. Small changes in applied stress will also cause some of the obstacles to be overcome and discontinuous motion results. The complicated nature of the discontinuous motion can be appreciated by consideration of an extension of the statistical theory of work-hardening presented by Kocks.

We expect this velocity relationship to hold for a dislocation in motion between obstacles in all stress ranges. The interesting point in the comparison between the damping constant found in the velocity measurements and in the internal friction measurements is that the losses are the same in both. This implies that the energy lost in large dislocation motions (200 μ) in high-purity, low-dislocation density ($<10^4$ cm^{-2}) copper crystals is the same as that in the much smaller motions which occur in the internal friction measurements ($\sim 10^{-2}\mu$).

DISLOCATION MOTION AND YIELD STRESS IN PURE COPPER AND ITS DILUTE ALLOY

T. Suzuki

Institute for Solid State Physics,
University of Tokyo

ABSTRACT

Dislocations move in pure copper and copper-nickel alloy quite nonuniformly, which is in marked contrast with the uniform motion of dislocations in LiF found by Johnston and Gilman. The velocity at yield point is 8×10^2 cm/sec in copper and 5×10^2 cm/sec in copper-nickel 0.16% alloy at room temperature. These velocities are in good agreement with those deduced from the phonon-drag coefficients obtained previously in an earlier report. The nonuniformity of dislocation motion is found to be due to the random distribution of impurities or solute atoms, similar to that suggested recently by Kocks, and Foreman and Makin. Taking these dynamical properties of dislocations, a theory of solid-solution hardening is worked out and compared with experiments made on copper-nickel alloys.

1   INTRODUCTION

Although some dislocations in copper can move by the application of an external stress of only a few grams per mm^2 as found by Young,[1] this stress has no relation to the yield stress, which was about 40 g/mm^2 in his crystals. According to our experiment on pure copper and copper-nickel alloy crystals, most dislocations at room temperature are trapped by obstacles after some movement and never move again unless an increased stress is applied. The dislocation motion is neither uniform nor steady, but quite random in nature. In addition, the randomness is found to be due to the random distribution of impurities or solute atoms. Such behaviors of dislocations are in marked contrast to those found in LiF crystals by Gilman and Johnston,[2] which are solution-hardened by Mg^{++} ions, and where a uniform motion of dislocations is found.

At present, it is difficult to estimate theoretically the effects of the random distribution of point obstacles on yield stress. Hence, the yield stress has been discussed by taking some kind of an average of the distribution. It should be noted however, that the yield stress τ_c for the regular distribution of point obstacles is different from σ_c for the random distribution of the same obstacles, i.e., if we put

$$\sigma_c = k\tau_c \tag{1}$$

k is generally not equal to unity.

Very recently, Foreman and Makin[3] made a computer experiment on the motion of a dislocation through the two-dimensional random distribution of point obstacles. Their experiment, which covered the whole range of the interaction strength of the point obstacle with the dislocation, is an extension of Kocks' experiment,[4] which dealt with the infinite strength of the interaction. Although these computer experiments are static in nature, their results are of great importance since they predict that a definite yield point exists even for the random distribution of point obstacles.

In this paper, the dislocation motion and yield stress in copper and copper-nickel alloys are studied experimentally. For the determination of k, the triangular lattice distribution of solute atoms is assumed, which can be adjusted as a function of the interaction energy in accordance with the flexibility of the dislocation at the yield point concerned.

From experimental knowledge of yield stress and dynamical properties, the interaction energy and, then k are determined as

0.10 eV and 0.11, respectively for Cu-Ni alloys. It will be shown that $k = 0.11$ holds for similar metallic solid-solutions as long as the interaction energy is less than 1 eV.

2   DISLOCATION MOTION IN COPPER AND COPPER-NICKEL ALLOYS

Details of experimental procedure and results obtained will be described elsewhere.[5,6,7] Here only the results necessary for the discussion as mentioned above are explained.

2.1   Number of Mobile Dislocations vs. Applied Shear Stress

The ratio of the number of mobile dislocations n_v to the number of grown-in dislocations n_0 increases with increasing stress starting at a very small reduced stress σ/σ_c, both in copper and copper-nickel alloys. All possible dislocations belonging to the primary slip-system move only after the yield point is passed for each crystal.

The fact that the dislocation motion is quite irregular is demonstrated by successive pulse-loading. The ratio of the number of dislocations moved by the first pulse to that by the second pulse of the same amplitude is only ten percent or less at a few g/mm^2 in copper and it decreases with increasing stress. Near the yield point it increases rapidly. Similar behavior is found in copper-nickel alloys, although the ratio is slightly larger than in copper. Most of the dislocations stopped after the first pulse do not move again unless the stress amplitude is increased, even if the pulse time is lengthened. Most barriers, therefore, are concluded to be athermal in copper at all temperatures and in copper-nickel alloys above a certain critical temperature.

Below a certain stress no increase of the number of dislocations was observed. Multiplication of dislocations occurred rapidly near yield point

2.2   Distance of Motion of Dislocations

The irregularity of the dislocation motion most dominantly appears in the distribution of the distance of motion. Figure 1 is an

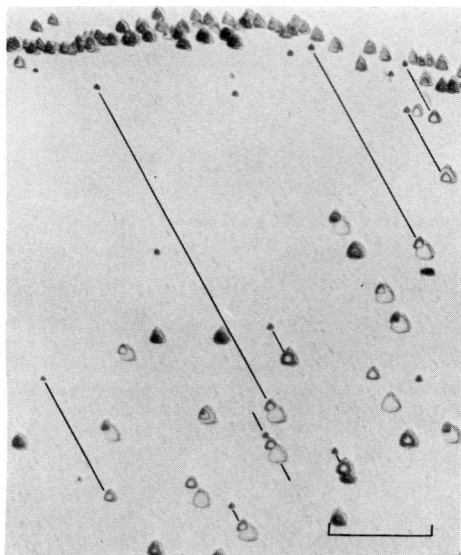

Fig. 1 Dislocation motion in copper by two successive pulse-loadings. The first pulse was 11.8 g/mm^2 - 1 sec and the second one 19.5 g/mm^2 - 1 sec. Irregular motion of dislocations can be seen from this photograph. The scale inserted denotes 100 μ.

example of this irregularity, where successive one-second pulses of 11.8 g/mm^2 and 19.5 g/mm^2 were applied. Some dislocations, moved for longer distances by the first pulse, moved for shorter distances by the second pulse; on the other hand, the distances of motion during a single pulse are quite different for different dislocations.

Figures 2 and 3 show the number of dislocations moved plotted against their distance of motion in copper and 0.16 at.% NI alloy, respectively. The distance varies from several microns to about 2000μ in both materials by the application of a stress pulse close to the yield stress.

The average distance of motion thus obtained $\overline{L}$ is independent of the length of pulse and testing temperature in copper. For all specimens of copper-nickel alloy in which the concentration of nickel varies from 0.011 to 5.4 at.%, it gives $70\mu \leq \overline{L} \leq 200\mu$ at yield point.

2.3   Multiplication of Dislocations

There seems to be a certain critical distance of motion L_c for the multiplication. Generally, it follows that $L_c > \overline{L}$ as long as $\sigma < \sigma_c$. However, we do not know what determines L_c. The Frank-Read source in crystals where the density of dislocations is lower

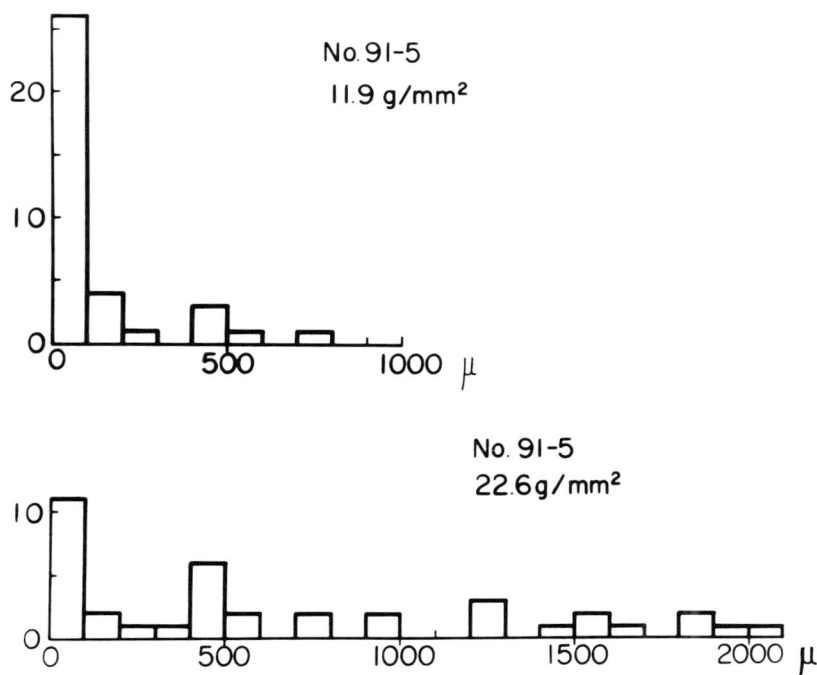

Fig. 2 Number of dislocations moved plotted against their distance of motion in copper at 77°K. The time of duration of a square-shaped pulse was 0.0005 sec.

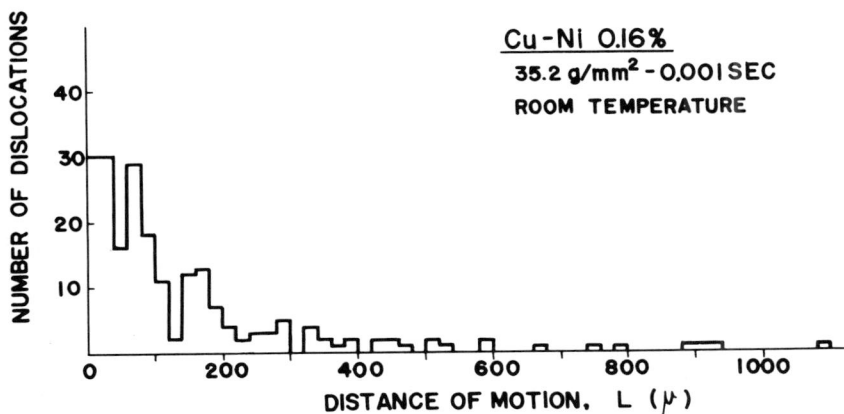

Fig. 3 Number of dislocations moved plotted against their distance of motion in copper-nickel 0.16 at. % alloy at room temperature. The pulse-time was 0.001 sec, and the stress was 35.2 g/mm^2 (= 0.78 σ_c).

than 10^4 cm^{-2} or so may be a segment of the line which lies in a single slip plane (neglecting the existence of a small number of jogs). The length of these segments L_s is probably a few hundred microns. The segments of the dislocations inside crystals, which do not end at the specimen surfaces, and which move for a longer distance than $L_s/2$ can multiply below or above the yield point. The critical stress for the multiplication, however, has no relation to the yield stress.

2.4　Dislocation Velocity

As explained in Sec. 2.2, the average distance of motion $\overline{L}(\sigma)$ is independent of pulse time Δt for copper and the distance of motion L at yield point distributes from several to about 2000 μ for copper and copper-nickel alloys. Therefore, to define the average velocity by $\overline{L}/\Delta t$ seems to be impractical. Hence, we define the

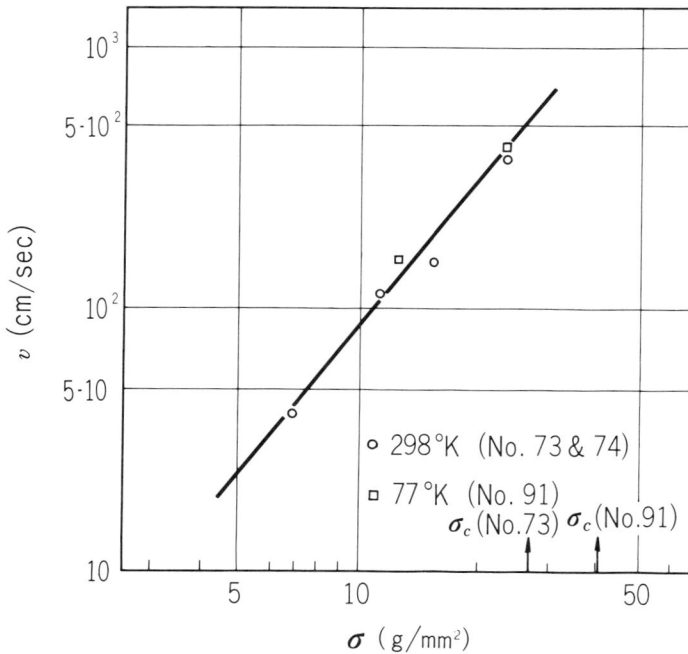

Fig. 4 Velocity of free flight of dislocations in copper. The velocity was obtained from $L_{max}/\Delta t$, that is, the longest distance of movement divided by the pulse-time equal to 0.0005 sec.

velocity of free flight (i.e., the intrinsic velocity) for the motion of dislocations between static barriers. It is definable for copper at all temperatures and for the alloys above a certain temperature well below room temperature.

To measure the velocity of free flight, we applied a 0.0005-sec pulse and observed the maximum distance of motion. Since a limitation exists for the observation, the observed maximum velocity may be the lower limit of the velocity of free flight.

Figure 4 shows the result for copper. In the figure it is seen that the velocity of dislocations reaches as high as 8×10^2 cm/sec at yield point, which is almost independent of temperature. For copper-nickel alloys, the velocity at room temperature is 3 to 5×10^2 cm/sec in the case of 0.16 at.% alloy. These velocities are, needless to say, quite large compared with other materials as seen in Table I.

If the observed velocity is truely the velocity of free flight, then it is expected that it should agree with that estimated from the

TABLE 1 Dislocation Velocity at Yield Point in Various Materials : $v = (\sigma / \sigma_0)^m$

Material	Temperature (°K)	σ_c (kg/mm²)	v (cm/sec)	m	σ_0 (kg/mm²)	Author
Si	873	33.4	1×10^{-4}			Suzuki and Kojima[8]
	973	17.4	5×10^{-4}	1.6	1660	
	1073	8.8	40×10^{-4}			
LiF	77	2.8	1×10^{-3}	16.5	0.54	Johnston and Gilman[2]
	300	0.9	1×10^{-3}			
W	77	58	15	9.3	49	Schadler[9]
	300	19	0.07	4.8	32	
Fe-Si 3.35%	78	29	1×10^{-3}	44	38	Stein and Low[10]
	198	18	1.2×10^{-4}			
	298	14	4×10^{-5}	35	20	
Cu	77	0.026	0.8×10^3	2	0.001	Marukawa[5]
	298	0.026	0.8×10^3			
Cu-Ni 0.16%	298	0.046	$0.3 - 0.5 \times 10^3$	–	–	Suzuki and Ishii[7]
0.35%	298	0.076	0.47×10^3	5.4	0.022	

overdamping experiment. Neglecting the terms of line tension and inertia, we obtain a relation from the equation of motion as follows:

$$\sigma = \frac{B v}{b} \tag{2}$$

where B is the overdamping constant, which is 1.2×10^{-4} cgs for copper[11] and 2.8×10^{-4} cgs for copper–nickel 0.1 at% alloy[12] at room temperature. For the external stress $\sigma = 22$ g/mm^2, Eq. (2) gives $v = 4.8 \times 10^2$ cm/sec. The etch-pit experiment gives $v = 4 \times 10^2$ cm/sec as seen in Fig. 4. From this figure we have

$$v = \left(\frac{\sigma}{\sigma_0}\right)^m \qquad m = 2 \quad \text{and} \quad \sigma_0 = 1 \text{ g/mm}^2$$

which, however, does not give a linear relation as Eq. (2). The above agreement is also satisfactory in copper–nickel alloys.

3   YIELD STRESS OF COPPER AND COPPER-NICKEL ALLOYS

3.1   Yield Stress vs. Initial Density of Dislocations

The yield stress of copper is independent of the initial density of dislocations n_0, in the range of the density of 1×10^3 to 10^5 cm^{-2}. Below 1×10^3, there is a minimum in yield stress at $n_0 = 4.4 \times 10^2$ cm^{-2}. The minimum stress is equal to 9.7 g/mm^2. The decrease of yield stress below 1×10^3 cm^{-2} is considered to be due to a decrease in the elastic interaction between the dislocations stopped by point obstacles and other moving dislocations. Above this density yield stress is determined by the stress needed to move the dislocations stopped by point obstacles and, therefore, it does not depend on n_0. Therefore, at least for copper, obstacles against moving dislocations are not only point obstacles, but also the dislocations stopped by these point obstacles.

Yield stresses described in the following section, for both copper and copper–nickel alloys, are those that are independent of the initial density of dislocations.*

---

*Copper crystals described here were not zone-refined. When the raw materials were zone-refined 20 times, the yield stress of the crystals was 13 g/mm^2.

3.2   Yield Stress vs. Concentration of Solute Atoms

It is important to study precisely the dependence of yield stress on the concentration of solute atoms c because information concerning the distribution of solute atoms as point obstacles is obtained. Figure 5 shows σ_c vs. c, in which the yield stress is proportional to $\sqrt{c}$, when c is smaller than a critical concentration c_0 (about 3 at.%) and increases linearly for $c > c_0$.

The $\sqrt{c}$-dependence of yield stress is undoubtedly due to the random distribution of solute atoms. It should be noted that no jump in stress is observed at c_0, as would be expected if the linear dependence above c_0 is due to the segregation of solute atoms along dislocation lines. Observations of the dislocation motion also give no evidence of the segregation.[7]

3.3   Yield Stress vs. Temperature

Yield stress of copper is independent of temperature except for that of the rigidity of modulus. As shown in Fig. 6, however,

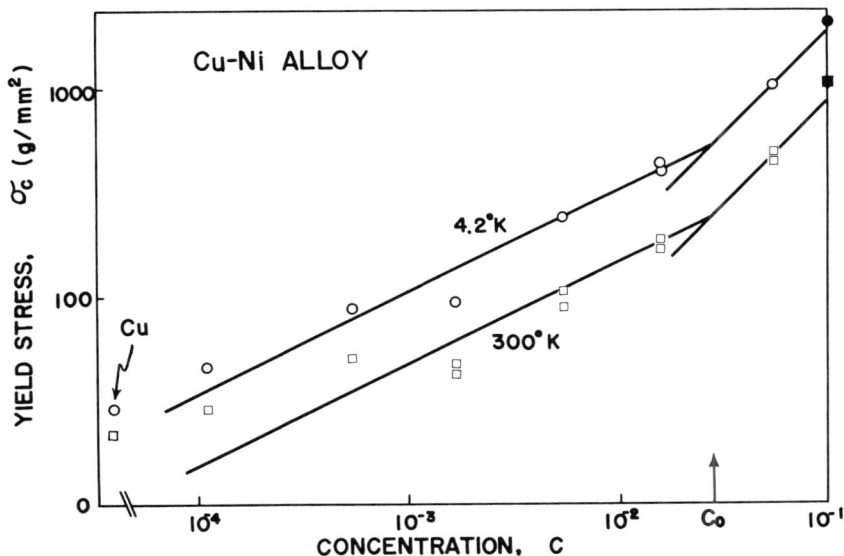

Fig. 5 Yield stress (resolved shear stress) of copper-nickel alloys as a function of the atomic concentration of nickel. Black points are taken from H. Suzuki's data.[13]

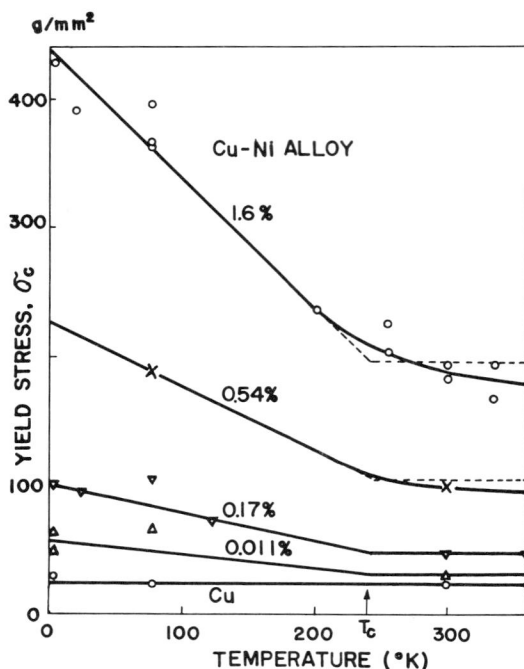

Fig. 6 Yield stress of copper-nickel alloys plotted
against testing temperature.

yield stress of copper-nickel alloys depends almost linearly on
temperature below T_c = 240°K and becomes constant above T_c at
least for dilute alloys.

4 YIELD STRESS AND DISLOCATION MOTION IN SOLID SOLUTION: THEORETICAL

4.1 Triangular Lattice Scheme as the Regular Distribution of Solute Atoms

To obtain the relation given by Eq. (1), first, we consider a
triangular lattice distribution as the regular distribution (see Fig.
7). Point obstacles are at A, B_1, C, B_2, etc., and the solid line is a
dislocation line to be considered. Although the point obstacles in

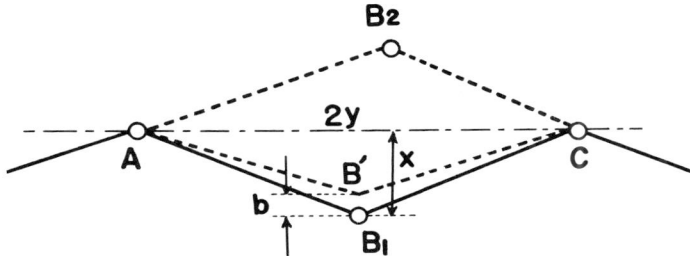

Fig. 7 Triangular lattice distribution of solute atoms. A, B_1, C, B_2, etc. indicate solute atoms and solid line is a dislocation line. Broken lines trace its successive movement. Attractive interaction between the solute atom and the dislocation was assumed for drawing. This is not essential, however, for the following discussion.

Fig. 7 are all attractive, there is no essential change in the following equations even if a part or all of them are repulsive. By the application of stress, the dislocation moves from AB_1C to AB_2C which is taken as a unit for the movement of the dislocations.

Although Fig. 7 is similar to the scheme proposed by Mott[14] and Friedel[15] for the discussion of solution hardening, this idea is physically different from that as follows.

The x and y of the triangular lattice are determined by the following two conditions. Since the interaction concerned is short range in nature, the solute atoms in the atomic planes next to the slip plane of the dislocation are considered in f.c.c. metals. Taking {110} planes as the slip plane in b.c.c. metals, interstitial atoms in two interstitial planes adjacent to the slip plane, are assumed to give an equal effect on the moving dislocation. Then the first condition gives

$$xy = \frac{b^2}{\alpha c} \tag{3}$$

and

$\alpha = 4.61$ for f.c.c. (substitutional solute atoms)

$\alpha = 6.36$ for b.c.c. (interstitials)

Next, we determine the triangular lattice distribution so as to give the maximum barrier to the moving dislocations. This is the second condition. It will be convenient to discuss the following two cases separately. The procedure of calculation is very similar to Friedel's.

(a) *Lower Concentration* $(c < c_0)$

The energy of the interaction per unit length is given by*

$$E_1 = \frac{|W_{max}|}{y} \tag{4}$$

The difference in the line energy per unit length of the dislocation at $B'(= x - b)$ and $B_1 (= x)$ will be given by

$$E_2 \simeq \frac{\mu b^3}{4y^2} (2x - b) \tag{5}$$

where we assumed $x \ll y$. Equation (5) can be rewritten, using
Eq. (3), as

$$E_2 = \frac{\mu b^2 \alpha^2}{4b^2} x^2 \left(\frac{2x}{b} - 1 \right) c^2 \tag{6}$$

The second condition is

$$\frac{\partial}{\partial x} (E_1 - E_2) = 0$$

Thus, it follows finally that

$$\left.\begin{array}{l} x = \left(\dfrac{2W}{3\alpha\mu b} \right)^{1/2} c^{-1/2} \\[4mm] y = \dfrac{b^2}{\alpha} \left(\dfrac{2W}{3\alpha\mu b} \right)^{-1/2} c^{-1/2} \end{array}\right\} \tag{7}$$

(b) *Higher Concentration* $(c > c_0)$

Since the relation $x \gg b$ is not valid in this case, it follows
that

$$E_2 \simeq \frac{\mu b^2}{4} \frac{x^2}{y^2}$$

---

*For the sake of brevity, the maximum binding energy $|W_{max}|$ will simply be designated
by W.

$$x = \frac{W^{1/3}}{\alpha^{1/3}\mu^{1/3}} c^{-1/3}$$

$$y = \frac{\alpha^{-2/3}\mu^{1/3}b^2}{W^{1/3}} c^{-2/3}$$

(8)

4.2 Critical Interaction Energy for Triangular Lattice Distribution

As explained in Sec. 4.1, it should hold that $x < y$. From this, we obtain the critical interaction energy as follows:

$$W_c = \frac{3}{2}\mu b^3$$

(9)

Bearing in mind that the lattice constant d of the square lattice for the same concentration should be smaller than $2y$, we will have another condition that $2y > d$. It gives

$$W_c = 0.98\,\mu b^3$$

(10)

Equation (10) gives a smaller value to W_c than Eq. (9); for example, for copper W_c is about 1 eV and for interstitials in iron it is 0.79 eV. It is to be noted that 0.79 eV is very close to the maximum binding energy of carbon in iron.

4.3 Yield Stress

To simplify the following argument, we take a triangular potential for the interaction of solute atom with a dislocation. Then, the yield stress will be given as follows:

(a) *Lower Concentration* $(c < c_0)$

Following Friedel,[15] we write the strain rate as

$$\dot{\epsilon}_{sp} = \frac{n}{2y}\,2xyb\,\frac{\nu b}{y}\,\exp\left(-\frac{V - (\tau - \tau_0)\,b^2 y}{kT}\right)$$

(11)

where $V = |E_1 - 2E_2|y = W/3$, n is the *effective number* of mobile dislocations, which will be discussed later, ν is the atomic vibrational frequency of the order of 10^{13} sec^{-1}, $\underline{k}$ is the Boltzmann constant and τ_0 is the stress independent of temperature which will be explained in (c). Taking the logarithm of Eq. (11), we have yield stress as follows:

$$\tau_c = \tau_0 + \frac{1}{3}\left(\frac{2}{3}\right)^{1/2} \alpha^{1/2} \frac{W^{3/2}}{\mu^{1/2} b^{9/2}} c^{1/2} \left[1 - \frac{T}{T_c}\right] \tag{12}$$

where

$$T_c = \frac{W}{3\underline{k}} \ln\left[\left(\frac{2W}{3\mu}\right)\left(\frac{n\nu}{b\dot{\epsilon}_{ext}}\right)\right] \tag{13}$$

It is assumed in the derivation of Eq. (12) that the dynamical condition $\dot{\epsilon}_{sp} = \dot{\epsilon}_{ext}$ holds at yield point, where $\dot{\epsilon}_{ext}$ is the strain rate forced by a hard testing machine and $\dot{\epsilon}_{sp}$ is that internally possible for a specimen.

(b) *Higher Concentration* $(c > c_0)$

Similarly, we obtain yield stress as follows:

$$\tau_c = \tau_0 + \frac{\alpha W}{2b^3} c \left[1 - \frac{T}{T_c}\right] \tag{14}*$$

where

$$T_c = \frac{W}{2\underline{k}} \ln\left[\left(\frac{2W}{3\mu}\right)\left(\frac{n\nu}{b\dot{\epsilon}_{ext}}\right)\right] \tag{15}$$

(c) *Stress* τ_0

Using the configuration illustrated in Fig. 7, the dislocation is forced by the stress τ_0, that is, $\tau_0 b(x - b)y = 2E_3 y$, where

---

*By increasing the concentration in the range $c > c_0$, the term τ_0 will disappear from Eq. (14). This is because the movement from AB_1C to AB_2C will be aided by thermal agitation.

$$E_3 = \frac{\mu b^2}{2y} \left(\left[(x - b)^2 + y^2 \right]^{1/2} - y \right) \simeq \frac{\mu b^2}{4} \frac{(x - b)^2}{y^2}$$

Accordingly, using Eq. (7) or (8), it follows that for $c < c_0$,

$$\tau_0 = \frac{1}{3} \left(\frac{2}{3} \right)^{1/2} \alpha^{1/2} \frac{W^{3/2}}{\mu^{1/2} b^{9/2}} c^{1/2} \tag{16}$$

for $c > c_0$,

$$\tau_0 = \frac{\alpha W}{2b^3} c \tag{17}$$

In conclusion, $\tau_c \propto \sqrt{c}$ for $c < c_0$ and $\tau_c \propto c$ for $c > c_0$. In both cases, yield stress is given by $\tau_c = 2\tau_0$ at $0\,^\circ$K. The above results all agree qualitatively with experimental results on copper-nickel alloys. A quantitiative examination will be given below.

4.4 Critical Concentration c_0

The critical concentration c_0 is given from Eqs. (7) and (8) by $x(c < c_0) = x(c > c_0)$, or from Eqs. (12) and (14) by

$$\tau_c(c < c_0) = \tau_c(c > c_0) \tag{18}$$

Both give the same result as

$$c_0 = \frac{1}{\alpha} \left(\frac{2}{3} \right)^3 \frac{W}{\mu b^3} \tag{19}$$

The concentration given by Eq. (19) is of course for the regular distribution of solute atoms, not for the random distribution as in real crystals considered. To estimate the latter from Eq. (19), it is necessary to know the value of k in Eq. (1). This is because k is different for the regular distribution for $c < c_0$ and $c > c_0$, respectively. The comparison, therefore, will be made after both values of k are determined.

4.5 Effective Number of Mobile Dislocations

As mentioned before, it must hold at yield point that $\dot{\epsilon}_{sp} = \dot{\epsilon}_{ext}$. When we write

$$\dot{\epsilon}_{sp} = bnv \tag{20}$$

v should be a steady velocity, in other words, Eq. (20) is valid for a uniform motion of dislocations.

In the metallic solid solutions considered, the motion of dislocations being nonuniform, the only significant velocity is that of free flight defined in Sec. 2.4, at least above a certain temperature. Therefore, to describe the macroscopic strain rate in terms of the dislocation motion, a physical translation of irregular motion into a regular motion is necessary.

Therefore, the effective number of dislocations in Eq. (11) is defined as*

$$n = \frac{\bar{A}}{A_m} n_v \tag{21}$$

where $\bar{A}$ is the average area swept by the dislocations, A_m is the maximum area covered by the uninterrupted motion of dislocations, n_v is the total number of mobile dislocations as defined before, and v is supposed to be given by the velocity of free flight at $T \geq T_c$.

At the critical temperature, we put $v = 5 \times 10^2$ cm/sec, $\dot{\epsilon}_{sp} = \dot{\epsilon}_{ext} = 4.2 \times 10^{-4}$ sec^{-1} and $b = 2.6 \times 10^{-8}$ cm, which are valid at the yield point of 0.16 at.% Ni alloy. Equation (20) gives

$$n = 32 \text{ cm}^{-2}$$

The effective number of mobile dislocations thus obtained is surprizingly small. However, this is verified from an independent argument as follows.

In Eq. (21), $\bar{A}$ and A_m are indeed functions of stress and related to the average distance $\bar{L}$ and L_m in Sec. 2.2, respectively. To determine the order of magnitude, it can be rewritten as

$$n = \left(\frac{\bar{L}}{L_m}\right)^2 n_v \tag{22}$$

Since $\bar{L}$ is approx. 100 μ, $L_m^2 = 0.1$ cm^2 and $n_v = 5 \times 10^4$ cm^{-2}, Eq. (22) gives $n = 50$ cm^{-2}.

4.6 Determination of W and k

The critical temperature T_c given by Eq. (13) or Eq. (15) is independent of the distribution of solute atoms. Unknown parameters

---

*The present author originally suggested a slightly different form for eq. (21). Eq. (21) is due to the suggestion given by U. F. Kocks at Argonne National Laboratory after the Colloquium.

in these equations are W and n. Therefore, if we insert $n = 32$ cm^{-2} obtained above into these equations, W can be estimated. Taking $T_c = 240$°K, $\mu = 4.9 \times 10^{11}$ dynes/cm^2, we then have

$$W = 0.10 \text{ eV}$$

Equations (12) and (14), into which $W = 0.01$ eV is inserted, give τ_c at 0°K and by comparison with experimental value σ_c, we have, finally,*

$$k = 0.11 \quad \text{for} \quad c < c_0$$
and

$$k = 0.04 \quad \text{for} \quad c > c_0$$

5 EXAMINATION OF RESULTS

5.1 Critical Concentration c_0

Equation (18) can be rewritten as $\sigma_c(c < c_0) = \sigma_c(c > c_0)$ using the k values above. Then, the critical concentration for the random distribution of solute atoms is found to be 1.1×10^{-2}. This is to be compared with the observed concentration 3×10^{-2}. The critical concentration for the regular distribution directly given by Eq. (19) is 1.4×10^{-3}.

5.2 Validity of $k = 0.11$ for Other Solid Solutions

Very few data are available unfortunately, for examining the validity of $k = 0.11$ for other solid solutions except for copper-aluminum alloys studied by Koppennal and Fine,[16] and iron-carbon alloys studied by Stein and Low.[17]

For copper-aluminum 5 at.% alloy, $\sigma_c(0°K) = 2.39 \times 10^8$ dynes/cm^2. Taking $k = 0.11$, it gives $W = 0.23$ eV. For iron-carbon 44 ppm alloy, if we take $\sigma(0°K) = \sigma_c - \sigma_{PN} = 8.96 \times 10^8$ dynes/cm^2, we have $W = 8.5$ eV. Here σ_{PN} is the yield stress at 0°K for the purest crystal, which is supposed to be due to the Peierls-Nabarro force.

In the following, we compare W obtained above with theory. If we assume that W is due to the size effect, the maximum binding

*A comparison of the present theory with computer experiments[3,4] is made in the reference (7).

energy to the edge dislocation is given by Friedel[15] as

$$| W_{\max} | = \frac{1}{2\pi} \frac{1 + \nu}{1 - \nu} \mu 3 \Omega \eta \tag{23}$$

where ν is Poisson's ratio, Ω is the atomic volume of a solvent atom and η is given by $\eta = \Delta\Omega/3\Omega$, $\Delta\Omega$ being the difference in atomic volume between solvent and solute atoms.

Since it is estimated as $\eta = -0.031$ for nickel in copper,[18] it gives $| W_{\max} | = 0.12$ eV. Similarly, $\eta = +0.064$ for aluminum in copper[18] it gives $| W_{\max} | = 0.24$ eV. These are in very good agreement with W already obtained.

On the other hand, carbon in iron generates a tetragonal strain $\Delta\epsilon$ and Fleisher[19] estimated the maximum binding energy as

$$| W_{\max} | \approx \frac{\sqrt{2}}{3\pi} \mu b^3 \Delta\epsilon \tag{24}$$

$\Delta\epsilon$ is estimated as $0.41 \sim 0.44$ by Stein and Low[17] and Cochardt et al[20] and, therefore, it gives $| W_{\max} | = 0.51 \sim 0.54$ eV. This is quite small compared with the obtained value that is 8.5 eV. As remarked in Sec. 4.2, the criticial interaction energy for an interstitial in b.c.c. iron is about 0.79 eV for the triangular lattice scheme, which is so close to the theoretical binding energy. Therefore, we should abandon an idea to apply the relation $k = 0.11$ for the triangular lattice scheme to this case.

The W observed is of course the sum of various interaction energies, i.e., elastic, electrical, chemical, etc. The agreement in copper-nickel and copper-aluminum alloys seems to be somewhat accidental. It is less doubtful, however, that the elastic interaction provides the most important contribution in which solid solutions as above, as Oren et al[21] and Fleischer[19] discussed. Although we have few data, $k = 0.11$ is thus likely to be valid for a fairly wide range of W and a variety of materials.

6   A REMARK ON THE DISLOCATION VELOCITY IN SOLID SOLUTIONS

Barriers against moving dislocations in the range of temperature $0 < T < T_c$ are not considered to be athermal, and we cannot define clearly the dislocation velocity in a random solid solution below the yield point. As far as $k = 0.11$ holds in this temperature range, however, the dislocation velocity at the yield point in real crystals must be expressed from Eq. (11) by

$$v = \frac{x}{y} \nu b \exp\left(-\frac{W}{3kT}\right) \exp\left[\frac{(\tau - \tau_0)\, b^2 y}{kT}\right]$$

or it can be written as

$$v = \frac{2}{3} \frac{W}{\mu b^2} \nu \exp\left(-\frac{W}{3\underline{k}T}\right) \exp\left[\frac{(\sigma - \sigma_0)(b^2 y/k)}{\underline{k}T}\right] \tag{25}$$

For copper-nickel 0.16 % alloy, using $W = 0.10$ eV, $k = 0.11$ and $\dot{\epsilon}_{ext} = 4.2 \times 10^{-4}$ sec^{-1}, Eq. (25) is estimated as shown in Fig. 8. The velocity increases linearly with $1/T$ and, accordingly, the random motion of dislocations is strengthened at lower temperatures. In other words, the effective number of mobile dislocations decreases rapidly with lower temperature compared with the total number of mobile dislocations.

The nonuniformity of the dislocation motion is due to the high intrinsic velocity and a small interaction energy W of randomly distributed solute atoms with dislocations. If the Peierls-Nabarro force or W is large, which may be the case for the dislocations in LiF (Mg^{++}) or in iron (carbon interstitials), then the dislocation velocity is so low that n in Eq. (20) will be taken without any important error as the total number of mobile dislocations, and v as the average velocity as assumed by Johnston.[22]

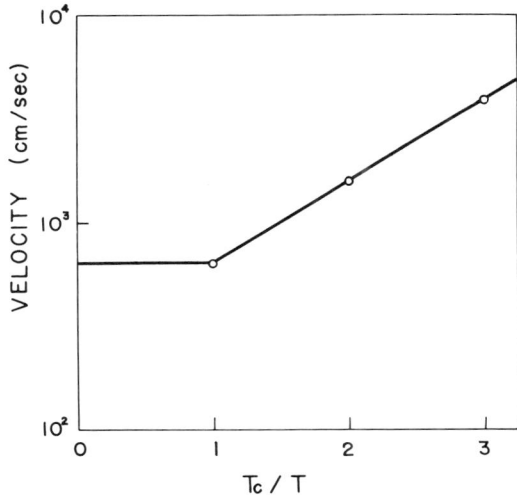

Fig. 8 Dislocation velocity as a function of temperature in copper-nickel 0.16 at. % alloy calculated from Eq. (25).

REFERENCES

1. Young, F. W. Jr.,: *J. Appl. Phys.* 33:963 (1962); ibid., 32:1815 (1961).
2. Johnston, W. G., and J. J. Gilman: *J. Appl. Phys.* 30:129 (1959).
3. Foreman, A. J. E., and M. J. Makin:*Phil. Mag.* 13:911 (1966).
4. Kocks, U. F.: *Phil. Mag.* 13:541 (1966).
5. Marukawa, K.: *J. Phys. Soc. Japan* 22:499 (1967).
6. Suzuki, T., K. Marukawa, and T. Imura: "Dislocation Motion and Yield Stress in Copper," to be published.
7. Suzuki, T., and T. Ishii: Proceedings of International Conference on the Strength of Metals and Alloys (Tokyo, Sept., 1967); see also Tech. Rep. of Inst. for Solid State Physics, Ser. A., No. 283 (Nov., 1967).
8. Suzuki, T., and H. Kojima: *Acta Met.* 14:913 (1966).
9. Schadler, H. W.: *Acta Met.* 12:861 (1964).
10. Stein, D. F., and J. R. Low, Jr.: *J. Appl. Phys.* 31:362 (1960).
11. Suzuki, T., A. Ikushima, and M. Aoki: *Acta Met.* 12:1231 (1964).
12. Ikushima, A., and T. Kaneda: private communication.
13. Suzuki, H.: "Dislocations and Mechanical Properties of Crystals," p. 361, John Wiley & Sons, Inc., 1957.
14. Mott, N. F.: "Imperfections in Nearly Perfect Crystals," p. 173, John Wiley & Sons, Inc., 1952.
15. Friedel, J.: "Dislocations," p. 379–393, Pergamon Press, New York, 1964.
16. Koppenaal, T. J., and M. E. Fine: *Trans. Met. Soc. AIME* 224:347 (1962).
17. Stein, D. F., and J. R. Low, Jr.: *Acta Met.* 14:1183 (1966).
18. Fleisher, R. L.: *Acta Met.* 11:203 (1963).
19. Fleisher, R. L.: "Strengthening Mechanism in Solids," p. 93, A.S.M., Metals Park, (1962).
20. Cochardt, A. W., G. Schoeck, and H. Wiedersich: *Acta Met.* 3:533 (1955).
21. Oren, E. C., N. F. Fiore, and C. L. Bauer: *Acta Met.* 14:245 (1966).
22. Johnston, W. G.: *J. Appl. Phys.* 33:2050 (1962).

DISCUSSION *on Paper by T. Suzuki*

P. HAASEN: A survey of solute solution hardening of copper single crystals[*][†][‡] results in a somewhat different picture than Dr. Suzuki has drawn so beautifully. The critical shear stress rises stronger than linearly from a plateau near 400°K toward low temperatures. The uncertain extrapolation to $T = 0$°K yields a stress τ_0 which depends linearly on solute concentration between 1 and 10 at % solute. The plateau stress τ_p, on the other hand, depends on concentration as $c^{1/2}$ for quite a number of solutes in the same range (zinc,[§] gallium, and germanium,[♦][▲]

---

[*]Haasen, P.: *Z. Metallkde* 55: 55 (1964).

[†]Haasen, P.: in "Physical Metallurgy," p. 821, R. W. Cahn (ed.), North Holland, 1965.

[‡]Haasen, P.: *Kristall und Technik*, 1967, in press.

[§]Suzuki, H.: in "Dislocations and Mechanical Properties," p. 361, John Wiley and Sons, N.Y., 1957.

[♦]Haasen, P., and A. H. King: *Z. Metallkde* 51, 722 (1960).

[▲]Peissker, E.: *Acta Met.* 13: 419 (1965).

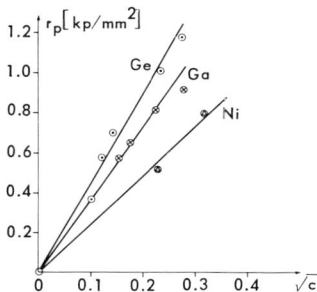

Fig. 1 Solid solution hardening of copper by Ge, Ga, and Ni.

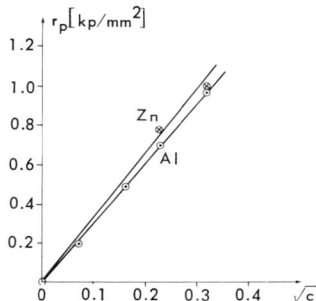

Fig. 2 Solid solution hardening of copper by Al and Zn.

aluminum* and nickel[†] (see Figs. 1 and 2). (Here the critical stress for pure Cu at 400°K was subtracted as 100 g/mm².) The slope of the straight lines in Figs. 1 and 2, $d\tau_p/dc$, is plotted vs. the combined misfit parameters, $|\eta' - 3\delta|$ and $|\eta' - 16\delta|$, respectively, in Fig. 3. Here $\delta = (1/a)(da/dc)$ is the size misfit, $a =$ lattice parameters, $\eta' = \eta/(1 + |\eta|/2)$, $= (1/\mu)(d\mu/dc)$ is the modulus misfit, $\mu =$ shear modulus. The factors 3 and 16 apply to screw and edge dislocations, respectively, according to Fleischer.[‡][§] A monotonic linear re-

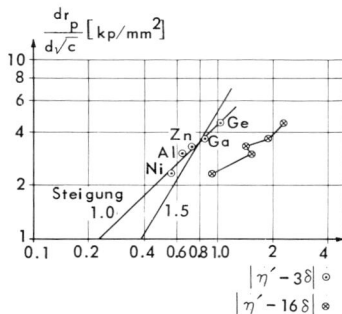

Fig. 3 Slope of strain hardening curves as a function of the misfit parameter.

lation obtains between $|\eta' - 3\delta|$ and $d\tau_p/dc$ according to Fig. 3 instead of a $|\eta - 3\delta|^{3/2}$ relation predicted by Fleischer. The variation of the combined misfit parameter is, however, not very large and the parameter itself is nearly one. For the absolute value of τ_p, Fleischer predicts in terms of $(G\sqrt{c}\,[\eta' - 3\delta]^{3/2})$ the value of $(1/760)$ while from Fig. 3 one reads $(1/950)$. Considering the assumptions the theory has to make, this must be regarded

*Koppenaal, T. J., and M. E. Fine: Trans. AIME 221, 1178 (1961).

†Suzuki, H.: in "Dislocations and Mechanical Properties," p. 361, John Wiley and Sons, N.Y., 1957.

‡Fleischer, R. L.: Acta Met. 9: 996 (1961); 11: 203 (1963).

§Fleischer, R. L.: in "Strengthening of Metals," p. 93, D. Peckner (ed.), Reinhold, N.Y., 1964.

as good agreement. It appears to be difficult to fit the above data to a linear $\tau_P(c)$ relations suggested by T. Suzuki in the $c = 1$ to 10 at % range.

T. SUZUKI: As regard to Fig. 1 in your comment, I hope you may compare it with Fig. 5 of the present paper (Cu-Ni alloys). In the latter, the data due to H. Suzuki are also cited as black points, which surely are not in the $\sqrt{c}$-dependence but in the c-dependence region. I have some doubt concerning your procedure of taking the extra yield stress value by subtracting the yield stress of pure copper, not only concerning the magnitude of 100 g/mm^2, but also due to its physical meaning involved. Concerning the dependence on the misfit parameter, my theory is exactly the same as Fleisher's, that is, $\overline{W}^{3/2}$ in my notation [Eq. (16)]. In the range $c > c_0$, however, it is linear to $\overline{W}$ [Eq. 17)]. Nevertheless, I thank you very much for your work which shows that the yield stress of many copper alloys may have the $\sqrt{c}$-dependence. Finally, in answer to your last question, I used a simple force-distance relation, which seems to suffice for such a system of Cu-Ni, probably because of the weak interaction. However, it is not difficult to use a much more realistic relation, which should give different temperature dependence and concentration dependence as regard to the yield stress at 0°K.

T. VREELAND, JR.: Suzuki's data shows a broad distribution of edge dislocation displacemnts in his stress pulse tests. Our data for mixed edge-screws (at somewhat higher stresses) show a more narrow distribution. In either case, we must conclude that ρ_m in the equation $\dot{\epsilon} = \rho_m b v$ is a small fraction of the total ρ in the usual type of stress-strain test.

T. SUZUKI: It is interesting to note that if I took average velocities properly, they seem to be almost equal to yours. This is because the dislocations moved for larger distance, say, more than several hundred microns (Figs. 2 and 3), are rather small in number compared to those moved for shorter distances. To compare the dislocation velocity obtained by the etch-pit method with that deduced from the overdamping data, however, we should not neglect the maximum displacement made by dislocations, no matter how small these are in number.

With respect to the equation $\dot{\epsilon} = \rho_m b v$, which is important for the discussion of such macroscopic, mechanical properties as yield stress, flow stress, etc., I completely agree with you, since $\bar{v}$ as you defined is still large, and ρ_m is only a fraction of the total mobile dislocations.

STRESS WAVE PROFILES
IN SEVERAL METALS

J. W. Taylor

University of California,
Los Alamos Scientific Laboratory

ABSTRACT

Under a wide range of impact loading conditions the motions of
dislocations appears to be the controlling factor in the structure of
shock fronts. This places an upper limit on the strain rate which
can be realized physically even for stress amplitudes of tens of
kilobars. In this paper, we present time-resolved measurements
of the detailed behavior of shocks in several metals in the stress
amplitude range roughly from ten to one hundred kilobars. It is
shown that single-crystal beryllium has an anisotropic dynamic
yield behavior from the known anisotropic mobility of dislocations
in this material. Experiments in which 2024-T4 aluminum alloy
specimens were subjected to dynamically induced tensile fracture
also show that the subsequent damping of mechanical vibrations
can be explained in terms of a simple linear relation between the
average dislocation velocity and the maximum resolved shear stress.

1   INTRODUCTION

In the study of plane stress waves in solids produced by shock loading, one traditionally refers to elastic and plastic waves. The terminology arises from the fact that a sufficiently weak plane impulse propagates in most solids at the velocity of longitudinal elastic acoustic waves. At some stress level which is typically of the order of a fraction of a kilobar to a few kilobars, the resolved shear stresses which follow a state of elastic plane strain are adequate to cause dislocations to move in such a direction that the shear stresses are at least partially relieved and the material approaches a hydrostatic state. The resulting stress time or stress distance profile which follows the ''elastic'' precursor is called the *plastic wave*.

We note that this plastic wave is not purely plastic because the material is also being subjected to ever-increasing compression, implying an increase in elastic potential energy. The compressibility of virtually all materials which do not undergo polymorphic phase changes is a decreasing function of compression. A strong tendency from this cause is noted for impulsively applied stress waves to propagate with wavefronts whose steepness is controlled only by such transport processes as heat conduction and viscosity as in fluid mechanics. For a wide range of stresses above the flow stress of solids, however, the creation and movement of dislocations appear to exert the greatest influence on the shock-front thickness.

The structure of rarefaction waves in shocked metals appears to be controlled by similar processes. The unloading is at first an essentially one-dimensional elastic process until the stress anisotropy becomes too great. Then the material again behaves somewhat like a fluid.

The study of elastic-plastic wave profiles in solids, so far, has largely been limited to time-resolved measurements of the free-surface velocity[1,2,3,4] of a shocked plate, or to piezoelectric transducer measurements[5] at the same surface. The interpretation of free-surface velocity measurements is complicated by the fact that in the last fraction of a microsecond, as the plastic wavefront approaches the free surface, many small signals can reverberate between the free surface and the advancing compression wave. Two consequences of this are that the material at the free surface is never actually subjected to the same stress history as the rest of the sample, and the free-surface velocity record is actually a record of complicated wave interactions. The piezoelectric transducer records are subject to similar difficulties if the sample under observation has an acoustic impedance different from that of the transducer. However, the difficulty is not really of major importance since to the extent that the relatively high-speed signals are truly elastic, they provide a faithful map of the stress front at which they arise.

This paper consists of a series of measurements, using the dc capacitor technique,[1,2] of the free-surface velocity profiles of various metals which were subjected to impact loads by gas and propellant fired projectiles. Usually, the collisions were between symmetric materials. Collision planarities were controlled so as to be simultaneous over the measuring area typically to within 0.02 μs or less. Further details of the technique were given in Refs. [1] and [2].

2   RESULTS

Figure 1 shows a comparison between the free-surface velocity profiles of polycrystalline Armco iron, annealed to Rockwell B 30, and an Fe-3% Si single crystal of comparable thickness which was impacted at a comparable velocity. The comparison is so good that one is strongly tempted to conclude that under impulsive loading conditions on this time scale, the properties having principal control over the stress-wave profile must not depend greatly on grain

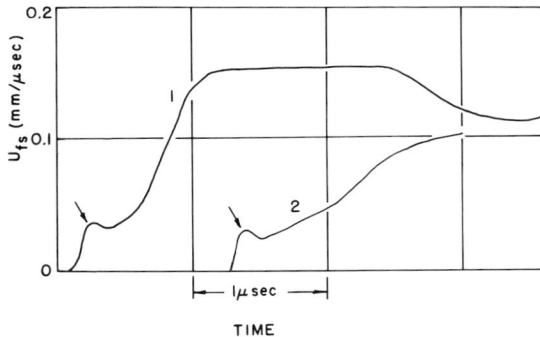

TIME

Fig. 1 Time-resolved free-surface velocity traces for annealed polycrystalline Armco iron (curve 1) and single crystal Fe-3% Si in the ⟨111⟩ orientation. Both samples were 12.7 mm thick and were impacted by Armco iron drivers. The similarities of the elastic precursors are evident. The elastic precursors are marked with arrows.

boundary effects. This assumption was employed by this author in a simple theory of the precursor decay in iron.[6] Further investigation of b.c.c. metals shows that, in general, polycrystalline samples also have fairly well-defined yield points, which suggests that the assumption remains fairly good for them. It is interesting to note, however, that apparently only a few polycrystalline b.c.c. metals exhibit the dynamic upper-lower yield-point effect. Figure 2 shows results for tantalum and niobium. Experiments by McQueen and Marsh using the optical lever technique[3] show that tungsten exhibits an upper-lower yield point, whereas chromium and molybdenum do not.[7]

An extreme test of the relation between single crystal properties and the properties of polycrystalline samples is afforded by beryllium, since in this metal only basal glide occurs easily at room temperature. Figure 3 shows time-resolved, free-surface velocities for two single crystals which were impacted parallel and perpendicular to the C-axis, and a typical result for a polycrystalline beryllium sample; the differences are very striking. Beryllium will apparently support a longitudinal stress of approximately 40 kb parallel to the C-axis before yielding; whereas, in the

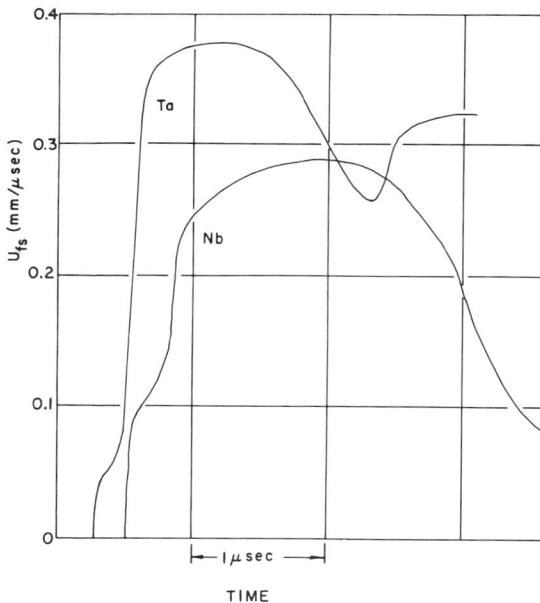

Fig. 2 Time-resolved free-surface velocity traces for annealed tantalum and niobium.

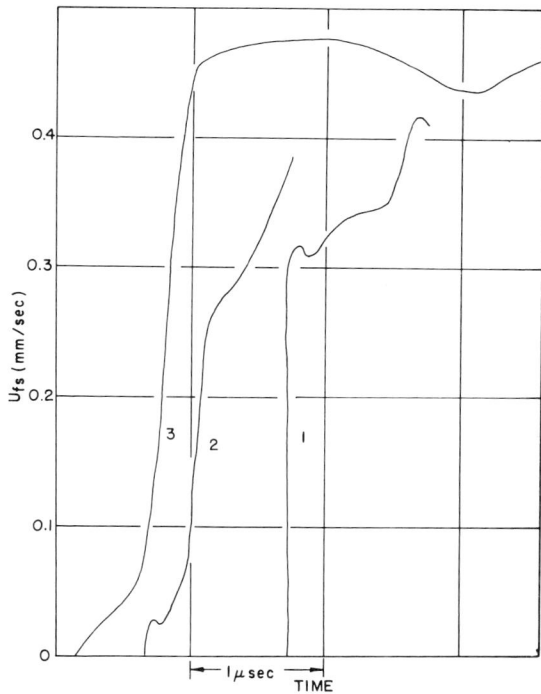

Fig. 3 Time-resolved free-surface velocities for (1) beryllium single crystal with the shock parallel to the C-axis; (2) beryllium single crystal with the shock perpendicular to the C-axis; and (3) polycrystalline sintered beryllium.

basal plane it will only support about 4 kb. This seems at least qualitatively reasonable when one considers that, to relieve shear stress in a plane shock, dislocations must move at some angle between the plane of the shock front and the shock normal. When the shock normal is parallel to the C-axis, there are no easy glide directions so oriented. The third trace shows that in polycrystalline samples there is considerable dispersion, and the sample actually appears to be even weaker than the basal plane orientation of a single crystal.

The impact properties of polycrystalline-annealed copper and uranium are rather similar in the low-velocity impact regime. Figures 4 and 5 show free-surface velocity profiles for several samples of each of these materials. The precursor waves start with essentially zero amplitude and rise gradually and continuously into the plastic wave-zone. It will be noted that in the case of

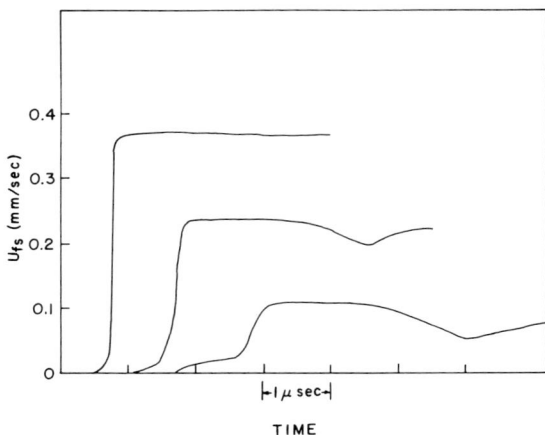

Fig. 4 Time-resolved free-surface velocity of annealed
polycrystalline copper, 12.7 mm thick. The targets were
impacted by 6.35 mm thick copper plates at three different velocities. The elastic precursor is barely visible, except where the plastic shock is weakest. This
is an accident since the precursor actually varies somewhat in appearance.

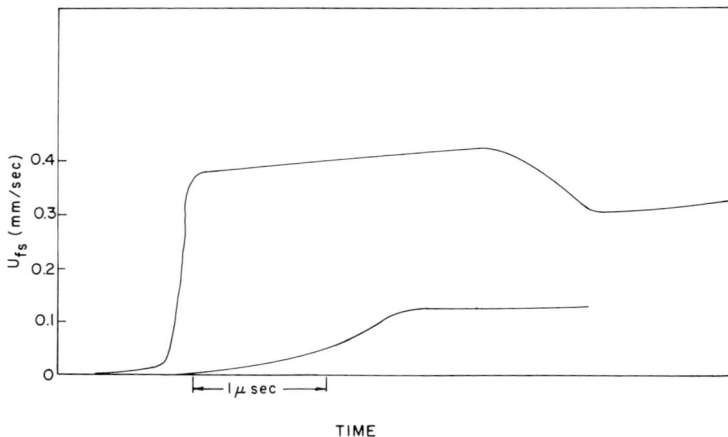

Fig. 5 Time-resolved free-surface velocity traces for 12.7 mm thick polycrystalline uranium resulting from symmetrical collisions at two different
velocities. The general similarity to annealed copper (Fig. 4) is evident.

copper, the plastic shock-front steepness decreases rapidly as the impact velocity is decreased. Thus for shock strengths of the order of 35 kb and greater (projectile impact velocities greater than about 0.025 cm/μs), the shock appears to propagate as a discontinuity. This effect is accompanied by another noteworthy phenomenon. The lower-velocity impacts produce free-surface velocities which do not achieve the projectile velocity (as they should in a symmetric collision) until many microseconds after the main front has arrived. If one asserts that the shock has "arrived" when the free-surface velocity vs. time curve first begins to be nearly constant, and plots the ratio of this amplitude to the projectile velocity against projectile velocity, the resulting curve passes through a minimum at approximately the 0.015 cm/μs projectile velocity (Fig. 6). This corresponds roughly to the shock strength of 30 kb at which Johari[8] found extensive microtwinning to occur in copper. It would seem that there may be some connection between these phenomena.

The polycrystalline uranium shock profiles also exhibit the qualitative feature of increasing steepness as the impact velocity is increased, but do not show the peculiar minimum in the free-surface-to-projectile velocity ratio. Polycrystalline uranium has a rather high ratio of longitudinal sound speed to "bulk sound

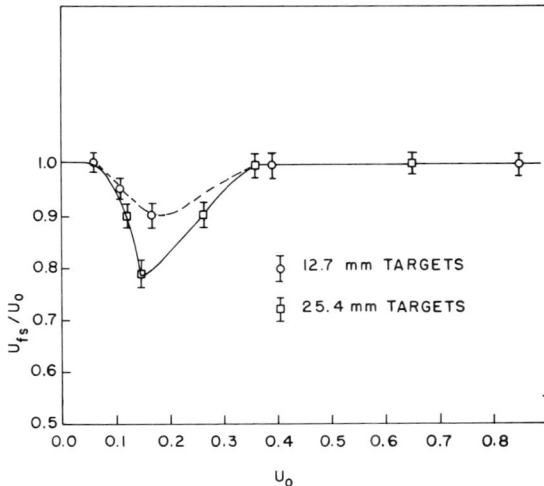

Fig. 6 The ratio U_{fs}/U_0 of the free-surface velocity behind the main shock in annealed copper to projectile velocity, plotted against projectile velocity. Two target thicknesses, 12.7 mm and 25.4 mm, were investigated.

speed'' (defined as $\sqrt{B/\rho}$) and this may be the reason why the uranium shock fronts never become discontinuities within the range of the present data, since plastic flow is always beginning so far in advance of the ''bulk wave.''

It has been noted by many investigators that the leading component of a rarefaction wave which is introduced into a preshocked metal travels with a high-pressure ''longitudinal sound speed'' rather than simply as a hydrodynamic disturbance.[2,9,10] The free-surface velocity traces in Figs. 4 and 5 implicitly contain such data, and a fairly extensive study of the effect was made for uranium. The resulting plot of longitudinal sound speed vs. pressure is shown in Fig. 7. The velocities measured at the lowest pressures scatter poorly because of the difficulty in defining signal arrivals in this material. One data point on this curve was obtained by using tungsten to reflect a shock rather than a rarefaction into the driver plate. The free-surface velocity profile resulting from this experiment is shown in Fig. 8. It will be seen that the reflected shock is considerably smeared out and shows every evidence that this further compressive disturbance into a material which is already under pressure, is again resulting in a comparatively slow dislocation motion. A similar effect was noted

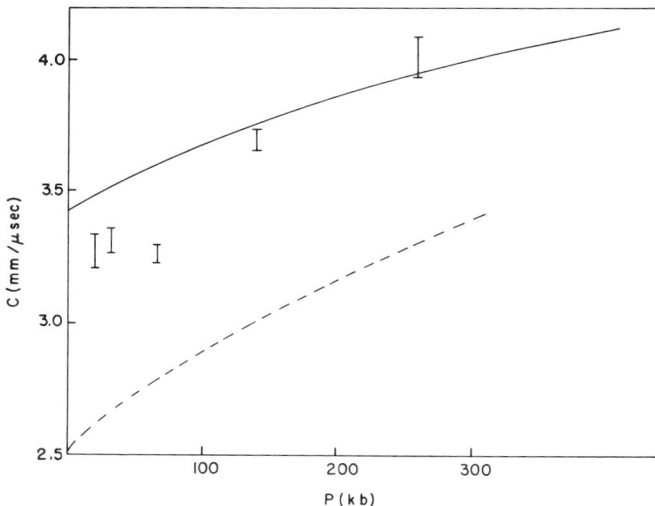

Fig. 7 Velocity of the first component of reflected signals in poly-crystalline uranium plotted as a function of the initial shock pressure. The point at zero pressure is the ultrasonically measured longitudinal sound speed. The dotted curve is the hydrodynamic sound speed, calculated from Hugoniot data.

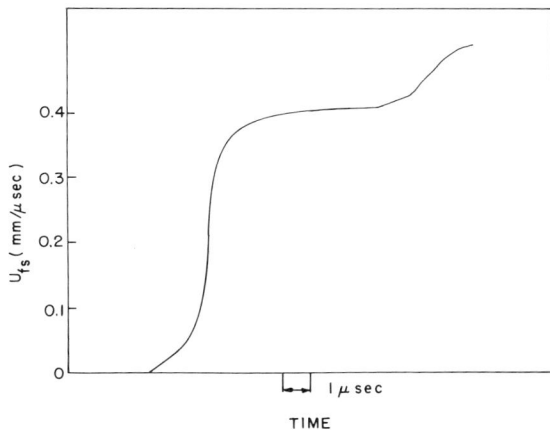

Fig. 8   Free-surface velocity versus time for a reflected shock experiment in polycrystalline uranium.

by Barker et al. in aluminum.[11] The tension waves arising from the rarefaction interactions arriving at the free surface are followed by further shocks and rarefactions which result from the fact that the target plate spalls under the tensile stresses, leaving a relatively thin sample in which relatively small amplitude signals resonate back and forth. This effect will be discussed in greater detail in connection with 2024-T4 Aluminum.

3   WORK-HARDENING

Samples of Armco iron were work-hardened by a 25% reduction in thickness, and samples of the pure copper whose properties in the annealed state have been reported above were rolled to a 50% reduction in thickness. The samples were impact-loaded, and the free-surface velocity profiles were recorded. A comparison between the two materials is shown in Fig. 9. The elastic-plastic profiles are qualitatively quite similar. In both metals, the first disturbance is several kb and is followed by a zone of gradually increasing stress amplitude. The yield-point effect characteristic of annealed iron has disappeared, which agrees with the supposition that it is caused by catastrophic multiplication of an initially fairly small number of mobile dislocations. The effect of cold work on copper is very striking since now several kb of stress are propagated at longitudinal sound speed. The effect of prior cold work

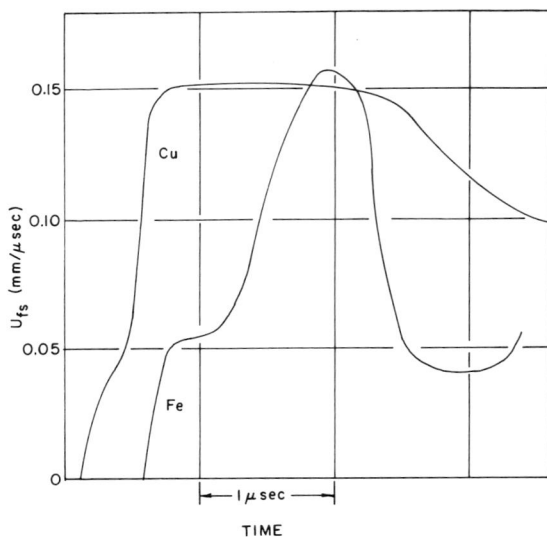

Fig. 9 Time-resolved free-surface velocities for cold
worked copper and Armco iron.

also shows up in the apparent tensile strength (spall stress) of
both materials under macroscopically plane strain conditions. The
spall stress can be correlated quantitatively with the magnitude
of the reduction of free-surface velocity through the relation
$\sigma_y = \rho C(\Delta U_{fs}/2)$. A comparison of Fig. 9 with Figs. 2 and 4 shows
that both iron and copper have their dynamic tensile strength increased by prior cold work. It is interesting to note from comparison of Figs. 4 and 9 that although dislocations are known to be produced in very large numbers by relatively weak shocks in copper,
the dislocations thus produced do not have nearly the effect on the
tensile strength that prior cold work does. The implication seems
to be that dislocations produced in shocks of moderate strength do
not cause as much work-hardening as one might expect, possibly
because they do not move far enough to interact strongly with each
other.

4   EXPERIMENTS WITH
2024-T4 ALUMINUM

The effect of the elastic overtaking wave and effect of spalling
and subsequent oscillations have been investigated extensively for

for 2024–T4 Aluminum in the T–4 state. The experiments consisted of symmetric collisions at two different driver velocities and with time-resolved free-surface velocity data are shown in Figs. 10 and 11, including an overtaking shock experiment.

From the experimental data one obtains the velocity of the leading edge of the rarefaction signal at material pressures of approximately 33 and 80 kb. The approximation involved is that the pressures can be calculated from the Rankine-Hugoniot relations which are strictly applicable only if the shock fronts are propagated as steady states, which is clearly not quite correct in this case. There is also a small uncertainty in the deduction of the actual velocity of the leading edge of the rarefaction signal. This arises from the fact that when it nears the free surface where measurements are made, it interacts with the

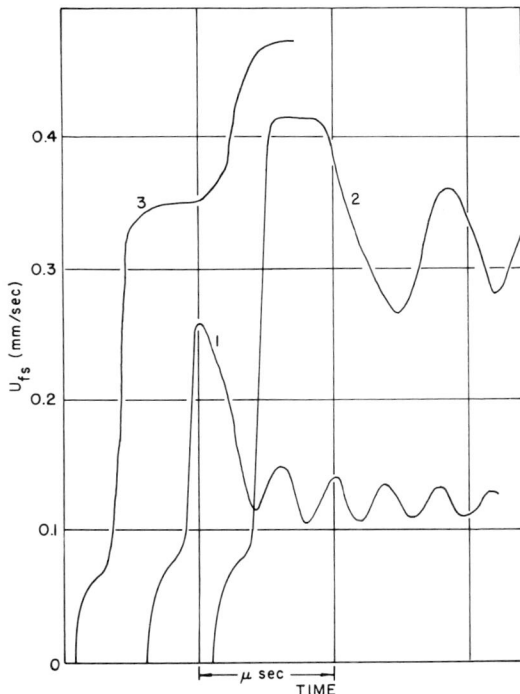

Fig. 10 Free-surface velocity data for samples of 2024-T4 Aluminum drivers at approximately 0. 4 mm/μsec. The targets were each 12.7 mm thick. The drivers were 1.0 and 3.0 mm thick for curves 1 and 2. Curve 3 was produced by a reflected shock from steel. The initial shocks were approximately 33 kb.

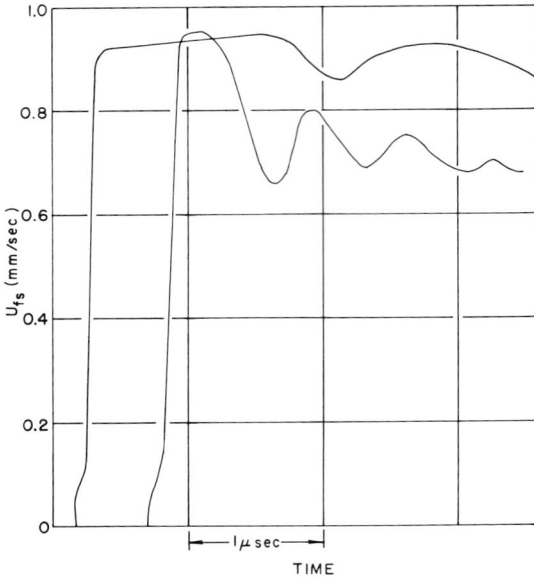

Fig. 11 Free-surface velocity data for 12.7 mm 2024-T-4 Aluminum targets impacted by 6.35 and 3.175 mm 2024-T-4 Aluminum drivers at 1.00 mm/μsec. The impact stress is approximately 80 kb.

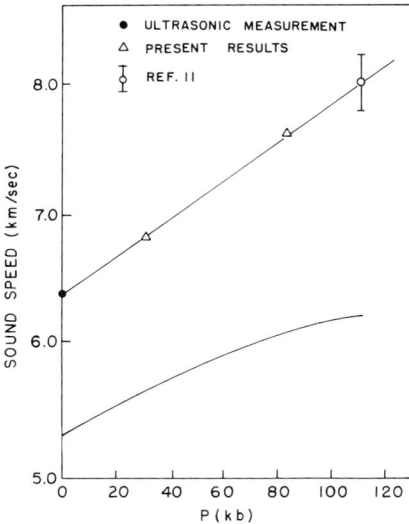

Fig. 12. Sound speed versus pressure for 2024 Aluminum. The present data (triangles) represent the results of three shots each. The scatter is within the size of the triangles, with the assumptions mentioned in the text. The lower curve, which is the calculated hydrodynamic sound speed, is taken from the work of Rice et al. [15]

rarefaction arising there and thereafter propagates into unstressed material. The chief uncertainty arises because of the interaction region between the two rarefaction signals. However, this is a very small percentage effect. Figure 12 shows these data together with the measured longitudinal sound speed (at zero pressure) and data measured at higher pressures by Erkman.[12] For additional comparison, the hydrodynamic sound speed calculated from high-pressure, shock-wave data is also shown.

The free-surface velocities shown in Figs. 10 and 11 each pass through minima, corresponding to the spall strength or maximum plane tension the material can withstand, and then oscillate. The frequency of the oscillations is of course determined by the thickness of the spalled layer, which is controlled largely by the driver thickness, but partly by such factors as the variation in spall strength with strain rate. This latter phenomenon has been investigated extensively by Breed et al.[13] using flash x-ray radiography. The periods of oscillation are quite constant, and in the case of the higher-frequency oscillations where many cycles can be followed, the data fit rather well to the simple functional form

$$(U - \bar{u}) = Ae^{-kt} \sin\omega t \tag{1}$$

where $\bar{u}$ is the average velocity of the metal surface. The constant k, for six experiments, lies between 0.6 and 0.8 μs^{-1}.

The stress amplitudes in the oscillating spalled samples are sufficiently small (a few kb) as might be expected from the amplitude of the elastic precursor to the first shock that the energy is oscillating at approximately longitudinal sound speed. This assumption is consistent with the thickness of recovered spalled samples.

The magnitude of the exponential decay constant can now be related to a simple linearized dislocation theory of the decay process. We assume that the stresses are never large enough during this part of the motion to cause the elastic moduli of the metal to deviate appreciably from their zero-pressure values. We then assume that over the range of stresses we are interested in, the average velocity of dislocations will be a linear function of the maximum resolved shear stress, which is equivalent to assuming that the intrinsic viscous damping fource on an element of dislocation line is directly proportional to its velocity and that the inertial term and restoring force (line tension) employed by Granato and Lucke[14] are unimportant in this problem. A further assumption in a linearized theory must be that the dislocation line density remains substantially constant during the oscillatory part of the motion. It is of course obvious that the present experiments represent the ultimate in large-strain damping.

It is readily shown that the above assumptions lead to a macroscopic constitutive relation of the form of the standard linear solid, in which the longitudinal stress and strain obey the relation

$$\tau \frac{\partial \sigma}{\partial t} + \sigma = \tau \left(\rho C_1^2 \right) \frac{\partial \epsilon}{\partial t} + \left(\rho C_2^2 \right) \epsilon \tag{2}$$

where the constants ρC_1^2 and ρC_2^2 are related to the Lame parameters λ and μ by the relations

$$\rho C_1^2 = \lambda + 2\mu \tag{3}$$

$$\rho C_2^2 = \lambda + \frac{2}{3}\mu \tag{4}$$

The relaxation time τ is related to the parameter (B) in the Granato-Lucke theory by the relation

$$\tau = \frac{B}{2Nb^2 \mu} \tag{5}$$

where B is the intrinsic damping parameter, b is the Burger's vector of the active dislocations, and N is the length of active dislocation line per unit volume. Of course, in this application, B may be expected to contain fairly large contributions from dislocation-dislocation interactions.

The constitutive relation (1) combined with the one-dimensional Lagrangian relations express the conservation of mass and momentum, viz.,

$$\frac{\partial u}{\partial x} = \frac{\partial \epsilon}{\partial t} \tag{6}$$

$$\rho \frac{\partial u}{\partial t} + \frac{\partial \sigma}{\partial t} = 0 \tag{7}$$

and suitable initial conditions define the motion of the system. In these expressions σ is longitudinal stress, ϵ is longitudinal strain and u is particle velocity.

As a reasonable approximation to the state of a spalled slab of thickness L, at the instant of spalling we assume that

$$u = U_0 \qquad 0 < X < \frac{L}{2}$$

$$U = 0 \qquad -\frac{L}{2} < X < 0$$

and that the stress and strain are everywhere zero. We wish to know the time-dependence of the velocity at $X = +L/2$ (the free surface) and are interested only in solutions which are periodic in time.

The problem is readily solved by the method of Laplace transforms with the result that

$$\frac{u\left(\frac{L}{2}, t\right)}{U_0} = 1 - \frac{1}{zi\pi} \int_{Br} \frac{e^{St} \, dS}{2S \cosh\left[\frac{LS}{2C_1} \sqrt{\frac{1 + \tau S}{\alpha + \tau S}}\right]} \tag{8}$$

where $\alpha = C_2^2/C_1^2$ is the ratio of the bulk modulus to the longitudinal modulus. We are interested only in oscillatory solutions which arise from the poles of the integrand, i.e., when

$$\frac{LS}{2C_1} \sqrt{\frac{1 + \tau S}{\alpha + \tau S}} = (2n + 1) \frac{\pi i}{2} \tag{9}$$

we put $Z = \tau S$ and $\zeta = (2n + 1)(\pi C_1/L)$

and

$$Z_n \sqrt{\frac{1 + Z_n}{\alpha + Z_n}} = i\zeta_n \tag{10}$$

Experimentally, we have found that the exponential damping time is essentially independent of the frequency of the oscillations. Therefore we want to look at solutions of Eq. (10) for which the real part of Z is negative and independent of ζ. This will occur if ζ_n is large (in which case Z_n will also be large). To first order in $1/Z$, Eq. (10) becomes

$$Z_n \left[1 + \frac{1}{2}(1 - \alpha) \frac{1}{Z_n}\right] = i\zeta_n$$

and
$$Z_n \approx -\frac{(1 - \alpha)}{2} + i\zeta_n$$

and the desired solution is of the form

$$\Delta U = e^{\frac{-(1 - \alpha) t}{2}} \sin\left(\frac{\pi C_1 t}{L}\right) \tag{11}$$

which is an acceptable form. Note that we have ignored the higher harmonics.

It remains to show that the quantity $C_1 \pi/L$ is at least greater than unity. For 2024-T4 Aluminum, $\alpha = 0.699$, and our measured time constant is approximately 1.5 μs. This means that the effective relaxation time is $\tau = 0.225$ μs, thus the parameter ζ_0 is given by $\zeta_0 = 3.61$ for our thinnest sample (0.125 cm) and $\zeta_0 = 1.51$ for the thickest sample. We conclude therefore that the approximate solutions are reasonably accurate.

The quotient of the damping parameter B and the dislocation density N can now be deduced from Eq. (4) and is given by $B/N = 72 \times 10^{-12}$, where we have assumed an average Burgers vector of 2 Å for this material. Some estimate of the same quotient can be obtained for the 33-kb experiments if we note from the experimental data that the plastic shock-front thickness is of the order of 0.2 μs. The total plastic shear strain involved in these compressions is approximately 4×10^{-2}; thus the strain rate is approximately 2×10^5/sec.

This strain rate is also given by

$$\frac{\partial \gamma}{\partial t} = \frac{Nb^2 \tau}{B} \tag{12}$$

where τ is now the maximum resolved shear stress during the flow, and which we take to be approximately 1.8 kb (the assumption being that the resolved shear stress which initiates plastic flow is maintained during the flow). We then find that from the estimated shock-front thickness that

$$\frac{B}{N} = \frac{b^2 \tau}{\partial \gamma/\partial t} = 3.6 \times 10^{-12}$$

which is only a factor of twenty lower than the estimate made from the damping of the lower amplitude oscillations. The difference is also probably in the correct direction since the lower stress amplitudes in the oscillations, combined with some work-hardening which must have occurred in the shock front, should reduce the number of active dislocations in the oscillatory phase.

5 CONCLUSION

It is clear that time-resolved, free-surface velocity experiments with metals, which are impact-loaded at stress levels of the order of a few tens of kilobars, may yield extensive information pertinent to the problems of dislocation dynamics. It is also clear that some of the experiments are currently capable of yielding information about the behavior of metals operating at the largest strain amplitudes which the metals will stand without rupture, while others pose new problems for theoretical consideration. Obviously, it would be of interest to perform experiments of the sort reported herein for 2024-T4 Aluminum on single crystals for which etch-pit counts can be made, preferably both before and after shock loading.

ACKNOWLEDGMENT

This work was performed under auspices of the U. S. Atomic Energy Commission.

REFERENCES

1. Rice, M. H.: *Rev. Sci. Inst.* 32:449 (1961).
2. Taylor, J. W., and M. H. Rice: *J. Appl. Phys.* 34:364 (1963).
3. Marsh, S. P., and R. G. McQueen: *Bull. Am. Phys. Soc.*, Ser. II, 5: No. 7 (Dec. 29, 1960).
4. Davis, W. C., and B. G. Craig:*Rev. Sci. Instr.* 32:579 (1961).
5. Jones, O. E., F. Nielson, and W. B. Benedict: *J. Appl. Phys.* 33:3224 (1963).
6. Taylor, J. W.: *J. Appl. Phys.* 36:3146 (1965).
7. R. G. McQueen, and S. P. Marsh: Private communication.
8. Om Johari: *UCRL* 10932 (September, 1963).
9. Al'tshuler, L. V., S. B. Kormer, L. A. Vladimirov, M. P. Speranspuya, and A. I. Funtikov: *Soviet Physic JETP* 11: 766 (1960).
10. Curran, D. R.: *J. Appl. Phys.* 34:2677 (1963).
11. Barker, L. M., C. D. Lundergan, and W. Herrmann: *J. Appl. Phys.* 35: 1203 (1964).
12. Erkman, J. O., A. B. Christensen, and G. R. Fowles: *AFWL-TR-66-12* (May 1966).
13. Breed, B. R., C. L. Mader, and D. Venable: *J. Appl. Phys.*, to be published.
14. Granato, A. V., and K. Lucke: "Physical Acoustics," Vol. IV, Part A, Warren P. Mason (ed.), Academic Press, New York and London, 1966.
15. Rice, M. H., J. M. Walsh, and R. G. McQueen: "Solid State Physics, Vol. 6, F. Seitz and D. Turnbull (eds.), Academic Press, New York and London, 1951.

THE APPLICATION OF DISLOCATION DYNAMICS TO IMPACT-INDUCED DEFORMATION UNDER UNIAXIAL STRAIN

B. M. Butcher and D. E. Munson

Sandia Corporation,
Albuquerque, New Mexico

ABSTRACT

Results from a series of plate impact experiments on annealed 1060 aluminum and annealed 4340 steel, which show the change in the plastic wave profile with distance of propagation, are presented and related to dislocation dynamics. First, functional forms for the dislocation velocity are examined with regard to the extent that they can be used to describe the observed elastic wave attenuation. The relation proposed by Gilman for the dislocation velocity in terms of the shear stress, $v = v_\infty e^{-\tau_0/\tau}$, describes the data quite well.

The response of the material above the elastic limit is considered in the second part of the discussion, which shows that dislocation multiplication is necessary to reproduce the plastic wavefronts in both metals. An approximate analysis of the steel data shows that the plastic strain at a given position in the material develops initially at a nearly constant stress, or constant dislocation velocity. A linear relation between dislocation density and plastic strain is also indicated.

1   INTRODUCTION

The stress waves which result from high velocity plate impacts of solids provide the most rapid loading rates available in the laboratory. For a material which deforms plastically, as do metals, this implies that the plastic deformation rate in the wavefront also attains large values. Adequate experimental resolution of these stress waves, with either time or propagation distance, permits demonstration of the influence of the high plastic strain rate on the flow stress of the solid. Fortunately, considerable improvements in time and displacement resolution through the use of thick quartz gages, optical interferometers, and rapic electronic devices now permit stress profiles to be measured in detail.

This paper describes a series of plate-impact experiments in which the wave profile is determined as a function of the distance of propagation. The materials, 1060 aluminum and fully annealed 4340 steel, represent the two common crystal structures of metals and produce typical elastic–plastic stress waves. These alloys are known to be sensitive to strain rate on the basis of previous studies.[1,2] Analysis of the elastic-wave attenuation and plastic waveform is based on various mathematical representations of dislocation multiplication and movement, and on numerical computer solutions of the equations used to describe the propagation of stress waves.

2   EXPERIMENTAL

The experiments were conducted on large-diameter, flat-plate targets impacted by gas-driven, four-inch diameter, flat-nosed projectiles which were normally the same material as the target. The impact produces a compressive stress wave which propagates through the target. The geometries were such that lateral rarefactions and projectile rear surface rarefactions did not interfere with the propagation of the planar wave during times of interest. Parameters of impact, projectile velocity, projectile tilt angle, and tilt orientation, were measured for the experiments. Precise experimental details have been described previously.[3]

The 4340 steel, fully annealed by heating to 1450°F and furnace cooling, was impacted at nominal velocities of 800 ft/sec using target thicknesses of 1/4, 1/2, 1, and 2 inches. The stress wave was observed as it arrived at the target free surface, which is the surface opposite the impact surface, by a guard-ring quartz gage.[4] The gage produced a current output proportional to the average stress over a 1/2-inch diameter area of the quartz gage at the quartz-specimen interface. Since the current was related to the stress in the quartz, the stress in the steel was obtained by an impedance calculation. A rather simple and quite small correction for average tilt (≈ 0.0007 radians) was made on the quartz gage records. The experimental stress data are plotted against a reduced time parameter in Fig. 1.

The fully annealed 1060 aluminum was impacted at a nominal velocity of 100 ft/sec using target thicknesses of 1/4, 1/2, and 1 inch. The reflection of the stress wave at the free surface was

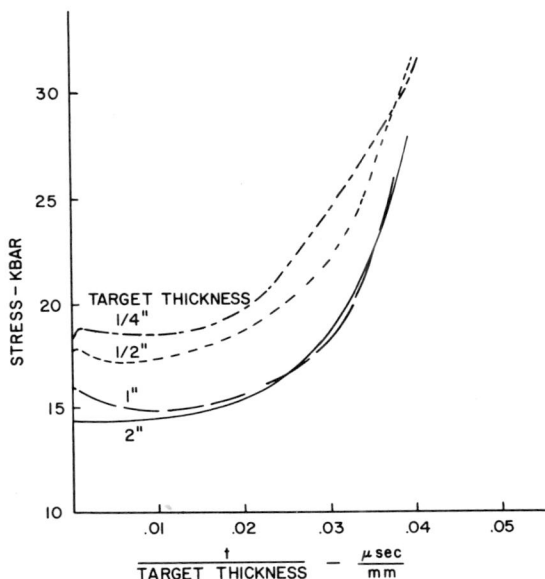

Fig. 1 Experimental stress versus reduced time curves for annealed 4340 steel impacted at 800 ft/sec. (The arrival time of the elastic wave front is taken as $t = 0$.)

observed by a laser-interferometer technique[5] which measures the surface displacement over a very small area; stress magnitudes may be deduced from the observed motion. Curves for the free-surface velocity as a function of reduced time are shown in Fig. 2.

The data of Figs. 1 and 2 have an uncertainty in stress which is caused by the unknown nature of the stress-strain release path of the material, i.e., the stress-strain curve which the material follows when the stress is decreased from the maximum compression. The only portion of this release path which is reasonably defined is the elastic portion, which in simple materials is about twice the stress amplitude of the compressive elastic wave ($2Y_0$). In materials which show Bauschinger strains, such as, steel and aluminum, true elastic release stress is probably of the same magnitude as Y_0.

In the steel experiments, a relief reflection is caused in the steel by the impedance mismatch between target and gage. The parts of the quartz records in which the unloading wave exceeds Y_0 are ignored in the analysis of the steel data, thereby eliminating

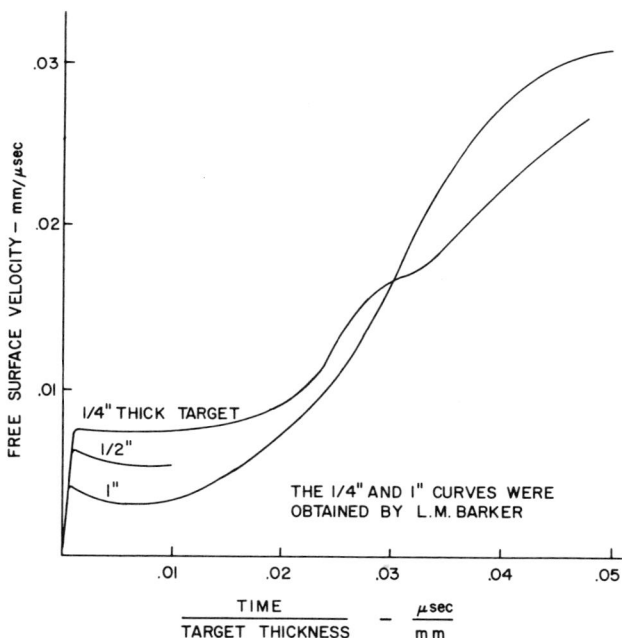

Fig. 2 Experimental free surface velocity-reduced time curves for 1060 aluminum impacted at 100 ft/sec. (The arrival time of the elastic wave front is taken as $t = 0$).

the unloading uncertainty. Also, the internal stress wave–inter-actions, which occur in the steel as a result of the reflection, are not considered in the data reduction. However, a detailed analysis of the problem suggests that this procedure introduces only a small error. Furthermore, the data are treated on a relative basis which further reduces the apparent error.

In the aluminum experiments, the compressive wave is reflected from the free surface in a way which reduces the stress to zero; therefore, the motion of the free surface is influenced by the entire elastic–plastic release path. This causes the interpretation of the data to be more difficult, as discussed in detail later in the paper.

3   DISCUSSION OF RESULTS

The approach to the interpretation of the results is necessarily parametric. We assume that functional equations for dislocation motion and multiplication derived from mathematically idealized models and confirmed by some measurements on single crystals can be extended to describe the behavior of polycrystalline alloy systems. In other words, the variables of stress, strain, and ma-terial velocity used in this study are assumed to reflect statistically events occurring on a microscopic scale.

During the passage of the stress front, two time intervals are of particular interest: the interval immediately at the elastic wave; and the interval encompassing all of the plastic wave. The former includes all events which occur as the elastic wave arrives. These are the dislocation multiplication and acceleration associated with the yield–point phenomena. Since time scales for these events are small compared to experimental resolution, the measurements re-ported here apply, in reality, to a state existing after some of these events are completed. Hence, the analysis ignores the fine detail of the elastic front. The second interval concerns the remainder of the wave where deformation events occur over resolvable time scales.

4   PARAMETRIC EQUATIONS REPRESENTING DISLOCATION DYNAMICS

A fundamental result from dislocation theory is that the plastic strain rate is given by

$$\dot{\gamma} = bN\upsilon \tag{1}$$

where $\dot{\gamma}$ is the plastic shear strain rate, b is Burgers vector, N is the mobile dislocation density, and υ is the dislocation velocity. To make Eq. (1) useful, functional forms of N and υ must be found which relate them to stress and strain.

Three typical equations for dislocation velocity will be considered. These functions are dependent only upon the maximum shear stress. Johnston and Gilman[6] have used an equation of the form

$$\upsilon = \upsilon_\infty e^{-\tau_0/\tau} \tag{2}$$

where υ_∞ is the limiting velocity (approximately the sear-wave velocity) of a dislocation, τ is a maximum shear stress, and τ_0 is a parameter. The second equation is

$$\upsilon = k_1 e^{\theta\tau} \tag{3}$$

where k_1 and θ are parameters. This is the form to be expected if the velocity is governed by thermal activation.[7] The parameter θ may contain a term in reciprocal temperature. Finally, a linear stress form based on athermal damping mechanisms[7] is considered:

$$\upsilon = k_2(\tau - k_3) \tag{4}$$

where k_2 and k_3 are parameters.

Dislocation multiplication during plastic strain is taken as

$$N = N_0(1 + \alpha\gamma) \cdot f \tag{5}$$

where α and N_0 are parameters, f is an exhaustion factor to prevent unlimited multiplication, and γ is the plastic shear strain. Gilman[8] gives

$$f = e^{-\phi'\gamma/\tau} \tag{6}$$

Johnston,[9] however, derives a form which does not explicitly include shear stress,

$$f = e^{-\phi\gamma} \tag{7}$$

Both ϕ and ϕ' are parameters. In all cases, the latter form will be used because of difficulties in experimental evaluation of ϕ'.

Specifically, this eliminates the need to consider stress in the multiplication equation.

5 THE CHARACTERISTIC EQUATIONS

For a material coordinate system, i.e., fixed with respect to the material, with x in the direction of propagation and normal to the plane of impact, the total stress (σ_{xx}),* total strain (ϵ_{xx}), and particle velocity (u_{xx}) which describe uniaxial strain conditions are related through the equations of motion

$$\frac{\partial \sigma_{xx}}{\partial x} = -\rho_0 \frac{\partial u_{xx}}{\partial t} \tag{8}$$

$$\frac{\partial \epsilon_{xx}}{\partial t} = -\frac{\partial u_{xx}}{\partial x} \tag{9}$$

$$\frac{\partial \epsilon_{xx}}{\partial t} = \frac{1}{\lambda + 2\mu} \frac{\partial \sigma_{xx}}{\partial t} + \frac{8}{3} \frac{\mu}{\lambda + 2\mu} \dot{\gamma} \tag{10}$$

where $\dot{\gamma}$ is the plastic strain rate or strain-rate function [Eq. (1)], and λ and μ are Lamé constants.

These partial differential equations may be reduced to ordinary differential equations by the method of characteristics,[10] and may be integrated directly along the characteristic lines. The characteristic equations† are

$$\frac{d\sigma}{\lambda + 2\mu} + \frac{du}{c} = -\frac{8\mu}{3(\lambda + 2\mu)} \dot{\gamma} dt; \quad \text{along } \frac{dx}{dt} = c \tag{11}$$

$$\frac{d\sigma}{\lambda + 2\mu} - \frac{du}{c} = -\frac{8\mu}{3(\lambda + 2\mu)} \dot{\gamma} dt; \quad \text{along } \frac{dx}{dt} = -c \tag{12}$$

$$\frac{d\sigma}{\lambda + 2\mu} - d\epsilon = -\frac{8\mu}{3(\lambda + 2\mu)} \dot{\gamma} dt; \quad \text{along } dx = 0 \tag{13}$$

$$c = \sqrt{\frac{\lambda + 2\mu}{\rho_0}} \tag{14}$$

*Compressive stress and strain are defined positive.
†Henceforth, subscripts will be dropped.

where c is the elastic dilational wave velocity and ρ_0 is the zero stress density of the material. The propagation solutions presented are obtained through numerical computer solutions of the characteristic equations[11] for suitable $\dot{\gamma}$ functions and appropriate initial and boundary conditions.

The lead characteristic defined by $dx/dt = c$ traces the propagation of the elastic wavefront. For elastic deformation,

$$d\sigma = \rho_0\, c\, du \tag{15}$$

and Eq. (11) may take one of the following forms:

$$\frac{d\sigma}{dt} = -\frac{4}{3}\mu\,\dot{\gamma} \tag{16}$$

$$\frac{d\sigma}{dx} = -\frac{4\mu}{3c}\,\dot{\gamma} \tag{17}$$

or an integral form:

$$\int_{\sigma_0}^{\sigma}\frac{d\sigma}{\dot{\gamma}} = -\frac{4}{3}\mu\,\Delta t \tag{18}$$

Most real material data do not reflect mathematically ideal elastic behavior. Real behavior may be described as nearly elastic within defined limits. Expansion of Eq. (15) about the velocity c gives

$$\Delta\sigma = \rho_0\, c\left(1 + \frac{\Delta c}{c}\right)\Delta u \tag{19}$$

where Δc is the deviation from c of the average propagation velocity of the stress increment $\Delta\sigma$. Equations (16), (17), and (18) are assumed to be applicable and the material is defined as elastic if $\Delta c/c < 0.02$.

6   ATTENUATION OF THE ELASTIC WAVE

The procedure used to interpret the elastic wave attenuation was similar to that used by Taylor.[12] The parameters of a given velocity equation [Eqs. (2)–(4)] are obtained by substituting $\dot{\gamma} = bN_0 v$ into either Eqs. (16), (17), or (18) and adjusting the values until agreement is obtained with the experimental data. A first approximation of the parameters is taken from a graphical representation of Eq. (17) plotted as $(\Delta\sigma/\Delta x)$ vs. either $\bar{\sigma}$ or $1/\bar{\sigma}$, depending upon the model, where $\bar{\sigma}$ represents the average stress in the gradient interval. If attenuation is rapid, the integral form [Eq. (18)] must be used in place of Eq. (17). Since a parametric determination requires one datum point per parameter, the redundancy of data permits independent checks.

The parameter-evaluation procedure is relatively straightforward; however a basic question about the stress condition of the impact surface has caused a dilemma. If the elastic stress at the impact surface precisely at the instant of impact $x = 0, t = 0$, is known, it can be used as an important additional datum point in the parameter evaluation. Assuming that the impact stress state is completely elastic, then regardless of any ensuing time-dependent decay, the elastic stress for like-material impacts, at the instant of impact, is given by

$$\sigma = \rho_0 c \frac{u_m}{2} \tag{20}$$

where u_m is the projectile velocity. The decision to include or omit the elastic impact stress becomes a critical step in the analysis.

If the elastic impact stress is not used, the parameters for Eqs. (2) and (3) are given in Table 1. Actually, for the range of experimental data, the activation equation [Eq. (3)] and the linear equation [Eq. (4)] produce a data fit nearly identical to that of Eq. (2). As shown in Fig. 3, the intercept of Eq. (2) fit (dashed line) gives a comparatively small value of impact elastic stress.

If the curves are forced through the elastic impact stress given by Eq. (20), then the parameters of Eqs. (2) and (4) are as given in the second column of Table 1. The Gilman parameters have changed markedly by including the impact elastic point. Although both the Gilman and linear equations give adequate data fit, the Gilman function seems better, as shown in Fig. 3. The fit of the activation function [Eq. (3)] had such substantial discrepancies under these conditions that it was not analyzed further. Indeed, the experimental results of Jones[13] suggest that thermal activation in iron alloys is not important. However, a variation of the elastic wave amplitude with temperature in copper, aluminum, and brass has been reported by Novikov et al.[14]

TABLE 1 Elastic Wave Attenuation Parameters

Material	Equation	Without elastic input stress data	With elastic input stress data
4340 Steel	$\dot{\gamma} = bN_0\,v_\infty\,e^{-\tau_0/\tau}$ (Gilman)	$N_0 = 10^7$ cm^{-2} $\tau_0 = 22$ kb	$N_0 = 2 \times 10^9$ cm^{-2} $\tau_0 = 41$ kb
$\rho_0 = 7.83$ gm/cc $c = 5.89$ mm/μs $K = 1700$ kb	$\dot{\gamma} = bN_0\,k_1\,e^{\theta\tau}$ (Activation)	$N_0\,k_1 = 2 \times 10^9$* $\theta = 1.04$ kb^{-1}	Solution does not agree with data
$b = 2.5 \times 10^{-8}$ cm	$\dot{\gamma} = bN_0\,k_2\,(\tau - k_3)$ (Athermal)	$N_0\,k_2 = 4 \times 10^{11}$* $k_3 = 3.9$ kb	$N_0\,k_2 = 8.2 \times 10^{10}$* $k_2 = 4.35$ kb
1060 Aluminum $\rho_0 = 2.71$ gm/cc $c = 6.395$ mm/μs $K = 754$ kb $b = 2.86 \times 10^{-8}$ cm	$\dot{\gamma} = bN_0\,v_\infty\,e^{-\tau_0/\tau}$ (Gilman)	Insufficient data	$N_0 = 3 \times 10^6$ cm^{-2} $\tau_0 = 0.41$ kb

*$N_0\,k_1$, $N_0\,k_2$ in cm^{-1} sec^{-1}

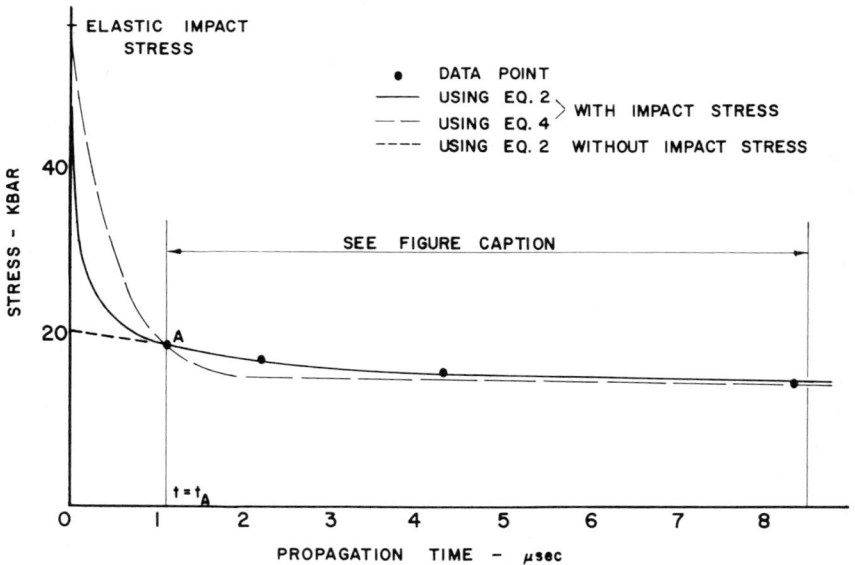

Fig. 3 Elastic wave attenuation in annealed 4340 steel: in the time range indicated the solid line approximates, within the maximum estimated experimental error of ±0.5 Kbar, the curves calculated from Eqs. 2, 3, or 4 using parameter values evaluated without consideration of the impact stress.

On the strength of the present experimental data, a clear separation between the Gilman and linear functions cannot be made. This result is substantiated by an empirical analysis of data obtained on 4340 steel at various velocities.[1] In that analysis, it was thought necessary, on the basis of uniaxial stress data, to use two equations to represent elastic wave attenuation. The first equation for $\dot{\gamma} < \dot{\gamma}_A$ and $t > t_A$ (see Fig. 3) was of the same form (exponential) as Eq. (3). The second equation for $\dot{\gamma} > \dot{\gamma}_A$ and $t < t_A$ was of the same form (linear) as Eq. (4). From previous discussions and Fig. 3, it is apparent that this separation could be made and still give reasonable approximation to the Gilman function.

Further experiments are being performed on thin targets ($t < 1\mu s$, Fig. 3); however, experimental problems of time resolution and projectile tilt have not been sufficiently resolved to give significant data.

The parameters of Eq. (2) are given in Table 1 for 1060 aluminum. Previous studies[2] of 1060 aluminum indicate that it has upper and lower yield phenomena, the stress values of which are not experimentally well defined. For this analysis, yield values close to the lower yield point have been chosen.

7 PLASTIC WAVEFRONT

Interpretation of the plastic wave requires iterative numerical computer solutions of the equations of motion.[11] This process is simplified if the materials are not strain-rate sensitive. Even though 4340 steel is known to be quite rate-sensitive, the analysis has been simplified by treating the propagation as rate independent over a small interval of travel. The difference between two adjacent target thicknesses has been taken as the interval of travel, e.g., between 1/4 and 1/2 inch is one increment. This assumes that the change in deformation rate is small in the region of interest and, thus, an average stress-strain curve is obtained for the stated interval. The averaging process, if the intervals are large, may obscure parts of the deformation process, hence care is necessary in this type of interpretation.

The stress vs. reduced time records for various target thicknesses (Fig. 1) are assumed to represent the same stress conditions that a plane at an equivalent location in the semiinfinite solid would experience. An average propagation velocity $\bar{u}_s$ can therefore be assigned to each part of the wave profile and used to estimate the change of strain for an increment of stress:

$$\bar{u}_s{}^2 = \frac{\rho_0}{\rho^2} \frac{\Delta\sigma}{\Delta\epsilon} \tag{21}$$

where ρ is the density ahead of the stress increment. Average stress–strain curves obtained in this manner are reduced to shear-stress-strain curves by using the relations derived by Taylor:[12]

$$\tau = 3/4 \, (\sigma - K\epsilon) \tag{22}$$

$$\gamma = \frac{1}{2} \left(\epsilon - \frac{\tau}{\mu} \right) \tag{23}$$

where K is the bulk modulus of the material. These curves are shown in Fig. 4.

It is evident that the curves of Fig. 4 are nearly elastic-perfectly plastic, with the yield point given by the average amplitude of the elastic precursor. They represent the deformation associated with the first part of the plastic wave in which the stress rate is still increasing and, for which, the unloading at the quartz interface is elastic. Eventually, the shear stress at a given plane will start to decrease as the stress rate begins to decrease and the peak of the wave is reached. The fact that the curves are not well spaced may be due to heat treatment or experimental error.

The assumption that the dislocation density is linearly related to plastic strain at small strains [Eq. (5)] can be checked. If the

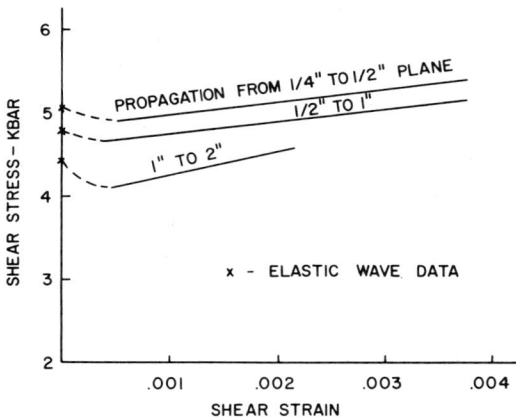

Fig. 4 Average shear stress–shear strain curves for annealed 4340 steel.

shear stress is approximately constant, as the shear stress-shear strain curves indicated, then the dislocation velocity is constant, and, according to Eq. (1), any increase in the plastic strain rate must be related to dislocation multiplication. If the dislocation density is a linear function of the plastic strain, a plot of average plastic strain rate vs. plastic strain should be a straight line. Plastic strain rates can be computed using stress rates derived from the curves of Fig. 1. Equation (13) is used, with the simplification $\dot{\sigma} = (\Delta\sigma/\Delta\epsilon)\dot{\epsilon}$ since the average stress is a function of strain only:

$$\dot{\gamma} = \frac{3}{8}\left[\frac{\lambda + 2\mu}{\Delta\sigma/\Delta\epsilon} - 1\right]\frac{\dot{\sigma}}{\mu} \tag{24}$$

Plastic strain rates have been evaluated at 1 kb steps from 16 kb to 30 kb. The plastic strain is found by using Eq. (23).

Curves derived in this manner are shown in Fig. 5. They do not superimpose because each corresponds to a slightly different, though nearly constant, shear-stress characteristic of a particular zone. The dislocation velocity corresponding to each shear-stress level (Fig. 4) may be found by using the relations in Table 1. In fact, the data can be replotted as $\dot{\gamma}/bv$ vs. γ for each relation in Table 1 to determine directly the dislocation multiplication. This has been attempted, and the curves come closer together, but not to the extent that they superimpose. The curves of Fig. 5 do show that at constant stress the dislocation density is initially nearly a linear function of plastic strain. This result is valid only for the deformation associated with the first part of the plastic wave.

Because the 1060 aluminum data is in terms of free surface motion and therefore involves the incompletely understood unloading path of the material, the analysis used for the steel does not seem warranted. Rather, a parametric study was made using Eqs. (2) and (5), and assuming only elastic unloading.

The dislocation multiplication is defined by assuming that the density increase is related to the maximum plastic strain γ_{max}. From Eqs. (5) and (7), the strain at which the dislocation density is a maximum is found from

$$\frac{dN}{d\gamma} = N_0\, e^{-\phi\gamma}[\alpha - \phi(1 + \alpha\gamma)] = 0 \tag{25}$$

The maximum strain, assuming no material shear strength, or $\tau = 0$, from Eqs. (20) and (23) will be

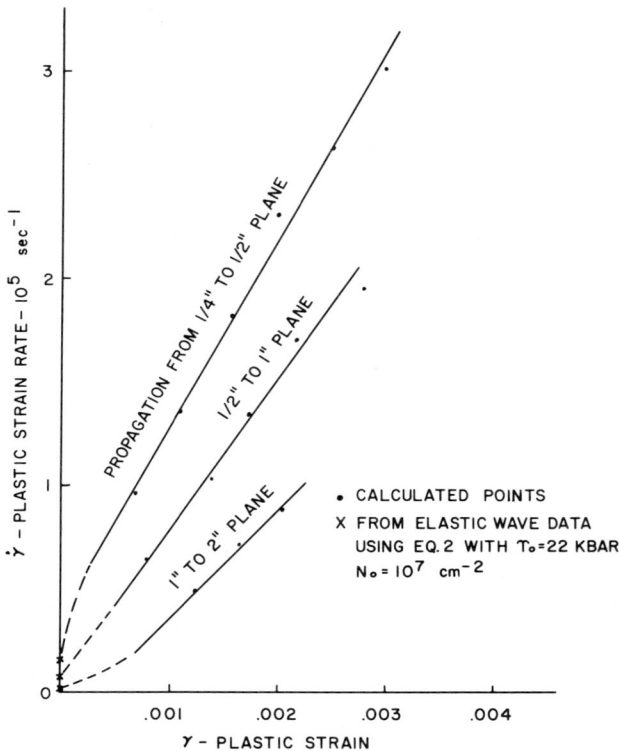

Fig. 5 The relation between plastic strain rate and plastic strain in annealed 4340 steel.

$$\gamma_{max} = \frac{\epsilon}{2} = \frac{\sigma}{2K} = \frac{\rho_0 \, c u_m}{4K} \qquad (26)$$

Now assume that the maximum dislocation density in terms of multiples of the initial density will be reached at some fraction of the total strain. Rather arbitrarily, $1/4 \, \gamma_{max}$ was used in these calculations. As discussed previously, the elastic attenuation is exponential.

In Figs. 6 and 7, calculated curves for zero and $5 N_0$ multiplication are compared with experimental data. Better agreement is obtained by using a multiplication factor. While the $5 N_0$ calculations give good agreement for the 1-inch thick target, the calculation for the 1/4-inch target does not reproduce the elastic wave amplitude. This is caused by the stress of the numerical solution relaxing very rapidly behind the elastic precursor, i.e., the dislocation function

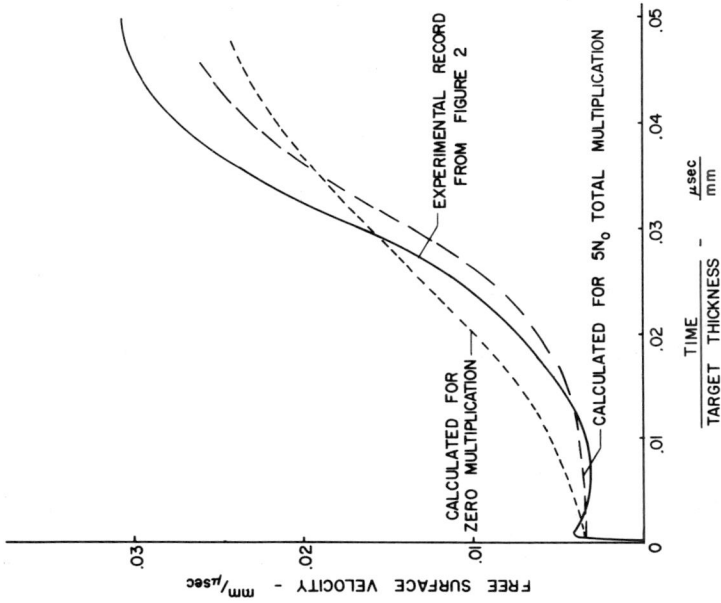

Fig. 7 Predicted and measured free surface motion for a 1-inch thick 1060 aluminum target impacted at 100 ft/sec.

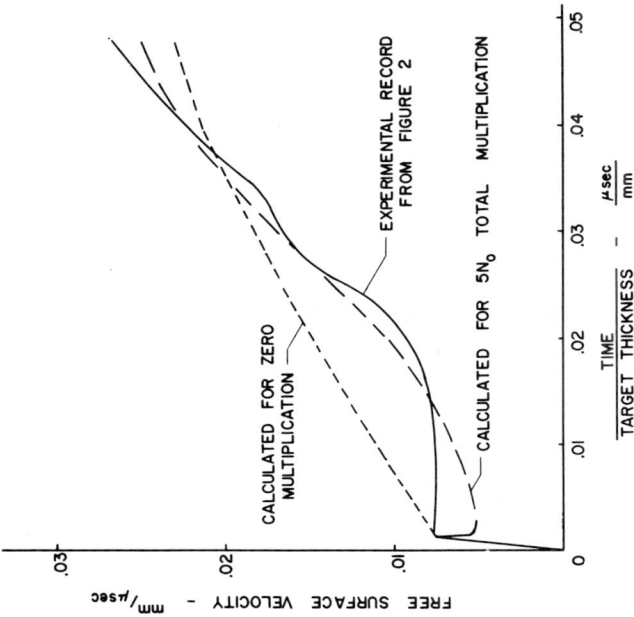

Fig. 6 Predicted and measured free surface motion for a 1/4-inch thick 1060 aluminum target impacted at 100 ft/sec.

provides too much relaxation. Although exact details are not reproduced, the multiplication model as treated here does give reasonable simulation of the measured plastic wave profile.

8 CONCLUSION

The extent to which several mathematical representations of dislocation motion can be used to describe elastic wave attenuation in 4340 steel was investigated. It was found that the Gilman relation described the experimental data quite well, although the values of the parameters were strongly influenced by the choice of the input stress condition. A linear velocity relationship also described the experimental data adequately, although not as precisely as the Gilman function. The thermal activation function showed marked discrepancies.

A simple analysis of the plastic wave in 4340 steel showed that the plastic strain at a given position in the material developed at a nearly constant value of the maximum shear stress. The magnitude of this stress was about the same as the amplitude of the elastic wave at that position.

Dislocation velocities appear to be small compared to the shear wave velocity. Thus, the large plastic strain rates calculated from the experimental records indicated considerable dislocation multiplication. A linear relation between plastic strain rate and plastic strain at constant stress was observed, but the data are not over a sufficient range to test the hypothesis that the dislocation density relation can be taken as a function of strain alone.

Parameters for the elastic wave attenuation in 1060 aluminum were computed using the Gilman function. This information and a dislocation multiplication model based only on plastic strain were used to make a parametric study of the plastic wave profile. Qualitative agreement was obtained for certain aspects of the wave. The model was apparently too simple to account for material deformation close to the impact plane; the calculations predicted much greater relaxation behind the elastic precursor than were experimentally observed.

ACKNOWLEDGMENT

We wish to thank T. C. Looby, R. G. Newman, and R. E. Hollenbach for their aid in the experiments, and to Emily G. Young for valuable help with the computations. L. M. Barker has kindly permitted use of his data.

REFERENCES

1. Butcher, B. M., and D. E. Munson: *The Fourth Symposium on Detonation*, U. S. Naval Ordnance Laboratory, Silver Spring, Maryland, October 1965.
2. Barker, L. M., B. M. Butcher, and C. H. Karnes: *J. Appl. Phys.* 37:1989 (1966).
3. Barker, L. M., and R. E. Holenbach: *Rev. Sci. Inst.* 35:742 (1964).
4. Graham, R. A., F. W. Neilson, and W. B. Benedick: *J. Appl. Phys.* 36:1776 (1965).
5. Barker, L. M., and R. E. Hollenbach: *Rev. Sci. Inst.* 36:1617 (1965).
6. Gilman, J. J.: *J. Appl. Phys.* 36:3195 (1965).
7. Dorn, J. E., J. Mitchell, and F. Hauser: *Dislocation Dynamics*, Document UCRL-11977, University of California, Berkeley, Calif., March 1965—Presented at the 1965 Spring Meeting of the Society for Experimental Stress Analysis.
8. Gilman, J. J.: *J. Appl. Phys.* 36:2772 (1965).
9. Johnson, J. N.: *Shock Waves in Stress-Relaxing Solids*, WSU SDL 66-01, Shock Dynamics Laboratory, Washington State University, Pullman, Washington, July 1966.
10. Courant, R., and K. O. Friedricks: "Supersonic Flow and Shock Waves, Interscience Publishers, Inc., New York, 1958.
11. Butcher, B. M.: *A Computer Program SRATE for the Study of Strain-Rate Sensitive Stress Wave Propagation*, SC-RR-65-298, Sandia Corporation, Albuquerque, New Mexico, September 1966.
12. Taylor, J. W.: *J. Appl. Phys.* 36:3146 (1965).
13. Jones, O. E.: Private communication.
14. Novikov, S. A., J. A. Sinitsyn, A. G. Ivanov, and L. V. Vasil'yev: *Fiz. Metal. Metalloved.* (translation) 21:452 (1966).

Agenda Discussion:

HIGH-SPEED DISLOCATIONS

J. Weertman*

Northwestern University

J. Krafft**

U. S. Naval Research Laboratory

At present, the subject of high-speed dislocations suffers from the lack of any experimental data whatsoever. (A fast dislocation velocity could be defined to be a velocity greater than about 0.9 times the transverse sound velocity.) Any discussion of this topic can only include theoretical calculations or speculation based on extrapolation of experimental results obtained on slowly moving dislocations. If no dislocation damping mechanism exists, theory predicts that fast dislocation velocities inevitably will occur in stressed crystals. Thus, our obvious question must be transformed into the problem of predicting the damping of fast-moving dislocations.

The damping of a dislocation is described formally by the following well-known equation

$$Bv = \sigma b \tag{1}$$

*Chairman
**Secretary

where σb is the force per unit length acting on a dislocation moving with velocity v. Here b is the Brugers vector, σ is the applied stress, and B is the damping constant. (It is now standard usage for the letter B to represent the damping constant.)

Theoretical attempts to predict the magnitude of B have been made in the past by Leibfried, Nabarro, Eshelby, Lothe, Weiner, T. Suzuki, and most recently by Mason and by Seeger (see their papers in this volume). These authors considered such mechanisms as thermoelastic dissipation, the scattering of phonons from the nonlinear elastic core region, the reradiation of elastic energy which occurs when thermal phonons cause a dislocation to oscillate about a mean position in a coordinate system that moves with the average dislocation velocity (called the flutter mechanism during the conference), and Mason's mechanism of electron and phonon viscosity which will damp out any sound wave. Most of these mechanisms or theories do predict values of B of the order of 10^{-4} to 10^{-3} cgs. They can account for the magnitude of B determined in internal friction experiments. The mechanisms involving phonons predict that B approaches zero as the temperature approaches absolute zero. The electron viscosity mechanism gives a B for a metal that increases with decreasing temperature. The minimum value of B for a metal occurs at an intermediate temperature.

The Agenda Discussion opened on the topic of the theoretical calculations of B. Nabarro stated that he no longer considered his old criticism of the Liebfried theory to be valid, and that he no longer believes one should introduce the Lorentz force into the theory. Both Nabarro and Lothe questioned the cut-off radius that Mason used in his calculation of B. They thought that the phonon (or electron) mean-free path should be used at all dislocation velocities. Mason felt his use of 3/4 b at low dislocation velocities and the phonon (or electron) mean-free path at high dislocation velocities was correct. Professor T. Suzuki, whose ideas are similar to Mason's, concurred. Mason did point out that the radius 3/4 b he now uses is larger than the one employed in his original paper. Professor Suzuki pointed out that impurities can have a large effect on B even at rather low concentration levels.

Gilman suggested a new damping mechanism. He pointed out that the flow of phonons around a moving dislocation may be turbulent rather than laminar. He estimated that the Reynolds number for turbulent phonon flow may be exceeded for dislocations in real crystals.

Seeger discussed damping found on a one-dimensional crystal model. He calls this model the *PDFK* (Prandtl-Dehlinger-Frenkel-Kontorova) rather than the usual name: Frenkel-Kontorova model.

A very interesting result of this model is that at high dislocation velocities, B decreases with increasing velocity and becomes zero when the transverse sound velocity is reached. Previously, Mason had predicted this behavior of B with his theory of damping. If this behavior of B were general for all damping mechanisms, a "window" in the damping constant would be available for obtaining fast dislocation velocities. However, the suggestion was made by Nabarro that the damping mechanism originally proposed by Orowan and analyzed by Hart could limit the dislocation velocity. This mechanism involves the energy radiated away when a dislocation moves over Peierls hills. Seeger pointed out that this energy loss is a constant. Thus, if the stress is increased, the effective value of B will decrease. Therefore, this mechanism will not limit the dislocation velocity to low velocities. Seeger thought that every damping mechanism except one leads to small or zero damping for dislocations moving with velocities close to the transverse sound velocity. The exceptional mechanism is the dispersion one (first suggested by Eshelby). The velocity of sound in real crystals is not independent of the wavelength of a sound wave (this is a nonlinear effect). The velocity decreases with decreasing wavelength. Therefore, when the dislocation moves near the transverse sound velocity, it will exceed the sound velocity of sound waves where wavelength is comparable to the core dimensions. Elastic energy will be radiated. Seeger, in an obvious analogy, calls this, Cherenkov radiation. The dispersion mechanism produces a large B at high dislocation velocities and thus may close the window mentioned earlier.

Seeger pointed out in response to a question that the $PDFK$ model could treat split dislocations containing a stacking fault merely by using the second harmonic of the particle motion.

The problem of the experimental determination of fast-dislocation velocities was considered by Gilman and Vreeland. Any measurement of dislocation velocity involves a measurement of the distance δx that a dislocation moves in a time interval δt. Thus, the measured velocity v is $\delta x/\delta t$. The time interval δt presents the most difficulty. In any method involving stress pulses in order for δt to be accurately found, a very short rise time is desired for the pulse. For specimens of finite geometry, a geometric dispersion sets simultaneous limits on δt and δx which can produce uncertainties in high velocity measurements of the order of 100%. The Vreeland torsional loading method (see his paper in this volume) minimizes these difficulties. Reflections of an applied stress pulse can also lead to erroneous results, since they will increase the length of time the dislocations are pushed. It is important to take account of or minimize these effects when the constant m in the equation

$$v = C \left(\frac{\tau}{\tau_0} \right)^m \tag{2}$$

is small and of the order of $m \approx 1$. This empirical equation is often used to describe the stress dependence of the dislocation velocity. In this equation, C and τ_0 are constants. If m is large, then the error in determining δt is greatly reduced, since the peak value of the stress pulse produces the major portion of the dislocation motion.

Vreeland pointed out an error that can easily be made in the determination of δx. Consider Fig. 1. Suppose an etch-pit technique is used to determine the velocity of the dislocation which originally is at position 1 and then moves to position 2. The velocity that will be determined will only be correct if the dislocation is perpendicular to the surface. If it forms an angle to the surface, and no account is taken of this fact, the apparent dislocation velocity will be larger than the true dislocation velocity.

Hahn raised the question of how well the internal friction determinations of the damping constant B compared with these determined by the etch-pit technique such as employed by Vreeland. This question was answered by Granato, who said that they agree very well with Vreeland's determination, but are higher than those determined by T. Suzuki. According to T. Suzuki, he, Ikushima, and Aoki, and later, Ikushima and his associates, determined B at room temperature to be 1.2×10^{-4} cgs for pure copper and 2.8×10^{-4} cgs for copper-0.1% nickel alloy. This value for copper is smaller than Granato's, which is about 7×10^{-4}, but very close to that obtained by Parè and Thompson (private communication). Furthermore, these values are in good agreement with the etch-pit determination by Suzuki and his associates, when

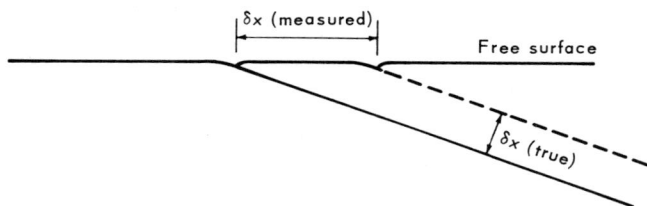

Fig. 1 Error introduced into etch pit measurements when dislocations move at an angle to the surface.

they take the maximum displacement of dislocations under a given stress-pulse (near the yield stress), which is considered to give the lower limit of "free-flight velocity." Accordingly, Vreeland and Suzuki both came to the same conclusion qualitatively, but not quantitatively.

Earlier in the conference Granato had pointed out that the damping constants B determined from internal friction are so large that stresses greater than the theoretical shear strength would be required to cause dislocation to move near the sound velocity, assuming that B was independent of velocity. This simple fact is the reason why the damping constant B is of such great interest for considerations whether dislocations can indeed move at higher velocities.

Granato suggested that a promising experiment to attempt to obtain high-speed dislocations is to stress a superconductor at temperatures below its critical temperature or a nonconductor at low temperatures. At low temperatures the phonon damping mechanisms give a small B. The electron viscosity damping mechanism is suppressed when a metal is in a superconducting state. This suppression can also be accomplished with the use of magnetic fields. Nabarro mentioned that Elbaum had found large changes in B, in going from the superconducting to the normal state in internal friction measurements. Thus, a promising window of low B does appear to exist in which high dislocation velocities can be produced.

Gilman considered a number of ingenious methods that might be used to measure and possibly produce fast-moving dislocations. Nonlinear scattering of laser light off dislocations could be used. He has tried this method, but without success thus far. Flash photography employing either light or x-rays might be used to take measurements of fast-moving dislocations during the motion (the term "in flight" was used during the conference for any possible measurement of dislocations while they are actually moving). Dislocations that move in luminescent crystals, such as doped KCl, will cause luminescence while they are moving. The radiated power produced could be used to measure or estimate the dislocation velocity. Spin resonance signals produced by moving dislocations or kinks is another possible method. Gilman suggested for short pulse-loading the use of pulsed electron or ion beams, pulsed magnetic fields, the laser pulses already mentioned, pulsed neutron beams, as well as the use of projectiles, explosive shocks, or electric sparks.

Young suggested that his x-ray method of observing dislocations in low-dislocation density metals could be used in measuring dislocation velocities. A crystal could be vibrated and a long dislocation loop set into vibration. A long x-ray exposure could then reveal the amplitude of dislocation motion.

The next topic considered was elastic radiation emitted from nonuniformly moving dislocations. Keh discussed work done at U. S. Steel by his colleagues, Fisher and Lally, who have measured acoustic pulses emitted from crystals pulled in tension.* It does not seem possible at present to make any reliable estimates of the velocities or accelerations of the dislocations that produce the acoustic pulses and discontinuous yielding.

Nabarro discussed a recent work of Laub and Eshelby on traveling or standing waves on a dislocation [*Phil. Mag.* 14:1285 (1966)]. In general, such waves on dislocations radiate elastic energy. Nabarro mentioned that Laub recently found a numerical mistake of a factor 2 near the end of this paper which, when corrected, makes the results more plausible. A difficulty of the string model of a dislocation when simple ideas of dislocation mass and line tension are used is that wave velocities down the dislocations are predicted that are greater than the transverse wave velocity. The corrected paper of Laub and Eshelby shows that the actual wave velocities are either less than the transverse velocity or are equal to it.

T. Suzuki presented evidence on point-defect production by fast-moving dislocations studied by Imura, Tanaka, and himself. Figure 2 shows a plot of the point-defect concentration vs. strain rate, which was deduced from electron-microscopic studies on aluminum. It can be seen that at fast strain rates (and presumably at faster dislocation velocities) the point-defect concentration is an order of magnitude larger. Figure 3 is an electron-micrograph of fast-strained aluminum, showing a number of small dislocation loops with stacking faults. These loops are thought to be produced by the condensation of point defects. These loops are characteristically small in size compared with dislocation dipoles predominantly observed in slowly strained specimens. It was pointed out in the discussion that heating effects at fast strain rates could have produced these point defects. The other explanation, of course, is that fast-moving dislocations should by theory be more efficient producers of point defects than slow-moving dislocations. It was also mentioned that vacancy- and interstitial-loops were almost equal in number.

*A report of this work is appended. pp. 617-620.

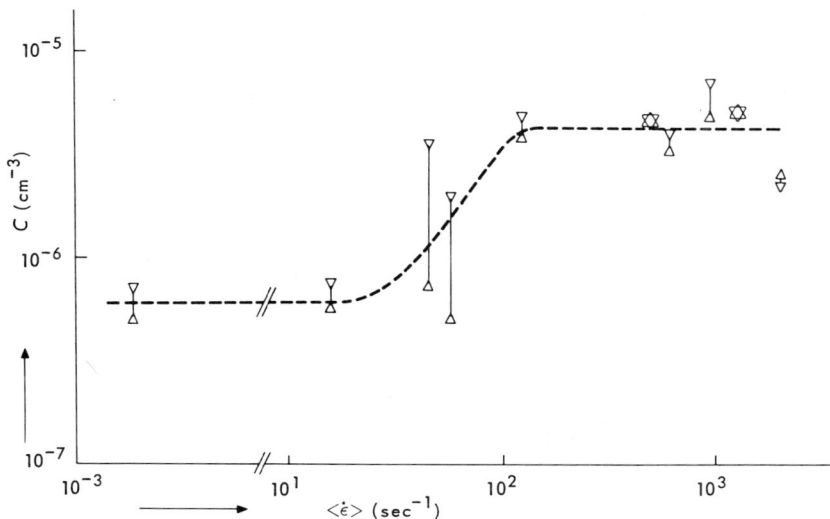

Fig. 2 Concentration of point defects C cm^{-3} vs. average strain rate $<\dot{\epsilon}>$ sec^{-1}.

In a communication received after the Colloquium, Brown commented on these results as follows:

"Suzuki's evidence is based on shocked aluminum which shows a vastly different structure than copper, nickel, and iron. The structure of aluminum after shock loading strongly indicates that much recovery takes place during or after shock loading. It contains a much smaller dislocation density. The behavior of nickel, whose recovery temperatures are very high compared to the temperature generated during shocking, is more typical. The latest results on point defects in shocked nickel are reported by Kressel and Brown [J. Appl. Phys., 38:1618 (1967)]. In connection with Fig. 2, the situation is really more complex than indicated. The dislocation density as a function of $\dot{\epsilon}$ for a given strain should be known. For a given strain, in nickel, it was found that the concentration of point defects (both interstitial and vacancies) was greater during shock loading than during conventional deformation. However, the dislocation density was also correspondingly greater so that the number of point defects per dislocation was nearly independent of strain rate for a given strain. However, not only the number of dislocations but the distance the dislocations moved should also be considered in order to understand the point-defect concentration. It is still not entirely clear as to whether fast-moving dislocations are more efficient generators of point defects than slower ones."

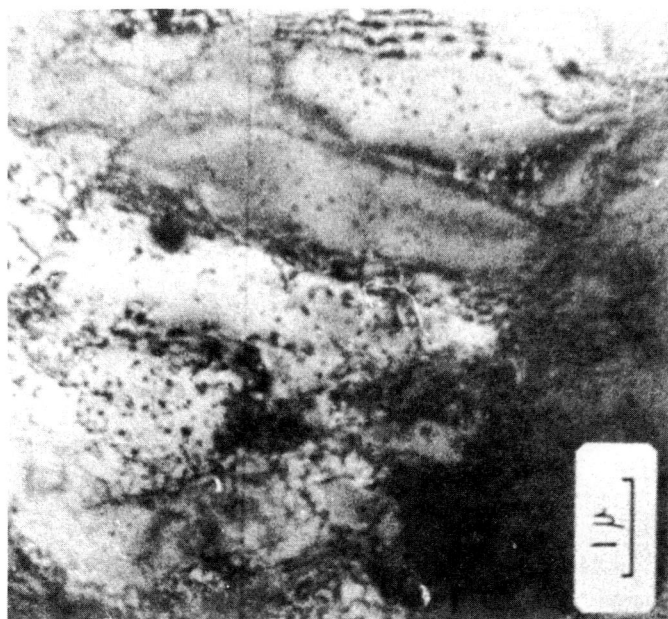

(b)

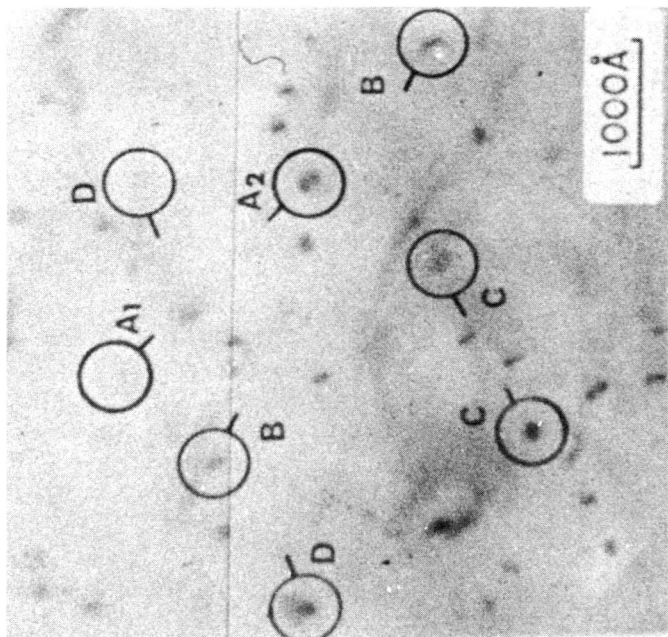

(a)

Fig. 3 Effect of strain rate on the deformation substructure of aluminum. (a) $\epsilon = 0.629$, $\langle \dot{\epsilon} \rangle = 1.26 \times 10^3 \cdot \sec^{-1}$ (500 kV). (b) $\epsilon = 0.69$, $\langle \dot{\epsilon} \rangle = 2.03 \times 10^3 \cdot \sec^{-1}$ (350 kV).

The topic of supersonic dislocations and dislocations moving on transformation interfaces provoked little discussion. It was pointed out by Hahn that twinning can take place at velocities near the sound velocity. Owen made a similar remark about martensitic transformation. Obviously, no progress will be made in this area until attempts are made to measure, however crudely, dislocation velocities on transformation interfaces.*

APPENDIX A DISCUSSION

Submitted by R. M. Fisher and J. S. Lally

U. S. Steel Corporation

(Presented by A. S. Keh)

In previous studies piezoelectric crystals were used to detect acoustic pulses emitted from single crystal specimens of Cu, Fe, Al and Mg which were pulled in tension [*Can. Journal of Physics* 45: 1147 (1967)]. Oscilloscope traces illustrating the appearance of these pulses from a Mg crystal are shown in Fig. A1 at different sweep rates. It may be seen that each pulse consists of a highly damped burst of sonic oscillations with strong frequency components at 100 and 20 kcps. These correspond to the natural frequencies of longitudinal vibrations of the piezoelectric detector and test specimens, respectively. The acoustic pulses were thought to be

---

*The following material was not discussed and is added to fill out the subject of supersonic dislocations: Frank and van der Merwe first pointed out the existence of supersonic dislocations for a one-dimensional crystal model. Eshelby pointed out that supersonic dislocations could exist on any interface that can give up energy. His observation sets the supersonic dislocation on a very plausible physical basis and makes it a subject that dislocation theory and experiment eventually must consider seriously. Eshelby pointed out that the supersonic screw dislocation will radiate energy away from it by two sound waves that move at an oblique angle to its slip plane. Stroh [*J. Mathematics and Physics* 41: 77 (1962)] worked out a completely general theory for a dislocation moving in anisotropic media at both supersonic and subsonic speeds. The radiated sound waves from supersonic dislocations that he assumed were step functions of the displacements (shock waves). Weertman (paper in preparation) has shown that Stroh's supersonic solution is only applicable in the interface upon which the dislocation moves gives up infinite energy per distance moved. When the transformation plane gives up finite energy, the shock wave is spread out into a sound wave of finite widths and the supersonic dislocation is spread out into a localized distribution of infinitesimal dislocations. The stress at any point on the transformation plane is proportional to the dislocation density at that point. This result is in contrast to the case for subsonic dislocations. For these, the stress on a point on the slip plane is produced by all dislocations other than the dislocation density at the particular point. When dislocations do move on an interface that gives up energy, it should always move faster than the transverse sound-wave velocity if it is a screw dislocation, and faster than the longitudinal wave velocity if it is an edge dislocation.

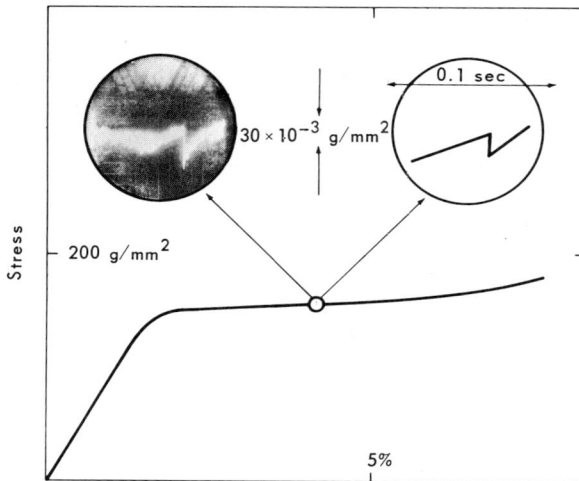

Fig. A1 Microscopic yield drops producing acoustic pulses.
Magnesium single crystal.

induced by microscopic load drops which occur during slip
avalanches but it was not possible to determine an absolute
calibration for the equipment. However, it was deduced that
the pulses shown in Fig. A1 represent plastic strain increments of the order 10^{-6} to 10^{-7}. The repetition rate, amplitude
and rise time of the pulses suggest that the dislocation velocities
in slip avalanches are of the order of 10^3 cm/sec.

Recently, a miniature quartz load cell has been used to detect
and obtain a direct measurement of the size of the microscopic
load drops produced by such slip avalanches. Figure A2 shows
a typical tracing on a storage oscilloscope of a microscopic yield
point during straining of a magnesium single crystal. Although
the sensitivity and noise rejection of this load cell is inferior
to the acoustic pickup, it is adequate to record the larger pulses
and the voltage amplitude is directly related to the stress change.
Use of a dual beam oscilloscope has verified that the acoustic
pulses and load drops occur simultaneously.

These preliminary results confirm the conclusions reached
previously, namely: (1) that plastic deformation occurs discontinuously with dislocation movement occurring only during very
brief intervals; (2) that the microscopic strain increments occur

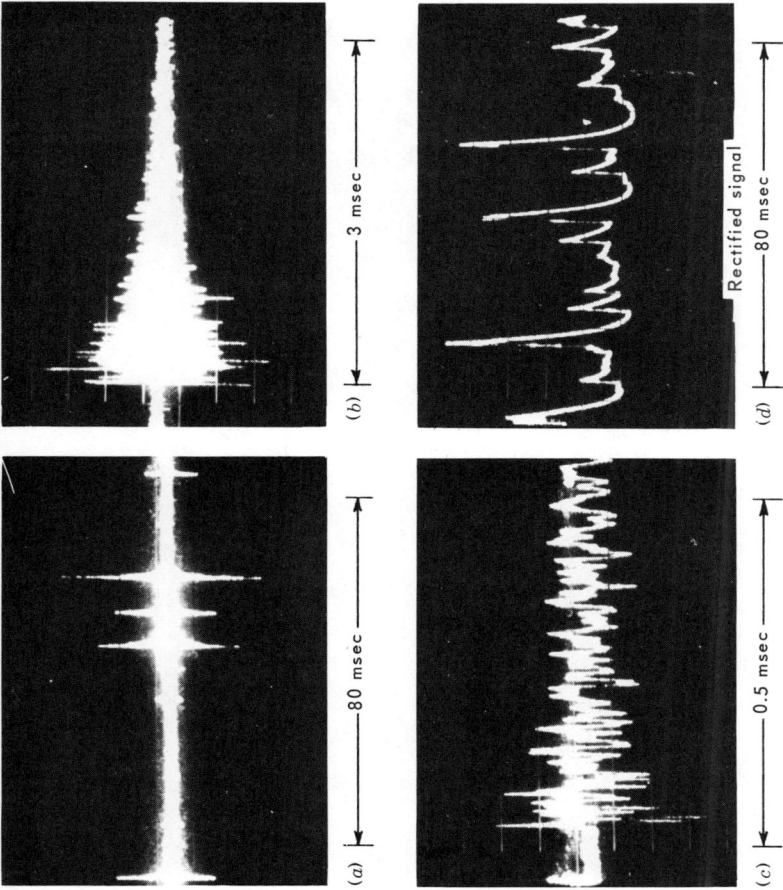

Fig. A2 Oscilloscope traces of acoustic pulses from Mg single crystal
$$\epsilon \approx 5\% \qquad \dot{\epsilon} \approx 10^{-4}\ \text{sec}^{-1}$$

in a few microseconds and result from a cooperative occurrence of numerous slip events across the full section of the crystal specimen; and (3) that the dislocation velocity during the slip avalanches is of the order of 10^3 cm/sec.

_____ *Part Five*

CRITICAL PROBLEMS

THEORY OF KINK
MOBILITIES AT
LOW TEMPERATURES

Alfred Seeger and Helmut Engelke

*Max-Planck-Institut für Metallforschung,
Stuttgart, Germany*

ABSTRACT

The low-temperature mobility of dislocations is best discussed in terms of the frictional force experienced by moving kinks, since due to the action of the Peierls barrier, the dislocation movement at low temperatures takes place through the displacement of kinks, and since by considering kinks the complications due to the flexibility of dislocation lines are avoided. The various mechanisms for kink friction will be considered. It will be shown that provided extremely low temperatures are excluded, the dominant effect is the scattering of phonons from the elastic inhomogeneities caused by uniformly moving kinks. The scattering of ordinary ("light," rest mass zero) phonons from an inhomogeneity is calculated by using nonlinear elasticity theory. The calculation of the scattering of heavy phonons (finite rest mass, corresponding to the quantized oscillation of a dislocation in a Peierls valley) are carried out within the framework of the Prandtl-Dehlinger-Frenkel-Kontorova model. The quantization procedures and the formulation of the master equation for the phonon distribution follow the techniques developed in the theory of heat conductivity. The calculations show that depending on the temperature range, either the light or the heavy phonon scattering may dominate. Numerical calculations are performed for kinks in screw dislocations in copper. Preliminary comparisons with experimental results show quite good agreement. In the Appendix, the "relativistic" effects on the phonon scattering are discussed within the framework of the Prandtl-Dehlinger-Frenkel-Kontorova model.

1 INTRODUCTION

It may be considered as experimentally established that the Bordoni relaxation (including the so-called Niblett−and−Wilks peak) is due to the thermally activated, stress-assisted formation of kink-pairs in dislocations that run approximately parallel to one of the close-packed directions of the glide plane.[1] This means that below the temperature of the Bordoni peak the shapes of the dislocation lines giving rise to the relaxation effect are to a considerable extent determined by the existence of Peierls barriers, i.e., the periodic variation of the dislocation energy as a function of the dislocation position in the crystal. At low temperatures a dislocation that is anchored in two Peierls valleys not too far apart from each other, may be decomposed into a series of straight sections (lying in Peierls valleys) that are connected by kinks leading from one valley to a neighboring one. This is quite different from the predictions of the string model, which neglects the existence of the Peierls barriers and is therefore applicable only at temperatures well above the Bordoni peak.* Nevertheless, as long as the dislocation response to an applied stress is linear, the predictions of the kink model and of the string model of a dislocation line are quite similar.[3,4] This means that in this domain the results obtained by fitting the parameters of the string model to experimental data may be reinterpreted in terms of the kink model, and vice

---

*A detailed investigation into the justification of the string model[2] indicates that even at high temperatures (where the existence of the Peierls barrier may rightly be neglected), the string equation for the dislocation motion is not obtainable by a first step in a series of well-defined approximations. The string model must thus be considered as a heuristic model rather than an exact description of a limiting case.

versa. On the other hand, as discussed in detail by Alefeld,[5] the low-temperature nonlinear behavior of the kink-model is quite different from that of the string model; this fact may serve to differentiate between these two models empirically.

From a theoretical viewpoint, the kink-model is superior to the string model with regard to first-principle calculations of the model parameters, e.g., those describing the frictional forces exerted on a dislocation. This is due to the fact that because of their flexibility, dislocations change their shapes in the course of their motion. The dynamic behavior of dislocations is determined not only by their instantaneous shapes and velocities, but also by their previous histories. This has the consequence that the difficulties experienced in defining such quantitites as the energy per unit length of an arbitrarily curved dislocation line are aggravated, and that it is not possible to define unambiguously a viscosity coefficient per unit length of a dislocation line. The state of motion of a kink, however, is to a fairly good approximation characterized by a single quantity, e.g., the velocity of the movement along the dislocation line. We should therefore be able to define a velocity-dependent viscosity coefficient for a kink.

The main friction mechanisms related to the interactions of dislocations with the crystal structure and the lattice vibrations have been discussed elsewhere.[1,2,6] All of them can be formulated in the kink picture. In a nonconducting perfect crystal, the two most important friction mechanisms are the "flutter mechanism"[2,6-8] and the "nonlinearity mechanism.[2,11] In the first mechanism, which has been discussed in a preliminary form by Leibfried[7] and in more detail by Eshelby[8] and Lothe,[6] the drag exerted on moving kinks is attributed to the elastic radiation due to kink oscillations caused by the thermal shear stresses. The second mechanism considers the scattering of lattice vibrations from the high-strain region in the environment of the kink, in which the superposition principle is violated. This leads to a dragging force on moving kinks, the calculation of which is the main object of this paper.

In metals, in addition to the phonon mechanisms, the motion of dislocations (and kinks) is damped by the interaction between the strain fields and the conduction electrons. In good electrical conductors this mechanism will dominate over the above-mentioned phonon mechanisms at very low temperatures[9,10] (e.g., He temperatures), since the phonon mechanisms become inefficient due to the small density of thermally excited phonons.

Our main interest lies in calculating the phonon contribution to the kink viscosity for temperatures up to the temperature of the Bordoni peak, i.e., well below the Debye temperature. In this temperature range, the dispersion of the lattice vibrations may be disregarded, and the nonlinear elasticity theory may be used to calculate the scattering of phonons from the strain field of a moving kink.

First, we shall calculate in general terms the dragging force exerted by the scattering of Bose-type excitations (quasi-particles) from a perturbation moving with the uniform velocity $\bar{v}$, and apply the result to a phonon gas only slightly perturbed from its equilibrium distribution (Sec. 2). In Sec. 3, the specific interaction of moving dislocations with "light" and "heavy" phonons will be treated. "Light" phonons are the ordinary phonons propagating in a three-dimensional elastic solid (rest mass zero); "heavy" phonons[8] possess finite rest mass and are obtained by quantizing the (one-dimensional) vibrations of a dislocation line lying in a Peierls valley.

Sections 3.1 and 3.2 give the classical treatment; Secs. 3.3 and 3.4 contain the extension to quantum mechanics. For light phonons (Secs. 3.2 and 3.3) our treatment will follow closely that developed elsewhere[11-18] for the effect of the phonon scattering from dislocations on the heat conductivity of insulators. The essential simplifying features of this approach are that phonon Umklapp processes are neglected and that the dispersion relations of the continuum theory are used. The calculation of the dragging force due to the heavy phonons will be based on the Prandtl-Dehlinger-Frenkel-Kontorova model (for a detailed discussion see Ref. [2]). This model can be used both for studying kinks in dislocations and as a simple one-dimensonal model of a dislocation. The results pertaining to the latter aspect will be briefly outlined in the Appendix. Section 4 gives an explicit treatment for a uniformly moving single kink. The numerical results for copper are discussed in Sec. 5.

2   THE DRAGGING FORCE ON A UNIFORMLY MOVING PERTURBATION

In this section we calculate the dragging force exerted by a gas of elementary excitations on a uniformly moving perturbation in a crystal. We characterize the excitations by their wave-vectors k and their energies $\hbar\omega_k^j$; j is a polarization index. Quantizing the classical fields (see Sec. 3) in the usual way leads to a Hamiltonian of the form (see Sec. 4)

$$H = H_0 + H_1(t) \tag{2.1}$$

For phonons (more generally: for Bose-Einstein-type excitations) H_0 takes the form

$$H_0 = \sum_{k,j} \hbar \omega_k{}^j \left(a_k^{+j} a_k{}^j + \frac{1}{2} \right) \tag{2.2}$$

which represents an infinite set of uncoupled quantized linear harmonic oscillators. Also, a_k^{+j} and $a_k{}^j$ are the Bose-Einstein creation and annihilation operators. The operator

$$H_1(t) = \sum_{kk'} \left\{ H_1' \exp[i(k_z + k_z') \cdot \bar{v}t] + H_1'' \right\} \tag{2.3}$$

gives rise to an interaction (coupling) between the oscillators. Both H_1' and H_1'' are time-independent operators, explicit forms of which will be derived below. H_1'' represents the effect of the time-independent ("nonmoving") part of the perturbation. The N-processes (three-phonon-processes) could also be included in H_1'', but will be shown to be unimportant in the present treatment. The time-dependent sum in Eq. (2.3) describes the interactions between the excitations due to the presence of a perturbation moving with the constant (average) velocity $\bar{v}$ in z-direction (i.e., the kink in a dislocation line). Here k_z and k_z' are the z-components of the wave-vectors k and k' of the excitation created and destroyed during the interaction described by the term $a_k^{+j} a_k{}^{j'}$ in H_1'. Since we neglect Umklapp processes, we may identify the quasi-momentum of a phonon with the ordinary momentum. This means that the change of the momentum p of the phonon gas in such a process is given by $\hbar(k - k')$. According to actio = reactio the perturbation causing this transition undergoes an opposite change in its momentum, and the force f exerted on the perturbation is given by

$$f = -\frac{\partial p}{\partial t} \tag{2.4}$$

If $N_k{}^j$ denotes the occupation of the oscillator $\{j, k\}$, then the total momentum of the phonon gas is given by

$$p = \sum_{j,k} \hbar k N_k{}^j \tag{2.6}$$

The calculation of the occupation numbers $N_k{}^j$ as a function of time would require the solution of a complicated many-body problem. It is expedient to introduce the average occupation numbers $\bar{N}_k{}^j$ according to

$$\bar{N}_k{}^j = \sum_\nu N_k{}^j P(\nu;\widetilde{\nu};t) \tag{2.6}$$

where the summation extends over all eigenstates ν of the unperturbed Hamiltonian H_0. The conditional probability

$$P(\nu;\widetilde{\nu};t) \equiv P\left(\ldots, N_k{}^j, \ldots, \ldots, \widetilde{N}_k{}^j, \ldots, t\right) \tag{2.7}$$

gives the probability of finding at time t the occupation numbers $N_k{}^j$, if the perturbation was switched on at $t = 0$ and if the occupation numbers at that time were $\widetilde{N}_k{}^j$. By averaging Eqs. (2.4) and (2.5) and inserting Eq. (2.6), the following expression for the z-component of the mean force on the perturbation is obtained.

$$\bar{f}_z = \sum_{j,k} \sum_\nu \hbar k_z N_k{}^j \dot{P}(\nu;\widetilde{\nu};t) \tag{2.8}$$

The conditional probability $P(\nu;\widetilde{\nu};t)$ obeys the master equation

$$\dot{P}(\nu;\widetilde{\nu};t) = \sum_{\nu''}\{W(\nu,\nu'')P(\nu'',\widetilde{\nu}) - W(\nu'',\nu)P(\nu,\widetilde{\nu})\} \tag{2.9}$$

where the sum extends over all intermediate states ν''.

According to the semiclassical derivation of Eq. (2.9), $W(\nu,\nu'')$ is Dirac's transition probability for a transition from state ν'' to state ν. Van Hove[19] has given a rigorous derivation of Eq. (2.9), leading to

$$W(\nu'',\nu) = \frac{2\pi}{\hbar^2}\left|<\nu''|H_1'|\nu>\right|^2 \delta(\alpha'_{\nu''}) + \frac{2\pi}{\hbar^2}\left|<\nu''|H_1''|\nu>\right|^2 \delta(\alpha''_{\nu''}) \tag{2.10}$$

where

$$\alpha'_{\nu''} = \frac{(E_{\nu''} - E_\nu)}{\hbar} + (k_z + k'_z)\bar{v} \;\;;\;\; \alpha''_{\nu''} = \frac{(E_{\nu''} - E_\nu)}{\hbar} \tag{2.10a}$$

$$E_\nu = \sum_{j,k} \hbar\omega_k{}^j\left(N_k{}^j + \frac{1}{2}\right)$$

Inserting Eqs. (2.10) and (2.9) into Eq. (2.8) gives the final expression for the mean force $\bar{f}_z$ on the perturbation in terms of the transition probabilities $W(\nu, \nu'')$. Although in the derivation of this result we had in mind the drag exerted by phonons on kinks in dislocations, the same derivation holds with small changes also for other problems, e.g., for the drag exerted by spin waves on a moving domain wall.

For phonons the operators H_1' and H_2'' take the following forms:

$$\mathrm{H}_1' = -\frac{(2\pi)^3}{V_0} \cdot \frac{\hbar}{2\rho_0} \sum_{jj'} \frac{1}{\sqrt{\omega_k^{\ j} \omega_k^{\ j'}}} \left\{ \left(V_{1\,kk'}^{\prime\ jj'} - V_{2\,kk'}^{\prime\ jj'} \right) a_k^{\ j} a_{k'}^{\ j'} \right.$$

$$- 2 \left(V_{1-kk'}^{\prime\ jj'} + V_{2-kk'}^{\prime\ jj'} \right) \epsilon^j a_k^{+j} a_{k'}^{\ j'}$$

$$\left. + \left(V_{1-k-k'}^{\prime\ jj'} - V_{2-k-k'}^{\prime\ jj'} \right) \epsilon^j \epsilon^{j'} a_k^{+j} a_{k'}^{+j'} \right\} \qquad (2.11)$$

H_2'' is obtained from H_1' by replacing the matrix elements $V_{i\,kk'}^{\prime\ jj'}$ by $V_{i\,kk'}^{\prime\prime\ jj'}$. These matrix elements will be derived in Sec. 3. The basic volume of the crystal, outside which the crystal is assumed to repeat itself periodically is denoted by V_0; ρ_0 is the density of the ideal crystal. The symbol ϵ^j, which assumes the values ± 1, will be explained in Sec. 3.

Inserting the operator H_1 into Eq. (2.10) gives

$$W(\nu'', \nu) = W_{-kk'}^{jj'} \left(N_k^{\ j} + 1 \right) N_{k'}^{\ j'} \qquad (2.12)$$

where

$$W_{-kk'}^{jj'} = 2\pi \left[\frac{(2\pi)^3}{V_0 \rho_0} \right]^2 \frac{1}{\omega_k^{\ j} \omega_{k'}^{\ j'}} \left\{ \left| V_{1-kk'}^{\prime\ jj'} + V_{2-kk'}^{\prime\ jj'} \right|^2 \delta(\alpha') \right.$$

$$\left. + \left| V_{1-kk'}^{\prime\prime\ jj'} + V_{2-kk'}^{\prime\prime\ jj'} \right|^2 \delta(\alpha'') \right\} \qquad (2.13)$$

and

$$\alpha' \equiv \omega_k^{\ j} - \omega_{k'}^{\ j'} - (k_z - k_z')\bar{v}; \qquad \alpha'' \equiv \omega_k^{\ j} - \omega_{k'}^{\ j'}$$

In Eq. (2.13) we have omitted terms containing $\left(\omega_k^{\ j} + \omega_{k'}^{\ j'} - (k_z + k_z')\bar{v} \right)$ or $\left(\omega_k^{\ j} + \omega_{k'}^{\ j'} \right)$ as factors, since due to energy conservation these factors are identically zero as long as $\bar{v} < C_j$, where C_j are the velocities of the waves with polarization j. (This holds both for the "light" phonons, $\omega_k^{\ j} = C_j|\,k\,|$, and for the "heavy" phonons.) With Eqs. (2.12) and (2.13) the intermediate states ν'' correspond to the creation of one additional phonon $\binom{j}{k}$ and the annihilation of one phonon $\binom{j'}{k'}$. The summation over ν necessary in order to obtain the average force $\bar{f}_z$ [Eq. (2.8)] requires averages over triple the products of the occupation numbers. Within the framework of the present approximation, we may replace these averages by making the products of the averages of the occupation numbers. Making use of the symmetry properties of $V_{i\mathbf{k}\mathbf{k}'}^{'jj'}$ and $V_{i\mathbf{k}\mathbf{k}'}^{''jj'}$, we obtain for the average dragging force

$$
\bar{f}_z = -\pi \left[\frac{(2\pi)^3}{V_0 \rho_0}\right]^2 \sum_{\substack{jj'\\ \mathbf{k}\mathbf{k}'}} \frac{\hbar(k_z - k_z')}{\omega_k^{\ j}\omega_{k'}^{\ j'}} \Biggl\{ \left| V_{1-\mathbf{k}\mathbf{k}'}^{'\ jj'} + V_{2-\mathbf{k}\mathbf{k}'}^{'\ jj'} \right|^2 \delta(\alpha')
$$

$$
+ \left| V_{1-\mathbf{k}\mathbf{k}'}^{''\ jj'} + V_{2-\mathbf{k}\mathbf{k}'}^{''\ jj'} \right|^2 \delta(\alpha'') \Biggr\} \left(\bar{N}_k^{\ j'} - \bar{N}_k^{\ j} \right)
$$

$$(2.14)$$

So far we have not restricted the occupation numbers of the phonon gas. Henceforth we assume that before switching on the perturbation the phonons were in thermal equilibrium, i.e., that the average occupation numbers were (K = Boltzmann's constant)

$$
\bar{N}_k^{\ j} = \bar{\bar{N}}_k^{\ j} \equiv \left[\exp \frac{\hbar\omega_k^{\ j}}{KT - 1} \right]^{-1} \tag{2.15}
$$

In contrast to the theory of heat conductivity by phonons, we obtain a physically significant result without calculating the deviation of $\bar{N}_k^{\ j}$ from $\bar{\bar{N}}_k^{\ j}$ caused by a small perturbation H_1 (which would require the solution of a Boltzmann-type integral equation), since a finite result for $\bar{f}_z$ is obtained already by inserting the equilibrium distribution (2.15) into (2.14). If the perturbation H_1 is small (which is necessary anyway for our perturbation theory approach to be valid), allowing for the deviation $\bar{N}_k^{\ j}$ from $\bar{\bar{N}}_k^{\ j}$ would give rise to a small correction to $\bar{f}_z$ only. In the present approximation we are permitted to neglect [as we have done from Eq. (2.11) on] the three-phonon collisions (normal processes in the theory of heat conductivity), since these do not have an influence on $\bar{\bar{N}}_k^{\ j}$.

Replacing the summations in k-space by integrations according to the rule of correspondence

$$\sum_{k} g(k) = \frac{V_0}{(2\pi)^3} \int g(k) \, d\tau_k \tag{2.16}$$

we obtain as the final result for the average force

$$\bar{f}_z = -\frac{\pi}{\rho_0^2} \sum_{j,j'} \int \frac{\hbar (k_z - k_z')}{\omega_k^j \omega_k^{j'}} \left| V_1' {}_{-kk'}^{jj'} + V_2' {}_{-kk'}^{jj'} \right|^2 \left(\bar{\bar{N}}_k^{j'} - \bar{\bar{N}}_k^{j} \right)$$

$$\times \delta \left(\omega_k^j - \omega_k^{j'} - (k_z - k_z') \bar{v} \right) d\tau_k d\tau_{k'} \tag{2.17}$$

The static perturbation H_1'' no longer contributes to Eq. (2.17). In the present approximation the dragging force is due entirely to the acoustic Doppler shift [see the δ-function in Eq. (2.17)].

3   THE INTERACTION OF LATTICE VIBRATIONS WITH MOVING DISLOCATIONS AND KINKS

3.1   The Prandtl-Dehlinger-Frenkel-Kontorova-Model— Classical Treatment

We shall employ here the continuum approximation to the PDFK model.[2] This describes a one-dimensional continuum ("string") with line tension S and mass per unit length γ. The displacement of the continuum as a function of position x and time t is denoted by $u(x, t)$. (In the application to kinks in dislocations and as a one-dimensional model of a screw dislocation, the displacement $u(x, t)$ is perpendicular to the x-direction; if used as a one-dimensional model for an edge dislocation, the u- and x- coordinates are parallel.) The essential feature of the model is that the potential energy of the string is a periodic function $U(u)$ of the displacement, with a period a equal to the periodicity of the crystal.

In this model, the kinetic energy is given by

$$E_{\text{kin}} = \frac{1}{2} \gamma \int_{-\infty}^{+\infty} \left(\frac{\partial u}{\partial t} \right)^2 dx \tag{3.1a}$$

the "elastic" energy by

$$E_S = \frac{1}{2} S \int_{-\infty}^{+\infty} \left(\frac{\partial u}{\partial x} \right)^2 dx \tag{3.1b}$$

and the "potential" energy by

$$E_U = \frac{1}{a} \int_{-\infty}^{+\infty} U(u)\, dx \qquad (3.1c)$$

Our calculations will be based on the following expression for the potential energy

$$U(u) = \left(\frac{U_0}{2}\right)\left(1 - \cos\frac{2\pi u}{a}\right) \qquad (3.2)$$

Other forms[2] for $U(u)$ may be treated in a similar manner. Since the displacements $u_1(x, t)$ due to the "heavy" phonons are small compared to the lattice constant a, it is expedient to write the total displacement as

$$u = u_0(x, t) + u_1(x, t) \qquad (3.3)$$

where $u_0(x, t)$ is the displacement of the uniformly moving defect, and where powers of u_1/a higher than the first will be neglected in the equations of motion. Similarly, we subdivide the Lagrangian density $\mathcal{L}$ according to

$$\mathcal{L} = \mathcal{L}_0 + \mathcal{L}_1 \qquad (3.4)$$

with

$$\mathcal{L}^0 = \frac{\gamma}{2}\dot{u}_0^2 - \frac{S}{2}u_0'^2 - \frac{U_0}{2a}\left[1 - \cos\frac{2\pi u_0}{a}\right] \qquad (3.5a)$$

$$\mathcal{L}^1 = \frac{\gamma}{2}\dot{u}_1^2 - \frac{S}{2}u_1'^2 + \left(-\gamma\ddot{u}_0 + Su_0'' - \frac{U_0\,\pi}{a^2}\sin\frac{2\pi}{a}u_0\right)u_1$$

$$-\left(\frac{U_0}{a}\cdot\frac{\pi^2}{a^2}\cos\frac{2\pi}{a}u_0\right)u_1^2 \qquad (3.5b)$$

$\mathcal{L}^1$ is the Lagrangian density for the elastic vibrations of the string in the presence of the "defect" described by u_0.

In the usual manner we obtain as the equations of motion

$$Su_0'' - \gamma\ddot{u}_0 - \frac{\pi U_0}{a^2}\sin\left(\frac{2\pi}{a}\cdot u_0\right) = 0 \qquad (3.6a)$$

$$Su_1'' - \gamma \ddot{u}_1 - \left(\frac{2\pi^2}{a^2} \cdot \frac{U_0}{a} \cos \frac{2\pi}{a} u_0\right) u_1 = 0 \qquad (3.6b)$$

The Hamiltonian of the elastic vibrations is

$$H = \int_{-\infty}^{+\infty} \left\{\frac{p_1{}^2}{2\gamma} + \frac{S}{2} u_1'{}^2 + \left(\frac{U_0}{a} \cdot \frac{\pi^2}{a^2} \cos \frac{2\pi}{a} u_0\right) u_1{}^2\right\} dx \qquad (3.7)$$

where p_1 is the momentum canonically conjugate to u_1. Introducing the Fourier decompositions

$$u_1 = \frac{1}{\sqrt{2\pi}} \int_{-\infty}^{+\infty} Q(k, t) e^{ikx} dk \qquad p_1 = \frac{1}{\sqrt{2\pi}} \int_{-\infty}^{+\infty} P(k, t) e^{-ikx} dk$$

$$(3.8)$$

we obtain (asterisks denoting the complex conjugate)

$$H = \int_{k=-\infty}^{+\infty} \left\{\frac{1}{2\gamma} P(k, t) P^*(k, t) + \frac{S}{2} k^2 Q(k, t) Q^*(k, t)\right\} dk$$

$$+ \iint_{-\infty}^{+\infty} V(k, k', t) Q(k, t) Q(k', t) dk dk'$$

where

$$V(k, k', t) \equiv -\frac{U_0 \pi}{a^3} \int_{-\infty}^{+\infty} \sin^2 \frac{\pi}{a} u_0 \cdot e^{i(k+k')x} dx = V(k', k, t) \qquad (3.9)$$

$$\omega^2(k) = c^2 (k^2 + k_0{}^2) \qquad (3.10)$$

with

$$c^2 = \frac{S}{\gamma}, \quad k_0{}^2 = \pi^2 \frac{2U_0}{Sa^3} \qquad (3.11)$$

The dispersion relation (3.10) is analogous to that of Yukawa's meson theory rather than Debye's theory of lattice vibrations; this has prompted Eshelby[8] to coin the expression "heavy" phonons.

3.2 Three-dimensional Elastic Continuum— Classical Treatment*

We consider an isotropic elastic medium with an elastic energy density

$$\Phi(y) = A_1 y_I^2 + A_2 y_{II} + A_3 y_I^3 + A_4 y_I y_{II} + A_5 y_{III} \qquad (3.12)$$

y_I, y_{II}, y_{III} are the invariants of the strain tensor y (referred to the undeformed state); A_1 and A_2 are related to the usual (second-order) elastic constants, A_3, A_4, A_5 to the third-order elastic constants. The total displacement s is subdivided according to

$$s = s_0{}^{el} + s_0{}^{pl} + s_1 \equiv s_0 + s_1 \qquad (3.13)$$

where $s_0{}^{el}$ and $s_0{}^{pl}$ denote the elastic and plastic part of the displacement due to lattice defects, and s_1 is the additional displacement due to the lattice vibrations. We may again confine ourselves to first powers in s_1 in the equations of motion. In this approximation the Lagrangian density may be written as

$$\mathcal{L} = \frac{1}{2} \bar{\rho} \left[\dot{s}_0(x, t) + \dot{s}_1(x, t) \right]^2 - \frac{\bar{\rho}}{\rho_0} \Phi(y) \qquad (3.14)$$

where $\bar{\rho}$ is the density and x is the position vector in the self-strained state. Analogous to Eq. (3.4) we can introduce a Lagrangian density $\mathcal{L}_1$ for the elastic vibrations in the presence of crystal defects

$$\mathcal{L}_1 = \frac{1}{2} \bar{\rho} \dot{s}_1^2 - \Psi(\bar{y}_1, \epsilon^0) + \sum_{k, l=1}^{3} \sigma_{kl}^0 \frac{\partial s_{1l}}{\partial x_k} \qquad (3.15)$$

In Eq. (3.15) the last term arises from the product $\dot{s}_0$, $\dot{s}_1$ in Eq. (3.14); σ_{ik}^0 are the components of the stress tensor of the defects in the self-stressed state without lattice vibrations), ϵ^0 is the corresponding strain tensor. The energy density of the elastic vibrations is defined as the difference between energy density in the state containing the defects plus the elastic vibrations and that in the state containing the defects alone (self-strained state), namely,

$$\Psi(\bar{y}_1, \epsilon^0) = \frac{\bar{\rho}}{\rho_0} \left\{ \Phi(y) - \Phi(y^0) \right\} \qquad (3.16)$$

*For more details see Refs. [11] and [12].

$\bar{\gamma}_1$ is the strain tensor of the elastic vibrations, referred to the self-strained state. The corresponding Hamiltonian is

$$H = \int \left\{ \frac{1}{2\bar{\rho}} \pi_1 \pi_1 + \Psi(\bar{\gamma}_1, \varepsilon^0) - \sum_{k,l} \sigma^0_{kl} \cdot \frac{\partial s_{1l}}{\partial x_k} \right\} d\tau_x \qquad (3.17)$$

where π_1 denotes the canonically conjugate momentum to s_1.

We introduce the Fourier decompositions

$$s_1(x,t) = \frac{1}{(2\pi)^{3/2}} \sum_j \int e_k{}^j Q_k{}^j(t) e^{ikx} d\tau_k \qquad (3.18a)$$

$$\pi_1(x,t) = \frac{1}{(2\pi)^{3/2}} \sum_j \int e_k{}^j P_k{}^j(t) e^{-ikx} d\tau_k \qquad (3.18b)$$

$$\varepsilon^0(x,t) = \frac{1}{(2\pi)^{3/2}} \int \varepsilon^0(k,t) \cdot e^{ikx} d\tau_k \qquad (3.18c)$$

where $e_k{}^j$ (longitudinal waves $j = 1$, transversal waves $j = 2, 3$) are the unit vectors (mutually orthogonal to each other) of the plane waves of wave vector k.

Since we are not interested in the effects of retaining the third-order (nonlinear) terms in Eq. (3.17) of the strain-fields of the defects described by s_0, but in calculating the scattering of the elastic vibrations by the defects in the lowest non-vanishing order of the perturbation theory, we may linearize the equations of motions with regard to ε^0. Inserting the Fourier representations (3.18) into Eq. (3.17) gives us by calculation analogous to that of Ref. [12] the following expression for the Hamiltonian:

$$H = \sum_j \int \left\{ \frac{1}{2\rho_0} P_k{}^j P_k{}^{j*} + \frac{\rho_0 (\omega_k{}^j)^2}{2} Q_k{}^j Q_k{}^{j*} \right\} d\tau_k$$

$$+ \sum_{j,j'} \iint \left\{ \frac{1}{\rho_0{}^2 \omega_k{}^j \omega_{k'}{}^{j'}} V_1{}^{jj'}_{kk'} P_k{}^j P_{k'}{}^{j'} + V_2{}^{jj'}_{kk'} Q_k{}^j Q_{k'}{}^{j'} \right\} d\tau_k d\tau_{k'} \qquad (3.19)$$

$$\omega_k{}^1 = \sqrt{\frac{2A_1}{\rho_0}} \cdot |k| \equiv C_l \cdot |k| \quad \text{and} \quad \omega_k{}^2 = \omega_k{}^3 = \sqrt{\frac{-2A_2}{\rho_0}} \cdot |k| \equiv C_t |k|$$

$$(3.20)$$

$$(2\pi)^{3/2} V_1{}^{jj'}_{kk'} = (2\pi)^{3/2} V_1{}^{j'j}_{k'k} = \frac{\omega_k{}^j \omega_k{}^{j'}}{2} \rho_0 \; e_k{}^j \cdot e_k{}^{j'} \; \epsilon_I{}^0(k'', t) \qquad (3.21)$$

$$(2\pi)^{3/2} V_2{}^{jj'}_{kk'} = (2\pi)^{3/2} V_2{}^{j'j}_{k'k} = (A_1 - 1/2 A_2 - 3 A_3 - 3/2 A_4)(e_k{}^j \cdot k)(e_k{}^{j'} \cdot k') \epsilon_I{}^0(k'', t)$$

$$+ \; (-A_1 - 1/4 A_2 + 1/4 A_4)(e_k{}^j \cdot e_k{}^{j'})(k \cdot k') \epsilon_I{}^0(k'', t)$$

$$+ \; 1/4(A_2 + A_4)(e_k{}^j \cdot k')(e_k{}^{j'} \cdot k) \epsilon_I{}^0(k'', t)$$

$$+ \; 1/2 A_2(e_k{}^j \cdot e_k{}^{j'})(k \cdot \epsilon^0(k'', t) \cdot k')$$

$$+ \; (-2A_1 + 1/2 A_4)\Big\{(e_k{}^j \cdot k)(e_k{}^{j'} \cdot \epsilon^0(k'', t) \cdot k')$$

$$+ \; (e_k{}^{j'} \cdot k')(e_k{}^j \cdot \epsilon^0(k'', t) \cdot k)\Big\}$$

$$+ \; (1/2 A_2 - 1/4 A_5)\Big\{[e_k{}^j \times e_k{}^{j'}] \cdot \epsilon^0(k'', t) \cdot [k \times k']$$

$$+ \; [e_k{}^j \times k'] \cdot \epsilon^0(k'', t) \cdot [k \times e_k{}^{j'}]\Big\}$$

where $k'' = -(k + k')$.

3.3   Three-dimensional Model— Quantum-theory

The integrations in k-space in Eq. (3.19) are replaced by summations according to Eq. (2.16). The canonically conjugate variables $P_k{}^j$ and $Q_k{}^j$ are quantized by postulating the commutation relations

$$[P_k{}^j(t), Q_k{}^{j'}(t)] = \frac{V_0}{(2\pi)^3} \cdot \frac{\hbar}{i} \delta^{jj'} \delta_{kk'} \qquad (3.22a)$$

$$[P_k{}^j, P_k{}^{j'}] = 0 \; ; \quad [Q_k{}^j, Q_k{}^{j'}] = 0 \qquad (3.22b)$$

Creation operators a_k^{+j} and annihilation operators $a_k{}^j$ are introduced through the relations

$$Q_k{}^j(t) = \sqrt{\frac{V_0}{(2\pi)^3} \frac{\hbar}{2\rho_0 \omega_k{}^j}} \; \left(\epsilon^j a_{-k}^{+j}(t) + a_k{}^j(t)\right) \qquad (3.23a)$$

$$P_k^{j*}(t) = i \sqrt{\frac{V_0}{(2\pi)^3} \frac{\hbar \rho_0 \omega_k^{j}}{2}} \left(\epsilon^j a_{-k}^{+j}(t) - a_k^{j}(t) \right) \qquad (3.23b)$$

The symbol ϵ^j is defined by $e_{-k}^{j} = \epsilon^j e_k^{j}$ and is capable of assuming the values ± 1. Introducing Eq. (3.23) into (3.19) leads to Eq. (2.11).

3.4   PDFK-Model— Quantum Theory

The quantization of the classical theory developed in Sec. 3.1 follows closely the treatment outlined in Sec. 3.3. Instead of carrying this out explicitly, we give a table of correspondence of the quantities and operations occurring in the two models (Table 1). The final result for the average force [comp. Eq. (2.14)] is

$$\bar{f}_z = - \frac{\pi}{\gamma^2} \int \frac{\hbar(k - k')}{\omega_k \omega_{k'}} |V'(-k, k')|^2 (\overline{\overline{N}}_{k'} - \overline{\overline{N}}_k) \delta[\omega_k - \omega_{k'} - (k - k') \bar{v}] \, dk dk'$$

$$(3.24)$$

TABLE 1 Correspondences Between "Light" and "Heavy" Phonons

Three-dimensional model	PDFK – model
wave vector k	k
j	no polarization index
ϵ^j	1
$\iiint g(k) \, dk_x dk_y dk_z$	$\int g(k) \, dk$
$\sum_j$	no sum
$\Delta \tau_k = \dfrac{(2\pi)^3}{V_0}$	$\Delta \tau_k = \dfrac{2\pi}{L_0}$
$\omega_k^j = c^j |k|$	$\omega_k = c \sqrt{k^2 + k_0^2}$
ρ_0	γ
$V_{1 \, kk'}^{jj'}$	0
$V_{2 \, kk'}^{jj'} = V_{2 \, kk'}'^{jj'} \exp i(k_z + k_z') \bar{v} t$	$V(k, k', t) = V'(k, k') \cdot \exp i(k + k') \bar{v} t$

4   CALCULATION OF THE FORCE EXERTED ON A KINK IN A DISLOCATION

4.1   PDFK-Model

The dynamic solutions of Eq. (3.6a) have been discussed in detail elsewhere.[2,20] The displacement function corresponding to a single kink moving uniformly with velocity $\bar{v}$ is

$$u_0(x, t) = \frac{a}{\pi} \arcsin \left[\cosh \pi \frac{x^*}{w_0} \right]^{-1} ; \quad x^* = \frac{x - \bar{v} t}{(1 - \beta^2)^{1/2}} ; \quad \beta = \frac{\bar{v}}{c}$$

$$(4.1)$$

The "rest-width" w_0 of the kink is related to the quantities introduced in Sec. 3.1 according to

$$w_0 = a \cdot \sqrt{\frac{aS}{2U_0}} = \frac{\pi}{k_0} \tag{4.2}$$

Using Eq. (4.1) we may now work out Eq. (3.9) explicitly:

$$V(k, k', t) = -\frac{U_0 \pi}{a^3} \int_{-\infty}^{+\infty} \frac{e^{i(k + k') x}}{\cosh^2 k_0 x^*} \, dx = V(k, k') \cdot e^{i(k + k')\bar{v} t}$$

$$(4.3)$$

where

$$V(k, k') = -\frac{2\pi U_0}{k_0 a^3} \cdot \sqrt{1 - \beta^2} \cdot \frac{\left[\dfrac{\pi}{2} \left(\dfrac{k}{k_0} + \dfrac{k'}{k_0} \right) \sqrt{1 - \beta^2} \right]}{\sinh \left[\dfrac{\pi}{2} \left(\dfrac{k}{k_0} + \dfrac{k'}{k_0} \right) \cdot \sqrt{1 - \beta^2} \right]}$$

$$(4.4)$$

If we insert Eq. (4.3) into Eq. (3.24), one of the integrations is taken care of by the δ-function, giving us the final result

$$\bar{f}_z(\bar{v}, T) = -\frac{4\hbar c k_0^2}{\pi} \int_0^\infty \frac{\xi(1 - \beta^2)\, d\xi}{(\xi + \zeta) \cdot \sqrt{1 + (\xi - \zeta)^2/4} + (\xi - \zeta)\sqrt{1 - (\xi + \zeta)^2/4}}$$

$$\times \frac{[(\pi/2)\xi\sqrt{1 - \beta^2}]^2}{\sinh^2[(\pi/2)\xi\sqrt{1 - \beta^2}]} \left(\frac{1}{\exp(T_c/T)[1 + (\xi - \zeta)^2/4]^{1/2} - 1} \right.$$

$$\left. - \frac{1}{\exp(T_c/T)[1 + (\xi + \zeta)^2/4]^{1/2} - 1} \right) \quad (4.5)$$

where

$$\zeta = \beta \left[\xi^2 + \frac{4}{1 - \beta^2} \right]^{1/2} \quad (4.6)$$

$$T_c = \hbar c \frac{k_0}{K} = \frac{\hbar c \pi}{w_0 K} = \frac{\hbar \pi^2}{2K} \cdot \frac{W_k}{a^2 \sqrt{S\gamma}} \quad (4.7)$$

$(K = $ Boltzmann's constant, $W_k = $ kink energy$)$

Since the computations in Sec. 4.2 will be carried out only to first powers in $\bar{v}/c$, it is natural to confine the further discussion of Eq. (4.5) to the same approximation.

Retaining only the first power in β in the expansion of Eq. (4.5), we obtain the following expression for the kink viscosity:

$$\eta_h(T) \equiv -\frac{\bar{f}_z}{\bar{v}} = \frac{2\hbar K_0^2 T_c}{\pi T} \int_{\xi=0}^\infty \frac{\xi}{\Lambda} \frac{[\pi\xi/2]^2}{\sinh^2(\pi\xi/2)} \frac{\exp(\Lambda T_c/T)}{[\exp(\Lambda T_c/T) - 1]^2}\, d\xi \quad (4.8)$$

with $\Lambda = (1 + \frac{1}{4}\xi^2)^{1/2}$. Figure 1 shows the numerically calculated function $\eta_h(T)$.

At low $(T \ll T_c)$ and high $(T \gg T_c)$ temperatures, Eq. (4.8) may be approximated as follows:

$$\eta_h^{LT} = \frac{8\hbar K_0^2}{\pi} \left\{ 1 - \frac{2}{3}\pi^2 \frac{T}{T_c} + \left(\frac{8}{15}\pi^4 - \frac{2}{3} \right) \frac{T^2}{T_c^2} + \cdots \right\} \exp\frac{-T_c}{T}$$

$$\frac{\bar{v}}{c} \ll 1 \ , \quad \frac{T}{T_c} \ll 1 \quad (4.9)$$

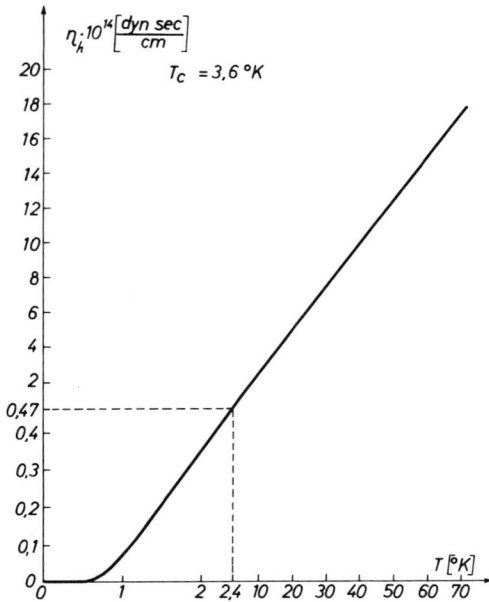

Fig. 1 Temperature dependence of viscosity η_h due to scattering of heavy phonons by kinks (limiting case of slow kinks). Numerical values for Cu (T_c = 3.6°K). Note the break-in.

$$\eta_h^{HT} = \frac{2\hbar K_0^2}{\pi} \left(0.488 \cdot \frac{T}{T_c} - 4.4 \cdot 10^{-2} \frac{T_c}{T} + \cdots \right), \quad \frac{\bar{v}}{c} \ll 1, \frac{T}{T_c} \gg 1 \tag{4.10}$$

The temperature dependences in Eqs. (4.9), (4.10) can be understood by assuming an approximate proportionality between η_h and the energy density of the lattice vibrations. The leading term of Eq. (4.9) is related to the exponential temperature variation of the energy density in an Einstein solid and reflects the fact that—according to the dispersion relation (3.10)—a minimum energy $\hbar\omega = c\hbar k_0$ is necessary to excite any heavy phonons. In the high-temperature approximation, the force is proportional to the absolute temperature for $T > 2T_c$.

Inserting into T_c [Eq. (4.7)] numerical results ($S \approx Gb^2/2$; $\gamma = \pi\rho_0 b^2$; Cu: ρ_0 = 8.9 g/cm³; shear modulus G = 4.68 · 10¹¹ dyne/cm²; W_k = 0.1 eV) gives us T_c = 3.6°K. It should be noted that T_c is not related to the Debye-temperature θ_h of the model, which would be defined by $k\theta_h = \hbar\omega_{max}$ (ω_{max} = cut-off frequency) and which for Cu would be $\theta_h \approx 86°K$. It can be shown that in Eq. (4.5) the high-frequency cut-off is unimportant, so that Eq. (4.5) remains valid even for $T > \theta_h$.

4.2   Three-dimensional Model

The partial differential equations for the strain field of an arbitrarily moving dislocation may be derived[21] from the equilibrium condition

$$\text{Div}\sigma_0 = \rho_0 \left(s_0^{el} + s_0^{pl} \right) \tag{4.11}$$

as follows:

$$\mu\Delta\epsilon_{ij}^0 + (\lambda + \mu)\nabla_i\nabla_j\epsilon_I^0 - \rho_0\ddot{\epsilon}_{ij}^0 = \rho_0\ddot{\epsilon}_{ij}^{0p} + \mu(\eta_{ij} - \delta_{ij}\eta_I) \tag{4.12}$$

Here λ and μ are Lamé's constants, ϵ_{ij}^{0p} and η_{ij} the components of the tensors of the plastic strain and of the incompatibility. Equations (4.12) can be written as inhomogeneous wave equations for the trace $\epsilon_I^0 \equiv \sum_i \epsilon_{ii}^0$ and for the deviator $\epsilon_{ij}'^0 = \epsilon_{ij}^0 - \frac{1}{3}\epsilon_I^0\delta_{ij}$.

The matrix elements $V_{kk'}'^{jj'}$, $V_{kk'}''^{jj'}$ depend only on the Fourier transforms of the strain ϵ^0 (V' contains the time-dependent, V'' the time-independent strain). It is therefore convenient to Fourier-transform the spatial parts of Eqs. (4.12) and to solve the resulting ordinary differential equations with time as independent variable.

We have done this for a kink in a screw dislocation moving uniformly in z-direction. The shape of the kink given by Eq. (4.1) was approximated by pieces of straight lines as follows:

$$u(z, t) = \begin{cases} -\dfrac{a}{2} & \text{for} \quad z < -\dfrac{w}{2} + \bar{v}t \\[2ex] \dfrac{a}{w}(z - \bar{v}t) & \text{for} \quad -\dfrac{w}{2} + vt < z < \dfrac{w}{2} + \bar{v}t \\[2ex] +\dfrac{a}{2} & \text{for} \quad z > \dfrac{w}{2} + \bar{v}t \end{cases} \tag{4.13}$$

where a = distance of neighboring Peierls' valleys, and w = kink width. We obtained the following expressions for the Fourier transforms of the strain:

$$\epsilon_1'^{0}(\mathbf{k},t) = -\frac{1-2\nu}{1-\nu}\frac{ab}{(2\pi)^{3/2}}C_l^2 \cdot F \cdot \frac{ik_y}{k^2C_l^2 - k_z^2\bar{v}^2} \cdot e^{-ik_z\bar{v}t}$$

$$\epsilon_{11}'^{0}(\mathbf{k},t) = \frac{ab}{(2\pi)^{3/2}}\frac{ik_yC_t^2F}{k^2C_t^2 - k_z^2\bar{v}^2} \cdot e^{-ik_z\bar{v}t}\left(\frac{2}{3} + \frac{(k_x^2 - \frac{1}{3}k^2)}{1-2\nu} \cdot \frac{2C_t^2}{k^2C_l^2 - k_z^2\bar{v}^2}\right)$$

$$\epsilon_{22}'^{0}(\mathbf{k},t) = \frac{ab}{(2\pi)^{3/2}}\frac{ik_yC_t^2F}{k^2C_t^2 - k_z^2\bar{v}^2} \cdot e^{-ik_z\bar{v}t}\left(-\frac{1}{3} + \frac{(k_y^2 - \frac{1}{3}k^2)}{1-2\nu} \cdot \frac{2C_t^2}{k^2C_l^2 - k_z^2\bar{v}^2}\right)$$

$$\epsilon_{33}'^{0}(\mathbf{k},t) = \frac{ab}{(2\pi)^{3/2}}\frac{ik_yC_t^2F}{k^2C_t^2 - k_z^2\bar{v}^2} \cdot e^{-ik_z\bar{v}t}\left(-\frac{1}{3} + \frac{(k_z^2 - \frac{1}{3}k^2)}{1-2\nu} \cdot \frac{2C_t^2}{k^2C_l^2 - k_z^2\bar{v}^2}\right)$$

$$\epsilon_{12}'^{0}(\mathbf{k},t) = -\frac{ab}{(2\pi)^{3/2}}\frac{ik_xC_t^2F}{k^2C_t^2 - k_z^2\bar{v}^2} \cdot e^{-ik_z\bar{v}t}\left(\frac{1}{2} - \frac{k_y^2}{1-2\nu} \cdot \frac{2C_t^2}{k^2C_l^2 - k_z^2\bar{v}^2}\right)$$

$$\epsilon_{13}'^{0}(\mathbf{k},t) = \frac{b}{(2\pi)^{3/2}}\frac{ik_y}{k^2}\frac{\sin[\frac{1}{2}(k_xa + k_zL)]}{k_z}$$

$$+\frac{ab}{(2\pi)^{3/2}}\frac{ik_yC_t^2F}{k^2C_t^2 - k_z^2\bar{v}^2} \cdot e^{-ik_z\bar{v}t}\left(-\frac{1}{2}\frac{k_x}{k_z} + \frac{k_xk_z}{1-2\nu}\frac{2C_t^2}{k^2C_l^2 - k_z^2\bar{v}^2}\right)$$

$$\epsilon_{23}'^{0}(\mathbf{k},t) = -\frac{b}{(2\pi)^{3/2}}\frac{ik_x}{k^2}\frac{\sin[\frac{1}{2}(k_xa + k_zL)]}{k_z}$$

$$+\frac{ab}{(2\pi)^{3/2}}\frac{C_t^2Fe^{-ik_z\bar{v}t}}{k^2C_t^2 - k_z^2\bar{v}^2}\left(\frac{1}{2}\frac{ik_x^2}{k_z} + \frac{1}{2}\frac{a}{w}\frac{\bar{v}^2}{C_t^2}ik_x + \frac{2C_t^2}{1-2\nu}\frac{ik_y^2k_z}{k^2C_l^2 - k_z^2\bar{v}^2}\right)$$

Here ν is Poisson's ratio, b is the modulus of the Burgers vector, L is a temporarily introduced crystal length in z-direction, and F is given by

$$F = \frac{\sin[\frac{1}{2}(k_xa + k_zw)]}{\frac{1}{2}(k_xa + k_zw)} \tag{4.14a}$$

Inserting of the strain field equation (4.14) into the matrix elements equation (3.19) leads to a sixfold integral for the force $\bar{f}_z$. This integral has been evaluated under the following simplifying assumptions:

1) Limitation on longitudinal-longitudinal-transitions of phonons $(j = 1, j' = 1)$. If $\bar{f}_z = \sum_{jj'} \bar{f}_z^{(jj')}$, we restrict ourselves to f_z.[11]
Other contributions $f_z^{(jj')}$ could be evaluated similarly.

2) The kink velocity $\bar{v}$ is small compared to the longitudinal sound velocity C_l; it suffices to retain the first power in v/C_l

3) The kinks are flat, i.e., $a/w \ll 1$.

Under these conditions all integrations except one could be carried out in closed form.

Numerical results were obtained for copper using the third-order elastic constants as tabulated by Seeger and Buck:[22]

$$A_1 = 0.994 \cdot 10^{12} \text{ dyn/cm}^2$$
$$A_2 = -0.936 \cdot 10^{12} \text{ dyn/cm}^2$$
$$A_3 = -4.58 \cdot 10^{12} \text{ dyn/cm}^2$$
$$A_4 = 12.16 \cdot 10^{12} \text{ dyn/cm}^2$$
$$A_5 = -15.60 \cdot 10^{12} \text{ dyn/cm}^2$$

$$C_l = 4.726 \cdot 10^5 \text{ cm/sec}$$
$$b = 2.56 \cdot 10^{-8} \text{ cm}$$
$$\rho_0 = 8.9 \text{ g/cm}^3$$
$$\theta = 317°\text{K} = \text{Debye temperature}$$

Introducing the energy density of the longitudinal phonons, $E_l = (\pi^2/30) \cdot (kT)^4/(\hbar C_l)^3$, we may write the kink viscosity as follows:

$$\eta^{(11)} \equiv -\frac{\bar{f}_z^{(11)}}{\bar{v}} = A\left(\frac{T}{\theta}, \frac{a}{w}\right) ab \frac{E_l}{C_l} \tag{4.15}$$

where

$$A\left(\frac{T}{\theta}, \frac{a}{w}\right) = \begin{cases} 0.422 \cdot 10^5 \left(\frac{T}{\theta}\right)^2 - 0.880 \cdot 10^6 \left(\frac{w}{a}\right)^2 \left(\frac{T}{\theta}\right)^4 & \text{for} \quad p < 1 \\[2ex] 2.68 \cdot 10^3 \left(\frac{a}{w}\right)^{1,4} \cdot \left(\frac{T}{\theta}\right)^{0,6} & \text{for} \quad 2 \le p \le 10 \\[2ex] 1.10 \cdot 10^3 \left(\frac{a}{w}\right)^2 + 0.625 \cdot 10^2 \left(\frac{a}{w}\right)^2 \ln\left(\frac{wT}{a\theta}\right) & \\[2ex] \qquad\qquad + 1.55 \cdot 10^3 \left(\frac{a}{w}\right)\left(\frac{T}{\theta}\right) & \text{for} \quad p > 10 \end{cases}$$

$$\tag{4.16}$$

$$p = 12.22 \frac{wT}{a\theta}$$

The function $A(T/\theta, a/w)$ represents an expansion in powers of p at $p = 0$ for $p < 1$. In the region $2 \le p \le 10$ it is an interpolation curve

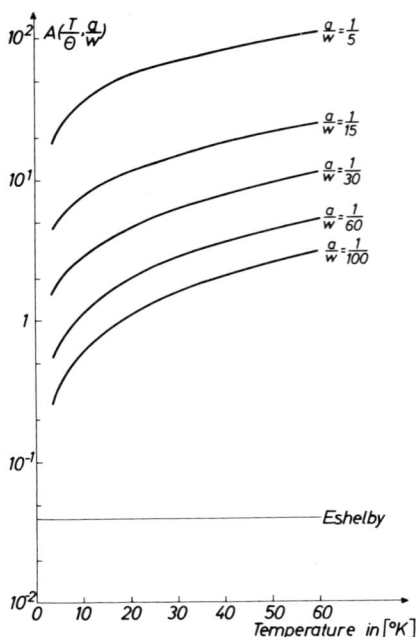

Fig. 2 Fraction $A(T/\theta, a/w)$ as given by Eq. (4.16) as a function of temperature (for Cu, $\theta = 317\degree$K). The line denoted by "Eshelby" is the corresponding value predicted by Eshelby's (ref. 8) treatment of the flutter mechanism for longitudinal waves only.

(relative error <5%). For $p > 10$ it is an asymptotic approximation. Figure 2 shows the factor $A(T/\theta, a/w)$ as a function of temperature for kinks of different widths.

5 DISCUSSION

The results obtained in the *three-dimensional model* are somewhat restricted by the approximations made in Sec. 4. The assumption of low kink velocities is permissible above $T \simeq 10\degree$K for a driving force of $10^{-6}\,abG$, which is a typical magnitude in internal friction experiments. The expansion into powers of a/w restricts us to crystals with not too high Peierls stresses; it is justified for f.c.c. metals, where the kink width is about 30 Burgers vectors. Furthermore, we have restricted ourselves to the scattering of the longitudinal lattice waves only. We may expect the contribution of each of the transverse modes to be similar to or slightly larger than that calculated for the longitudinal modes, although a direct comparison with thermal conductivity is not possible.

The two main restrictions in the treatment of the *heavy phonons* is the retention of only one harmonic in the potential $U(u)$ and the use of a line tension. If a second harmonic were included in $U(u)$,

T_c could vary between zero and $\sqrt{2}\, T_c^{(l)}$, where $T_c^{(l)}$ is calculated from Eq. (4.7) for a given kink energy W_k. The use of a line tension has already been discussed (see footnote on p. 624). In a more exact treatment it should be replaced by a variational approach which determines the shape of a kink allowing both for the potential energy E_U and the elastic energy as calculated from the elastic strain field of the kinked dislocation. Such a treatment would lead to the same kink shape to be used for both heavy and light phonon scattering and would permit including in a consistent way the "relativistic" contraction of a fast moving kink. This cannot be done in the present approach, in which different kink shapes were used for the calculation of the scattering of the light and heavy phonons. Therefore, we were forced to confine ourselves in the final evaluation to slowly moving kinks.

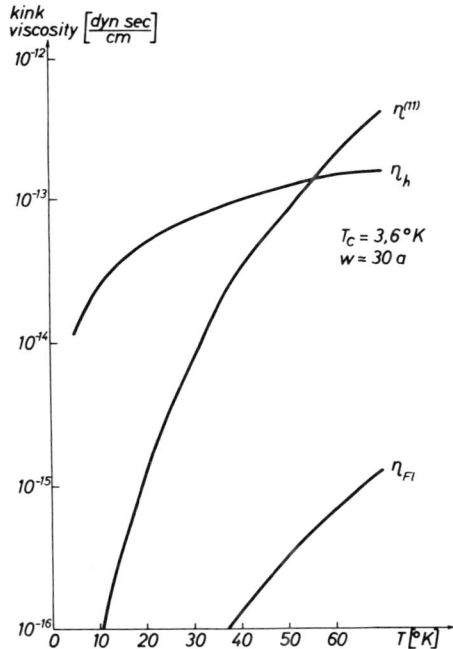

Fig. 3 Kink viscosities (low velocity limits) as a function of temperature for Cu according to various models. $\eta^{(11)}$: kink in screw dislocation, scattering of longitudinal branch of "light" phonons into longitudinal branch only. η_h = kink viscosity due to scattering of "heavy phonons". ("Einstein" temperature T_c = 3.6°K). η_{Fl}: Viscosity due to flutter mechanism according to Eshelby[8], taking into account longitudinal phonons only.

Figure 3 gives a comparison between our results for the kink viscosities [Eqs. (4.8) and (4.15)] and calculations of Eshelby[8] obtained for the flutter mechanism.

One sees that the kink viscosities for the PDFK-model (η_h) and the longitudinal contribution to the three-dimensional model $(\eta^{(11)})$ are equal at $T = 56°K$, but that $\eta^{(11)}$ rises more quickly with temperature than η_h. Figure 3 also includes Eshelby's result (taking into account longitudinal waves only)

$$\eta_{fl} = \frac{1}{25} \cdot \frac{ab}{C_l} E_l \tag{5.1}$$

We see that compared with the nonlinearity mechanism, the flutter mechanism can be neglected at least for the temperature range in which the phonon processes are dominant in metals.

The magnitude of the kink viscosity $\eta^{(11)}$ may be illustrated by introducing the stopping path l_{st}, which is defined as the distance over which a kink with the kinetic energy $KT/2$ comes to rest under the sole action of the frictional force. The stopping path is given by

$$l_{st} = \frac{(m_k KT)^{1/2}}{\eta} \approx \frac{ab}{\eta} \left(\frac{2\rho_0 KT}{w} \right)^{1/2} \tag{5.2}$$

In Fig. 4 the stopping path has been plotted for Cu in units of the kink width w as a function of temperature. In these computations, the kink viscosity η in Eq. (5.2) was obtained as follows: η was assumed to be the sum of the contributions from heavy phonon scattering (η_h), from light phonons (η_l), and from the flutter mechanism (η_{Fl}). The contribution of the transversal waves to η_{fl} is larger than that of the longitudinal waves by a factor of the order of magnitude $2(C_l/C_t)^4$ [for Cu appr. 36]. This suggests that the contribution of the transversal waves to η_l is also much larger than that of the longitudinal waves. We have arbitrarily assumed $\eta_l = 10\,\eta^{(11)}$.

A comparison of the calculated kink-viscosities is possible with the megacycle attenuation measurements on Cu by Alers and Thompson.[23] These authors interpret their results in terms of the string-model and find for the string-model damping constant B (see Ref. [2]) at 60°K and 20°K the following experimental values:

$$B(20°K) = 1.4 \cdot 10^{-5} \text{ dyn-sec-cm}^{-2} \qquad B(60°K) = 1.15 \cdot 10^{-4} \text{ dyn-sec-cm}^{-2} \tag{5.3a}$$

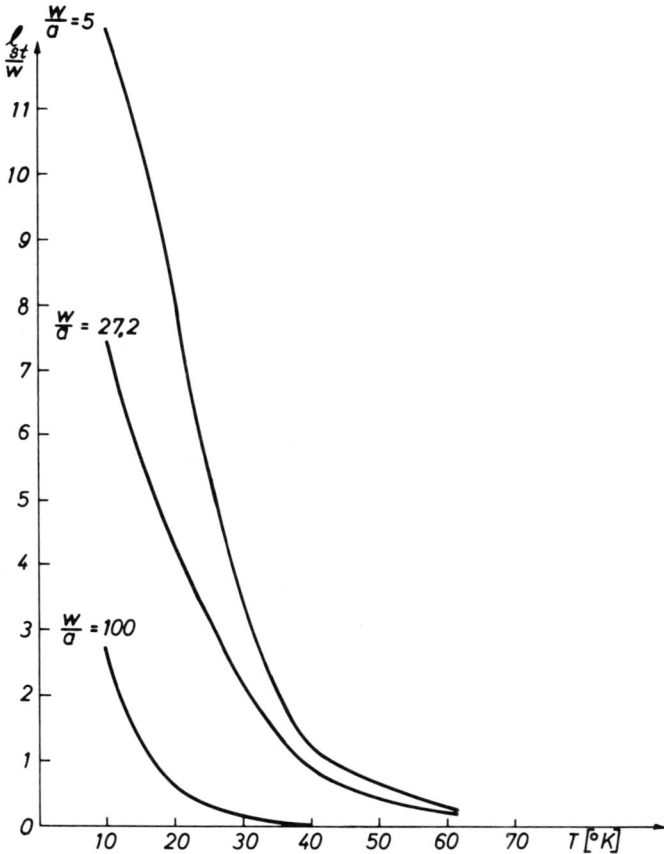

Fig. 4 Stopping path l_{st} of a kink (in units of the kink width) in a screw dislocation in Cu as a function of temperature for different kink widths (for details see text).

Comparison between the equations of the string model[24] and the kink-model[3] for the high-frequency range gives us the following relation between B and η:

$$\eta = \frac{\pi}{8} a \overline{|\sin \phi|} B \qquad (5.4)$$

where $\overline{|\sin \phi|}$ is an orientation factor which will be assumed to be 1/2. Inserting the result of Alers and Thompson into Eq. (5.4) gives us for Cu at 20°K and at 60°K:

$$\eta(20°\text{K}) = 1.23 \cdot 10^{-13} \text{ dyn-sec-cm}^{-1} \, , \quad \eta(60°\text{K}) = 1.01 \cdot 10^{-12} \text{ dyn-sec-cm}^{-1}$$

$$(5.3b)$$

Comparison of the 60°K value with Fig. 3 shows that the experimental value is about five times the calculated value of $\eta^{(11)}$. This suggests that about 4/5 of the total effect comes from the interaction with transversal waves. This is not an unreasonable order of magnitude. According to the theory, the dominant term at 20°K is η_h. The theoretical value of η_h is about 1/2 of the experimental value equation (5.3b). Inclusion of η_l and η_{fl} would decrease this discrepancy but not remove it completely. In view of the theoretical simplifications (elastic isotropy, guessing of the importance of transversal waves) and the experimental uncertainties (absolute value and temperature dependence of the density of dislocations contributing to the internal friction), the comparison between experiment and theory is considered to be satisfactory and to indicate that the principal damping processes in the temperature range between 10°K and 80°K have been included.

The theory predicts that in the temperature range of the Bordoni relaxation (in copper: from 45°K upwards, depending on the measuring frequency) the stopping path of a kink is comparable with or smaller than the kink width. This means that in this temperature range the usual transition state theory for the thermally activated formation for kink pairs should be replaced by a diffusion theory of thermal activation.[25] This has been done by Stenzel;[26] a brief account of this work has been given elsewhere.[2]

6 APPENDIX

6.1 Dislocation Interpretation
of the PDFK-Model

Before it was used as a model for kinks in dislocation lines, the PDFK-model was introduced as a simple model for a dislocation in a crystal (see Ref. [2]). In this interpretation the force $\bar{f}_z$ of Eq. (4.5) is the dragging force on a unit length of a straight dislocation moving uniformly with velocity $\bar{v}$. The restriction to small velocities is no longer necessary. Figure 5 shows a plot of the numerically evaluated dragging force as a function of the dislocation velocity $\bar{v}$ for various temperatures. The force is proportional to the velocity up to $\bar{v} \approx 1/3\, c$, where $c \equiv C_l$ for an edge dislocation, $c \equiv C_t$ for a screw dislocation; $\bar{f}_z$ reaches a maximum for

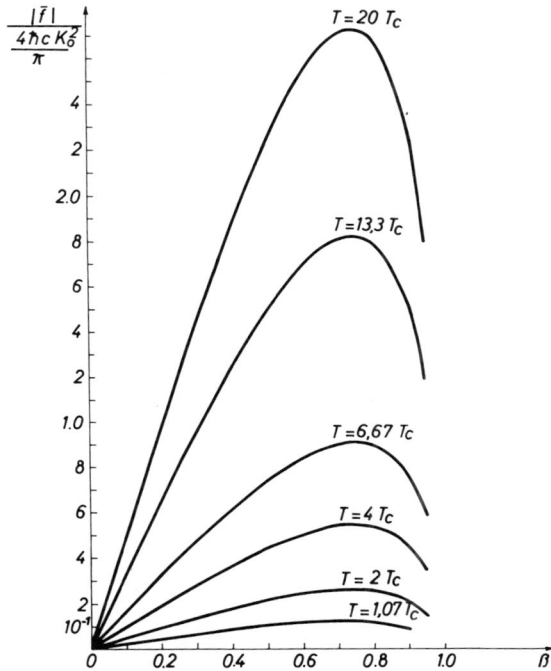

Fig. 5 Dragging force $\bar{f}$ on a dislocation in the Prandtl-Dehlinger-Frenkel-Kontorova model as a function of the dislocation velocity $\bar{v}$ ($\beta = \bar{v}/c$; c = velocity of sound) for various temperatures.

$\bar{v} \approx \tfrac{3}{4} c$ and decreases for higher $\bar{v}$. The decrease of $\bar{f}_z$ to higher velocities is due to the Lorentz contraction of the dislocation width. If one keeps the displacement field of the dislocation independent of the velocity $\bar{v}$ [and thus violates Eq. (3.6a)], the force becomes a monotonically increasing function of $\bar{v}$.

The reduction of a three-dimensional crystal with an isotropic spectrum of sound waves to the glide plane of a dislocation with a spectrum restricted to one propagation direction represents a strong simplification of the situation in a real crystal. Therefore only the order of magnitude in the results of Sec. 4.1 is of interest in the dislocation interpretation. Assuming a dislocation rest width $w_0 = 2b$ leads to a temperature $T_c = 70°K$ and to a room-temperature viscosity per dislocation unit length $\eta = 1.9 \cdot 10^{-4}$ dyn-sec/cm^2. Various authors [25,27-29] estimate from megacycle measurements in Cu at $T = 300°K$ an approximate value $\eta = 7.5 \cdot 10^{-4}$ dyn-sec/cm^2. The theoretical result is too low by a factor of 4. This is not unreasonable in view of the simplicity of the model.

ACKNOWLEDGMENT

The authors are very grateful to Professor H. Bross, now at the University of Munich, for his advice and help in the early part of this work, particularly, in formulating the general theory of Sec. 2.

REFERENCES

1. Seeger, A.: In S. Flügge (ed.), "Handbuch der Physik," Vol. 7, Part 1, p. 383, Springer, Berlin-Göttingen-Heidelberg, 1955.
2. Seeger, A., and P. Schiller: In W. P. Mason (ed.), "Physical Acoustics," Vol. IIIA, p. 361, Academic Press, New York, 1966.
3. Seeger, A., and P. Schiller: Acta Met. 10:348 (1962).
4. Suzuki, T., and C. Elbaum: J. Appl. Phys. 35:1539 (1964).
5. Alefeld, G.: J. Appl. Phys. 36:2642 (1965).
6. Lothe, J.: J. Appl. Phys. 33:2116 (1962).
7. Leibfried, G.: Z. f. Phys. 127:344 (1949).
8. Eshelby, J. D.: Proc. Roy. Soc. A266:222 (1962).
9. Mason, W. P.: Phys. Rev. 143:229 (1966).
10. Tittmann, B. R., and H. E. Bömmel: Phys. Rev. 151:178 (1966).
11. Bross, H.: Phys. Stat. Sol. 2:481 (1962).
12. Bross, H., A. Seeger, and P. Gruner: Ann. Phys. Leipzig (7), 11:230 (1963).
13. Bross, H., A. Seeger, and R. Haberkorn: Phys. Stat. Sol. 3:1126 (1963).
14. Seeger, A., H. Bross, and P. Gruner: Faraday Soc. 38:69 (1964).
15. Gruner, P.: Phys. Stat. Sol. 12:679 (1965).
16. Bross, H., P. Gruner, and P. Kirschenmann: Z. Naturforschg. 20a:1611 (1965).
17. Gruner, P.: Z. Naturforschg. 20a:1626 (1965).
18. Bross, H.: Z. f. Phys. 189:33 (1966).
19. van Hove, L.: Physica 21:517 (1955).
20. Seeger, A., and A. Kochendörfer: Z.f. Phys. 130:321 (1951).
21. Bross, H.: Phys. Stat. Sol. 5:329 (1964).
22. Seeger, A., and O. Buck: Z. Naturforschg. 15a:1056 (1960).
23. Alers, G. A., and D. O. Thompson: J. Appl. Phys. 32:283 (1961).
24. Granato, A., and K. Lücke: J. Appl. Phys. 27:583 (1956).
25. Kramers, H. A.: Physica 7:284 (1940).
26. Stenzel, G.: Diplomarbeit, T. H. Stuttgart (1965).
27. Stern, R. M., and A. V. Granato: Acta Met. 10:358 (1962).
28. Suzuki, T., A. Ikushima, and M. Aoki: Acta Met. 12:1231 (1964).
29. Greenman, W. F., T. Vreeland, Jr., and D. S. Wood: J. Appl. Phys. 38:3595 (1967).

DISCUSSION on Paper Presented by A. Seeger

G. ALEFELD: I think that your calculations for the damping constant, as determined by considering a free moving kink, are applicable only to a small fraction of dislocations. The majority of dislocations, at least in f.c.c. and b.c.c. metals, is made up of strongly interacting geometric kinks. For such kink chains the damping constant B will have to be determined by considering the interaction of phonons with the lowest vibrational modes of the kink system.

A. SEEGER: In our paper we were mainly interested in relatively low temperatures, and our analytical approximations confine our numerical results to the temperatures at about or below the Bordoni maximum. Our main intention was to obtain theoretical results on kink properties which can be used in further refinements of the theory of the Bordoni relaxation.

I agree that for a treatment of a situation with high kink density and comparison with the "background" dislocation friction it would be of considerable interest to treat the vibrational modes of kink-chains in the manner indicated by Alefeld. For a chain of geometrical kinks this could indeed be done by a slight extension of our treatment.

J. P. HIRTH: Seeger has analyzed his data in terms of the diffusion theory of Kramers. In fact, the original theory by Lothe and myself [*Phys. Rev.* 115:543 (1959)] is a kink-diffusion theory for dislocation motion. Later work by Seeger and Schiller [*Acta Met.* 10:348 (1962)] showed that the activation energy should be amended to include the roughly 10% correction associated with the kink-kink interaction energy. In his present paper, Seeger suggests that the kink diffusivity should be lower than the Leibfried estimate that we used originally because of his additional kink-damping term. With these modifications of our original theory, I would anticipate that the predictions of our kink diffusion theory and that of Kramers would be similar, differing only by a small factor in the preexponential term.

Specifically, the above discussion emphasizes the point that the Farkas-Zeldovich treatment of transition state theory envisions a diffusion-type process, not an Umklapp-type process; see the criticism by Lothe [*Z. Phys.* 157:457 (1960)] of Donth's work. Indeed, while it is generally presented as a "correction factor" to quasi-equilibrium nucleation theory, the so-called Zeldovich factor is related to a diffusion coefficient in energy space [Farkas, L.: *Z. Phys. Chem.* 125:236 (1927); Zeldovich, J. B.: *Acta Physicochem.* USSR 18:1 (1943); Feder, J., et al.: *Advan. Phys.* 15:111 (1966)]. Random walk in energy space physically corresponds to random motion in coordinate space, thus the two types of diffusion treatments should be closely related.

A. SEEGER: I have compared the diffusion-type formula of Lothe and Hirth [*Phys. Rev.* 115:543 (1959)] with those derived from Kramers' theory of chemical reactions ("Physical Acoustics," vol. IIIA, chap. 8, IV F 3, Academic Press, New York—London, 1966) and found that the preexponential factors are quite different, even if the long-range interaction between the kinks is disregarded. I doubt therefore that Hirth's first statement is correct. Furthermore, the main point of my argument was that it is just the increased magnitude of the kink viscosity (or the smallness of the kink diffusion coefficient) that leads us to consider the diffusion theory rather than the transition state theory of double-kink formation. With the previous theoretical estimates for the order of magnitude of the kink-diffusion coefficient the use of a diffusion theory could hardly be justified.

Also, I cannot agree with Hirth's second comment. The last one of the quoted references has indeed enabled me to better appreciate the origin of the so-called Zeldovich factor. As stated by Hirth, the Farkas-Zeldovich approach considers *diffusion in energy space*. Such a diffusion is inherent in the transition state theory. It is true that in most presentations of the transition state theory it is not treated explicitly. However, it has been considered explicitly in a number of papers [e.g., Donth, H.: *Z. Physik* 149: 111 (1957); Zener, C.: Suppl. al *Nuovo Cimento* VII /X/, 544 (1958); Geszti, T.: *Phys. Stat. Sol.* 20: 165 (1967)]. By contrast, the Kramers theory considers *diffusion in coordinate space*; it is applicable to the overcoming of a potential barrier by a particle which, due to Brownian movement, changes many times its direction of motion on its way to the saddle point. Accordingly, the Zeldovich factors appears to be different from the correction introduced by the Kramers-type treatment in the usual formulation of the transition rate.

Final judgment on that point, however, should await the application of the two theories to the same problem.

J. LOTHE: How do the kink-mobility calculations relate to mobility calculated from the scattering in the strain field of the infinite straight dislocations?

A. SEEGER: My general attitude is, as stated repeatedly, that the standard calculation of mobilities of infinitely straight dislocations are not fully satisfactory from the theoretical viewpoint, since they do not adequately take into account the fact that a dislocation line is at the same time flexible and continuous. This is quite clear from Leibfried's basic 1950 paper, in which the assumption is made that adjacent sections of a dislocation line move independently from each other. An adequate treatment within the framework of the string-model would have to consider the scattering of the phonons from the quantized normal modes of the dislocations, as has been done in a different context by S. Ishioka and H. Suzuki [*J. Phys. Soc.* Japan 18: Suppl. II, 93 (1963)] and as has been suggested for the kink-chain by Alefeld in his comments to our contribution.

The physical reason why in the kink picture the scattering from the strain field (treated in our paper) is, under most conditions, so much more effective than the radiation damping due to the oscillations of the kink positions is thought to be as follows: Thermal fluctuation with too short wavelengths do not contribute to the radiation damping (flutter) mechanism, since they will only distort the shape of the kink but not lead to an oscillation of the kink position. The phonon-spectrum therefore has to be cut off at wavelengths comparable with the kink-width, which except for very low temperatures is larger than the cut-off in the Planck distribution. As is well known, the treatments by Eshelby and Lothe differ in the details of this cut-off. No such cut-off is required in the scattering mechanism, since phonons, even of very short wavelengths, are scattered effectively by the elastic inhomogeneities due to the kinks.

G. ALERS: You have presented a very general approach to the calculation of the interaction between dislocations and phonons. Is your approach sufficiently general to use it with little modification for the calculation of the electron-dislocation interaction which should be the dominant source of dislocation damping at very low temperatures?

A. SEEGER: I agree that at very low temperatures (presumably below 4.2°K) the interaction between electrons and dislocations (or kinks) will be the dominant damping process for dislocation motion in pure metals. As stated in our paper our approach to the problem is general in the sense that it is also valid for the interaction with other Bose particles, and that small modifications will allow us to extend the theory to electrons, at least within the framework of perturbation theory. This has not yet been done, however.

A. GRANATO: Do you plan to work out your formulas in terms of anistropic third-order elastic constants as well as in the isotropic approximation?

A. SEEGER: Yes. The whole work was done when there were no measurements of third-order elastic constants on metal single crystals, so we used the compilation that Buck and I made in 1960, which was the best thing then available.

As better single crystal data become available, we shall do these calculations using them. In fact, we are applying anisotropic elastic constants to our calculations of heat conductivity. I am very happy to see that George Alers has produced low temperature elastic constant data. These data are quite important because: (a) the temperature dependence is quite significant; and (b) because they enter into our formulas as squares and errors by a factor of 3 can very easily be produced by using room temperature elastic constant data.

A. GRANATO: Does the calculation you have made throw any light on the problem of the anomalously high dislocation thermal resistivity found in alkali halides?

A. SEEGER: We have produced a series of theoretical papers on this topic and also have done some experiments on copper-gallium alloys, where we know the dislocation density fairly well from x-ray and from electron microscopy studies. In the case of this alloy, the discrepancy can be almost completely removed if we use the low temperature constants that Alers has produced instead of the room temperature constants which we have used previously. (The agreement is now to within a factor of 2.) We have studied the temperature dependence of the thermal resistance in great detail including the effect of piled-up dislocations and the changes in dislocation configurations going from Stage I to Stage II of the stress-strain curve. Theory and experiment agree well. This means that now heat conductivity measurements can be used to study dislocation densities in materials with very high dislocation density and also dislocation arrangements. It would appear that some of the previous comparisons of this type have grossly underestimated the dislocation density. We are thinking of experimental measurements in alkali halides to check further this theory.

A. GRANATO: Is not the discrepancy in lithium floride too large to start with? Many authors claim it is on the order of 1000.

A. SEEGER: In fact, the worst disagreements are on the order of 10^3, but in the best case it is as low as a factor of 24, and a factor of 24 is probably not too serious in the light of the improved experimental and theoretical possibilities. You must remember that this factor of 10^3 depends on etch-pit measurements of dislocation densities and that this technique has serious disadvantages in that it reveals neither dislocation dipoles nor dislocations lying at a shallow angle to the surface. Thus, we have tentatively condluded that the fault is more likely to lie with the experiments than with the theory.

CROSS-SLIPPING PROCESS
IN THE F.C.C. STRUCTURE

B. Escaig

Lab. de Physique des Solides,
Associé au CNRS,
Faculté des Sciences,
91-Orsay-France

ABSTRACT

A cross-slip model for the extended screw dislocations, as suggested earlier by Friedel, is theoretically studied, and its stress dependence is derived. In this model the cross-slipping loop can split in the cross-slip plane as soon as it begins to form. It is found that, in addition to the compressive force on the faulted ribbon in the primary glide plane, the driving force for cross-slip is the widening of the cross-slipped ribbon in the cross-slip plane, and not its bending as assumed by most investigators. Therefore, the cross-slipping process and the motion of the cross-slipped loops can be two quite separate processes.

The cross-slip energy is derived, and asymptotic forms are given in the limiting cases of low and high stresses, compared with the stacking fault energy. At low stresses, a linear relation in stress is obtained.

On this basis, one can account for the onset of State III of pure copper with a cross-slip mechanism initiated at preexisting constrictions, neglecting any possible stress concentration factor. An order of magnitude of the stacking fault energy f can thus be deduced, without the need of a more-or-less known parameter. For pure copper the value $\mu b/f \simeq 200$ or 250 has been obtained.

INTRODUCTION

The cross-slipping process is known as a low-temperature process of dynamic recovery in cold-worked metals; it determines the onset of Stage III of deformation of f.c.c. metals; it probably occurs in low-temperature anneals of strongly strained f.c.c. metals; it should also be of interest to high-temperature creep tests.

However, the exact laws of cross-slipping are still not well known. It is usually described by the theory initially proposed by Schoek and Seeger,[1] then refined at Stuttgart.[2] In this model, cross-slip requires the recombination of the extended screw dislocation along a certain length long enough to be unstable in the cross-slip plane, and to be able to expand under the action of the effective stress acting in that plane. Such a recombination implies two main difficulties as detailed in a paper of Thornton et al.:[3]

i) the high value of the cross-slip energy, mainly at low stresses (infinite under zero stress); and

ii) the necessity of strong piled-up groups to ensure an effective stress high enough to recombine the head dislocation. Particularly, it is hard to accept large, unrelaxed piled-up groups, and also barriers strong engough to withstand such high stresses. As a result, the stacking fault energies deduced by the theory from τ_{III} experiments for Cu, and Cu-Zn alloys are clearly unrealistic,[3] e.g., 175 ergs/cm^2 for Cu.

The purpose of this paper is to develop quantitatively a model for the cross-slip of screw dislocations, as suggested earlier by Friedel.[4] Let us start from an initial constriction in an extended screw dislocation which is held up by some obstacle in the initial glide plane. As the two halves A and B of this constrictions are separating, due to the stress-aided, thermal activation, we assume following Friedel, that the loop AB can split in the cross-slip plane as soon as it begins to form (Fig. 1). A critical separation is reached, so that further separation occurs spontaneously under the action of the stress in the cross-slip plane until the cross-slip is complete, with two completely separated constrictions A and B.

In Friedel analysis, the applied stress pushed both partials of the cross-slipped portion AB into a loop, and moved it to give cross slip. On the other hand, we establish in this paper that the cross-slipping process, and the motion of the cross-slipped loops

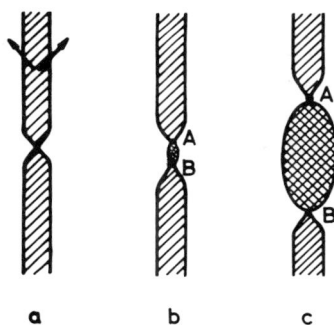

a   b   c

Fig. 1 Successive events leading to cross-slipping.

can be two quite separate processes, obtained under different stresses. It is mainly the widening of the faulted ribbon in the cross-slip plane, not its bending, that induces cross slip, as the ribbon is compressed in the initial plane. More precisely, we are concerned in the cross-slip plane with quite different shear stress components. One of these σ_t acts identically on both partials, through their similar screw character, and gives them some curvature; the other σ_s pushes the two partials in opposite directions, through their opposite edge character, and changes only their separation. The contribution of σ_t to the cross-slip energy is shown to be always small, compared with the σ_s one. Therefore the driving force for cross slip is the widening of the faulted ribbon in the slip-plane; the stress on the total dislocation σ_t is only active for further glide on the cross-slip plane, as, for example, for mutual annihilation with an opposite dislocation.

This is the main result of the first part of this paper, devoted to the calculation of the cross-slip energy. We first compute the energy of an extended cross-slipping loop, at its equilibrium position (Secs. 1.1 and 1.2) for a given length AB between two constrictions A and B. This computation is based on two assumptions:

i) the classical Stroh assumption in constriction calculations[5] for local elastic interactions; and

ii) the mean line of the extended loop AB fits the circular one into which the perfect equivalent dislocation would be bent under the action of σ_t.

Once the energy has been obtained for each given length AB (Sec. 1.3), the calculation continues with the determination of the saddle-point energy for the critical length AB (Sec. 1.4). Allowance is made for the stress σ' acting in the initial plane. The cross-slip energy is found to be only a function of the ratio of the stresses to the stacking fault energy f. This function can be computed numerically. Asymptotic forms give analytic formula valid in the limiting cases

$$\frac{\sigma b}{f} > 1 \quad \text{or} \quad \frac{\sigma b}{f} \ll 1 \qquad \text{(Sec. 1.5)}$$

In the second part, we have applied these results to some phenomena usually linked to cross slip, together with two additional assumptions on the occurrence of cross slip. (i) Cross slipping is more often initiated at preexisting constrictions such as jogs or nodes of attractive junctions; thus we are concerned only with the pinching energy in the cross-slip plane. (ii) Cross slipping takes place under the action of the one externally applied stresses σ' and σ_s; therefore, we can express σ_s, σ' and the cross-slip energy in terms of the maximum resolved shear stress (Sec. 2.1). A few remarks are made concerning high-temperature creep (Sec. 2.2), while the main interest is given to the onset of Stage III of deformation of f.c.c. metals (Sec. 2.3). At low stresses, $\sigma b/f \ll 1$, the theory predicts a linear dependence of the τ_{III} upon the temperature and the logarithm of the strain rate, depending on the orientation of specimen.

Experimental verification of such a relation is tested in the third part, with data on pure copper single crystals reported in the literature. Part of the orientation effect is qualitatively understood, mainly near the (100) and (111) orientations (Sec. 3.2). Strain rate and temperature dependence are approximately verified (Sec. 3.1) and both of them yield a reasonable value of the stacking fault energy in Cu, about 50 ergs/cm^2. No more than an order of magnitude should be expected, for one is comparing crystals at different strains, thus with different dislocation densities. Differential measurements would be of interest in this field.

The main results obtained are discussed in the fourth part.

1   CALCULATION OF THE CROSS-SLIP ENERGY

Here we compute the activation energy to develop in the cross-slip plane a cross-slipping loop *from a preexisting constriction on* an extended screw dislocation. Thus we are concerned only with the pinching in the cross-slip plane. Taking account of the constrictions lying in the initial plane would consist of adding to the energy computed in Eq. (14), the Stroh constriction energy, Eq. (14a), $W_s = 1.1 a\mu b^2 \, l'/16\pi$, in the notations of Eq. (14), l' being the dissociation width in the initial plane.

First the equilibrium position (Secs. 1.1 and 1.2) and its energy (Sec. 1.3) are obtained for each given length AB, then the critical length AB which makes this energy maximum, and this maximum itself is derived (Sec. 1.4). Asymptotic formulas are obtained in the limiting cases of high and low stresses (Sec. 1.5).

The following notations are used for the stresses to be considered:

a) σ' is the stress which compresses in the initial plane one partial against the other. Attention must be paid that throughout this paper, the *total* force per unit length is written $\sigma' b/(2\sqrt{3})$;

b) σ_s and σ_t are the stresses in the cross-slip plane; σ_t is the component which acts on the total cross-slipping dislocation, giving it some curvature; on the other hand, σ_s acts only on its width.

Further, general notations are used: b is the lattice period, f is the stacking fault energy, A is $\mu b^2/2\pi$, and μ is the shear modulus.

1.1 Equilibrium Equations

The equilibrium equations are set up under the following assumptions.

a) Local elastic interactions between the two partials are treated as interactions between two concentric circular loops, an adaptation for circular geometry of the Stroh assumption in constriction calculations.[5]

b) These interactions are developed to the first order in (h/r); h is the local half separation of the two partials and r is the local curvature of the mean line AOB of the extended loop (Fig. 2).

One can show[6] that the determination of the equilibrium position turns into two problems that are solved *independently* of one another: (i) the determination of the mean line AOB, that is, the function $r(M)$; and (ii) the determination of the relative positions of the partials to this mean line, that is, the function $h(M)$.

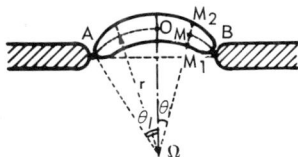

Fig. 2 Equilibrium of the cross-slipping loop; $r = \Omega M$, $h = MM_1 = MM_2$.

i) The mean line is found to fit the circular one into which the perfect equivalent dislocation would be bent under the action of σ_t. This is the physical content of the first-order approximation in (h/r).

ii) The relative position of the partials to this mean line is found[6] to be independent of its curvature, that is, the same one as if the mean line was straight on the whole length. Thus the curvature perturbs the differential equation for $h(M)$ only through a term in $(h/r)^2$, which cancels in a first-order approximation. Therefore, $h(M)$ is a solution of a Stroh-type equation:

$$2 \frac{d^2 H}{dZ^2} = 1 - \frac{1}{H} \tag{1}$$

Here H is the separation of the partials at M, taken perpendicular at the mean line, and related to l, the dissociation width of a straight screw dislocation in the cross-slip plane:

$$H = 2 \frac{h}{l} \tag{2}$$

Z is the dimensionless coordinate for the curve length OM

$$Z = \left(\frac{r\theta}{l}\right)\left(\frac{A}{2C}\right)^{1/2} \tag{3}$$

θ is the angle shown in Fig. 2, and C is the effective line tension of the partials to be used in their restoring force.

As a result of the independence of the partial separation $H(Z)$ on the curvature of the whole r, the energy of the equilibrium configuration, splits into two additive terms:

$$W = W_1 + W_2 \tag{4}$$

corresponding to the contributions of the pinching W_1 and of the curvature W_2. We shall return to this relation in Sec. 1.3.

Finally, the validity of the first-order approximation used here, implies the condition, which would easily be fulfilled over a wide range of stress values of physical interest:

$$\left(\frac{h}{r}\right)^2 \lesssim \left(\frac{l}{2r}\right)^2 \simeq \left[\frac{\sigma_t b}{20(f - \sigma_s b/2\sqrt{3})}\right]^2 \ll 1 \tag{5}$$

1.2   Boundary Conditions. Integration
of Equilibrium Equations

We must add boundary conditions on the local width $H(Z)$ to obtain it from the integration of Eq. (1); $H(Z)$ and its derivatives must be continuous and symmetrical relative to the middle point 0, that is, for $Z = 0$:

$$Z = \theta = 0 \; , \quad \frac{dH}{dZ} = 0 \; , \quad H = H_0 \qquad (6)$$

Taking H_0 as a parameter, we obtain the two necessary boundary conditions to integrate Eq. (1). For each value $H_0 \leq 1$, there is one solution $H(Z)$, i.e., a well-defined configuration; $H(Z)$ is obtained more conveniently under the form $Z(H)$:

$$Z = H_0 \int_{t=H/H_0}^{1} [-\mathrm{Ln}t + H_0(t - 1)]^{-1/2} \, dt \qquad (7)$$

This configuration can also be characterized by another parameter, the half-length of the cross-slipped loop $D \propto AO = OB$, which is the Z value at $H = 0$, or at $\theta = \theta_l$, θ_l being the angular extension of the curve AO:

$$Z = D \; , \quad \theta = \theta_l \; , \quad H = 0 \qquad (8)$$

Using either of these two parameters, we can follow the energy variations during the successive events which lead to cross slip. Clearly, D increases with H_0, from a zero common value to $H_0 = 1$ when the central separation reaches the width of an extended straight dislocation. Then D is infinite and the cross slip is complete. As shown below, the energy W increases first with H_0, with an infinite slope at the origin, passes through a maximum value, then decreases due to the work of stresses. To obtain this maximum, we must first express the function $W(H_0)$.

Thus, by making $H = 0$ at $Z = D$, similar to Eq. (8), we neglect the necessary width ($\simeq b$) of the perfect dislocation; therefore, throughout this paper, our equations should be applied with caution as the dissociation width becomes of the order of b.

1.3   The Energy of Equilibrium Configurations

For a given H_0, the equilibrium energy $W(H_0)$ splits into two additive terms, as mentioned above: one from the pinching, and the other from the curvature of the whole,

$$W = W_1 + W_2 \tag{4}$$

where W_1 is the pinching-dependent term, which would be the same for a zero curvature, or a zero σ_t stress component:

$$W_1 = l \left(\frac{AC}{8}\right)^{1/2} (W_{11} + W_{12}) \tag{9}$$

$$W_{11} = D(H_0) \cdot E(H_0) \ , \quad E(H_0) = -\mathrm{Ln}\, H_0 + H_0 - 1 + \mathrm{Ln}\left(\frac{l'}{l}\right) \tag{9a}$$

$$W_{12} = \left(\frac{\tau}{C} + 1\right) U(H_0), \quad U(H_0) = H_0 \int_0^1 [-\mathrm{Ln}\, t + H_0 (t - 1)]^{1/2} \, dt \tag{9b}$$

Here, l and l' are the dissociation widths of a straight screw in the cross-slip plane, and in the initial glide plane, respectively; C and τ are the effective line tension, and the energy per unit length of the partials, respectively; for the mean character of the partials (30° Shockley partials), the relation $\tau \simeq 0.7\ C$ holds. Finally, the two components W_{11} and W_{12} represent two possible steps to make the pinching:

i) First, the straight partials separation is moved from l' in the initial plane to $(H_0\, l)$, the distance required in the cross-slip plane, at the middle point of the cross-slipping loop. Thus, one is varying the energy by $E(H_0)$ per unit length, by about a factor $A/8$, and by W_{11} for a length D. The half-length D of the cross-slipping loop is given by Eqs. (7) and (8):

$$D = H_0 \int_0^1 (-\mathrm{Ln}\, t + H_0 (t - 1))^{-1/2} \, dt \tag{9c}$$

ii) W_{12} is the energy supplement to pinch the preceding partials separated by $H_0\, l$. This is the true pinching term, the value of which Stroh has computed[5] for $H_0 = 1$. It is the sum of a lengthening term $\tau U/C$, and an interaction term between the partials, which equals U if the loop is at its equilibrium position between A and B.

Thus W_2 is the curvature-dependent term arising when an extended dislocation is bent under the action of σ_t. Considering the line lengthening and the work of the σ_t stress, and assuming the curvature to be equal to that of the perfect equivalent dislocation, we obtain

$$W_2 = -l\left(\frac{AC}{8}\right)^{1/2}\left(\frac{D^3}{60}\right)\left(\frac{\sigma_t b}{\gamma}\right)^2 , \quad \gamma = f - \left(\frac{\sigma_s b}{2\sqrt{3}}\right) \tag{10}$$

as the angular extension, $2\theta_l$ is small enough to develop $\sin\theta_l$ and $\cos\theta_l$ in the first order. This can be shown[6] to be consistent with $(h/r)^2 \ll 1$. We note that the one contribution of σ_t to the total energy is (through W_2) a term in $(\sigma_t b/\gamma)^2$. This result still holds for the saddle-point energy, which is calculated in the following section.

1.4 Saddle Point. Activation Energy

Here we determine the maximum value of the one-variable function $W(H_0)$, as defined by Eqs. (4), (9), and (10). This function represents the energy of the equilibrium configuration obtained for each H_0 or D value fixed by a boundary condition. The saddle point must be chosen from among all the equilibrium configurations for all possible D values, as that which makes zero the differential variation of the energy W, when D and H_0 are being varied, as seen by Eq. (9c).

First, we assume the equality between C, the effective line tension, and τ, the line energy per unit length of the partials. This postulates a character-independent line tension τ, i.e., the isotropy of τ relative to the lattice. In such a case, keeping D fixed, one must obtain the equilibrium condition $\partial W/\partial H_0 = \partial W_1/\partial H_0 = 0$. Thus the relation

$$2\,dU + D\,dE = 0 \tag{11}$$

holds between the H_0 functions U, D, E. The saddle-point equation $dW = 0$ then becomes $\partial W/\partial D = 0$, that is:

$$E(H_0) - \left(\frac{1}{20}\right)\left(\sigma_t b \frac{D}{\gamma}\right)^2 = 0$$

In the limiting case $\sigma' = \sigma_s = \sigma_t = 0$, the saddle-point occurs as $E(H_0) = 0$, i.e., at $H_0 = 1$, for an infinite length D. This result is obvious as one considers that without stresses and with an isotropic line tension, the saddle point cannot occur for any finite length D. On the other hand, this no longer holds if the initial screw position is energetically more favorable. Starting from some finite length D, one would gain some energy in spreading the configuration along the screw direction, thus putting further constrictions on A and B.

Since $\tau = 0.7\,C$, this is the main modification to be added to the above results. More precisely, writing $dW = 0$, dW_1 is no longer equal to EdD, but DdE is also to be introduced. This introduces a new function:

$$\delta(H_0) = D \frac{dE}{dD} = - \frac{\displaystyle\int_0^1 dt[-\mathrm{Ln}\,t + H_0(t-1)]^{-1/2}}{\displaystyle\int_0^1 dt(1 - H_0 t)^{-2}[-\mathrm{Ln}\,t + H_0(t-1)]^{-1/2}} \tag{12}$$

Therefore, the final saddle-point relation is given by

$$-\mathrm{Ln}\,H_0 + H_0 - 1 + 0.15\,\delta(H_0) = \left(\frac{1}{20}\right)\left(\frac{\sigma_t\,bD}{f - \sigma_s b/2\sqrt{3}}\right)^2 - \mathrm{Ln}\left(\frac{1 - \sigma_s b/2\sqrt{3}f}{1 + \sigma' b/2\sqrt{3}f}\right) \tag{13}$$

When all the stresses are zero, the saddle point occurs at $H_0 \simeq 0.95$, and the corresponding D value is no longer infinite, but of the order of 5, that is, a nucleation critical length of about ten times the dissociation width under zero stress $(2r\theta_l \simeq 2\,lD)$.

Finally, the cross-slip energy is obtained from Eqs. (4), (9), (10), and (13). Eliminating $E(H_0)$, we find

$$W = a\left(\frac{\mu b^2}{16\,\pi}\right)\left\{ l\left[2U - 0.18\,D\delta + 16D\left(\frac{\sigma_t\,bD}{20(f - \sigma_s b/2\sqrt{3})}\right)^2\right]\right\} \tag{14}$$

$$a = \left(\frac{8\tau}{A}\right)^{1/2} = \left[1.5\,\mathrm{Ln}\left(\frac{lD\sqrt{3}}{b}\right)\right]^{1/2}, \quad l = \frac{\mu b^2}{16\,\pi(f - \sigma_s b/2\sqrt{3})}$$

The value of U, D, δ, in Eq. (14) must be taken for the H_0 value obtained from Eq. (13), depending on the stresses $\sigma', \sigma_s, \sigma_t$. Thus the energy can only be computed numerically for each particular case. However, asymptotic forms are developed in the next section for $H_0 \to 1$ (low stresses) and $H_0 \to 0$ (high stresses).

First, an important simplification of Eq. (14) can be made. The contribution of the σ_t component occurs in Eqs. (13) and (14) through such terms as $(D^2/20)(\sigma_t b/\gamma)^2$. The following are always negligible:

— at low stress, for this is a second-order term in $(\sigma_t b/\gamma)$, D^2 being of the order of 20;

— at high stress, for $D \to 0$, and the contribution becomes of the order of $(\sigma_t b/20\gamma)^2$, a term which has been neglected throughout [cf. Eq. (5)].

— Therefore, writing $\sigma_t = 0$, we can use Eqs. (13) and (14). Thus *the driving force for cross slip is the widening of the extended cross-slipped loop, not its bending*, as emphasized in the introduction.

Finally, we obtained the following simplified expression, instead of Eqs. (13) and (14):

$$W \simeq a\left(\frac{\mu b^2 l}{16\pi}\right) w(s) \quad , \quad s = \mathrm{Ln}\left(\frac{1 + \sigma' b/2\sqrt{3}f}{1 - \sigma_s b/2\sqrt{3}f}\right) \tag{14b}$$

where $w(s)$ is the function $2U - 0.18\,D\delta$ in Eq. (14) related to $-\mathrm{Ln}H_0 + H_0 - 1 + 0.15\,\delta(H_0)$ in Eq. (13); $w(s)$ has been computed and is shown in Fig. 3.

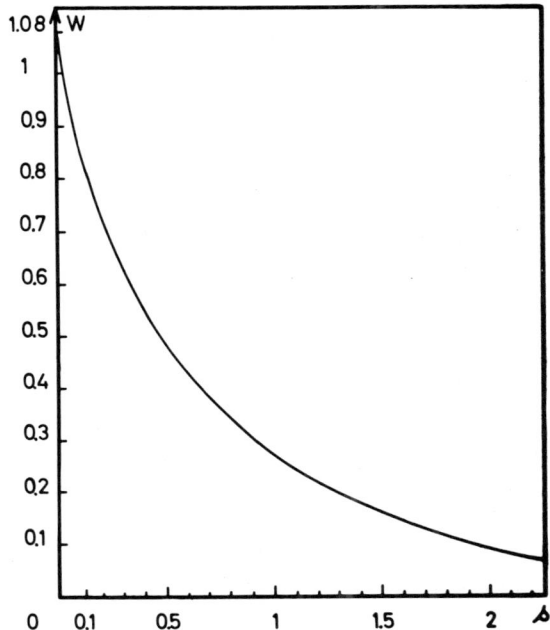

Fig. 3 Function $w(s)$.

1.5   Asymptotic Formula

From the saddle-point relation (13) it can be seen that for zero stress $H_0 \simeq 1$, while for increasing stress $H_0 \to 0$. In both cases we can develop asymptotic forms. We have neglected the σ_t terms in them.

1.5.1   Low Stress: $H_0 \to 1$. We have obtained the cross-slip energy to the first order in $\sigma_s b/f$ and $\sigma' b/f$. Such a development may be of physical interest in many cases (creep, onset of Stage III). The saddle-point equation shows that it is consistent with a development in $(1 - H_0)$ at the second order. Therefore, we obtained at this order Eqs. (13) and (14) to calculate the stress derivative of the function W at zero stress. Further, we have used a line tension of the partials in $\mathrm{Ln}(lD\sqrt{3}/b)$, thus weakly dependent on stress because D is itself logarithmically stress-dependent in that range; typically we have chosen $D \simeq 2$, corresponding to $\sigma_s b/f \simeq 0.1$. Hence

$$W \simeq W_0 \left(1 - 1.2 \, \frac{\sigma_s b}{f} - 1.5 \, \frac{\sigma' b}{f} \right)$$

$$W_0 \simeq \mu b^2 \frac{l_0 \, (\mathrm{Ln} \, 2\sqrt{3} \, l_0/b)^{1/2}}{37.3} \quad , \quad l_0 \simeq \frac{\mu b^2}{16 \pi f}$$

(15)

The value of W_0 is consistent with other classical values computed previously for the energy of one constriction.

1.5.2   High Stress: $H_0 \to 0$. The development of the saddle-point equation (13) at $H_0 \to 0$ gives the order relative to stresses of the approximation; this turns into

$$H_0 l \simeq \frac{l'}{\sqrt{10}} \quad \text{or} \quad H_0 \simeq 0.3 \, \frac{1 - \sigma_s b/2\sqrt{3}f}{1 + \sigma' b/2\sqrt{3}f} \to 0$$

(l, l' are the dissociation widths in the cross slip or initial plane). Thus, the first order in H_0 is the first order in $l'/l\sqrt{10}$ in terms of stresses; typically for $\sigma b/f \simeq 1$, $H_0 \simeq 0.18$. Hence this order is valid for $\sigma b/f > 1$. Let us obtain Eqs. (13) and (14) at the first order in H_0. Then, eliminating H_0 and taking into account the line tension τ of the stress dependence of D [here $D \ll 1$ is linearly dependent on (l'/l)], we obtain

$$W \simeq W_\infty \left(1 + 0.5 \frac{1 - \sigma_s b / 2\sqrt{3} f}{1 + \sigma' b / 2\sqrt{3} f}\right)$$

$$W_\infty \simeq \mu b^2 \, l' \, \frac{(\text{Ln}\, l'/b)^{1/2}}{61} \ , \quad l' \simeq \frac{\mu b^2}{16 \pi (f + \sigma' b / 2\sqrt{3})}$$

(16)

2 APPLICATIONS

Now we apply the preceding results to some phenomena linked to cross slip, on the occurrence of which, two additional assumptions are made.

i) *Cross slipping is more often initiated at a preexisting constriction than at a place where new constrictions are required.* Such constrictions should be numerous as a result of the cutting through the forest by moving dislocations. Thus cross–slip nucleation has been considered to occur preferentially at jogs[7] or at nodes of attractive junctions.[8] Here we are only concerned with the pinching in the cross–slip plane.

ii) *Cross slipping takes place under the action of the applied stresses σ' and σ_s.*

First we express the cross–slip energy in terms of the maximum resolved shear stress (Sec. 2.1); then after a few remarks about high–temperature creep (Sec. 2.2), we consider the onset of Stage III (Sec. 2.3).

2.1 Cross-slip Energy in Terms of the Applied Maximum Resolved Shear Stress

Let λ be the angle of the tensile axis with the glide direction, χ that with the initial glide plane, and ω the oriented one from the glide direction to the normal projection of the tensile axis on the glide plane (for $\lambda = \chi$, $\omega = 0$). Here λ, χ, ω refer to the sample orientation at the cross–slip occurrence. The subscript zero on these parameters refers to the initial sample orientation. Finally, τ is the maximum resolved shear stress on the initial glide plane.

We have first described the stress components σ_s, σ' in terms of λ, χ, ω, and τ:

$$\sigma_s = \alpha_s \tau , \quad \alpha_s = \left[\frac{2\sqrt{2}(\cos^2\lambda - \cos 2\chi)}{9 \sin\chi \cos\lambda} \right] - 7 \frac{\tan\omega}{9} \tag{17a}$$

$$\sigma' = \alpha'\tau , \quad \alpha' = \sqrt{3} + \tan\omega$$

Then, under the assumption of single slip, we have related λ, χ to the initial orientation λ_0, χ_0 and to the deformation ϵ at the cross-slip occurrence from the classical relations[9] $\epsilon = (\cos\lambda/\sin\chi) - (\cos\lambda_0/\sin\chi_0)$ and $\sin\chi \sin\lambda_0 = \sin\lambda \sin\chi_0$. Hence:

$$\alpha_s = \frac{\alpha_{s0}}{1 + (\epsilon \sin\chi_0/\cos\lambda_0)} \qquad \alpha' = \sqrt{3} + \frac{\tan\omega_0}{1 + (\epsilon \sin\chi_0/\cos\lambda_0)} \tag{17b}$$

These equations together with Eqs. (13) and (14) define the cross-slip energy as a function of orientation. Particular attention is given to the asymptotic form at low stresses, $\sigma b/f \ll 1$; hence Eq. (15) can be written as

$$W \simeq W_0 \left[1 - \alpha \left(\frac{\tau b}{f} \right) \right], \quad \alpha \simeq 1.2\alpha_s + 1.5\alpha' \tag{15a}$$

α is weakly orientation dependent, and is usually on the order of 15%. For most usual orientations, $\alpha \simeq 3$, as seen from Eq. (17a) with, for example $\lambda = \chi = \pi/4$. We can deduce a constant activation volume at low stress

$$V = -\frac{dW}{d\tau} \simeq b^3 \left(\frac{\mu b}{f} \right)^2 \frac{[\text{Ln}(\mu b/15 f)]^{1/2}}{600} \tag{15b}$$

which can vary from some b^3 for high $f(\mu b/f \simeq 50)$ to several hundreds for low $f(\mu b/f \simeq 500)$.

We have also considered the orientation dependence of σ_t, the contribution of which to W is through σ_t^2:

$$\sigma_t^2 = \beta\tau^2 \qquad \beta = \frac{8}{9} \left(\frac{\tan\omega \cos\lambda}{\sin\chi} - \frac{1}{2\sqrt{2}} \right)^2$$

where β is found to be independent of ϵ, $\beta = \beta_0$; but the most important fact is that for most of the orientations [except near (100) and (111) orientations] β is clearly smaller than $|\alpha_s|$ and α', often by one order of magnitude. Thus the σ_t contribution is still weakened as far as applied stresses are concerned.

2.2 Creep

Here only a few remarks will be made concerning creep. An activation energy and an activation volume such as Eqs. (15a) and (15b) should be of interest for high-temperature creep experiments; it was previously suggested by Dorn in particular—see, for example, Refs. [10] and [11]—that cross slipping could behave independently of the diffusion processes occurring at higher temperature. Thus, one could expect that for aluminum in which the width $l_0 \simeq 1.5\,b$ can tentatively be ascribed to the screws (that gives $f \simeq \mu b/75$), the zero-stress energy $W_0 \simeq 0.2$ eV (with $\mu b^3 = 4$ eV) and the activation volume $V \simeq 10\,b^3$; for copper, in which the stacking fault energy is estimated (see below, Sec. 3) at about $\mu b/250$, i.e., a screw width $l_0 \simeq 5\,b$, the zero-stress energy would be $W_0 \simeq 1.1$ eV, and the activation volume $V \simeq 150\,b^3$.

However, the activation energy in this temperature (below the diffusion range) when it can be isolated, is found experimentally to depend weakly on the stacking fault energy: about 1.2 eV for Al[11] and Cu[12], 1.9 eV in Ni-Co system,[13] whatever the Co-content. To date, no definitive experimental evidence exists for predominance of cross slip in these experiments.

2.3 The Onset of Stage III

Cross slipping seems to predominate for the activated slip observed at the onset of Stage III. We assume that once the cross slip is over, the glide of cross-slipped loops occurs easily across some area A. The strain rate is then given by the usual condition $\dot{\epsilon} = \dot{\epsilon}_0 \exp(-W/kT)$, where W is the cross-slip energy for the shear stress $\tau = \tau_{\mathrm{III}}$ at the onset of Stage III. Here we can use the linear asymptotic form (15a) for $\tau_{\mathrm{III}}\, b/f < 1$ if the dislocations are not too widely dissociated, a condition which holds for usual pure metals, and if the temperature is not too low [in Cu, $\tau_{\mathrm{III}}\, b/f < 0.3$ for $T > 100°$K, cf. Sec. 3]. Hence the linear relation

$$\frac{\alpha \tau_{\mathrm{III}}}{\mu} \simeq \frac{kT}{300}\, \mathrm{Ln}\,\frac{\dot{\epsilon}}{\dot{\epsilon}_0} + \frac{f}{\mu b}$$

$$K \simeq 10\left(\frac{f}{\mu b}\right)^2 \left(\mathrm{Ln}\,\frac{\mu b}{15 f}\right)^{-1/2} \quad , \quad \frac{\tau_{\mathrm{III}}\, b}{f} < 1$$

(18)

$\dot{\epsilon}_0$ is the frequency factor of the rate equation. It is given by $\dot{\epsilon}_0 = (\nu b/l_c)\, Np\, Ab$, where ν is an atomic frequency, l_c is the critical nucleation length, N is the number of nucleation sites per unit length along the dislocation line, here the preexisting constriction density, and ρ is the dislocation density. For $l_c \simeq 4 l_0$, l_0 being given in Eq. (15), and for reasonable values of the other parameters, $\rho \simeq 10^{10}$ cm^2, $N \simeq \rho^{1/2} \simeq 10^5$, $b \simeq 3.10^{-8}$ cm, $\nu \simeq 10^{13}$ sec^{-1}

$$\mathrm{Ln}\,\dot\epsilon_0 \simeq 20 + \mathrm{Ln}\,\frac{10^3 f}{\mu b} + \mathrm{Ln}\,\rho A$$

For not too widely dissociated metals, $\mathrm{Ln}(10^3 f/\mu b)$ is of the order of unity, and ρA can vary from $\rho A = 1$ (the cross-slipped loop glides from one tree to another) to $\rho A \simeq 100$ (the cross-slipped loop glide across one cell of the cell structure, of mean size 1 μ). Therefore, $\mathrm{Ln}\,\dot\epsilon_0$ should be a constant equal to about 24 ± 2.

3   COMPARISON WITH THE EXPERIMENTAL VALUES OF τ_{III} OF COPPER

Generally speaking, here crystals are compared at different strains, thus with different dislocation densities. Therefore, no more than an order of magnitude should be predicted. Data on single crystals of pure copper published in the literature are considered which report stress (τ_{III}) and strain (ϵ_{III}) values, allowing a comparison with the predicted ϵ_{III} value from Eqs. (15a), (17), and (18).

First (Sec. 3.1), the strain rate and the temperature dependence are considered, having regard to two possibilities: (i) the cross slipping is initiated at a preexisting constriction, as assumed for Eqs. (15), (17), and (18); (ii) the cross slipping is initiated at any place, thus taking account of the necessary constrictions lying in the initial plane. Then one deals with the orientation effect (Sec. 3.2), a somewhat more refined comparison.

3.1 Strain Rate and Temperature Dependence

Strain-rate dependence of $\alpha\tau_{III}$, measured by Mitchell et al.[15] at room temperature (cryst. no. 17) is shown in Fig. 4 where α has been calculated from Eqs. (17a) and (17b). Similar results have been obtained by other investigators.[16,17] It can be seen that a linear dependence on $\mathrm{Ln}\,\dot\epsilon$, as predicted by Eq. (18) is approximately verified. The measured slope, $p \simeq 10^{-4}$ cgs, gives the value of the

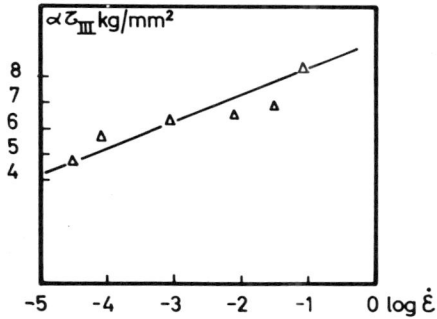

Fig. 4 Strain rate dependence of $\alpha\tau_{III}$, in kg/mm^2, following Mitchell et al.[15] Cryst. no. 17. Room temperature. (Decimal logarithms are used in abscissa).

constant K in Eq. (18), a function only of the stacking fault energy f. Thus f can be deduced directly without using other approximately known parameters. For pure copper, f is thus evaluated at about $\mu b/250$, i.e., 50 ergs/cm^2.

Figure 5 shows the temperature dependence measured by Mitchell[15] in the same work, at a strain rate $\dot{\epsilon} \simeq 10^{-3}$ sec^{-1}, for four

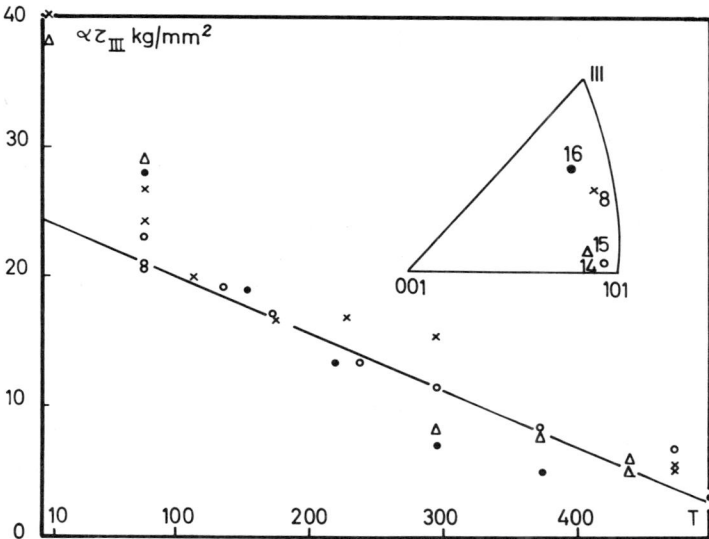

Fig. 5 Temperature dependence of $\alpha\tau_{III}$, in kg/mm^2, following Mitchell et al.[15] for four different orientations. $\epsilon \simeq 10^{-3}$ sec^{-1}. Cryst. no. 14, 15, 16, 8. (T in degrees Kelvin).

different orientations [cryst. no. 16, 8, 15, 14]. A linear relation such as Eq. (18) is still obeyed at least for not too low temperature: with the preceding estimation of f, $\tau_{III} b/f < 0.3$ if $T > 100°K$. From such data, two parameters can be deduced: (i) the energy f is obtained from the extrapolated $\alpha \tau_{III}$ at $T = 0$, that gives f/b following Eq. (18). Therefore, $f/b \simeq 2.3 \ 10^9$ dynes, i.e., $f \simeq \mu b/210$, which is consistent with the preceding one; (ii) the $Ln \dot\epsilon_0$ value [see Eq. (18a)] is obtained from the measured slope, $p' \simeq 10^{-5}$ cgs, the ratio of which to the preceding slope p is, following Eq. (18),

$$\frac{p'}{p} = \frac{Ln(\dot\epsilon/\dot\epsilon_0)}{300} \simeq 0.1$$

Thus, $Ln \dot\epsilon_0 \simeq 23$, yielding a ρA value [see Eq. (18a)] of about 10. Such a determination must only be considered as a test for the consistency of the theory, reasonable value of ρA lying from 1 to 1000; for the ratio of $Ln \rho A$ to $Ln \dot\epsilon_0$, about 20%, falls into the scatter of experimental errors in the slope determination. Hence no reliable determination of ρA will be obtained here.

The above experiments are explained on the basis of a cross-slip mechanism initiated at a preexisting constriction. It is interesting to compare the results with the predictions which would follow from a cross-slip mechanism which occurs anyplace. Then, the activation energy consists of the sum of the cross-slipping loop energy W, as in Eq. (15) or Eq. (15a), and the constriction energy W_s, as in Eq. (14a). As a result α changes into $\alpha/2$, and W_0 into $2W_0$, thus leaving unchanged the activation volume $\alpha W_0 b/f$; the predicted slope $d\tau_{III}/dT$ or $d\tau_{III}/d(Ln\dot\epsilon)$, leads to the same stacking-fault energy estimation. But the predicted absolute τ_{III} values which are as $1/\alpha$ [see Eq. (18)], have now become greater than the experimental ones by a factor two if this same stacking-fault energy is used. Thus the cross slip is considered to be initiated at preexisting constrictions.

3.2 The Orientation Effect

Here we used the measurements of Diehl[14] made at room temperature for various orientations. For such a temperature, τ_{III}/μ is sufficiently weak (smaller than 10^{-3}) to allow the use of the asymptotic formula at low stress [Eqs. (15) and (18)]. Thus the quantity $\alpha \tau_{III}$ should be a constant, α being calculated from Eqs. (17a) and (17b), τ_{III} and ϵ_{III} being taken from experimental values.

Clearly, the slight orientation dependence of α [α varies from 2.65 near (100) to 3.14 near (110)] is not sufficient to explain quantitatively the large variation observed to τ_{III}: from 4.9 kg/mm^2 near (111) to 2.5 near the center of the unit stereographic triangle. However, qualitatively, α shows a minimum near (111) [$\alpha = 2.87$] and near (100) [$\alpha = 2.65$] which is consistent with the maximum τ_{III} values often observed near these orientations. But the maximum of α near (110) [$\alpha = 3.14$] does not seem to be followed by a corresponding minimum in τ_{III} observed at such orientation, as it seems to occur in Al. Therefore, only qualitative agreement is obtained in this area.

4 CONCLUSIONS

The cross-slip model discussed in this paper and its stress dependence are based on the following two points:

i) the cross-slipping loop can split in the cross-slip plane as soon as it begins to form;

ii) the driving force for cross slip is the widening of the faulted ribbon in the cross-slip plane.

On this basis, assuming further that cross slipping occurs preferentially at the preexisting constrictions, and neglecting any possible stress-concentration factor, we can explain quantitatively the onset of Stage III, as shown on pure copper, except perhaps for orientation effect, which is only qualitatively understood. An order of magnitude of the stacking fault energy can be deduced, which yields a reasonable value for copper, 50 ergs/cm^2. Differential measurements in Stage III would be of interest for further verification of the theory.

REFERENCES

1. Schoek, G., and A. Seeger: *Rep. Conf. Defects Solids*, p. 340, Phil. Soc., London, 1955.
2. Wolf, H.: *Zeit. Nat.* 15A:180 (1960).
3. Thornton, P. R., T. E. Mitchell, and P. B. Hirsch: *Phil. Mag.* 7:1349 (1962).
4. Friedel, J.: "Dislocations and Mechanical Properties of Crystal," p. 330, John Wiley & Sons, Inc., New York, 1957.
5. Stroh, A. N.: *Proc. Phys. Soc.* B67:427 (1954).
6. Escaig, B.: *J. Phys.* In press.
7. Hirsch, P. B.: *Phil. Mag.* 7:67 (1962).
8. Washburn, J.: *Appl. Phys. Letters* 7:183 (1965).
9. Jaoul, B.: "Etude de la Plasticite et Application aux Metaux," pp. 192, 194, Dunod, Paris, 1965.

10. Friedel, J.: "Dislocations," p. 305, Pergamon Press, New York, 1964.
11. Jaffee, N., and J. E. Dorn: *Trans. Met. Soc. AIME* 224:1167 (1962).
12. Barrett, C. R., and O. D. Sherby: *Trans. Met. Soc. AIME* 230:1322 (1964).
13. Davies, C. K. L., P. W. Davies, and B. Wilshire: *Phil. Mag.* 12:827 (1965).
14. Diehl, J.: *Zeit. Met.* 47:331 (1956); see also B. Jaoul: "Etude de la Plasticite et Application aux Metaux," p. 315, Dunod, Paris, 1965.
15. Mitchell, T. E., and P. R. Thornton: *Phil. Mag.* 8:1127 (1963).
16. Peissker, E.: *Acta Met.* 13:419 (1965).
17. Berner, R.: *Zeit. Nat.* 15A:689 (1960).

DISCUSSION *on Paper by B. Escaig*

Z. BASINSKI: Mader's work on slip lines in Stage II indicates that the slip distance for edge dislocations is almost twice that for screws. One would therefore expect that screw dislocation density in deformed crystals should be about twice higher than density of edges. Electron microscopy results obtained in Cambridge, Ottawa, and Stuttgart, on the other hand, show that the majority of dislocations in crystals deformed into State II are either edge or reacted 60° dislocations. It appears therefore that there must be a mechanism of annihilation of screw dislocations operating in Stage II. Our work (unpublished) shows in fact that a large amount of very fine slip takes place on the cross-slip plane in Cu crystals deformed in Stage II at temperature as low as 4.2°K. If this is correct, it is difficult to see how cross slip can explain the transition from Stage II to Stage III.

B. ESCAIG: From a theoretical point of view, the occurrence of cross slipping during Stage II can be quite expected. I have only described here a stress-induced cross-slip mechanism. But other factors can induce it before the stress reaches the τ_{III} value. Particularly, line tension can be a quite efficient driving force at nodes of attractive junctions for drawing the dislocation in the cross-slip plane, as Washburn [*Appl. Phys. Letters* 7: F 183 (1965)] has convincingly suggested. In the case of favorable configuration, Washburn has shown that no activation energy is required.

I have intended to show that, as the applied stresses reached the τ_{III} level, cross slipping can become a much more frequent process to relax the internal stresses built up during the Stage II, and when screw parts appear during the stress-induced bending of dislocation lengths.

P. HAASEN: I must strongly object to your sole use of the early τ_{III} measurements by Thornton et al. to support your theory which itself I will not discuss at this stage. It was pointed out already by Peissker [*Acta Met.* 13:419 (1965)] that Thornton et al. neglected some experimental precautions necessary to obtain meaningful τ_{III} data. As a result, some of their claims of experimental evidence against the Schoek-Seeger-Wolf model of cross slip are clearly unrealistic. Neither do the six measured points in your Fig. 4 support in my opinion a logarithmic relation between τ_{III} and the strain rate $\dot{\epsilon}$ with any statistical significance. On the contrary, Peissker showed by measurements on some 400 single crystals of copper and copper alloys that $\log \tau_{III}$ was proportional to $\log \dot{\epsilon}$ with a mean-square deviation of 5 to 10%. The latter relation follows from the S-S-W theory which allows the stacking-fault energy of copper to be measured as (50 ± 6) ergs/cm^2 without significant influence of ill-known parameters (Peissker, loc. cit.). The value of 175 ergs/cm^2 you quote from Berner's early measurements evidently is typical for copper with oxide inclusions as was discussed in detail also for pure and oxidized silver (Ahlers, M.: *Z. Metallkde.* 1966). The stacking-fault energies obtained for pure Cu and Ag by Peissker and Ahlers, respectively, using the S-S-W theory are in good agreement with extended node, tetrahedra, and extended dipole measurements. Their data do not fit the relation $\tau_{III}(\epsilon_1 T)$ you propose with any accuracy.

B. *ESCAIG*: I would point out that the τ_{III} theory as based on cross-slip mechanism is still a difficult problem. The proposed cross-slip models, including mine, are very crude. Particularly, a realistic description of obstacles to be overcome by cross slip is still missing, despite its possible importance for the laws of cross slip. In Stage III the dislocation structure is very complex, mainly in the cell walls where cross slip is thought to occur. Finally, the predominance of cross slipping itself at the onset of Stage III is still a controversial subject as the question of Dr. Basinski shows. Concerning the Schoek-Seeger-Wolf model, the theoretical criticisms of Thornton et al [*Phil. Mag.* 7:1349 (1962)] strongly object against this model.

The experimental situation is not better. As quoted in the paper of Peissker, mentioned by Haasen, the case for copper is not clear. I do not know if oxide inclusions can account for the discrepancy in stacking-fault energy measurement between the value of 175 ergs/cm^2 found in the Berner's work and in Thornton's work, previously mentioned, and the value of 50 ergs/cm^2 found in Peissker's paper. In regards to this, it seems a little surprising that some results, for instance, the temperature dependence, is found, however, to be quite similar in the two papers (cf. Peissker). In Ahler's paper, mentioned by Haasen, it is shown that, as well as I can remember, oxide inclusions in silver can change the energy measurement by at most a factor of two. Finally, in Peissker's paper, some constants of the theory are found to vary experimentally in an appreciable manner.

Mainly, I wish to point out that, as the strain-rate range, or the temperature range, is extensively explored, one is comparing crystals at different strains, thus with different dislocation densities.

As a result of all these comments, I do not think that a good agreement with experimental data is to be expected at this time. One can only verify the order of magnitude of the parameters used in the theory and the stacking-fault energy. Therefore, I have only used an asymptotic formula in the experimental comparison. A use of my full law would lead to a law nearer the Peissker one. But I do not think that a distinction between a straight law and a logarithmic law is very significant because of the experimental scatter in strain-rate measurements, and nearly flat range of the logarithmic function when $\tau_{III}/\mu \simeq 10^{-3}$. (See Ref. [6] in addition to the text.)

P. HIRSCH: (1) I would like to support what Prof. Haasen has said; you should use the more extensive results on τ_{III} now available to test your theory over as wide a range of stress, strain-rate, and temperature as possible.

(2) The essence of your theory appears to be that the main driving force for cross slip comes from the widening of the dislocation stacking fault ribbon on the cross-slip plane. This effect is due to the component of the applied stress normal to the Burgers vector. In the face-centered cubic structure, however, because the partials occur in a definite sequence so that the fault between them is intrinsic, this component of stress will either tend to widen the ribbon, or to constrict it, depending on the sense of the stress. Thus, for a given orientation, if the applied stress acts in such a direction that the ribbon becomes wider after cross-slip, then on reversing the stress, the opposite case should obtain. Thus one would expect the magnitude of τ_{III} to depend on whether tests are made in tension or compression. I do not know of any relevant experimental evidence on this point.

(3) It seems a little surprising that the orientation dependence of τ_{III} on your model is relatively small. It appears to me that in the [00$\bar{1}$]–[011]–[$\bar{1}$11] triangle the stress component normal to the Burgers vector [$\bar{1}$01] in the primary glide plane (111) changes sign across the [1$\bar{2}$1] zone, while the same component in the cross-slip plane changes sign in the [1$\bar{1}$1] zone. Thus, there are two regions in the triangle where the ribbons widen in the cross-slip plane, and one region where the ribbons become narrower

in the cross-slip plane, or vice versa, depending on the sense of the stress. One might expect therefore a striking orientation dependence, which should reverse on reversing the stress direction. The orientation dependence τ_{III} should be examined in detail to test this point.

B. *ESCAIG* : I agree with the comments of Prof. Hirsch on the orientation effect on the stress component normal to the Burgers vector, as one can deduce it from Eq. (17a). But, in addition to the stress component normal to the Burgers vector b, one has to take into account the stress component parallel to the Burgers vector acting in the primary glide plane on one partial against the other. This force arises when one of the partials is being held up by some obstacle, the other is forced against it under the action of the force inducing the partial dislocation to move, i.e., acting on its screw character $b/2$. Thus, this contribution $\tau b/2$, where τ is the maximum resolved shear stress, is responsible for the term $\sqrt{3}$ in Eq. (17a), i.e., for $1.5\sqrt{3} \simeq 2.6$ in the value of α involved in the cross-slip energy as given by Eq. (15a).

Clearly, this contribution is independent of the orientation. Thus, the orientation dependence of α is screened by this constant background, for a mean value of α is about 3, and so the orientation dependent part is only 20%.

Such a compressive force in the primary glide plane, $\tau b/2$, is only a very crude approximation. As a matter of fact, it is the stress gradient which is to be considered, involving a specification of the obstacle holding up the dislocation in the initial glide plane, and of the stress field which is acting. But such a contribution would not be orientation-dependent, so experimental studies on the orientation dependence of τ_{III}, and particularly on reversing the stress, should be of great interest to estimate at what extent τ_{III}, and the cross-slip mechanism, is dependent on it. In this connection, the Diehl measurements, quoted in my paper, would not be reliable (Seeger, private communication). Therefore, more experimental data are needed in the orientation dependence of τ_{III}; it seems to me that it would be the better way to check the τ_{III} theories.

MOTION OF DISLOCATIONS IN BODY-CENTERED CUBIC CRYSTALS

Hideji Suzuki

Department of Physics, Faculty of Science,
University of Tokyo,
Bunkyo-ku, Tokyo, Japan

ABSTRACT

The frictional stress of a screw dislocation in a body-centered cubic crystal is calculated using a simple model. The Peierls stress derived from potential energy of an undissociated screw dislocation is about 0.05 μ. The effect of zero-point vibration on the Peierls Stress is also discussed. In the case of Lithium, the Peierls stress disappears, and in the transition metals, it decreases by a few tens percent due to zero-point vibration.

1 INTRODUCTION

It seems to be established that the Peierls-Nabarro-type frictional stress in a body-centered cubic transition metal is considerably larger than a face-centered cubic metal.[1] Conrad[2] analyzed the strain-rate sensitivity and the temperature dependence of the yield stress extensively. He obtained activation energies and activation volumes which are favorable for the concept of high Peierls stress. According to his result, the Peierls stresses of body-centered cubic metals are around 0.005 μ, where μ is the shear modulus. As shown elsewhere[3] there is no reason to suppose that the Peierls stress of an edge dislocation in a body-centered cubic metal is greatly different from that in a face-centered cubic metal.

An origin of frictional stress of a screw dislocation in a body-centered cubic metal was suggested by Mitchell, Foxall, and Hirsch.[1] A screw dislocation can dissociate into three partials on three or two {112} planes. If such dissociations occur in screw dislocation, either constrictions must be formed or high-energy stacking faults must be created before the dislocations can move.

The extended screw model, however, does not seem to be justified, because the separation between partials is less than the period of atomic arrangement along the stacking-fault plane due to the high stacking fault energy in a body-centered cubic crystal, as discussed elsewhere.[3] Then the atomic arrangement in the core of dislocation cannot be described appropriately in terms of stacking faults. In this paper it is pointed out that the core energy of a screw dislocation must change significantly in a periodic manner during its motion due to the crystal structure itself. It is also pointed out that the change in zero-point energy is quite large compared to the change in potential energy during the motion of dislocation over one atomic distance.

2 ATOMIC ROW MODEL OF A
BODY-CENTERED CUBIC CRYSTAL

To treat the Peierls-Nabarro-type frictional stress of a screw dislocation in a body-centered cubic crystal, we will use an atomic row model of the crystal. A body-centered cubic crystal may be considered as an assembly of atomic rows parallel to the [111] direction. Since we are interested in the problem in which each atom on a [111] atomic row moves simultaneously, we consider

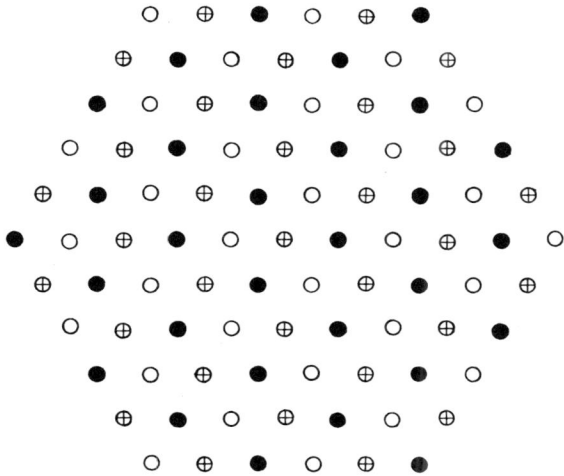

Fig. 1 Arrangement of [111] atomic rows in the body-centered cubic crystal. The paper is parallel to (111) plane.

each [111] atomic row as a rigid one, and also each atomic row has only one degree of freedom of motion parallel to the direction of the atomic row. The energy change of the crystal due to the displacement of atomic rows may be denoted by the summation of the potential energy between atomic rows.

The atomic rows are arranged as shown in Fig. 1, where each is surrounded by the six nearest neighbors at the same distance. If we consider an atom on an atomic row, the second nearest neighboring atoms, as well as the nearest neighboring atoms, of this atom are situated on the atomic row or on the nearest neighbors. Therefore, to calculate the energy change, we consider only the potential energy between the nearest neighboring atomic rows. Let us take the z-axis parallel to [111], denote the z-coordinate of an atom on the ith atomic row by Z_i, and on the jth atomic row by Z_j. Then the interaction potential between ith and jth atomic rows is denoted as a periodic function of $Z_i - Z_j$, namely,

$$\Phi_{ij} = \sum_{n=0}^{\infty} A_n \cos\left(2\pi n \frac{Z_i - Z_j}{b}\right) \qquad n = 0, 1, 2, 3, \cdots \cdots \qquad (1)$$

where A_ns are constants, and b is the distance between neighboring atoms on a [111] atomic row.

For simplicity, neglecting the terms except $n = 1$ in Eq. (1), we have

$$\Phi_{ij} = A \cos\left(2\pi \, \frac{Z_i - Z_j}{b}\right) \tag{2}$$

This potential is the simplest one indicating that the interatomic force becomes repulsive when two atoms approach too close to each other, and attractive if the distance between them increases beyond a certain value. Of course, the absolute value of cohesive energy obtained by Eq. (2) is not correct, but we are interested only in the difference in energy between two different configurations of the crystal. Even if we substitute $Z_i + nb$ for Z_i in either Eqs. (1) or (2), Φ_{ij} keeps the same value for any integer n. Therefore, we may assume

$$0 < Z_i \le b \tag{3}$$

where Z_i/b is the phase of ith atomic row. In a perfect body-centered cubic crystal, the phase of each atomic row has any of three values $1/3$, $1/2$, 1. The phase of each atomic row changes periodically as shown in Fig. 2.

The constant A in Eq. (2) is easily determined by shear modulus in the direction of [111]. The shear modulus of this atomic row crystal is $2\sqrt{3}\pi^2 (A/b^2)$ independent of the shear plane. On the other hand, the shear modulus of a cubic crystal in the direction of [111] is given approximately by

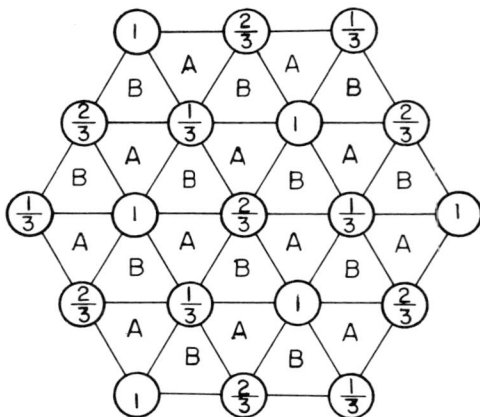

Fig. 2 Two kinds of triangles of neighboring atomic rows. A is the screw triad axis 3_1, and B is 3_2 axis.

$$\mu = \frac{1}{3}(C_{11} - C_{12} + C_{44})$$

and, therefore,

$$A = \frac{b^2(C_{11} - C_{12} + C_{44})}{6\sqrt{3}\,\pi^2} \tag{4}$$

3 ENERGY OF A SCREW DISLOCATION

First, we will illustrate qualitatively that the Peierls stress of a screw dislocation is very large in a body-centered cubic crystal. It is evident that the energy of a screw dislocation becomes high when the center of the screw dislocation gets near to an atomic row. Therefore, a screw dislocation may be assumed to move along the path connecting the center of each triangle of neighboring atomic rows, as shown in Fig. 2. There are two kinds of triangles as shown in this figure. Around A the phase of atomic rows increases, while around B the phase decreases provided that the atomic rows were seen counterclockwise. That is, A is the screw triad axis 3_1, while B is 3_2 axis. Meanwhile, around a screw dislocation, the displacement of atomic rows changes in proportion to the angle around the dislocation, and it is just b after 2π rotation. If the displacement around the screw dislocation increases with increasing angle around the dislocation, the resultant phases of three atomic rows in the core of the dislocation are as follows:

A position	$\frac{1}{3} + \frac{1}{3} \rightarrow \frac{2}{3}$	$\frac{2}{3} + \frac{2}{3} \rightarrow \frac{1}{3}$	$1 + 1 \rightarrow 1$
B position	$\frac{1}{3} + \frac{1}{3} \rightarrow \frac{2}{3}$	$1 + \frac{2}{3} \rightarrow \frac{2}{3}$	$\frac{2}{3} + 1 \rightarrow \frac{2}{3}$

The relative phases between three atomic rows in the core of screw dislocation at A position is exactly the same as those of perfect crystals, namely, the core energy is zero. Meanwhile, in the core of dislocation at B position the phase of each atomic row coincides with the others, namely, the core energy is highest at this position. If the energy of the outside of the core were constant in both positions, the difference of the energy might be about $4.5A \simeq 0.13\,\mu b^2$. Of course, this value is too high. To obtain the energy of dislocation, the sum of potential energy of atomic rows must be taken over a wide region. The calculation was made using

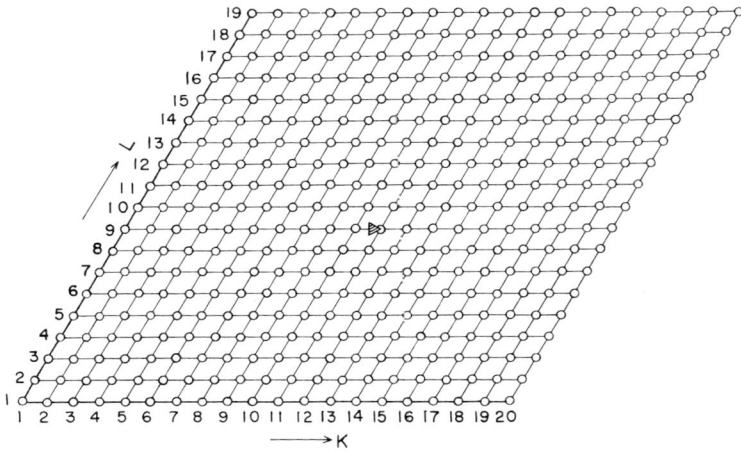

Fig. 3 A model crystal used in the calculation of Peierls potential. The screw dislocation is placed at 15 points in the triangle around the center of the crystal.

a model crystal which is made of $19 \times 20 = 380$ atomic rows, as shown in Fig. 3. The total energy of this crystal is:

$$U = \frac{1}{2} \sum_i \sum_{j(i)}^{6} \Phi_{ij} = \frac{A}{2} \sum_i \sum_{j(i)}^{6} \cos\left(2\pi \frac{Z_i - Z_j}{b}\right) \qquad (6)$$

where $j(i)$ is the nearest neighboring atomic row of ith atomic rows. Equation (6) is written as

$$U = \frac{A}{2} \sum_i \sum_{j(i)}^{6} \left[\cos\left(2\pi \frac{Z_i}{b}\right)\cos\left(2\pi \frac{Z_j}{b}\right) + \sin\left(2\pi \frac{Z_i}{b}\right)\sin\left(2\pi \frac{Z_j}{b}\right)\right] \qquad (7)$$

The force acting on ith atomic row is exerted only by the nearest neighbors and given by

$$f_i = \pi \frac{A}{b} \sum_{j(i)} \left[\cos\left(2\pi \frac{Z_i}{b}\right)\sin\left(2\pi \frac{Z_j}{b}\right) - \sin\left(2\pi \frac{Z_i}{b}\right)\cos\left(2\pi \frac{Z_j}{b}\right)\right] \qquad (8)$$

Putting $\cos(2\pi Z/b) = X$, $\sin(2\pi Z/b) = Y$, and denoting the quantities of the atomic row at the coordinates K and L by $X[K,L]$, $Y[K,L]$, etc., and also

$$G[K,L] = X[K - 1,L] + X[K - 1,L + 1] + X[K,L - 1] + X[K,L + 1]$$
$$+ X[K + 1,L - 1] + X[K + 1,L]$$
$$H[K,L] = Y[K - 1,L] + Y[K - 1,L + 1] + Y[K,L - 1] + Y[K,L + 1] \qquad (9)$$
$$+ Y[K + 1,L - 1] + Y[K + 1,L]$$

Then Eqs. (7) and (8) become

$$U = \frac{A}{2} \sum_L \sum_K (X[K,L] G[K,L] + Y[K,L] H[K,L]) \qquad (10)$$

$$f[K,L] = \frac{\pi A}{b} (X[K,L] H[K,L] - Y[K,L] G[K,L]) \qquad (11)$$

where

$$X^2[K,L] + Y^2[K,L] = 1$$
$$K = 1, 2, \ldots, 20$$
$$L = 1, 2, \ldots, 19$$

When the mechanical equilibrium is established,

$$f[K,L] \equiv 0 \qquad (12)$$

In principle, we can obtain the strain around the dislocation by solving Eqs. (11) and (12) under suitable boundary conditions. But these equations are quadratic of 760 variables of X and Y, and impossible to solve directly. Therefore, we must solve them by successive approximations.

The simplest calculation of the Peierls potential is to use the values of $X[K,L]$, $Y[K,L]$ obtained by isotropic elasticity, instead of the solution of Eq. (12). That is, denoting the coordinate of an atom on the ith atomic row in the perfect crystal by Z_{io}, and using a polar coordinate with the origin at the center of the dislocation, we assume

$$Z_i - Z_{io} = \frac{\theta b}{2\pi} \qquad (13)$$

It must be mentioned that this assumption is equivalent to that of Peierls and Nabarro in the case of a screw dislocation.

Figure 4 shows calculated equipotential energy curves. The filled circles in the figure indicate positions of atomic rows. The

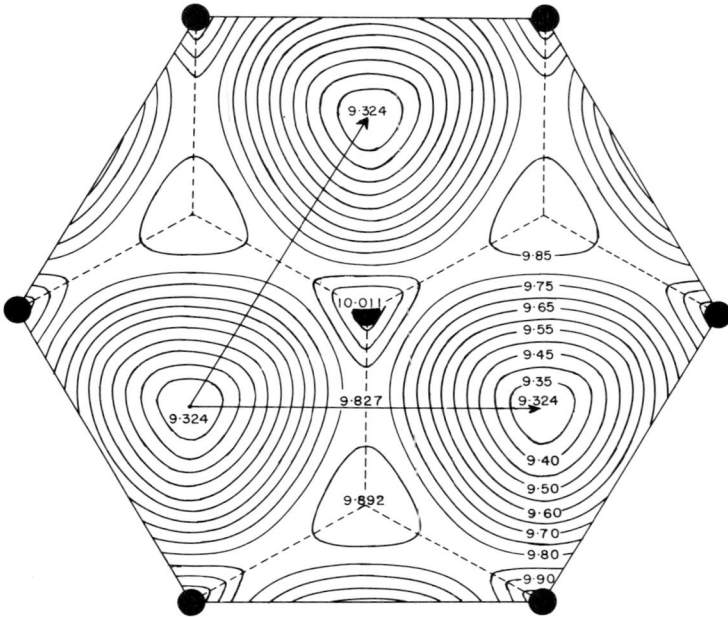

Fig. 4 Potential energy of a screw dislocation in a body-centered cubic crystal. Filled circles indicate the positions of atomic rows. The numerical quantities in the figure indicate energy in the unit of $A = \mu b^2/2\sqrt{3}\,\pi^2$.

potential energy of screw dislocation at the center of each triangle of atomic rows is the maximum or minimum, alternatively, as already noted by preliminary consideration. The screw dislocation has the highest energy at the position of each atomic row. A screw dislocation may move along the lines with arrows, and must overcome the saddle points. The difference between the saddle point and the lowest energy is given by

$$\Delta U = 0.503\,A \qquad A = \frac{\mu b^2}{2\sqrt{3}\,\pi^2}$$

Peierls stress τ_p is then

$$\tau_p = 0.049\,\mu$$

This value is 6–12 times higher than those obtained by Conrad.[2]

Some of the discrepancy is, of course, due to rough approximation used in the above calculation. That is, we have assumed too simple a potential, and employed too simple a method of computation.

To compare the relation between the potential assumed here and the usually employed interatomic potential, the atomic row potential was calculated using Johnson's potential[4] for iron. Figure 5 shows the potential curve J, obtained from Johnson's potential and the curve S, given by Eq. (2). The curvature at the bottom of each potential curve agrees with the others. This is because it gives shear modulus. Curve J, however, differs significantly from Curve S at the top of the hill. Since Peierls stress seems to increase with increasing sharpness of the hill, it may increase if Curve J in Fig. 5 is used in the calculation instead of Curve S. However, actually, the neighboring atomic rows increase the distance between them when their phases are nearly equal to each other, and the potential energy between them decreases.

We have use the solution of elasticity as the displacement of atomic rows. But the solution does not satisfy Eq. (12). The discrepancy is most significant in the core of dislocation and the solution approaches asymptotically to the exact one with the increase of the distance from the center of dislocation.

To obtain a more exact solution a successive approximation was used. Comparing Eqs. (9), (11), and (12), we obtain

$$X[K,L] = \pm\left(1 + \frac{H^2[K,L]}{G^2[K,L]}\right)^{-1/2} \tag{14}$$

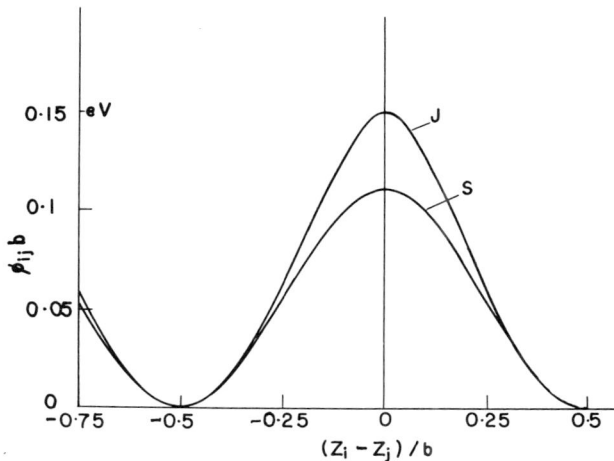

Fig. 5 Comparison of the potential energies between atomic rows calculated from Johnson's potential and that used in this calculation.

$$Y[K,L] = \frac{X[K,L]H[K,L]}{G[K,L]} \qquad (15)$$

At first the elasticity solution was used as starting values of X and Y. The atomic rows at this extreme outside of the model crystal were fixed at the elasticity solution. The improved values of X and Y of an atomic row were determined from the corresponding values of its nearest neighbors using Eq. (14). This procedure was repeated successively for each atomic row, in the order of K and L, using the improved values of X and Y of the nearest neighbors.

Since the change in core energy is very large, the boundary condition of the strain field in the extreme outside atomic rows cannot restrict the position of the core of the dislocation. Thus, for any starting values of X and Y corresponding to various dislocation positions, the dislocation core moves to the lowest position A after repetition of successive approximation. Therefore, we cannot calculate the equipotential curves as shown in Fig. 4 using the successive approximation. The successive approximation was used only for the calculation of the potential energy at the position A. Table 1 shows the change of this energy by repeating the successive approximation. Figure 6 shows the difference between Z values obtained by successive approximation after 100 repetitions and that of elasticity solution.

TABLE 1.   Decrease in Potential Energy of a Dislocation by Repeating Successive Approximation. Unit of Energy is $A = \mu b^2/2\sqrt{3}\pi^2$.

Number of repeat	Energy of model crystal
0	9.3210379
50	9.2224896
70	9.2224652
90	9.2224597
100	9.2224578

4   ZERO POINT ENERGY

The calculations in the previous sections were carried out only for the potential energy. The energy of a dislocation, however, must include energy of zero-point vibration which necessarily exists at absolute zero temperature. Denoting the frequency of the nth harmonic oscillator in the crystal by ν_n, the zero-point energy

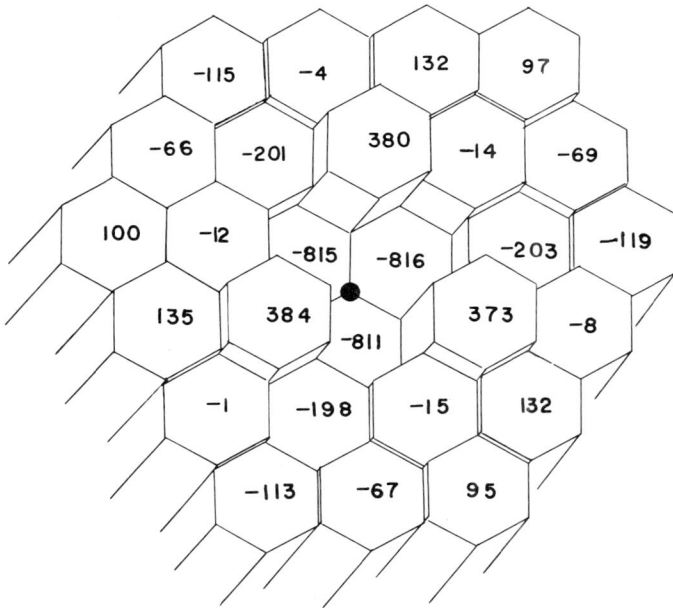

Fig. 6 Differences in displacement of atomic rows around a screw dislocation between elasticity solution and solution of (12). Numerals in the figure indicate those differences in the unit of 0.0001 b.

is given by the relation

$$E_0 = \sum_{n=1}^{N} \frac{1}{2} h \nu_n \tag{16}$$

The summation is taken over all of the harmonic oscillators. It is, however, almost impossible to calculate exactly the frequency spectrum of an imperfect crystal including a dislocation.

Ishioka and Suzuki[5] have discussed the vibration of a dislocation in a valley of Peierls potential using the string model. If we assume that the vibration spectrum denoted in terms of vibration of dislocation line depends on the position of the dislocation, but another vibration spectrum is assumed to be independent of the position of dislocation, we can obtain the approximate value of the difference in zero-point energy between the valley and hill of the Peierls potential, using their calculations.

Consider a part of screw dislocation which lies along the z-axis between $Z = -L/2$ and $Z = L/2$, and vibrates in the valley of

Peierls-Nabarro potential. The dislocation is regarded as a string which has a mass M per unit length, a line tension T, and is subjected to a Peierls potential U. For small vibration U is given by $P\phi^2/2$ where P is a constant and ϕ is the displacement of the part of dislocation from the position of minimum potential energy. The Lagrangian function of the vibrating dislocation is then

$$\mathcal{L} = \int_{-L/2}^{L/2} \frac{1}{2}\left[M\left(\frac{\partial \phi}{\partial t}\right)^2 - P\phi^2 - T\left(\frac{\partial \phi}{\partial z}\right)^2 \right] dz \tag{17}$$

This is simply the Lagrangian for a one-dimensional scalar field. Using the standard quantization procedure, we have the eigenvalue of the vibrating dislocation*

$$H = \sum_n E_n \left(a_n^* a_n + \frac{1}{2} \right) \tag{18}$$

$$E_n = 4\,\hbar c\,(m^2 + k_n{}^2)^{1/2} \tag{19}$$

$$k_n = \frac{n\pi}{L} \qquad (n = \pm 1, \pm 2, \cdots) \tag{20}$$

where a^*, a are a creation and an annihilation operator of the quantum of the vibrating dislocation, c is the velocity of shear wave, and m is the mass of the quantum which is given by $(P/T)^{1/2}$.

From Eq. (16), the zero-point energy of the nth vibrational mode is given by $E_n/2$, and the zero-point energy of the dislocation per unit length at the bottom of the potential energy is then given by

$$E_0 = \frac{4}{L}\sum_{n=1}^{N} \hbar c \left[\left(\frac{n\pi}{L}\right)^2 + \frac{P}{T} \right]^{1/2} \tag{21}$$

Next, we consider the vibration of a dislocation at the top of the Peierls potential. For simplicity, the Peierls potential is assumed to be denoted by a sinusoidal function. Then the potential energy of a dislocation in the vicinity of the top of hill is denoted by $-P\phi^2/2$ using the same constant P as in the bottom of the potential valley.

---

*See Addendum following the Discussion on this paper, p. 700.

It is easily seen that the vibrations of a dislocation with longer wavelength than a certain value are forbidden at the top of the Peierls potential. The vibration is possible only for the values of n satisfying the relation

$$\frac{P}{T} < \left(\frac{n\pi}{L}\right)^2 \tag{22}$$

The vibrations with longer wavelength change into translational motions and the corresponding zero-point energy disappears. The zero-point energy at the bottom of the Peierls potential is, therefore, higher than that at the top of the hill. The difference per unit length is given by the equation

$$\Delta E_0 = \frac{4}{L}\left[\sum_{n=1}^{N} \hbar c \left\{\left(\frac{n\pi}{L}\right)^2 + \frac{P}{T}\right\}^{1/2} - \sum_{n=n_0}^{N} \hbar c \left\{\left(\frac{n\pi}{L}\right)^2 - \frac{P}{T}\right\}^{1/2}\right] \tag{23}$$

where

$$n_0 = \frac{L}{\pi}\sqrt{\frac{P}{T}}$$

$$N = \frac{L}{2b}$$

Putting $n\pi/L = X$, Eq. (23) can be replaced by integral

$$\Delta E_0 = \frac{4\hbar c}{L}\left[\int \sqrt{X^2 + \frac{P}{T}} \, \frac{L}{\pi} \, dX - \int \sqrt{X^2 - \frac{P}{T}} \, \frac{L}{\pi} \, dX\right] \tag{24}$$

$$= \frac{hc}{4b^2} \, g(\alpha) \tag{25}$$

$$g(\alpha) = \frac{1}{\sqrt{\alpha}}(\sqrt{\alpha + 1} - \sqrt{\alpha - 1}) + \frac{1}{\alpha} \log(\sqrt{\alpha} + \sqrt{\alpha + 1})(\sqrt{\alpha} + \sqrt{\alpha - 1}) \tag{26}$$

and

$$\alpha = \frac{\pi^2}{4b^2} \frac{T}{P} \tag{27}$$

Meanwhile, the height of the Peierls potential is

$$\Delta U = \frac{b^2}{18\,\alpha} \tag{28}$$

Then the ratio of ΔE_0 to ΔU is given by the relation

$$\frac{\Delta E_0}{\Delta U} = \frac{9h}{2b^4 \sqrt{\mu\rho}} \, \alpha g(\alpha) \tag{29}$$

where ρ is the density. Tables 2 and 3 show the values of $\alpha g(\alpha)$ and $9h/2b^4\sqrt{\mu\rho}$, respectively. Inserting appropriate values into Eq. (29), we can easily see that the zero-point energy correction is of the order of the potential energy difference in alkali metals, and the correction is of the order of a few tens percent in transition metals. Of course, the above-mentioned calculation is not applicable if the zero-point energy correction ΔE exceeds the height of potential ΔU, but it may be correct to say that the Peierls potential is overcome by zero-point motion.

We must discuss the activation energy for the kink-pair formation in order to compare the calculation with observed Peierls stress. To determine the shape of dislocation at the saddle point, the summation of potential energy and zero-point energy must be minimized because the state of zero-point vibration is the lowest energy state of the dislocation. It is evident that the zero-point energy of dislocation is not a unique function of the position of the

TABLE 2. Values of $\alpha g(\alpha)$

α	$\alpha g(\alpha)$	α	$\alpha g(\alpha)$
10	4.69	500	8.60
20	5.38	1,000	9.29
50	6.30	2,000	9.99
100	6.99	5,000	10.90
200	7.69	10,000	12.29

TABLE 3. Constants Appearing in the Zero-Point Energy Correction

Metal	$\mu[111]$ 10^{11} dyne/cm^2	ρ g/cm^3	b 10^{-8} cm	$\dfrac{9h}{2b^4\sqrt{\mu\rho}}$
Li	0.393	0.53	3.039	0.242
Na	0.261	0.97	3.715	0.098
K	0.117	0.86	4.627	0.065
Fe	7.07	7.87	2.481	0.0334
Mo	13.1	10.2	2.725	0.0148
W	15.1	19.3	2.739	0.0098

dislocation because it depends on the restoring force over a half of the wavelength. The length of dislocation segment passing over the energy barrier at the saddle-point configuration is always longer than a half of the maximum wavelength which is stable at the top of the energy hill under the applied stress. It is evident that zero-point energy of a dislocation segment longer than a half of the maximum stable wavelength is nearly equal to that calculated for a straight dislocation. Thus we may suppose that the Peierls barrier for kink-pair formation is equal to the summation of potential energy and zero-point energy of straight dislocation.

5 DISCUSSION

A simple theory of the Peierls stress for a screw dislocation in a body-centered cubic crystal was given. The high Peierls stress in this crystal was attributed entirely to the geometry of the crystal. The frictional stress for an edge dislocation has not been calculated explicitly, but it seems to have rather low values.

According to Johnston and Gilman[6] the ratio of the number of edge dislocations n_e to that of screw dislocations n_s is related to the ratio of the velocities as follows:

$$\frac{n_e}{n_s} = \frac{v_s}{v_e}$$

where v_s and v_e are velocities of screw and edge dislocations, respectively. It has been reported repeatedly in transmission electron microscopic observations that screw dislocations predominate in deformed body-centered cubic metals and alloys.[7] These observations support the assumption that edge dislocations can move more easily than screw dislocations in a body-centered cubic crystal.

If the mobility of an edge dislocation differs considerably from that of a screw dislocation, then the strain rate is given by the relation approximately

$$\dot{\epsilon} = N v_s b$$

as already discussed by Johnston and Gilman, where N is the total dislocation length per unit volume, v_s is the velocity of the slower dislocation, and b is the magnitude of the Burgers vector. Thus the apparent velocity of dislocations appearing in macroscopic quantity denotes the velocity of the slower dislocations. The Peierls-Nabarro-type frictional force for the screw dislocation, therefore, characterizes the mechanical properties of a body-centered cubic crystal.

Although the effects of solute atoms on the mobility of dislocations have not been discussed in this paper, it is expected that combined effects of Peierls stress and interaction with interstitial solute atoms may explain the complicated characteristics of body-centered cubic alloys.

REFERENCES

1. Mitchell, T. E., R. A. Foxall, and P. B. Hirsch: *Phil. Mag.* 8:1895 (1963).
2. Conrad, H.: "The Relation between the Structure and Mechanical Properties of Metals," p. 476, Her Majesty's Stationery Office, London, 1963.
3. Suzuki, H.: *J. Phys. Soc. Japan* (to be published).
4. Johnson, R. A.: *Phys. Rev.* 134:A 1329 (1964).
5. Ishioka, S., and H. Suzuki: *J. Phys. Soc. Japan* 18: Suppl. II, 93 (1963).
6. Johnston, W. G., and J. J. Gilman: *J. Appl. Phys.* 30:129 (1959).
7. Low, J. R., Jr., and A. M. Turkalo: *Acta Met.* 10:215 (1962).

DISCUSSION *on Paper by H. Suzuki*

A. SEEGER: As I understand your paper, the zero-point energy of a dislocation lying in its Peierls valley will be so large that it exceeds the potential barrier of the Peierls hills and therefore you cannot confine a dislocation line to a valley. This would result in there being no Peierls barrier at all. This is a very attractive idea since we know that if you confine a system to a very small volume, its zero-point energy becomes very large. However, I do not see how you can make this calculation leaving out the zero-point energy of the saddle-point configuration. This seems somewhat arbitrary. I would have thought that it would be better to work out the quantum mechanical partition function in both configurations (as we have done in another context) in both the valley and the saddle-point configurations; this would automatically include the zero-point energy of the saddle point. I would like to see your conclusion result from this type of calculation, rather than be simply an *ad hoc* assumption.

H. SUZUKI: One point we have to consider is the configuration of the dislocation at the saddle point. To determine this configuration we must use the sum of potential energy and zero-point energy instead of potential energy, because the former is the lowest energy. In this paper a rough estimation of the zero-point energy required for this calculation was given. I think that the configuration of dislocation must be determined in this way before carrying out the calculation according to your procedure.

A. SEEGER: Schiller and I discussed this point in Volume IIIA of Mason's series on Physical Acoustics, and we show indeed that the energy barrier is lowered at low temperatures if you take into account zero-point energies. Our calculaions show that these effects are not very large. Certainly they are not as large as you suggest, and I suspect that the truth lies somewhere in between.

F. NABARRO: Is it not true that the body-centered cubic alkali metals also become very hard at low temperatures? One can hardly believe that interstitial impurities will have very large effects in such open structures, which suggests that these metals also have a large Peierls stress. Could it be that the roles of the "valley" and "hill" positions are inverted, the geometrical "valley" having the higher energy?

H. SUZUKI: The high-potential energy positions have low zero-point energy. Therefore, the sum of potential and zero-point energy may sometime become lower at the potential hill than at the potential valley. The highest potential-energy state, however, is not the stable configuration, so I wonder if we could make such a conclusion.

A. SLEESWYK: It would seem that in a calculation of this type it makes quite a difference whether one considers interactions between 1st, 2nd, and 3rd or more neighbors. An objection to the method employed here is that the b.c.c. lattice, as such, is not stable if only 1st neighbor central forces are considered. The number of neighbors to be considered if one wants a b.c.c. lattice that may be stable in principle relative to f.c.c. involves interactions between 1st, 2nd, and 3rd neighbors for the central force model. The way the lattice is stabilized in Professor Suzuki's treatment involves fixing the positions of the $<111>$ close-packed rows. Lattice symmetry would require this to be done for all four $<111>$ orientations, but then one could not even introduce a dislocation in the lattice; in other words, the present treatment does not take into account the lattice symmetry.

H. SUZUKI: The model crystal used in my calculation does not have the same lattice symmetry as a body-centered cubic crystal, but it has the same atomic arrangement as the crystal. A body-centered cubic crystal loses some of its lattice symmetry when it contains a lattice defect, such as a screw dislocation, a stacking fault, or a twin boundary. The lattice symmetry of the model crystal is sufficient to treat the properties of these defects in the crystal. Thus, since we are concerned with the three kinds of defects of the same displacement vector, the lattice symmetry neglected in the model crystal has no effects on the final results, but the calculation becomes extremely simple.

N. BROWN: I would like to make a statement about the multiplicity of deformation mechanisms. About a year and a half ago, Kossowsky and I wrote a paper on microstrain down to 4.2°K, and we found that we could move dislocations at a fairly low stress, compared to the macroscopic yield point. We concluded that the Peierls stress was a very complicated problem. We faced the problem of trying to explain three different stages of yielding. One has to consider the motion of kinks in edge dislocations, the existence of long lengths of screw dislocations, and other phenomena. One phenomenon that probably enters in is that discussed in this colloquium by Hirsch and Sleeswyk involving constriction and cross slip of screw dislocations. We disagree with the suggestion put forth by a number of people, including, for example, Conrad, that a single thermally activated process is controlling dislocation motion through the microstrain region and up into the macrostrain yielding.

H. SUZUKI: I agree to the multiplicity of deformation mechanisms in b.c.c. metals, but we have not really considered this problem in detail.

R. BULLOUGH: In your model, the <111> rows of atoms parallel to the dislocations are rigid rows of atoms which only interact with the atoms in the nearest neighbor parallel <111> rows by a shear law of force. Do you allow any relaxation other than parallel to the dislocation and if not, do you think this could possible explain why your core configuration is symmetrical, whereas our results apparently indicate an asymmetrical configuration.

H. SUZUKI: An asymmetrical configuration may be possible if my model is modified to allow relaxations other than parallel to the dislocations.

A. GRANATO: I wonder if I have misunderstood you. I thought you said that you did consider all the frequencies including the translations at the saddle-point configuration. But Professor Seeger's remarks suggest that you neglected them entirely. Would you clarify this?

H. SUZUKI: I have not considered the zero-point energy of a dislocation at the saddle point (double-kink position) explicitly. I considered translations at the top of the potential hill for a nearly straight dislocation. At the top of the potential hill long wave length vibrations are unstable and turned into independent translational motions.

A. SEEGER: But not the low-frequency vibrations. Physically, there are only two translational modes, one of kinks moving sideways or one if you consider rigid dislocations. At best you should lose two of these modes and perhaps one in Suzuki's model, but he left out quite a few.

A. GRANATO: You have the same number of total modes in the saddle point?

H. SUZUKI: Yes.

A. SEEGER: So how many translational modes did you take out?

H. SUZUKI: The number of translational modes in a nearly straight dislocation at the top of hill per unit length is the inverse of a half of the shortest unstable wavelength (which is slightly longer than the longest stable wavelength). The distance between the kinks at the saddle point is slightly longer than a half of the longest stable wavelength. Therefore, there are only one or two translational degrees of freedom at the saddle point.

A. SEEGER: I would have thought choosing a unit length to be somewhat arbitrary and that the only thing that there is physically is the separation between the kinks, but in this case the treatment boils down to the treatment I was proposing, and this is the type of calculation we have made where we find a logarithmic variation of the activation energy with temperature.

H. SUZUKI: (Written answer to Professor Seeger after the Colloquium.) The most important difference between Professor Seeger and myself is as follows: Seeger determined the configuration of dislocation at the saddle point and of kink only by potential energy, while I thought that the configuration of dislocation is determined by the sum of potential and zero-point energy. Therefore, I think that only a part of zero-point energy is taken into account in the calculation by Seeger and Schiller. If their calculation is modified as follows, the effect of zero-point energy may be included thoroughly: the free energy of dislocation at the saddle point is calculated as a function of configuration of dislocation using their quantum mechanical partition function, then the free energy is minimized by changing the configuration using the calculus of variation. This procedure, however, seems to give almost the same saddle-point free energy or kink energy as those calculated by the usual method using the sum of Peierls potentials and zero-point energy calculated by the procedure in this paper instead of Peierls potential.

P. HIRSCH: (1) Your potential appears to be rather similar to that assumed by Bullough, yet your dislocation seems to have a different structure. Can you comment on this.

(2) Have you calculated the Peierls stress for slip in the "hard" and "soft" directions on (112) planes and for slip on (110) planes?

H. SUZUKI: (1) In my model relaxations in directions other than parallel to the dislocation line were not allowed, while in Bullough's calculation relaxations were allowed in any direction. This difference may give different core configurations. Another possibility is that Bullough has shown the atomic arrangement on three atomic planes which might not pass through the center of dislocation.

(2) I have calculated the potential energy map of a screw dislocation by a crude approximation. The result shows anisotropy of Peierls stress, but we cannot say quantitatively about the anisotropy, because of very crude approximation.

R. BULLOUGH: I just wanted to say that I do not think they will be the same. I have to check to see if my dislocation did move. However, I do think that neglect of relaxation along the rows would be important.

H. CONRAD: I would like to mention one additional piece of experimental evidence which is not in accord with your model in addition to the two that have been mentioned, one by yourself that the Peierls stress that you calculated is higher than the experimental value observed in b.c.c. metals; second that the alkaline metals do show a high Peierls stress similar to the b.c.c. transition metals; third that the edge dislocations as determined by dislocation velocity measurements show a high Peierls stress as well as the screw dislocations as you might deduce from deformation experiments.

H. SUZUKI: I would like to comment on your second and last points: Dr. Li reported in his talk that Peierls stress in potassium is about 0.002μ, where μ is the shear modulus. Meanwhile, Peierls stress in iron is about 0.005μ. My calculation indicates that the decrease of Peierls stress due to zero-point motion is about 40% in potassium and about 20% in iron. Since we have no reliable measurements of Peierls stress in sodium and lithium the calculation in my paper does not contradict the experiments.

Velocities of edge dislocations must somehow be coordinated with screw dislocations, and I wonder how the velocity of one type of dislocation effects the velocities of the other type.

Maybe the curvature at the point where edge converts to screw may hinder the velocity of edge dislocations and the curvature depends upon the velocity of screw dislocations.

D. STEIN: In examination of dislocations produced in b.c.c. metals, one would expect that if only the screw dislocation were responsible for the temperature dependence of the yield stress that the dislocation loops would become much more asymmetrical at low temperatures. This is not observed.

H. SUZUKI: The asymmetry of dislocation loop does not seem to increase very much at low temperatures due to the interactions with obstacles such as impurities and intersecting dislocations. For example, we consider an extreme case in which the frictional stress of edge dislocations can be neglected in comparison with that of screw dislocations. In this case, the work done by the applied stress to overcome the obstacle is given only through the motion of edge component of the dislocation segment between two neighboring obstacles. Denoting the edge component by l_e and the screw component by l_s, the condition to overcome the obstacle is given by

$$\tau_a l_e b \geq F$$

where τ_a is the applied stress, b is the magnitude of the Burgers vector, and F is the maximum stress required to overcome the obstacle. Meanwhile, the density of the obstacle C must satisfy the relation

$$l_e l_s \simeq \frac{1}{C}$$

then we have

$$\frac{l_e}{l_s} = \left(\frac{F}{b\tau_a}\right)^2 C$$

This quantity which may indicate the degree of asymmetry of dislocation loop, does not change very much with temperature.

ADDENDUM

After the colloquium the author was made aware that the numerical factor in the expression of zero-point energy is not correct. If we assume the relation $T = Mc^2$, the numerical factor 4 in Eq. (19) must be removed. Here T is the line tension of dislocation, M the effective mass, and c the shear wave velocity. The zero-point energy correction, however, does not seem to become so small as a quarter of the value in this paper because of the following two reasons.

(1) Ninomiya and Ishioka* discussed the vibration of a dislocation using the continuum approximation and found that both the tension and mass of a dislocation depend on the frequency and wave number. Over a wide range of frequency and wave number these quantities are approximately connected by the relation

$$T \simeq 2Mc^2$$

(2) A screw dislocation can also vibrate in the plane normal to the arrow in Fig. 4. The contribution from this vibration is about forty percent of that from the vibration in the direction of arrow.

Thus Eq. (29) and also values in the last column of Table 3 should be reduced to half. The conclusion of this paper must be modified to some extent, but the effect of the zero-point energy is still significant.

*T. Ninomiya and S. Ishioka: *J. Phys. Soc. Japan*, **23**: 361 (1967).

DISLOCATION DYNAMICS IN THE DIAMOND STRUCTURE

P. Haasen

Institut für Metallphysik,
Universität Göttingen

ABSTRACT

Dislocation dynamics in germanium and other diamond structure crystals is studied in two ways: (1) by measuring dislocation velocities by double etching at various temperatures and stresses; and (2) by analyzing macroscopic creep curves in a model assuming exponential dislocation multiplication and elastic dislocation interaction. The two methods are in quantitative agreement with each other and, to a certain extent, also with first principle dislocation theory. Recent work at Göttingen on these points is reviewed.

1 INTRODUCTION

Crystals with diamond cubic structure (DS), particularly germanium and silicon, and the related III-V compounds with sphalerite structure like InSb are semiconductors with essentially covalent bonding. Mechanically, they are characterized by complete brittleness at room temperature. At higher temperatures they become increasingly ductile until above 2/3 of the melting temperature they are as ductile as f.c.c. metals. Since the Bravais lattice of the DS is f.c.c., the crystal geometry of plastic glide, that is, the dislocation geometry, is identical in both structures (there are, however, no extended dislocations in the DS, see Ref. [1]). DS crystals offer several advantages in the study of dislocation dynamics:

1) dislocations move relatively slowly in the DS;

2) dislocation velocity is clearly governed by the intrinsic friction of the lattice, i.e., the Peierls force;

3) Peierls force calculations are tractable to good approximations in the DS, see Refs. [2, 3]. A recent calculation by Teichler[4] will be described in Sec. 2;

4) high-quality crystals of Ge and Si are available;

5) dislocations can be made visible in the DS particularly easy by etching and by electron transmission. One can ensure by cooling under load to R.T., that these techniques of observation do not disturb the deformation-produced arrangement of dislocations.[5] A comparison of the results of both methods has been performed recently on deformed germanium[6] (see Sec. 3).

Dislocation dynamics in the DS has been studied in Göttingen for the past 10 years basically in two ways, by:

a) microscopic measurements by the Johnston–Gilman technique of the velocity v of individual dislocations piercing a surface; and

b) macroscopic measurements of v via the well-known geometrical relation for the strain rate

$$\dot{\epsilon} = Nbv \qquad (1)$$

where N, the mobile dislocation density, is determined separately (b = Burgers vector).

Recent results obtained by both methods will be reported below. The general state of our knowledge of the deformation of the DS has been reviewed previously.[7,8,9,10]

2 DIRECT MEASUREMENTS OF DISLOCATION VELOCITIES

The double etching method for measuring $v(\tau, T)$ where τ is the applied stress and T is the temperature was first applied to germanium by Chaudhuri et al.[11] Dislocations were introduced by scratching a specimen and subsequent heating. This created a rather dense array of etch pits in several slip systems which was subsequently dispersed by a stress pulse of height τ and duration Δt at a temperature T. A range of stresses τ was created in a single specimen by three-point bending, and the corresponding distances of dislocation movement $\Delta x(\tau)$ were used to calculate $v = \Delta x / \Delta t$ as a function of τ. It must be said, however, that the stress acting on any particular dislocation in such an experiment is rather uncertain. This is partly inherent in inhomogeneous plastic bending, partly due to the interaction between the dislocations either piling up behind a moving dislocation on its own slip plane or interfering with it on intersecting slip planes. Chaudhuri et al. expressed the stress-temperature dependence of v by

$$v = B_0 \tau^m \exp\left(\frac{-U}{kT}\right) \equiv B\tau^m \tag{2}$$

where, for Ge, $m \approx 1.3$ to 1.7 and $U = 1.5$ to 1.7 eV. Kabler[12] repeated these measurements at somewhat lower temperatures also in three-point bending. He did, however, etch away most of the initial dislocations and established the Burgers vectors and line directions of his dislocation half loops. Kabler's results were represented by an inverse exponential dependence of v on τ with the numbers depending somewhat on dislocation character. Such a stress dependence has been taken as support for theoretical models of Celli et al.[13] and Gilman.[14] It does not fit, however, the macroscopic plastic properties of germanium which are rather in line with Eq. (2), though m is found nearer to one.[7]

In view of the above discrepancies, the problem has been taken up again by Schäfer[15] who avoided a number of sources of errors and came to much better-founded results than previous investigators. Schäfer scratched a germanium compression specimen *perpendicular to its expected main slip plane*. On heating, only half loops on this one set of slip planes appear (Fig. 1). Most of these loops are etched away except the outermost ones which are subsequently

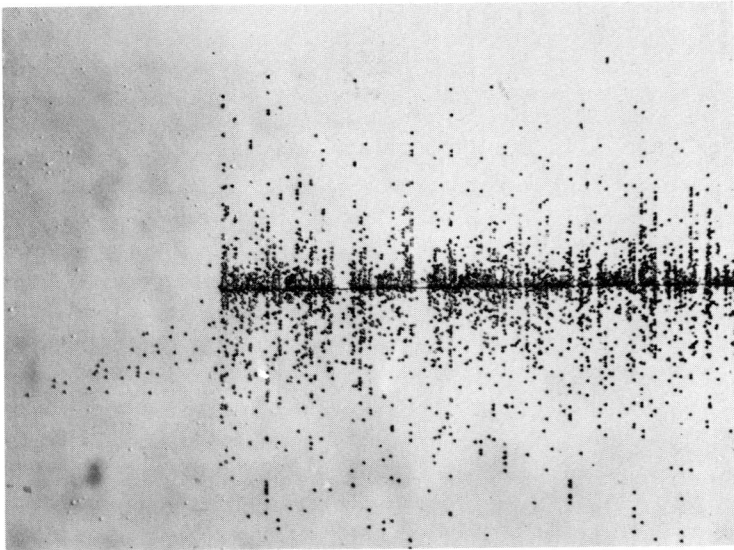

Fig. 1 Dislocation etch pits after scratching a (1$\bar{1}$1) plane of Ge in [121] direction and heating to 440°C under a stress τ = 2 kg/mm^2, magn. 160× [15].

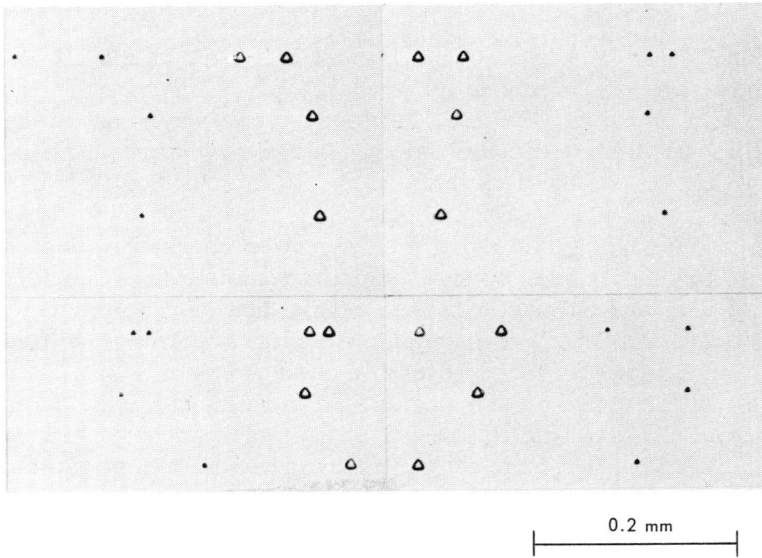

0.2 mm

Fig. 2 Dislocation half loops before (large pits) and after (small pits) deformation at τ = 1.5 kg/mm^2, T = 440°C. Right side = screws, left side = 60° dislocations [15], magn. 130×.

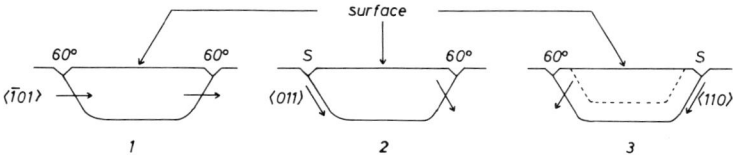

Fig. 3 Dislocation half loops in (1̄11) slip plane with 3 possible Burgers vectors.

moved by a homogeneous shear stress applied to the crystal in uniaxial compression. Schäfer noticed that the loops almost always expanded asymmetrically (Fig. 2). This means that of the three possible Burgers vectors (Fig. 3) only that of case 3 is realized. (The outer stress component in case 2 is very small.) Thus the velocities of screw and 60° dislocations are measured separately as the semihexagonal shape of the loops was ascertained by repeated polishing and etching. Interference microscopy also showed the two etch pits of a half loop to be of different depth (Fig. 4). Another important observation of Schäfer's shows that the line tension permits the half loops to shrink during the heating and cooling periods of the test. Unless this line tension is compensated by a stabilizing stress it will result in erroneous velocity data, particularly for small loops (Fig. 5). Schäfer could show that v measured on unstabilized loops appeared to depend much stronger on τ than v for stabilized loops. We think now that Kabler's[12] results are strongly influenced by this artifact. Chaudhuri et al.'s[11] were probably less affected since they did not polish away the innermost loops and these did somewhat stabilize the outer ones.

Considering all of the above precautions and measuring the velocities of 150 to 1000 loops for each combination (τ, T) in the range

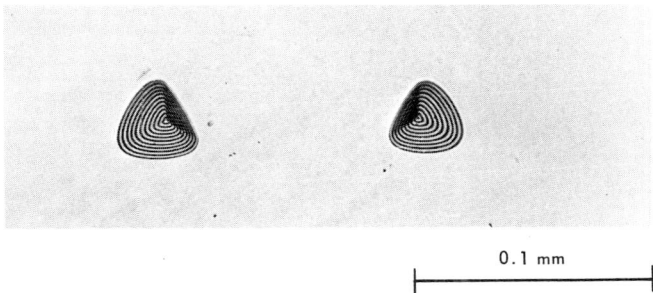

0.1 mm

Fig. 4 Interference photograph of dislocation loop pits ($\lambda = 0.54$ μm); magn. 350×[15].

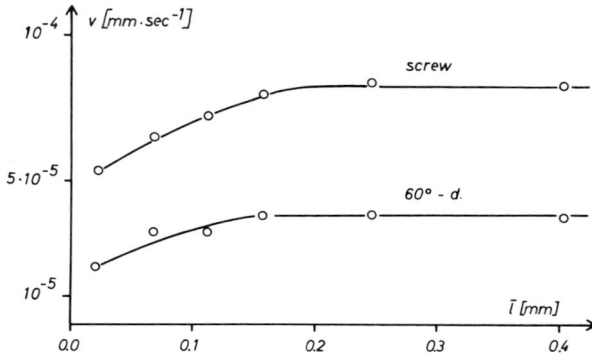

Fig. 5 Dislocation velocity vs. loop diameter for unstabilized loops. $\tau = 1.5$ kg/mm^2, $T = 440\,°C$ [15].

220 to 580°C, 0.1 to 15 kg/mm^2, Schäfer confirms relation (2) as is shown in Figs. 6 and 7. He finds the same activation energy for the velocity of screws and 60° dislocations $U = (1.62 \pm 0.1)$ eV independent of stress, but slightly different stress exponents $m_{\text{screw}} = 1.0 \pm 0.1$; $m_{60°} = 1.2 \pm 0.1$ which are definitely lower than Chaudhuri et al.'s and are in fact in very good agreement with those obtained from macroscopic experiments (see Sec. 3). Schäfer also

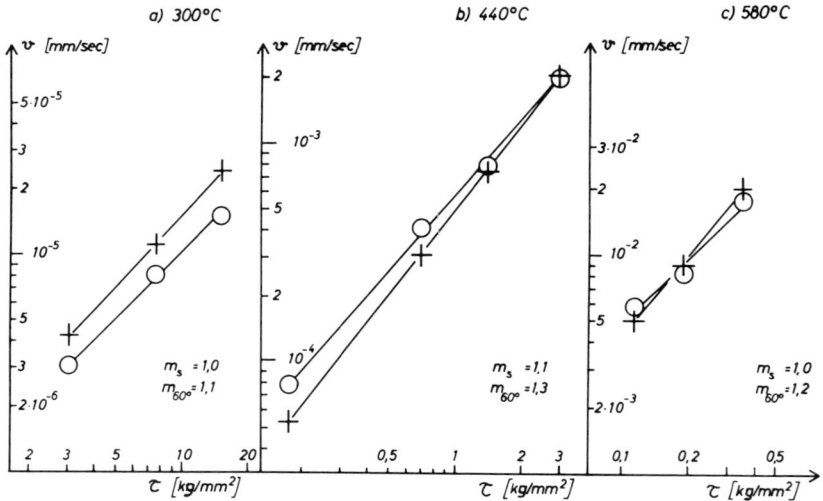

Fig. 6 Dislocation velocity vs. stress at various temperatures. ○ = screws, + = 60° dislocations [15].

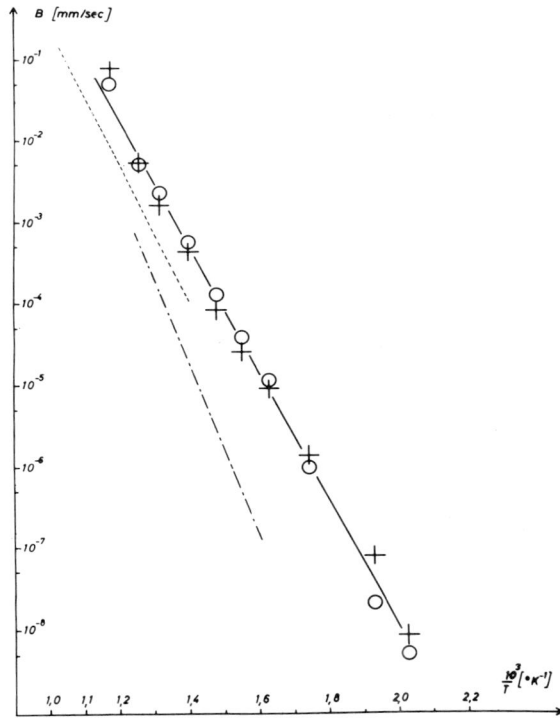

Fig. 7 Dislocation mobility $v/\tau^m = B$ vs. reciprocal temperature for screws (○) and 60° dislocations (+). (—— According to [15], ---- according to [11], ·—·— according to [12])

measures $B_0 = (1 \pm 0.5)\ 10^7 \text{cm/sec}\!/\!(\text{kg/mm}^2)^{1.1}$. Since $m_{60°} > m_{\text{screw}}$ the ratio of the velocities of the two dislocation types may be larger or smaller than one (see Fig. 6).

A velocity-stress exponent of essentially *one*, that is, viscous dislocation movement in Ge (and according to macroscopic experiments [16,17] also in Si and InSb[21]) is plausible but still unexplained by any rigorous theory. The activation energy U can be identified with the calculated formation energy of a double kink in the Peierls potential.[3,4] Labusch[3] calculates the structure of a screw dislocation in Ge and Si by Peierls' variational method as worked out by Dietze.[18] The atomic interaction over the slip plane is determined by a force constant which in turn is obtained from the known dispersion spectrum of lattice vibrations. Labusch obtains for the Peierls potential of Si 0.266 eV, Ge 0.225 eV, for the Peierls "forces" 275 and 212 kg/mm², for the double-kink formation energy 2.14 and 1.75 eV, respectively. The latter value agrees within the scatter with Schäfer's activation energy of the dislocation velocity in Ge. Celli's[2] Peierls potential for a screw dislocation in Ge, calculated for a harmonic interaction model crystal, is more than a factor of two smaller than Labusch's. Teichler[4] uses Bennemann's Green's functions method of treating the covalent bond to calculate the structure and energy of dislocations in the DS. Interaction between nearest neighbors all over the dislocated crystal is expressed in terms of the known pseudopotential of Ge. Teichler obtains nearly the same Peierls potential as Labusch which is somewhat surprising considering the different approximations involved. On the other hand, the identification of the velocity determining rate process with kink formation appears justified. However, it is not possible to fit the velocity data to an activation energy which is inversely proportional to stress as is implied by the theories of Gilman[14] and Celli et al.[13]

3   DISLOCATION VELOCITIES FROM MACROSCOPIC EXPERIMENTS

A quantitative theory of the initial stage of plastic deformation of DS crystals was developed some years ago by the present author (see Refs. [7–10]. It is based on Johnston and Gilman's idea of the multiplication yield point[19] and on the concept of an effective stress τ_{eff} acting on a dislocation in a uniform dislocation array of density N, that is,

$$\tau_{eff} = \tau - \alpha G b \sqrt{N}, \qquad \alpha \approx 0.2 \tag{3}$$

The dislocation velocity in the presence of N other dislocations is then written with Eqs. (2) and (3) as

$$v = B(\tau - A\sqrt{N})^m \tag{3a}$$

More accurately, but without change in the following conclusions, one should calculate the mean dislocation velocity $\bar{v}$ in a sinusoidal internal stress field of amplitude $A\sqrt{N}$. The result is for $1 \leq m \leq 2$ (Haasen and Labusch, see Ref. [10])

$$\bar{v} = B \frac{(\tau^2 - A^2 N)^{m-1/2}}{\tau^{m-1}} \tag{3b}$$

As observed by electron transmission microscopy,[5] dislocations in the DS become immobilized by forming dipoles which interact more weakly than monopoles according to Eq. (3). Therefore it is sensible to identify N in Eq. (3) with the *mobile* dislocation density in Eq. (1). With this assumption, which appears to be justified to a good approximation[11,20] for small strains, i.e., in the yield point and incubation creep range, the lower yield stress τ_{ly} and the stationary creep rate $\dot{\epsilon}_s$ of DS crystals can be expressed straightforwardly[7-10] in terms of the velocity and interaction determining parameters as

$$\tau_{ly} = \left(\frac{A^2 C_m}{bB_0}\right)^{1/(2 + m)} \dot{\epsilon}^{1/(2 + m)} \exp\left[\frac{U}{(2 + m) kT}\right] \qquad (4a)$$

$$\dot{\epsilon}_s = \left(\frac{bB_0}{A^2 C_m}\right) \tau^{(m + 2)} \exp\left[-\frac{U}{kT}\right] \qquad (4b)$$

Both expressions follow from an equation of state and are stationary values reached for a certain critical dislocation density N_s as a compromise between multiplication and interaction of dislocations. In the case of Eq. (4a) $\dot{\epsilon}$ and T are kept constant experimentally; for Eq. (4b) τ and T are kept constant, $A = \alpha Gb$, $C_m = [1 + (2/m)]^{[m + 2]}(m/2)^2$ $(= 6.75$ for $m = 1)$. It is obvious from Eq. (4) that by measurement of τ_{ly} and $\dot{\epsilon}_s$ as a function of the experimental variables τ, $\dot{\epsilon}$, T the theoretical parameters (A^2/B_0), m and U may be determined. Before reviewing such data on Ge we would like to report on a direct experimental confirmation of the important Eq. (3) which yields the interaction parameter A independently.

Berner and Alexander[20] have counted dislocation etch pits* as a function of creep strain under various stresses on germanium at 582°C. The creep curve has S-shape (Fig. 8) and displays an initial multiplication and a final work-hardening stage superimposed in an intermediate stationary stage. According to Eq. (3), $N_\infty = \tau^2/A^2$ should be the final dislocation density reached for vanishing $\dot{\epsilon}$ at any τ. Figure 9 shows the experimental results[20] in a double log plot for (a) local areas of maximum dislocation density, and (b) the average N over the specimen. The results strictly confirm Eq. (3) with $A_{\text{local}} = 6.6 \cdot 10^{-3}$ kg/cm, i.e., $\alpha = 0.29$. The authors also

*Springer[6] has recently compared dislocation lengths per cm³ N as measured by electron transmission with etch-pit densities N_p counted on the surface of Ge crystals deformed in single slip. Theoretically, one expects $N \approx 1.5 N_p$. Springer finds, however, in the range $10^7 \leq N_p \leq 10^8$ cm^{-2} and for the CP4 etch $N \approx 9 N_p$ independent of N_p. Although these results need confirmation and extension to a larger range of N_p they could indicate that dipoles are not etched which are the predominant dislocation arrangement in deformed Ge crystals.[5] Further experiments along these lines are underway.

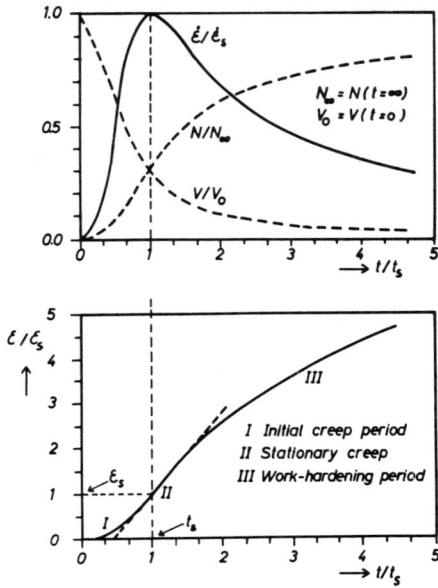

Fig. 8 Calculated creep curve normalized to the inflection point coordinates and dependence of dislocation velocity and density on creep time[21].

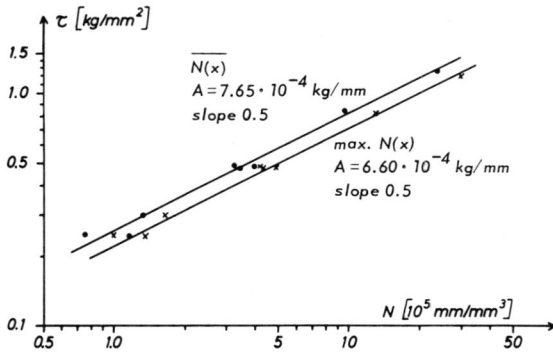

Fig. 9 Limiting dislocation density vs. stress, average (·) and maximum (×) values.[20]

investigated dislocation multiplication leading from the grown-in dislocation density N_0 up to N_∞. The distribution of strain and of dislocations over the length of a compression specimen was found to be rather inhomogeneous, particularly for Ge crystals with grown-in dislocation densities $N_0 < 3000$ cm^{-2} (Fig. 10). Therefore a correlation was established between local strain and local etch-pit densities for various stresses. The results for $\tau = 0.5$ kg/mm^2 and $T = 582°$C are shown in Fig. 11 together with a theoretical curve derived from the following assumptions:

a) dislocations multiply exponentially as first proposed by Gilman

$$dN = N \cdot v dt \cdot \delta = d\epsilon \frac{\delta}{b} \tag{5}$$

b) the mean free path between multiplications (δ^{-1}) depends on stress as proposed by Peissker et al.[21]

$$\delta^{-1} = K\tau_{\text{eff}} \tag{6}$$

c) stress in Eq. (6) means effective stress according to Eq. (3).

For $\tau = $ const, $m = 1$, Eqs. (5), (6), and (3) integrate for the initial condition $N = N_0$ for $\epsilon = 0$ to

$$\epsilon = \frac{2b\tau}{KA^2} \left[\frac{A\sqrt{N_0} - A\sqrt{N}}{\tau} + \ln\left(\frac{1 - A\sqrt{N_0}/\tau}{1 - A\sqrt{N}/\tau}\right) \right] \tag{7a}$$

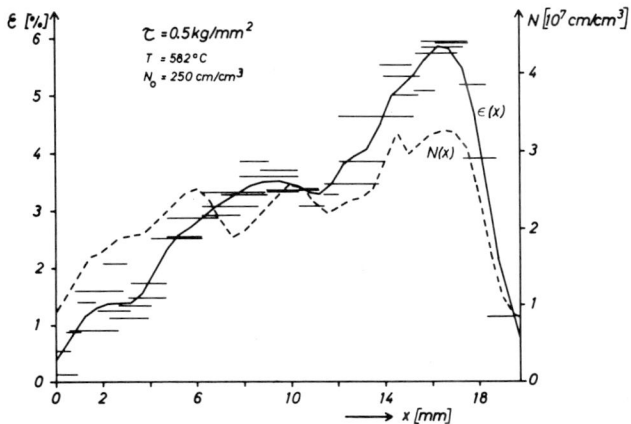

Fig. 10 Local shear strain and local dislocation density along compressed crystal[20].

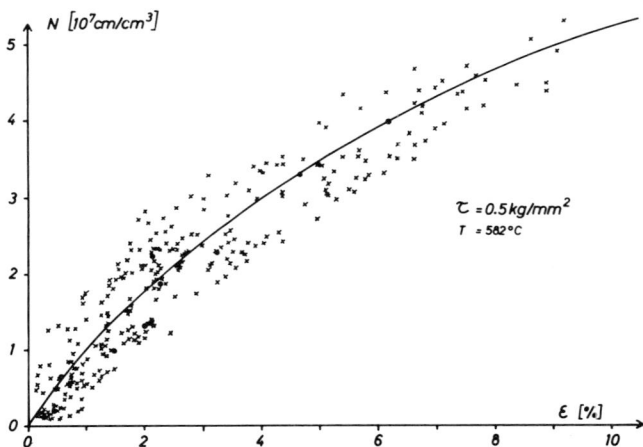

Fig. 11 Local ($\times$) and average ($\cdot$) dislocation density vs. strain.[20]

For $N \gg N_0$, $A\sqrt{N_0} \ll \tau$ this can be simplified to

$$\epsilon = \frac{2b\tau}{KA^2}\left[-\frac{A\sqrt{N}}{\tau} - \ln\left(1 - \frac{A\sqrt{N}}{\tau}\right)\right] \tag{7b}$$

which is the theoretical curve drawn in Fig. 11. One notices that, contrary to the model of dislocation multiplication by a fixed density M of Frank-Read sources (of length l_0), the relation between ϵ and N depends explicitly on stress. This τ-dependence is supported by experiment as Fig. 12 shows (Berner and Alexander[20]). The Frank-Read model gives, on the other hand,

$$N = N_0 + \left(\frac{9M}{l_0 b^2}\right)^{1/3} \epsilon^{2/3} \tag{8}$$

which is also a nonlinear relation as is the exponential multiplication law, Eq. (7b), although Eq. (8) is τ-independent without further assumptions.

Comparing Eq. (7b) to the experiments of Berner and Alexander[20] on Ge at 582°C, one obtains $A = (5.5 \pm 0.3)\,10^{-3}$ kg/cm, $K = (10.5 \pm 0.8)$ mm/kg. The latter figure means that a dislocation travels a distance $\delta^{-1} = 0.2$ mm in between multiplications. It should be noticed that former but less accurate experimental determinations[22,23] of the law of dislocation multiplication at constant τ yield

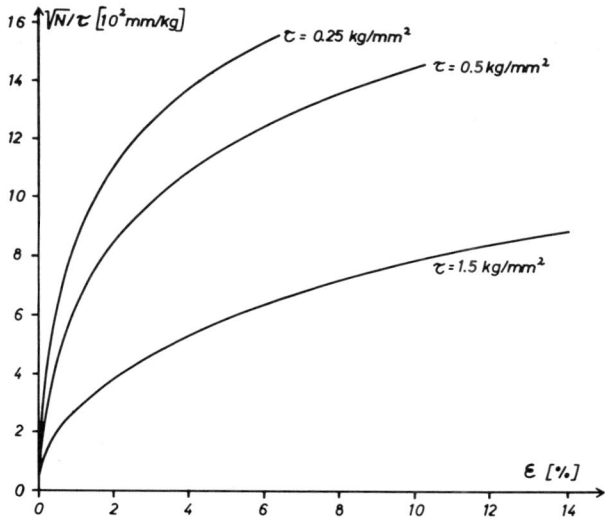

Fig. 12 Dislocation density vs. strain at various stresses[20].

$N \sim \epsilon^{3/2}$, that is, a curvature opposite to that shown in Figs. 11 and 12. In these figures the inflection point of the creep curve does not show up in any particular way. The dislocation density at this point N_s can easily be obtained by differentiation of $\dot{\epsilon}$ with respect to N as $d\dot{\epsilon}/d\epsilon = (d\dot{\epsilon}/dN)(dN/d\epsilon) = 0$, Eqs. (1), (2), and (3). One obtains[7]

$$N_s = N(t_s) = \left[\left(\frac{2}{m+2}\right)\left(\frac{\tau}{A}\right)\right]^2 \tag{9}$$

which is plotted in Fig. 13. This relation was also shown to hold for silicon.[16] Putting $m = 1.1$ the data in Fig. 13 result in $A = 8.6 \cdot 10^{-3}$ kg/cm in good agreement with the above evaluation of the limiting dislocation density. Inserting Eq. (9) into Eq. (7b) one obtains further the strain at the inflection point

$$\epsilon_S = 0.38 \frac{b\tau}{KA^2} \qquad \text{for} \qquad m = 1.1 \tag{10}$$

Experimental results on Ge[20] to check Eq. (10) are plotted in Fig. 14. They yield $KA^2 = 7.5 \cdot 10^{-5}$ kg/cm again in good agreement with the above-determined values of K and A. Peissker et al.[21] and Reppich et al.[16] have shown for InSb and Si, respectively, that ϵ_s is indeed independent of temperature as well as linear in stress.

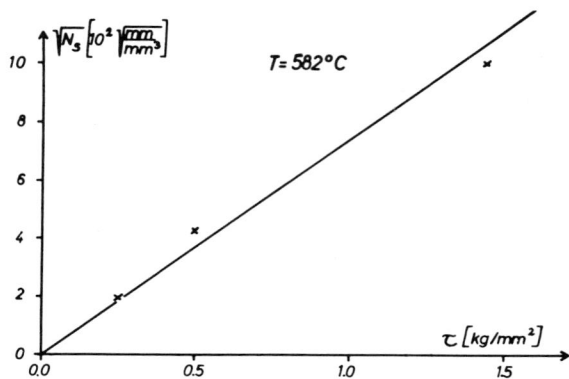

Fig. 13 Dislocation density at inflection point of creep curve vs. stress[20].

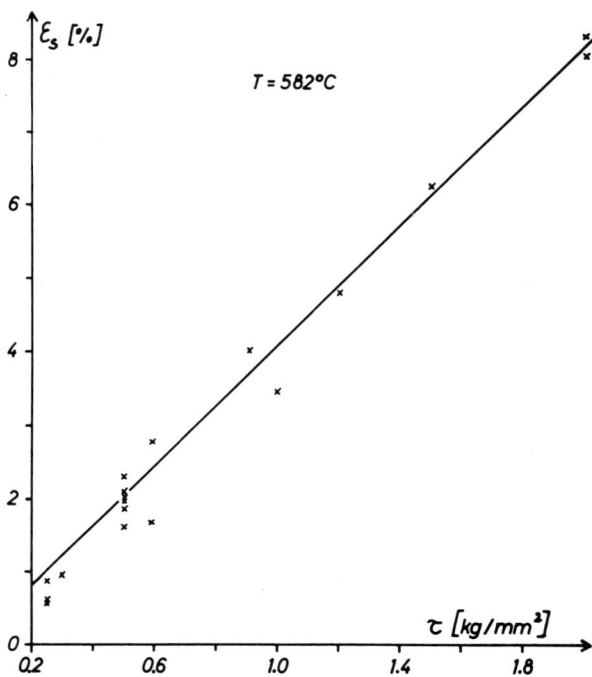

Fig. 14 Creep strain to inflection point vs. stress[20].

Finally, the creep time to the inflection point t_s is obtained from Eq. (5) together with Eqs. (6), (3), and (9),

$$t_s = \frac{\exp(U/kT)}{B_0 K \tau^{(m+1)}} \left(2.7 + \ln \frac{\tau}{A\sqrt{N_0}}\right) \qquad (11)$$

Its stress dependence is characteristically different from that of $\dot{\epsilon}_s$ and ϵ_s; it is plotted in Fig. 15 according to unpublished results of K. Berner on Ge. One obtains $(m + 1) = 2.04$ in good agreement with the results of direct velocity measurement (Sec. 2, above). Also, the temperature dependence of t_s as well as its absolute value fit independent measurements of U, B_0, and K as shown by Hewing[26] for Ge, Reppich et al.[16] for Si, and Peissker et al.[21] for InSb. The same is true for the stationary creep velocity [Eq. (4b)], or the corresponding lower yield stress [Eq. (4a)]. Figure 16 shows results for Ge at 580°C: $\dot{\epsilon}_s(\tau)$ and $\tau_{ly}(\dot{\epsilon})$ which yield $m = 1.12 \pm 0.05$. One recognizes the validity of an equation of state for the range of stationary deformation as dynamic and static results agree in this plot. They also have the same temperature dependence which for the stationary creep rate $\dot{\epsilon}_s$ in compression is shown in Fig. 17. The activation energy obtained from this plot $U = (1.75 \pm 0.25)$ eV is to be compared directly with the activation energy of the dislocation velocity in Ge [Eq. (4b)]. The activation energy of the lower yield stress of Ge (0.53 ± 0.02) eV obtained by Schäfer et al.[24] according to Eq. (4a) is to be identified with $U/(2 + m)$, where U is the activation energy of dislocation velocity. With $m = 1.1$ obtained from the strain-rate dependence of τ_{ly} one

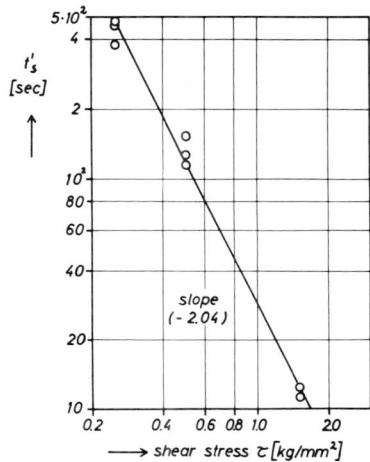

Fig. 15 Creep time t'_s to inflection point vs. stress, logarithmic scales. $t'_s = t_s/\ln(2.7 + \tau/A\sqrt{N_0})$, N_0 = grown-in dislocation density $\approx$ 1000 cm^{-2}, according to measurements of [20].

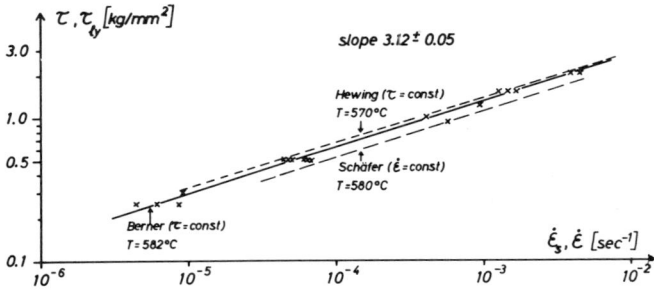

Fig. 16 Stationary creep rate $\dot{\epsilon}_s$ and lower yield stress τ_{ly} vs. stress and strain rate, respectively[20]. Included (– – –) are data of Hewing[26] and Schäfer et al.[24]

obtains $U = (3.1)(0.53)$ eV $= 1.64$ eV in very good agreement with direct determinations. The same is true for the absolute values of τ_{ly} and $\dot{\epsilon}_s$ making use of the constants A and B_0 determined by independent experiments.

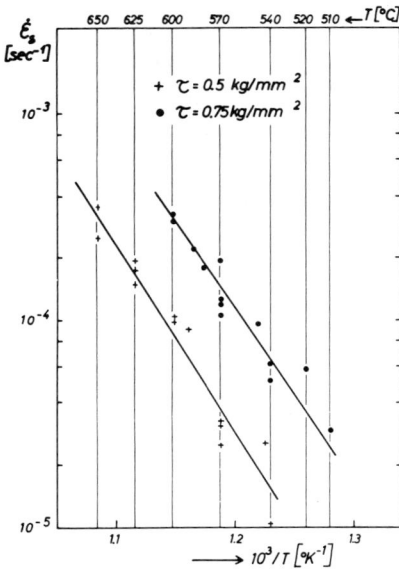

Fig. 17 Stationary creep rate vs. reciprocal temperature at two stresses in compression[26].

4 CONCLUSIONS

It can therefore be concluded from the creep experiments on germanium described above as well as from the dynamic experiments[24] that by means of a simple theory, details of the dislocation dynamics can be extracted from macroscopic experiments. The accuracy of microscopic and macroscopic determinations of the dislocation velocity is comparable in favorable cases, that is, for the stationary states of deformation which are rather independent of grown-in dislocation density above $N_0 \approx 3000$ cm^{-2}. It is interesting to note that for Ge, Si, and InSb, the velocity stress exponent is close to one, that is, dislocations move viscously, a fact which is not well understood at present. The activation energy of this motion, on the other hand, can safely be identified with that of the formation of double kinks in the Peierls potential. Dislocations appear to multiply continuously as they move [Eq. (5)]. A dipole crossing mechanism has been proposed[21,7] for this which accounts for the stress dependence of multiplication [Eq. (6)].

The interaction of dislocations in the DS is described best by a temperature-independent, long-range elastic stress [Eq. (3)], as originally proposed by Taylor.[25] DS crystals thus appear well suited as a model substance for the microdynamical approach to crystal plasticity.

ACKNOWLEDGMENT

The author wishes to thank his co-workers in Göttingen, particularly, Dr. H. Alexander, Dr. S. Schäfer, Dr. W. Schröter, and Dipl.-Phys. K. Berner, J. Hewing, and E. Springer for their contributions and help.

REFERENCES

1. Haasen, P., and A. Seeger: *Halbleiterprobleme* 4:68 (1958) [Vieweg, Braunschweig].
2. Celli, V.: *J. Phys. Chem. Sol.* 9:1 (1959).
3. Labusch, R.: *Phys. Stat. Sol.* 10:645 (1965).
4. Teichler, H.:*Phys. Stat. Sol.* 23:341 (1967).
5. Alexander, H., and P. Haasen: *Canadian J. Phys.* 45: 1209 (1967).
6. Springer, E.: Thesis, Göttingen, 1967.
7. Haasen, P.: *Festkörperprobleme* 3:167 (1964) [Vieweg, Braunschweig].
8. Haasen, P.: *Disc. Faraday Soc.* 38:191 (1965).
9. Haasen, P.: *J. de Phys.* 27:C 30 (1966).
10. Haasen, P., and H. Alexander: "Sol. State Physics," F. Seitz and D. Turnbull (eds.) [in press].

11. Chaudhuri, A. R., J. R. Patel, and L. G. Rubin: *J. Appl. Phys.* **33**: 2736 (1962).
12. Kabler, M. N.: *Phys. Rev.* 131:54 (1963).
13. Celli, V., M. Kabler, T. Ninomiya, and R. Thomson: *Phys. Rev.* 131:58 (1963).
14. Gilman, J. J.: *J. Appl. Phys.* 36:3195 (1965).
15. Schäfer, S.: *Phys. Stat. Sol.* 19: 297 (1967).
16. Reppich, B., P. Haasen, and B. Ilschner: *Phys. Stat. Sol.* 5:247 (1964).
17. Siethoff, H., and P. Haasen: *Tokyo Conf. Defects in Semicond.* (1966) [to be published].
18. Dietze, H. D.: *Z. Phys.* 131:113, 156 (1951).
19. Johnston, W. G., and J. J. Gilman: *J. Appl. Phys.* 30:129 (1959).
20. Berner, K., and H. Alexander: *Acta Met.* 15:933 (1967).
21. Peissker, E., P. Haasen, and H. Alexander: *Phil. Mag.* 7:1279 (1962).
22. Patel, J. R., and B. H. Alexander: *Acta Met.* 4:385 (1956).
23. Dew-Hughes, D.: *IBM J.* 5:279 (1961).
24. Schäfer, S., H. Alexander, and P. Haasen: *Phys. Stat. Sol.* 5:247 (1963).
25. Taylor, G. I.: *Proc. Roy. Soc.* A145:362 (1934).
26. Hewing, J.: Thesis, Göttingen, 1963.

DISCUSSION *on Paper by P. Haasen*

A. SEEGER: For the two types of dislocations investigated, 0° and 60° dislocations, the same activation energy was found. This would, in turn, mean that the Peierls barrier is not the same because the line tensions would be expected to vary by a factor of 2 or 3. Has any theoretical work been done in your laboratory to determine whether the Peierls energy is orientation-dependent?

P. HAASEN: We have not been able to do so because of mathematical difficulties. Both the Labusch and Teichler calculations become very involved if the 60° dislocation is considered. However, I agree that there must be a difference in Peierls potential between the two types of dislocations according to our experiments.

T. SUZUKI: In the calculation of the formation of double kinks, what was the width of the kink? Was it smooth or sharp? According to our experiment on Si, it was concluded to be very sharp, the half-width of the double kink being equal to (1.5 ± 0.3) *b*.

P. HAASEN: The kink width calculated by Labusch is 0.4 *b*. Thus, the kink is indeed sharp. Also, the kink potential came out very small, ~ 0.04 eV. A relaxation peak with such an activation energy was recently observed by Kaipnazarov et al. (*Phys. Stat. Sol.*, 21:805, 811 (1967) working at some hundred cps and 100°K on silicon and germanium. The peak increased with deformation and decreased with γ-irradiation although its nonlinear characteristics indicating kink motion have not yet been investigated.

G. ALEFELD: Because of the varying internal stress field there will also be regions where the internal stress will aid the nucleation of kinks and one should add the magnitude of the internal stress field to the applied stress. However, perhaps instead of using the peak value of the internal stress, one should use Schoeck's idea, incorporating some mean value of the internal stress.

P. HAASEN: We have considered this in the calculation of the average dislocation velocity which incorporated the internal stress field. We found the time it takes a dislocation to move a certain distance in a periodic field of internal stress.

G. ALEFELD: Yes, but the question is, what does one put in for the maximum height of the periodic internal stresses. If one takes the attitude that the double kink can be generated every-where along the line, they will be generated preferentially at those particular spots of the dislocation line where the forward stress is *the highest*. If there are no forward stresses the kinks will be generated where the back stresses are *the smallest*. That means that kinks never must be generated in regions where large back stresses exist. So the large back stresses will not appear in the averaging procedure whereas the largest forward stresses do. I think that for the amplitude of the periodic internal stresses one should not put in the maximum internal stresses but only some fraction of it.

P. HAASEN: One can take care of this by integrating

$$\bar{v} = \frac{L}{\int_0^L \partial x / v [\tau - \tau_i(x)]}$$

Then one automatically takes into account both positive and negative internal stresses $\tau_i(x)$, and produces the average velocity $\bar{v}$ over a distance L (τ = applied stress).

N. BROWN: In the generation of a double kink, the activation energy would be stress-dependent and therefore a stress term would appear in the exponential. However, you do not have a stress term in your exponential. Therefore, why do you talk about double-kink generation in light of your experimental results?

P. HAASEN: The activation energy measured this way is not de-pendent upon stress. Certainly one would expect the energy for the formation of double kinks to be stress-dependent with a small activation volume. Now, if you consider both forward and back-ward jumps, the stress term is removed from the exponential. Possibly a square root of the stress before the exponential is the first term of a series expansion which one would expect at low stresses. We have no theory for the linear variation of velocity with stress at the moment.

P. B. HIRSCH: Assuming that the interaction energy between the kinks is*

$$\sim \frac{Gb^2}{8\pi} \frac{(1 + \nu)}{(1 - \nu)} \frac{a^2}{l}$$

the critical length of the double kink is

$$l_c = \left[\frac{Gb}{8\pi} \frac{(1 + \nu)}{(1 - \nu)} \frac{a}{\tau} \right]^{1/2}$$

in an obvious notation (*a* is the distance between neighboring valleys), and the activation energy is

$$U = U_{DK} - \left[\frac{G}{2\pi} \frac{(1 + \nu)}{(1 - \nu)} a^3 b^3 \tau \right]^{1/2}$$

where U_{DK} is the double-kink energy. The dislocation velocity v taking the first term of the series expansion

$$v = 2\nu_{eff} \frac{L}{l_c} \left[\frac{G}{2\pi} \frac{(1 + \nu)}{(1 - \nu)} a^3 b^3 \tau \right]^{1/2} \frac{\exp(-U_{DK}/kT)}{kT}$$

$$v = 4\nu_{eff} \frac{Lab\tau}{kT} \exp\left(\frac{-U_{DK}}{kT} \right)$$

where ν_{eff} is the effective frequency, and L is the free length of dislocation. Note that the velocity depends on L which itself may vary and depend on the velocity of other parts of the loop. Assuming L to be reasonably constant $v \sim \tau$, as observed; provided $\nu_{eff} \sim \nu_0 (b/l_c)$, where ν_0 is the Debye frequency $v \sim \tau^{3/2}$.

P. HAASEN: I agree with you that it is possible to explain the proportionality of v and τ by a product of the kink-interaction energy (as the first term in the kink-energy expansion) and the number of sites being proportional to the inverse of the critical kink spacing.

*Kroupa, F., L. M. Brown: *Phil. Mag.* **6**: 1267 (1961).
Seeger, A., P. Schiller: *Acta Met.* **10**: 348 (1962).

P. B. HIRSCH: Is the multiplication parameter δ temperature-dependent?

P. HAASEN: The multiplication parameter δ is independent of temperature according to our experiments.

J. LI: First, I would like to agree with Dr. Haasen that a calculation of the average velocity through both the positive and negative internal stress regions is all you need to take care of inhomogeneous internal stresses. Now can I ask Dr. Haasen to comment on the density of mobile dislocations during steady-state creep as a function of stress and temperature.

P. HAASEN: According to our experiments, almost all the dislocations are mobile in the initial range of deformation (up to a few percent shear strain) in diamond structure crystals. The stress dependence of the dislocation density-strain relation is given by Eq. (7) of my paper. This relation is independent of temperature according to our experiments.

A. SEEGER: I would like to comment on the question of the linear dependence of velocity on stress. Such an experimental finding does not appear too surprising as long as the experimental results are linear when plotted as stress divided by temperature instead of simply as stress, as was done in the paper. In the present experiments, one has fairly high stresses and narrow kinks and a small activation volume. One could then expand the complete activation energy in powers of stress and the leading term can be taken out of the exponential as τ/kT. Whether this makes sense could be determined by examination of the magnitude of the preexponential factor. One would have to be consistent with the dislocation density determinations.

P. HAASEN: I do not think the dislocation density enters because this is a velocity measurement on one dislocation. I would have expected a square root of stress over kT in the preexponential.

A. SEEGER: This would only be expected if the kinks were very far apart, so that the Coulomb-type interaction dominates.

J. HIRTH: Haasen has noted that his results agree well with a double-kink theory except for the stress dependence of the strain rate. I would propose that his results, including the stress dependence, might be rationalized by a revision of the double-kink theory developed by Lothe and myself [*Phys. Rev.* 115: 543 (1959)]. Seeger and Schiller [*Acta Met.* 10: 348 (1962)] extended our result to include the kink-kink interaction in the activation energy. Including this factor, our equation for kink nucleation in finite segments of length L becomes

$$\dot{\epsilon} = \frac{A\tau}{T} \exp\left[-\left(\frac{2U_{DK} - C\sqrt{\tau}}{kT}\right)\right] \tag{1}$$

where τ is stress, $2U_{DK}$ is the double-kink formation energy, and A and C are essentially independent of stress and temperature.

In a range of stresses and temperature, the $\sqrt{\tau}$ part of the exponential can be expanded in a series of development, giving

$$\dot{\epsilon} \simeq \frac{A\tau}{T} \left(1 - \frac{C\sqrt{\tau}}{kT}\right) \exp\left(-\frac{2U_{DK}}{kT}\right) \tag{2}$$

To first order in stress, Eq. (2) becomes

$$\dot{\epsilon} \simeq \frac{A\tau}{T} \exp\left(-\frac{2U_{DK}}{kT}\right) \tag{3}$$

This expression is of the form found by Haasen. The range of parameters where the approximation of Eq. (3) is valid can be determined from the explicit expression for A and C given in the papers of Lothe and Hirth and of Seeger and Schiller, respectively.

THE DYNAMICS OF
NONUNIFORM PLASTIC
FLOW IN LOW-CARBON STEEL

J. D. Campbell, R. H. Cooper, and T. J. Fischhof

University of Oxford,
United Kingdom

ABSTRACT

The development of macroscopically nonuniform flow is discussed and related to the magnification of microscopic irregularities in the test specimen. Experimental results are given for the upper and lower yield stresses, and the fracture elongation, of low-carbon steel deformed at mean plastic strain rates between 2×10^{-4} and 2.6×10^{3} sec^{-1}, and temperatures between 77 and 293°K.

The dependence of the upper yield stress on the time to yield, and on the temperature, is analyzed in terms of a dislocation model for the growth of plastic heterogeneity. The analysis is used to derive the temperature dependence of the "drag stress" controlling the motion of dislocations, in the temperature range 180 to 293°K.

This temperature dependence is related to the activation energy for dislocation motion.

The dependence of the lower yield stress on temperature and strain rate is discussed and it is shown that increasing the plastic strain rate, or the grain size, has a considerable effect on the temperature at which the fracture changes from ductile to brittle.

1   INTRODUCTION

It is well known that under certain circumstances the plastic deformation of a test-piece may become macroscopically nonuniform although the applied stress is macroscopically uniform. This is related to the fact that any real specimen is microscopically imperfect in its geometry, and any real material is microscopically heterogeneous because of variations in its composition and structure. Two types of nonuniform deformation may therefore be distinguished: the first due to the magnification of initial imperfections in the geometry of the specimen; and the second to the magnification of initial heterogeneities in the material. In each case, the result is a localization of plastic flow.

The most common occurrence of the first type of nonuniform deformation is the necking of a tension specimen; at the ultimate tensile stress, the variations in cross-sectional area of the specimen are greatly magnified, so that flow continues only in the neighborhood of the section of minimum area.

The equally familiar phenomenon of the yield drop in certain materials, notably steels, is an example of the second type of nonuniform deformation. In tension or compression tests using a stiff testing machine, this type of nonuniform deformation is accompanied by a sudden drop in the applied load. Johnston and Gilman,[1] Hahn,[2] and others have shown that in crystalline materials a drop in load is to be expected if the density of mobile dislocations increases rapidly with plastic strain, and their velocity depends strongly on the stress. According to this treatment, the upper yield stress is defined by the load at which the plastic strain rate becomes large enough to accommodate the imposed rate of deformation. This load is found to be a function of the initial mobile dislocation density, which is assumed to be uniform. However, as pointed out by Hahn,[2] in real materials this density will vary within a given specimen; there is therefore a difficulty in applying the above definition of the upper yield stress to real materials. A further difficulty arises in applying it to a constant-stress test, such as those of Wood and Clark[3] and others. In these tests, the

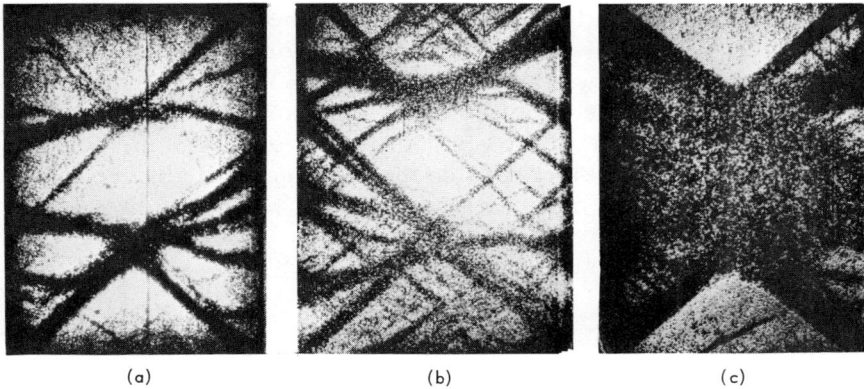

(a)   (b)   (c)

Fig. 1 Macrophotographs showing non-uniform yield in compression specimens deformed at constant stress. Longitudinal sections through axes of specimens $3/8$ in. diam., $1/2$ in. long. Amount of deformation increases from (a) to (c).

imposed deformation rate is essentially zero for a finite time (the delay time) after the initial application of the load. It can be shown experimentally that the macroscopic yielding which takes place after the delay period is nonuniform, even though there is no drop of load. Thus Fig. 1 shows sections of mild steel specimens subjected to a constant compressive load for a short time and rapidly unloaded; the regions in which macroscopic flow occurred have been exhibited by etching the sectioned surfaces with Fry's reagent.

It seems from the above discussion that a full description of the yield phenomenon must take account of the growth of microscopic heterogeneities of the material during the period preceding gross yielding. An analysis of this growth has been given in a previous paper,[4] and this analysis is used in the present paper to interpret experimental results obtained for a wide range of temperatures and strain rates. The effect of these variables on the post-yield flow and fracture is also discussed.

2   DISLOCATION THEORY OF NONUNIFORM YIELDING

For the sake of completeness, an outline will be given here of the analysis given in the earlier paper.[4] The heterogeneity of the material is defined in terms of the gradient of the density of mobile dislocations. Experimental observations[5,6] have shown that the total dislocation density is a function of the plastic strain, and it may be assumed that for small strains and in the absence of strain aging, the fraction of mobile dislocations is constant or varies only with the plastic strain; thus we take the density of mobile dislocations ρ to be a function of the plastic strain ϵ_p. Experiments[1,7] have also shown that, at least for low strains, the velocity of dislocations varies strongly with the stress on the slip plane; we assume therefore that the mean dislocation speed v is a function of the local stress σ.

The local plastic strain rate is given by

$$\dot{\epsilon}_p = ab\rho v \tag{1}$$

where α is an orientation factor, and b is the magnitude of the Burgers vector. Thus if v remains constant, $\dot{\epsilon}_p$ will be greatest in regions where ρ (and hence ϵ_p) is greatest. This would seem to indicate that heterogeneity of plastic strain will increase for any material, which is not the case. In fact, however, the local stress will be relaxed in regions in which the plastic strain is above the average, so that the dislocation velocity is reduced. This introduces a stabilizing factor which may be large enough to eliminate the tendency for the initial variations in dislocation density to be magnified.

The simplest basis upon which to treat the problem is to assume that the total (elastic plus plastic) strain is constant at any given time, being determined by the imposed displacements at the ends of the specimen. It then follows[4] that the gradient λ of (mobile) dislocation density at a given point in the material varies with time according to the equation

$$\frac{\partial}{\partial t}(\ln \lambda) = ab\left[\frac{v}{\rho'}\frac{d}{d\epsilon_p}(\rho\rho') - E\rho\frac{dv}{d\sigma}\right] \tag{2}$$

where $\rho' \equiv d\rho/d\epsilon_p$ and E is Young's modulus. Thus the condition for increasing heterogeneity of dislocation distribution is

$$\frac{d}{d\epsilon_p}[\ln(\rho\rho')] > E\frac{d}{d\sigma}(\ln v) \tag{3}$$

This condition is easily satisfied in materials in which the dependence of mobile dislocation density on plastic strain is strong and the dependence of dislocation velocity on stress is relatively weak. In these circumstances, the first term on the right-hand side of Eq. (2) is much greater than the second. Then we may write

$$\frac{\partial}{\partial t} (\ln \lambda) = \frac{\alpha b v}{\rho'} \frac{d}{d\epsilon_p} (\rho \rho')$$

and hence

$$\ln \frac{\lambda}{\lambda_0} = \alpha b \int_0^t \frac{v}{\rho'} \frac{d}{d\epsilon_p} (\rho \rho') dt \qquad (4)$$

where λ_0 is the initial value of λ.

In the above treatment the local stress is taken to be $E(\epsilon - \epsilon_p)$, where ϵ is the imposed strain. Thus its value in an unyielded region is $E\epsilon$. However, if this region is a grain adjacent to a yielded grain, there will be additional local stresses in the unyielded grain caused by dislocations held up at the boundary between the two grains. These stresses will be determined by the number of mobile dislocations in the yielded grain, and by the grain size. It is reasonable to assume that when the dislocation density gradient reaches a critical value λ_c, these local stresses in the unyielded material become large enough to permit the generation of new dislocations in this material, or the penetration of dislocations through the grain boundary, thus enabling the yielded zone to propagate much more rapidly as a Lüders band. According to this hypothesis, the upper yield stress σ_{uy} is reached when the integral in Eq. (4) reaches the value $(\alpha b)^{-1} \ln(\lambda_c/\lambda_0)$. It is shown in Ref. [4] that if the dislocation velocity is given by the formula proposed by Gilman,[8] i.e.,

$$v = v_0 \exp\left(\frac{-D'}{\tau}\right) = v_0 \exp\left(\frac{-D}{\sigma}\right) \qquad (5)$$

where τ is the shear stress on the slip plane and v_0 is a constant, the time to yield t_y, in a test at constant stress rate, is given by

$$t_y = (\alpha b \rho' v_0)^{-1} \left(\frac{D}{\sigma_{uy}}\right) \exp\left(\frac{-D}{\sigma_{uy}}\right) \ln\left(\frac{\lambda_c}{\lambda_0}\right) \qquad (6)$$

if ρ' is assumed to be independent of ϵ_p.

The previous experiments[4] showed that for an annealed low-carbon steel at room temperature, the measured values of σ_{uy} and t_y were in good agreement with Eq. (6), in the range of 0.5 msec $< t_y < 0.2$ sec. The work described here provides data within the same range of yield times, at temperatures down to 77°K. These data were obtained with the special hydraulic test apparatus that was used for the previous investigation. In addition, results are included which had been obtained during earlier work[9] carried out on a very similar steel using a drop-weight impact machine[10] with which yield times down to 25 μs are obtainable. From the combined results, the temperature-dependence of the drag stress D is derived.

3   THE LOWER YIELD STRESS

In tensile tests using a stiff machine, deformation continues at a lower load by spreading of one or more Lüders bands along the specimen. However, in a soft machine, yielding takes place without any appreciable reduction of load.[11] It seems certain, moreover, that the applied load during nonuniform flow is influenced by the number and distribution of Lüders bands, as well as by the rate of deformation. For these reasons, the "lower yield stress," if it exists, cannot be considered to be an intrinsic property of the material. For a given specimen geometry, however, the variation of the lower yield stress with rate of deformation and temperature should be significant, since it gives some indication of the extent to which these factors are equivalent. In this paper, values of lower yield stress are given for nominal strain rates in the range 2×10^{-4} to 2.6×10^{3} sec^{-1} and temperatures in the range 77 to 293°K. These data were obtained using conventional screw-driven machines, in addition to the hydraulic machine and the impact machine.

4   EXPERIMENTAL RESULTS

The specimen material used for the medium strain-rate tests was an En2A steel containing 0.045% C and 0.45% Mn, and small amounts of other impurities.[4] The specimens were 0.125 in. in diameter and had a gage length of 0.35 in.; they were vacuum-annealed after machining, and the mean ferrite grain size was 0.028 mm. Since strain measurements were not required for this investigation, the two oscilloscope traces were in many cases used to record the applied load at two different time-base speeds, as shown in Figs. 2a and 2b. This was done to improve the accuracy in determining the stress-time curve before yield.

Temperatures down to that of liquid nitrogen were obtained by means of a special cryostat surrounding the specimen. This is

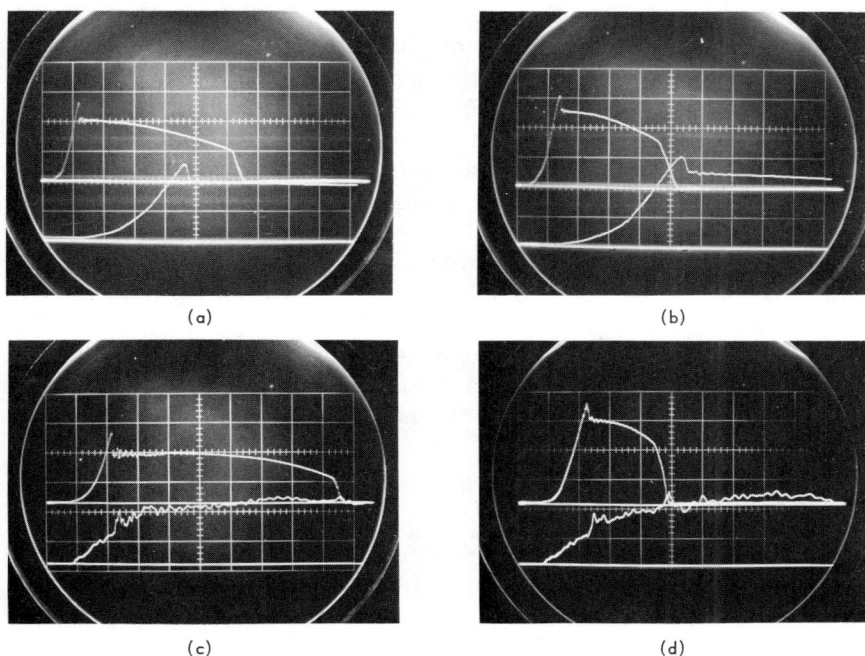

(a)

(b)

(c)

(d)

Fig. 2 Tensile test oscillograms obtained at medium strain rates. (a) Stress-time (both traces); total time base for upper trace 20 msec, for lower trace 5 msec.; temperature, 200°K. (b) Trace data as for (a); temperature, 163°K. (c) Stress-time (upper trace), strain rate-time (lower trace); total time base for both traces 10 msec.; temperature, 250°K. (d) Trace data as for (c); temperature, 100°K. Stress trace shows evidence of twinning before yield.

supplied with liquid or gaseous nitrogen at a rate controlled by a valve electromagnetically operated by a signal from a potentiometer controller. This, in turn, is operated by the output of a thermocouple mounted close to the specimen. The potentiometer can be preset to hold the measured temperature at any desired value to an accuracy of ± 0.25°C. The maximum temperature variation along the gage length of the specimen was found to be 3°C.

Figure 2 shows typical oscillograms obtained at low temperatures, and in Figs. 3b, c, and d, the upper and lower yield stresses are plotted against temperature for each of three nominal plastic strain rates $\dot{\epsilon}_p$: 0.20, 10 and 55 sec^{-1}. These rates were obtained by using different orifice sizes in a special throttle valve which regulates the flow of oil by which the test machine is operated.[12]

Figure 3a shows results obtained at $\dot{\epsilon}_p = 4 \times 10^{-4}$ sec^{-1} in a screw-driven machine. These data were derived in an earlier investigation[9] using an En2 steel containing 0.1% C, heat-treated to give mean ferrite grain sizes of 0.0154 and 0.051 mm. In Figs. 3e and f values are plotted for the upper and lower yield stresses of this steel at $\dot{\epsilon}_p$ = 460 and 2600 sec⁻¹; these values were obtained from tests using the drop-weight tensile impact machine. At the higher of these two rates, there is very considerable scatter in the results, so that separate curves have not been drawn for the two grain sizes.

In Fig. 4, curves are drawn showing the dependence of the lower yield stress on the mean plastic strain rate $\dot{\epsilon}_p$ at six different temperatures. The points plotted in Fig. 4 were obtained from the curves of Fig. 3, together with similar curves obtained at $\dot{\epsilon}_p$ = 0.7, 1.8, and 20 sec⁻¹, and values obtained for the En2A steel at $\dot{\epsilon}_p = 10^{-2}$ sec⁻¹, T = 77°K and 293°K. In deriving the values

Fig. 3 Variation of yield stresses and fracture strain with temperature, at constant mean strain rates $\dot{\epsilon}_p$. (a) 0.1% C

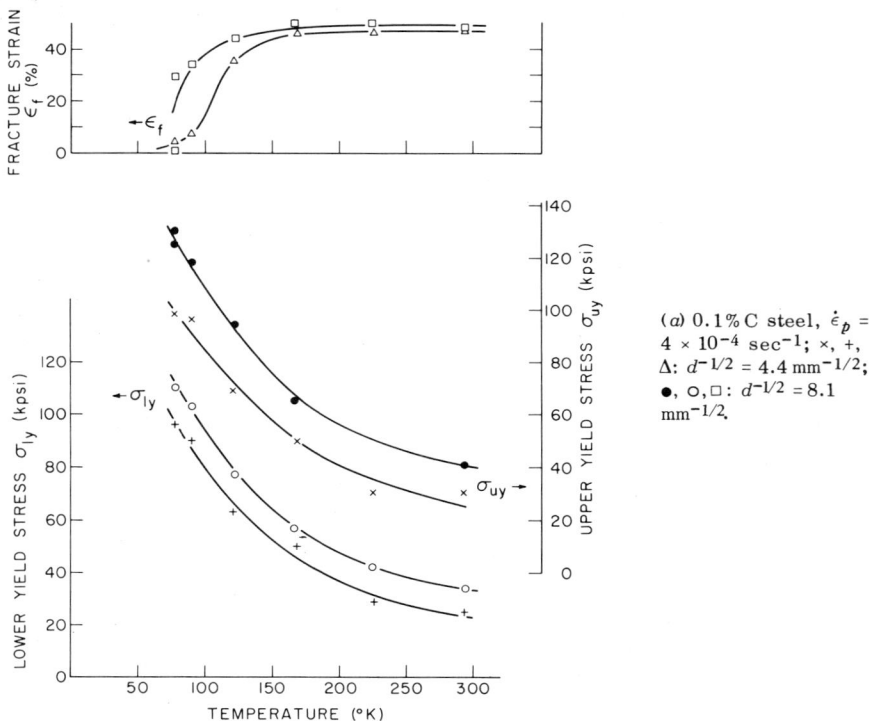

(a) 0.1% C steel, $\dot{\epsilon}_p$ = 4 × 10⁻⁴ sec⁻¹; ×, +, Δ: $d^{-1/2}$ = 4.4 mm⁻¹/²; ●, ○, □: $d^{-1/2}$ = 8.1 mm⁻¹/².

(b) 0.045 % C steel, $\dot{\epsilon}_p = 0.20$ sec^{-1}, $d^{-1/2} = 6.0$ mm$^{-1/2}$.

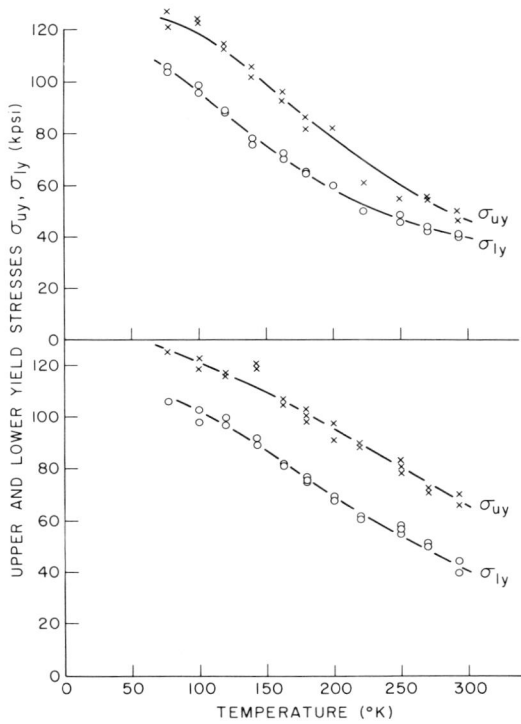

(c) 0.045 % C steel, $\dot{\epsilon}_p = 10$ sec^{-1}, $d = 6.0$ mm$^{-1/2}$.

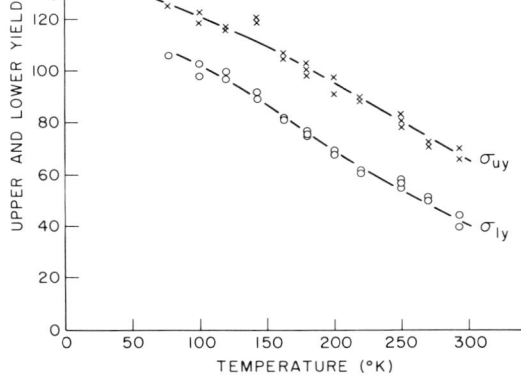

(d) 0.045 % C steel, $\dot{\epsilon}_p = 55$ sec^{-1}, $d^{-1/2} = 6.0$ mm$^{-1/2}$.

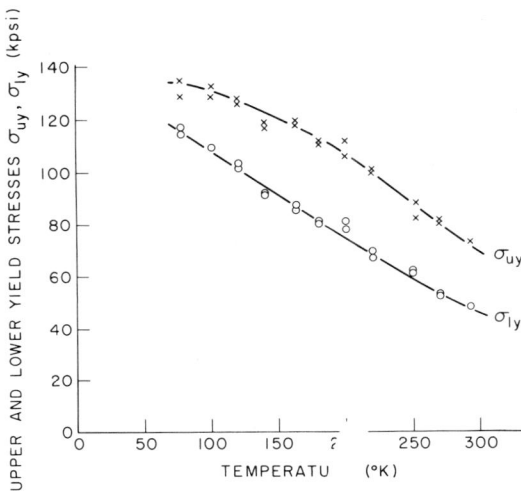

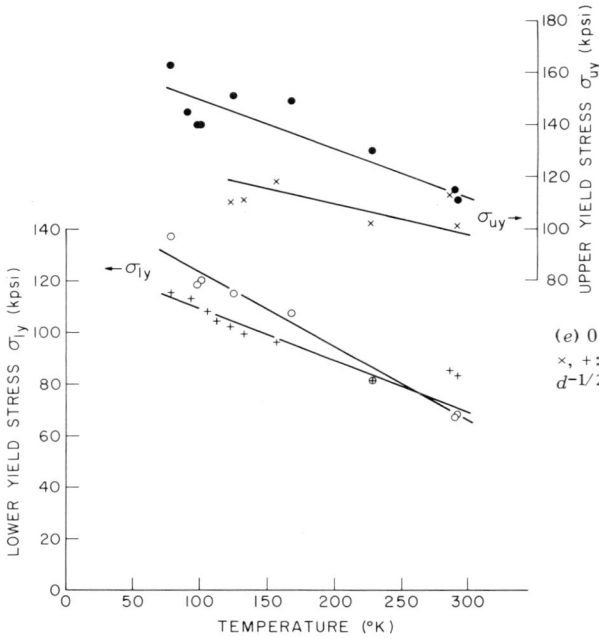

(e) 0.1% C steel, $\dot{\epsilon}_p = 460$ sec^{-1}; ×, +: $d^{-1/2} = 4.4$ mm$^{-1/2}$; ●, ○: $d^{-1/2} = 8.1$ mm$^{-1/2}$.

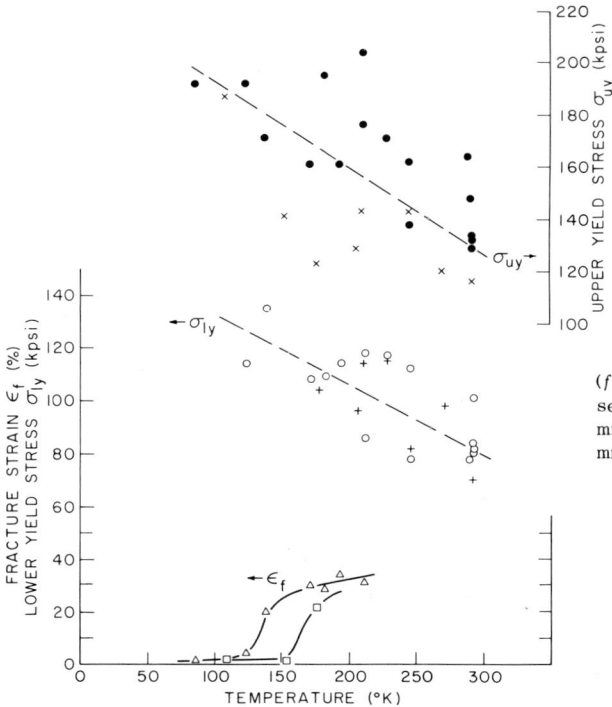

(f) 0.1% C steel, $\dot{\epsilon}_p = 2600$ sec^{-1}; ×, +, □: $d^{-1/2} = 4.4$ mm$^{-1/2}$; ●, ○, Δ: $d^{-1/2} = 8.1$ mm$^{-1/2}$.

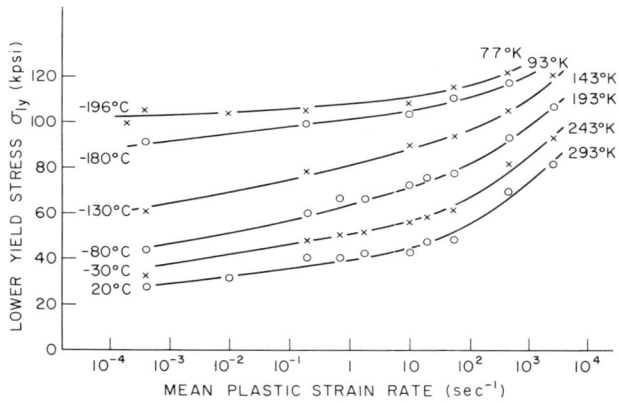

Fig. 4 Variation of lower yield stress with mean plastic strain rate, at constant temperatures.

for the En2 steel, an interpolation was made between the curves relating to the two different grain sizes; this interpolation was based on the assumption that σ_{ly} varies linearly with the inverse square root of the grain diameter. At the highest rate, $\dot\epsilon_p = 2600$ sec^{-1}, however, the scatter is such that separate curves for the two grain sizes were not drawn; the mean line shown in Fig. 3f was used to determine the values for plotting in Fig. 4.

5 DISCUSSION

The upper yield stress results have been compared with Eq. (6) as follows. From the curves of Figs. 3b, c, and d, together with similar curves obtained at $\dot\epsilon_p = 0.7$, 1.8, and 20 sec^{-1}, values of σ_{uy} were obtained for six temperatures: 180; 200; 220; 250; 270; and 293°K. (Lower temperatures were not included, as twinning becomes prevalent at these temperatures.) For each nominal strain rate, the preyield loading rate was determined from a number of oscillograms;* these loading rates were then used to calculate the yield times t_y corresponding to the various values of σ_{uy}. The resulting points are plotted in Fig. 5, which also shows data obtained with the drop-weight machine at a mean plastic strain rate $\dot\epsilon_p = 460$ sec^{-1}, and temperatures between 180 and 293°K. The

---

*In determining this rate, the linear part of the stress-time curve was used, since the shape of the curve at low stresses has a negligible influence on yielding.

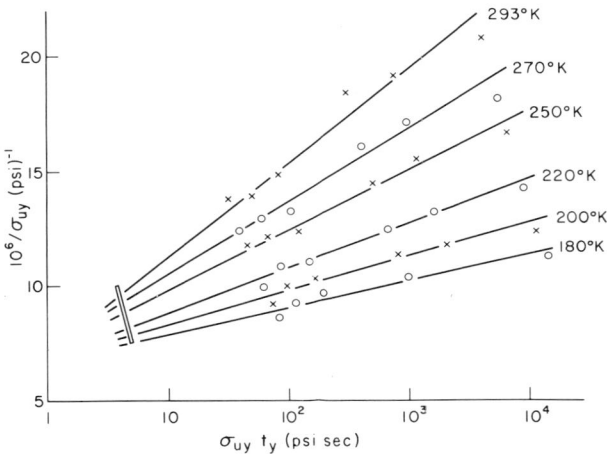

Fig. 5 Plot of $(\sigma_{uy})^{-1}$ vs. $\log(\sigma_{uy} t_y)$ at constant temperatures. Points derived from medium strain-rate tests; band gives range of values obtained from impact tests at $\dot{\epsilon}_p$ = 460 sec^{-1}.

latter results are plotted as a continuous band, because of the considerable scatter in the data. The variations in the values of σ_{uy} obtained at $\dot{\epsilon}_p$ = 2600 sec^{-1} are too large to permit a significant comparison with Eq. (6).

It is seen that the points plotted in Fig. 5, while showing a certain amount of scatter, may reasonably be fitted by straight lines, as indicated by Eq. (6). The maximum error in σ_{uy}, between the points and the lines drawn, is about 5%. From the slopes of these lines, the (tensile) drag stress D was derived for each temperature. The values obtained in this way are given in Table 1.

It follows from the values in Table 1 that as the absolute temperature decreases by a factor of 1.63, the value of D increases by

TABLE 1. Effect of Temperature on Parameters in Eq. (6)

Temperature, T (°K)	Drag Stress D (kpsi, tensile)	Product DT (10^8 psi °K)	$(ab\rho' v_0)^{-1} \ln(\lambda_c/\lambda_0)$ (sec)
293	567	1.66	2.9×10^{-8}
270	728	1.97	6.4×10^{-9}
250	883	2.21	1.8×10^{-9}
220	1185	2.61	5.2×10^{-10}
200	1565	3.13	1.3×10^{-11}
180	2020	3.64	5.0×10^{-13}

a factor of 3.56. The temperature dependence of the drag stress is therefore greater than that corresponding to a temperature-independent activation energy for dislocation motion. Gilman[13] has given an analysis of the motion of a dislocation randomly held up by interaction with point defects or impurities; the analysis leads to the result

$$D' = \frac{\omega \chi C^{1/2}}{bkT} = \frac{D}{2} \tag{7}$$

In Eq. (7), ω and χ are measures of the strength of the interaction, C is the concentration of defects, and k is Boltzmann's constant. Gilman has remarked that ω and χ are expected to increase with decreasing temperature because of a reduction in the effective width of the dislocation core; he has suggested that the increase may be described by the equation

$$\omega \chi = (\omega \chi)_0 \exp(-\beta T) \tag{8}$$

where β is a constant coefficient.

To compare the present results with Eq. (8), the values of DT from Table 1 are plotted logarithmically against the absolute temperature in Fig. 6, and it is seen that the points are well fitted by a straight line; from this line the constants $(\omega \chi)_0 C^{1/2}/bk$ and β are

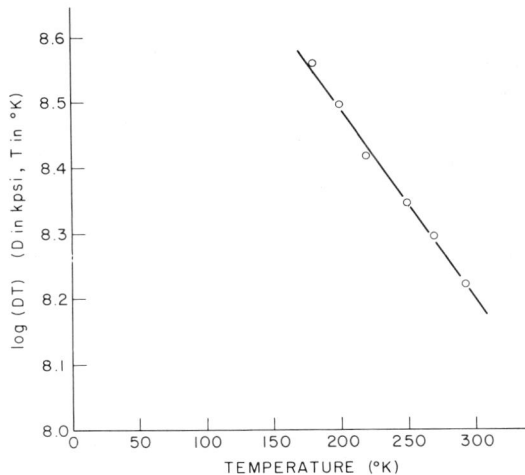

Fig. 6 Plot of log(DT) vs. T.

found, and the value of D at a temperature $T°K$ is then determined
by

$$D = \frac{1.16 \times 10^6}{T} \exp \frac{-T}{151} \text{ kpsi} \tag{9}$$

The value obtained for the drag stress D at room temperature
(567 kpsi) may be compared with several values derived by various
methods for iron alloys. Thus Gilman,[13] using the results of Man-
joine[14] for a commercial mild steel, has plotted $(\sigma_{ly})^{-1}$ against $\log \dot\epsilon_p$;
this plot indicates a value $D \simeq 700$ kpsi. Taylor[15] has analyzed re-
sults obtained by Taylor and Rice[16] in high-velocity impact tests
on Armco iron plates; two values of D are derived from these results by different methods, namely 490 and 400 kpsi. A value for D
can also be obtained from the measurements of dislocation velocities
in silicon-iron made by Stein and Low;[7] this value is about 1400 kpsi,
but its significance is questionable since, as discussed by Gilman,[13]
the velocities do not extrapolate to the shear wave velocity at
infinite stress.

In many investigations, data have been interpreted in terms of
a power relationship between stress and yield time or strain rate.
This relationship corresponds to the use of the equations

$$v = A'\tau^n = A\sigma^n \tag{10}$$

for the dislocation velocity in terms of the applied stress, instead
of Eq. (5). The values of n and D may be related by equating the
stress-sensitivities of the dislocation velocity as given by the two
equations. These are

$$\frac{dv}{d\sigma} = \frac{Dv}{\sigma^2} \tag{11}$$

and

$$\frac{dv}{d\sigma} = \frac{nv}{\sigma} \tag{12}$$

from Eqs. (5) and (10), respectively. Thus we obtain the relation
between D and n as

$$D = n\sigma \tag{13}$$

As was shown in Ref. [4], there is very little difference in the
curves derived from Eqs. (5) and (10) within normally attainable

ranges of stress. The most extensive range available for steel is that of the combined delay-time data of Wood and Clark[3] and Krafft and Sullivan[17] for a 0.17% C steel, in which the stress varies from 40 to 100 kpsi at room temperature. In this range, the deviation from a straight-line logarithmic plot, if Eq. (5) is obeyed instead of Eq. (10), is only about $\pm 5\%$ in stress, which is within the scatter of the data. We may therefore take D as the geometric mean stress, 63 kpsi, multiplied by the mean slope $n = 12.5$,* giving $D \simeq 790$ kpsi. Similarly, from the data of Campbell and Marsh[11] for a 0.085 % C steel, we obtain $D \simeq 460$ kpsi.

It is encouraging that these various values of D, derived in very different ways for similar materials, do not differ greatly. Moreover, the values appear generally to increase with increasing impurity content, which is consistent with the random pinning model on which Eq. (7) is based.

The straight lines of Fig. 5 defined by their intercepts the values of the quantity $(ab\rho'v_0)^{-1} \ln(\lambda_c/\lambda_0)$ in Eq. (6). These values are given in Table 1, from which it is seen that they decrease by a factor of some 6×10^4 within the temperature range indicated. It seems clear that the only term capable of varying by this amount is the dislocation multiplication rate ρ'. There appear to be no published data concerning the effect of temperature on this parameter in iron. Hull et al.[18] have, however, shown by electron microscopy that in tantalum there is a large increase in the dislocation multiplication rate as the temperature decreases from 293 to 240°K; they suggest that this is due to the more uniform distribution of dislocations at the lower temperature. While no accurate numerical comparison with the present results is possible because of the difference in material and the fact that no dislocation density measurements are available at low strains, it would seem that the increase in the dislocation multiplication rate at low temperatures may be sufficiently great to account for the variation in the values given in the last column of Table 1.

The results for the lower yield stresses (Fig. 4) show the same general features as those noted by Rosenfield and Hahn.[19] Thus the rate dependence, as characterized by the derivative $(\partial\sigma_{ly}/\partial \log\dot\epsilon_p)_T$, is relatively small at rates below about 10 sec^{-1}; it shows a rapid rise at higher rates, the increase being greater, the greater the temperature. The curves of Fig. 4 also show that the increase in σ_{ly} due to a decrease of temperature from 293 to 93°K is

---

*The value 9.6 quoted in Ref. [4] corresponds to a temperature of 65°C, not room temperature as stated there.

approximately independent of $\dot{\epsilon}_p$ at strain rates below 10 sec⁻¹; at higher rates the effect of temperature on the yield stress is smaller.

The curves of (nominal) fracture strain presented in Figs. 3a and 3f show that the ductile-to-brittle transition temperature of the En2 steel is raised about 30°C by the increase in mean grain size from 0.0154 mm to 0.051 mm; they also show that the transition temperature is raised nearly 60°C by the increase in mean plastic strain rate from 4×10^{-4} to 2.6×10^3 sec⁻¹. These temperature rises are sensibly independent of each other.

6   CONCLUSIONS

a) The occurrence of nonuniform plastic flow in materials has been discussed and it is concluded that such flow may be caused by the magnification of initial irregularities in the geometry or material of the test specimen.

b) The rate- and temperature-dependence of the upper and lower yield stresses of low-carbon steel have been determined at plastic strain rates and temperatures in the ranges 2×10^{-4} to 2.6×10^3 sec⁻¹, 77 to 293°K, respectively.

c) The dependence of the upper yield stress on yield time at temperatures between 180 and 293°K has been shown to be consistent with the predictions of a dislocation model for the magnification of initial heterogeneities in the material.

d) On the basis of the model, the "drag stress" opposing the motion of dislocations has been derived as a function of temperature; it is found to increase by a factor of 3.56 in the temperature range given in (c) above. It is shown that the increase can be explained in terms of an activation energy which increases as the temperature falls.

e) The dependence of the lower yield stress on the logarithm of the mean plastic strain rate shows an increase as the rate is raised above 10 sec⁻¹, the increase being less marked the lower the temperature.

f) The ductile-to-brittle transition temperature is raised about 30°C by increasing the mean ferrite grain size from 0.0154 mm to 0.051 mm.

g) The transition temperature is raised about 60°C by increasing the mean plastic strain rate from 4×10^{-4} sec⁻¹ to 2.6×10^3 sec⁻¹.

ACKNOWLEDGMENTS

Financial support for part of the work reported in the paper was received from the U. K. Ministry of Defence. One of us (T. J. Fischhof) was awarded a scholarship by the Colonial Sugar Refining Company of Australia. The final preparation of the paper was undertaken while the first author held a Visiting Professorship in the Division of Engineering, Brown University, partially supported by the Advanced Research Projects Agency under the Materials Research Program.

The authors are grateful to Dr. T. L. Briggs, who completed the final series of tests at medium strain rates.

REFERENCES

1. Johnston, W. G., and J. J. Gilman: *J. Appl. Phys.* **30**: 129 (1959).
2. Hahn, G. T.: *Acta Met.* **10**:727 (1962).
3. Wood, D. S., and D. S. Clark: *Trans. Amer. Soc. Metals* **43**:571 (1951).
4. Campbell, J. D., and R. H. Cooper: *Proc. Conf. on the Physical Basis of Yield and Fracture*, p. 77, Inst. Phys. and Phys. Soc., London, 1966.
5. Carrington, W. E., K. F. Hale, and D. McLean: *Proc. Roy. Soc.* A., **259**:203 (1960).
6. Keh, A. S., and S. Weissmann: "Electron Microscopy and Strength of Crystals," p. 231, Interscience Publishers, New York, 1963.
7. Stein, D. F., and J. R. Low: *J. Appl. Phys.* **31**: 362 (1960).
8. Gilman, J. J.: *Aust. J. Phys.* **13**:327 (1960).
9. Fischhof, T. J.: D. Phil. Dissertation, Oxford University, 1963.
10. Harding, J., E. O. Wood, and J. D. Campbell: *J. Mech. Eng. Sci.* **2**:88 (1960).
11. Campbell, J. D., and K. J. Marsh: *Phil. Mag.* **7**:933 (1962).
12. Cooper, R. H., and J. D. Campbell: *J. Mech. Eng. Sci.* **9**: 278 (1962).
13. Gilman, J. J.: *J. Appl. Phys.* **36**:3195 (1965).
14. Manjoine, M. J.: *J. Appl. Mech.* **11**:211 (1944).
15. Taylor, J. W.: *J. Appl. Phys.* **36**:3146 (1965).
16. Taylor, J. W., and M. H. Rice: *J. Appl. Phys.* **34**:364 (1963).
17. Krafft, J. M., and A. M. Sullivan: *Trans. Amer. Soc. Metals* **51**:643 (1959).
18. Hull, D., I. D. McIvor, and W. S. Owen: "The Relation Between the Structure and Mechanical Properties of Metals," p. 596, *Nat. Phys. Lab. Symp.*, No. 15, Her Majesty's Stationery Office, London, 1963.
19. Rosenfield, A. R., and G. T. Hahn: *Trans. Amer. Soc. Metals* **59**:962 (1966).

DISCUSSION *on Paper Presented by* *J. D. Campbell*

J. TAYLOR: It would seem that it is dangerous to compare impact experiments at $\dot{\epsilon} = 10^3$ sec^{-1} with lower strain-rate data because these experiments involve times in which only ten or so acoustic signals have passed through the specimen. This means that the approximation that total strain is homogeneous at a given time becomes very poor. One therefore expects large scatter in the data because one is essentially sampling from a small number of quantized events.

J. CAMPBELL: I agree that the wave transit time sets an upper limit to the rate of loading at which valid results can be obtained. In the present work, the elastic wave transit time through the gage length is 1.75 μs; thus, if yield occurs at 25 μs, the maximum uncertainty in the upper yield stress due to wave effects is 1.75/25, i.e., 7%. The scatter in the values of upper yield stress obtained at the highest rate is considerably greater than this, however; thus it appears that other factors, such as imperfect alignment of the specimen, must be significant.

T. VREELAND: Is it possible to extract from your application of this theory to your constant strain-rate tests, the parameters which can be used to quantitatively predict the delay time for yielding under constant stress? The delay times could be taken as the time to build up the critical λ in such a calculation.

J. CAMPBELL: Yes, the calculation was performed in a previous paper (Ref. [4] of the present paper); it was shown that if yield occurs at a stress σ, the ratio of the yield time in a constant stress-rate test to that in a constant stress test is given by $(n + 1)$ or D/σ, where n and D are the parameters occurring in the dislocation velocity vs. stress relations. A direct comparison of this kind has not yet been made, however, due to a variation in grain size between the specimens used for the two types of tests.

A. ROSENFIELD: Our experiments (Fig. D1) indicate that the semilogarithmic dependence of lower yield stress on strain rate

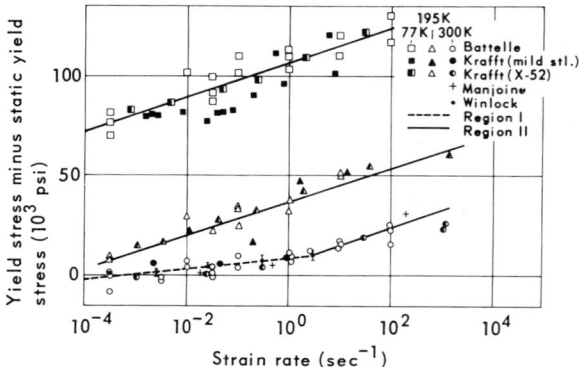

Fig. D1 Influence of strain rate and temperature on the lower yield stress of steel. The graph reports the difference between the yield stress at the indicated strain rate and the room temperature value at 10^{-3} sec^{-1} (the static yield stress). Source: A. R. Rosenfield and G. T. Hahn: *Trans. ASM* **59**: 962 (1966).

$(\partial Y/\partial \ln \dot{\varepsilon})$ is temperature independent (except at temperatures near the ambient when the strain rate is low). Do you have evidence to support this observation?

J. CAMPBELL: The curves of Fig. 4 of our paper show general agreement with the observation you make, except at temperatures below 140°K; at these temperatures twinning occurs, especially at high strain rates, and this has the effect of reducing the apparent rate dependence of the lower yield stress.

G. HAHN: I think the suggestion we put forward is that in our steels, which were very fine grained, there was no twinning at liquid nitrogen temperature. Accompanying twinning there is usually a diminution in the temperature and strain-rate dependence which we avoided perhaps by using a fine grain size which other people have not avoided. In covering this strain rate-temperature spectrum which you covered in your paper you may have encountered several deformation processes, perhaps several kinds of slip as Professor Hirsch has suggested, and perhaps at the lower temperature some twinning in addition to slip.

J. CAMPBELL: As noted in this paper, in our material twinning becomes prevalent at temperatures below about 180°K, and it is therefore true that at these temperatures the yield stress values given in Fig. 4 are governed by more than a single deformation process. In considering the upper yield stress, however, the low-temperature results were excluded so that an interpretation based on a single process should be valid.

J. SPREADBOROUGH: We have studied various aspects of the propagation of Lüders bands in Armco iron; we normally found that the D values were inversely proportional to T, except in cases where twinning intervened. Given your high strain rates, I wonder whether twinning might account for the product TD increasing with decreasing T in your work.

J. CAMPBELL: In deriving the variation of D with temperature, only results obtained at temperatures above 180°K were used, because of twinning at lower temperatures. I therefore do not think that twinning is the cause of the variation in DT. Possibly some difference in the behavior found in your work and ours is attributable to the fact that D was derived from lower and upper yield points, respectively.

J. SPREADBOROUGH: By measuring the profile of a Lüders front in a cylindrical specimen, we were able to derive the plastic strain rate as a function of distance from the front and also the dislocation density gradient. This might give some idea of the value for the critical gradient λ_c, referred to in your paper. We have made etch-pit studies across the Lüders front and these could also be used to estimate λ_c and λ_0.

J. CAMPBELL: I am very interested to hear of your work on Lüders band propagation, and I look forward to seeing the details of your observations. The measurements of dislocation density gradients may enable one to obtain an independent estimate of D or the multiplication parameter C.

W. OWEN: In your model you consider the elastic constraints on a small volume of material located within the specimen. I think it is found quite generally that Lüders bands start near a free surface and usually in the shoulder of the specimen. Do you agree that this is the experimental observation and, if you do, how do you take this into account in your model?

J. CAMPBELL: I agree that Lüders bands are often found to have originated at the shoulder of the specimen. One factor at least partly responsible for this is clearly the local stress concentration in this region. A small element of material situated at a free surface is still constrained elastically by the material adjacent to it in the interior of the specimen, but it is possible that this constraint is appreciably lower than that for an element with no free surface. If so , this could be taken into account by a reduction in the modulus in Eq. (2) of the paper. Such a reduction would increase the rate of growth of dislocation density gradient, and it appears quite likely that this is a contributory factor in determining the position at which a Lüders band is initiated in static tests. For dynamic yielding, however, the term involving the modulus has only a small influence so that the effect should be negligible.

J. KRAFFT: An understanding of dislocation processes relevant to the strain-hardening rate is applicable to several practical problems. Professor Dorn has emphasized the importance of such applications in another discussion of this Colloquium. As a specific example, I would note here the connection of strain hardening to brittle fracture, in rate-sensitive mild steels. Within the stress rates covered by Prof. Campbell's work, a marked speed sensitivity of the strain-hardening rate has been observed, and reasonably associated with dislocation mobility. To observe this, however, the specimen temperature must be low, the thermal condition of deformation controlled. It is the purpose of this discussion to suggest the modifications in Prof. Campbell's technique which would permit him to observe this, as well as to draw the attention of this most eminently qualified assembly to the use of low-temperature strain-rate sensitivity to probe the nature of dislocation dynamics.

Two kinds of test are considered as a model of the fracture problem, an edge-cracked tensile specimen provides a measure of the plane strain fracture toughness over a wide range of speeds: fracture times from 10^3 to 10^{-3} sec. The second experiment is a test for the plastic flow behavior: compression tests of small uniaxial plugs. The compression test records can be analyzed for the strain for (triaxial) tensile instability. The direct correspondence of this, a direct function of the strain-hardening rate, to the fracture toughness indicates control of fracture instability by tensile instability at the crack tip.

Now, unfortunately the plastic flow test may not be carried out in a straightforward stroke as can the fracture test. Since the process zone of the crack is small, it is deformed and ruptured in an isothermal state; its deformation heat is quenched to the nearby cold elastic material. The compression specimen, on the other hand, is orders of magnitude larger. It begins to retain its heat for straining times shorter than about 10 sec. One can correct for the consequent thermal softening by assuming an equation of state $S(\epsilon, \dot{\epsilon}_1, T)$; alternatively, the high strain rate may be applied in small fast steps, interrupted by time to quench back to the initial temperature. The point of yielding at the resumption of plastic flow in each step can be taken as a point on an isothermal though rapid stress-strain curve. The arrest of the testing machine must be as abrupt as possible. Confrontation with a mechanical stop is possible in that of our own design by its low inertia of driving piston and oil. Possibly Prof. Campbell's machine could be similarly programmed.

The result of all this is quite astonishing.* At temperatures below 160°K, strong troughs in fracture toughness are found in direct correspondence to the tensile instability strain. The major depression, some factor of two in depth, is found at a strain rate of about 3.5×10^{-1} sec^{-1} (i.e., fracture process zone position as well as in plastic flow test). There is another, some factor of 100 faster. Their connection to a lattice resonance effect is taken from an apparent harmonic distribution of subsidiary troughs to either side of these major depressions. If they do indeed reflect dislocation motion at the stress-wave velocity, where the resonance would ease the path, then the mobile dislocation density would have to be rather low, the order of 10^{-2} cm for the lowest trough. The low values reported by T. Suzuki at this conference are encouraging in this respect. That the speed

*Krafft, J. M.: Report of NRL Progress, Jan. 1966, pp 6-16, and also Mar. 1967, pp 6-18.

structure appears only at low temperature is consistent with occurrence of low damping here of which Prof. Granato and Prof. Mason have reminded us at this meeting.

I think that the study of these effects may eventually prove quite helpful to our understanding of dislocation dynamics. I would suggest then that Prof. Campbell review his technique with a view to controlling temperature and to simulating isothermal deformation to enable him to pursue this promising field of study.

J. CAMPBELL: The adiabatic heating of a specimen deformed at a high rate, to which Dr. Krafft draws attention, is a very important factor in relation to the post-yield flow. It seems likely, however, that its effect is negligible during the period preceding Lüders band initiation, because of the very small plastic work involved. The effect during the Lüders elongation is less certain, and may well be significant. If so, the curves of Fig. 4 cannot be considered as isothermals: the true isothermal rate sensitivity $(\partial\sigma_{ly}/\partial \log\dot\epsilon_p)_T$ would be larger than the slope shown.

The technique of incremental deformation suggested by Dr. Krafft is a valuable one, and its use may give a close approximation to the isothermal stress-strain relation at a given strain rate. It is, however, necessary to assume that no significant rearrangement of dislocations occurs during the unloading and reloading processes, and this assumption may not be valid if unloading occurs during the period of nonuniform flow.

F. NABARRO: Is it valid to take the dislocation velocity to be a function of the applied stress alone, without subtraction of an internal stress field? In your model, the regions of high-dislocation density deform most rapidly. Might not one find that the applied stress exceeded the internal stress only in the regions of low-dislocation density, so that these, and not the regions of high density, deformed most rapidly? Is it that your applied stresses are so large that the internal stresses are negligible in comparison with them?

J. CAMPBELL: I believe that it is valid to assume that the mean dislocation velocity is a function of the applied stress during the period preceding macroscopic yielding in steel. If the effective internal stress is estimated by extrapolating the work-hardening part of the stress-strain curve back to the origin, the value of this stress is negligible at strains up to the upper yield point strain.

The theory given in the paper could, however, be modified to take account of an effective internal stress which varies with dislocation density, so that the mean dislocation velocity is a function of the applied stress and the dislocation density. If this is done, the condition (3) for the growth of dislocation density gradient is replaced by

$$\frac{d}{d\epsilon_p}[\ln(\rho\rho')] > \left(E\frac{\partial}{\partial\sigma} - \rho'\frac{\partial}{\partial\rho}\right)(\ln v)$$

Since $\partial(\ln v)/\partial\rho < 0$, the effect of the additional term is to reduce the rate of increase of dislocation density gradient, possibly making the rate negative so that heterogeneous yielding would not occur.

R. ARMSTRONG: I should like to point out that the Rosenfield-Hahn data (Fig. D1) are consistent with the data reported in my paper, where one plots, at constant stress, log strain rate vs. the reciprocal square root of grain size. At larger thermal component of stress the activation volume is constant and that is the internal check that Campbell, too, has made on the plot that I have given. At large strain rate you do get a constant activation volume, which is essentially what Rosenfield and Hahn are saying. My second comment is related to what Spreadborough was saying and that is that the stress levels at high strain rate are very close to the twinning stresses which one observes at very low temperatures. The twinning stress appears to be essentially independent of temperature and strain rate. A preliminary analysis based on fitting the relationship

$$\sigma = B \exp(-\beta T)$$

to the data obtained at high strain rates indicates, when compared with the thermal activation rate equation, that β values too small to be explained by a slip process are obtained [cf. Armstrong, R. W.: *Acta Met.* **15**: 667 (1967)].

J. CAMPBELL: Certainly there is twinning at high rates and low temperatures. This gives rise to rapid fluctuations in the recorded load, which may occur before the upper yield point is reached (see Fig. 2*d*). Tests in which this occurred were, however, not taken into account in discussing the dependence of the upper yield stress on temperature and rate of loading.

P. HAASEN: I would like to report some observations on Lüders band propagation in Ge and Si [Schröter, Alexander, and Haasen, *Phys. Stat. Sol.* 7:983 (1964); Siethoff and Haasen, *Proceed. Conf. on Defects in Semiconductors*, Tokyo (1966)]. We observe deformation to occur by Lüders band propagation in crystals with less than 3000 cm^{-2} grown-in dislocations. The Lüders bands proceed from the compression faces of the specimens. The band front is aligned perpendicular to the main active slip planes. The velocity of the band front V_B is somewhat higher than the dislocation velocity V at the same applied stress. Actually, one could identify $V_B = V(\tau + A\sqrt{N})$ as if the dislocations behind the front, in the band, helped pushing the front. This phenomenon may be related to the proposal of Rosenfield and Hahn, (this Colloquium, p. 255).

J. *CAMPBELL*: In our earlier paper we have provided an analysis and it would be interesting to put your results into this formulation.

D. *STEIN*: If one makes a calculation using the Johnston* equation developed from dislocation dynamics, one finds that the yield point increases as one increases the drag stress keeping the other parameters fixed. Therefore, in these experiments one would expect the upper yield point to increase more rapidly with decreasing temperature (increasing drag stress) than would the lower yield point.

J. *CAMPBELL*: As pointed out by Johnston, his calculations do not apply to the problem of a material deforming by the propagation of Lüders bands, since uniform plastic strain in the specimen is assumed. Qualitatively, one may assume that the effect of non-uniform flow is to increase the plastic strain rate, thereby increasing the applied load. Thus, the smooth curve giving a minimum after the yield point is to be replaced by a higher, approximately horizontal, line. It is not possible from this treatment to predict quantitatively the variation in yield drop with temperature.

*Johnston, W. G.: *J. Appl. Phys.* **33**: 2716 (1962).

Agenda Discussion
CRITICAL PROBLEMS

---

*A. Seeger**

Max-Planck-Institut für Metallforschung,
Stuttgart, West Germany

*D. F. Stein*****†

Metallurgy and Ceramics Research Department,
General Electric Company,
Schenectady, New York

*G. T. Hahn*****

Metal Science Group,
Battelle Memorial Institute,
Columbus, Ohio

The discussion touched on a large number of topics:

1. the interpretation of macroscopic strain, microstrain, internal friction, and dislplacement measurements;

2. the identity of rate-controlling processes in b.c.c. metals;

*Chairman.
**Secretary.
†Present Address—University of Minnesota, Minneapolis, Minn.

3. promising experimental techniques, including model materials for studying events on the atomic scale; and

4. critical experimental and theoretical problems.

The report presented here is not a faithful transcript, but summarizes the substance of the discussions under a few main headings. To accomplish this some liberties were taken in organizing the material, and some of the material now appears out of their original context.

INTERPRETATION OF MEASUREMENTS

There was general agreement that macroscopic strain measurements—ordinary tensile tests, creep tests, impact tests, etc.— are difficult to interpret in mechanistic terms. This is because both the number of moving dislocations and their average velocity can change during an experiment. Since the strain rate is proportional to the product of number and velocity, strain rates will only reflect velocity changes when the density of moving dislocations is relatively constant. In some cases a constant-mobile-density assumption is valid. Dorn described impact tests on polycrystalline aluminum samples which show that the deformation at the highest strain rates was athermal. Furthermore, the drag coefficient value obtained in this way agrees with internal friction experiments. Li also pointed out that values of the velocity exponent m, deduced from stress relaxation measurements, are in good agreement with etch-pit measurements. On the other hand, the work on germanium reported by Haasen in this conference shows that only the initial parts of stress-strain and creep curves can be accounted for by dislocation velocities and dislocation (etch-pit) densities measured directly. At higher strain the dislocation density thus measured no longer agrees with the mobile dislocation density derived from $\dot{\epsilon}$ and v.

Basinski pointed out that measurements of the "reversible" change in flow stress are often misinterpreted. Figure 1 shows that stress-strain curves obtained by changing the strain rate can display transients (see arrows). These transients could result from adjustments in the dislocation structure accompanying the change in strain rate. If the adjustments occur in a relatively short time, careful analysis is needed to interpret the data.[1] However, if the transients are caused by relaxations within the testing machine, they should be ignored. Basinski feels that a standard method of dealing with these transients should be adopted. Other

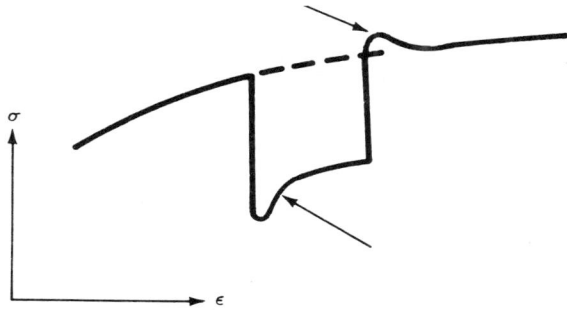

Fig. 1 Transient behavior of a stress-strain curve resulting from imposed changes in strain-rate.

evidence that the "reversible" change in flow stress is strain- and strain rate-history dependent was cited including the recent work of Galligan and Davidson.[2]

Dorn argued that macroscopic stress–strain measurements are important in their own right, serving as part of the all-important link between engineering properties and the processes on the atomic scale. Macroscopic properties may also distinguish between well-defined microscopic processsses provided they are quite different, but the identity of a given process and the detailed mechanism is difficult to establish in this way. In general, it is useful to apply two or more techniques to the same material. For example, Granato mentioned that internal friction measurements concurrent with straining are now being made in his laboratory.

Microstrain measurements suffer the same lack of resolution attributed to macroscopic studies. With current methods capable of detecting strains $\sim 10^{-6}$, the following interpretations of σ_A are equivalent for a total dislocation density $\sim 10^9$ cm^{-2}: (a) every dislocation advances ~ 1 atomic spacing; or (b) 0.1% of the dislocations advance 10^3 atomic spacings. For this reason the nature of the obstacles is difficult to interpret and it is not clear that the width of the hysteresis loop at $\sigma = \sigma_E$ is a function of the extent of the internal stress field. Since the fraction of dislocation line moved may vary with temperature, the temperature dependence of mobility does not follow directly from the dependence of σ_A. Brown agreed that further increases in sensitivity might lead to a reduction of σ_A and σ_E. However, he also argued that any resultant decrease in σ_E would arise from its dependence on strain, since one cannot measure the friction stress below the strain sensitivity of the experiments. Thus, even more sensitive detection methods are needed to support the view that σ_E values now being quoted correspond to a friction stress. This does not mean that a significant friction stress does not exist. Young expressed the view that it has a finite value even in the case of copper crystals where the macroscopic yield stress is 15–20 gm mm^{-2}. Vreeland mentioned that etch-pit studies together with microstrain measurements on the same material are in progress at Cal. Tech., and this combination should provide useful insights. Brown also feels that microstrain measurements are useful because their interpretation should be less complicated than the values derived from the macrostrain region. It also appears to be the only method for studying the relaxations that occur when the stress is removed.

More direct measurements of dislocation velocity have been performed with two techniques: (1) internal friction; and (2) displacement measurements. The internal friction technique measures the damping of the dislocations and this is analogous to the hysteresis loop obtained with the microstrain techniques. However, the frequency dependence of the damping provides additional information, which when coupled with measurement of the elastic modulus defect and estimates of the total line length and line tension, offers a means of separating the number of loops, the distance moved, and the mobility.

Displacement techniques involve measuring the position of a dislocation before and after it is subjected to a well-defined stress pulse. So far, both the etch pitting and the Berg-Barrett technique have been used to define dislocation positions. The major difference is that the distance a dislocation moves in the displacement technique is very large and one observes the dislocation only in the static condition. In contrast, the internal friction technique measures only very short distances of motion and the measurement is made while the dislocation is in motion. It should be realized that the two techniques are complementary. The fact that Vreeland's displacement technique and the Stern and Granato[3] and Alers and Thompson[4] internal friction studies give identical results in the appropriate range of velocity, is one of the most encouraging results of recent work. Agreement between both techniques was also reported by T. Suzuki in his paper at this Colloquium. In light of these advances, it appears that the factors influencing multiplication and the magnitude of the mobile fraction are now less well understood than dislocation velocity considerations.

RATE-CONTROLLING PROCESSES

Seeger noted that the processes controlling the mobility of a dislocation can conveniently be divided into two categories—intrinsic and extrinsic. The intrinsic processes include the various Peierls models, constriction and recombination (pseudo-Peierls), and jogging of screw dislocations. The extrinsic processes would include impurity and dislocation interaction models. Seeger pointed out that the mathematics of some of these models is so similar that they obey much the same functional relationships. For this reason, it will be difficult to distinguish between the various possibilities on the basis of the temperature or rate dependence at the flow stress. Hirsch suggested even "simple" cases such as dislocation velocity in germanium may involve several processes whose effects may not be additive. This will further complicate the interpretation. Therefore, there was some pessimism over prospects of identifying the specific mechanisms in the near future.

There was not general agreement on the specific mechanisms operating in b.c.c. metals at low temperature, although the predominant feeling was that an intrinsic mechanism dominates the thermal part of the yield stress. One school of thought represented by Dorn, Conrad, and others believes that interstitials make no contribution to the temperature dependence of the yield stress. Their models do not predict the gross failure of the Schmid law at low temperatures, especially in high-purity materials. The other school, represented by Haasen,[5] Stein, Seeger, and others, still leaves room for an impurity contribution to thermal activation, as do the results reported by Stein for molybdenum crystals. Certainly, more experimentation on the connection between b.c.c. crystal orientation and impurity effects may help to distinguish between various possible mechanisms. In contrast to the "high-purity-approach," Owen suggested that a larger excess of defects be introduced, so that the extrinsic mechanism completely overshadows the intrinsic process. However, some reservations were voiced since an extrinsic mechanism may depend on the defect concentration.

The "polarity" of dislocation mobility—differences between the tension and compression flow stress—is another key feature of the intrinsic mechanism. Although the "pseudo-Peierls" model, treated in the paper presented by Duesbery and Hirsch in this conference, can account for this effect, it deserves more attention. Hirsch suggested that direct measurements of edge and screw dislocation mobility in both directions should be attempted. Hirth noted that elastic anisotropy was another complication which could cause strong effects in many ways, such as stabilizing or destabilizing pileups and jogs on dislocations, and altering image stresses at grain boundaries. It was agreed that these effects may be important and that they are not well understood.

ATOMIC MECHANISMS

In Gilman's view, velocity measurements, although still fruitful, are not sufficient. Further progress requires the additional step that will bring atomic or molecular motions into focus. The spin resonance technique may be of value here, but complications from conduction electrons must be avoided. Gilman suggests that this can be accomplished by adopting insulators as models, and that these could prove as useful as LiF did a decade ago. According to Haasen, an electron spin resonance signal from dislocations in silicon has been identified in Göttingen.[6] The signal appears to depend critically on dislocation character rather than on kink density, but such correlations are difficult to establish.

Keh gave a brief report of electron drag calculations performed in his laboratory by Huffman and Louat.* Consistent with previous findings, their results show that the electron drag coefficient is proportional to conductivity. Seeger noted that unusual effects might be observed by suppressing the electron drag effect, either by going to the superconducting state, by means of high magnetic fields, or by alloying. However, interest in kinks is not confined to low temperatures. Seeger noted that kinks have definite properties and relevance at elevated temperatures.

EXPERIMENTAL AND THEORETICAL PROBLEMS

Deformation studies can benefit from a variety of measurements. Systematic studies of b.c.c. crystal orientation and slip polarity effects at various purity levels are needed. Existing techniques may prove adequate here. More velocity measurements should be performed and this can be done either with internal friction, with etch pits, with the Berg-Barrett technique or with several other methods described in a following paragraph. The

---

*These calculations appear as Appendix A of this Agenda Discussion.

internal stress field, its intensity and extent, must be characterized; at least four ways of approaching this problem are suggested below. On the other hand, experimentation to define the structure and processes at the dislocation core has virtually no established techniques to draw on. Hirth proposed that field ion mocroscopy may be useful in this respect, and the possibility of nuclear spin resonance was mentioned. Finally, some way of measuring the instantaneous number of moving dislocations is needed. At present, this can only be done indirectly provided the stress-strain curve, the strain rate, and the dislocation velocity characteristics are known.

The discussion of velocity measuring techniques touched on a number of potentially interesting methods. Seeger pointed out that early cinematic studies of growing slip steps were of limited value because the microscopes had poor resolution. However, new methods such as the Nomarski technique can resolve slip steps as small as 10 Å. Sleeswyk noted that although the resolution was good, this technique suffers from poor depth of focus. The scanning electron microscope is another possibility with inherently good depth of focus. Seeger also proposed direct observation of moving dislocations by transmission microscopy, especially with the new high voltage units. According to Suzuki, this is already being attempted by Fujita and Yamada of the National Research Institute for Metals in Japan using a 500-kV electron microscope. Rosenfield noted one difficulty with techniques relying on high magnification: the relatively small field of view restricts the experiment to very slow moving dislocations such as those obtained under creep conditions. Young suggested that it might be possible to couple internal friction and x-ray topography, if a dislocation can be kept oscillatory in the same mode for a long period of time. Since the dislocation will spend more time in the extreme positions, enhanced contrast there might reveal the extent of the oscillation.

In response to a question by Rosenfield, Seeger outlined a number of techniques for measuring internal stress fields. Transmission microscopy affords one opportunity. Following Essmann[7] and Ramsteiner,[8] a crystal in the as-deformed but unloaded state is irradiated with fast neutrons to pin the dislocations completely. Thin foils are then prepared for transmission microscopy, say, parallel to the main glide plane. By observing the local curvature of dislocations it is possible to deduce the magnitude of internal stresses as well as their wavelength. This approach has been successfully applied to copper and gold crystals. Dislocations pinned by irradiation can also be studied by Berg-Barrett x-ray topography and this has been demonstrated by Wilkens.[9] The magnitude and axis of lattice rotations (averaged over a few microns) can be determined from reflections from differently oriented crystal surfaces. The magnitude of internal stresses and new dislocations arrangements responsible has been inferred in this way for copper crystals. In germanium, Alexander[10] has observed dislocations frozen in position after deformation and cooling *under stress* and in this way measured internal stress.

Young suggested that new insights to the processes controlling dislocation movement can be obtained by observing dislocations pinned *while the crystal is still under load*. Young presented a prepared discussion* of preliminary findings based on this technique.

Another approach involves phonon heat conductivity at low temperatures. The phonon contributions κ_{ph} to heat conductivity involves the scattering of phonons from various imperfections including internal strains.[11] If the internal strains are due entirely to randomly distributed dislocations, $\kappa_{ph} \propto T^2$ (T is the temperature). Gruner[12] and Bross[13] have calculated the deviation from this law due to dislocation arrangements that produce long-range stress fields. During Stage I, deformation of Cu-4.6 at % Ga crystals, the T^2 law is accurately obeyed[14] (indicating the absence of long-range internal stresses). In Stage II, deviations from the T^2 relation are found. These have the right sign and magnitude to be compatible with other evidence of the long-range internal stresses. Similar deviations can be found in the literature on heat conductivity of deformed alkali halide crystals.

In materials with high magnetostriction (Ni, Fe, etc.), magnetic measurements can be used to measure the magnitude and the wavelength of internal stresses. The experiments that may be interpreted in the most direct way are those involving the rotation of magnetization against internal stresses. Where applicable, such measurements are very powerful if carried out over a sufficiently wide range of orientation and magnetic fields since they allow a nondestructive Fourier-type analysis of the internal stresses according to magnitude and wavelength (for a recent summary, see Kronmüller[15]).

It is also apparent that a good many theoretical problems remain unresolved. Those that were touched upon in the course of the discussion are discussed below.

Our theoretical knowledge of the various dislocation multiplication mechanisms, particularly in their quantitative aspects, is

*Appendix B of this Agenda Discussion.

rather rudimentary. Further theoretical work on these questions and on the magnitude of the mobile fraction of dislocations appears desirable.

Another field in which theoretical work is required is the structure of the dislocation core and its effects on dislocation interactions and on dynamical dislocation properties. At present, the main uncertainty in the quantitative analysis of such dislocation configurations as nodes and stacking-fault tetrahedra is the uncertainty in the core energies.

Although considerable work has recently been done on problems involving elastic anisotropy, further theoretical studies of the effect of the elastic anisotropy on dislocation interactions, the internal stresses created by dislocation arrangements, etc., are desirable.

An important and, so far rather neglected, theoretical field is the nonadditivity of closely linked thermally activated processes. Examples are the effect of alloying on dislocation recombination or intersection processes, or the effect of a Peierls barrier on the characteristics of other thermally activated processes. It appears likely that theoretical treatments of such "combined" processes will help to alleviate some of the present difficulties in analyzing thermally activated dislocation processes. For example, such studies might lead to a better understanding of the critical flow stress in b.c.c. metals and the observed "polarity effects" and gross deviations from Schmid's law.

Since many of the thermally activated processes are observed at rather low temperatures, the overcoming of dislocation barriers by the combined effects of thermal and quantum mechanical fluctuations should be studied by theory.

Considerable work has recently been done on calculations on phonon and electron drag mechanisms. There is nevertheless room for substantial improvements. In the phonon calculations the elastic anisotropy should be taken into account. The electron drag calculations should be extended to realistic electron energy surfaces and wave-functions and to fuller treatments of the electron transport problem at low temperatures.

REFERENCES

1. Mecking, H., and K. Lücke: Mater. Sci. Engg. 1:349 (1967).
2. Galligan, J. M., and J. L. Davidson: J. Appl. Phys. 38:3420 (1967).
3. Stern, R. M., and A. V. Granato: Acta Met. 10:358 (1952).
4. Alers, G. A., and D. O. Thompson: J. Appl. Phys. 32:283 (1961).
5. Haasen, P.: "Realstruktr und Eigenschaften von Reinstoffen," E. Rexer (ed.), Akademie-Verlag, Berlin, 1967, p. 387.

6. Alexander, H., R. Labusch, and W. Sander: *Sol. State Comm.* 3:357 (1965); Wöhler,
   F. D., H. Alexander, and W. Sander: To be published.
7. Essmann, U.:*Phys. Status Solidi* 12:723 (1965).
8. Ramsteiner, F.:*Mater. Sci. Eng.* 1:281 (1966/67).
9. Wilkens, M.:*Can. J. Phys.* 45:567 (1967).
10. Alexander, H.: *Can. J. Phys.* 45:1209 (1967); *Phys. Stat. Sol.* (in press).
11. Klemens, P. G.: "Solid State Physics," Vol. 7, Academic Press, New York, 1958.
12. Gruner, P.:*Z. Naturforschg.* 20a:1626 (1965).
13. Bross, H.:*Z. Physik* 189:33 (1966).
14. Zeyfang, R.:*Phys. Stat. Sol.* 24:221 (1967).
15. Kronmüller, H.: *Can. J. Phys.* 45:631 (1967).

APPENDIX A   DISCUSSION

Submitted by G. P. Huffman and N. P. Louat

U. S. Steel Corporation

(Presented by A. S. Keh)

When a dislocation moves through a metal, it generates electric
fields which produce currents in the conduction electron gas. The
energy dissipated by such currents can be calculated in a manner
analogous to that used in theories of acoustic attenuation. One first
solves the Boltzmann equation to obtain the electron distribution
function, from which the electron current can be calculated. Then,
using Maxwell's equations and the charge-current continuity equation, the electric field can be derived from the current. In the final
result both the current and field are given entirely in terms of the
time rate of change of the atomic displacement function.

The energy dissipated by currents in the electron gas when the
dislocation jumps from one position in the lattice to the adjacent
position is

$$W = \int_{j\tau_D}^{(j+1)\tau_D} dt \int_V d^3 r \, \vec{J}_e(\vec{r}, t) \cdot \vec{E}(\vec{r}, t) = S b^2 L \tag{1}$$

where $\vec{J}_e$ is the electron current, $\vec{E}$ is the electric field, τ_D is the
time required to make the jump, S is the stress causing the motion,
b is the Burgers vector, and L is the dislocation length.

The calculation is most easily performed by transforming
Eq. (1) into momentum space and introducing a Debye cutoff in
momentum. In analogy to acoustic attenuation theories, it is convenient to divide the stress into two parts: that associated with
longitudinal currents and fields; and that associated with transverse
currents and fields. An edge dislocation has both longitudinal and
transverse components, and for the longitudinal part, one finds:

$$S_{\text{long.}}^{(\text{edge})} \approx \frac{Nmbv_d}{16\pi\tau}\left[1 + 2\left(\frac{v_2}{v_1}\right)^2\right]^2 \left\{\frac{(q_D\Lambda)^2}{3}\,\mathcal{F}\left(\frac{q_D}{q_{TF}}\right) + \frac{\pi}{6}(q_D\Lambda)\,\mathcal{L}\left(\frac{q_D}{q_{TF}}\right) + .2\right\} \quad (2)$$

where N, m, τ and Λ are the electronic density, mass, relaxation time and mean free path, respectively, b is the Burgers vector, v_d is the dislocation velocity, v_1 and v_2 are the longitudinal and transverse sound wave velocities, q_D and q_{TF} are the Debye and Thomas-Fermi wave vectors, and $\mathcal{F}$ and $\mathcal{L}$ are simple functions of order .1 and 1, respectively. In obtaining Eq. (3), Eshelby's nonrelativistic atomic displacement functions for a dislocation moving through a continuum have been used; $q_D\Lambda$ is of the order 10^5 at helium and 10^2 at room temperatures, so the first term is dominant. Using well-known relations between Λ, the Fermi energy, the relaxation time and the dc electrical resistivity, we find, keeping only the first term,*

$$S = \frac{mE_F\, b\, q_D^2 \left\{1 + 2(v_2/v_1)^2\right\}^2 \mathcal{F}}{24\pi e^2}\,\frac{v_d}{\rho} \quad (3)$$

where ρ is the resistivity, e is the electron charge, and E_F is the Fermi energy. Inserting values appropriate for a typical metal and assuming a dislocation velocity of 10^3 cm/sec, Eq. (3) gives about 10^{10} dyne/cm^2 at liquid helium temperature and 10^6 dyne/cm^2 at room temperature.

The evaluation of the integrals giving the part of the stress for an edge dislocation due to transverse fields, and the stress for a screw dislocation, which involves only transverse fields and current, is more difficult, and has not yet been done analytically. However, we have been able to put upper and lower bounds on these stresses and find them to be less than 10^3 dyne/cm^2 for all temperatures of interest. These terms are, therefore, readily neglected.

Since a moving dislocation can be pictured as a phonon wave packet, we expect its velocity to be a few percent of the speed of sound in the metal in question. Moreover, since the velocity of sound changes by only a few percent from liquid helium to room temperature, it seems reasonable to assume that the dislocation velocity is approximately constant with respect to temperature. The temperature dependence of the stress should, therefore, be about the same as that of the inverse resistivity, or the conductivity. To compare Eq. (3) with existing experimental yield stress

*$S/v_d = B$, where B is the damping constant.

data, it is necessary to have some estimate of the residual resistivity of the samples tested, which is primarily a function of sample purity, and of the dislocation density. These numbers can be found by plotting $1/S$ vs. the ideal resistivity, ρ_i. As Eq. (4) shows, this should give a straight line whose slope determines v_d and whose intercept then gives the residual resistivity ρ_0.

$$\frac{1}{S} = \frac{24\pi e^2}{mE_F\, b\, q_D{}^2 \left[1 + 2(v_2/v_1)^2\right]^2 \mathcal{F}} \left(\frac{\rho_i(T) + \rho_0}{v_d}\right) \tag{4}$$

In Fig. A1 we show a plot of $1/S$ vs. ρ_i for vanadium, obtained from the yield stress data of Christian and Masters[1] and the resistivity data of White and Woods.[4] Using a free electron model with one conduction electron per atom to calculate the Fermi energy, we find from the slope a dislocation velocity of 9.0×10^4 cm/sec and a residual resistivity of $3.85\,\mu\Omega$-cm from the intercept, both of which are reasonable. The same procedure for the potassium yield stress data of Bernstein[3] and the resistivity values of Dugdale and Gugan[2] gives a dislocation velocity of 34.5 cm/sec and a residual resistivity of $0.54\,\mu\Omega$-cm. When these values are inserted into the previous equations, one obtains the yield stress

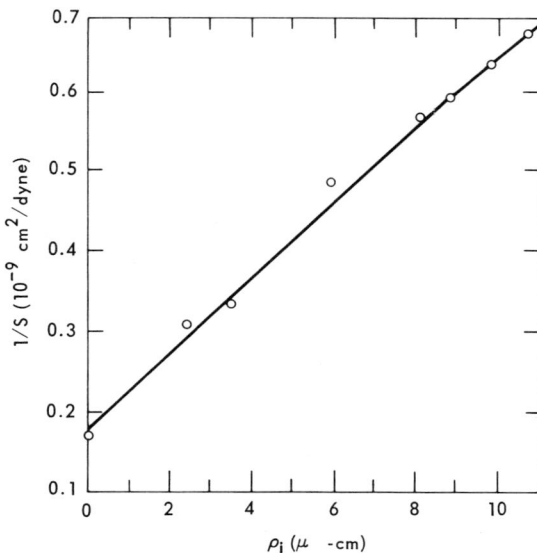

Fig. A1 Inverse yield stress vs. ideal resistivity for vanadium.

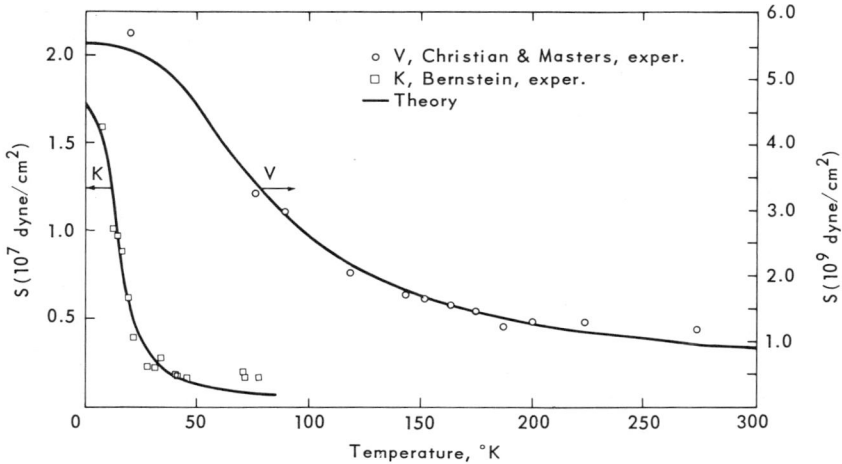

Fig. A2 Temperature dependence of yield stress of potassium and vanadium.

vs. temperature curves shown in Fig. A2. Here, the left-hand scale refers to potassium and the right to vanadium. The smooth curves are the theoretical results, while the squares are the experimental values of Bernstein[3] for K and the circles are the data of Christian and Masters[1] for V. The agreement is quite satisfactory. Similar analyses have been carried out for Fe, Mo, W, and Cr with satisfactory results

The physical picture of a moving dislocation as a wave packet of phonons makes dislocation velocities of the order of 10^2 to 10^4 cm/sec seem quite reasonable.

References

1. Christian, J. W., and B. C. Masters: *Proc. Roy. Soc.* **A281**:223 (1964).
2. White, G. K., and S. B. Woods: *Phil. Trans. Roy. Soc.* (London) Ser. **A251**:272–302 (1959).
3. Bernstein, I. M.: Ph.D. Thesis, Columbia Univ., 1965.
4. Dugdale, J. S., and D. Gugan: *Proc. Roy. Soc.* **270**: 186 (1962).

APPENDIX B   DISCUSSION

Submitted by J. C. Crump and F. W. Young, Jr.

(Presented by F. W. Young, Jr.)

The results of recent investigations by Young and Sherrill (1966) on processes occurring in the preyield range of copper single crystals using Borrmann x-ray topography techniques have clearly indicated that back motion of dislocations occurs when stressed crystals are relaxed and that the dislocation configurations in the stress relaxed condition are different from those in the stress-applied condition. Similar results have also been reported by Brydges (1966), who has made surface-etching studies on the early stages of plastic deformation in copper. This study has been extended into higher stress ranges using transmission electron microscopy techniques.

Three copper crystals having dimensions 1 cm × 1 cm × 2 cm with one of their large faces parallel to the $(1\bar{1}1)$ planes and with their long axis several degrees off of either $[2\bar{1}1]$ or $[1\bar{1}0]$ were spring loaded under compression and were irradiated at ambient temperature with 2×10^{17} fast neutrons with the stress applied to pin dislocations in their as-stressed position. The first crystal was compressed to 500 gm/mm^2 resolved shear stress, the second to 100 gm/mm^2 and the third to 40 gm/mm^2. The yield points for similar crystals compressed in an Instron were less than 20 gm/mm^2 resolved shear stress. Lamellae parallel to the primary slip plane were cut from the center of the crystals with an acid saw and electropolished for electron microscope studies.

The dislocation configurations observed in the primary slip planes in the crystal compressed to 500 gm/mm^2 were identical to those which have been observed previously in copper samples similarly stressed and then relaxed. The most prominent feature in this crystal was the presence of dislocation bundles which tended to be aligned along the traces of the conjugate and critical planes and which were composed of primary edge dislocation dipoles and some secondary edge dislocations. No long single dislocations and only a few short jogged screw dislocations have been observed in this crystal.

The dislocation arrangements in the crystal compressed to 10 gm/mm^2 were different from those observed previously in similarly stressed crystals studied after the stress was removed and also from those observed in the crystal compressed to 500 gm/mm^2. Whereas previous studies have indicated that large numbers of edge dislocation bundles were present in crystals stressed to this range and then relaxed, very few bundles have been observed during the present investigation. A number of long dislocations, some of which were in the screw orientation and others which were in a mixed orientation, have been observed in this crystal, however, and these have not previously been reported in crystals similarly stressed and relaxed.

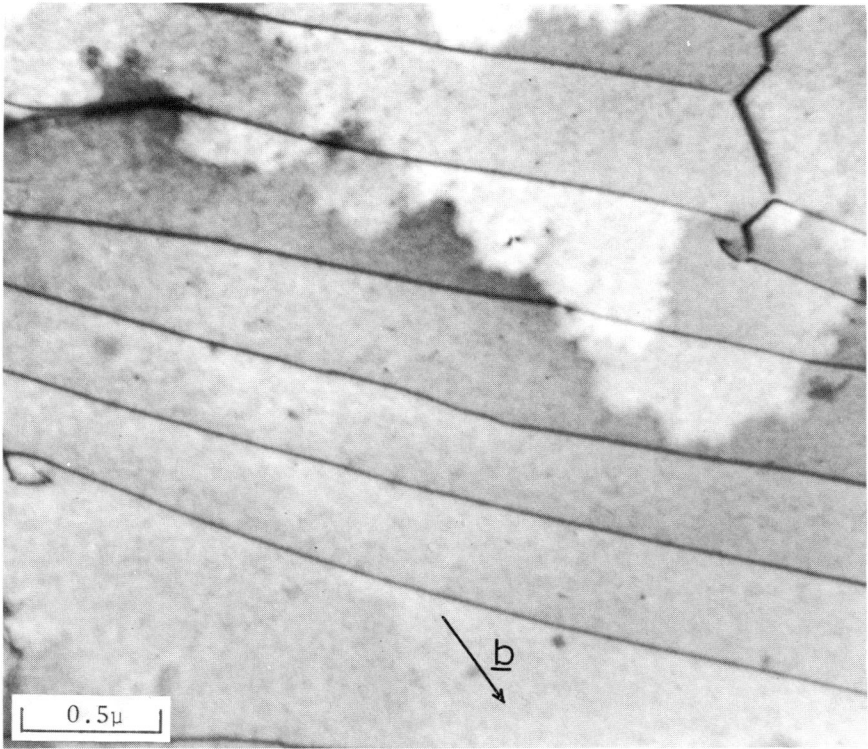

Fig. B1 Electron micrograph of a lamella cut from a copper crystal loaded to a resolved shear stress of 40 g/mm². The dislocations were pinned by irradiation with the stress applied.

The results of deformation in the crystal which was compressed to 40 gm/mm² (Fig. B1) indicate the presence in this crystal of numerous long dislocations, most of which are in a mixed orientation, which lie on the primary slip planes and intersect to form networks. A few single edge dipoles have been seen but no bundles of edge dipoles.

Apparently little or no relaxation occurs when a crystal stressed to 500 gm/mm² is relaxed, and the dislocation arrangements in such a crystal in the stress-relaxed condition are, therefore, similar to those in the crystal in the stress-applied condition. For lower stresses, however, i.e., 100 gm/mm² or less, relaxation does occur and the dislocation configurations observed in the stress-relaxed condition very likely do not represent the true dynamic phenomena in the stressed crystal.

INDEX